W0106289

The Electronic Structure of Complex Systems

NATO ASI Series

Advanced Science Institutes Series

A series presenting the results of activities sponsored by the NATO Science Committee, which aims at the dissemination of advanced scientific and technological knowledge, with a view to strengthening links between scientific communities.

The series is published by an international board of publishers in conjunction with the NATO Scientific Affairs Division

A Life Sciences **B Physics**	Plenum Publishing Corporation New York and London
C Mathematical and Physical Sciences	D. Reidel Publishing Company Dordrecht, Boston, and Lancaster
D Behavioral and Social Sciences **E Engineering and Materials Sciences**	Martinus Nijhoff Publishers The Hague, Boston, and Lancaster
F Computer and Systems Sciences **G Ecological Sciences**	Springer-Verlag Berlin, Heidelberg, New York, and Tokyo

Recent Volumes in this Series

Volume 111—Monopole '83
edited by James L. Stone

Volume 112—Condensed Matter Research Using Neutrons: Today and Tomorrow
edited by Stephen W. Lovesey and Reinhard Scherm

Volume 113—The Electronic Structure of Complex Systems
edited by P. Phariseau and W. M. Temmerman

Volume 114—Energy Transfer Processes in Condensed Matter
edited by Baldassare Di Bartolo

Volume 115—Progress in Gauge Field Theory
edited by G. t'Hooft, A. Jaffe, G. Lehman, P. K. Mitter, I. Singer and R. Storra

Volume 116—Nonequilibrium Cooperative Phenomena in Physics and Related Fields
edited by Manuel G. Velarde

Volume 117—Moment Formation in Solids
edited by W. J. L. Buyers

Series B: Physics

The Electronic Structure of Complex Systems

Edited by

P. Phariseau

Rijksuniversiteit
Ghent, Belgium

and

W. M. Temmerman

SERC Daresbury Laboratory
Daresbury, United Kingdom

Plenum Press
New York and London
Published in cooperation with NATO Scientific Affairs Division

Proceedings of a NATO Advanced Study Institute on Electronic
Structure of Complex Systems, held July 12–23, 1982, at the
State University of Ghent, Belgium

Library of Congress Cataloging in Publication Data

NATO Advanced Study Institute on Electronic Structure of Complex Systems (1982: State University of Ghent)
The electronic structure of complex systems.

(NATO ASI series. Series B, Physics; v. 113)
"Published in cooperation with NATO Scientific Affairs Division."
"Proceedings of a NATO Advanced Study Institute on Electronic Structure of Complex Systems, held July 12–23, 1982, at the State University of Ghent, Belgium"—Verso t.p.
Includes bibliographical references and index.
1. Electronic structure—Congresses. 2. Free electron theory of metals—Congresses. I. Phariseau, P. II. Temmerman, W. M. III. Title. IV. Series.
QC176.8.E4N336 1982 530.4′1 84-17856

ISBN-13: 978-1-4612-9466-5 e-ISBN-13: 978-1-4613-2405-8
DOI: 10.1007/978-1-4613-2405-8

©1984 Plenum Press, New York
Softcover reprint of the hardcover 1st edition 1984

A Division of Plenum Publishing Corporation
233 Spring Street, New York, N.Y. 10013

All rights reserved. No part of this book may be reproduced, stored in a retrieval system, or transmitted, in any form or by any means, electronic, mechanical, photocopying, microfilming, recording, or otherwise, without written permission from the Publisher

PREFACE

We present here the transcripts of lectures and talks which were delivered at the NATO ADVANCED STUDY INSTITUTE "Electronic Structure of Complex Systems" held at the State University of Ghent, Belgium during the period July 12-23, 1982.

The aim of these lectures was to highlight some of the current progress in our understanding of the electronic structure of complex systems. A massive leap forward is obtained in bandstructure calculations with the advent of linear methods. The bandtheory also profitted tremendously from the recent developments in the density functional theories for the properties of the interacting electron gas in the presence of an external field of ions. The means of performing fast bandstructure calculations and the confidence in the underlying potential functions have led in the past five years or so to a wealth of investigations into the electronic properties of elemental solids and compounds. The study of the trends of the electronic structure through families of materials provided invaluable insights for the prediction of new materials.

The detailed study of the electronic structure of specific solids was not neglected and our present knowledge of d- and f-metals and metal hydrides was reviewed. For those systems we also investigated the accuracy of the one electron potentials in fine detail and we complemented this with the study of small clusters of atoms where our calculations are amenable to comparison with the frontiers of quantum chemistry calculations.

The bandtheory of random solids based on KKR-CPA calculations has also seen a rapid developement over the past years. Those methods have not only led to the study of the electronic structure of random substitutional alloys but the technique can also be used to study metallic magnetism at finite temperature.

We also devoted some time to the techniques of first principles lattice dynamics which has seen steady progress and where the present techniques will lead to first principles calculations of phonon spectra in the near future.

Fermi surface studies were crucial to the development of bandtheory in the 1960's and 1970's and is now complemented by spectroscopy calculations which can probe the occupied and unoccupied bands and is developing as an invaluable way of analyzing the calculated bands.

We set aside some time to the study of materials which demand more theoretical investigations and where in the near future the bandstructure calculations should contribute.

In short, this institute was dedicated to the coming-of-age of bandtheory techniques in the study of the electronic properties of metals and this will lead in the 1980's to the prediction of new materials and the understanding from first principles of such outstanding issues in solid state physics as metallic magnetism at finite temperature and electron-ion interactions.

This institute was timely both in the presentation of the methods developed and to give the flavor of new things to come.

While the individual contributions are self-contained accounts of the relevant topics, and no effort has been made to standardize the notations all through the text, cross references are frequent and each is written with evident awareness of the unity of the subject.

The Advanced Study Institute was financially sponsored by the NATO Scientific Affairs Division (Brussels, Belgium). Co-sponsors were the National Science Foundation (Washington, D.C., U.S.A.) and the State University of Ghent. In particular we are indebted to Prof. Dr. Ir. A. Cottenie, Rector of the University of Ghent, who made it possible that the institute took place in ideal circumstances.

We are grateful to all lecturers for their most valuable contribution and their collaboration in preparing the manuscripts. The Institute itself could not have been realized without the enormous enthusiasm of all participants and lecturers and without the untiring efforts of our co-workers at the "Seminarie voor Theoretische Vaste Stof- en Lage Energie Kernfysica". Also Mrs. A. Goossens-De Paepe's help in typing the manuscripts is gratefully acknowledged.

P. Phariseau
W. Temmerman

Ghent and Daresbury, December 1983

CONTENTS

THE ROLE OF ELECTRONIC STRUCTURE COMPUTATIONS IN THE ADVANCEMENT OF SOLID STATE PHYSICS

Volker Heine

Cavendish Laboratory
Madingley Road
Cambridge CB3 OHE
England

ABSTRACT

The present state of electronic structure computations is reviewed, and the contribution it can make to the main stream of solid state physics is analysed with examples.

At the opening of this Nato Advanced Study Institute I would like to thank the University of Gent and Prof. Phariseau and Dr. Temmermann for bringing it about : I know from recent personal experience what a lot of work is required. The subject of electronic structure calculations is a very timely one. We are often reminded how the power of computers is growing exponentially. Is our subject of electronic structure calculations also growing exponentially in its contribution to the advancement of main-stream solid state physics ? What is its role ?

Much of the following two weeks will be devoted to the technical intricacies of making such calculations, and before we get wholly immersed in those I want to set the work in the wider context : what do we hope to achieve ? In all of science, achieving a real advance in understanding often depends on selecting just the right experiment or calculation to shed light on a phenomenon and not merely in heaping up more facts.

Before answering those questions we need to focus on what are the present-day challenges in solid state physics. After all, superconductivity and semiconductors have been around a long time. Broadly speaking I see three main frontiers.

1. There is what one could call solid state chemistry : understan-

ding the differences between similar elements or compounds. Why are Cu, Ag and Au different in various respects ? Why do Nb compounds or certain crystal structures figure prominently among high T_c superconductors ?

2. Complex materials. Amorphous metals, metal cluster compounds, layer compounds, one-dimensional materials etc. etc. with their fascinating and unexpected variety of properties.

3. Complex phenomena : such as defects with their surrounding atomic displacements and their interactions, and our old friends corrosion and catalysis.

I want to emphasize that in all these fields there are interesting principles to be discovered and understood. It is not just a matter of dotting detailed i's or crossing t's. But coming to grips with the details is an essential part of the process : we cannot say that we fully understand invar alloys until we understand why a few alloys show the phenomenon and others do not.

Clearly such understanding is not going to come without intelligent computation, which establishes an essential role for our subject at the core of modern solid state research. But our analysis also shows a difficulty of the computational approach. I believe Rutherford warned that the occupational hazard of experimental physics is to degenerate into what he called mere "stamp collecting". Similarly a computation can easily boil down to mere retesting of the Schrödinger equation. If one computes the conductivity or magnetic susceptibility of some material and gets agreement with experiment, one has achieved two things. Firstly one has shown that one's approximations (and all electronic structure calculations involve approximations) are valid, but in most cases this will already have been established by similar calculations of related materials or properties. Secondly one has demonstrated that God believes in the same Schrödinger equation governing the real material on the laboratory bench as one has programmed into the computer. I exaggerate slightly. Nevertheless the truth remains that the essential difficulty of research lies in getting beyond this point. Analogous comments apply to experimental work, as already indicated, and of course to theoretical physics too. So I do not make my remarks in order to denigrate computational physics, but as a spur to formulating its proper positive role. There is a further difficulty : having invested several man-years in developping a computational method, one cannot toss it aside and develop a different one just to respond to a fashionable hot discovery or controversy. One is tied to one's soft-ware, and in order to participate in the most active research front one has to sift literally hundreds or thousands of experimental what-is-going-on in order to see what is most significant and where one can best jump in. But again the same applies to experimentalists and theorists.

Before discussing the role of computation I want to review briefly the present state of electronic structure calculations : what can the computations currently achieve and where can rapid development be expected in the near future ? The first item is the potential to be used in the Schrödinger equation, its self-consistency and the inclusion of exchange and correlation. My drawing of Mr. Smiley (Fig. 1) indicates general satisfaction and great advance, mostly based on the density functional approach (which I want to emphasize includes in principle an exact treatment of correlation) and on the local density approximation. But amongst the general euphoria I want to mention a few limitations of the local approximation which are also being researched. Firstly no single calculation can be relied on to the last decimal place (or even the first decimal place in some instances) for, say, the band gap in a band structure, remarkable though the results of the local approximation are. However chemical trends among a series of compounds are even better established. For example a recent band structure calculation of TiS_2 caused considerable controversy with the experimentalists about whether it was a semiconductor or a semimetal. The comparison with analogous calculations for $TiSe_2$ and for both materials under high pressure helped to establish the degree of reliability and a wider comparison with a range of experimental data. Just as a surveyer establishes the position and height of a peak by triangulating from as many directions as possible, so all computations should include trends wherever possible.

Fig. 1

Secondely the Mott metal-insulator transition would appear at first sight to be in principle outside the local density approximation. However the Mott insulator MnO with the d^5 ^6S state can be treated in band theory by filling all of the up-spin band to represent the ferromagnetic configuration. In other cases one has to use the calculated bands to form Wannier functions. The point is that by expanding one's conceptual horizons one can extract crystal field splittings and sometimes super-exchange interactions from the bands. I mention it because I know such types of extension offend some people's mathematical purity so that they won't do them, even when correct physically. For example I have met people who are happier doing straight Hartree-Fock calculations, even though this leads to infinite Fermi velocity in a metal, than using the local density approximation. (There are of course situations where one might argue that Hartree-Fock would give better some off-diagonal exchange interaction).

My next example concerns identifying non-local effects beyond any local potential approximation. If one calculates, say, just a band bap, how can one say what precisely any remaining discrepancy from experiment is due to ? In silicon the effective mass sum rule says that the effective mass $m^\star$ in the conduction band is approximately inversely proportional to the related band gap, and Kane some years ago showed that the quotient (Gap/$m^\star$) is very nearly the same calculated for a whole range of local potentials. This value differs appreciably from the experimental one which must be due to genuine non-local many-electron effects. We have therefore a real handle on these effects. In magnetic transition metals I wonder how well the local density approximation gives the sd, pd and $e_g t_{2g}$ exchange : in Hartree-Fock theory I believe the off-diagonal interactions are noticeably smaller than the diagonal ones, and it may not be good to treat both on the same footing as the local density approximation does. I have not investigated the matter.

My second area of assessment concerns the treatment of relevant materials and atomic configurations such as surfaces, defects and amorphous substances. Here progress has been quite good as I am sure will be shown by many examples given by the other lecturers. The smile is not as broad as it might be, but there is a definite up-turn at the edges indicating rapid progress now, partly in England due to the Cray computer. For example in Cambridge in our tight-binding calculations for transition metals with arbitrary magnetic configurations, we use clusters of about 1000 atoms to demonstrate convergence, with 10 d-electron spinorbitals per atom and 8 more sp-spinorbitals when required.

Fig. 2

The third area concerns calculation of relevant physical properties beyong the one-electron-like energy levels and wave functions. Here the smile is not very wide (Fig. 3) because we still have a long way to go, although again there is a definite up-turn, for example in calculation of photo-emission and other response functions. I want to refer to just two aspects. Most physical properties of solids relate not just to the ground state but involve excitations due to prodding the material with photons, electric or magnetic fields, temperature, etc. These Landau quasi-particle excitations are rigorously definable in many-body theory, with energy not precisely equal to the one-electron-like energy bands usually calculated. Although the relationship between them was already discussed by Kohn and Sham in their original paper, for many years electronic structure calculators didn't seem to want to know. The differences are relatively small, except for Ni, but are now being given the fundamental attention they deserve.

Fig. 3

Of greater practical significance is the calculation of total energy for arbitrary atomic positions with an accuracy sufficient to give force constants and equilibrium atomic positions. This promises to open up whole new fields. The modern so-called "norm-conserving" pseudopotentials have been very valuable here. Of course the variational principle guarantees convergence of the energy faster than anything else, but I learnt recently from Bob Jones that some people have been missing out on this for the following reason which may be of interest to some of you. Suppose one has just completed an iteration of the Schrödinger one-particle-like equation, yielding energy levels $\varepsilon_{n,new}$ and wave functions ψ_{new} giving a new electron charge density ρ_{new}. Ignoring for simplicity all except the Hartree terms, we have the total energy

$$E = \Sigma\, \varepsilon_n - \frac{1}{2} \iint d\underline{r}\, d\underline{r}'\ \rho(\underline{r})\rho(\underline{r}')\ e^2/|\underline{r} - \underline{r}'| \qquad (1)$$

Does one substitute the new ε_n and ρ into this ? No : it is necessary to calculate correctly the variational quantity

$$<\psi_{new}|T + V_N + e^2/r_{ij}|\psi_{new}> \qquad (2)$$

to obtain a rapidly convergent sequence. The ψ_{new} of $\varepsilon_{n,new}$ derive from a Schrödinger equation whose Hartree potential was calculated from the previous iteration, i.e. from the previous electron density ρ_{old}. The energy (2) is equal to

$$\Sigma\, \varepsilon_{n,new} - \iint d\underline{r}\, d\underline{r}'\ \frac{\rho_{old}\rho_{new}}{|\underline{r} - \underline{r}'|} + \frac{1}{2} \iint d\underline{r}\, d\underline{r}'\ \frac{\rho_{new}\rho_{new}}{|\underline{r} - \underline{r}'|} \qquad (3)$$

which of course converges to (1) for the self-consistent limit $\rho_{new} = \rho_{old}$.

My fourth area of assessment concerns status : an unhappy state of affairs. This can be documented for example by the almost vanishing number of senior positions, or even junior ones, in university physics and theoretical physics departments for computational physicists. By and large more of the lead over the last 20 years has come from industrial and government research labs than from universities, on both sides of the Atlantic, though some countries are exceptional. I have often heard comments from experimental and theoretical physicists, including the high priests of the subject in universities, which make me think "He does rather despise computational physics". Perhaps I can talk about such things because these remarks are not usually directed at me and I do not take them

Fig. 4

personally since I am not primarily a computational physicist. However I am saddened by such attitudes because I am a great customer of computational physics in my research and I try to support its development when I can. What to do about it ? A bit of self-criticism is already sprinkled through these paragraphs. Much more important is positive consciousness raising, by which I mean clarifying the role of computational research in solid state physics. I invite you to join me in this excercise because sharpening up one's thinking can contribute discrimination and thus benefit to the quality of one's research. Finally we are not very good at image building. We don't need to oversell our subject as one sometimes feels certain areas of science may tend to do, but we could articulate our achievements, insights and role better.

Let me therefore return in my remaining remarks to talking about the role of computations. I have already defined the frontier areas, or some of the frontier areas of solid state physics in terms of understanding the difference between related materials, and understanding complex materials and complex phenomena. Part of the fun of solid state physics is the richness of phenomena and their subtlety. Even at the most elementary level, we have that one element is a ferromagnet, a neighbouring one an antiferromagnet, and another one superconducting. This does not mean that everything in the subject is a "dirt effect", i.e. the result of more or less random variation in the accidental small differences between cancelling large effects. After all, there are well defined trends, once you see them ! And I have been around long enough to have witnessed how phenomena once puzzling and inexplicable have become "obvious" to any well-educated graduate student of today. However, it remains true that throughout solid state physics we are concerned with approximations and relative magnitudes : what is the dominant factor in a given situation ? Clearly computation has here an important role and a unique contribution. If we return to our earlier question why niobium compounds tend to be high T_c superconductors, we see that experiment cannot ultimately answer that question. Experiment can amass enough data to test whether the assertion is really true, and it could bring evidence that another element might be better if its even stronger electron-phonon coupling didn't lead to lattice instabilities. But experiment cannot in principle establish the logical connection to atomic properties of the element, ultimately to its atomic number. Similarly theory is impotent beyond a certain point. After providing the general explanation for superconductivity and the formula for T_c, it can suggest arguments about trends etc., but a loss of credibility sets in when ad hoc approximation is piled on ad hoc assumption. One ultimately has to test these by some computations where one can follow if necessary the details of the electron-phonon coupling at any point on the Fermi surface, and one can arbitrarily vary individual aspects of the calculation to disentangle competing effects. Thus I have always regarded computational physics as a third pillar of wisdom, in its role logically separate from theory or experiment though closely related to both ; in a sense often the interface between them (Figure 5).

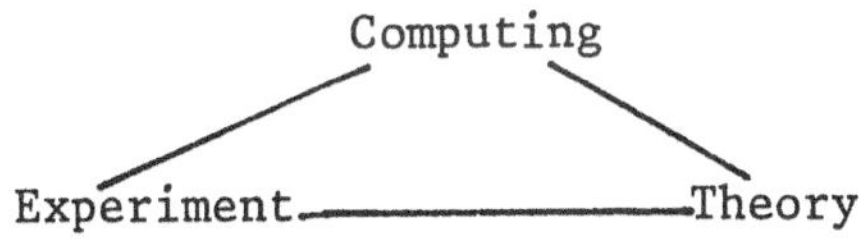

Fig. 5

I want to interpolate here some remarks about electronic structure determination, interpreted in its wide sense to include all the properties such as magnetism and interatomic forces etc. etc. that can be computed from electronic structure. In the Physics Abstracts Classification System (PACS) condensed matter physics as ordinarily understood comprises about 40% of the column-centimeters of the whole classification list. That indicates the scope of the field in relation to high energy, nuclear, astro-, bio-, geo-, atomospheric and space physics etc. Within condensed matter physics, practically all properties of solids and liquids result from the behaviour of the outer shell of valence or/and conduction electrons. Thus I can give examples of electronic structure computations, in the wide sense including derived properties, which have contributed to almost every subject in the field : magnetism, lattice dynamics, alloy structures and phase diagrams, semiconductors, lattice defects, mechanical properties, electrical and thermal conductivity, etc. etc., even polymers. Indeed electronic structure is the link, for example, from the electrical and optical properties of silicon to its lattice vibrations and crystal structure under pressure. We conclude that electronic structure calculators are not off on some ego-trip of their own, but stand strongly linked in the mainstream of experimental and theoretical solid state physics.

I have already in my survey mentioned some examples of computations properly focussed to yield understanding. I want therefore to continue analysing the relation to theory and experiment a bit further. At one extreme are situations of data processing. For example Low Energy Electron Diffraction measurements cannot be analysed directly to yield atomic positions at surfaces but require comparison with the computed diffraction intensities for various atomic configurations.

Secondly I see an enormous role for what I call computer experiments. A corner-stone of the scientific method is to create in an experiment a controlled situation to focus on some features and exclude others. We study the onset of corrosion, at the present stage of knowledge, not by tossing a hunk of steel out into the weather but by taking a single crystal with a carefully prepared surface in ultrahigh vacuum in a system baked to high temperature to remove just abour every atom of possible contaminant. It is only a small step further to build the crystal and attacking oxygen molecule in the computer, programmed to exactly the same laws of physics as apply in the laboratory. The computer has the advantage that one has much better probes to observe the course of the reaction. This is not fanciful. Much of our modern understanding of liquids has been derived this way. A recent computation 'measures' the entropy of a Lennard-Jones glass by taking the crystal from low temperature up through the melting point and down into the glassy state. This experiment can only be done in the laboratory on much more complicated materials.

The situation is analogous in the field of electronic structure computations. The experimental study of the more common invar alloys has gone about as far as it can but we still lack microscopic probes of what is happening at the atomic level. This gap is beginning to be filled by computations. Already the behaviour under uniform volume changes have been calculated and one can put in Boltzmann excitation from the high spin to the low spin state. With some more computing power will come the study of magnetically fluctuating regions and coupling to lattice vibrations. Indeed one can see an exciting future on our horizon when soon a host of traditional phenomena can be cleaned up by computational study at the microscopic level.

In this role computation is closely akin to experiment in showing what happens, generating data that are then grist for the mill of further theoretical analysis. For example the IBM group have computed the electronic structure of many binary alloys, real and thermodynamically unstable ones. The computer generates just the kind of data the crystallographer delights in measuring, heat of formation and electron density map, plus band structure and everything decomposed according to s,p,d angular momenta. Current theoretical research is taking this mass of 'experimental' data, finding the significant trends and providing a theoretical explanation, something which had to some degree eluded us previously.

This example also illustrates my third point, the role of computation in making, testing and extending simplified models. For instance magnetism in metals was once the province of many-body mystique, but it seems increasingly that all can be understood within the ordinary framework of electronic structure : even simpler, apart from quantitative corrections, within the model of a tight-binding d-band and an intra-atomic Stoner-like exchange interaction I which itself agrees with local density functional calculations. Computations and models in solid state physics extend on various levels and hierarchies. For example the Stoner I has been computed self-consistently for Fe in a ferromagnetic and an antiferromagnetic arrangement, and in the ferromagnetic state under pressure. In all cases there was little change. This valides the simple model of a constant intra-atomic parameter I, which can then be used in calculations with arbitrary magnetic configurations over a cluster of 1000 atoms, calculations that could not possibly be carried out at the original level of complete self-consistent ab initio band structure computations. In this way models and computations climb on each others backs to reach up to the higher levels of representing complex real situations.

Fourthly I want to return to the matter of exploring trends already mentioned earlier. This need not be restricted to simple chemical substitutions or moderate compressions. Just as in theoretical physics, one can gain insight by going to the very dilute limit, e.g. for a system of hydrogen atoms, or to extremely small separations. Similarly the element hydrogen plays a unique role in studying atomic vibrations experimentally because it has isotopes with such a large mass ratio : but in the computer any atom can have its mass varied at will as a way of probing the whole vibrational system. Such changes are not intended to correspond to real experiments : they represent the kind of controlled changes which experimentalists would make if they could in order to establish the atomic behaviour of the system. I am currently involved in a collaboration to try to understand the crystal structures of the sp bonded elements up to valence 4. When one narrows the 92 elements down to this group there are not very many, and even if one models the pseudopotential by a single Ashcroft radius R_c one has also the valence Z and atomic radius R_a, making almost every element unique with not enough data to extract clear trends. (Though we are not in as bad a pickle as the geophysicists and atmospheric physicists who have only one earth to play with : hence their interest in the planets). However by treating Z as a continuous variable we found in the computer that the interatomic interaction Φ depended strongly on the electron density parameter $R_s = R_a/Z^{1/3}$ as expected but very little on the orthogonal variable $R_a Z^{1/3}$. This eliminates one variable. Plotting as a function of $x = 2k_F R/\pi$ instead of distance R nails down the Friedel oscillations at integer values of x which eliminates that aspect of R_s. Taking out the dominant amplitude factor Z^2/R_a finally allows one to see the main effects of the pseudopotential and the screening at around the nearest neighbour distance, while numerical double differentiation in the computer enables one to follow the oscillations of Φ when they disappear into the strongly repulsive region at smaller R. This illustrates the concept of trends in a somewhat wider sense than we set out with.

I want to end with some remarks on courage, cunning and serendipity. Firstly about courage. I think computational physics has for too long been bringing up the rear, happy to obtain agreement with experiment and leaving it at that. That is not the way great discoveries are made. Partly it has been dictated by the limitations of computer hardware, but I detect also to some degree a frame of mind. In retrospect the calculations with a square-well model pseudopotential that I was involved with, could have been done on a desk calculator in the 1930's, involving much less computations than Hartree's self-consistent ones and using ideas all present in Hellmann's papers and the book by Mott and Jones. The fact that Pb was found experimentally to have a nearly-free-electron Fermi surface need not have come as a surprise in the 1950's. I have been involved in a similar situation this last week : new experimental data have kicked me into focussing my mind better and starting some calculations that we ought to have had the cunning to do two years ago, one jump ahead of the experimentalists ! I have painted a present, and even more a future, in which computation of electronic structure is a versatile tool for research at various levels in a whole range of solid state physics. There is room for different modes of research, some more akin to experiment and others more an extension of theory, some relating to a depth of understanding on canonical simple systems and others modelling complexicities. Who, for example, will become the Berndt Mathias of electronic structure calculations ?! Mathias made important discoveries by making thousands of compounds in his chemical preparation room, guided by experience and intuition, but also a strong believer in inspired serendipity and just mixing a bit of this and that which no-one seemed to have tried before.

I want to thank all those on whose work I have drawn for examples. I have omitted all references and almost all names because this is not a review article to give balanced credit where credit is due. I have concentrated on presenting some points of view, and a clutter of references would have detracted from that. I hope I will be forgiven.

LINEAR METHODS IN BAND THEORY

O.K. Andersen

Max-Planck-Institut für Festkörperforschung
D-7000 Stuttgart-80
Federal Rep. of Germany

I. INTRODUCTION

We would like to improve our understanding of the behaviour of real solids. Why do atoms combine into molecules and solids in the way they do and what is the response of these aggregates of atoms to external fields, stress, temperature, etc. ? Answers to such questions would, for instance, be a useful guide in our, so far mostly empirical, search for new and better materials : Stronger materials, materials which corrode less, better catalysts, better magnets, better semi- or superconductors, and, materials with properties unknown today. I expect that future advances in this field will be aided by insight obtained from calculations of the electronic structure.

The properties of matter under normal conditions are, at the microscopic level, governed by the behaviour of the electrons because these light particles in their motion almost immediately follow the far heavier nuclei. The basic problem is therefore to calculate the stationary states for the system of electrons moving in the electrostatic field of the fixed nuclei ; that is, the electronic structure. At a later stage, the energy of the electronic ground state obtained as a function of the nuclear positions can serve as potential energy for the motion of the nuclei.

The electronic-structure problem for a solid is one of infinitely-many interacting fermions, but for realistic materials we can, as a first step, at most attempt to solve Schrödinger's equation for one electron moving in the mean field of the other electrons plus the field of the nuclei. This mean field is composed of the electrostatic field from the electronic charge cloud plus corrections for exchange and correlation and it is usually determined by the following self-consistency procedure. For a supposed total field, V, one-electron energies, E_j, and wave-functions ψ_j, are obtained by solving the one-electron Schrödinger equation[1]:

$$(-\nabla^2 + V - E_j)\psi_j = 0 \tag{1.1}$$

The one-electron wavefunctions are then occupied according to the Pauli principle, and a new field is obtained by solving Poisson's equation for the nuclear point-charges screened by the electronic charge cloud of density

$$n(\underline{r}) = \sum_j^{occ} |\psi_j(\underline{r})|^2 \tag{1.2}$$

and by addition of the corrections for exchange and correlation. With this new field, or rather a weighted sum of the old and the new field, the calculation is now repeated and the cycle iterated until the input and output fields are consistent. From the one-electron energies and wavefunctions the total electronic energy may be estimated while the one-electron energies themselves are approximations to the one-electron ionization energies.

The classical <u>self-consistent field method</u> is the Hartree-Fock which, by definition, neglects correlation. For a solid, the method would not only be laborious because the exchange field depends in a complicated way on the one-electron state in question, but it would also yield too large a separation between occupied and unoccupied one-electron levels and hence make a subsequent treatment of correlation necessary.

The mean field used in calculations for solids is a much simpler, state-independent potential, $V(\underline{r})$, that can be constructed from the electron <u>density</u> alone, and which includes important effects of both exchange and correlation through the use of accurate many-body calculations for a simple model system. The model system is taken to be the so-called jellium, in which the nuclei are smeared out to a uniform background. Taking the contribution from exchange and correlation to the total energy to be $\varepsilon_{xc}(n)$ per electron, in a jellium of density n, the contribution in the real system can be expressed as

$$\int \varepsilon_{xc}(n(\underline{r}))n(\underline{r})d^3r \tag{1.3}$$

This so-called <u>local</u> approximation to <u>density functional theory</u> works surprisingly well and examples are given elsewhere in this book. Even for isolated atoms and small molecules, the method yields energies of the ground state and low-lying excited states which are more accurate than those obtained with the Hartree-Fock method. So much we have learned during the past ten years, partly with the aid of the one-electron techniques that I shall describe below. Nevertheless, before self-consistent calculations of total energies became feasible, one-electron calculations with the so-called Slater exchange potential had been highly successful in describing the band structures of the elemental semiconductors and the Fermi surfaces of the elemental metals.

The motivation therefore existed to devise methods for computing one-electron energies and wavefunctions with the speed and accuracy needed in self-consistent calculations. Such a method should apply to atomic constituents from all parts of the periodic table and potentially be applicable to realistic structures : crystals with many atoms per cell, crystals with surfaces, interfaces and impurities, or even clusters of impurities, amorphous systems, molecules and so on. In addition to being numerically accurate, fast and reliable, such a one-electron method should be physically transparent and thereby have the flexibility to be used at various levels of approximation. It could then form the basis for analytical modes which would make the computed results intelligible and hence open the way for further progress.

None of the classical one-electron methods satisfies the abovementioned requirements. The augmented-plane-wave (APW) and Korringa-Kohn-Rostocker (KKR) methods based on scattering theory, have proven highly accurate for Fermi surface calculations but they are numerically cumbersome and lack transparency. The linear combinations of Atomic Orbitals (LCAO) method is cumbersome when used as a first-principles method and, when parametrized, it has either too many parameters or the wavefunctions are ill-defined. The LCAO schemes using Gaussian orbitals have some computational advantages but they need at least twice as many basis functions as, for instance, the KKR method and they are not formulated in terms of physically significant parameters. The modern first-principles pseudopotential method meets the requirements but it is limited to treating sp-like valence and conduction electrons. Computationally this can be remedied by the addition of localized orbitals to the plane-wave basis set but such a hybrid scheme is neither elegant nor in accord with chemical and physical intuition based up on the smooth trends observed for the band structures through the Periodic Table : s-, p-, d- and f-electrons ought to be treated on the same footing.

The socalled <u>linear</u> methods of band theory satisfy the requirements rather well. This is particularly true for the Linear-Muffin-Tin Orbitals (LMTO) method in the atomic spheres approximation (ASA). The linear APW (LAPW) method is less transparent than the LMTO method because it is based on a plane-wave representation and therefore needs more basis functions. On the other hand, the LAPW method is the most accurate band structure method in use. Normally, the linear methods are used to solve Schrödinger's equation for the full one-electron potential but they do, of course apply to pseudopotentials as well.

In this chapter I shall explain the idea of the linear methods and present the formalism of the LAPW and, in particular, of the LMTO-ASA method. The presentation will be based upon recent, mostly unpublished developments. In addition, the concepts of canonical bands and potential parameters will be discussed.

II. LINEAR METHODS

Before I get into the detailed formulation of specific linear methods I will try to explain their basic idea and compare them with the traditional one-electron methods.

We want to solve Schrödinger's equation (in atomic Rydberg units)

$$[-\nabla^2 + V(\underline{r})]\psi_j(\underline{r}) = E_j\psi_j(\underline{r}) \tag{2.1}$$

for the bound states of a potential $V(\underline{r})$ characteristic for a system of atoms. A part of such a potential is shown in Fig. 1a where, specifically, we show the behaviour near the surface of a crystal, in a plane perpendicular to the surface. For an infinite crystal the potential is simpler than this because it has three-dimensional periodicity. The one-electron states relevant for most physical and chemical properties are those with energies in the neighborhood of 1 Ry around the Fermi level, i.e. from about - 2 to 0 Ry in Fig. 1, and the contribution to the electron density from the lower-lying states is approximately equal to the corresponding contribution in the isolated atom (frozen core approximation). The relevant energy range begins where the electron has sufficient energy to move from one atom to the other and, hence, when its energy reaches the level of the potential between the atoms.

The traditional one-electron methods may be classified according to whether they seek the wavefunctions in (2.1) as an expansion in some set of fixed basisfunctions, like atomic orbitals, Gaussians or plane waves, or they expand the wavefunctions in the set of energy - and potential - dependent partial waves, as done in the KKR and APW methods.

A. Methods using fixed basisfunctions

In order to be specific we consider the LCAO method. Here one makes the Ansatz

$$\sum_{RLn} \chi_{Rn\ell m}(\underline{r}_R)a_{Rn\ell m,j} \approx \psi_j(\underline{r}) \tag{2.2}$$

where

$$\chi_{Rn\ell m}(\underline{r}_R) \equiv \chi_{Rn\ell}(r_R)Y_{\ell m}(\hat{r}_R) \tag{2.3}$$

is an atomic orbital with radial quantum number n and angular momentum quantum numbers ℓ and m. (We shall use the notation $\ell m \equiv L$). The atomic orbital is centered at the atomic site $\underline{R}$ with the local coordinates

$$\underline{r}_R = \underline{r} - \underline{R} \tag{2.4}$$

The unknown coefficients, a in (2.2), and the one-electron energies E, are now obtained by use of the Rayleigh-Ritz variational principle for the Hamiltonian (2.1). As a result, they are respectively the eigenvectors and eigenvalues of the algebraic eigenvalue problem

$$(H - E\mathcal{O})a = 0 \tag{2.5}$$

with the Hamiltonian matrix

$$H_{R'n'L',RnL} \equiv \langle\chi_{R'n'L'}|-\nabla^2 + V|\chi_{RnL}\rangle^{\infty} \tag{2.6}$$

and the overlap matrix

$$\mathcal{O}_{R'n'L',RnL} \equiv \langle\chi_{R'n'L'}|\chi_{RnL}\rangle^{\infty} \tag{2.7}$$

Here, and in the following, $\langle|\rangle^{\infty}$ means an integral over all space. The important advantage of using a set of fixed basisfunctions is that the one-electron problem reduces to an algebraic eigenvalue problem. The essential disadvantage is that the set must be large in order to be reasonably complete. The LCAO method has the specific advantage of using a local description based on the atomic constituents. A drawback lies in the necessity to compute a large number of integrals involving orbitals and atomic potentials centered at two and three different sites. As an empirical approach it is customary to neglect all but the two-centre integrals between the nearest and next-nearest neighbors and to treat these integrals as adjustable parameters.

B. Methods using partial waves

An important simplifying feature of the problem (2.1) with a potential like the one shown in Fig. 1a is that in those regions of space where the kinetic energy, $E - V(\underline{r})$, is numerically large, and the wavefunction therefore varies rapidly, the potential essentially varies in only one dimension : Near the nuclei it is spherically symmetric and near the surface it varies mostly in the perpendicular direction. In the partial-wave approach space is therefore divided into regions inside which only the one-dimensional part of the potential is retained such that Schrödinger's equation becomes separable. As an example we show in Fig. 1b the socalled muffin-tin (MT) average of the potential in Fig. 1a : The MT-potential is spherically symmetric inside touching spheres surrounding the atoms, it only depends on the perpendicular direction outside the half-plane touching the outermost layer of spheres, and it is flat in the remaining; socalled interstitial region. For open structures like the diamond structure it is customary to describe not only the potential around the atomic sites but also the (repulsive) potential around the interstitial sites as spherically symmetric. Let us now, for simplicity, just consider a bulk crystal where space is divided into spherical (R) regions and the interstitial (i) region.

For the approximate MT potential

$$V(\underline{r}) \approx \sum_R \theta_R(r_R) v_R(r_R) + \theta_i(\underline{r}) V^i \tag{2.8}$$

it is now possible to solve Schrödingers equation exactly in terms of energy-dependent partial-wave expansions

$$\sum_{RL} \theta_R(r_R) \phi_{RL}(E,\underline{r}_R) b_{RL} = \psi(E,\underline{r}) \tag{2.9}$$

Here θ is one inside the region and zero outside. Moreover, L labels the angular momentum as in (2.2-3) and the partial wave

$$\phi_{RL}(E,\underline{r}_R) \equiv \phi_{R\ell}(E,r_R) Y_L(\hat{r}_R) \tag{2.10}$$

has a radial part which is obtained by integrating (numerically) the radial Schrödinger equation

$$[r\phi_{R\ell}(E,r)]'' = [v_R(r) + \ell(\ell+1)r^{-2} - E]\, r\phi_{R\ell}(E,r) \tag{2.11}$$

for the potential v and energy E outwards from the centre to the sphere radius, s.

The one-electron energies, E_j, are now those values of E for which b-coefficients can be found such that the partial-wave expansions join continuously and differentiably together at the boundaries between the regions. The way in which this matching condition is formulated algebraically, differs for the various partial-wave methods, but in general, the result is a set of linear, homogeneous equations

$$M(E)b = 0 \tag{2.12}$$

with a secular matrix, M, which in contrast to the matrix $H - EO$ in (2.5) has a complicated, non-linear energy dependence. As a consequence, the one-electron energies, E_j, and the corresponding solutions $b_{RL,j}$ must be found individually by first tracing a root

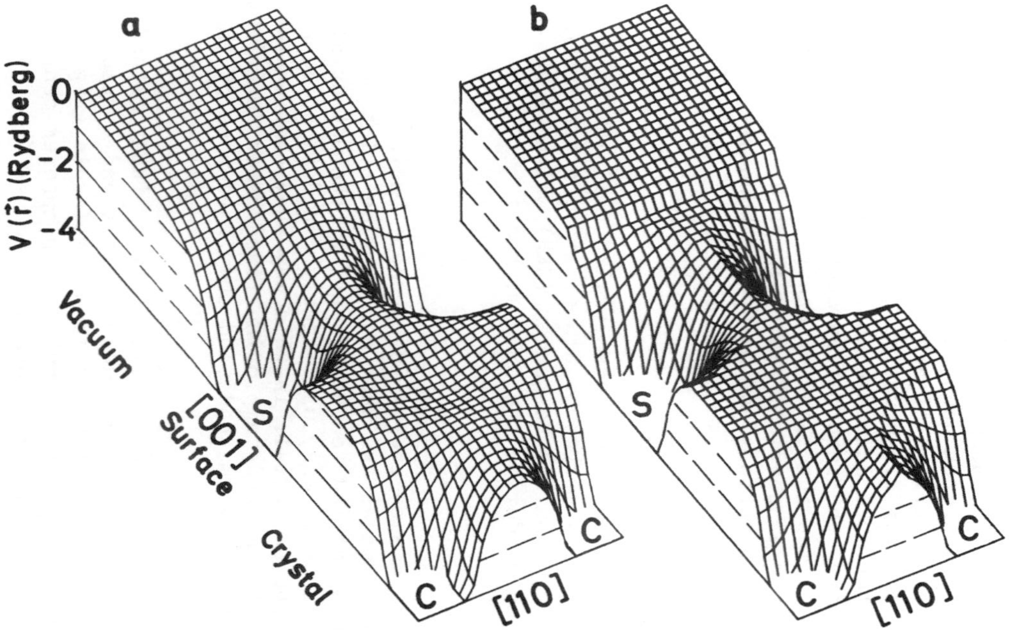

Fig. 1.a : One-electron potential in a plane perpendicular to the surface of a nickel crystal. S and C denote atoms in, respectively, the surface plane and the third crystal plane below the surface. The potential is self-consistent and was obtained with the linear augmented-plane-wave method.

b : The muffin-tin average of the potential shown in Fig.1a : Inside the touching spheres surrounding the atoms, the muffin-tin potential is spherically symmetric, in the vacuum region it only depends on the distance from the surface plane and in the remaining interstitial region it is constant. (Jepsen, Madsen and Andersen 1982).

of the determinant of M, and then solving the linear equations.

Even for moderately sized matrices this requires orders of magnitude more computation than does the solution of the eigenvalue problem (2.5). Further disadvantages of the partial-wave methods are that their formalisms are complicated and that perturbations, such as the remaining, non-muffin-tin part of the potential, are difficult to include in (2.12) because this condition does not derive from the variational principle for the one-electron Hamiltonian. The partial-wave approach does, however, have two distinct advantages : it provides solutions of arbitrary accuracy for the muffin-tin potential, and for close-packed crystals this makes the partial-wave methodes for more accurate than any traditional fixed basis method. Furthermore, the information about the potential enters (2.12) only via a few functions of energy, the logarithmic derivative functions. For a spherical region these are essentially the slopes of the radial functions at the sphere boundary, or precisely

$$D_{R\ell}(E) \equiv s_R \phi'_{R\ell}(E,s_R)/\phi_{R\ell}(E,s_R) \tag{2.13}$$

Among the partial-wave methods are the so-called cellular method of Wigner and Seitz, the APW method of Slater and the scattered-wave method of Korringa, Kohn and Rostocker (KKR) which we shall consider in detail when we derive the LMTO method.

C. Linear Methods

The linear methods employ energy-independent basis functions derived from the partial waves and they can provide solutions of arbitrary accuracy when applied to the MT-part of the potential. The linear methods, moreover, only use logarithmic derivative parameters to describe the potential. In this way they combine the desirable features of the methods employing fixed basis functions with those of the partial-wave methods and they have none of the drawbacks.

In order to obtain the basis functions for a linear method we now compare the LCAO and partial-wave descriptions and consider the simplest model of a homo-nuclear, diatomic molecule (Fig. 2). The potential is described by the two muffin-tin spheres centered at the nuclei at O and at R. We only consider s-waves ($\ell = 0$) (or assume that the problem is one -rather than three-dimensional). Due to the centre of inversion between the atoms the lowest, bonding state with energy B is even and the highest, anti-bonding state with energy A is odd.

In the LCAO description the two states are

$$\psi^B_A \approx \chi(r) \pm \chi(r - R) \tag{2.14}$$

and their energies may be found from (2.5-7).

In the partial-wave description, which is the exact one for the model considered, we must solve the radial Schrödinger equation (2.11) for the potential in the MT-sphere for a number of energies E. As the energy increases the radial function contracts more and more and pulls one node after the other into the sphere. The logarithmic derivative function (2.13) diverges when a node enters and D(E) is, in general, a never increasing, cot-like function of energy. We shall only be interested in a narrow energy range corresponding to one value of the radial quantum number. The energy for which the radial wave-function has zero slope at the sphere is then the energy of the bonding state, i.e.

$$\psi_B = \phi(B,r) \quad , \quad \text{for } r \leqslant s \tag{2.15}$$

if

$$D(B) = 0 \tag{2.16}$$

Similarly, the energy for which there is node at the sphere is the energy of the antibonding state, i.e.

$$\psi_A = \phi(A,r) \tag{2.17}$$

if

$$D(A) = \infty \tag{2.18}$$

By comparison of the description (2.14) in terms of the energy-independent, overlapping atomic orbitals with the exact description (2.15 and 17) in terms of the energy-dependent, non-overlapping partial waves we see that the LCAOs have the proper symmetry and, hence, the proper slope and amplitude (i.e. logarithmic derivative) at the sphere. However, inside the sphere at the origin the tail of the atomic orbital centered at R should equal $[\phi(B,r) - \phi(A,r)]/2$ but there is of course no reason why the tail generated by an isolated atom-potential should do so ; the tail merely matches onto $[\phi(B,r) - \phi(A,r)]/2$ at the boundary of the sphere. In a <u>linear</u> LCAO method, however, the basisfunctions are <u>chosen</u> to be <u>atomic</u> orbitals whose tails are <u>augmented</u> by essentially $[\phi(B,r) - \phi(A,r)]/2$ inside the other spheres. In this way they can form the proper solutions for the MT-part of the potential inside those spheres. Similarly, inside its own sphere the atomic orbital should equal $[\phi(B,r) + \phi(A,r)]/2$, and essentially this is the case in the linear LCAO method.

Now, for an elemental solid there usually exists a continuous band of states between the energies B and A defined in (2.16 and 18). In this case, the entire energy dependence of $\phi(E,r)$ should be

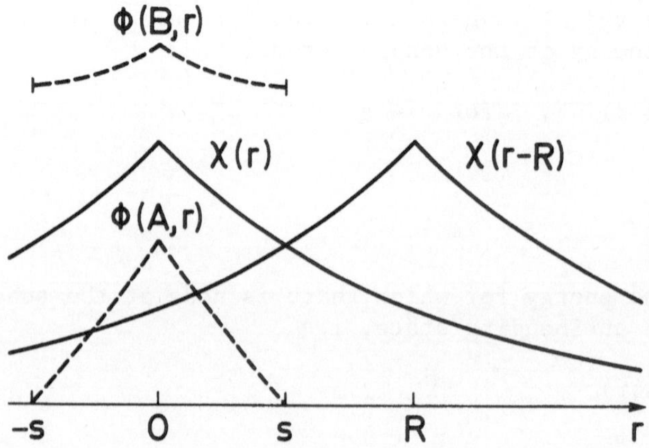

Fig. 2 : Atomic orbital, χ, and paritial waves, ϕ, for a diatomic molecule (schematic). The atoms are centered at 0 and R, and the radius of the muffin-tin sphere at 0 is s. Inside the atom at the origin the bonding state is $\chi(r) - \chi(x-r) \approx \phi(A,r)$.

supplied by the tails of the orbitals from the other sites and this can only be accomplished in a practical way if instead of using $\phi(B,r) - \phi(A,r)$ the tails are augmented by the energy-derivative function

$$\dot{\phi}(r) \equiv [\partial\phi(E,r)/\partial E]_{E=E_\nu} \tag{2.19}$$

evaluated at some energy, E_ν, at the centre of interest. This augmentation usually works very well because the Taylor series

$$\phi(E,r) = \phi(r) + (E - E_\nu)\dot{\phi}(r) + o(E - E_\nu) \tag{2.20}$$

with

$$\phi(r) \equiv \phi(E_\nu,r) \quad \text{and} \quad \dot{\phi}(r) \equiv \partial\phi(E,r)/\partial E|_{E_\nu} \tag{2.21}$$

is well converged in the energy range of interest after the two first, linear terms. For our MT-model of the diatomic molecule this linear method which employs augmentation by phi and phi-dot functions will provide trial functions (2.14) with errors of order $(B - E_\nu)^2$ and $(A - E_\nu)^2$, respectively. The energies obtained using the variational principle with (2.5-7) will consequently have errors of order $(B - E_\nu)^4$ and $(A - E_\nu)^4$, respectively.

Sofar the normalizations of the partial waves (2.15) and (2.17) have, according to (2.14), been such that they match onto $\chi(r) \pm \chi(r - R)$ at the sphere. As a consequence, the augmentation of χ is by $[\phi(B,r) + \phi(A,r)]/2$ inside its own sphere and by $[\phi(B,r) - \phi(A,r)]/2$ inside the other sphere. Had we chosen a different for instance a structure- and χ-independent, normalization of the two partial waves the rule for augmentation would simply have been that χ should be augmented by that linear combination of $\phi(B,r)$ and $\phi(A,r)$ which matches continuously and differentiably onto χ at the sphere in question. Similarly, the argumentation of the tail by the energy-derivative fuction $\dot{\phi}(r) \propto \phi(B,r) - \phi(A,r)$ only holds for the appropriate energy-dependent normalization of $\phi(E,r)$. With an unspecified normalization the rule for augmentation is that χ must be augmented by that linear combination of $\phi(r)$ and $\dot{\phi}(r)$ which matches continuously and differentiably onto χ at the sphere in question. This is so because the energy-derivative function corresponding to an arbitrary normalization, specified by I(E), is

$$\partial I(E)\phi(E,r)/\partial E|_{E_\nu} = \dot{I}(E_\nu)\phi(r) + I(E_\nu)\dot{\phi}(r) \tag{2.22}$$

i.e., an arbitrary linear combination of phi and phi-dot.

From now on we shall assume that the partial waves (2.10-11) are normalized to unity in their sphere, i.e. that

$$\int_0^{S_R} \phi^2_{R\ell}(E,r)r^2dr \equiv \langle\phi^2_{RL}(E)\rangle = 1 \tag{2.23}$$

With this choice phi and phi-dot are orthogonal in the sphere, i.e.

$$\int_0^{s_R} \phi_{R\ell}(E,r)\dot{\phi}_{R\ell}(E,r)r^2 dr \equiv \langle\phi_{RL}|\dot{\phi}_{RL}\rangle = 0 \qquad (2.24)$$

as seen by differentiation of (2.23) with respect to energy. Here and in the following a bracket indicates an integral in the sphere. (For simplicity of notation we shall not explicitly treat other kinds of regions such as planar surface regions, but the formalism is quite analogous).

In Fig. 3 we show the radial phi, phi-dot, and phi-double-dot functions for $\ell = 0,1$ and 2 in Yttrium metal. The values of E_ν were chosen such that $D_\ell(E_\nu) = -\ell - 1$ in all cases. The good convergence of the Taylor series (2.20) is evident from the relative magnitudes of the energy-derivative functions.

We are now ready to set up a linear method for solving Schrödinger's equation for a real structure. We first choose a set of basis functions which, after the continuous and differentiable augmentation by phi and phi-dot inside the MT-spheres, will equal the original functions, χ^i, in the interstitial region only. The atomic orbitals used in our discussion of the diatomic molecule are not very practical for this purpose because they are difficult to expand about other sites and because their kinetic energy (curvature) in the interstitial region is too low for solid-state potentials. It should, however, be obvious that any smooth set of functions , χ^i_G, which is reasonably complete in the interstitial region can in principle be used. We shall be using plane waves or muffin-tin orbitals (MTOs) but, for the moment, this need not be specified and we may quite generally express the linear basis function as

$$\sum_{RL} [\phi_{RL}(\underline{r}_R)\pi_{RL,G} + \dot{\phi}_{RL}(\underline{r}_R)\Omega_{RL,G}] + \chi^i_G(\underline{r}) = \chi_G(\underline{r}) \qquad (2.25a)$$

For convenience of notation we have here dropped the step functions used in (2.8-9). We thus assume that phi and phi-dot vanish outside their respective spheres and that χ^i equals χ in the interstitial region but vanishes in the spheres. The matrices π and Ω are such that the function (2.25) is everywhere continuous and differentiable, and they contain the coefficients in the expansion of χ^i_G in spherical harmonics, $Y_L(\hat{r}_R)$, about site R. Moreover, they contain the potential parameters which are the values of phi and phi-dot and their logarithmic derivatives at the sphere boundary. This will be explicit when we come to consider the LAPW and LMTO methods.

In the following it will often be convenient to regard phi, phi-dot, χ^i and χ as row-vectors with indices R and G, respectively, and therefore to use the matrix notation

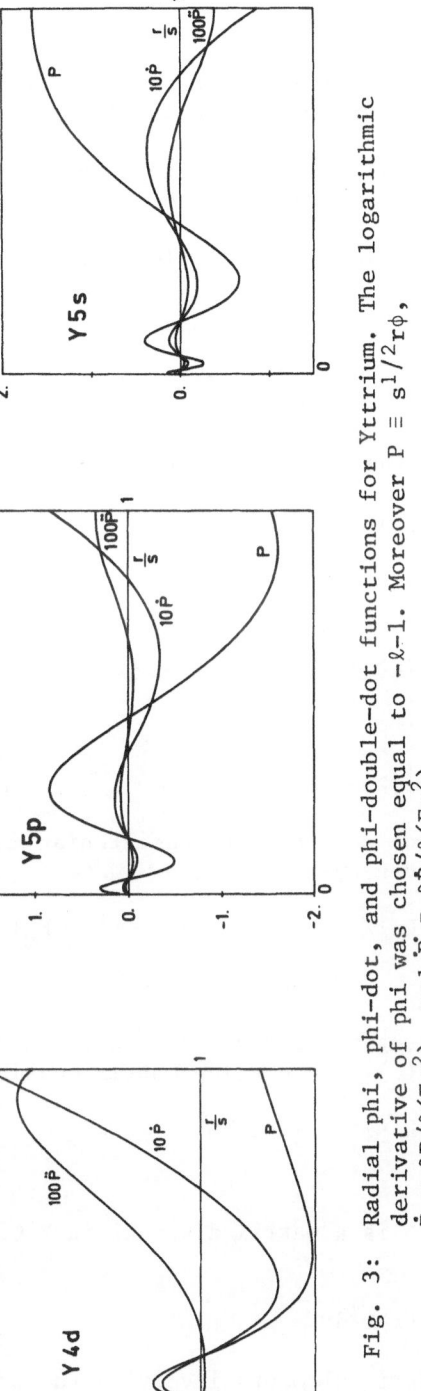

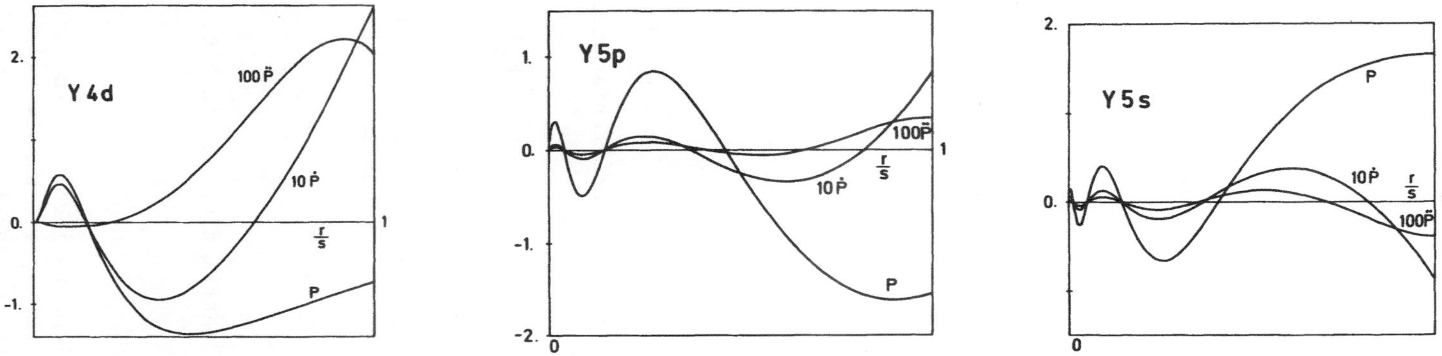

Fig. 3: Radial phi, phi-dot, and phi-double-dot functions for Yttrium. The logarithmic derivative of phi was chosen equal to $-\ell-1$. Moreover $P \equiv s^{1/2} r\phi$, $\dot{P} \equiv \partial P/\partial(Es^2)$, and $\ddot{P} \equiv \partial \dot{P}/\partial(Es^2)$.

$$|\phi>\pi + |\dot{\phi}>\Omega + |\chi>^{i} = |\chi>^{\infty} \tag{2.25b}$$

Here $|>$, $|>^{i}$ and $|>^{\infty}$ denote functions defined in respectively a sphere, the interstitial region, and in all space.

It is now a simple matter to write down the Hamiltonian and overlap matrices for use in (2.5). From (2.23) and 24) we have

$$<\phi_{RL}|\phi_{R'L'}> = \delta_{RR'}\delta_{LL'} \quad \text{or} \quad <\phi|\phi> = 1 \tag{2.26}$$

and

$$<\phi_{RL}|\dot{\phi}_{R'L'}> = 0 \quad \text{or} \quad <\phi|\dot{\phi}> = 0 \tag{2.27}$$

Furthermore, since the partial wave is a solution of Schrödinger's equation for the MT-part of the potential in the sphere at R, i.e.

$$[-\nabla^2 + v_R(r)]\phi_{RL}(E,\underline{r}) = E\phi_{RL}(E,\underline{r}) \tag{2.28}$$

we obtain by differentiation with respect to energy and by setting $E = E_\nu$:

$$(-\nabla^2 + v - E_\nu)|\phi> = 0 \tag{2.29}$$

and

$$(-\nabla^2 + v - E_\nu)|\dot{\phi}> = |\phi> \tag{2.30}$$

Using now (2.25-30) we see that the Hamiltonian matrix with the elements $H_{GG'}$, is given by

$$<\chi| -\nabla^2 + V - E_\nu|\chi>^{\infty} = \pi^{+}\Omega + <\chi| -\nabla^2 - E_\nu|\chi>^{i} + <\chi|V^{nmt}|\chi>^{\infty}$$
$$= H - E_\nu O \tag{2.31}$$

In the interstitial region we have defined the non-MT potential, V^{nmt}, as the total $V(\underline{v})$. The overlap matrix equals

$$<\chi|\chi>^{\infty} = \pi^{+}\pi + \Omega^{+}<\dot{\phi}^2>\Omega + <\chi|\chi>^{i}$$
$$= O \tag{2.32}$$

In this equation $<\dot{\phi}^2>$ is a matrix diagonal in R,ℓ, and m with the elements

$$<\dot{\phi}^2_{RL}> = \int_0^{s_R} \dot{\phi}^2_{R\ell}(r)r^2dr \tag{2.33}$$

In (2.31 - 32) the matrix elements involving the strong parts of the potential, i.e. the MT-parts, are simply expressed in terms of π, Ω and $<\dot{\phi}^2>$. These parts we shall study in detail later. The remaining integrals involve smooth functions, but the region of integration may be complicated. The ease with which we can perform these remaining integrals of $-\nabla^2 - E_\nu$ over the interstitial region and of the non-muffin-tin part of the potential

$$V^{nmt}(\underline{r}) \equiv V(\underline{r}) - \sum_R \theta_R(r_R)v_R(r_R) \tag{2.34}$$

over the entire space depends on the basis set chosen. For this purpose the set of plane waves is convenient and accurate and the potential shown in Fig. 1a is, in fact, the result of a self-consistent LAPW calculation. A drawback of the plane-wave set is that it needs 10 - 50 plane waves per atom. For the same accuracy this is still less than needed in current first principles pseudopotential calculations but, it is significantly more than the $(\ell_{max} + 1)^2 = 4 - 16$ orbitals needed per atom in an LMTO calculation. The MTOs, on the other hand, have difficulty in describing the kinetic energy in the interstitial region with high accuracy and the integrals in this region are cumbersome.

If the basis set is overcomplete in the interstitial region, as is for instance the set of plane waves, and if the potential has the muffin-tin form, the only source of error in a linear method is the first-order expansion (2.20). Consequently the wave functions obtained by solving the eigenvalue problem (2.5) have errors of second and higher order in $E_i - E_\nu$ and the eigenvalues, being variational estimates, have errors of fourth and higher order in $E_i - E_\nu$. This means that a linear method can be viewed as a fixed-basis method which needs only one (arbitrary and non-integer) value ν, of the radial quantum number (per atom and angular momentum).

In Fig. 4a we show the energy-band structure of palladium metal in the wide range from about 0.5 Rydberg below the Fermi level to 4 Rydberg above. In order to cover this large range, where the narrow 4d-bands and the wide 5s-, 5p-, 5d- and 4f-bands are situated, three LAPW calculations with three different values of E_ν were performed. The good overlap between the three densities of states obtained indicates the smallness of the fourth-order errors. In Fig. 4b we show the corresponding $\ell = 1$ projected densities of states and the quality of the overlap is a measure of the second-order errors in the wave functions.

D. The atomic spheres approximation (ASA) and the orthogonal theta orbitals

In closely packed solids each atoms has 8 - 12 nearest neighbors and the MT-approximation (Fig. 1b) is a good one. More accurate in general and far more convenient is an approximation in which the interstitial region is "annihilated" through expansion of the MT-spheres and neglect of the slight overlap. The MT-spheres in this approximation thus become the Wigner-Seitz atomic spheres

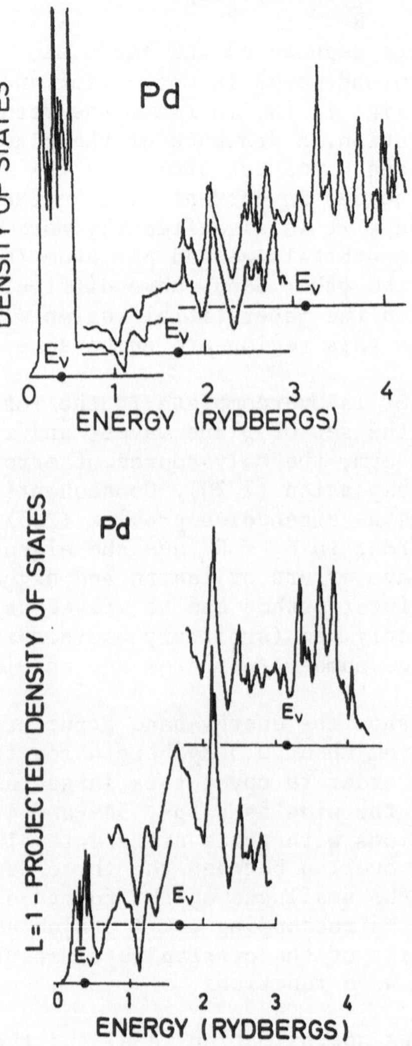

Fig. 4 : Calculated density of one-electron states in face-centered cubic palladium metal. The energy range of over 4 Rydberg, starting at the bottom of the 5s-band, was covered by three linear bandstructure calculations using the values of E_ν shown. The Fermi level is at 0.5 Rydberg. (Müller, Jepsen, Wilkins and Andersen 1978).

which are supposed to fill space. This atomic spheres approximation is more accurate than the MT-approximation, except for free electrons and for elemental metals, because it has no flat "muffin-tin floor" extending throughout the interstitial region of the entire solid. This is important, for instance, in the treatment of impurities.

In open structures the interstitial positions often have so high symmetry that not only the atomic but also the repulsive interstitial potential can be approximated by a spherically symmetric one and the atomic and interstitial spheres, taken together, form a close packing. In such cases, the atomic spheres approximation (ASA) is also a good starting point. We shall now use this approximation to derive a simple description of the electronic structure.

In the ASA the terms in (2.25, 31 and 32) referring to the interstitial region and to the non-muffin-tin part of the potential drop out.

We now combine the set of basisfunctions (2.25) into a new set. The new basisfunction, $\Theta_{RL}(\underline{r}_R)$, is an orbital having the property of being augmented by phi-dot functions only, except inside its own sphere and for its own angular momentum. The new set of basisfunctions thus comprises the linear combination of the old set specified by the application of the matrix π^{-1}, i.e.

$$|\phi\rangle + |\dot{\phi}\rangle\Omega\pi^{-1} = |\chi\rangle^{\infty}\pi^{-1} \equiv |\Theta\rangle^{\infty} \qquad (2.35a)$$

where

$$(\Omega\pi^{-1})_{RL,R'L'} = \sum_G \Omega_{RL,G}(\pi^{-1})_{G,R'L'} \qquad (2.36)$$

it may be shown that $\Omega\pi^{-1}$, in contrast to π and Ω, is Hermitian. If we now operate with $-\nabla^2 + V - E_\nu$ on (2.35a) using (2.26-30) we find that the Hamiltonian $H_{RL,R'L'}$ in the theta-representation is given by

$$h \equiv H - E_\nu O \equiv \langle\Theta|-\nabla^2 + V - E_\nu|\Theta\rangle^{\infty} = \Omega\pi^{-1} \qquad (2.37)$$

The theta-orbital (2.35a) may therefore be expressed as

$$\phi_{RL}(\underline{r}_R) + \sum_{R'L'} \dot{\phi}_{R'L'}(\underline{r}_{R'})\, h_{R'L',RL} = \Theta_{RL}(\underline{r}_R) \qquad (2.35b)$$

and the overlap matrix $O_{RL,R'L'}$ may be expressed as

$$O \equiv \langle\Theta|\Theta\rangle^{\infty} = 1 + h\langle\dot{\phi}^2\rangle h \qquad (2.38)$$

The second, non-diagonal term of the overlap matrix (2.38) is of second order in $h \equiv H - E_\nu O$. Moreover, a measure of the smallness of the parameter $<\dot{\phi}^2>$ is given by (Andersen 1975)

$$<\dot{\phi}^2> = (-1/3)[\ddot{\phi}(s)/\phi(s)] \tag{2.39}$$

which relates the integral in the sphere to the values of $\phi(r)$ and $\ddot{\phi}(r)$ at the sphere. Realistic values of $<\dot{\phi}^2>^{-1/2}$ are several Rydbergs. (see also Fig. 3).

We have thus constructed a set of orbitals which are orthogonal to first order in h. Moreover, within the ASA, these theta-orbitals are independent of the basis set, χ, chosen for their construction. In Sect. IV we shall use muffin-tin orbitals. With the theta-orbitals it is now a simple matter to solve Schrödinger's equation. We first neglect the second-order term of the overlap matrix (2.38) and therefore diagonalize $h \approx H - E_\nu 1$ (2.37) :

$$b^+ h b = \omega \tag{2.40}$$

obtaining the eigenvectors $b_{RL,j}$ and the eigenvalues $\omega_j \approx E_j - E_\nu$. The corresponding wavefunctions obtained from (2.35) are

$$\psi_j(\underline{r}) = \sum_{RL} \Theta_{RL}(\underline{r}_R)\, b_{RL,j}$$

$$= \sum_{RL} [\phi_{RL}(\underline{r}_R) + \omega_j \dot{\phi}_{RL}(\underline{r}_R)]\, b_{RL,j} \tag{2.41}$$

because b diagonalizes h. This is nothing but the partial-wave expansion (2.9) to first order in energy. We have thus seen explicitly how the cumbersome partial-wave problem (2.12) may be cast into the simplest possible eigenvalue problem (2.40) with a small matrix (dimension ≈ 9 per atom). Of course, only those of the solutions (2.41) which have eigenvalues $E_\nu + \omega_j$ in the neighborhood of E_ν are physically significant but, if solutions over a broader range are needed, we just have to cover that range with "energy-panels" each one having a different E_ν, and perform the calculation for each panel at a time. This was illustrated in Fig. 4.

The eigenvalues ω_j of (2.40) have errors of third and higher order because, so far, we neglected the second-order term in the overlap matrix (2.38). This term is most simply taken into account by first-order perturbation theory. From (2.5)

$$0 = b_j^+ [h - (E_j - E_\nu)\mathcal{O}] b_j$$

$$= \omega_j - (E_j - E_\nu)(1 + \omega_j b_j^+ <\dot{\phi}^2> b_j \omega_j) \tag{2.42}$$

The one-electron energy correct to third order, in $E_j - E_\nu$ is therefore given by

$$E_j = E_\nu + \omega_j [1 + \omega_j^2 <\dot{\phi}^2>_j]^{-1} \tag{2.43}$$

where

$$<\dot{\phi}^2>_j \equiv \sum_{RL} <\dot{\phi}^2_{RL}>|b_{RL,j}|^2 \tag{2.44}$$

It may sometimes be advantageous to use different E_ν's for different partial waves. In order to generalize our formalism accordingly we just remember that E_ν is no longer a number, and therefore diagonal in all representations, but a matrix which is diagonal in the RL-representation and has the elements $E_{\nu R\ell}$. In (2.37) $E_\nu O$ should therefore be substituted by $E_\nu + hE_\nu<\dot{\phi}^2>h$. The Hamiltonian that we diagonalize is

$$\tilde{H} \equiv h + E_\nu = \Omega\pi^{-1} + E_\nu \tag{2.45}$$

and, in terms of the eigenvectors $b_{RL,j}$ and the eigenvalues $\tilde{E}_j$,

$$b^+\tilde{H}b = \tilde{E} \tag{2.46}$$

The corresponding wavefunction is

$$\psi_j(\underline{r}) = \sum_{RL}[\phi_{RL}(\underline{r}_R) + (\tilde{E}_j - E_{\nu R\ell})\dot{\phi}_{RL}(\underline{r}_R)]b_{RL,j} \tag{2.47}$$

Finally, the formula including the third-order correction of the energies is found to be

$$E_j = \frac{\tilde{E}_j + \sum_{RL} E_{\nu R\ell}<\dot{\phi}^2_R>(\tilde{E}_j - E_{\nu R\ell})^2|b_{RL,j}|^2}{1 + \sum_{RL}<\dot{\phi}^2_{R\ell}>(\tilde{E}_j - E_{\nu R\ell})^2|b_{RL,j}|^2} \tag{2.48}$$

In order to perform selfconsistent calculations in the ASA we need the electron density spherically averaged in each sphere. As seen from (2.9-10), the density (per spin) is in general given by

$$n_R(r) = (4\pi)^{-1}\sum_\ell \int^{E_F}\phi^2_{R\ell}(E,r)N_{R\ell}(E)dE \tag{2.49}$$

in terms of the projected densities of states

$$N_{R\ell}(E) \equiv \sum_j \delta(E - E_j)\sum_m|b_{RL,j}|^2 \tag{2.50}$$

If we now expand the partial wave in a Taylor series as in (2.20) and define the projected number of states

$$n_{R\ell} \equiv \int^{E_F} N_{R\ell}(E)dE = \sum_j^{\text{occ.}} |b_{RL,j}|^2 \tag{2.51}$$

and the energy moments

$$\varepsilon^q_{R\ell} \equiv (n_{R\ell})^{-1}\int^{E_F}(E - E_{\nu R\ell})^q N_{R\ell}(E)dE \tag{2.52}$$

the density may conveniently be expressed as

$$n_R(r) = (4\pi)^{-1} \sum_\ell n_{R\ell} \times [\phi^2(r) + 2\varepsilon\phi(r)\dot{\phi}(r) + \varepsilon^2\{\dot{\phi}^2(r) + \phi(r)\ddot{\phi}(r)\}] \quad (2.53)$$

This is valid to second order and we have dropped the subscripts $R\ell$ inside the square paranthesis. Only the first term of (2.53) contributes net charge to the sphere and the remaining terms merely redistribute the charge within that sphere. This is so because integration in the sphere yields

$$n_R = \sum_\ell n_{R\ell}[1 + 2\varepsilon<\phi\dot{\phi}> + \varepsilon^2\{<\dot{\phi}^2> + <\phi\ddot{\phi}>\}] = \sum_\ell n_{R\ell} \quad (2.54)$$

due to the orthogonality of phi and phi-dot, and because

$$0 = (\partial/\partial E)<\phi(E)\dot{\phi}(E)> = <\dot{\phi}^2> + <\phi\ddot{\phi}> \quad (2.55)$$

An important consequence of this is that we need not truncate (2.53) after the first two terms even if we use a linear method (2.46 and 48) to compute the projected densities of states. Since b is simply found as the eigenvectors of a Hermitean matrix it is strictly unitary and the integrated density $n \equiv \sum_R n_R$ therefore exactly equals the number of occupied states.

For a crystal with lattice translation T and sphere-positions Q in the primitive cell the atomic-sphere potentials only depend on Q, and not on

$$\underline{R} = \underline{Q} + \underline{T} \quad (2.56)$$

The Hamiltonian is therefore diagonal in the Bloch representation and it is only necessary to consider basisfunctions with the property that

$$\chi^{\underline{k}}(\underline{r} + \underline{T}) = e^{i\underline{k}.\underline{T}} \chi^{\underline{k}}(\underline{r}) \quad (2.57)$$

where $\underline{k}$ is the Bloch vector. From (2.25) we see that

$$\pi^{\underline{k}}_{(\underline{Q}+\underline{T})L,G} = e^{i\underline{k}.\underline{T}} \pi^{\underline{k}}_{\underline{Q}L,G} \quad (2.58)$$

and analogously for Ω. The number of values taken by the indices QL is only $(\ell_{max} + 1)^2$ times the number of spheres in the primitive cell and it is therefore possible to invert $\pi_{QL,G}$ and form

$$h^{\underline{k}}_{Q'L',QL} = (\Omega\pi^{-1})^{\underline{k}}_{Q'L',QL} \quad (2.59)$$

for each Bloch vector. In order to obtain the theta-orbital (2.35) we finally transform from the $\underline{k}$- to the $\underline{T}$-representation, i.e.

$$h_{Q'L',(Q+T)L} = (BZV)^{-1} \int e^{i\underline{k}\cdot\underline{T}}\, h^{\underline{k}}_{Q'L',QL}\, d^3k \tag{2.60}$$

Here, $BZV \equiv (2\pi)^3/v$ and v are, respectively, the volumes of the primitive cells in the reciprocal and direct lattices.

E. The linear method for a single sphere

In this final part of the section dealing with general aspects of the linear methods we shall derive some formulas involving the radial phi and phi-dot functions. These will be useful when deriving the LAPW and LMTO methods in the following. We shall also parametrize the logarithmic derivative function (2.13) in terms of phi, phi-dot parameters. This will further illustrate the relation between the partial-wave and the linear methods, and it will be particularly useful for comparing the KKR- with the LMTO-ASA method. In the following the subscripts $R\ell$ will be dropped.

In order to derive explicit expressions for π and Ω in (2.25) we must find that linear combination of phi and phi-dot which matches continuously and differentiably onto a given radial function at the radius, s, of the sphere. Let us assume that the given radial function has the logarithmic derivative D. Then we require that the linear combination

$$\Phi(D,r) \equiv \phi(r) + \omega(D)\dot{\phi}(r) \tag{2.61}$$

has the same logarithmic deviative and find that

$$\omega(D) = -\frac{\phi(s)}{\dot{\phi}(s)}\,\frac{D - D\{\phi\}}{D - D\{\dot{\phi}\}} \tag{2.62}$$

Here,

$$D\{\phi\} = s\phi'(s)/\phi(s) \tag{2.63}$$

and

$$D\{\dot{\phi}\} = s\dot{\phi}'(s)/\dot{\phi}(s) \tag{2.64}$$

Eq. (2.62) apparently contains four potential parameters, namely the values and the slopes of phi and phi-dot at the sphere. However, only three of these parameters are independent because (2.29-30) and Green's second identity give the following Wronskian relation between them

$$1 = \langle\dot{\phi}^2\rangle = \langle\dot{\phi}| - \nabla^2 + v - E_\nu|\dot{\phi}\rangle$$
$$= \langle\dot{\phi}| - \nabla^2 + v - E_\nu|\dot{\phi}\rangle + [\dot{\phi}'(s)\dot{\phi}(s) - \dot{\phi}'(s)\dot{\phi}(s)]\,s^2$$
$$= s\phi(s)\dot{\phi}(s)[D\{\phi\} - D\{\dot{\phi}\}] \quad (2.65)$$

In order to match also the amplitude at the sphere we need the value

$$\Phi(D,s) = \phi(s) + \omega(D)\dot{\phi}(s) = [s\dot{\phi}(s)(D - D\{\dot{\phi}\})]^{-1} \quad (2.66)$$

Instead of using the four phi, phi-dot parameters we can equally well choose two numbers D_+ and D_- and then use

$$\omega_+ \equiv \omega(D_+) \qquad \omega_- \equiv \omega(D_-)$$
$$\Phi_+ \equiv \Phi(D_+,s) \quad \text{and} \quad \Phi_- \equiv \Phi(D_-,s) \quad (2.67)$$

as potential parameters. The Wronskian relation between them is

$$(D_+ - D_-)s\Phi_+\Phi_- = \omega_- - \omega_+ \quad (2.68)$$

and eqs. (2.62) and (2.66) become respectively

$$\omega(D) = \omega_+ + (D_+ - D_-)s\Phi_+^2 + \left[\frac{\Phi_+}{\Phi_-} - \frac{D - D_-}{D - D_+}\right]^{-1}$$
$$= \omega_- + (D_+ - D_-)s\Phi_-^2 \,\frac{D - D_-}{D - D_+}\left[1 - \frac{\Phi_-}{\Phi_+}\,\frac{D - D_-}{D - D_+}\right]^{-1} \quad (2.69)$$

and

$$\Phi(D,s) = \frac{(D_+ - D_-)\Phi_+\Phi_-}{(D - D_-)\Phi_- - (D - D_+)\Phi_+} \quad (2.70)$$

A useful parametrization of the logarithmic derivative function, D(E), defined in (2.13), may be found if we use $\Phi(D(E),r)$ as trial-function for $\phi(E,r)$ and employ the variational principle. In this way we obtain an estimate, E(D), of the function inverse of D(E), i.e.

$$E(D) = \frac{\langle\Phi(D)| - \nabla^2 + v|\Phi(D)\rangle}{\langle\Phi(D)|\Phi(D)\rangle}$$
$$= E_\nu + \omega(D)[1 + \omega^2(D)\langle\dot{\phi}^2\rangle]^{-1} \quad (2.71)$$

This is the single-sphere analogue to (2.43) and it is correct to third order. The estimate

$$\tilde{E}(D) \equiv E_\nu + \omega(D) \quad (2.72)$$

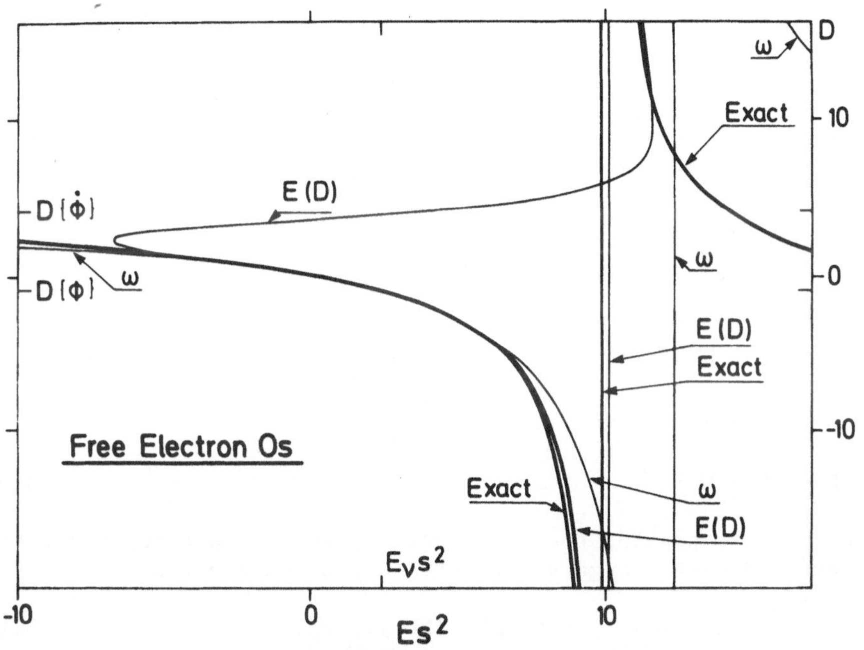

Fig. 5 : Estimates of the first branch ($\nu = 0$) of the logarithmic derivative function for a free s-electron. The estimate ω is correct to second order in $E - E_\nu$, while $E(D)$ is correct to third order. The potential parameters are given in the second column of Table V.

is consequently correct to second order.

The approximations (2.71) and (2.72) to the logarithmic derivative function are illustrated in Fig. 5.

III. LINEAR AUGMENTED-PLANE-WAVE METHOD

For simplicity of notation we consider a crystal with lattice translations T, reciprocal lattice vectors G satisfying

$$\underline{T} \cdot \underline{G} = 2\pi \times \text{integer} \tag{3.1}$$

and sphere-positions, Q, in the primitive cell. The cell volume is v.

A. Basisfunctions

In the LAPW method the envelope basisfunctions are the plane waves

$$\chi_G^k(\underline{r}) = v^{-1/2}\, e^{i(\underline{k} + \underline{G}).\underline{r}} \tag{3.2}$$

and in the following we shall use the notation

$$\underline{K}_G = \underline{k} + \underline{G} \equiv \underline{K} \tag{3.3}$$

We now wish to augment the plane wave inside the spheres and therefore write down the angular-momentum expansion about site Q :

$$e^{i\underline{K}.\underline{r}} = e^{i\underline{K}.\underline{Q}}\, 4\pi \sum_L i^\ell Y_L^\star(\hat{K}) j_\ell(Kr_Q) Y_L(\hat{r}_Q) \tag{3.4}$$

The augmentation in the sphere at Q is

$$\begin{aligned} j_\ell(Kr_Q) \rightarrow j_{Q\ell K}(r_Q) &\equiv j_\ell(Ks_Q)\Phi_{Q\ell}(D_{Q\ell K}, r_Q)/\Phi_{Q\ell}(D_{Q\ell K}, s_Q) \\ &\equiv j_\ell(Ks_Q)\Phi_{Q\ell K}(r_Q)/\Phi_{Q\ell K} \\ &\equiv j_{Q\ell K}\, [\phi_{Q\ell}(r) + \omega_{Q\ell K}\dot{\phi}_{Q\ell}(r)]/\Phi_{Q\ell K} \end{aligned} \tag{3.5}$$

In these equations we have used the definitions of Sect. IIE and have defined further short-hand notations. Moreover,

$$D_{Q\ell K} \equiv D\{j_\ell(Ks_Q)\} = [x j_\ell'(x)/j_\ell(x)]_{x=Ks_Q} \tag{3.6}$$

and

$$\omega_{Q\ell K} \equiv \omega_{Q\ell}(D_{Q\ell K}) \tag{3.7}$$

By comparison with (2.25) and using (2.57 and 61) we see that

$$\pi_{QL,K} = v^{-1/2}\, e^{i\underline{K}\cdot\underline{Q}}\, i^{\ell}\, Y_L^{\star}(\hat{K})\, 4\pi s_Q^2$$
$$\times\,[\, Kj_{\ell}'(Ks_Q)\dot{\phi}_{Q\ell}(s_Q) - j_{\ell}(Ks_Q)\dot{\phi}_{Q\ell}'(s_Q)] \qquad (3.8)$$

and

$$\Omega_{QL,K} = \pi_{QL,K}\, \omega_{Q\ell K}$$
$$= -v^{-1/2}\, e^{i\underline{K}\cdot\underline{Q}}\, i^{\ell} Y_L^{\star}(\hat{K})\, 4\pi s_Q^2$$
$$\times\,[\, Kj_{\ell}'(Ks_Q)\phi_{Q\ell}(s_Q) - j_{\ell}(Ks_Q)\phi_{Q\ell}'(s_Q)] \qquad (3.9)$$

The major advantage of the LAPW method is the high accuracy and relative ease with which it can treat a general potential. In the present notes we shall concentrate on this feature and will therefore use the formulation (2.31-32) of the Hamiltonian and overlap matrices rather than the formulation in terms of theta orbitals. The method that we shall now sketch is the crystalline version of the one employed by Jepsen et al. for the surface calculation mentioned in Fig. 1a. A presumably more accurate LAPW technique has been devised by Wimmer et al. (1981).

It is convenient to let the plane waves extend throughout space, because there they are orthogonal, and we therefore regroup the terms in (2.25) as follows

$$\chi_K(\underline{r}) = v^{-1/2}[\, e^{i\underline{K}\cdot\underline{r}} + \sigma_K(\underline{r})] \qquad (3.10)$$

Each of the two terms is now continuous and differentiable in all space and the function sigma vanishes at and outside the spheres. Inside the sphere at Q

$$\sigma_K(\underline{r}) \equiv e^{i\underline{K}\cdot\underline{Q}}\, 4\pi\sum_L i^{\ell} Y_L^{\star}(\hat{K}) Y_L(\hat{r}_Q)[\, j_{Q\ell K}(r_Q) - j_{\ell}(Kr_Q)] \qquad (3.11)$$

B. Electron density and potential

Let $a^{k}_{G,j}$ be the solutions of the LAPW equations (2.5) orthonormalized according to

$$a^{+}Oa = 1 \qquad (3.12)$$

Here, and in the following, we drop the superscript k. The expressions for the Hamiltonian and overlap matrices will be written down later.

The electron density for the state $|\underline{k}j\rangle$ is

$$|\psi_j^k(\underline{r})|^2 = v^{-1}\sum_{G'}\sum_{G}[e^{i(\underline{G}-\underline{G}')\cdot\underline{r}} + e^{-i\underline{K}'\cdot\underline{r}}\,\sigma_K(\underline{r})$$

$$+ \sigma_{K'}^{\star}(\underline{r})e^{i\underline{K}\cdot\underline{r}} + \sigma_{K'}^{\star}(\underline{r})\sigma_K(\underline{r})]\,a_{G',j}^{k\star}\,a_{G,j}^{k} \qquad (3.13)$$

The last three terms in the square bracket vanish outside- and at the spheres, and their gradients vanish at the spheres. Inside the spheres they must therefore be very nearly spherically symmetric (except for an atom with an open, localized shell) and we shall, in fact, neglect their non-spherical components. With this approximation

$$|\psi_j^k(\underline{r})|^2 \approx v^{-1}\sum_{G}\sum_{G'}\,[e^{i(\underline{G}-\underline{G}')\cdot\underline{r}} + \sum_{T}\sum_{Q} e^{i(\underline{G}-\underline{G}')\cdot\underline{Q}}$$

$$\times\ \theta_Q(r_Q)\sum_{\ell}(2\ell+1)P_\ell(\hat{K}'\cdot\hat{K})\{j_{Q\ell K'}(r_Q)j_{Q\ell K}(r_Q) - j_\ell(K'r_Q)j_\ell(Kr_Q)\}]$$

$$\times\ a_{G',j}^{k\star}\,a_{G,j}^{k} \qquad (3.14)$$

Here we have used the addition theorem

$$4\pi\sum_{m} Y_L(\hat{K}')Y_L^{\star}(\hat{K}) = (2\ell+1)P_\ell(\hat{K}'\cdot\hat{K}) \qquad (3.15)$$

We have thus expressed the valence-electron density as a plane-wave series plus spherical contributions which vanish continuously and differentiably outside the spheres. With this form it is a simple matter to solve Poissons equation as needed in self-consistent calculations :

To the spherical valence density we first add the spherical density from the core electrons, the point charge of the nucleus, and a fictitious, neutralizing, smooth and spherical charge. The Coulomb potential from this spherical density now vanishes continuously and differentiably outside the sphere and, inside each sphere, it is found by solving the radial Poisson equation. From the plane-wave part of the density we first subtract the Fourier transform of the fictitious charge. The Coulomb potential from the resulting, neutral charge is then found simply by multiplying the plane-wave components by $8\pi G^{-2}$ and omitting the G = 0 term.

The exchange-correlation potential is a non-linear but local function of the density in $\underline{r}$-space. It must therefore be found point by point and then expressed as the sum of plane-wave and spherical parts.

The total one-electron potential is now expressed as the sum of a plane-wave and a spherical part :

$$V(\underline{r}) = \sum_{G} e^{i\underline{G}.\underline{r}} V(\underline{G})$$
$$+ \sum_{T}\sum_{Q} [v_Q(r_Q) - \sum_{G} j_0(Gr_Q)e^{i\underline{G}.\underline{Q}} V(\underline{G})] \theta_Q(r_Q) \tag{3.16}$$

where $v_Q(r_Q)$ is the MT-potential in the sphere at Q. This potential, of course, includes the spherically symmetric part of the plane-waves.

Various contributions to the density are illustrated in Fig. 6.

C. Hamiltonian and overlap matrices

The solutions of the radial Schrödinger equation for the MT-potentials $v_Q(r_Q)$ provide the phi, phi-dot functions and hence the potential parameters as described in Sect. IIE.

The overlap matrix may be obtained by integrating (3.14) over all space. Equivalently we may use (2.32), employing (3.15) to the matrix products, $\pi^+\pi$ and $\Omega^+\langle\dot\phi^2\rangle\Omega$, as well as the result

$$\langle\chi|\chi\rangle^i = v^{-1}\langle e^{i(\underline{G}-\underline{G}').\underline{r}}\rangle^\infty - v^{-1}\langle e^{i\underline{K}'.\underline{r}}|e^{i\underline{K}.\underline{r}}\rangle$$
$$= \delta_{G'G} - \sum_{Q} e^{i(\underline{G}-\underline{G}').\underline{Q}} \frac{4\pi s_Q^3}{v} \frac{j_1(|\underline{G}-\underline{G}'|s_Q)}{|\underline{G}-\underline{G}'|s_Q} \tag{3.17}$$

We now express the Hamiltonian matrix (2.31) for the potential (3.16). The non-MT part of the potential inside a sphere, $V^{nmt} = V(\underline{r}) - v_Q(r_Q)$, vanishes at the centre of the sphere, and the part $\sigma_K(\underline{r})$, of the LAPW vanishes at and near the surface of the sphere. We shall neglect the overlap between these two functions and therefore use

$$\langle\chi_{K'}|V^{nmt}|\chi_K\rangle \approx v^{-1}\langle e^{i\underline{K}'.\underline{r}}|V^{nmt}|e^{i\underline{K}.\underline{r}}\rangle \tag{3.18}$$

in a sphere. Since in (2.31) we defined V^{nmt} as the non-MT part of the potential inside the spheres and as the total potential between the spheres we have

$$\langle\chi_{K'}|V^{nmt}|\chi_K\rangle^\infty \approx v^{-1}\langle e^{i\underline{K}'.\underline{r}}|V^{nmt}|e^{i\underline{K}.\underline{r}}\rangle^\infty$$
$$= v^{-1}\langle e^{i\underline{K}'.\underline{r}}|\sum_{G''} [e^{i\underline{G}''.\underline{r}} - \sum_{T}\sum_{Q} \theta_Q(r_Q)j_0(G''r_Q)e^{i\underline{G}''.\underline{Q}}]V(\underline{G}'')|e^{i\underline{K}.\underline{r}}\rangle^\infty$$
$$= V(\underline{G}'-\underline{G}) - \sum_{Q} e^{i(\underline{G}-\underline{G}').\underline{Q}} \frac{4\pi s_Q^3}{v} \sum_{G''} e^{i\underline{G}''.\underline{Q}} V(\underline{G}'') \times$$

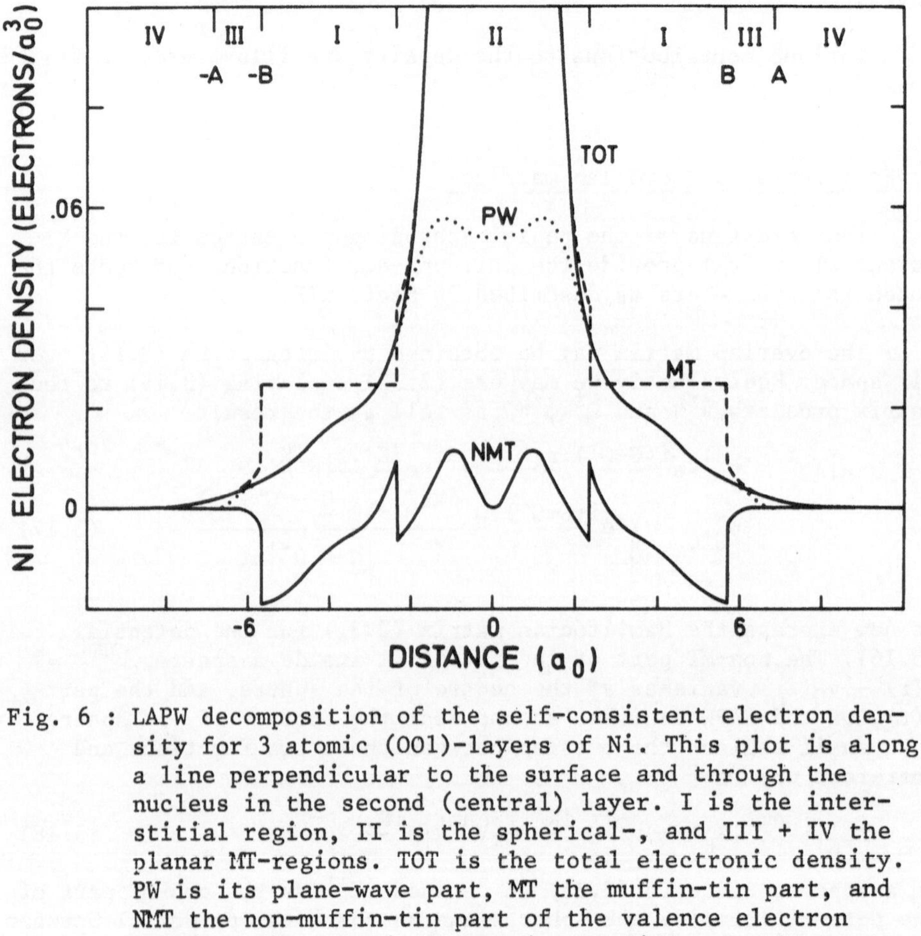

Fig. 6 : LAPW decomposition of the self-consistent electron density for 3 atomic (001)-layers of Ni. This plot is along a line perpendicular to the surface and through the nucleus in the second (central) layer. I is the interstitial region, II is the spherical-, and III + IV the planar MT-regions. TOT is the total electronic density. PW is its plane-wave part, MT the muffin-tin part, and NMT the non-muffin-tin part of the valence electron density. The spherical part is the difference between TOT and PW. (Jepsen, Madsen and Andersen 1982).

$$\left[\frac{\cos x'' \dfrac{\sin x}{x} - \cos x \dfrac{\sin x''}{x''}}{x^2 - x''^2}\right]_{\substack{x \equiv |\underline{G} - \underline{G}'| s_Q \\ x'' \equiv G'' s_Q}} \tag{3.19}$$

The final term in (2.31) is

$$\langle \chi | -\nabla^2 - E_\nu | \chi \rangle^i = (K^2 - E_\nu) \langle \chi | \chi \rangle^i \tag{3.20}$$

An expression for the Hamiltonian matrix in which each term is Hermitean is

$$\begin{aligned} H^k_{G'G} &= K^2 \delta_{G'G} + \langle \chi_{K'} | V^{nmt} | \chi_K \rangle^\infty \\ &+ \sum_Q e^{i(\underline{G}-\underline{G}')\cdot\underline{Q}} \sum_\ell W_{Q\ell}(\underline{K}',\underline{K}) [\Gamma_{Q\ell}(K',K) + E_{\nu Q\ell} \Delta_{Q\ell}(K',K)] \end{aligned} \tag{3.21}$$

where Γ and Δ are given in (Andersen 1975). In that same reference one finds

$$\mathcal{O}^k_{G'G} = \delta_{G'G} + \sum_Q e^{i(\underline{G}-\underline{G}')\cdot\underline{Q}} \sum_\ell W_{Q\ell}(\underline{K}',\underline{K}) \Delta_{Q\ell}(K',K) \tag{3.22}$$

IV. LINEAR MUFFIN-TIN-ORBITAL METHOD

The LMTO-ASA method is the simplest and most transparent of the linear methods and in the present chapter the main emphasis will be on this version of the LMTO method.

For the treatment of non-MT, or non-ASA, potentials using muffin-tin orbitals we refer to the papers by Andersen, Woolley, Jones, Harris, Gunnarsson, Müller, Kasowski and Herman.

A. Muffin-tin orbitals

In the interstitial region (see Fig. 1) the potential is rather flat and the absolute value of the kinetic energy, $|E - V(\underline{r})|$, is fairly small. If, in addition, the interstitial region has a relatively small volume then it is reasonable to choose basisfunctions whose envelopes are solutions of the wave equation

$$[\nabla^2 + \kappa^2] \chi^i(\kappa, \underline{r}) = 0 \tag{4.1}$$

with a constant, numerically small kinetic energy, κ^2. We emphasize that kappa must be independent of energy in order that the basisfunctions are independent of energy and hence, that the eigenvalue equation (2.5) is linear in energy. This is not so in the KKR partial-wave method where the choice $\kappa^2 = E - V^{mtz}$ is made.

The muffin-tin orbitals (MTOs) are angular-momentum eigenfunctions centered at the sphere-sites R, i.e.

$$\chi_{RL}(\underline{r}) = \chi_{R\ell}(r)\, Y_L(\hat{r}) \tag{4.2}$$

and the envelope functions, χ^i_{RL} satisfy (4.1). The radial envelope function, χ^i_ℓ , is singular at the origin and well-behaved at infinity. This means that, for $\kappa^2 < 0$, chi must be that solution of (4.1) which decays asymptotically as $r^{-1}\exp(-|\kappa|r)$ and, for $\kappa^2 = 0$, chi must be proportional to $r^{-\ell-1}$. Finally, for $\kappa^2 > 0$, chi can be any linear combination of spherical Bessel, $j_\ell(\kappa r)$, and Neumann, $n_\ell(\kappa r)$, functions and these decay asymptotically as $r^{-1}\exp(i\kappa r + i\eta)$.

The linear MTO is now obtained by augmenting the envelope function inside its own and inside all other spheres. For the latter purpose we need to expand chi about the other sites. Since chi is regular, except at its own site, R, it must have an expansion about site R' of the form

$$\chi^i_{RL}(\kappa,\underline{r}_R) = \sum_{L'} j_{\ell'}(\kappa r_{R'})\, i^{\ell'}\, Y_{L'}(\hat{r}_{R'})\, S_{R'L',RL}(\kappa) \tag{4.3}$$

where j is that solution of the radial wave equation which is regular at the origin. The expansion is valid for $r_{R'} < |\underline{R} - \underline{R}'|$. The coefficients, S, are the socalled KKR structure constants whose traditional normalization is such that (4.3) is correct when we choose

$$\chi^i_{R\ell}(\kappa,r) = \kappa n_\ell(\kappa r)\, i^\ell \tag{4.4}$$

It may be shown that

$$S_{R'L',RL}(\kappa) = 4\pi \sum_{\ell''} C_{LL'L''}\, \kappa n_{\ell''}(\kappa R'')\, i^{-\ell''}\, Y^{\star}_{L''}(\hat{R}'') \tag{4.5}$$

where

$$\underline{R}'' \equiv \underline{R} - \underline{R}'$$

and

$$C_{LL'L''} \equiv \int Y_L(\hat{r})\, Y^{\star}_{L'}(\hat{r})\, Y_{L''}(\hat{r})\, d^2\hat{r} \tag{4.6}$$

are the socalled Gaunt coefficients. The matrix (4.5) is Hermitian. The major advantage of choosing the orbitals to be solutions of the translationally invariant wave equation (4.1) rather than, for instance atomic orbitals, is the simplicity of the expansion (4.3). The radial behaviour is $j_{\ell'}(\kappa r)$ regardless of the distance $\underline{R} - \underline{R}'$ and, for a crystal, we may therefore perform the lattice-summations on the structure constants only, i.e.

$$S^{\underline{k}}_{Q'L',QL} = \sum_{T\neq 0} e^{i\underline{k}.\underline{T}} S_{Q'L',(Q+T)L} \tag{4.7}$$

At this place we shall not carry the formalism with a general kappa further because the traditional normalization (4.3-5) is unsuitable for imaginary kappa (exponentially decaying MTOs) and impossible for vanishing kappa. A "renormalized" formalism may be found in (Andersen and Woolley 1973). We shall now make the simplest choice

$$\kappa^2 = 0 \tag{4.8}$$

which is accurate in connection with the ASA where the sphere packing is close. The wave equation (4.1) thus reduces to the Laplace equation and the expansion (4.3) becomes the well-known expansion

$$\left[\frac{|\underline{r}-\underline{R}|}{a}\right]^{-\ell-1} Y_L(\widehat{\underline{r}-\underline{R}}) = -\sum_{L'} \left(\frac{r}{a}\right)^{\ell'} \frac{Y_{L'}(\hat{r})}{2(2\ell'+1)} S_{OL',RL} \tag{4.9}$$

of a static multipole-potential. The solutions of the Laplace equation are harmonic functions and this allows us to measure all distances in units of some arbitrary scale constant, a,say the lattice constant. Defined in this way, the structure constants become <u>independent of the scale of the structure</u>. They are, of course, independent of the potential and the sphere radii, s_R, and their explicit expression is

$$S_{R'\ell'm',R\ell m} = (4\pi)^{1/2} g_{\ell'm',\ell m} \left[\frac{|\underline{R}-\underline{R}'|}{a}\right]^{-\ell'-\ell-1} Y^{\star}_{\ell'+\ell,m'-m}(\widehat{\underline{R}-\underline{R}'}) \tag{4.10}$$

with

$$g_{\ell'm',\ell m} = (-)^{\ell+m+1}\, 2\left[\frac{(2\ell'+1)(2\ell+1)(\ell'+\ell+m'-m)!(\ell'+\ell-m'+m)!}{(2\ell'+2\ell+1)(\ell'+m')!(\ell'-m')!(\ell+m)!(\ell-m)!}\right]^{1/2} \tag{4.11}$$

and the relation to the traditional normalization is

$$S_{R'L',RL} = \lim_{\kappa^2\to 0} [(a/2)\kappa^2 n_{\ell'}(\kappa a) n_\ell(\kappa a)]^{-1} i^{\ell'-\ell} S_{R'L',RL}(\kappa) \tag{4.12}$$

The radial part of the LMTO can now be defined as

$$\chi_{R\ell}(r) = \begin{cases} \Phi_{R\ell}(-\ell-1,r) & \text{for} \quad r \leqslant s_R \\ (\bar{r}/s_R)^{-\ell-1}\, \Phi_{R\ell-} & \text{for} \quad r \geqslant s_R \end{cases} \tag{4.13}$$

where the augmentation inside the own sphere has been specified and we have used the notation (2.67) choosing

$$D_- \equiv -\ell-1 \tag{4.14}$$

which is the logarithmic derivative of $r^{-\ell-1}$. The augmentation inside all other spheres follows from (4.9) and it involves that linear combination of phi and phi-dot which has the logarithmic derivative

$$D_+ \equiv \ell' \tag{4.15}$$

Specifically,

$$\chi_{RL}(\underline{r}_R) = -\sum_{L'} \frac{\Phi_{R'L'}(\ell',\underline{r}_{R'})}{2(2\ell'+1)\Phi_{R'\ell'+}} \left(\frac{s_{R'}}{a}\right)^{\ell'} S_{R'L',RL} \left(\frac{a}{s_R}\right)^{-\ell-1} \Phi_{R\ell-} \tag{4.16}$$

inside the sphere at R'.

In order to simplify the notation it is useful to renormalize the potential parameters Φ_+ and Φ_- defined in (2.67) such that they appear as if they were evaluated at the common distance, a, rather than at the radius, s_R, of the sphere in question. The new potential parameters are

$$\begin{aligned}\sqrt{\Delta} &\equiv (a/s)^{-\ell-1}\,(a/2)^{1/2}\Phi_- \\ &\approx (a/2)^{1/2}\,\Phi(-\ell-1,a)\end{aligned} \tag{4.17}$$

and

$$\begin{aligned}\sqrt{\Gamma} &\equiv 2(2\ell+1)(a/s)^{\ell}(a/2)^{1/2}\,\Phi_+ \equiv Q\Delta^{1/2} \\ &\approx 2(2\ell+1)(a/2)^{1/2}\,\Phi(\ell,a)\end{aligned} \tag{4.18}$$

where, for reasons to be seen later, a factor $(a/2)^{1/2}$ has been included, too. In this way Δ and Γ become band-width parameters with the dimension of energy. The energy parameters ω_- and ω_+ remain as defined in (2.67) and the relation (2.68) between the four potential parameters becomes

$$\sqrt{\Delta}\,\sqrt{\Gamma} = \omega_- - \omega_+ \text{ , or}$$

$$\Delta = (\omega_- - \omega_+)/Q \quad \text{and} \quad \Gamma = (\omega_- - \omega_+)Q \tag{4.19}$$

We shall, furthermore, use the notation introduced in (2.25) and therefore write

$$\Phi(-\ell-1,r) \equiv |\Phi_-\rangle \quad \text{and} \quad \Phi(\ell,r) \equiv |\Phi_+\rangle \tag{4.20}$$

These functions will be regarded as vectors with indices R and L.

With these definitions we may now express (4.13) and (4.16) in the form (2.25b), i.e.

$$|\Phi_-\rangle - |\Phi_+\rangle \, \Gamma^{-1/2} \, S \, \Delta^{1/2} + |\chi\rangle^i = |\chi\rangle^\infty \tag{4.21}$$

Using for $|\Phi_-\rangle$ and $|\Phi_+\rangle$ the definitions

$$|\Phi\rangle \equiv |\phi\rangle + \omega|\dot{\phi}\rangle \tag{2.61}$$

we can finally identify the π and Ω matrices of the LMTO representation

$$\pi = 1 - \Gamma^{-1/2} \, S \, \Delta^{1/2} \tag{4.22}$$

and

$$\Omega = \omega_- - \omega_+ \, \Gamma^{-1/2} \, S \, \Delta^{1/2} \tag{4.23}$$

B. The LMTO method

The LMTO Hamiltonian and overlap matrices may now be obtained from (2.31-32) and they are seen to contain one-centre terms proportional to S^0, two-centre terms proportional to S, and three-centre terms proportional to S^2.

The integrals over the interstitial region and those involving the non-MT terms may be performed in angular-momentum representation using the tail-expansion (4.16), which is valid inside the sphere centered at R' passing through R (Fig. 7). With this procedure we must divide the region of integration into Wigner-Seitz cells, and then expand these in angular-momentum series as well. (Andersen and Woolley 1973, Gunnarsson, Harris and Jones 1976, Casula and Herman 1983).

Alternatively, we may proceed as we did in the LAPW method (Chapt. III). We first Fourier-transform χ^1, augmented by some smooth pseudo-phi inside the spheres, and then use this pseudo-LMTO to evaluate all the remaining integrals in K-space. For crystals, this procedure (Andersen 1975) has the advantage that the integration region offers no problem because the plane waves are orthonormalized in the cell.

The pseudo-LMTO is conveniently taken as the LMTO for a potential which is flat throughout space, and with E_ν equal to the value of this potential. This pseudo-LMTO is thus proportional to $r^{-\ell-1}$ outside its own sphere and therefore easy to Fourier-transform. The pseudo-phi is

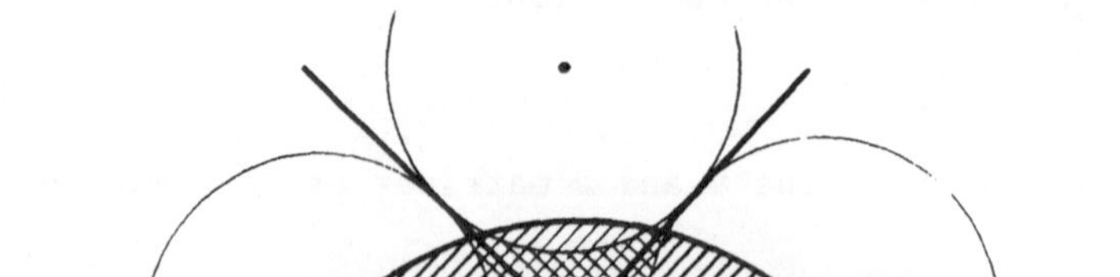

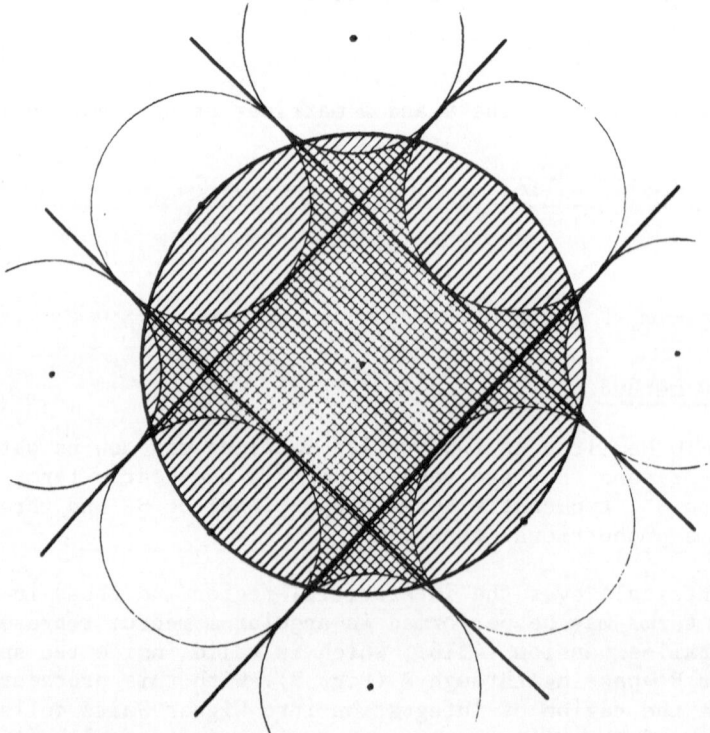

Fig. 7 : Radius of convergence for the one-centre expansions (4.3), (4.9) and (4.16).

$$\phi_\ell^{ps}(r) = [(2\ell + 3)s^{-3}]^{1/2} (r/s)^\ell = \Phi_+^{ps}(r) \tag{4.24}$$

where the prefactor comes from the normalization in the sphere. The unnormalized radial wavefunction for the flat potential is the spherical Bessel function, and the pseudo phi-dot is therefore found by orthogonalizing the energy derivative

$$\partial\kappa^{-\ell} j_\ell(\kappa r)/\partial\kappa^2|_{\kappa=0} \propto r^{\ell+2} \tag{4.25}$$

to phi in (4.24). This leads to

$$\dot\phi_\ell^{ps}(r) \propto (\frac{r}{s})^\ell - \frac{2\ell + 5}{2\ell + 3} (\frac{r}{s})^{\ell+2} \tag{4.26}$$

and therefore

$$D\{\dot\phi^{ps}\} = 3\ell + 5 \tag{4.27}$$

The normalization of phi-dot (4.26) may now be found from the Wronskian relation (2.65) and, as a result,

$$\dot\phi_\ell^{ps}(r) = \frac{[(2\ell + 3)s]^{1/2}}{2}[\frac{1}{2\ell + 5} (\frac{r}{s})^\ell - \frac{1}{2\ell + 3} (\frac{r}{s})^{\ell+2}] \tag{4.28}$$

Having evaluated phi and phi-dot all the pseudo-potential parameters may easily be found.

The Fourier-transform may be evaluated if we observe that $\nabla^2\exp(i\underline{K}.\underline{r}) = - K^2\exp(i\underline{K}.\underline{r})$ and $\nabla^2\chi^{ps} = 0$ outside its own sphere. We may then use Greens second identity to express a volume integral involving ∇^2 into a surface integral involving logarithmic derivatives. For a crystal the result is

$$\chi_{\underline{Q}L}^{\underline{k}}(\underline{r})^{ps} = \Phi_- \sum_G e^{i\underline{K}.\underline{r}} F_L(\underline{K})\, e^{-i\underline{K}.\underline{Q}} \tag{4.29}$$

with $\underline{K} \equiv \underline{k} + \underline{G}$, and

$$F_L(\underline{K}) = (2\ell + 1)(2\ell + 3) \frac{4\pi s^3}{V} \frac{j_{\ell+1}(Ks)}{(Ks)^3} Y_L(\hat{K}) \tag{4.30}$$

The real potential parameter Φ_- appears in (4.29) because the pseudo LMTO must match the real LMTO at its sphere. The integral over the interstitial region may now be performed as an integral over all space using (4.29) minus the integrals in the spheres using the one-centre representation

$$[\,|\phi^{ps}\rangle\pi^{ps} + |\dot\phi^{ps}\rangle\Omega^{ps}](\Phi_-/\Phi_-^{ps}) + |\chi\rangle^i = |\chi^{ps}\rangle^\infty \tag{4.31}$$

C. The LMTO-ASA method in the theta representation

We now use the atomic-sphere approximation introduced in Sect. IID. With π and Ω expressed by (4.22-23) we seek $h \equiv \Omega\pi^{-1}$. From (4.22) we obtain

$$\pi^{-1} = \Delta^{-1/2}\,(Q - S)^{-1}\,\Gamma^{1/2} \tag{4.32}$$

and, from (4.23)

$$\begin{aligned}\Omega &= \omega_- - \omega_+\,\Gamma^{-1/2}(Q - \Gamma^{1/2}\Delta^{-1/2} + S)\Delta^{1/2}\\ &= (\omega_- - \omega_+) + \omega_+\Gamma^{-1/2}(Q - S)\Delta^{1/2}\end{aligned} \tag{4.33}$$

Using then the Wronskian relation (4.19) the result is

$$h = \Gamma^{1/2}(Q - S)^{-1}\,\Gamma^{1/2} + \omega_+ \tag{4.34a}$$

$$\begin{aligned}&\Gamma^{1/2}(1 - Q^{-1}S)^{-1}\Delta^{1/2} + (\omega_- - \Gamma^{1/2}\Delta^{1/2})\\ &\Gamma^{1/2}[\,(1 - Q^{-1}S)^{-1} - 1]\,\Delta^{1/2} + \omega_-\\ &\Delta^{1/2}S(1 - Q^{-1}S)^{-1}\Delta^{1/2} + \omega_-\end{aligned} \tag{4.34b}$$

The second-order Hamiltonian (2.45) in the theta-orbital representation (2.35) is thus given by the two equivalent expressions

$$\tilde{H}_{RL,R'L'} = V_{R\ell}\,\delta_{RR'}\,\delta_{LL'} + \Gamma^{1/2}_{R\ell}(Q - S)^{-1}_{RL,R'L'}\Gamma^{1/2}_{R'\ell'} \tag{4.35a}$$

and

$$\tilde{H}_{RL,R'L'} = C_{R\ell}\,\delta_{RR'}\,\delta_{LL'} + \Delta^{1/2}_{R\ell}\,\tilde{S}_{RL,R'L'}\Delta^{1/2}_{R'\ell'} \tag{4.35b}$$

Here, we have defined the "screened" structure constants

$$\tilde{S} \equiv S(1 - Q^{-1}S)^{-1} = Q(Q - S)^{-1}\,Q - Q \tag{4.36}$$

as well as the energies

$$V_{R\ell} \equiv \omega_{R\ell}(\ell) + E_{\nu R\ell} = \tilde{E}_{R\ell}(\ell) \tag{4.37}$$

and

$$C_{R\ell} \equiv \omega_{R\ell}(-\ell-1) + E_{\nu R\ell} = \tilde{E}_{R\ell}(-\ell-1) \tag{4.38}$$

which to second order correspond to the logarithmic-derivative values ℓ and $-\ell-1$, respectively. Eq. (4.35) is the most important single result of the LMTO formalism and we shall now discuss it in some detail. For simplicity we consider the example of elemental metals.

D. Canonical Bands

The potential parameter defining the "screening" of the structure constants depends, according to (4.17), (4.18) and (2.66), only on $D\{\dot{\phi}\}$ and not on phi

$$Q = 2(2\ell + 1)\ \left(\frac{a}{s}\right)^{2\ell+1} \frac{\Phi(\ell,s)}{\Phi(-\ell-1,s)}$$

$$= 2(2\ell + 1)\ \left(\frac{a}{s}\right)^{2\ell+1} \frac{D\{\dot{\phi}\} + \ell+1}{D\{\dot{\phi}\} - \ell} \qquad (4.39)$$

If we therefore construct LMTOs using the correct phi functions but incorrect phi-dot functions, whose logarithmic derivatives are set to the value ℓ, and whose values at the sphere are in accordance with the Wronskian relation (2.65) with $D\{\dot{\phi}\} \equiv \ell$, then our trial functions are correct to zeroth order and we obtain eigenvalues correct to first order. This proves that the Hamiltonian $C + \sqrt{\Delta}S\sqrt{\Delta}$ obtained from (4.35b) by substituting S for $\tilde{S}$ is correct to first order. As a consequence, we can start discussing (4.35b) by considering the properties of the unscreened structure constants. According to (2.65) the above argument only holds when $D\{\phi\} \neq \ell$. If $D\{\phi\} \approx \ell$ we may set $D\{\dot{\phi}\}$ equal to $-\ell-1$ and, hence, use (4.35a) in the form $V - \sqrt{\Gamma}S^{-1}\sqrt{\Gamma}$. Again, this means that we can start discussing (4.35a) by considering the properties of the unscreened structure constants.

The Hamiltonian (4.35b) has the two-centre form. The on-site, "atomic" energies are C and the hopping integrals are $\sqrt{\Delta}\ S\ \sqrt{\Delta}$, which are factorized into products of on-site potential parameters, $\sqrt{\Delta}$, and the structure matrix, S. The latter was given by (4.10-11) and it may be expressed in the usual two-centre form by choosing the z-direction of the spherical harmonics in the direction of the inter-atomic vector $\underline{R} - \underline{R}'$. The result is

$$S_{ss\sigma} = -\ 2(a/d)$$

$$S_{sp\sigma} = 2\sqrt{3}(a/d)^2$$

$$S_{sd\sigma} = -\ 2\ \sqrt{5}(a/d)^3$$

$$S_{pp(\sigma,\pi)} = 6(a/d)^3(2,-1)$$

$$S_{pd(\sigma,\pi)} = 6\ \sqrt{5}(a/d)^4(-\ \sqrt{3},1)$$

$$S_{dd(\sigma,\pi,\delta)} = 10(a/d)^5(-6,4,-1) \text{ etc.} \qquad (4.40)$$

In general

$$S_{\ell'\ell M} = (-)^{\ell+M+1}(\ell'+\ell)!2\left[\frac{(2\ell'+1)(2\ell+1)}{(\ell'+M)!(\ell'-M)!(\ell+M)!(\ell-M)!}\right]^{1/2}\left(\frac{a}{d}\right)^{\ell'+\ell+1} \qquad (4.41)$$

and $d \equiv |\underline{R} - \underline{R}'|$ is the inter-atomic distance. In Table I we list the structure matrix as functions of the orbitals involved (the Y_L's are here chosen as real cubic harmonics) and the directional cosines, ℓ,m, and n, of the interatomic vector ($\underline{R} - \underline{R}' \equiv d\{\ell,m,n\}$). These expressions are very simple but, for crystals, only the pd- and dd-interactions converge sufficiently rapidly in real space to be useful. In general, the Ewald procedure must be used to perform the lattice summations (4.7).

For finite clusters the long range of the interaction makes it necessary to embed the cluster in an outer, concave sphere whose LMTO's have tails sticking into the cluster. We shall not go into the details of this formalism but refer to the papers by Andersen and Woolley (1973) and Gunnarsson, Harris and Jones (1976). At present we only state a simple result obtained by including only the s-waves on the outer sphere and by folding down the corresponding block of the Hamiltonian. The effect of the outer sphere is essentially to add $2a/s_0$, where s_0 is the radius of the outer sphere, to all $\ell = \ell' = 0$ - elements of the structure matrix. This simple correction has been included in Table I.

We now consider a crystal and use the notation (2.56-60). The Hamiltonian (4.35) now depends on the Bloch-vector, k, but R and R' only run over the atoms in the primitive cell. The eigenvalues E^k_i are the energy bands which are periodic in reciprocal space and therefore only need to be calculated in its primitive cell, the Brillouin zone. Only the structure constants (4.7), and not the potential parameters, depend on k and, writing $\underline{k}.\underline{T} = (\underline{ka}).(\underline{T/a})$ in (4.7), we realize that the structure constants are canonical in the sence that they only depend on the crystal structure and not on the lattice constant. For elemental solids we always chose

a = s = atomic Wigner-Seitz radius.

Since we have chosen the potential to be spherically symmetric the potential parameters only depend on ℓ but not on m. If we therefore neglect the $\ell\ell'$-hybridization by neglecting the off diagonal blocks $S^k_{\ell m,\ell' m'}$ with $\ell \neq \ell'$ of the structure constants then the correspondingly unhybridized energy bands $E^k_{\ell i}$ may be obtained simply by energy scalings of socalled canonical bands $S^k_{\ell i}$. The canonical bands consist of the eigenvalues of the diagonal blocks $S^k_{\ell m,\ell' m'}\delta_{\ell\ell'}$ of the structure constants and the corresponding eigenvalues $U^k_{\ell m,\ell' i}\delta_{\ell\ell'}$ define a unitary transformation which applied to the Hamiltonian (4.35) yields

$$U^{-1}\tilde{H}U \equiv \tilde{E}^k_{\ell i}\delta_{\ell\ell'}$$

$$= [V_\ell + \frac{\Gamma_\ell}{Q_\ell - S^k_{\ell i}}]\,\delta_{\ell\ell'} \qquad (4.42a)$$

$$= [C_\ell + \frac{\Delta_\ell S^k_{\ell i}}{1 - Q_\ell^{-1} S^k_{\ell i}}] \delta_{\ell\ell'} \tag{4.42b}$$

if in (4.35-36) the off-diagonal blocks of S are neglected. Defined in this way the unhybridized ℓ-bands $\tilde{E}^k_{\ell i}$ thus equal the canonical ℓ-bands $S^k_{\ell i}$ positioned at the energy C_ℓ scaled by the band-width parameter Δ and distorted by the dimensionless parameter Q^{-1}.

The energy scaling (4.42) of the canonical bands to unhybridized energy bands is only correct to second order in $E - E_\nu$. The expression correct to third order is according to (2.43-44)

$$E^k_{\ell i} = E_{\nu\ell} + \omega^k_{\ell i} \; [1 + (\omega^k_{\ell i})^2 \langle \dot{\phi}^2_\ell \rangle]^{-1} \tag{4.42c}$$

Where $\omega^k_{\ell i} = \tilde{E}^k_{\ell i} - E_{\nu\ell}$. The exact energy scaling will in Sect. III G (eqs. 4.58 and 61) be shown to be given by the implicit equation

$$S^k_{\ell i} = 2(2\ell + 1)(\frac{a}{s})^{2\ell+1} \frac{D_\ell(E) + \ell + 1}{D_\ell(E) - \ell} \tag{4.43}$$

involving the logarithmic derivative function (2.13). For the moment we can see that (4.43) is correct, at least to third order in $E - E_\nu$, by comparing (2.69 and 71) with (4.42).

The canonical s-, p-, and d-bands for the bcc structure are shown in Fig. 8 and they look just like real energy bands. The canonical bands for the fcc and hcp structures may be found in Andersen and Jepsen (1977) and in Mackintosh and Andersen (1980). For certain high-symmetry k-points bands of different ℓ-character belong to different irreducible representations and hybridization between them is therefore forbidden by symmetry. In those cases the transformation from the canonical- to the real energy bands is exact. The values of the fcc, bcc, and hcp canonical bands at the high-symmetry points have been listed in Tables II - IV.

The canonical d-bands can with reasonable accuracy be calculated by hand using Table I but, for the p- and s-bands the Ewald procedure must be used as was mentioned above. The distortion parameter Q^{-1} is small for transition-metal d-waves and the unhybridized d-bands therefore look like the canonical d-bands. For the s- and p-waves $Q_\ell^{-1} S^k_{\ell i}$ is only slightly less than unity near the top of the band and the distortion is therefore larger. As we shall see below, the canonical s-band is such that the form (4.42a) rather than (4.42b) is appropriate.

From (4.7) and (4.10) we may derive the following general properties of the structure constants. The centre of gravity of a canonical band is zero. This holds at each k-point except for canonical p-bands at k = 0 for canonical s-bands throughout the Bril-

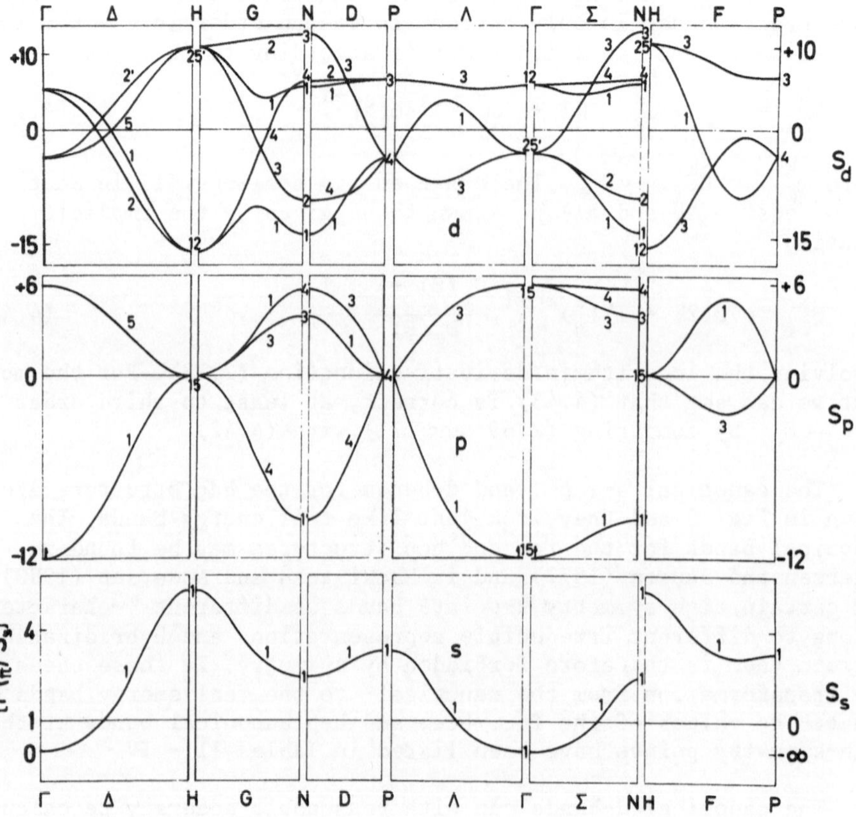

Fig. 8 : The canonical bands for the bcc structure.

Table I. Canonical Structure Constants.
The notation follows that of Slater and Koster. The vector from the first to the second orbital has the length d and the direction cosines ℓ, m, and n. The distance a, which also enters the definition of the potential functions, is arbitrary. The entries not given in the table may be found by cyclically permuting the coordinates and direction cosines. Moreover, $S(\ell'm',\ell m) = (-)^{\ell'+\ell} S(\ell m,\ell'm')$, where ℓm refers to the angular momentum. The present, real structure constants equal those defined in Andersen (1975) times $i^{\ell'-\ell}(a/s_{R'})^{\ell'+1/2}(a/s_R)^{\ell+1/2}$.
For clusters, s_0 is the radius of the outer sphere.
For crystals, $s_0 = \infty$.

$S(s,s)$	$2(a/s_0) - 2(a/d)$
$S(s,x)$	$\ell \; 2\sqrt{3}(a/d)^2$
$S(x,x)$	$(3\ell^2-1) \; 6(a/d)^3$
$S(x,y)$	$3\ell m \; 6(a/d)^3$
$S(s,xy)$	$-\sqrt{3}\ell m \; 2\sqrt{5}(a/d)^3$
$S(s,x^2-y^2)$	$-\sqrt{3}(\ell^2-m^2)/2 \; 2\sqrt{5}(a/d)^3$
$S(s,3z^2-r^2)$	$(1-3n^2)/2 \; 2\sqrt{5}(a/d)^3$
$S(x,xy)$	$(1-5\ell^2)m \; 6\sqrt{5}(a/d)^4$
$S(x,x^2-y^2)$	$[1-5(\ell^2-m^2)/2]\ell \; 6\sqrt{5}(a/d)^4$
$S(x,yz)$	$-5\ell mn \; 6\sqrt{5}(a/d)^4$
$S(z,x^2-y^2)$	$-5n(\ell^2-m^2)/2 \; 6\sqrt{5}(a/d)^4$
$S(x,3z^2-r^2)$	$(\sqrt{3}/2)(1-5n^2)\ell \; 6\sqrt{5}(a/d)^4$
$S(z,3z^2-r^2)$	$(\sqrt{3}/2)(3-5n^2)n \; 6\sqrt{5}(a/d)^4$
$S(xy,xy)$	$[-35\ell^2m^2-5n^2+4] \; 10(a/d)^5$
$S(x^2-y^2,x^2-y^2)$	$[-35(\ell^2-m^2)^2/4-5n^2+4] \; 10(a/d)^5$
$S(3z^2-r^2,3z^2-r^2)$	$(-3/4)[35n^4-30n^2+3] \; 10(a/d)^5$
$S(xy,x^2-y^2)$	$-35\ell m(\ell^2-m^2)/2 \; 10(a/d)^5$
$S(zx,x^2-y^2)$	$-5[7(\ell^2-m^2)/2-1]\ell n \; 10(a/d)^5$
$S(yz,x^2-y^2)$	$5[7(m^2-\ell^2)/2-1]mn \; 10(a/d)^5$
$S(yz,zx)$	$-5(7n^2-1)\ell m \; 10(a/d)^5$
$S(x^2-y^2,3z^2-r^2)$	$(-\sqrt{3}/2)5(7n^2-1)(\ell^2-m^2)/2 \; 10(a/d)^5$
$S(xy,3z^2-r^2)$	$(-\sqrt{3}/2)5(7n^2-1)\ell m \; 10(a/d)^5$
$S(xz,3z^2-r^2)$	$(-\sqrt{3}/2)5(7n^2-3)\ell n \; 10(a/d)^5$

Table II. BCC Canonical Bands, $S^{k}_{\ell i}$, at Points of High Symmetry

	Γ		H		N		P	
	12	5.394	25'		3	12.292	3	6.214
	12		25'	10.792	4	6.064	3	
d	25'		25'		1	5.394	4	
	25'	- 3.596	12	- 16.187	2	- 9.603	4	- 4.142
	25'		12		1	- 14.146	4	
	15		15		4'	5.689	4	
p	15	6.000	15	0.000	3'	3.834	4	0.000
	15		15		1'	- 9.524	4	
s	1	- ∞	1	2.004	1	1.458	1	1.721

Table III. Fcc Canonical Bands, $S^{k}_{\ell i}$, at Points of High Symmetry

	Γ		X		W		L		K	
	12	4.116	5	10.936	1'	10.940	3	9.516	2	9.243
	12		5		1	4.887	3		4	5.970
d	25'		2	9.127	3	- 3.258	3	- 3.058	3	4.395
	25'	- 2.744	3	- 14.595	3		3		1	- 7.662
	25'		1	- 16.404	2'	- 9.311	1	- 12.915	1	- 11.946
	15		5'	3.104	2'	2.258	3	5.178	4	3.839
p	15	6.000	5'		3	- 1.129	3		1	- 0.596
	15		4'	- 6.208	3		2'	- 10.357	3	- 3.243
s	1	- ∞	1	1.762	1	1.809	1	1.366	1	1.769

Table IV. HCP Canonical Bands, $S^{k}_{\ell i}$, at Points of High Symmetry

	Γ		K		M		L		A		H	
	5^-	8.61	4	8.41	4^+	10.36	2	10.06	3		3	9.11
	5^-		5	7.93	2^+	9.65	2		3	6.51	3	
	6^+	1.88	5		3^-	9.53	1	9.09	3		1	2.68
	6^+		3	3.50	1^+	7.09	1		3		1	
d	5^+	− 0.47	6	− 1.80	2^-	2.06	2	3.17	3		2	0.60
	5^+		6		4^-	0.52	2		3	− 2.63	2	
	6^-	− 2.14	1	− 2.60	1^-	− 2.99	1	− 8.88	3		1	− 4.48
	6^-		5	− 4.78	2^-	− 9.12	1		3		1	
	1^+	− 2.86	5		3^+	− 12.68	1	− 13.44	1	− 7.76	3	− 7.91
	4^-	− 12.91	2	− 12.01	1^+	− 15.01	1		1		3	
	6^-	6.000	6	2.546	4^+	5.180	2	4.149	3		2	4.807
	6^-		6		3^+	3.928	2		3	5.588	2	
p	2^-	6.000	3	1.871	3^-	3.107	1	3.815	3		1	− 1.082
	5^+	5.180	5	− 1.273	4^-	0.437	1		3		1	
	5^+		5		2^-	− 3.544	1	− 7.964	1	− 11.175	2	− 3.725
	3^+	− 10.362	1	− 4.417	1^+	− 9.109	1		1		2	
s	4^-	1.356	5	1.694	2^-	1.731	1	− 1.580	1	− 1.907	3	1.718
	1^+	$-\infty$	5		1^+	1.334	1		1		3	

louin zone. It always holds for the average over the Brillouin zone. The meaning of the potenital parameter C_ℓ is thus that it would be the centre of the unhybridized ℓ-band if the distortion parameter Q^{-1} were zero.

Due to the infinite range of the ss-interaction, (4.40) which behaves like the Coulomb interaction, the canonical s-band diverges for k = 0. By Fourier-transformation we realize that

$$S_s^k \to - 6(ks)^{-2} + \text{const.}, \text{ for } k \to 0 \tag{4.44}$$

and from (4.43) we see that this corresponds to $D \to 0 = \ell$. We therefore use (4.42a) instead of (4.42b) and find the well-known parabolic behaviour

$$E_s^k \to V_s + (ks)^2\Gamma_s/6 \equiv V_s + k^2/\tau_s \tag{4.45}$$

The potential parameter V_s is thus the bottom of the s-band and the parameter

$$6[\Gamma_s s^2]^{-1} = 3[s^3\Phi_s^2(0,s)]^{-1} = \tau_s \tag{4.46}$$

is the mass of the s-band.

One subband of the canonical p-band is discontinuous for $k \to 0$ and the continuity of the corresponding energy band arises only through hybridization with an s-band. This behaviour of the longitudinal (i.e. the p-orbital points in the direction of the k-vector) canonical p-band arises from the requirement that the structure matrix be independent of the lattice constant. We shall see below that the screened canonical p-band $\tilde{S}_p^k$ is perfectly continuous and differentiable.

The width of and the average hybridization between canonical bands may conveniently be estimated from the following expression for the second moment

$$|S_{t'\ell'}^{t\ell}|^2 \equiv \sum_{R'm'} \sum_{Rm} |S_{R'\ell'm',R\ell m}|^2$$

$$= \frac{4(2\ell'+1)(2\ell+1)(2\ell'+2\ell)!}{(2\ell')!(2\ell)!} \sum_{\underline{R}} \sum_{\underline{R}'} (\frac{a}{d})^{2(\ell'+\ell+1)} \tag{4.47}$$

This expression is quite general and it only depends on the numbers of atoms in the various shells and on the distance between them. R runs over all atoms of type t and R' runds over all atoms of type t'. For t = t' the terms with R = R' are excluded from the double sum. For a crystal R runs over all type t atoms in the cell and R' runs over all type t' atoms in the entire crystal. If we put t = t' and ℓ = ℓ' the second moment (4.47) equals

$$|S^{t\ell}_{t\ell}|^2 = \sum_i (BZV)^{-1}\int(S^{k}_{t\ell\, i})^2 d^3k \qquad (4.48)$$

and this provides the measure

$$[12|S^{t\ell}_{t\ell}|^2/n_{t\ell}]^{1/2} \equiv W_{t\ell} \qquad (4.49)$$

for the width of the canonical tℓ-band. Here $n_{t\ell}$ is the number of tℓ-bands, which equals 2ℓ + 1 times the number of type t atoms in the primitive cell. The factor 12 in (4.49) arises from our definition that W must be the exact width when the canonical band has a rectangular density of states (i.e. when the eigenvalues $S^{k}_{t\ell\, i}$ are uniformly distributed from -W/2 to W/2).

For the elemental bcc, fcc and hcp structures (4.47) yields

$$W_p = 18.8,\ 18.7, \text{ and } 18.6$$

and

$$W_d = 23.8,\ 23.5, \text{ and } 23.5 \qquad (4.50)$$

respectively. Due to the long range of the canonical s-band its second moment diverges and the scaling (4.42b) from the canonical band to the energy band is so non-linear that the assumption of a rectangular density of states does not hold. In (4.45) we did, however, already determine the position of the bottom of the s-band and we now estimate the position of the top. We simply assume that (4.44) holds throughout the Brillouin zone and determine the value of the constant from the requirement that the integral of (4.44) in the Brillouin zone vanishes. For this purpose we approximate the Brillouin zone by a sphere of the same volume, that is, with the radius

$$k_{BZ} = (9\pi/2)^{1/3}\ s^{-1} \qquad (4.51)$$

Wit this simple model the value of the constant in (4.44) is $18(9\pi/2)^{-2/3} = 3.08$ and the result for the top of the canonical s-band becomes

$$S^{k_{BZ}}_{s} = 12(9\pi/2)^{-2/3} = 2.05 \qquad (4.52)$$

The simple estimates (4.44), (4.52) and (4.50) of the canonical band widths may now be compared with the computed values given in Tables II-IV and shown in Fig. 8. The agreement is quite satisfactory. We may also check the quality of the simple Wigner-Seitz rule considered in (2.15-18). Translated into the S-scale (4.43) the WS-rule states that the extents of the canonical s-, p- and d-bands should be $(-\infty,2)$, $(-12,6)$ and $(-15,10)$, respectively. Compared with Tables II - IV the WS-rule is seen to hold within 10 per cent of the band widths for the closely packed solids considered. At the bottom of the s-band the rule is, of course, exact if the symmetry is cubic and if the partial waves with $\ell \geqslant 4$ are neglected. For a general ℓ-band where the WS-rule is merely intuitive its accuracy is surprisingly good in view of the complicated three-dimensional nature of the WS-matching problem. Had the WS-rule instead been defined for the logarithmic derivative at half the nearest neighbor distance (i.e. at the MT-radius), which is a very reasonable choice, the estimates of d-band widths in transition metals would have been about 50 per cent too large !

E. Potential parameters and screened canonical bands

For each type of atom and for each value of ℓ the energy bands are described by 4 independent potential parameters : three of the 4 numbers, C, V, $\sqrt{\Delta}$ and $\sqrt{\Gamma}$, plus $<\phi^2>$. To second order only, that is in (4.35), $<\phi^2>$ is not needed. The potential parameters were defined in terms of phi and phi-dot in (4.14-15), (4.17-18), (4.37-39) and in Sect. IIE. The physical significance of the parameters should be apparent from the form of the Hamiltonian (4.35) and from our discussion of the canonical bands in the previous section. The energy C_ℓ is called the centre of the band and it is the energy of the LMTO inside its own sphere. To second order it is the energy for which the logarithmic derivative takes the value $-\ell-1$.

The energy V_ℓ is called the square-well pseudo-potential because if the potential in the atomic sphere were flat and equal to v we would have $V_\ell = v$, to second order. For free electrons $(C_s - v)s^2 = \pi^2/4$ and $(C_p - v)s^2 = \pi^2$, if we choose $E_\nu = C$. The value of the remaining potential parameters for a flat potential both for $E_\nu = C$ and for $E_\nu = V$ may be found in Table V.

The energy Δ_ℓ translates the width of the canonical band into the width of the energy band and $\Delta_\ell^{1/2}\,\Delta_{\ell'}^{1/2}$ gives the strength of the hybridization between the ℓ- and ℓ'-bands. It is sometimes convenient to transform Δ into a relative band-mass μ according to

$$\Delta \equiv (a/s)^{-2\ell-1}\,(s/2)\Phi_-^2$$
$$\equiv (a/s)^{-2\ell-1}\,(\mu s^2)^{-1} \tag{4.53}$$

For a flat potential $\mu_\ell = 1$ to first order in $E_{\nu\ell} - C_\ell$. This may be shown from the properties of the spherical Bessel functions. Since the distortion Q^{-1} of transition-metal d-bands is small the d-band width in an elemental transition metals is approximately $25\Delta_d \approx 25(\mu_d s^2)^{-1}$ as seen from (4.50).

The energy Γ_ℓ gives the scale of the band near V_ℓ. As for Δ

we may transform Γ into a relative band mass T

$$\Gamma \equiv 4(2\ell + 1)^2 (a/s)^{2\ell+1} (s/2)\Phi_+^2$$
$$\equiv (a/s)^{2\ell+1}\; 2(2\ell + 1)^2 (2\ell + 3)(Ts^2)^{-1} \tag{4.54}$$

In (4.24) we already proved that $\tau_\ell = 1$ to first order for free electrons and this result was anticipated in (4.45 - 46).

While the potential parameters for free electrons are listed in Table V we list in Table VI the selfconsistently calculated (Glötzel, Andersen and Jepsen 1984) potential parameters for 33 elemental metals. Here we used a ≡ s and the $E_{\nu\ell}$'s were chosen such that the first moment, ε_ℓ^1 (2.52), of the projected density of states vanishes, i.e. the $E_{\nu\ell}$ listed are the centers of gravity for the occupied parts of the bands.

The values of the distortion parameters Q_ℓ^{-1} for a given ℓ exhibit only a weak variation through the part of the periodic table considered, and the values for ℓ = 0 and 1 are not far from those listed for free electrons in Table V. Moreover, according to the discussion around (4.39), the use of "standard-values" for Q^{-1} need only lead to small second-order errors of the energy bands. To this degree of accuracy we may therefore consider the <u>screened</u> structure constants (4.36) as being properties of the structure and not of the potential.

The properly defined <u>unhybridized energy bands</u> are, of course, those for which the hybridization between theta-orbitals of different ℓ's are neglected, rather than those obtained by setting the unscreened structure constants with ℓ ≠ ℓ' equal to zero. The unhybridized energy bands are therefore to second order

$$\tilde{H}^k_{\ell i} = C_\ell + \Delta_\ell \tilde{S}^k_{\ell i} \tag{4.55}$$

in terms of the eigenvalues of the diagonal blocks $\tilde{S}^k_{\ell m,\ell m'}$ of the <u>screened</u> structure constants. The latter we might call <u>screened canonical bands</u>. In Fig. 9 we show the unhybridized and hybridized energy bands of bcc Vanadium (Andersen and Jepsen 1984). The unhybridized bands are according to (4.55) simply the screened canonical bands scaled by Δ and positioned at C (which need no longer be the centre of gravity). We may now compare the bcc screened canonical bands with the unscreened bands in Fig. 8 and we see that the pathological behaviour of the <u>unscreened</u> p-band arising from the long-range interaction of the LMTOs has disappeared. The Fourier-transform of the smooth $\tilde{S}^k_{\ell i}$-bands in Fig. 9 into $\tilde{S}_{R\ell i}$ seems to converge reasonably fast in R-space and this means that the generalized Wannier functions defined by transforming the theta-orbitals $\Theta_{R\ell m}$ (2.35) into $\Theta_{R\ell i}$ are reasonably well localized.

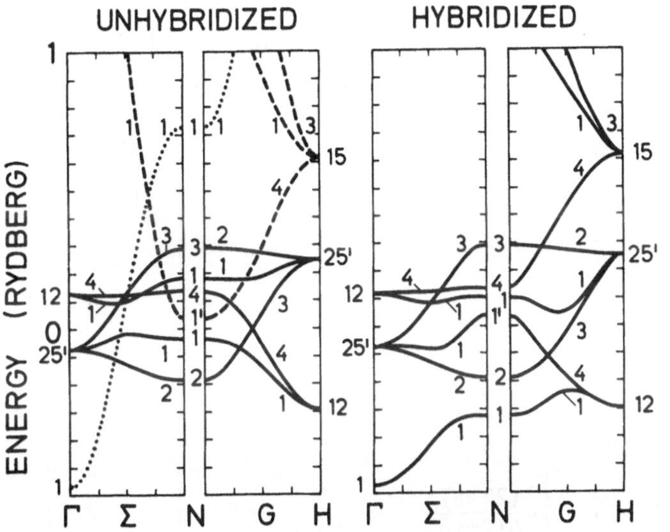

Fig. 9 : The unhybridized screened energy bands $\tilde{H}^k_{\ell i}$ and the hybridized energy bands E^k_j of bcc vanadium metal. Φ_-/Φ_+ = 0.869(s), 0.544(p), 0.095(d).

Table V. Potential Parameters for Free Electrons

s^2E_ν	s	0		$\pi^2/4$	2.47
	p	0		π^2	9.87
	d	0			20.19
$s^2\omega(-\ell-1)$	s		2.5	0	
	p	$\frac{1}{2}(2\ell+1)(2\ell+5)$	10.5	0	
	d		22.5	0	
$s^3\Phi^2(-\ell-1)$	s		2.08	2	
	p	$\frac{(2\ell+5)^2}{4(2\ell+3)}$	2.45	2	
	d		2.89	2	
$\frac{\Phi(-\ell-1)}{\Phi(\ell)}$	s		.83	$1-4/2\pi^2$	.80
	p	$\frac{2\ell+5}{2(2\ell+3)}$	.70	$1-9/2\pi^2$	.54
	d		.64		.38
$s^2\langle\dot{\phi}^2\rangle^{-1/2}$	s	$\frac{(2\ell+5)\times}{\sqrt{(2\ell+3)(2\ell+7)}}$	23.	$\sqrt{\frac{3\pi^2}{1-9/\pi^2}}$	18.
	p		47.	$\sqrt{\frac{12\pi^2}{1-33/4\pi^2}}$	27.
	d		79.		35.

Table VIa : Potential Parameters at experimental equilibrium atomic volume

		K Rb Cs	Ca Sr Ba	Sc Y Lu	Ti Zr Hf	V Nb Ta	Cr Mo W	Mn Tc Re	Fe Ru Os	Co Rh Ir	Ni Pd Pt	Cu Ag Au
E_ν (m Ry)	s	-270	-340	-351	-351	-350	-374	-438	-474	-497	-521	-545
		-260	-320	-337	-337	-329	-352	-381	-424	-485	-537	-545
		-243	-271	-377	-397	-399	-433	-464	-504	-565	-620	-638
	p	-247	-286	-264	-233	-223	-227	-290	-325	-346	-370	-404
		-241	-282	-260	-226	-208	-209	-235	-277	-348	-422	-450
		-229	-246	-271	-242	-223	-225	-246	-285	-354	-435	-491
	d	-233	-257	-222	-207	-198	-215	-248	-263	-272	-282	-357
		-226	-250	-229	-219	-211	-240	-245	-277	-330	-387	-509
		-214	-222	-230	-224	-209	-242	-242	-269	-320	-383	-471
	f	-214	-251	-251	-184	-177	-185	-233	-256	-264	-281	-359
		-163	-220	-217	-190	-184	-197	-216	-242	-302	-375	-496
		-212	-219	-223	-193	-180	-193	-206	-220	-274	-357	-465
ω_- (m Ry)	s	31	16	74	136	188	215	184	182	174	161	105
		21	2	50	106	161	189	200	205	192	171	89
		18	27	10	45	80	95	98	99	92	76	27
	p	347	418	642	837	1011	1099	1038	1053	1067	1063	961
		334	400	607	804	1005	1092	1140	1160	1116	1047	878
		308	473	645	804	982	1052	1094	1118	1086	1020	911
	d	468	306	238	215	189	157	104	71	50	30	11
		392	300	276	266	251	214	155	107	60	23	0
		266	236	337	319	302	264	197	143	84	34	2
	f	1160	1747	2422	3011	3552	3733	3999	3946	4367	4494	4415
		920	1181	1634	2012	2378	2656	2814	2967	3112	3061	2998
		771	989	1913	2225	2533	2707	2827	2908	2925	2991	2899
$10 s \phi_-^2$ (m Ry)	s	1031	1468	2318	2036	3611	3958	3743	3744	3772	3719	3325
		910	1256	1981	2629	3207	3542	3701	3755	3561	3223	2702
		797	1323	2009	2544	3040	3302	3433	3480	3317	2988	2562
	p	1045	1423	2171	2805	3320	3624	3446	3466	3507	3482	3162
		970	1286	1957	2567	3129	3425	3566	3615	3447	3165	2705
		868	1386	2099	2617	3117	3371	3507	3569	3441	3167	2796
	d	621	411	394	414	418	397	319	280	256	230	174
		544	469	541	614	667	651	601	539	440	337	220
		386	449	646	715	783	773	728	672	569	448	325
	f	831	1630	2300	2867	3366	3292	3740	3416	4046	4146	4022
		640	851	1278	1649	1980	2254	2325	2429	2587	2385	2221
		549	829	1603	1923	2220	2363	2451	2498	2443	2523	2280
$10^3 \phi_- / \phi_+$	s	846	846	860	866	869	868	863	860	857	853	845
		847	848	863	871	875	876	874	872	868	862	850
		852	865	854	861	866	867	866	864	860	854	843
	p	694	685	702	711	715	711	697	691	687	682	671
		711	706	728	746	758	753	744	733	720	709	693
		727	758	722	736	749	746	740	734	723	712	700
	d	469	283	173	128	94	67	33	10	- 6	- 21	- 29
		472	354	285	251	225	195	159	129	97	69	56
		431	351	314	285	265	240	207	179	146	113	88
	f	402	561	553	550	550	508	556	514	562	565	571
		331	360	405	424	434	449	441	444	471	455	458
		405	448	466	467	469	464	461	458	454	483	471

Table VIb : Band Parameters at experimental equilibrium atomic volume

		K Rb Cs	Ca Sr Ba	Sc Y Lu	Ti Zr Hf	V Nb Ta	Cr Mo W	Mn Tc Re	Fe Ru Os	Co Rh Ir	Ni Pd Pt	Cu Ag Au
$\langle \dot{\phi}_\nu^2 \rangle^{-\frac{1}{2}}$ (m Ry)	s	1300	1840	3100	4200	5070	5560	5120	5060	5030	5870	4190
		1150	1590	2710	3750	4690	5200	5420	5460	5080	4490	3560
		1040	1840	2640	3490	4280	4680	4850	4990	4580	4030	3300
	p	2000	2670	4000	5020	5900	6550	6470	6570	6660	6600	5920
		1880	2470	3500	4250	4890	5560	6130	6580	6640	6280	5330
		1680	2280	3960	4700	5320	5880	6330	6690	6730	6360	5620
	d	1190	810	850	950	1000	1000	860	800	770	730	610
		1120	960	1140	1320	1460	1460	1390	1290	1110	930	710
		610	960	1360	1530	1700	1710	1640	1540	1360	1140	910
s_{exp} (a.u.)		4.862	4.122	3.427	3.052	2.818	2.684	2.699	2.662	2.621	2.602	2.669
		5.197	4.494	3.761	3.347	3.071	2.922	2.840	2.791	2.809	2.873	3.005
		5.656	4.652	3.624	3.301	3.069	2.945	2.872	2.825	2.835	2.897	3.002
n $\left(\frac{\text{electrons}}{\text{atom}}\right)$	s	0.616	0.849	0.757	0.685	0.637	0.624	0.646	0.633	0.644	0.651	0.699
		0.633	0.905	0.795	0.715	0.657	0.649	0.650	0.641	0.631	0.616	0.687
		0.634	0.799	0.886	0.818	0.770	0.760	0.766	0.764	0.759	0.756	0.801
	p	0.309	0.656	0.674	0.722	0.693	0.776	0.785	0.751	0.740	0.721	0.733
		0.273	0.526	0.587	0.659	0.649	0.795	0.791	0.788	0.746	0.653	0.667
		0.237	0.334	0.645	0.768	0.776	0.937	0.958	0.962	0.928	0.839	0.782
	d	0.071	0.480	1.542	2.539	3.605	4.518	5.489	6.528	7.533	8.551	9.505
		0.084	0.544	1.575	2.536	3.579	4.407	5.399	6.400	7.464	8.601	9.556
		0.125	0.841	1.426	2.324	3.332	4.138	5.087	6.055	7.095	8.217	9.276
	f	0.004	0.015	0.027	0.054	0.065	0.082	0.080	0.088	0.083	0.077	0.063
		0.010	0.025	0.043	0.090	0.115	0.149	0.160	0.171	0.159	0.130	0.090
		0.004	0.026	0.043	0.090	0.122	0.165	0.189	0.219	0.218	0.188	0.141
	tot	1	2	3	4	5	6	7	8	9	10	11
q $\left(\frac{\text{electrons}}{\text{atom}}\right)$		1.14	2.16	2.76	3.13	3.26	3.40	3.22	3.06	2.93	2.78	2.54
		1.15	2.20	3.04	3.66	4.02	4.36	4.27	4.14	3.76	3.23	2.75
		1.16	2.24	3.12	3.88	4.40	4.86	4.92	4.89	4.58	4.04	3.45
v (m Ry)		-363	-501	-630	-724	-785	-830	-812	-810	-811	-803	-766
		-345	-468	-598	-699	-774	-828	-844	-849	-821	-770	-706
		-321	-457	-622	-719	-795	-850	-871	-883	-863	-817	-755
E_F (m Ry)		-199	-200	-166	-123	-111	- 62	-119	-125	-134	-150	-145
		-197	-190	-166	-119	- 98	- 33	- 42	- 57	-109	-202	-186
		-193	-187	-153	-115	- 90	- 15	- 6	- 4	- 36	-133	-165
rms($E-E_\nu$) (m Ry)	s	43	74	86	89	94	105	112	113	123	128	154
		40	66	74	77	85	104	114	121	118	108	155
		33	49	96	102	111	128	142	155	153	144	171
	p	35	66	82	90	93	103	108	107	115	118	136
		32	59	75	84	88	104	115	122	118	103	146
		27	42	102	113	119	133	147	156	156	143	153
	d	28	46	50	58	62	74	82	87	88	88	75
		25	41	49	63	72	92	117	127	124	111	83
		20	28	67	79	88	105	138	153	156	146	121
	f	38	43	48	52	57	70	71	75	79	78	72
		20	39	46	58	68	93	108	119	115	99	89
		9	83	65	72	80	101	121	141	144	131	108

F. Ghost bands

As defined in (2.35) the theta orbitals are not always localized. If we choose $E_{\nu\ell}$ between the $n\ell$- and the $(n+1)\ \ell$-band, say we choose $E_{\nu d}$ in Vanadium between the 3d- and the 4d-band at about 1 Ry, then a linear method using one non-integer radial quantum number ν does not know which one of the two bands to pick. As a result of this schizophrenia the method might choose to deliver the top of the 3d-band together with the bottom of the 4d-band and to connect the two pieces by a steep "ghost-band" which crosses the region of interest or diverges. The corresponding theta orbital is of course not localized.

What happens in the formalism is that since the logarithmic derivative is positive between two bands we may have $D\{\phi\} > 0$ and $D\{\dot{\phi}\} < 0$. With such a phi-dot it may not be possible to construct theta orbitals with decreasing tails. The parameter Q (4.39) becomes small and π^{-1} (4.32) develops a pole. To cure this disease we may, for such a partial wave which is not contributing a band in the region of interest but merely hybridizes weakly with the other (4s- and 4p- in Fig. 9) bands, set $D\{\dot{\phi}\}$ equal to a standard value which does not cause a pole. As discussed above, this merely leads to second-order errors for the weak ℓ-component in question.

G. The KKR method in the ASA

In this final section we shall shift the emphasis from the linear methods to the ASA and derive a partial-wave method in the ASA without the additional approximations arising from the linearization of the wavefunctions. This will simply be the KKR-method with kappa set equal to zero and the MT-spheres substituted by the overlapping atomic Wigner-Seitz spheres. Apart from being illuminating this formalism is useful in connection with the LMTO method for treating localized impurities (Gunnarsson, Jepsen and Andersen 1983). Instead of the LMTO defined in (4.13) we now define the radial part of the energy-dependent MTO as follows

$$\chi(E,r) = \begin{cases} \phi(E,r) - \dfrac{D(E)+\ell+1}{2\ell+1}\left(\dfrac{r}{s}\right)^{\ell}\phi(E,s) \\[2ex] \dfrac{\ell-D(E)}{2\ell+1}\left(\dfrac{r}{s}\right)^{-\ell-1}\phi(E,s) \end{cases} \tag{4.56}$$

Here, the upper part is valid for $r \leqslant s$ and the lower part is valid for $r \geqslant s$. The MTO (4.56) is everywhere continuous and differentiable and for the energy C, where $D = -\ell-1$, it equals the LMTO (4.13) with $E_\nu = C$. The MTO is not augmented inside the other spheres and we may use (4.9) to expand its tail. If we define

$$|r_+> \equiv (\frac{r_R}{a})^\ell \frac{Y_L(\hat{r}_R)}{2(2\ell + 1)} \tag{4.57}$$

and

$$P(E) \equiv 2(2\ell + 1) (\frac{a}{s})^{2\ell+1} \frac{D(E) + \ell + 1}{D(E) - \ell} \tag{4.58}$$

which is the so-called potential function, we may write the MTO in the one-centre form (4.21) as

$$|\phi(E)> + |r_+>[P(E) - S]\chi(E,a) = |\chi(E)>^\infty \tag{4.59}$$

Here, S is the structure matrix and P(E) and $\chi(E,a)$ are matrices diagonal in R and L.

We may now place a linear combination of MTOs at each site of the structure and ask whether we can determine the coefficients in such a way that

$$\sum_{RL} \chi_{RL}(E,\underline{r}_R)b_{RL} \tag{4.60}$$

is a solution of Schrödingers equation at energy E. The condition is, of course, that the one-centre expansion (2.9) in terms of the partial waves is valid. This means that inside any sphere the sum of the MTO tails coming from all other sites must cancel the unphysical r_+-parts (4.56) of the MTO's on that site. From (4.59) we realize that this matching condition may be expressed analytically as

$$\sum_{RL} [P_{R\ell}(E)\delta_{R'R}\delta_{L'L} - S_{R'L',RL}]\dot{P}_{R\ell}(E)^{-1/2}b_{RL} = 0 \tag{4.61}$$

for all R' and L'. We have here used that

$$\dot{P}(E)^{-1/2} = (a/2)^{1/2} \chi(E,a)$$
$$= \frac{\ell - D(E)}{2\ell + 1} (\frac{s}{a})^{\ell+1/2} (\frac{s}{2})^{1/2} \phi(E,s) \tag{4.62}$$

This may be proved from the definition (4.58) and the fact that

$$\dot{D}(E) = - [s\phi^2(E,s)]^{-1} \tag{4.63}$$

which follows from (2.62) for $D \approx D\{\phi\} + \dot{D}\{\phi\}dE$ with the help of (2.6). Eqs. (4.61) are then the KKR equations in the ASA and they have the form (2.12). These equations only have a non-trivial solution for those energies where the determinant of the matrix in the square bracket vanishes, i.e.

$$\det|P(E) - S| = 0 \tag{4.64}$$

and these are then the one-electron energies. Eq. (4.64) is seen to be the generalization of (4.43).

We may finally close the circle and "linearize" (4.61). For this purpose we note that (2.69) allows us to write

$$P(E) = \frac{\Gamma}{V - E} + Q \qquad (4.65)$$

to second order, and hence

$$\dot{P}(E)^{-1/2} = (E - V)\Gamma^{-1/2} \qquad (4.66)$$

We now multiply (4.61) on the left by the matrix $\Gamma^{1/2}(Q - S)^{-1}$ and obtain

$$\begin{aligned} 0 &= \Gamma^{1/2}(Q - S)^{-1} \left(Q - S + \frac{\Gamma}{V - E}\right) \frac{E - V}{\Gamma^{1/2}} b \\ &= (E - \widetilde{H})b \end{aligned} \qquad (4.67)$$

with $\widetilde{H}$ given by (4.35a), Q.E.D.

REFERENCES

Linear Methods in general

Andersen, O.K., Mont Tremblant Lectures 1973 (unpublished), Phys. Rev. B12, 3060 (1975) and Europhysics News 12, 5, 1 (1981).

LMTO Method

Andersen, O.K., and Woolley, G., Mol.Phys. 26, 905 (1973).
Andersen, O.K., Solid St. Commun. 13, 133 (1973).
Andersen, O.K., and Jepsen, O., Physica, 91B, 317 (1977).
Andersen, O.K., Klose, W., and Nohl, H., Phys.Rev. B17, 1209 (1978).
Casula, F., and Herman, F., J.Chem.Phys., Jan 15 (1983) and references therein.
Glötzel, D., Segall, B., and Andersen, O.K., Solid St. Commun. 36, 403 (1980).
Glötzel, D., Andersen, O.K., and Jepsen, O., Adv.Phys. (1984).
Gunnarsson, O., Harris, J. and Jones, R.O., Phys.Rev. B15, 3027 (1977).
Gunnarsson, O., Jepsen, O., and Andersen, O.K., Phys.Rev. B27, 7144 (1983).
Harris, J., Chapter of the present volume, and references therein.
Jarlborg, T., J.Phys. F9, 283 (1979).
Kasowski, R.V., Phys.Rev. B25, 4189 (1982) and references therein.
Mackintosh, A.R., and Andersen, O.K., in Electrons at the Fermi Surface. Ed. Springford (Cambridge Univ. Press 1980).
Pettifor, D.G., J.Phys. F7, 613 (1977).

Skriver, H.L., The LMTO-method (Springer 1983) and references therein.
Williams, A.R., Kübler, J. and Gelatt, C.D.,Jr., Phys.Rev. B19, 6096 (1979)

LAPW Method

Hamann, D.R., Mattheiss, L.F., and Greenside, H.S., Phys.Rev. B24, 6151 (1981).
Jepsen, O., Madsen, J. and Andersen, O.K., Phys.Rev. B18, 605 (1978) and ibid B26, 2790 (1982).
Koelling, D.D., and Arbman, G., J.Phys. F5, 2041 (1975).
Müller, J.E., Jepsen, O., Andersen, O.K., and Wilkins, J.W., Phys. Rev.Lett. 40, 720 (1978).
Wimmer, E., Krakauer, H., Weinert, M., and Freeman, A.J., Phys.Rev. B24, 864 (1981).

DENSITY FUNCTIONAL THEORY FOR SOLIDS

Ulf von Barth

IBM Thomas J. Watson Research Center
Yorktown Heights, New York 10598

and

Department of Theoretical Physics
University of Lund, 22362 Lund, Sweden

1. DENSITY-FUNCTIONAL THEORY

1.1. Introduction

A considerable part of all efforts in electronic structure theory is directed towards the understanding of the energetics of atoms, molecules and solids. Problems of interest are for instance the binding energy of molecules and solids, the compressibility of solids, the heat of formation of compounds, the energy associated with structural changes in solids and, more recently, the determination of the spatial arrangement of atoms on the surfaces of solids. These problems are often of considerable interest as far as practical applications are concerned. With a few exceptions, notably the energetics of atoms and small molecules for which there exist sophisticated and quantitavely accurate methods, density-functional theory is presently our only means of attacking these complicated problems. Fortunately, as will become evident from the different lectures of this school, density-functional theory is capable of yielding not only insight into the qualitative aspects of these problems but can also provide rather accurate quantitative answers to them. This is the basis for the large and growing interest in density-functional methods in the last decade. This interest is reflected in several recent review articles on the subject and for further information we refer the reader, e.g., to Ref.1 and Ref. 2. The present notes rely heavily on Ref. 1 and they can be described as an abbreviated and edited version of Ref. 1 but with new material added.

1.2. Explicit Functionals

Density-functional theory in its earlier versions such as the statistical theory of the atom dates back almost as far as quantum mechanics[3,4,5]. The modern version of the theory which underlies the successes of the last 5-10 years was introduced by Hohenberg, Kohn and Sham[6,7,8] in their pioneering work from the mid-sixties. The theory is based on two theorems which center on the particle density as the fundamental variable for the description of any many-electron system. The first theorem states that the total ground state energy E of any many-electron system is a functional of the one-particle density $n(\vec{r})$. In this context different many-electron systems differ only by the local external potential felt by the electrons. Furthermore, splitting off from the total energy the explicit interaction with the external potential $w(\vec{r})$, the theorem also states that the rest is a universal functional of $n(\vec{r})$ i.e., independent of the external potential. Thus, if

$$E[n] = F[n] + \int w(\vec{r})n(\vec{r})d^3r \tag{1}$$

then the functional F depends only on n and not on w. The second theorem states that for any system (any w) the functional E[n] for the total energy has a minimum equal to the ground-state energy at the physical ground state density of that system. These theorems although rather abstract in nature were of immense importance to the rapid development of density-functional theory. They were proven in the paper by Hohenberg and Kohn[6] under certain simplifying assumptions such as for instance the non-degeneracy of the ground state and the so-called w-representability† of the particle density. The proofs are abstract and often rather difficult to grasp to a beginner in this field. Therefore, we will here follow a different route due to Levy[9] which I believe is much easier to follow in that it displays the functionals much more explicitly. Let us pick a density which is N-representable i.e., a density which is the expectation value of the density operator for some N-particle wave-function. Let us then consider the set of wave-functions M(n) which all yield this density. We can then define several functional of interest by means of the following construction

$$O[n] = \inf_{|\psi\rangle \in M(n)} \langle\psi|\hat{O}|\psi\rangle \tag{2}$$

where $\hat{O}$ is the operator corresponding to some physical observable as for instance the kinetic energy $\hat{T} = -1/2\sum_i \nabla_i^2$, the Coulomb interaction energy $\hat{U} = 1/2\sum_{ij} v(\vec{r}_i - \vec{r}_j)$ where $v(\vec{r}) = 1/r$, or their

† A particle density is called w-representable if it is the ground-state density of a many-electron system subject to some external local potential w.

sum $\hat{F} = \hat{T} + \hat{U}$. Thus, for each chosen density $n(\vec{r})$ we search among all wave functions that yield this density in order to find the smallest (infimum) expectation value of the chosen operator. Clearly this minimum value defines a functional of n for that operator. In our three examples above the sum of the kinetic energy and the intraction energy was deliberately labeled $\hat{F}$ because, for physical densities, we will now establish the identity of the corresponding functional F n and that introduced by Hohenberg and Kohn (see Eq. 1). Consider N electrons subject to the external potential $w(\vec{r})$ and let E_0 and $n_0(\vec{r})$ be their ground-state energy and density. The ground state could be degenerate in which case $n_0(\vec{r})$ would be one of the possible ground-state densities. By the explicit construction of Levy the first Hohenberg-Kohn theorem becomes trivially satisfied by defining the total energy functional according to Eq. 1 where $F[n]$ is defined in analogy with Eq. 2. It remains to be shown that $E[n]$ has a minimum at n_0 equal to the ground state energy E_0, $E[n_0] = E_0$. Let $|\psi_0\rangle$ be the ground state and let $|\psi_n\rangle$ be a state which yields a density $n(\vec{r})$ and which minimizes the expectation value of $\hat{F} = \hat{T} + \hat{U}$. Then by definition

$$E[n] = F[n] + \int n(\vec{r})w(\vec{r})d^3r = \langle\psi_n|\hat{T} + \hat{U} + \hat{W}|\psi_n\rangle \tag{3}$$

where $\hat{W} = \Sigma_i w(\vec{r}_i)$. From the variational principle for the Hamiltonian $\hat{H} = \hat{T} + \hat{U} + \hat{W}$ it follows immediately that

$$E[n] \geqslant E_0 \tag{4}$$

for all N representable densities $n(\vec{r})$. Furthermore from the definition of $F[n]$ we have

$$F[n_0] \leqslant \langle\psi_0|\hat{T} + \hat{U}|\psi_0\rangle \tag{5}$$

Adding $\int w(\vec{r})n_0(\vec{r})d^3r$ to this gives

$$E[n_0] \leqslant E_0 \tag{6}$$

which, together with Eq. (4), gives the desired result $E[n_0] = E_0$. This result also shows that equality must hold in Eq. (5) which demonstrates the equivalence of the "Levy" functional F and the "Hohenberg-Kohn" functional F for physical (w-representable) densities.

In contrast to the original work by Hohenberg and Kohn[6] which establishes a one-to-one correspondence between the ground state density and the external potential the Levy[9] construction is readily generalized to spin-polarized systems[10] where no such correspondence exist between the spin-density matrix and the external spin-dependent potential. In the spin-polarized case we simply search for the smallest expectation value of our operator of interest (e.g. $\hat{F} = \hat{T} + \hat{U}$) among those wave functions which yield a particular density matrix. The latter is defined by

$$n_{\sigma\sigma'}(\vec{r}) = \langle\psi|\psi^{\dagger}_{\sigma'}(\vec{r})\psi_{\sigma}(\vec{r})|\psi\rangle \tag{7}$$

where $\psi_{\sigma}(\vec{r})$ and $\psi^{\dagger}_{\sigma}(\vec{r})$ are the field operators for the annihilation and creation of an electron with spin σ at $\vec{r}$. Clearly this minimum is a functional of the density matrix and we can then proceed to show that the total energy including a spin-dependent external potential has a minimum equal to the ground state energy at the physical density matrix. The proof is completely analogous to that demonstrated above. Other extensions of density-functional theory are equally simple to construct. Gunnarson and Lundqvist[11] suggested a density-functional theory within each symmetry channel of a problem as a means of obtaining the energies of those lowest excited states which have a symmetry different from that of the ground state. In other words, consider a Hamiltonian which possesses some symmetry, i.e. which has a commuting symmetry operator $\hat{S}$. Then clearly the energy eigenstates can be labeled by the eigenvalues of $\hat{S}$ and for each eigenvalue S we can construct an energy functional by searching for the smallest expectation value of the Hamiltonian among wave functions which all are eigenfunctions of $\hat{S}$ with eigenvalue S and which all give the same specified particle density. Again these functionals can for each S be minimized with respect to the density to yield the lowest energy of each symmetry. In a similar way functionals can be constructed using many kinds of restrictions on the wave functions. We can, for instance, consider a density-functional theory in a world without correlation[7,9], i.e. with only exchange, be restricting the search to wave functions which are single Slater determinants giving a particular density. The generalization to finite temperatures[12] is also straightforward using the language above. In this case expectation values are replaced by ensemble averages and Eq. 2 becomes

$$O[n] = \inf \mathrm{Tr}(\rho.\hat{O}) \tag{8}$$

where Tr denotes the operation of taking the trace. Here, the search is to be carried out not over wave functions but over density matrices ρ which all yield the same particle density $n(\vec{r})$ according to

$$n(\vec{r}) = \mathrm{Tr}[\rho\psi^{\dagger}(\vec{r})\psi(\vec{r})] \tag{9}$$

As an example we can consider the analog of the energy in the case of the grand canonical ensemble, i.e. the grand potential defined by

$$\Omega(\rho) = \mathrm{Tr}[\rho(\hat{H} - \mu\hat{N} + \frac{1}{\beta}\ell n\ \rho)] \tag{10}$$

where μ is the chemical potential, $\hat{N}$ is the operator for the number of particles $[\hat{N} = \int\psi^{\dagger}(\vec{r})\psi(\vec{r})d^3r]$, and β is the invers temperature measured in units of the Boltzmann constant. We know from statistical mechanics (see e.g. Ref. 12) that the grand potential is minimized by the equilibrium density matrix given by

$$\rho_0 = e^{-\beta(\hat{H} - \mu\hat{N})}/\mathrm{Tr}\; e^{-\beta(\hat{H} - \mu\hat{N})} \tag{11}$$

so that $\Omega(\rho) \geqslant \Omega(\rho_0)$. The grand potential is readily made a functional of $n(\vec{r})$ by the prescription

$$\Omega[n] = \inf_{\mathrm{Tr}(\rho\hat{n}) = n} \Omega(\rho) \tag{12}$$

and if ρ_n is that density matrix which minimizes $\Omega(\rho)$ under the condition that it also yields $n(\vec{r})$ through Eq. (9) then $\Omega[n] = \Omega(\rho_n) \geqslant \Omega(\rho_0) \geqslant \Omega[n_0]$ where n_0 is the equilibrium density. The first inequality follows from the minimal property of the grand potential with respect to density matrices and the second inequality is a consequence of the "infinium" in the definition of $\Omega[n]$ (notice that ρ_0 produces n_0). Furthermore, if ρ' is a density matrix which both minimizes the grand potential and yields the equilibrium density n_0 then $\Omega[n_0] = \Omega(\rho') \geqslant \Omega(\rho_0)$. Thus, $\Omega[n_0] = \Omega(\rho_0)$ and it follows that the functional Ω n is minimized by the equilibrium density.

1.3. One-particle Equations

The theorems discussed above are obviously rather abstract and their utility might appear obscure at this point. Their importance stems, however, from the fact that they allow us to construct an equivalent one-particle formulation of the complicated many-body problem at hand. Consider first the simple case of N electrons moving in the external local potential $w(\vec{r})$. As prescribed above, we are to search for the minimum of the total energy functional $E[n]$ among all N-representable densities $n(\vec{r})$. Using μ as a Lagrange parameter to ensure that the density $n(\vec{r})$ integrates to N electrons.

$$N = \int n(\vec{r})d^3r \tag{13}$$

we obtain the Euler equation

$$\frac{\delta E}{\delta n(\vec{r})} = \mu \tag{14}$$

Since the explicit form of the function $E[n]$ is unknown, we have not come much further. Let us rewrite the total energy as

$$E[n] = T[n] + \tfrac{1}{2}\int n(\vec{r})v(\vec{r}-\vec{r}\,')n(\vec{r}\,')d^3r d^3r' + \int w(\vec{r})n(\vec{r})d^3r + E_{xc}[n] \tag{15}$$

where $v(\vec{r}) = \frac{1}{r}$ is the Coulomb interaction and $T[n]$ is the functional for the kinetic energy defined by replacing the operator $\hat{O}$ in Eq. 2 by the operator for the kinetic energy, $-\frac{1}{2}\Sigma_i\nabla_i^2$. The equation (15) is actually a definition of the functional $E_{xc}[n]$, called the exchange-correlation functional, and we have only performed a reshuffling of terms. The Euler equation becomes

$$\frac{\delta T}{\delta n(\vec{r})} + V_{eff}(\vec{r}) = \mu \tag{16}$$

where

$$V_{eff}(\vec{r}) = w(\vec{r}) + \int v(\vec{r} - \vec{r}\,')n(\vec{r}\,')d^3r' + v_{xc}(\vec{r}) \tag{17}$$

and

$$v_{xc}(\vec{r}) = \frac{\delta E_{xc}}{\delta n(\vec{r})} \tag{18}$$

At this point Kohn and Sham[7] made the crucial observation that the Euler equation (16) is the Euler equation of non-interacting particles subject to the external potential $V_{eff}(\vec{r})$. Thus, given V_{eff}, we now precisely know how to compute $n(\vec{r})$ and $T[n]$. We just solve the one-particle Schrödinger equation with the potential $V_{eff}(\vec{r})$ to get a set of one-particle orbitals $\varphi_i(\vec{r})$ and then we compute n and T from

$$n(\vec{r}) = \sum_{i=1}^{N} |\varphi_i(\vec{r})|^2 \tag{19}$$

$$T[n] = -\frac{1}{2}\sum_i \int \varphi_i^{\star}(\vec{r})\nabla^2\varphi_i(\vec{r})d^3r \tag{20}$$

The problem, of course, is that we do not know $E_{xc}[n]$ and thus not v_{xc} into which the many-body problem has been transferred. Fortunately, keeping only the first two terms in Eq. 18, i.e. completely neglecting $v_{xc}(\vec{r})$, accounts for a large part of the truth and with a rather simple-minded approximation to E_{xc} and v_{xc} one can actually obtain rather accurate answers as we shall see later. One of the main reasons for this is that the large kinetic energy is treated exactly through Eq. 20 whereas earlier density-functional theories such as the Thomas-Fermi theory[4,5] used an approximate local form for the kinetic energy. Note that the charge density $n(\vec{r})$ appears explicitly in the second term in the effective potential (Eq. 18) and we have thus transformed the many-body problem to a one-particle problem with self-consistency. Note also that what we have really done is to demonstrate how to construct a non-interacting system such that the functional $E[n]$ is stationary at the ground-state density of the non-interacting system. We have, however, not shown that this procedure leads to the absolute minimum of $E[n]$. The self-consistency requirement poses a non-linear problem which could sometimes have more than one solution[13] for the density $n(\vec{r})$ in which case one would have to choose the solution which yields the lowest value of the functional for the total energy (Eq. 15).

In the case we are interested in the lowest energy within a particular symmetry channel S we can proceed as follows. The functional T_S for the kinetic energy now becomes

$$T_S[n_S] = \inf_{\substack{\hat{S}|\psi\rangle = S|\psi\rangle \\ \langle\psi|\hat{n}|\psi\rangle = n_S}} \langle\psi|\hat{T}|\psi\rangle \qquad (21)$$

where we have put a subscript S on n_S to indicate that only a limited set of densities can be obtained from wave functions of symmetry S. The functional for the total energy $E_S[n_S]$ is defined similarly and will also be symmetry dependent. Let us for a moment define the functional $V_S[n_S]$ through the equation

$$E_S[n_S] = T_S[n_S] + V_S[n_S] \qquad (22)$$

and let us assume that there exists a functional derivative of V_S with respect to n_S. This might not be the case in general but V_S corresponds to the last three terms of Eq. 15 and for the known approximations to E_{xc} it is certainly true. Let us then try and find a density n_S such that E_S becomes stationery, i.e. such that

$$\delta T_S + \int \frac{\delta V_S}{\delta n_S} \cdot \delta n_S(\vec{r}) d^3r = 0 \qquad (23)$$

This would be achieved if we could find the density that minimized the functional

$$H_S^0[n_S] = T_S[n_S] + \int w_S(\vec{r}) n_S(\vec{r}) d^3r \qquad (24)$$

with respect to $n_S(\vec{r})$. Here, the potential

$$w_S(\vec{r}) = \frac{\delta V_S}{\delta n_S(\vec{r})} \qquad (25)$$

should be considered as an external potential independent of n_S. From the discussion above we realize that this later problem is equivalent to finding the minimum of the expectation value of the one-particle Hamiltonian $\hat{H}_S^0 = \hat{T} + \int w_S(\vec{r})\hat{n}(\vec{r}) d^3r$ with respect to wave functions $|\psi\rangle$ of symmetry S, i.e. $\hat{S}|\psi\rangle = S|\psi\rangle$ ($\hat{n}(\vec{r}) = \psi^\dagger(\vec{r})\psi(\vec{r})$ is the density operator). If $|\psi_S^0\rangle$ is the wave function which achieves this minimum and if n_S is the corresponding density then, obviously, $T_S[n_S] \leq \langle\psi_S^0|\hat{T}|\psi_S^0\rangle$ by the definition of T_S. If further $|\psi_{ns}\rangle$ is the state with density n_S which yields $T_S[n_S]$

then $T_S[n_S] = \langle\psi_{n_S}|\hat{T}|\psi_{n_S}\rangle \geq \langle\psi^0_S|\hat{T}|\psi^0_S\rangle$ by the fact that $|\psi^0_S\rangle$ minimizes $\langle\psi|\hat{H}^0_S|\psi\rangle$. Thus we have $T_S[n_S] = \langle\psi^0_S|\hat{T}|\psi^0_S\rangle$. It is also easy to see that $H^0_S[n'_S] \geq H^0_S[n_S]$ for any other density n'_S and therefore the first-order variation of $H^0_S[n_S]$ must vanish. The state $|\psi^0_S\rangle$ and the corresponding density n_S is found by diagonalizing the one-particle Hamiltonian $\hat{H}^0_S$ in the subspace of symmetry S, i.e. by solving for the lowest eigenvalue of the Hamiltonian $P_S\hat{H}^0_S P_S$ where P_S is the projector on to this symmetry channel. The importance of the projector P_S can be appreciated from the following example. Consider a free atom with a singlet S ground state such as for instance a rare gas atom. The lowest state of P symmetry is obtained by exciting one of the p-electrons in the outermost shell to a higher lying s-orbital. The charge density of this P state is not spherically symmetric which means that our effective one-particle potential $w_S(\vec{r})$ above would not be spherical, at least not for the most common approximations to the exchange-correlation energy in which this energy is written as an integral over a simple function of the density. Consequently our effective one-particle Hamiltonian $\hat{H}^0_S$ does not commute with the square of angular momentum ($\vec{L}^2$). Still the projection operators force us to minimize $\hat{H}^0_S$ with respect to wave functions of P (L = 1) symmetry.

We end this section by writing down the effective one-particle problem for the case of a spin-polarized system where the external field couples only to the charge and the spin of the electrons and not to their angular momenta (see Ref. 10). In this case we obtain a Pauli-like theory in which the effective one-particle orbitals become two-component spinors with components $\varphi_{i\sigma}(\vec{r})$ obeying the coupled Schrödinger equations

$$\sum_{\sigma'} \{-\frac{1}{2}\delta_{\sigma\sigma'}\nabla^2 + \delta_{\sigma\sigma'}\int v(\vec{r}-\vec{r}\,')n(\vec{r}\,')d^3r' + w_{\sigma\sigma'}(\vec{r}) + v^{xc}_{\sigma\sigma'}(\vec{r})\}\varphi_{i\sigma'}(\vec{r}) = \varepsilon_i\varphi_{i\sigma}(\vec{r}) \tag{26}$$

The exchange-correlation potential is given as a functional derivative of the exchange-correlation energy with respect to the density matrix

$$v^{xc}_{\sigma\sigma'}(\vec{r}) = \frac{\delta E_{xc}}{\delta n_{\sigma\sigma'}(\vec{r})} \tag{27}$$

The density matrix is obtained from the one-particle spinors $\varphi_{i\sigma}(\vec{r})$ by means of the following sum over the N lowest states

$$n_{\sigma\sigma'}(\vec{r}) = \sum_{\varepsilon_i<\varepsilon_{max}}^{occ} \varphi_{i\sigma}(\vec{r})\varphi^{\star}_{i\sigma'}(\vec{r}) \tag{28}$$

and the total energy E[n] is obtained from

$$E[n] = \sum_{\varepsilon_i<\varepsilon_{max}}^{occ} \varepsilon_i - \frac{1}{2}\int n(\vec{r})n(\vec{r}')v(\vec{r}-\vec{r}')d^3r' -$$

$$- \sum_{\sigma\sigma'} \int v^{xc}_{\sigma\sigma'}(\vec{r})n_{\sigma'\sigma}(\vec{r})d^3r + E_{xc}[n] \qquad (29)$$

The above equations [(26)-(29)] represent the density-functional description of systems in which only the charge and the spin of the electrons couple to external electric and magnetic fields. Several interesting physical phenomena lie outside this framework. For instance, the spin-orbit interaction in heavier elements and magnetism in materials such as the rare-earth metals, where the coupling of the orbital angular momentum of the electrons to the magnetic field can play an important role, cannot be described by the above equations. Gunnarsson and Lundqvist[11] have shown how the spin-orbit interaction can be readily incorporated into the above theory by simply adding this interaction to the functional for the kinetic energy. This is due to the fact that both the kinetic-energy operator and the spin-orbit operator are one-particle operators and do not cause any difficulties in the effective one-particle Schrödinger problem. (The many-body problem with spin-orbit interaction is transferred to a non-interacting problem with spin-orbit interaction). The drawback of this somewhat ad hoc procedure is that it does not provide a recipe for the coefficient in front of the spin-orbit term. In the non-interacting case, the strength of the spin-orbit coupling is proportional to the gradient of the external potential but the coupling strength is certainly modified by exchange and correlation in the interacting case. Since spin-orbit coupling really is a relativistic effect, a relativistic generalization of density-functional theory becomes necessary to properly deal with the problem. This has been achieved by Rajagopal and Callaway[14] who showed that, in this case, the fundamental variable is not the density but the four-current density. In analogy with the non-relativistic theory their generalization leads to an effective one-particle Dirac equation with effective scalar and vector potentials determined by the current density. The non-relativistic limit of the theory leads naturally to a Pauli-like two-component theory with the proper form for the spin-orbit term. A further advantage of the relativistic formulation is that it shows how the coupling of the orbital angular momentum of the electrons to the external magnetic field is to be treated in a density-functional theory.

2. THE LOCAL-DENSITY APPROXIMATION : INTERPRETATION AND EXTENSIONS

2.1. The Local-Density Approximation

In order to solve the self-consistent set of equations (26)-(29) we must introduce some approximation to the functional $E_{xc}[n]$

for the exchange-correlation energy and to the corresponding potential $v^{xc}_{\sigma\sigma'}(\vec{r})$. A very simple choice is the local-density (LD) approximation[6,7] which nowadays is the most extensively used technique for the calculation of the electronic properties of real materials. The idea is simply to consider the inhomogeneous system as consisting of small boxes each containing a homogeneous but interacting electron gas. Such a gas is completely characterized by two numbers, namely its spin-up and spin-down densities which we take to be the eigenvalues of the local density matrix in each box. In other words, in each box we have, in general, a density matrix $n_{\sigma\sigma'}(\vec{r})$ which we first diagonalize to obtain the local spin direction which is then taken as the direction of polarization of the local homogeneous gas. The total exchange-correlation energy is finally taken to be the sum of the contributions from all boxes. Thus,

$$E_{xc}[n_{\sigma\sigma'}] = \int n(\vec{r})\varepsilon_{xc}(n_{\uparrow}(\vec{r});n_{\downarrow}(\vec{r}))d^3r \tag{30}$$

where $n_{\uparrow}(\vec{r})$ and $n_{\downarrow}(\vec{r})$ are the eigenvalues of the one-particle density matrix $n_{\sigma\sigma'}(\vec{r})$, $n(\vec{r}) = n_{\uparrow}(\vec{r}) + n_{\downarrow}(\vec{r})$ is the electron density and $\varepsilon_{xc}(n_{\uparrow},n_{\downarrow})$ is the exchange-correlation energy per particle of a homogeneous spin-polarized electron gas with the spin densities $n_{\uparrow}$ and $n_{\downarrow}$. The general formulation presented above can, in principle, be used to treat cases in which the local spin vector $\vec{S}(\vec{r})$ defined by

$$\vec{S}(\vec{r}) = \frac{1}{2}\,\mathrm{Tr}(n(\vec{r}).\vec{\sigma}) \tag{31}$$

changes not only its magnitude but also its direction as a function of position ($n(\vec{r})$ is short for the density matrix and $\vec{\sigma}$ is the Pauli vector operator for the spin). This would be the case with the complicated spiral-shaped magnetic structures found in some of the rare-earth metals. In most systems, however, the direction of $\vec{S}(\vec{r})$ is independent of position, only its magnitude might vary, and a direction for spin quantization can always be found such that the density matrix becomes diagonal. This is the case when spin-orbit coupling can be neglected in finite systems such as atoms and molecules. The ground state can then be chosen to be an eigenstate of the z-component of the total spin yielding a vanishing matrix element of the operator $\psi^{\dagger}_{\uparrow}\psi_{\downarrow}$ responsible for an off-diagonal matrix element of the density matrix (Eq. 7) (the operator $\psi^{\dagger}_{\uparrow}\psi_{\downarrow}$ flips a spin). In a non-polarized system or in ferromagnetic as well as in anti-ferromagnetic systems the density matrix can also be chosen to be diagonal which has the simplifying consequence that the coupled equations (26) become two independent equations each with an exchange-correlation potential $v^{xc}_{\sigma}(\vec{r})$ given by[10]

$$v^{xc}_{\sigma}(\vec{r}) = \frac{\delta E_{xc}}{\delta n_{\sigma}(\vec{r})} \tag{32}$$

where $n_{\sigma}(\vec{r})$, for $\sigma = \uparrow$ or $\downarrow$, is a diagonal element of the density matrix.

Through the local-density approximation (Eq. 30) the problem of exchange and correlation in an inhomogeneous system has been transferred to the problem of calculating the exchange-correlation energy $\varepsilon_{xc}(n\zeta)$ of the homogeneous electron gas. Successively more accurate results have recently been obtained for this quantity as a function not only of density n but also of spin density $\zeta = (n_\uparrow - n_\downarrow)/n$. These results are usually presented in the form of tables and have to be parametrized in order to facilitate computations. Von Barth and Hedin[10] proposed the following approximate interpolation formula for the electron-gas data required by the formalism

$$\varepsilon_{xc}(n.\zeta) = \varepsilon^P_{xc}(n) + [\varepsilon^F_{xc}(n) - \varepsilon^P_{xc}(n)].f(\zeta) \tag{33}$$

where the function $f(\zeta)$ gives the exact spin dependence of the exchange energy

$$f(\zeta) = \frac{1}{2}(2^{1/3} - 1)^{-1}\{(1 + \zeta)^{4/3} + (1 - \zeta)^{4/3} - 2\} \tag{34}$$

The exchange-correlation energies $\varepsilon^P_{xc}(n)$ and $\varepsilon^F_{xc}(n)$ appearing in Eq. 33 are those of the unpolarized (paramagnetic) and completely-polarized (ferromagnetic) homogeneous electron gas. This simple formula (Eq. 34) accurately approximates the spin dependence of the exchange-correlation energy given by both the random-phase approximation and by more sophisticated electron-gas calculations[15]. The density dependence of both $\varepsilon^P_{xc}(n)$ and $\varepsilon^F_{xc}(n)$ is conveniently parametrized in terms of the function

$$F(x) \equiv (1 + x^3)\ell n(1 + \frac{1}{x}) + \frac{x}{2} - x^2 - \frac{1}{3} \tag{35}$$

introduced by Hedin and Lundqvist[16]. In their work, the correlation energy was written simply as $- 0.045.F(x)$ Ry, where $x \equiv r_s/21.0$ (r_s is the usual electron-gas density parameter). This description of the electron-gas data cannot, however reproduce the known density dependence in the high- and low-density limits. This is an unfortunate shortcoming, because the electron densities encountered in an atom, for example, span the entire high-density/low-density range. Although it would certainly be desirable not to have to worry about the parametrization of the electron-gas data, when studying atoms, molecules and solids, e.g. it is noteworthy that such errors are fortunately systematic (as opposed to random) and tend therefore to subtract out of many calculated properties of interest. Arbman and von Barth[17] have shown that a much improved description of the electron-gas data results if the function F(x) defined in Eq. 35 is used twice in the description of $\varepsilon^P_{xc}(n)$ and $\varepsilon^F_{xc}(n)$. That is

$$\varepsilon^P_{xc}(n) = \varepsilon^P_x(n) - c^P_1\, F(r_s/r^P_1) - c^P_2\, F(r_s/r^P_2) \tag{36}$$

and similarly for $\varepsilon^F_{xc}(n)$. The eight parameters introduced in this way $(r_1^P, c_1^P, r_2^P, c_2^P, r_1^F, c_1^F, r_2^F, c_2^F)$ can be set so that Eq. 36 reproduces the high-density result obtained by Gell-Mann and Bruckner[18], the low density result of Wigner[19] while also providing a good fit to the results of numerical calculations (see e.g. Refs. 20-23) in the metallic-density range ($2 \leqslant r_s \leqslant 5$). A particular set of numerical values for the eight parameters which yields the correct limiting behavior, a good fit to the correlation energies of Singwi et al[20] and which obeys approximate RPA scaling[10] between $\varepsilon_{xc}^P(n)$ and $\varepsilon_{xc}^F(n)$ is $r_1^P, r_2^P, r_1^F, r_2^F$ = 32.5, 0.8, 30.6, 0.1 Bohr radii and $c_1^P, c_2^P, c_1^F, c_2^F$ = 0.0352, 0.0270, 0.0278, 0.0034 Ry. The exchange energy $\varepsilon_x^P(n)$ appearing in Eq. 36 is given by $\varepsilon_x^P(n) = -\,0.9163/r_s$ Ry ; the analogous result for polarized systems is $\varepsilon_x^F(n) = -\,1.1545/r_s$ Ry. Note that, while the errors introduced by the use of the local-density approximation are usually larger than those due to the imperfect description of the electron-gas data, there are important exceptions[24,25,26] that make it desirable to improve the description, particularly as improved electron-gas results become available. Such improved descriptions have recently been proposed by Vosko et al[27].

Once the density dependence of the electron-gas data is represented analytically by a form, such as that in Eq. 36, the density derivative required by the exchange-correlation potential (recall that $v_{xc}(n) \equiv d/dn[n\varepsilon_{xc}(n)]$) can be taken easily. Note that, quite independent of the accuracy of representations such as Eq. 36, properties such as, e.g. the virial theorem for the inhomogeneous system[1] depend crucially on the consistency which results when $\varepsilon_{xc}(n)$ and $v_{xc}(n)$ are obtained from the same representation of the electron-gas data.

In order to be able to apply the local-density approximation to relativistic systems involving heavy atoms we must, of course, use some approximation to the relativistic analog of ε_{xc}, i.e. of the exchange-correlation energy per particle of the relativistic electron gas. An expression for the exchange contribution has recently been obtained by Rajagopal[28] and by Mc Donald and Vosko[29]. In the relativistic case the exchange energy can be thought of as a sum of two contributions ; one which amounts to a minor rescaling of the non-relativistic exchange energy, and one, called the transvers exchange energy, which has no non-relativistic analog and which is large and varies rapidly with the electron density. This latter term actually reverses the sign of the total exchange potential at very high densities. However, in actual calculations[28,30] on heavy atoms this term only contributes of the order of 10% to the total exchange energy because relatively few of the electrons are relativistic even in a heavy element like uranium.

Also the correlation energy of the relativistic electron gas has recently been calculated by Ramana and Rajagopal[31]. It is shown that this energy can be twice as large as the corresponding non-relativistic energy at densities found in the vicinity of the nucleus in heavy atoms. Again, however, actual calculations on heavy atoms by Ramana et al[32] show that the relativistic correction to the total correlation energy is only a few percent because such a small fraction of the electrons are rally relativistic. In this context it seems appropriate to note that exchange is by far the dominant effect at densities where relativistic effects become important. At these densities correlation energies are typically a factor of 50 smaller than exchange energies. It should also be noted that the local-density approximation to the non-relativistic part of the exchange energy is particularly bad for the tightly bound relativistic electrons close to the nucleus. For these electrons the error in the non-relativistic exchange energy is typically 14%[33].

The finite-temperature extension[12] of density-functional theory presented in Sec. 1.2 also leads to an effective one-particle Schrödinger equation for obtaining the grand potential and the particle density. The two main differences in this case as compared to the zero-temperature case is that, (i) each term in the equations (19) and (20) is preceded by an occupation number n_i appropriate to the Fermi-Dirac distribution

$$n_i = \frac{1}{e^{\beta(\varepsilon_i - \mu)} + 1} \tag{37}$$

and, (ii) in the local-density approximation, the free energy per particle of the homogeneous but interacting electron gas becomes a function not only of density but also of temperature. For a wide range of temperatures and densities this free energy has recently been calculated by Gupta and Rajagopal[34] from the finite-temperature RPA. As might have been expected correlation plays a much larger role at elevated temperatures and can actually become more important than exchange. Their calculation now enable us to perform local-density calculations for matter under extreme conditions.

By construction, the local-density approximation is exact in the limit of slowly varying spin densities[6,7] and one would expect it to work well only for systems in which the densities do not change appreciably over a distance corresponding to an inverse Fermi wave vector. Unfortunately, most real systems such as atoms, molecules, or solids show strong variations in the particle density over this distance. Still, the local-density approximation has been applied to many of these systems and we think it is fair to say that the results have been far more accurate than one had any reason to expect. Numerous examples of this success are given in other lectures of this school. The next sections will be devoted to a discussion of the physics behind the success of the local-density approximation, but we will also point out some difficulties associated with it.

2.2. Gradient Corrections

A very natural way to try to understand why a zeroth-order theory works is to estimate the first-order corrections in the hope that they will be small. Coming from the limit of a slowly varying density, these corrections are the so-called density-gradient terms. If only the lowest order correction is included, the exchange-correlation functional becomes

$$E_{xc}[n] = \int n(\vec{r})\varepsilon_{xc}(n(\vec{r}))d^3r + \int B_{xc}(n(\vec{r}))|\nabla n(\vec{r})|^2 d^3r \tag{38}$$

in the case of no spin polarization[6]. The function $B_{xc}(n)$ is determined by the density-density response function of the homogeneous electron gas, and it has been extracted from various approximate response functions[35-41]. The function $B_{xc}(n)$ has also recently been obtained by Geldart and Rasolt from a careful analysis of the diagrammatic expansion for the static dielectric function[42,43].

On dimensional grounds, we would expect a gradient expansion to be reasonable only if

$$\left|\frac{1}{k_F(n(\vec{r}))} \cdot \frac{\nabla n(\vec{r})}{n(\vec{r})}\right| \ll 1 \tag{39}$$

where the local Fermi wave-vector $k_F(n(\vec{r}))$ is given by $(3\pi^2 n(\vec{r}))^{1/3}$. The inequality (39) is a mathematical formulation of the requirement that the density variations be small over distances of the order of an inverse Fermi wave-vector. This criterion is, however, seldom satisfied in real systems, as illustrated by the two examples in Fig. 1 (from Ref. 44) which shows the relative density variations in bulk copper and in the surface region of semi-infinite jellium with a density corresponding to that of metallic sodium. It is therefore not surprising that the results for the total energy of atoms turns out to be better in the local-density (LD) approximation than if also the lowest order gradient correction is included. This is obvious from the work by Herman et al[45] who tried to improve the total energies of atoms by treating B_{xc} in Eq. 38 as an adjustable parameter. This optimal B_{xc} has a sign opposite that of the B_{xc} calculated from first principles[42]. As a next step, one can try to include higher-order gradient corrections of, e.g., the form $\int C_{xc}(n)|\nabla^2 n|^2 d^3r$ or $\int D_{xc}(n)\nabla^2 n|\nabla n|^2 d^3r$[6]. For a system in which the particle density deviates little from its average value $\bar{n}$, we can use linear-response theory which amounts to neglecting all gradient erms involving the density raised to powers greater than two. The remaining subclass of gradient terms can be summed exactly, to infinite order giving[6]

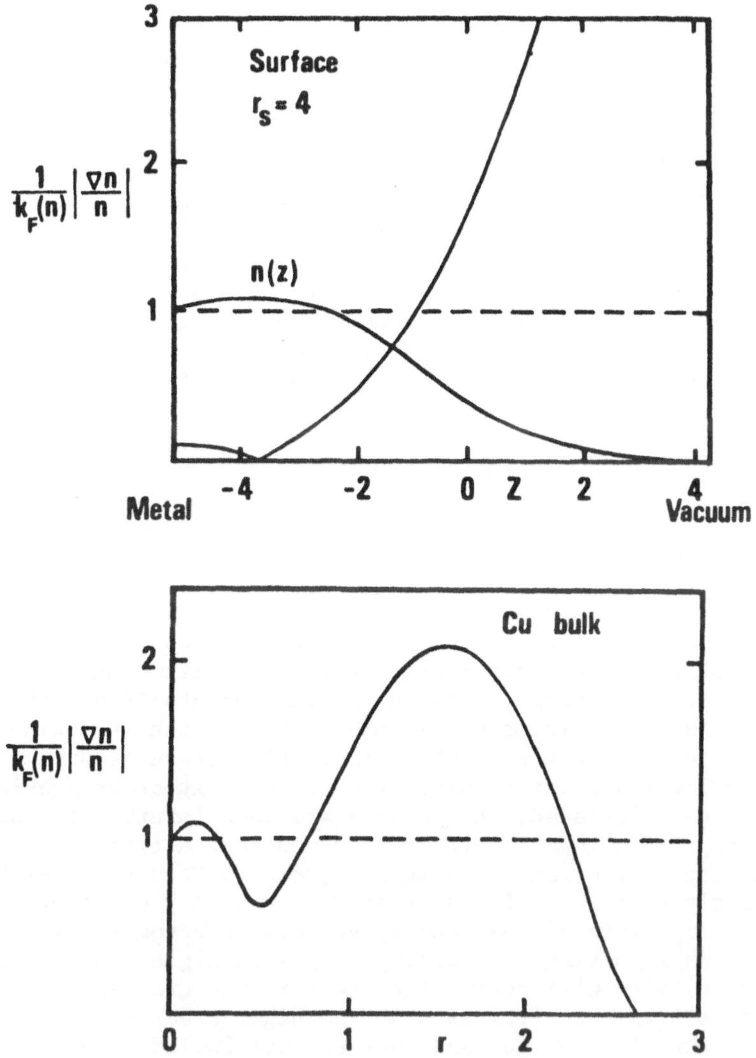

Fig. 1. The dimensionless quantity $|\vec{\nabla}n|/(nk_F(n))$ giving the change in the density n over the local Fermi wavelength $k_F^{-1} \equiv (3\pi^2 n)^{-3}$ is here displayed as a function of the perpendicular distance z from the surface of semi-infinite jellium ($r_s = 4$) and as a function of the distance r from the nucleus in bulk copper. It is this quantity that should be much less than unity in order that the use of gradient corrections be reliable. The curve labelled n(z) shows the surface density profile. Distances are measured in Bohr radii. The figure is taken from Ref. 44.

$$E_{xc}[n] = \int n(\vec{r})\varepsilon_{xc}(n(\vec{r}))d^3r -$$
$$-\frac{1}{4}\int\int K_{xc}(\vec{r}-\vec{r}';\bar{n}).(n(\vec{r})-n(\vec{r}'))^2 d^3r d^3r' \qquad (40)$$

The Fourier transform of the kernel $K_{xc}(\vec{r},n)$ is given by $-(4\pi/q^2).G(q,n)$ where the local-field factor G, as defined by Singwi et al[20], can be obtained from the static density-density response function of the homogeneous electron gas of density n. If we choose the density argument in Eq. 40 to be the average density, then Eq. 40 is valid for arbitrarily rapid density variations, as long as $n(\vec{r})$ deviates little from $\bar{n}$. It thus provides a possibility to test the quality of gradient corrections. These are proportional to the coefficients in the expansion of the Fourier transform $K_{xc}(q)$ of the kernel K_{xc} in powers of q. For instance, from Ref. 6 we have

$$K_{xc}(q) = K_{xc}(0) + 2B_{xc}q^2 + O(q^4) \qquad (41)$$

In Fig. 2, (from Ref. 44), we show $K_{xc}(q)$ obtained from the dielectric function of Geldart and Taylor[46] together with the results for $K_{xc}(q)$ obtained by truncating its series expansion in q to order q^0, q^2 and q^4 corresponding to the LD approximation and the lowest and next-higher-order gradient corrections, respectively. Looking at Fig. 2 and remembering that realistic density profiles have appreciable Fourier components with momenta larger than the Fermi momentum, it seems questionable whether the inclusion of the lowest-order gradient correction (Eq. 38) represents an improvement over the local-density result. If also the next-order gradient correction is included, the results are much inferior to those of the LD approximation. The argument given here against the relevance of gradient corrections is originally due to Geldart et al[36]. It is strictly valid only for systems with a nearly constant density. When the amplitude of the density variations becomes larger, gradient terms involving the density to powers higher than two, e.g. $\int D_{xc}\nabla^2 n|\nabla n|^2 d^3r$, will become important in the gradient expansion. Such terms originate from second- and higer-order response functions of the electron gas and they are not included in the infinite sum given by Eq. 40. There is, however, no reason to believe that gradient corrections would become more appropriate when the amplitude of the density variations increase.

The convergence of the gradient expansion has also been studied by Gunnarsson, Jonson, and Lundqvist[44]. They applied Eq. 40, representing an infinite summation of a subclass of gradient terms to the calculation of the total energies of atoms and surfaces. In contrast to the case of an almost constant density, discussed previously, the choice of density argument in the kernel K_{xc} of Eq. 40 is not unique for these strongly inhomogeneous systems. There are several possibilities which are physically reasonable and which reduce to the correct limit when the density variations are

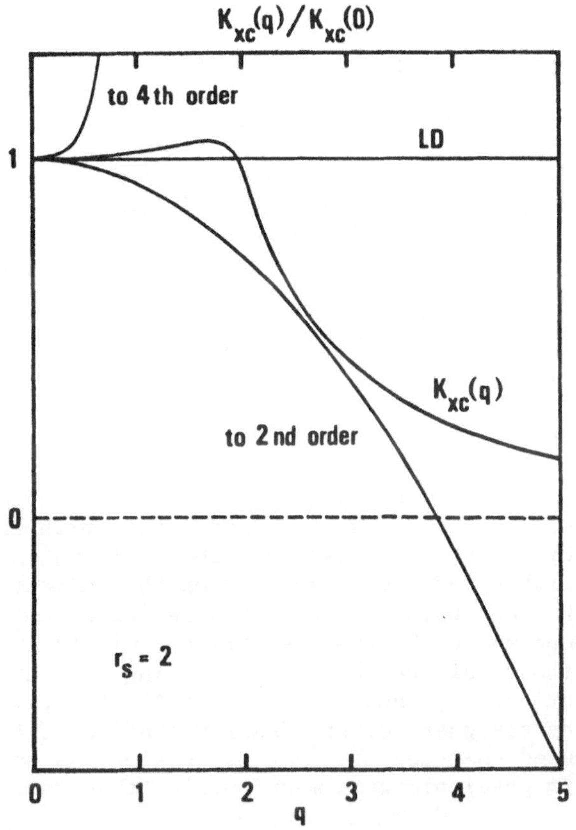

Fig. 2. The local interaction K_{xc} describing the effects of exchange and correlation on the static density response function of the homogeneous electron gas (r_s=2, see Eqs. 40,41,90) is here displayed as a function of wave vector q given in units of the Fermi wave vector. Also displayed is the same quantity obtained within the local-density approximation and obtained by also including the lowest (2nd) order and next higher (4th) order gradient corrections. The figure demonstrates the futility of low-order gradient corrections. The figure is based on the dielectric function of Geldart and Taylor (Ref. 46) and appears in Ref. 44.

made small. Gunnarsson et al discovered that the exchange-correlation energies of both atoms and surfaces are infinite for some of these choices, thus demonstrating the divergence of this particular, infinite subseries of gradient terms.

Perhaps the most conclusive argument against the relevance of gradient corrections has recently been presented by Langreth and Perdew[47]. In a previous paper[48], they had analyzed both the exact exchange-correlation energy of a metallic surface and the LD approximation to it in terms of the wave vectors of the density fluctuations. They found that the LD approximation was quite accurate for large wave vectors but failed badly for small wave vectors. Using the surface-density profile of the infinite-barrier model, and treating correlation within the random-phase approximation, they also showed that a simple interpolation between the exact limiting behavior (found by them) at small and large wave vectors gave an exchange-correlation energy very close to the exact result[49]. In Ref. 47 a similar analysis is carried out for the lowest-order gradient correction (Eq. 38). It is shown that the wave-vector-dependent gradient correction has the wrong analytical behaviour for small wave vectors. It is also shown that the largest contribution to the integrated wave-vector-dependent gradient correction comes from this region, resulting in a much too large correction to the LD approximation. This result is consistent with an earlier investigation by Perdew, Langreth and Sahni[50] in which accurate results for the exchange-correlation part of the surface energy, obtained using the same wave-vector interpolation scheme mentioned above, were compared to the results obtained within the LD approximation and to those obtained by also including the lowest-order gradient correction. By means of an external potential, they were able to vary the steepness of the density profile of the surface, and they concluded that the gradient correction was appropriate only when the density gradients were much smaller than those of realistic surfaces.

It is clear from the discussion given above that a strong case can be made against gradient corrections as a way to improve the LD approximation for the total energies of systems in which the particle density varies as rapidly as in, e.g., atoms and metallic surfaces. The gradient expansion is, at best, asymptotically convergent ; keeping only the lowest-order gradient term is a doubtful approximation. There are, however, other situations in which a gradient expansion might be more appropriate. This could be the case when there is hope for a cancellation of the errors due to strong inhomogeneities. The total-energy differences between different crystal structures or the energies of chemisorption might be examples. The discussion above was not primarily intended to demonstrate the inadequacy of gradient corrections, but rather to point out that the LD approximation gives quite reasonable answers even when gradient corrections are not appropriate. This means that the

LD approximation cannot be justified as the zeroth-order approximation in a rapidly convergent expansion. It is rather an approximation in its own right and the reason for its success must be sought elsewhere.

2.3. Pair-Correlation Hole and Sum Rules

The most extensive investigation of the factors governing the success of the LD approximation is due to Gunnarsson et al[11,33,44]. Their discussion is based on an exact expression, due to Almbladh[51] relating the exchange-correlation energy $E_{xc}[n]$ to a quantity $\tilde{g}(\vec{r},\vec{r}')$ which is the pair-correlation function of the system, integrated over the strength, λ, of the electron-electron Coulomb interaction,

$$E_{xc}[n] = \frac{1}{2}\int n(\vec{r})n(\vec{r}')\{\tilde{g}(\vec{r},\vec{r}') - 1\}v(\vec{r} - \vec{r}')d^3rd^3r' \tag{42}$$

Here, $v(\vec{r}) = 1/r$ is the full-strength Coulomb interaction and $n(\vec{r})$ is the particle density. (For simplicity we have assumed the system to be non-spin-polarized). The integral of the interaction-strength parameter λ has to be performed in the presence of an additional λ-dependent local external potential which vanishes at the physical value of λ but, which is otherwise designed so as to make the particle density independent of λ. That it is possible to define such an external potential is postulated in the original derivation of density-functional theory[6]. A derivation of Eq. 42 can be found in Ref. 11.

Gunnarsson and Lundqvist[11] then defined the exchange-correlation hole to be

$$n_{xc}(\vec{r},\vec{r}') = n(\vec{r}')\{\tilde{g}(\vec{r},\vec{r}') - 1\} \tag{43}$$

in terms of which the exchange-correlation energy is given by

$$E_{xc}[n] = \frac{1}{2}\int n(\vec{r})v(\vec{r} - \vec{r}')n_{xc}(\vec{r},\vec{r}')d^3rd^3r' \tag{44}$$

According to this equation, the exchange-correlation energy arises from the Coulomb interaction of each electron (e.g. the one at $\vec{r}$) with a charge distribution $n_{xc}(\vec{r},\vec{r}')$ i.e. the exchange-correlation hole surrounding that electron. The hole is a consequence of the exchange and Coulomb interactions which cause a depletion of charge in the vicinity of each electron. The fact that the total amount of displaced charge corresponds to one unit of charge is expressed by the sum rule

$$\int n_{xc}(\vec{r},\vec{r}')d^3r' = -1 \tag{45}$$

valid for all $\vec{r}$. The language of the present section can be used to give a definition of the LD approximation which is completely equivalent to Eq. 30. This equation results from a particular approximation for the exchange-correlation hole

$$n_{xc}^{LD}(\vec{r},\vec{r}\,') = n(\vec{r})\{\tilde{g}_h(\vec{r} - \vec{r}\,' \; ; n(\vec{r})) - 1\} \tag{46}$$

Here, $\tilde{g}_h$ (r,n) is the pair-correlation function of the homogeneous electron gas of density n integrated with respect to the strength of the interaction. We note immediately that the exchange-correlation hole in the LD approximation fulfills the sum rule (45) because this sum rule is valid also in the homogeneous case

$$\int n_{xc}^{LD}(\vec{r},\vec{r}\,')d^3r' = -1 \tag{47}$$

Due to the long range of the Coulomb interaction, we know that the interaction energy between two charge distributions only depends weakly on the shape of the distributions. The most important factor is the total amount of charge in each distribution. We therefore expect that a misrepresentation of the exchange-correlation hole will not result in a large error in the exchange-correlation energy as long as the sum rule (45) is fulfilled. This is probably the most important single factor underlying the success of the LD approximation. We can carry the analysis one step further[11] and define a spherical average $\bar{n}(\vec{r},R)$ of the hole $n_{xc}(\vec{r},\vec{r}\,')$ with respect to the point $\vec{r}$,

$$\bar{n}_{xc}(\vec{r},R) = \int n_{xc}(\vec{r},\vec{r} + \vec{R}) \frac{d\Omega_{\vec{R}}}{4\pi} \tag{48}$$

It is then easy to see that

$$E_{xc}[n] = \frac{1}{2} \int n(\vec{r})d^3r \int v(R)\bar{n}_{xc}(\vec{r},R)d^3R \tag{49}$$

which means that the exchange-correlation energy only depends on the spherical average of the exchange-correlation hole[11]. This exact result is also very important for the quantitative success of the LD approximation as can be appreciated from Figs. 3 and 4, both taken from Ref. 44. In Fig. 3, the exact exchange hole in the neon atom is compared to the corresponding quantity in the LD approximation for different positions of the electron considered (different $\vec{r}$). We have chosen to consider exchange only because in this case we have exact results to compare with. Exchange is also the dominant effect in an atom and the exchange hole has the same qualitative features as the full hole, e.g. it obeys the sum rule (45). The exchange-only version of the LD approximation is obtained by using the Hartree-Fock approximation for $\tilde{g}_h$ in Eq. 46. As seen in Fig. 3, the exact hole is always centered close to the nucleus whereas the LD-hole is always spherically symmetric and centered on the electron, as is evident from Eq. 46. Furthermore, the exact hole is rather localized whereas the LD-hole is much more extended. Thus, the LD approximation gives a rather poor

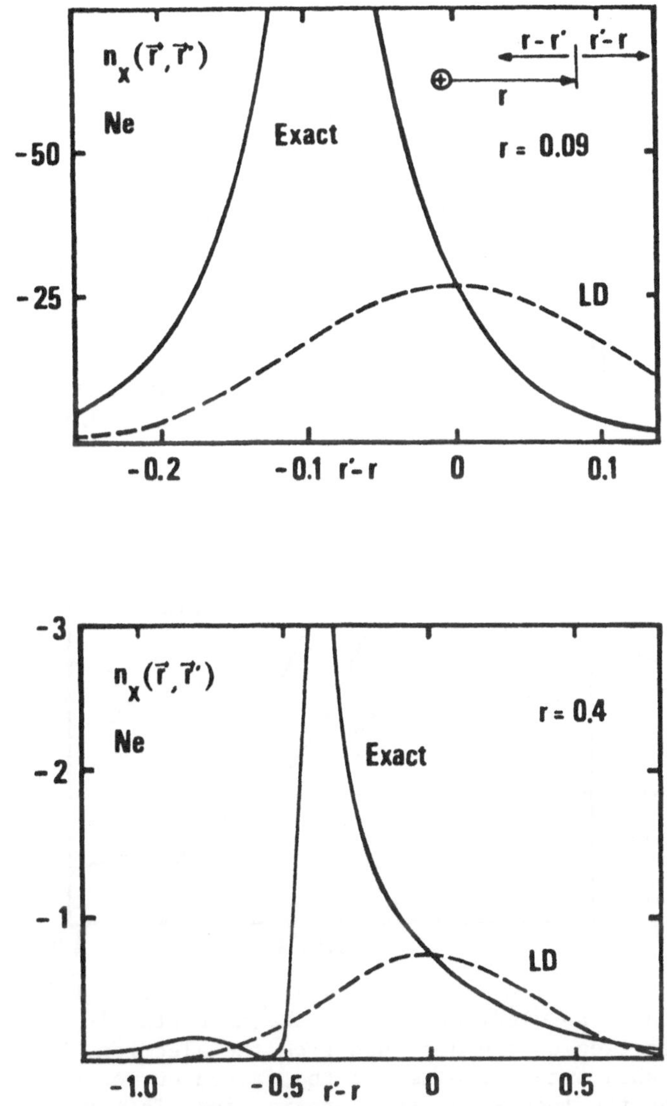

Fig. 3. The exchange hole around an electron at r in the neon atom is here shown as a function of distance from the electron along a line connecting the electron and the nucleus. The full curves show the exact hole (Eq. 43) and the dashed curves show the results obtained within the local-density approximation (Eq. 46) which is seen to give a rather poor description of the hole. The two sets of curves correspond to the electron being at two different distances from the nucleus. Distances are measured in Bohr radii. The figure is taken from Ref. 44.

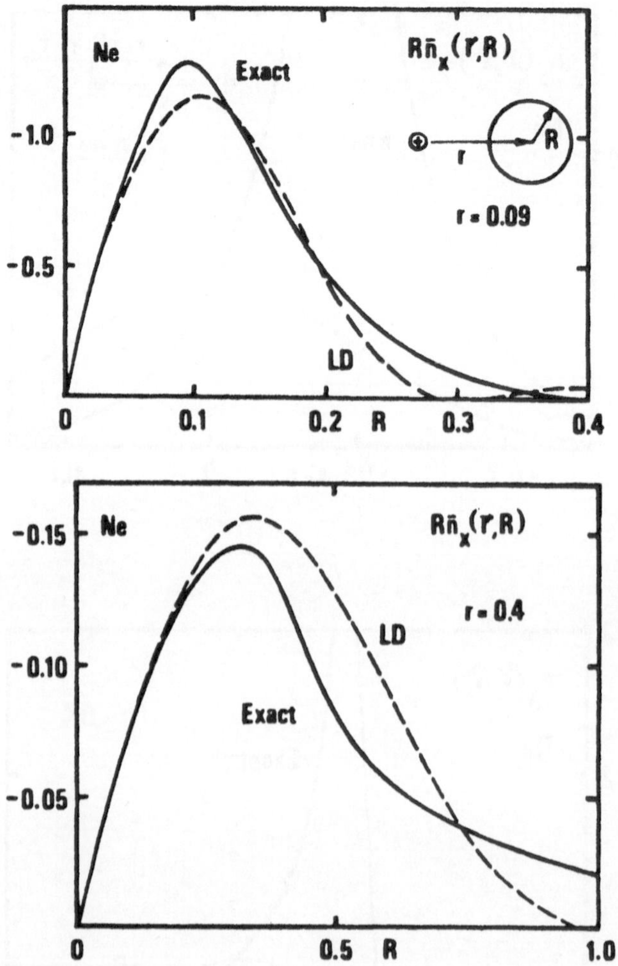

Fig. 4. The spherical average (Eq. 48) of the exchange hole surrounding an electron at r in the neon atom is here shown as a function of the distance from the electron. The full curves show exact results and the dashed curves were obtained using the local-density approximation. The hole is shown multiplied by R so that the area under each curve is directly proportional to the exchange-energy density (Eq. 50). When the curves shown here are compared with the corresponding curves in Fig. 3, it is seen that the local-density approximation gives a much more accurate description of the spherical average of the exchange hole than it gives of the hole itself. It is important to note in this context that the exchange energy depends only on the spherical average. Distances are measured in Bohr radii (From Ref. 44).

representation of the exchange hole. Looking instead at the more relevant spherically averaged hole in Fig. 4, we see that the picture has improved considerably. The LD approximation gives a rather accurate description of the spherical average of the hole and therefore also of the exchange energy according to Eq. 49. Since we have plotted $R.\bar{n}_x(\vec{r},R)$ the area under each curve is directly proportional to the corresponding exchange-energy density $\varepsilon_x(\vec{r})$, which is defined by

$$\varepsilon_x(\vec{r}) = 2\pi \int_0^\infty \bar{n}_x(\vec{r},R)\ R\ dR \tag{50}$$

and which determines the exchange energy E_x according to

$$E_x[n] = \int n(\vec{r})\varepsilon_x(\vec{r})d^3r \tag{51}$$

We note the strong cancellation of errors that occurs for large and intermediate distances from the center of the hole. This cancellation is a consequence of the sum rule (45) which implies that the area under the curve $-4\pi R^2\bar{n}_x(\vec{r},R)$ is 1. Thus, multiplying the curves in Fig. 4 by R would give two other curves of equal area which would intercept at the same place as the curves already shown. For these new curves we would have exact cancellation which is only partly destroyed by dividing by R.

The arguments given above will only undergo minor quantitative modifications if correlation effects are included. The basic conclusions remain the same. Correlations mainly reduce the spatial extent of the electron-gas pair-correlation function. As a consequence, the exchange-correlation hole of the local-density approximation will be more localized, thereby effectively improving this approximation for strongly inhomogeneous systems. This is the basic argument in favor of treating exchange and correlation effects together. At this point, we would like to draw the reader's attention to the fact that obtaining correlation energies from a density-functional formalism is a much more difficult problem than obtaining exchange energies. As stressed by Tong[52], correlation energies are connected also to the excitation spectrum and not just to the wave functions, as is the case for exchange energies. However, the difficulty can also be appreciated in a nice way using the language of the present section. Let us imagine the exchange-correlation hole as a sum of an exchange hole and a correlation hole. Since both the exchange hole and the exchange-correlation hole obey the sum rule (45) it folloxs that the correlation hole contains zero charge. Thus, the exchange energy is a consequence of the interaction of each electron with a charge distribution containing one unit of charge. The correlation energy, on the other hand, comes from the interaction of each electron with a neutral charge distribution. From this it follows that exchange energies are much larger than correlation energies although the difference becomes less pronounced when the electron density decreases and the exchange-correlation hole becomes more extended. Still, the exchange energy per particle is a factor of four larger than the correlation energy even in an electron gas with a density corresponding to metallic sodium. In atoms, the ratio between exchange and correlation energies can easily be as large as thirty. A further consequence is that the percentage error in the exchange energy (or in the exchange-correlation energy) is substantially reduced, due to the sum rule (45), even though there is a misrepresentation of the exchange hole (or exchange-correlation hole). No such reduction can occur for the correlation energy since it is entirely due to the shape of the correlation hole.

The arguments given by Gunnarsson and Lundqvist[11] to explain the quantitative success of the LD approximation for exchange-correlation energies of inhomogeneous systems can be summarized as follows : (i) the exchange-correlation energy is determined by the spherical average of the exchange-correlation hole, (ii) the LD approximation gives a relatively accurate description of this spherically averaged hole, and (iii) in addition, there is a systematic cancellation of errors due to the sum rule that expresses conservation of charge.

Before ending this section, we would like to stress that the arguments given above represent one possible way in which to explain the success of the LD approximation. There may be other ways that are equally valuable and informative. The discussion above is based on a description of electronic properties in real space. As mentioned in Sec. 2.2, Langreth and Perdew[47,48] have reformulated the problem of exchange and correlation in reciprocal space by decomposing the exchange-correlation energy into contributions from different wave vectors of the density fluctuations. In this language, the success of the LD approximation is due to its ability to accurately describe the density fluctuations with large wave vectors. As will be discussed in Sec. 2.5 there is now some evidence in favor[44,53,54] of the reciprocal-space approach as a starting point for improving on the local-density approximation.

2.4. Non-Local Functionals in Real Space

An implication of the previous section is that the many-body problem can be transformed into the problem of modeling the exchange correlation hole of the system at hand. This is a rather difficult task for any specific system. The fact that we would like the model to be able to carry us continuously from the electron-gas limit to the atomic limit does not make the problem any easier. The knowledge of why the LD approximation works so well should, however, be useful in trying to improve it. Based on their earlier analysis, Gunnarsson and co-workers[55,56] have recently presented two "non-local" schemes to improve the LD approximation, both of which hold some promise. The schemes are "non-local" in the sense that the local exchange-correlation-energy density, as defined by an equation analogous to Eq. 50, is made to depend on some average of the surrounding density, rather than just on the local density as in the LD approximation. Since exchange and correlation effects are inherently non-local, there is hope for an improved description of these effects in the new schemes. Note that they will still give rise to a local effective one-body potential as will any density-functional scheme.

The proposed methods have one feature in common with the LD approximation : the pair-correlation function of the homogeneous electron gas is used to model the exchange-correlation hole. The density argument and the density prefactor in Eq. 46 are, however, altered.

In the average-density (AD) approximation[55] both the argument and the prefactor if chosen to be an average of the density, $\bar{n}(\vec{r})$ defined in terms of a density-dependent weight factor, $W(\vec{r},n)$,

$$\bar{n}(\vec{r}) = \int W(\vec{r} - \vec{r}\,' \; ; \bar{n}(\vec{r})).n(\vec{r}\,')d^3r' \tag{52}$$

The AD-hole is given by

$$n_{xc}^{AD}(\vec{r},\vec{r}\,') = \bar{n}(\vec{r}) \; \{\tilde{g}_h(\vec{r} - \vec{r}\,' \; ; \bar{n}(\vec{r})) - 1\} \tag{53}$$

and like the LD-hole and unlike the true hole, it is spherically symmetric and centered on the electron. The corresponding exchange-correlation energy is easily seen to be

$$E_{xc}^{AD}[n] = \int n(\vec{r})\varepsilon_{xc}(\bar{n}(\vec{r}))d^3r \tag{54}$$

from which the exchange-correlation potential $v_{xc}^{AD}(\vec{r})$ is obtained by taking the functional derivative with respect to the density $n(\vec{r})$. This derivative will contain the functional derivative $\delta\bar{n}(\vec{r})/\delta n(\vec{r}\,')$ which is easily obtained from the definition of the average density (Eq. 52). By construction, the AD approximation obeys the sum rule (45) irrespective of the choice of weight function. This follows from the corresponding sum rule for the homogeneous gas. The weight function $W(\vec{r},n)$ can thus be used to further improve the description of the exchange-correlation hole. Gunnarsson et al[55] choose W such that the AD approximation becomes exact when the density is almost constant everywhere, i.e. such that the exchange-correlation energy of the AD approximation, in the linear-response limit, is given by Eq. 40. Consequently, in this limit, the AD approximation is exact, even for arbitrarily rapid density variations. It turns out that the chosen requirement uniquely determines the weight function W which is given by a first-order non-linear differential equation involving coefficients determined by the wave-vector- and density-dependent static dielectric function of the homogeneous electron gas. This equation can be solved numerically and tabulated once and for all. A table can be found in Ref. 44.

The second scheme proposed by Gunnarsson et al[56], the weighted density (WD) approximation, has also been suggested by Alonso and Girifalco[57] for the exchange energy. In this scheme, the correct density prefactor $n(\vec{r}')$ form Eq. 43 is retained and the sum rule (45) is satisfied for each $\vec{r}$ by a proper choice of density argument $\tilde{n}(\vec{r})$ in the pair-correlation function of the homogeneous gas. The exchange-correlation hole becomes

$$n_{xc}^{WD}(\vec{r},\vec{r}') = n(\vec{r}')\ \{\tilde{g}_h(\vec{r}-\vec{r}'\ ;\ \tilde{n}(\vec{r})) - 1\} \tag{55}$$

which, when inserted into Eq. 44, gives the corresponding exchange-correlation energy E_{xc}^{WD}. By taking the functional derivative of E_{xc}^{WD} with respect to the density $n(\vec{r})$, we obtain the exchange-correlation potential $v_{xc}^{WD}(\vec{r})$ of the WD approximation. The potential will contain the functional derivative $\delta\tilde{n}(\vec{r})/\delta n(\vec{r}')$ for which an explicit expression can be obtained from the sum rule (45). Thus, the exchange-correlation hole of the WD approximation is obtained by digging a hole in the true density. Consequently, this hole, like the true hole, is not spherically symmetric around the electron as is the case for the LD and AD approximations. Due to this feature the WD approximation is exact for any one-electron system. The only way in which the sum rule can be satisfied in a one-electron system is by choosing $\tilde{n}$ so small that $\tilde{g}_h$ vanishes almost everywhere and $n_{xc}^{WD}(\vec{r},\vec{r}')$ becomes $-n(\vec{r})$ which, in this case, is the exact result. The WD approximation also gives the exact exchange energy for a two-electron system. The reason is that the Hartree-Fock approximation $\tilde{g}_h^{HF}$ to $\tilde{g}_h$ is always greater than 1/2 and, when there are only two electrons of opposite spin, one must choose $\tilde{n}$ very small in order to fulfill the sum rule. The exchange hole $n_x^{WD}(\vec{r},\vec{r}')$ then becomes $-1/2\ n(\vec{r}')$ which is the exact result for this case.

The AD approximation was designed to be exact for a gas of almost constant density, and we can assess the accuracy of the WD approximation in this limit by using it to calculate the static density-response function of the homogeneous electron gas (see Ref. 2). One then obtains a response function of a quality similar to that obtained by e.g. the approach by Singwi et al[20]. Thus, to within our present knowledge of the density-response function of the electron gas, the AD and WD approximations are equivalent in the limit of an almost constant density. Note that they become identical in this limit, if the response function of the WD approximation is used as input to the AD approximation. Since the WD approximation is exact in the "opposite" limit of the hydrogen atom, we could expect the overall performance of the WD approximation to be superior to that of the AD approximation. It is therefore rather surprising that actual tests on atoms show the reverse to be true[44]. The error in the exchange-correlation energy in the usual LD approximation ranges from 8% to 5% for light to heavy atoms. The corresponding numbers for the AD approximation are 3% and 1%. The AD approximation thus represents a clear improvement over the LD approximation. The WD approximation, on the other hand, giving errors ranging from 15% to 6%, is inferior to the LD approximation. If only exchange energies are considered, the AD and WD approximations are of comparable accuracy and both are superior to the LD approximation. The former approximations give exchange energies which are too large (~ 5%), whereas the exchange energy is too small (~ 10%) in the LD approximation. The correlation energy of the WD approximation is larger than that of the LD approximation which is already about a factor of two too large[58]. As a consequence, the error in the total energy of the WD approximation is enhanced when correlation is included. In the case of the AD approximation, the error is instead reduced when the correlation energy is added. This is because the correlation energy of the AD approximation is much too small and actually has the wrong sign for some atoms. This fact and the fact that the WD approximation should, by construction, be superior to the AD approximation suggest that the good performance of the latter might be fortuitous. In the LD, AD and WD schemes, one tries to model the exchange-correlation hole by means of the pair-correlation function of the homogeneous electron gas. It is tempting to guess that this procedure is too simple-minded in a strongly inhomogeneous system like an atom. It could be that the errors inherent to this procedure are the same order of magnitude as the energy differences between the three schemes.

Gunnarsson and Jones[59] have recently suggested a modification of the WD scheme in which the true pair-correlation function is modeled by an analytic function of the form

$$\tilde{g}_m(r,n) - 1 = A(n).\{1 - \exp[-B^5(n)/r^5]\} \quad (56)$$

The density-dependent parameters A(n) and B(n) are chosen so that $\tilde{g}_m$ obeys the sum rule imposed by charge conservation $[n\int(\tilde{g}_m - 1)d^3r = -1]$ and gives the correct exchange-correlation energy of the homogeneous electron gas. The exchange-correlation hole of the modified scheme is then given by Eq. 55, with $\tilde{g}_h$ replaced by $\tilde{g}_m$. The advantage of this scheme is that the long-range oscillatory tail of the electron-gas pair-correlation function, which certainly has no relevance for localized systems such as, e.g., atoms, is replaced by a rapidly decaying (r^{-5}) tail without destroying the basic correctness of the theory in the slowly-varying limit. The scheme has been tested[59] in several atoms and the resulting exchange-correlation energies are in error by only a few percent. The new scheme thus represents a definitie improvement over both the LD approximation and the original WD approximation. The choice of formula (56) is, however, somewhat ad hoc. It would therefore be valuable to investigate whether the results obtained using this formula are indeed due to the short-range nature of the parametrization and are not a fortuitous consequence of the particular analytical form employed in formula (56).

Gunnarsson, Jonson and Lundqvist[44] have also tested the AD and WD schemes in a calculation of the exchange-correlation part of the surface energy of semi-infinite jellium. The results are quite disappointing. It is uncertain whether the AD approximation gives a convergent result for this energy and in any case, the surface energy is much too large. The WD approximation gives a divergent exchange energy, but the total exchange-correlation energy is convergent although a factor two smaller than the exact result within the RPA approximation applied to the infinite-barrier model[49]. Consequently, the AD and WD approximations are much inferior to the LD approximation for this particular problem. The difficulty seems to stem from a slight misrepresentation of the exchange-correlation hole on the bulk side of the surface. Since the surface energy is an energy difference between an infinite and semi-infinite system, these small errors seem to have a tendency to add up rather than cancel. This problem ought to be less severe in the modified WD scheme of Gunnarsson and Jones because the exchange-correlation hole in their scheme is less extended.

One should keep in mind that an atom and a surface represent very severe test cases for any density-functional theory based on data from the homogeneous electron gas. We still can hope that the AD and WD schemes represent a substantial improvement on the LD approximation in systems where the density variations are less rapid. Thus, it would be interesting to apply these non-local schemes to calculations of molecular binding energies or cohesive energies of solids. These energies are essentially determined by the other valence electrons[60] and one could hope that the new schemes would yield results for e.g. the 3d metals of comparable accuracy as the results obtained by Moruzzi et al[61] for the 4d metals.

Another case for which the AD and WD schemes may prove useful is the calculation of binding energies and equilibrium distances for atoms and molecules adsorbed on surfaces. The difficulties encountered with these schemes in connection with the surface energy will not influence the result for binding energies since these are energy differences for the semi-infinite system, with and without the adsorbed species. The AD and WD schemes will, however, give a more realistic exchange-correlation potential in the surface region as compared to the LD approximation. Far outside a metallic surface a classical unit charge will experience an image potential of the form $-1/(4z)$, z being the distance from the surface. Since classical arguments should apply also for an electron far enough outside a metallic surface, we would expect the asymptotic form of the exchange-correlation potential to be

$$v_{xc}(\vec{r}) = -\frac{1}{4z} \tag{57}$$

a conjecture which can be confirmed from microscopic theory [62]. Due to its density prefactor in Eq. 46, however, the LD approximation predicts an exponentially decaying potential outside the surface. Thus, the LD approximation cannot describe the van der Waal's interaction between e.g. an atom and a surface. On the other hand, the AD and WD schemes, as well as the modified WD scheme[59], all exhibit image-like behavior, although only the latter gives the correct coefficient (- 1/4) in Eq. 57. One can therefore hope that these schemes will produce more accurate potential surfaces than the LD approximation, at least for larger distances from the surface. A word of caution is appropriate here. The exact exchange potential decays exponentially away from the surface. Thus, the image potential (Eq. 57) is entirely a correlation effect. In the AD and WD schemes however, the image-like exchange-correlation potentials are a consequence of the large-r behavior of the pair-correlation function $\tilde{g}_h(r)$ of the homogeneous electron gas. Since this behavior is not very different[63] from that of the Hartree-Fock approximation to $\tilde{g}_h(r)$ both the AD and WD schemes would produce image potentials even if only exchange effects were considered. Thus, one can say that the desirable image-like behavior of these schemes is obtained for the wrong reason and their performance will have to be judged on the basis of numerical tests.

It is also of interest to discuss the asymptotic behaviour of the exchange-correlation potential v_{xc} far away from an atom or an ion. Classical arguments suggest the asymptotic form

$$v_{xc}(\vec{r}) = -\frac{1}{r} - \frac{\alpha}{2r^4} \tag{58}$$

which can also be rigorously derived from microscopic theory [62]. (See also Sec. 4.2). The quantity α is the static polarizability of the corresponding ion. The first term in Eq. 58 is a pure exchange effect. The second term is a pure correlation effect and should therefore be as difficult to reproduce as the image potential outside a surface. In the LD approximation v_{xc} decays exponentially away from the atom and thus neither of the two terms is given correctly by this approximation. The AD and WD approximations, on the other hand, both give a 1/r - dependence in v_{xc} although the coefficient is not -1 as in Eq. 58 but rather -0.55[44]. In the case of the WD approximation this partial failure can be traced to the violation of the symmetry condition $g(\vec{r},\vec{r}') = g(\vec{r}',\vec{r})$ obeyed by the exact pair-correlation function (see Eq. 55). In the AD scheme, the origin of the erroneous coefficient seems to be a more subtle numerical issue. Since the charge density of an atom is very low in the asymptotic region, an erroneous coefficient cannot affect the total energy in any significant way. In the case of negative ions (e.g. H^-, Na^-, Cl^-), however, this deficiency in the AD and WD schemes is of crucial importance. The potential from the nucleus and from all the electrons will here have a positive Coulomb tail

(+ 1/r) which will be present, although with a reduced coefficient, in the total effective one-electron potential unless the exchange-correlation potential has a Coulomb tail with a coefficent less than or equal to -1. It seems likely that a positive Coulomb tail will produce a positive eigenvalue for the outermost one-electron orbital, which explains the instability of the negative ions in the LD approximation[64,65]. This argument suggests that the AD and WD approximations also suffer from the same deficiency. A slight modification of the WD scheme (not to be confused with the modification mentioned above) that retains the above mentioned symmetry of the pair-correlation function has, however, recently been suggested (see Ref. 2). The tail from the total Coulomb potential will then be exactly cancelled by a term in the asymptotic exchange-correlation potential which ensures a total potential that decays faster than 1/r. Thus, there is some hope that the new, symmetrized scheme will predict stable negative ions.

At this point we would like to issue a warning. Even though there are cases, such as that described above, in which it is essential for our approximations to have a proper asymptotic behavior outside atoms and surfaces this requirement has, in our opinion, often been overemphasized in the litterature. It is, e.g., often pointed out that the LD approximation fails to provide a correct van der Waals interaction at large distances. The encouraging results Jones[66] obtained, within the LD approximation, for the binding energy and the equilibrium distance of the supposedly typical van der Waals molecule Be_2 indicate that this deficiency of the LD approximation is of minor importance. A lucid discussion of why this is to be expected has recently been given by Lang[67]. Using the LD approximation Lang also obtained rather accurate values for the binding energy and the equilibrium distance of rare-gas atoms adsorbed on metal surfaces.

Having discussed the virtues and drawbacks of the new, non-local schemes, we now add a few comments on their computational feasibility. We have already described how to construct the local exchange-correlation potentials ($v_{xc}(\vec{r})$) of the AD and WD schemes and in both cases it involves doing a small number of three-dimensional integrals for each point in space where the potential v_{xc} is needed. In all integrals the particle density appears as a weight factor. In the case of spherical symmetry, often encountered in practice, these integrals become one-dimensional and both schemes become quite feasible, although the computational effort is larger than for the LD scheme. Any band-structure or atomic program based on the LD scheme is, however, easily converted to the new non-local schemes by merely replacing the subroutine that computes $v_{xc}(r)$. The WD scheme requires a table of the electron-gas pair-correlation function for different densities and different inter-particle separations. In the case of exchange only, there is a single analytical formula available[68]. If correlation effects are of interest, the parameterized pair-correlation function recently suggested by Rajagopal et al[69] might be useful. In the modified WD scheme[59], the necessary pair-correlation function is instead given by a simple analytical formula (Eq. 56), which is a clear computational advantage. In the AD scheme, one needs to know the weight function which defines the average density discussed above and which is given in a table in Ref. 44.

By making use of the following two facts, a substantial amount of computational work can, however, be avoided with only a minor loss of accuracy. i) The charge density is relatively insensitive to difference choices of the exchange-correlation functional and the corresponding potential. For instance, the charge density of an atom in the LD approximation including exchange only differs by only a few per cent from the full Hartree-Fock result. ii) Due to the variational principle, the total energy is insensitive to errors in the charge density. Thus, we recommend that the total energy of the non-local schemes be calculated by inserting the charge density of a self-consistent LD calculation into the non-local energy functionals.

In this section we have discussed several methods for going beyong the LD approximation and they were all based on a real-space description of exchange and correlation. In the next section we will advocate a new and promising reciprocal-space approach.

2.5. Non-Local Functionals in Reciprocal Space

So far we have discussed the advantages and disadvantages of the LD approximation in terms of a description of the exchange-correlation hole in real space. The work by Langreth et al[47,48,53] mentioned previously seems, however, to suggest that it might be even more advantageous to consider the exchange-correlation energy as a sum of contributions from different wave vectors in reciprocal space. Defining the quantity $\tilde{S}(\vec{q})$ as

$$\tilde{S}(\vec{q}) = 1 + \frac{1}{N} \int n(\vec{r})n(\vec{r}\,')\{\tilde{g}(\vec{r},\vec{r}\,') - 1\}\, e^{i\vec{q}(\vec{r}-\vec{r}\,')} d^3r d^3r' \tag{59}$$

Langreth and Perdew[47,48] rewrote Eq. 42 as

$$E_{xc}[n] = \frac{N}{2} \int \frac{d^3q}{(2\pi)^3}\, v(\vec{q})\{\tilde{S}(\vec{q}) - 1\} \tag{60}$$

where N is the number of particles in the system and $v(\vec{q}) = 4\pi/q^2$ is the Fourier transform of the Coulomb potential. The purpose of the notation $\tilde{S}(\vec{q})$ is to suggest that this quantity is the ordinary static structure factor only integrated over the strength of the electron-electron interaction. Obviously, $\tilde{S}(\vec{q})$ tends to 1 for large q and for small q it tends to zero by virtue of the sum rule (45). Because of the Coulomb potential in Eq. 60, large wave vectors will give a relatively small contribution to the exchange-correlation energy whereas small and intermediate wave vectors contribute considerably. This was clearly demonstrated by Langreth and Perdew[48] in the case of a jellium surface for which the correlation effects were treated in the RPA. Concentrating only on that part $\delta \underset{\sim}{S}(\vec{q})$ of the structure factor which is due to the presence of the surface, they were able to show that $\delta \underset{\sim}{S}$ behaved as $\alpha . q$ for small q and they also gave an exact expression for the proportionality constant α which is valid to all orders in the electron-electron interaction. As discussed previously (see Eq. 46), the $\underset{\sim}{LD}$ approximation consists of replacing the pair-correlation function g by the corresponding quantity of the homogeneous electron gas and by replacing $\vec{r}'$ by $\vec{r}$ in the second density in Eq. 59. Langreth and Perdew then found that the corresponding $\delta \underset{\sim}{S}_{LD}$ has the wrong small-q behavior (it is proportional to q^2). For larger wave vectors, however, the LD approximation gave a rather satisfactory description of $\delta \underset{\sim}{S}$ and by constructing a simple scheme that interpolated between the known exact result for small q and the LD result for intermediate and large q, Langreth and Perdew were actually able to accurately reproduce the results of the full RPA for the exchange-correlation energy of the jeliium surface treated within the infinite-barier model[49]. When the interpolation scheme was applied to surfaces with more realistic density profiles such as those calculated by Lang and Kohn[70] from self-consistent LD calculations, the interpolation scheme gave corrections to the Lang-Kohn values of the order of 5-10% depending on the density. In atoms the LD approximation is known to give errors in the total exchange-correlation energies which are of the same order of magnitude[58]. Thus, the LD approximation is surprisingly accurate for these very inhomogeneous systems and therefore Langreth and Perdew took a different approach in later work. They left the LD approximation untouched and added a gradient correction. As discussed at length in Sec. 2.2, a straightforward lowest-order gradient term is not a good idea and actually leads to worse results for atoms. However, by carrying out the wave-vector analysis discussed above also for the gradient term, Langreth and Perdew[47] discovered the cause of the difficulty with the gradient correction. In the case of a jellium surface the straightforward gradient term tends to a large constant in the limit of small wave vectors thus violating the previously discussed exact behavior of the total wave-vector decomposed exchanged-correlation energy. This results in a large overestimate of the correction to the LD approximation, the latter giving a too-small exchange-correlation energy. Langreth and Perdew then argued that it is unphysical to use the local slope of the density in the gradient term in the case of long wave-length fluctuations which certainly do not feel finer details in the density. Instead they proposed the use of a slope that is averaged over the "size" ($\sim 1/q$) of the fluctuations. The resulting gradient correction has the correct analytical behavior ($\beta . q$) at small wave vectors although the coefficient β is slightly off and at large wave vectors the new correction becomes equal to the bare gradient correction which is rather accurate and small for large q. The new scheme yields surface exchange-correlation energies very similar to the interpolation scheme discussed above. In contrast to the interpolation scheme, however, the average-gradient scheme lends itself to relatively easy parametrization such that is can be applied to arbitrary inhomogeneous systems. This was recently demonstrated by Langreth and Mehl[54]. The idea is simply to neglect the correlation part of the wave-vector decomposed gradient correction below a certain cut-off wave vector proportional to the local density gradient and to retain the full correction at large wave vectors. (The exchange part of the wave-vector decomposed gradient correction has the correct analytical behavior at small wave vectors). The resulting expression for the exchange-correlation functional reads as follows

$$E_{xc}[n] = E_{xc}^{LD}[n] + C \int |\nabla n|^2 \, n^{-4/3} \, \{2e^{-F} - \tfrac{7}{9}\} d^3r \tag{61}$$

where $E_{xc}^{LD}[n]$ is the ordinary LD approximation, C is a constant equal to $(3/\pi)^{2/3}/(72\pi)$ in Rydbergs and $F = 0.262.|\nabla n|.n^{-7/6}$. The numerical constant in F determines the precise value of the cut-off wave vector and it is chosen such that the parametrized gradient correction (Eq. 61) rather accurately reproduces the results of the average-gradient approximation for metallic surfaces (a slightly smaller constant gives better results for the surfaces but somewhat worse results for small atoms). An important advantage of the parametrized scheme over the full scheme is that the expression (61) can be functionally differentiated to yield the exchange-correlation potential

$$v_{xc}(\vec{r}) = v_{xc}^{LD}(\vec{r}) + 2C\, n^{-7/3} \{\tfrac{7}{9}(n\nabla\vec{K} - \tfrac{2}{3}K^2) - e^{-F}[(2-F)n\nabla\vec{K} - (\tfrac{4}{3} - \tfrac{11}{3}F + \tfrac{7}{6}F^2)K^2 + nF(F-3)\frac{\vec{K}.\nabla|\vec{K}|}{|\vec{K}|}]\} \tag{62}$$

where $\vec{K} = \nabla n$. A convenient expression for the potential is of course crucial in order to do self-consistent calculations. Such calculations have been carried out by Langreth and Mehl[53,54] for the total energy of several atoms, and they found that the parametrized gradient scheme reduces the errors in the LD approximation by typically a factor of three. They also computed ionization energies of atoms and found even larger reductions of the error. The self-consistent charge densities of the atoms obtained with the potential (62) were consistently better than the Hartree-Fock (HF) densities when compared with more accurate results from configuration-interaction calculations. From this we conclude that the potential given by Eq. 62 contains much of the correct physics and is rather close to the unique exact density-functional (DF) potential for exchange and correlation. Unfortunately, the Langreth-Mehl potential gives eigenvalues which are rather far from the exact DF eigenvalues. The fact that the densities and also the eigenvalue differences obtained[54] from this potential are close to the correct results suggests, however, that the error in the potential (62) is almost constant over most of the atom. We will return to this point in Sec. 4.3.

Langreth and Mehl[54] also managed to show that the largest error in their energy functional (Eq. 61) is due to their treatment of the exchange energy which may be easier to improve on than the correlation energy. By constructing a scheme for separating exchange and correlation effects they obtained an energy functional for the correlation energy to be added to the energy of an exact HF calculation. In this way, the managed to calculate correlation energies of atoms with an accuracy which was better than 0.01 Ry. This is truly a remarkable achievment considering the simplicity of their scheme.

The Langreth-Mehl approximation has recently been tested by von Barth and Car[71] in a calculation of the ground-state properties of bulk silicon. According to their preliminary results the cohesive energy is much better than that obtained within the LD approximation and the equilibrium volume and the bulk modulus are, within the numerical accuracy, equal to the corresponding experimental results. On the other hand, the band gap of silicon is still a factor of two too small in the new scheme[71] as it is also in the LD approximation[72].

These results are of course very encouraging in particular because the parametrized gradient scheme substantially reduces the errors of the LD approximation in such dissimilar systems as atoms and surfaces for which the errors actually are of opposite sign. This is a clear advantage over all other presently available methods for improving on the LD approximation. The major drawback of the new scheme is its present inability to treat spin-polarized systems, but this will probably be rectified in the near future.

We end this section by noting that the non-local corrections represented by the second terms in Eqs. 61 and 62 has been computed from an RPA-like theory of the electron gas. Since there are strong cancellations between local and non-local effects for inhomogeneous systems, the local terms E_{xc}^{LD} and v_{xc}^{LD} should also be computed from the RPA of the gas. Thus, local energy functionals based on more sophisticated electron gas calculations should not be used in conjunction with the non-local corrections mentioned above.

3. ENERGIES OF EXCITED STATES

3.1. States with a Different Number of Particles

Clearly, density-functional (DF) theory as discussed previously in these notes is a theory for the ground-state properties of electronic systems. There are however several obvious cases in which the theory can be used to obtain excitation energies. The ionization potential (I) of any localized system such as an atom or a molecule is the difference between two ground-state energies - that of the original system, $E_0(N)$, and that of the corresponding ion, $E_0(N - 1)$

$$I = E_0(N - 1) - E_0(N) \tag{63}$$

A similar case is the electron affinity A, given by

$$A = E_0(N) - E_0(N + 1) \tag{64}$$

in obvious notation. Since this way of obtaining an excitation energy involves the energy difference between two self-consistent calculations the procedure is often referred to as a SCF calculation and we will return to it in connection with core-electron binding energies.

The ΔSCF procedure is also applicable to infinite solids because these can be considered as limiting cases of very large molecules. In the case of metals the ionization potential and the electron affinity is one and the same quantity, namely the chemical potential relative to the vacuum level which, therefore, is given correctly by the ΔSCF procedure. This fact was pointed out already by Sham and Kohn in their original work on DF theory[8]. In the case of semi-conductors and insulators the ionization potential and the electron affinity correspond respectively to the top of the valence band and the bottom of the conduction band - both measured relative to the vacuum level. Thus. These two important quantities are also given correctly be the ΔSCF construction. We postpone for the moment the discussion of how this leads to the fact that these two quantities are given by the corresponding exact DF eigenvalues. In the next subsection we will consider energies of states of different symmetries.

3.2. Pure-Symmetry States

Another rather obvious case in which the SCF procedure discussed above can be used to obtain an exact excitation energy is when the excited state has a symmetry S different from that of the ground state. In addition, the excited state has to be the lowest state within the subspace of symmetry S. We can then obtain the excited state energy by minimizing the density functional for the total energy within that subspace as discussed in section 1.3. This possibility was first pointed out by Gunnarsson and Lundqvist[11]. Clearly, the functional for the total energy does, in principle, depend on the symmetry S. This must also be the case for the exchange-correlation part, $E^S_{xc}[n]$, of that functional and, unfortunately, there is, as yet, no theory of how to incorporate this symmetry dependence into $E^S_{xc}[n]$. As we shall see, our limited knowledge of $E^S_{xc}[n]$ can be a much more severe problem when the lowest state of symmetry S is not the absolute ground state. For the lack of a better alternative we may consider the consequences of using the local-density (LD) approximation for $E^S_{xc}[n]$ (Eq. 30). Two specific examples will illustrate the inadequacy of this approximation in connection with the procedure outlined in section 1.3. Consider first the excited 1s2s configuration of atomic helium which gives rise to four states, one ^1S-state denoted $|0,0\rangle$ and three ^3S-states denoted $|1,M_S\rangle$, $M_S = -1,0,1$. It can easily be seen that the density matrices corresponding to these states are diagonal and spherically symmetric. They have the form

$$\begin{pmatrix} \frac{1}{2}\,n(r) & 0 \\ 0 & \frac{1}{2}\,n(r) \end{pmatrix} \tag{65}$$

for the states $|0,0\rangle$ and $|1,0\rangle$ and

$$\begin{pmatrix} n(r) & 0 \\ 0 & 0 \end{pmatrix} \tag{66}$$

for the state $|1,1\rangle$. Because there is no explicit symmetry dependence in the LD approximation to $E^S_{xc}[n]$, only an implicit dependence through the density matrix, states with the same form of the density matrix will acquire the same energy and the same density-matrix upon minimization. Consequently, the scheme predicts the same energy for the states $|0,0\rangle$ and $|1,0\rangle$ and, as a matter of fact[73], a different energy for the state $|1,1\rangle$ which should have been degenerate with the state $|1,0\rangle$. This clearly does not make sense. The difficulty stems from the fact that the LD approximation is not invariant under rotations in spin-space thus producing different energies for different M_S quantum numbers. Another example is furnished by the p^2 configuration of the carbon atom which gives rise to the three terms ^3P, ^1D and ^1S. The states belonging to the ^1S and the ^1D terms are singlets and therefore have no net spin-density. Furthermore, the ^1S state has a spherical charge density, and it is possible to form a linear combination of the five degenerate ^1D states that, according to the prescription given in section 1.3, also will have a spherical (and therefore ultimately the same) charge density. Consequently, the present scheme has the undesirable feature of giving the same energy for the ^1S and the ^1D. It will also predict energy differences between the degenerate ^3P states which are of the same order of magnitude as the whole ^3P - ^1D splitting[73]. These errors reflect the fact that the LD approximation is not invariant under rotations in real space. Thus we have to conclude that the DF theory of states of pure symmetry can not be used as a practical tool for calculations unless these deficiencies of the LD approximation are removed.

3.3. Mixed-Symmetry States

The failure of the straightforward generalization of the density-functional theory outlined in Sec. 3.2 suggests that the symmetry dependence of the exchange-correlation functional $E^S_{xc}[n]$ must be accounted for. As mentioned, there is, however, presently no available prescription for a symmetry-dependent $E^S_{xc}[n]$. Fortunately, there exists an alternate procedure based on the notion of mixed symmetry states[73] that allows us to keep the simple LD approximation for $E_{xc}[n]$. As observed by Ziegler et al[74] one obtains rather accurate total energies for those particular symmetry states that would reduce to single Slater determinants if there were no Coulomb interaction between the electrons. The energies of the states consisting of several determinants, on the other hand, often turn out to be quite poor. The reason for this can be understood from the exact expression for the exchange-energy functional obtained by using the Hartree-Fock pair-correlation function in Eq. 42. Knowing that the pair-correlation function has the four spin components $g_{\sigma\sigma'}$, where $\sigma,\sigma' = \uparrow,\downarrow$, this expression allows us to consider the exchange energy as a sum of four spin components $E^X_{\sigma\sigma'}$. Now, the crucial point to realize is that any single-determinantal state will always have $g_{\uparrow\downarrow} = g_{\downarrow\uparrow} = 1$, and consequently the unequal-spin exchange energies $E^X_{\uparrow\downarrow}$ and $E^X_{\downarrow\uparrow}$ vanish for such a state. In the LD approximation one replaces the true pair-correlation function with that of the homogeneous electron gas which, in the Hartree-Fock approximation, is obtained from a single Slater determinant. Therefore, the unequal-spin exchange energies $E^{LDX}_{\uparrow\downarrow}$ and $E^{LDX}_{\downarrow\uparrow}$ are always zero in the LD approximation. As a result, states which reduce to single Slater determinants in the absence of interactions will have this part of their exchange energies given exactly by the LD approximation. A state consisting of several determinants can, however, have large unequal-spin exchange energies, and this contribution to the total energy is completely lost in the LD approximation. To overcome this difficulty von Barth[73] constructed a rigorous density-functional theory of mixed-symmetry states. These states are defined as linear combinations of those states of pure symmetry which are lowest in energy

$$|D_i\rangle = \sum_j \alpha_{ji}\, |S_j,0\rangle \tag{67}$$

Here, $|S_i,0\rangle$ is the "ground-state" of the S_i symmetry channel, and the significance of the method lies in choosing the coefficients α_{ij} such that the mixed-symmetry state $|D_i\rangle$ reduces to a single Slater determinant in the non-interacting case. The energy expectation value $E(D_i)$ of this state is given by

$$E(D_i) = \sum_j |\alpha_{ji}|^2\, E_0(S_j) \tag{68}$$

where $E_0(S_i)$ is the lowest energy in the S_i symmetry channel. The quantity $E(D_i)$ can, however, also be estimated from the LD approximation by minimizing the functional for the total energy (Eq. 15) with respect to density matrices of a form appropriate to $|D_i\rangle$. If we can find as many mixed-symmetry states as there are pure-state energies, we can invert the system of linear equations (Eq. 68) to get the pure-state energies $E_0(S_i)$. In cases of high symmetry, such as an atom, we generally get an overdetermined set of equations which can be solved by a min-max procedure. There are, however, low symmetry cases in molecular physics for which the number of single-determinantal states are too few to enable us to determine all pure state energies[74].

We end this section by noting an interesting consequence of the DF theory of mixed-symmetry states with regard to the statistical average of the term energies of many configurations. This average is, in general, not given by the energy of a spherical non-spin-polarized calculation as is often assumed in the literature. (See also Sec. 3.5 below).

3.4. Valence Exchange and Correlation Energies

As reiterated several times in these notes, the LD approximation was designed for systems with a slowly varying electronic density. We would therefore expect the approximation to work much better for the more spread-out valence electrons than for the tightly bound core electrons. Fortunately, many interesting physical properties such as the cohesive energy of a solid and the binding energy and the equilibrium distance of a molecule are mainly a consequence of rearrangements among the valence electrons[60]. Still, the density of these electrons can have a substantial spatial variation and it would be nice if we, in some way could monitor the accuracy of the approximation as we go from localized to more extended electron orbitals. Since density-functional theory is a theory of the total energy and the total charge density, this would perhaps seem to be a difficult task, but the multiplet splittings of atoms and molecules offer an interesting possibility in this regard. These splittings are almost entirely due to exchange and correlation effects among the outer valence electrons. By comparing the splittings obtained from the LD approximation with the experimental splittings for cases in which the valence electrons are more or less localized, we can get a handle on the accuracy of the approximation as a function of the degree of localization.

Several people[24,73-76] have carried out calculations of multiplet splittings for many different atoms using the method of mixed-symmetry states which was described in Sec. 3.3. Typical results are shown in Table 1, in this case for the p^2 configuration of the carbon and silicon atoms and for the p^3 configuration of the nitrogen atom[73]. As can be seen from the table, the LD

Table 1. Multiplet splittings in eV for the carbon, silicon, and nitrogen atoms. The configuration is indicated. HF is Hartree-Fock. LDX is the local-density approximation with exchange only. LD is the local-density approximation, present theory. Exp is spectral data from Ref. 77. LDC is local-density correlation, LDX-LC.C is the correlation contribution, HF-exp.

		HF	LDX	LD	Exp	LDC	C
C p^2	$E(^1D)-E(^3P)$	1.56	1.59	1.33	1.26	0.26	0.30
	$E(^1S)-E(^3P)$	3.90	3.97	3.30	2.68	0.67	1.22
Si p^2	$E(^1D)-E(^3P)$	1.07	1.09	0.85	0.76	0.24	0.30
	$E(^1S)-D(^3P)$	2.67	2.71	2.11	1.89	0.60	0.78
N p^3	$E(^2D)-D(^4S)$	2.81	2.90	2.46	2.38	0.44	0.42
	$E(^2P)-E(^4S)$	4.68	4.83	4.09	3.58	0.74	1.10

approximation including only exchange reproduces the full Hartree-Fock results with a remarkable accuracy, the error being typically of the order of 0.1 eV. These results permit us to conclude that the charge density of the outer valence electrons in atoms and molecules (and in solids ?) is sufficiently slowly varying to allow for a suantitative description of their exchange energies within the LD approximation. More interesting, however, is the fact that when correlations are included in the LD approximation, one also obtains reasonable correlation-energy contributions to the multiplet splittings. As mentioned in Sec. 2.4, the total correlation energy of an atom is typically a factor of two too large in the LD approximation[58]. In the case of the multiplet splittings, the corresponding error is only of the order of 20%-50% depending on the diffuseness of the orbital. Furthermore, the correlation energy is usually too small. There are two things that indicate that this encouraging result is significant and not fortuitous : (i) The larger correlation-energy error in carbon relative to silicon is consistent with the relative localization of the valence p-orbitals in the two atoms (2p in C, 3p in Si). This is the behavior one would expect from a beginning convergence of the LD approximation with respect to the rapidness of the density variations. (ii) Within perturbation theory one can show[73] that the LD results for the multiplet splittings can be described in terms of reduced Slater integrals (usually denoted F_k). The formula for the reduction term only involves quantities associated with the correlation part of the LD energy functional. This is in accord with what is usually expected from a theory of correlations. It is also consistent with an observation by Wood[75] that the correlation energy cannot be obtained by choosing a particular α in the Xα energy functional. On the other hand, the perturbation treatment also points to a limitation of the LD theory of multiplet splittings. The theory no longer applies when the correlations become so strong that the splittings cannot be described in terms of reduced Slater integrals. Another deficiency has been pointed by Gunnarsson and Jones[76] who show that there is a peculiar asymmetry between particles and holes in the theory. For example, the results for several sp configurations are consistently better than those for the sp^5 configurations. They also found a recipe for restoring the symmetry which considerably improved the results for the sp^5 configurations but there is, as yet, no formal justification for this procedure. In summary of this section, we would like to point out that the high accuracy with which both exchange and correlation energies of the outer valence electrons of atoms can be obtained from the LD approximation, as demonstrated by the accuracy of the atomic multiplet splittings, strongly supprts the use of this approximation for obtaining those physical properties of molecules and solids which primarily depend on rearrangements of the outermost electrons.

3.5. Density-Functional Theory of Subspaces

In the original density-functional (DF) theory as formulated by Hohenberg, Kohn, and Sham[6,7] the ground-state density and the total ground-state energy are the fundamental variables for the description of a many-body system. Theophilou[78] has, however, shown that there exist subspaces of the eigenstates of the Hamiltonian such that the corresponding subspace energies are functionals of the subspace densities. Consider, for instance, the subspace spanned by all the eigenstates belonging to the two lowest eigenenergies. The subspace energy is defined as the trace of the Hamiltonian with respect to the subspace, in this case, simply the sum of the two lowest eigenvalues weighted by their corresponding degeneracies. Similarly, the subspace density is defined to be the trace of the density operator over the subspace, i.e. the sum of all the particle densities of the individual states of the subspace. Theophilou also showed that the functional for the subspace energy is minimized by the exact subspace density. This theory allows us, in principle, to obtain the subspace energy and therefore, by subtraction, the energy of the first excited state provided that the ground-state energy is known. Further advantages in the theory of subspaces over the original formulation of DF theory is that a degenerate ground state does not cause a problem and that the subspace densities usually have the full symmetry of the Hamiltonian.

As usual, the difficulty with the subspace procedure lies in our inadequate knowledge of the exchange-correlation part of the functional for the subspace energy. Clearly, this must depend explicitly on the chosen subspace and it is not difficult to realize that the LD approximation is inadequate for the exchange-correlation energy of the subspace. Consider, for instance, again the p^2 configuration of the carbon atom and the subspace spanned by its 15 states (see Sec. 3.2). Clearly, the subspace energy is the configurational average and the subspace density is spherically symmetric (the spin-up and spin-down densities of the subspace are also the same, i.e. the subspace is not spin polarized). Thus, within the LD approximation, the subspace method predicts that the configurationally averaged energy is given by the energy of the spherical non-spin-polarized calculation which is not correct as discussed in Sec. 3.3. We thus conclude that some way of incorporating the subspace dependence of $E_{xc}[n]$ must be found in order to render the subspace method useful for practical calculations.

3.6. Core-Level Excitations

At first thought, it appears to be a particularly unreliable procedure to apply the DF ground-state formalism to the calculation of core-level excitation energies - especially using the LD approximation. In the first place, such an excitation involves a highly excited state of the atom, molecule, or solid which certainly is well outside the formal validity of the ground-state theory. Secondly, the highly localized character of the core-hole state casts doubt on the validity of its description in terms of the LD approximation. Nevertheless, the LD approximation has been used successfully to calculate core-electron binding energies in both simple metals and transition metals - the accuracy being typically of the order of 1 eV. By calculating the valence-electron response as a function of core-level occupation, it was possible of either one or two core holes. In this way, the difference between screening in the context of photoemission (a single core hole)[79,80] and of Auger (two core holes)[79,81] experiments to be understood. Thus, in these calculations the difficulties with the LD approximation in connection with very localized orbitals were essentially avoided by considering only that part of the core-hole binding energy which is due to the rearrangement of the outer valence electrons, i.e., due to the valence relaxation effects.

We have, however, yet to provide any justification for using even the exact ground-state formalism to describe such highly excited states as systems containing core holes. We offer in this regard only our speculation that the crucial property of an electronic system that determines the appropriateness of the ground-state theory is the lifetime of the state to which the theory is applied. If the interactions that cause a core-hole state to decay (e.g. Auger) are neglected, then the core-hole and no-core-hole states become orthogonal subspaces of the full many-body Hamiltonian. If we think of such subspaces as possessing different symmetries, then we can appeal to the proposal by Gunnarsson and Lundqvist[11] to use the ground-state theory to describe the lowest-energy state of each symmetry subspace. In the discussion (Sec. 3.2) of atomic multiplets, which constitute an example of symmetry subspaces with different energies, we saw that a different energy functional is required for each symmetry subspace. In the present context, we believe that the distinguishing feature of the energy functional appropriate to the core-hole state is simply the fact that the core orbital is kept unoccupied. Note that occupying single-particle states of the energy functional other than those which lie lowest in energy represents a change in the functional, not the minimization of the ground-state functional in a restricted space of electron densities.

In Sec. 4.3 we will return to the core-electron binding energies and the interpretative power of the LD eigenvalues.

3.7. Time-Dependent Density-Functional Theory

In quantum mechanics the problem of finding the excited states of a system is intimately connected to the problem of describing the response of the system to time dependent external perturbations. The difficulties encountered in our present version of DF theory in connection with excited states orginiate from the fact that this theory is basically a theory of static phenomena. A time-dependent generalization of the DF theory of Hohenberg and Kohn[6] has, however, been proposed by Peuckert[82]. The variational principle for the ground-state energy is here replaced by the stationary property of the action integral under small variations of the time-dependent state vector around the true physical state of the system. From a strict mathematical point of view there are more difficulties associated with the time-dependent theory than with the static theory. In analogy with the static theory it is, for instance, necessary to assume that there exists a non-interacting time-dependent system which reproduces the time-dependent density of the interacting system. Furthermore, the one-to-one correspondence that exists between the density and the external potential in the theory of Hohenberg and Kohn is, through the assumed invertability of the density-density response function, replaced by a one-to-one correspondence between the infinitesimal deviations of these quantities from their physical values. These assumptions are, however, easily accepted on physical grounds. The new theory has a structure rather similar to that of Hohenberg, Kohn, and Sham[6]. The interacting N-body problem is again reduced to that of N non-interacting electrons moving in an effective time-dependent potential which is the sum of the external potential w, the Hartree potential V_H, and the now also time-dependent exchange-correlation potential v_{xc}. Thus,

$$\{-\frac{1}{2}\nabla^2 + w(\vec{r},t) + \int v(\vec{r}-\vec{r}')n(\vec{r}',t)d^3r' + v_{xc}(\vec{r},t)\}\psi_i(\vec{r},t) = i\frac{\partial}{\partial t}\psi_i(\vec{r},t) \tag{69}$$

and the density is obtained from

$$n(\vec{r},t) = \sum_{i=1}^{N} |\psi_i(\vec{r},t)|^2 \tag{70}$$

where N is the number of electrons. As usual the many-body problem has been transferred into v_{xc} for which Peuckert[82] gives an expression quite analogous to Eq. 42 and Eq. 18

$$v_{xc}(\vec{r},t) = \frac{\partial}{\partial n(\vec{r},t)} \int dt \int \frac{d\lambda}{2\lambda} \int d^3r d^3r' \times \{\langle\Psi_\lambda(t)|\psi^+(\vec{r})\psi^+(\vec{r}')\psi(\vec{r}')\psi(\vec{r})|\Psi_\lambda(t)\rangle - n(\vec{r},t)n(\vec{r}',t)\}v_\lambda(\vec{r}-\vec{r}') \tag{71}$$

Here, $|\Psi(t)\rangle$ is the state vector of the interacting system at time t, $\psi^+(\vec{r})$ and $\psi(\vec{r})$ are the field operators that create and annihilate an electron at $\vec{r}$, and an integration over the strength λ of the electron-electron Coulomb interaction is performed in analogy with Eq. 42 ($v_\lambda(r) = \lambda/r$).

It is well known that the interaction energy of a static system can be expressed in terms of its frequency-dependent density-density response function. As can be shown from, for instance, a many-body diagrammatic analysis this is also true for interacting electrons subject to a time-dependent external potential (see e.g. Ref. 83) and we have

$$v_{xc}(\vec{r},t) = \frac{\partial}{\partial n(\vec{r},t)} \int dt \int \frac{d\lambda}{2\lambda} \int d^3r d^3r' \times \{iX(\vec{r},t;\vec{r}',t) - n(\vec{r},t)\delta(\vec{r}-\vec{r}')\}v(\vec{r}-\vec{r}') \tag{72}$$

where X is the time-ordered response function. This expression offers an interesting possibility. In the static DF theory is, in principle, possible to calculate the exact static response function by considering non-interacting electrons moving in the effective potential given by Eq. 17. The density is calculated from Eq. 19 in the case of an external potential $w(\vec{r})$ and also in the case when this potential has been incremented by $\delta w(\vec{r})$. The density difference is then expanded in powers of δw, the coefficient of the leading term being the response function X ($\delta n = X.\delta w$, see also Ref. 1). The frequency or, equivalently, the time dependence of X is, however, beyond the static theory. On the other hand, using the time-dependent generalization of Peuckert[82] it becomes possible to compute $X(\vec{r},t;\vec{r}',t')$ from the equivalent one-particle scheme defined by Eqs. 69 and 70, provided that we have some approximation to $v_{xc}(\vec{r},t)$. The resulting X can then be inserted into Eq. 72 giving an improved v_{xc} which, in turn, can be used to construct an improved X. In principle, the exact v_{xc} could be constructed iteratively in this way but, in practice, already the second step involves, as yet, insurmountable difficulties. The first step, i.e., obtaining X from some known approximation to v_{xc} is, however, quite feasible as has been demonstrated by Zangwill and Soven[84]. They used the dynamical theory outlined above to calculate the frequency-dependent density-density response function $X(\vec{r},\vec{r}';\omega)$ for several atoms. Using the simple LD approximation for v_{xc} in Eq. 69 they obtained surprisingly accurate results for the atomic photoabsorption crossections which are simply related to the imaginary part of X. Even in such strongly correlated systems as Ba and Ce[85] their results compared well with results from quite sophisticated many-body perturbation expansions.

The application of time-dependent DF theory to dynamical phenomena is a new and exciting field of research and I would expect it to grow rapidly in the future. The new theory gives quite reasonable dynamical response functions of the homogeneous electron gas (see Ref. 1) and the encouraging results obtained for such localized systems as atoms induce hope that the theory will provide reliable results also for, e.g., optical absorption spectra of solids.

4. EXCITATION ENERGIES AND DENSITY-FUNCTIONAL AND LOCAL-DENSITY EIGENVALUES

4.1. Density-Functional Eigenvalues of Extended Systems

The exact DF eigenvalues are defined as those obtained from the solutions of the effective one-particle Schrödinger equation (Eq. 26) of DF theory with the exchange-correlation potential v_{xc} derived from Eq. 27. In the original formulation of the theory[6,7] these eigenvalues were considered as auxiliary quantities for obtaining the total energy (Eq. 29) and the charge density (Eq. 28). They were assumed to have no physical significance - except for the highest occupied eigenvalue which was shown to give the correct Fermi energy in a metal. Nevertheless, these eigenvalues as they are obtained from the LD approximation, have been used extensively and rather successfully to interpret excitation spectra such as photoemission spectra of both simple metals and transition metals. There are, however, also some cases in which the LD eigenvalues are clearly incompatible with the experimental evidence. Consequently, it would be quite useful to understand the connection between one-particle excitation energies and exact DF eigenvalues and how the latter are corrupted by the use of the LD approximation. A first clue to this connection is provided by the relation

$$\frac{\partial E}{\partial n_k} = \varepsilon_k \tag{73}$$

which is obtained by extending the functional for the total energy into the domain of densities containing an non-integral number of particles. We consider densities of the form

$$n(\vec{r}) = \sum_i n_i |\psi_i(\vec{r})|^2 \tag{74}$$

where $\psi_i(\vec{r})$ is an ordinary DF orbital, and take derivatives with respect to the fractional occupation numbers n_i. At the end we let the n_i:s resume their ordinary values for Fermions, i.e. either 0 or 1. The extension is straightforward within, e.g., the LD approximation and the relation (73) is quite easily proven[79,86,87]. In the exact case the difficulty lies in extending the exchange-correlation energy $E_{xc}[n]$ to a function of fractional occupation numbers. The following arguments suggest, however, that the relation (73) is correct also in the exact DF theory. If x is a variable that measures transfer of charge from orbital j to orbital i (i.e. $n_i = x$ and $n_j = 1 - x$), there is no difficulty in proving the relation

$$\frac{\partial E}{\partial x} = \varepsilon_i - \varepsilon_j \tag{75}$$

because the total charge is conserved. If we now think of orbital j as an extremely diffuse orbital at the vacuum level ($\varepsilon_j = 0$) and far away from the rest of the system it seems clear that this orbital has no self energy (either kinetic or potential) and no interaction with the rest of the system. We therefore identify the energy of the total system with that of the rest of the system and arrive at the relation (73). Thus the trick is to move charge far away from the system under consideration but to still consider it as part of the system. We also note that there was no restriction on the orbital labeled k in our discussion of Eq. 73. Hence this equation is valid for all DF orbitals - occupied as well as unoccupied.

If we now consider large systems in which the DF orbitals extend over the entire system it seems intuitively clear that the DF eigenvalues can not change upon removal of an orbital. Each DF orbital contains a factor one over the square root of the volume of the system and removing such an orbital can only produce a change in the effective potential which tends to zero as the inverse of the volume. Thus, in the limit of a very large system the DF eigenvalues do not depend on occupation. As a matter of fact it is quite easy to prove the exact relation

$$\frac{\partial \varepsilon_i}{\partial n_j} = \langle ij|(v + K_{xc})\tilde{\varepsilon}^{-1}|ij\rangle \tag{76}$$

where

$$<ij|v|ij> = \int |\psi_i(\vec{r})|^2 \, v(\vec{r} - \vec{r}') |\psi_j(\vec{r}')|^2 d^3r d^3r' \tag{77}$$

$v(r) = 1/r$ is the Coulomb interaction, and where $\tilde{\varepsilon}$ is the dielectric function for quasi-particles as defined in Ref. 16. The quantity

$$K_{xc}(\vec{r},\vec{r}') = \frac{\delta^2 E_{xc}}{\delta n(\vec{r}) \delta n(\vec{r}')} \tag{78}$$

is the exchange-correlation interaction discussed previously in Sec. 2.2. This interaction is short ranged[20] so that the long-range properties of the right-hand side of Eq. 76 are determined by v(r). The normalization factors in the orbitals together with two integrations over the volume Ω and the 1/r-dependence of the Coulomb interaction give, at least, rise to a $\Omega^{-1/3}$ decay of the right-hand side of Eq. 76 as the volume tends to infinity. Thus, $\partial\varepsilon_i/\partial n_j = 0$ for an extended system and we find that the eigenvalue equals a ΔSCF energy

$$\varepsilon_k = E(n_k = 1) - E(n_k = 0) \tag{79}$$

If k labels the highest occupied state in a metal then the right-hand side of this equation is that difference between two ground-state energies which corresponds to the energy of removing one particle, i.e. to the Fermi energy, which therefore is given correctly by DF theory[8]. Similarly, if k labels the state at the top of the valence band in an insulator the right-hand is the ΔSCF result for ionizing the system as discussed in Sec. 3.1. Equation 79 shows that this excitation energy is again given correctly by the corresponding DF eigenvalue. This is also the case for the energy gained by adding a particle to the system, i.e., per definition, the bottom of the conduction band. Consequently, DF theory can be used to determine the fundamental band gap of a semiconductor or an insulator. In this context it is important to stress that one can, in principle, not obtain the energies of the excitons from DF theory since these represent excited states of the N-particle system.

As we move away from the Fermi level in a metal or from the band gap in an insulator the right-hand side of Eq. 79 is no longer the difference between ground-state energies and it becomes doubtful whether or not the exact DF eigenvalues represent excitation energies. The formal similarity between Eq. 73 and a corresponding equation within an exact theory of low lying excitations, namely Fermi liquid theory, suggests that they are. It is tempting to draw the same conclusion from the strong similarity between measured valence photoemission spectra spectra of simple metals and the results of band structure calculations based on the LD approximation. Fortunately, there exists a system for which the relationship between the ground-state eigenenergies and the quasi-particle excitation spectrum can be explored in detail, and without the ambiguity usually introduced by use of the LD approximation. This system is the homogeneous, interacting, electron gas. On the one hand, the translational symmetry of this system allows its accurate description using the techniques of traditional many-body theory ; on the other hand, for this system, the LD approximation is not an approximation. This system therefore, allows us to compare ground-state eigenenergies which are uncorrupted by the LD approximation with the exact quasi-particle excitation spectrum.

The exact quasi-particle excitation spectrum, E(k), is given in terms of the momentum- and frequency-dependent self-energy operator $\Sigma(k,\omega)$ by the following implicit relationship[83]

$$E(k) = \frac{k^2}{2m} + \Sigma[k,E(k)] \tag{80}$$

The single-particle equations of the ground-state theory, on the other hand, contain a k-independent, translationally invariant potential ; the dispersion of the ground-state eigenenergies, $\varepsilon(k)$, is therefore simply proportional to k^2. As discussed above, however, the ground-state eigenenergies correctly describe the Fermi energy E_F, and are therefore given by

$$\varepsilon(k) = \frac{k^2}{2m} + \Sigma[k_F,E(k_F)] \tag{81}$$

As the comparison of Eqs. 80 and 81 clearly shows the ground-state eigenenergies $\varepsilon(k)$ and the quasi-particle spectrum E(k) differ as we move away from the Fermi energy, but they differ only to the extent that the self-energy varies with k and ω. In the immediate vicinity of the Fermi level, the quasi-particle energies and the DF eigenvalues can be expanded, to first order, in Taylor series giving

$$E(k) = E_F + \frac{k_F}{m^*}(k - k_F) \tag{82}$$

$$\varepsilon(k) = E_F + \frac{k_F}{m}(k - k_F) \tag{83}$$

where

$$\frac{m^*}{m} = (1 - \frac{\partial\Sigma}{\partial\omega}) \cdot (1 + \frac{m}{k_F} \cdot \frac{\partial\Sigma}{\partial k})^{-1} \tag{84}$$

Thus, close to E_F, the distinction between the ground-state eigenenergies and the quasi-particle spectrum is described by the ratio of the effective mass m^* to the bare-electron mass m.

The difference between m and m^* provides a clear demonstration that the ground-state eigenenergies are not formally equivalent to excitation energies. Thus, the total-energy functional of density-functional theory differs from that of Fermi-liquid theory, and the state that corresponds to adding a density-functional orbital with wave vector k is in general not equal to the many-body state described by a quasi-particle of the same wave vector. Furthermore, while the ground-state theory in principle, gives the correct Fermi energy, it remains an open question whether or not it gives the correct Fermi surface. Note also that the deviation of $m^\star$ from m implies that the state density at the Fermi energy, where lifetime effects are unimportant, is not given precisely by the distribution of density-functional ground-state eigenenergies. We can ask independently, as a practical matter, what is the magnitude of the difference between m and $m^\star$? While in some special cases, such as ^{3}He, the difference can be large ($m^\star \sim 3m$)[88], in the electron gas at metallic densities the difference is typically 5%[83].

Away from the immediate vicinity of the Fermi energy, the distinction between the ground-state eigenenergies and the excitation spectrum is most easily described by the graph of the k dependence of the self-energy shown in Fig. 5. The most qualitative way in which the self-energy changes as we move away from the Fermi energy is that it develops an imaginary part, reflecting the finite lifetime of the quasi-particle excitations. Beyond that, we see in Fig. 5 that the constant value of Σ appearing in the ground-state eigenenergies (Eq. 81) means that the ground-state eigenenergies reflect neither the rapid variation of the self-energy for $k \sim 2k_F$ which reflects the onset of quasi-particle decay via plasmon creation nor the gradual loss of exchange and correlation energy with increasing quasi-particle velocity. The latter reflects the relative inability of the electron gas to screen a rapidly moving disturbance. The picture of the difference between the quasi-particle spectrum and the ground-state eigenenergies provided by Fig. 5 is consistent with the known need for using an energy-dependent exchange-correlation potential in the analysis of LEED[89] and EXAFS[90] experiments because of the large energies involved. Figure 5 is also consistent with the good description of simple-metal valence-band spectra provided by the ground-state eigenenergies, for we see that the variation of Σ in the vicinity of the valence band is relatively weak. We remind the reader that, even for simple metals, the LD approximation is not exact and its use introduces an additional source of error which is quite independent of that due to the variation of the self-energy.

4.2. Density-Functional Eigenvalues of Localized Systems

In localized systems intuition suggests that the removal of one unit of charge must affect the DF eigenvalues and consequently we would expect $\partial\varepsilon_i/\partial n_j$ to be finite for a localized system. Thus,

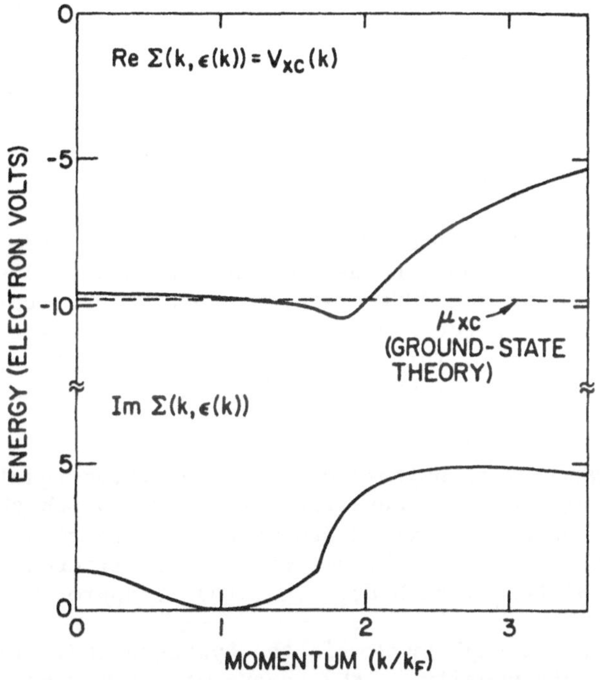

Fig. 5. Variation of the self-energy with quasi-particle velocity. The deviation of this curve from its value at $k/k_F = 1$ is the energy error made when the eigenenergies of the density-functional description of the ground state of jellium are used to approximate the quasi-particle excitation spectrum. The error is seen to be relatively small (but finite) over the velocity range corresponding to the valence band, but to become large in the range relevant to LEED and EXAFS experiments. The self-energy depends of the value of the uniform electron density ; the curve shown here is appropriate to most metals. The values of $\Sigma[k,E(k)]$ were taken from Ref. 83.

Eq. 79 has to be replaced by the integrated version of Eq. 73.

$$E(n_k = 1) - E(n_k = 0) = \int_0^1 \varepsilon_k(n)dn \tag{85}$$

In order to more clearly exhibit the difference between localized and extended systems we expand the DF eigenvalues in a Taylor series around $n_k = 1$ and, to first order, we obtain

$$E(n_k = 1) - E(n_k = 0) = \varepsilon_k + \Delta_k \tag{86}$$

where

$$\Delta_k = -\frac{1}{2} \left(\frac{\partial \varepsilon_k}{\partial n_k}\right)_{n_k=1} \tag{87}$$

Thus, assuming the second term on the right-hand side of Eq. 86 to be finite leads to the conclusion that for localized systems, the DF eigenvalues are not ΔSCF energies. Furthermore, relying on our experience from Hartree-Fock theory, we might guess that $\partial\varepsilon_k/\partial n_k < 0$ ($\Delta_k > 0$). This is because, within Hartree-Fock theory the difference Δ_k^{HF} between the ΔSCF energy and the eigenvalue is a physical relaxation shift which is always positive.

As we shall see, neither of these intuitive results are correct. There are several localized systems for which the exact DF eigenvalues can be determined and which thus serve as important test cases for assessing the performance of different approximate versions of DF theory such as, e.g., the LD approximation.

The simplest such system is the hydrogen atom. In this case the ground-state density is the square of exactly one DF orbital (Eq. 28) which thus, to within an irrelevant phase factor, has to be the hydrogen 1s-orbital. Consequently, the DF eigenvalue is -1 Rydberg and the exact exchange-correlation potential is the negative of the Coulomb potential from the 1s-electron (see Eq. 26). Helium is another atom in which the two identical DF orbitals (one for each spin) can be obtained from the square root of an accurately known density obtained from, for instance, a Hylleraas kind of wave function. By inverting the effective one-particle Schrödinger equation of DF theory (Eq. 26) one can then obtain the effective one-particle potential minus the DF eigenvalue. Far away from the atom this quantity approaches a constant. Since the effective potential vanishes outside the atom (see Eq. 94) this constant is the negative of the DF eigenvalue. Unfortunately, this procedure can not produce an accurate eigenvalue because it relies on having a charge density which is very accurate far from the atom. This poses a difficult problem since this region contributes very little to the total energy of the atom. It is quite possible to have a many-term variational wave function which gives a total energy correct to within one milli-Rydberg but which still gives a density which is not accurate enough outside the atom. The difficulty can also be appreaciated in the following way. Suppose we add to the nuclear potential of the atom a potential which is constant (V_0) inside some very large radius R and which vanishes outside R. Clearly this potential can essentially change the density of the atom only outside R where the density is very small. On the other hand, the eigenvalues all shift by the constant V_0 which can be quite large.

This same problem also occurs for many-electron atoms. In this case the charge density is a sum of squares of several DF orbitals (Eq. 28) and we can no longer obtain the one-particle DF potential by inverting the Schrödinger equation. To within a constant, the potential can still be obtained from the following procedure if we assume that the charge density is accurately known from, for instance, a large-scale configuration-interaction calculation. One constructs the charge density of the correct number of non-interacting electrons moving in a local one-particle potential with a large number of variational parameters. These are then varied until the resulting density agrees with the correct density. The fact that there exists a solution to this problem demonstrates the non-interacting w-representability of the density of the interacting system and, by the uniqueness theorem of Hohenberg and Kohn[6], the resulting potential and the exact effective DF potential only differ by a constant. The exact DF potential for exchange and correlation is finally obtained by subtracting the Coulomb potential from the nucleus and from the exact charge density. The feasibility of such calculations have been demonstrated by von Barth and Car[91] and by Almbladh and Pedroza[92]. In practice it is quite easy to reproduce the exact density to within one part in a thousand using only ~ 15 parameters in the one-particle potential. For the reasons discussed above this accuracy is far from sufficient for determining the DF eigenvalues. Note, however, that it is only the absolute position of the eigenvalues relative to the zero of the potential which is hard to determine and not their mutual separation (see the discussion of the previous paragraph).

In Hartree-Fock (HF) theory the asymptotic large-r behavior of the density is known analytically which offers a nice possibility for determining the exact DF eigenvalues. Note that a DF equivalent to HF theory was shown to exist already by Kohn and Sham[7] It is easy to see that, for large r, the highest occupied HF orbital decays exponentially as $\exp(-r\sqrt{|\varepsilon_{HF}|})$ where ε_{HF} is the corresponding eigenvalue in Rydbergs. All the HF orbitals are solutions to inhomogeneous Schrödinger equations and in the absence of the inhomogeneities all other orbitals would decay exponentially faster than the highest orbital owing to the fact that their eigenvalues are larger in absolute magnitude. The inhomogeneous term for one orbital, however, contains all other occupied orbitals and in particular the highest orbital which thus determines the decay of all orbitals to exponential accuracy. Therefore, the HF density being the sum of squares of the orbitals, decays as $\exp(-2r\sqrt{|\varepsilon_{HF}|})$. On the other hand, the DF orbitals are all solutions to the same homogeneous Schrödinger equation and their decays are, to exponential accuracy, determined by their corresponding eigenvalues. Thus, the decay of the density is determined by the slowest decaying DF orbital, i.e., the highest occupied DF orbital with eigenvalue ε_{DF}. Consequently, the density decays as $\exp(-2r\sqrt{|\varepsilon_{DF}|})$. Since, by definition, the DF version of HF exactly reproduces the HF density, we conclude that, within HF, the highest occupied DF eigenvalue is identical to the corresponding HF eigenvalue.

The picture that emerges within HF theory is illustrated in Table 2 comparing the DF eigenvalues with the corresponding excitation energies ΔE of the 1s- and 2s-levels of atomic beryllium. Within HF, the latter energies are simply given by the HF-ΔSCF results. Also shown are the eigenvalues obtained from the LD approximation which will be discussed in the next section. We see that the uppermost DF eigenvalue is close to the ionization potential whereas the DF eigenvalue is way above the excitation energy of the deeper lying level. In the event the exact DF eigenvalues are monotonic functions of occupation number we conclude that the "relaxation shift" of DF theory, as defined by Eq. 87, is negative and thereby loses its physical significance.

Table 2. Eigenvalues in the Be atom within Hartree-Fock (HF) theory. The excitation energy ΔE is the HF-ΔSCF result. DFX and LDX are the results of the exchange-only versions of the exact DF theory and the LD approximation respectively. Energies in eV. A minus sign is implied for all entries.

	ΔE	HF	DFX	LDX
2S	8.0	8.4	8.4[a]	4.6
1S	123.1	128.8	112.2[a]	103.2

[a]From Ref. 91.

In connection with our discussion of HF theory we would like to mention a theoretical framework which is closely related to HF. The HF ground state is that Slater determinant which minimized the expectation value of the full many-body Hamiltonian and this requirement leads to the well-known integro-differential equations for the one-particle orbitals making up the Slater determinant.

It is quite useful to restrict the variational freedom of the orbitals by demanding that they be solutions to the same Schrödinger equation with a local one-electron potential which is then considered as the variational parameter for minimizing the expectation value of the total Hamiltonian. This procedure was used by Talman and Shadwick[93] who shoved that it leads to an integral equation for the local potential which they solved for several atoms. They found, rather surprisingly, that their method reproduces the HF total energies to within some 50 parts in a million. Thus, for localized systems, such as e.g. atoms, very little is to be gained from the extra complication arising from non-local exchange. It is precisely the long-range properties of the non-local exchange which are responsible for the well known anomalies of HF theory of extended systems such as the vanishing density of states at the Fermi level, the much too large band width of metals and the singular response functions. It seems likely that many of these problems with HF theory will be cured by the Talman procedure. (The band width of jellium in this procedure, is e.g., the same as for non-interacting electrons which is much closer to the correct result than the HF band width). Consequently, for all systems, the Talman procedure is to be preferred over HF theory.

Our discussion of the Talman procedure has led us somewhat astray. For the purpose of our present discussion it is useful to notice that the densities of this procedure are almost as close to the HF densities as those obtained from the numerical fitting procedure which we discussed above and which were aimed at constructing the exact DF exchange potentials. Thus, for all practical purposes, we can say that the Talman potentials are rather accurate representations of the DF exchange potentials. (Note that, in principle, the Talman procedure is not the DF equivalent of HF even though it results in a local potential for all orbitals. The HF total energy is, for instance not obtained by inserting the DF orbitals into the usual expression for the HF total energy. In the Talman procedure, on the other hand the total energy is obtained by inserting the Talman orbitals into this expression. A lucid discussion of the distinction between the procedure of Talman et al.[93,94] and HF theory has recently been presented by Langreth and Mehl[54]).

Talman et al.[94] have published eigenvalues for many levels in several atoms and the picture that emerges is that described previously and illustrated in Table 2. The highest occupied DF eigenvalue is close to the excitation energy whereas the lower eigenvalues are way above.

We have so far discussed the DF eigenvalues within the framework of HF and we will now show that our major conclusions are not altered by the introduction of correlation effects. It is possible to show quite rigorously that the highest occupied DF eigenvalue of any localized system is, in fact, equal to the negative of the ionization potential[62]. Thus, the small discrepancy found in our discussion above is an artifact of HF theory. The proof relies on a rigorous asymptotic result for the density far away from the system and the above mentioned fact that the density decays as $\exp(-2r\sqrt{|\varepsilon_{DF}|})$. We will here briefly sketch that proof[62]. We write the ground-state density as

$$n(\vec{r}) = \langle N|\psi^+(\vec{r})\psi(\vec{r})|N\rangle \tag{88}$$

where $|N\rangle$ is the ground state and $\psi^+(\vec{r})$ and $\psi(\vec{r})$ are the operators for creating and annihilating an electron at $\vec{r}$. Inserting a complete set of eigenstates $|N-1,s\rangle$ of the ion and defining the quasi-particle amplitude $f_s(\vec{r})$ through

$$f_s(\vec{r}) = \langle N-1,s|\psi(\vec{r})|N\rangle \tag{89}$$

we have

$$n(\vec{r}) = \sum_s |f_s(\vec{r})|^2 \tag{90}$$

By commuting $\psi(\vec{r})$ with the full Hamiltonian one obtains a Schrödinger-like equation for $f_s(\vec{r})$ which far outside, e.g. an atom, has the form

$$\{-\frac{1}{2}\nabla^2 + \frac{N-Z-1}{r}\}.f_s(\vec{r}) + \frac{1}{r^2}.\sum_s d_{ss'}f_{s'}(\vec{r}) = \varepsilon_s f_s(\vec{r}) \tag{91}$$

Here, N is the number of electrons, Z is the nuclear charge, $d_{ss'}$ is a matrix element of the dipole operator between the ionic states s and s', and the "eigenvalue" ε_s is given by

$$\varepsilon_s = E_0(N) - E_s(N-1) \tag{92}$$

in obvious notation. By inspection, it follows from Eq. 91 that all quasi-particle amplitudes $f_s(\vec{r})$ which do not correspond to the ionic ground state decay faster at large r than the ground state amplitude $f_0(\vec{r})$ - but only by a power of r. Thus,

$$n(\vec{r}) = |f_0(\vec{r})|^2 \ , \ r \to \infty \tag{93}$$

It also follows that

$$\{-\frac{1}{2}\nabla^2 + \frac{N-Z-1}{r} - \frac{\alpha}{2r^4}\}.f_0(\vec{r}) = \varepsilon_0 f_0(\vec{r}) \tag{94}$$

where

$$\alpha = -2 \sum_{s\neq 0} \frac{|d_{os}|^2}{\varepsilon_s - \varepsilon_0} \tag{95}$$

is the static polarizability of the N-1 ion. The two Eqs. 93 and 94 prove the asymptotic result (Eq. 58) for the exchange-correlation potential $v_{xc}(\vec{r}) = \delta E_{xc}/\delta n(\vec{r})$ discussed in Sec. 2.4. These equations also prove that $\varepsilon_{DF} = \varepsilon_0 = E_0(N) - E_0(N-1) = -I$ as claimed (see Eq. 63).

We see that the inclusion of correlation effects shifts the uppermost occupied DF eigenvalue from the highest occupied HF eigenvalue to the negative of the exact ionization potential. This shift is, in most cases, small and the lower lying eigenvalues are shifted by similarly small amounts[92]. The lower eigenvalues thus remain above their corresponding excitation energies and we have the inequality

$$\frac{\partial \varepsilon_k}{\partial n_k} > 0 \tag{96}$$

with equality only for the highest occupied DF level. The final picture is illustrated in Table 3 comparing the exact DF eigenvalues to the exact excitation energies in the Be atom.

So far in this section we have only discussed DF eigenvalues of occupied levels. From the point of view of band gaps in solids it would be quite interesting to have some idea of where we can expect to find the unoccupied levels of DF theory in order to see later (in Sec. 4.3) how these are corrupted by the use of the LD approximation.

Table 3. The exact DF eigenvalues and those obtained from the LD approximation compared with the experimental results (ΔE) for atomic Be. Energies in eV. A minus sign is implied for all entries.

	ΔE	DF	LD
2S	9.3[a]	9.3[b]	5.7
1S	123.6[a]	116.0[c]	105.0

[a] From Ref. 95

[b] From Ref. 62

[c] From Ref. 92
According to the authors of Ref. 92 this value is somewhat uncertain but it is included in order to demonstrate the small effect of correlation on the DF eigenvalues (c.f. Table 2).

Using neutral atoms as illustrative examples, we note that the effective DF potentials of such atoms always have Coulomb tails (-1/r) which assure that they always have infinitly many bound states below the vacuum level. On the other hand many of these atoms, notably the closed-shell systems have no electron affinities. Thus, we conclude that, for these atoms and most likely for all atoms, the lowest unoccupied DF eigenvalue is below the affinity level. Note that this result consistent with the inequality above (Eq. 96) because, in this case, we are considering an unoccupied level and, in relating the eigenvalue to the affinity level, the former has to be expanded around zero occupation number (see Eqs. 64 and 85).

4.3. Local-Density Eigenvalues

In the previous sections we have shown where one can expect to find the exact DF eigenvalues, in both extended and localized systems. We will now discuss how these eigenvalues are corrupted by the LD approximation. In Sec. 4.1 we have already discussed some of the implications of the LD approximation in extended systems with relatively diffuse orbitals. As an example of a localized system we will again take the simple hydrogen atom in which the DF ground-state eigenvalue and the ionization potential are the same that is -1 Ry. The corresponding eigenvalue in the LD approximation is, however, only -0.56 Ry even when spin polarization is accounted for. The case of atomic hydrogen has the virtue of clearly indicating the source of the error. The effective potential entering the DF description of hydrogen contains the electrostatic interaction of the electron with its own charge density. In the exact DF theory this self-interaction is eliminated by the contribution of the exchange term to the single-particle potential. In the LD approximation the elimination of the self-interaction is incomplete leading to an elevated (~ 6 eV) eigenvalue. Thus, in this case, the DF eigenvalue gives the correct ionization potential while the use of the LD approximation introduces a large error.

As a second example we will consider the helium atom within the Hartree-Fock (HF) approximation*. Because He contains only one electron of each spin there is no inhomogeneous exchange term in the HF equations, only a one-particle potential which is local. Thus, the HF description of He and the DF description, in which correlation is ignored, are the same. A particularly illuminating comparison of the LD (exchange only) and DF (Hartree-Fock) results is contained in Fig. 6, where the spin-up and spin-down ground-state eigenvalues of He are shown as functions of the occupation of the spin-down orbital while the spin-up orbital is being kept fully

* Because of new finding made after the completion of Ref. 1 the discussion and the conclusions of this subsection differ from those of the corresponding section in Ref. 1.

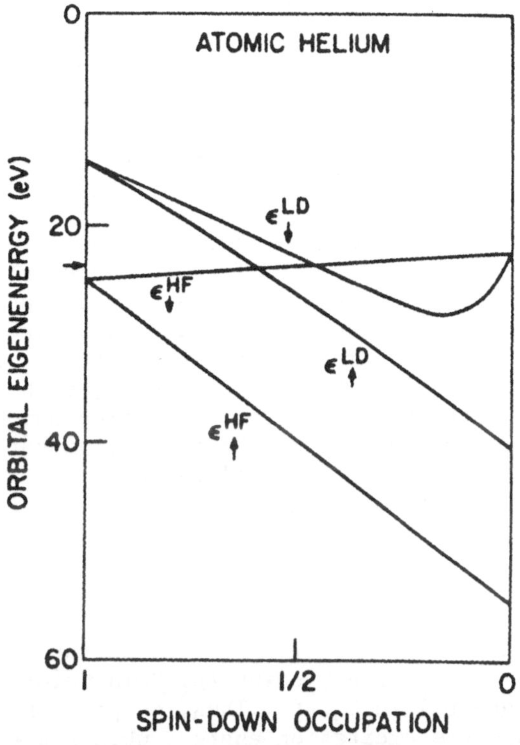

Fig. 6. Effect of the local-density approximation on the occupation-number dependence of the ground-state eigenenergies of atomic He. The spin-up orbital contains one electron. Because the He atom contains only one electron of each spin, the Hartree-Fock potential is local. The Hartree-Fock description of this system therefore constitutes the density-functional treatment in which correlation is ignored. The Hartree-Fock results are compared to the corresponding results in the local-density approximation (i.e., correlation is ignored in both the density-functional and local-density calculations). In the density-functional description, the spin-down level is seen to rise as charge is removed from the level, whereas in the local-density approximation, the level falls. The incorrect behavior of the approximate result is an artifact of the incomplete elimination of the electrostatic self-interaction in the local-density approximation. The arrow on the energy axis indicates the value of the ΔSCF total-energy difference. The similarity of the ΔSCF result to both the density-functional and local-density eigenvalues at half occupation illustrates the utility of the transition state approximation to the ΔSCF total-energy difference, even when the local-density approximation is used. The corresponding measured value is 24.6 eV, indicating the relative unimportance of correlation for this system.

occupied. The extension of the HF formalism to non-integral occupation numbers used here is analogous to that used in the LD calculations. (This extension may not be unique). Note first that the self-interaction has elevated the LD eigenvalues by ~ 11 eV. The difference, some 14 eV, between the spin-up eigenvalues when the spin-down orbital is empty is also purely and unambiguously a self-interaction artifact. This effect is much larger in He than in H because the greater nuclear charge leads to more localized orbitals. Next we note that, in the LD approximation, the spin-down eigenvalue falls when charge is removed from that orbital whereas in the DF (Hartree-Fock- case the level remains almost constant. If we had included the effect of correlation in the DF treatment we would have expected the level to remain precisely at its ground-state value. This is because, in the full DF theory, both the eigenvalue and its integral with respect to the occupation number give the same result i.e., the negative of the ionization potential (Eqs. 63, 85 and 94). Furthermore Fig. 6 illustrates another fact of great utility. We see that, despite the generally large differences between the DF and the LD eigenvalues, the spin-down eigenvalues of the two procedures are remarkably similar when the spin-down orbital is half occupied. The significance of this fact is that the eigenvalue at half occupation is a good approximation to the corresponding SCF total-energy difference. This is the basis for Slater's transition-state idea[96], i.e., using the "mid-point formula" for evaluating the integral in Eq. 85. Thus, despite the large errors in the eigenvalues the excitation energy, when represented as the ΔSCF total-energy difference is reasonably well described by even the LD approximation to the ground-state DF theory. Note that the ΔSCF value for the ionization potential is also indicated in Fig. 6. Note also that the transition-state concept allows the ΔSCF approximation to excitation energies to be constructed without the explicit evaluation and subtraction of total energies (Eq. 85).

In our two examples above we considered only the highest occupied orbital which is rather special within DF theory (see Sec. 4.2). As our final example we consider the Be atom which has also a deeper lying level. We see from Table 3 that, whereas the uppermost DF eigenvalue correctly predicts the true 2s-level, the corresponding LD level is 3.6 eV too high. This result is quite analogous to those discussed above for H and He although the error is somewhat smaller in Be due to the lesser localization of the Be 2s-orbital. The new information provided by Table 3 is that, for the deeper 1s-level, the DF eigenvalue is also above the experimental result (~ 7.6 eV). Since the Hartree-Fock eigenvalue, for which the self-interaction has been eliminated, is below the experimental 1s-level we can interpret this result as being due to an incomplete cancellation of the self-interaction in the full DF theory. As seen from Table 3, this effect is exaggerated in the LD approximation which places the 1s-level another 11 eV above the DF result. We can summarize the discussion as follows. In the case of the highest occupied level the corresponding DF eigenvalue gives the correct result whereas the LD approximation places the level too high, the error being proportional to the degree of localization of the corresponding orbital. For the deeper levels the difference between the LD eigenvalue and experiment becomes much more pronounced and the exact DF eigenvalue ends up approximately half way between the LD eigenvalue and the experimental excitation energy.

Our examples above also reveal another interesting fact with regard to the accuracy of the LD approximation. This approximation is much more accurate for total energies than for orbital eigenvalues. For H and He, for example, the total energy errors are only 0.3 eV and 1.7 eV, whereas the corresponding errors in the eigenvalues are 6.0 eV and 9.0 eV. We believe this difference in accuracy to be due to the fact that total-energy errors reflect the imperfections of the exchange-correlation functional whereas the eigenvalue errors are caused by the corresponding imperfections in the effective single-particle potential. Because the exchange-correlation energy represents an average over all the occupied states, it is relatively insensitive to the choice of exchange-correlation functional, as long as constraints such as the sum rule discussed in Sec. 2.3 are maintained. The potential, on the other hand, is obtained by functionally differentiating the energy with respect to the density. It therefore probes much finer details of the density dependence of the energy functional and an exceedingly accurate functional is needed in order to obtain a reasonably accurate potential. This idea is supported by the relatively poor eigenvalues obtained using the scheme by Langreth and Mehl[53,54]. As discussed in Sec. 2.5 this scheme is much superior to the LD approximation for the total energies of atoms and solids. Still the eigenvalues of the new scheme are almost the same as those of the LD approximation. In Be for instance, the 1s- and 2s-eigenvalues of the new scheme are - 105.8 eV and - 5.9 eV compared to - 105.0 eV and - 5.7 eV within the LD approximation (see Table 3). In the Si atom the 3s and 3p eigenvalues of the Langreth scheme only differ by a few tenths of an eV from the corresponding LD eigenvalues and as mentioned in Sec. 2.5, the new scheme gives only half the correct bandgap of bulk silicon[71] as does the LD approximation[72]. This idea also suggests a way to obtain an improved potential and improved eigenvalues. By differentiating the exact expression (Eq. 42) for the exchange-correlation energy one eliminates the weakest link in constructing the exchange-correlation potential. One obtains

$$v_{xc}(\vec{r}) = \int n(\vec{r}')\{\tilde{g}(\vec{r},\vec{r}') - 1\}v(\vec{r} - \vec{r}')d^3r' + \int n(\vec{r}_1)n(\vec{r}_2)\,\frac{\delta\tilde{g}(\vec{r}_1,\vec{r}_2)}{\delta n(\vec{r})}\,v(\vec{r}_1 - \vec{r}_2)d^3r_1 d^3r_2 \qquad (97)$$

In the hydrogen atom $\delta\tilde{g}/\delta n$ vanishes identically and by neglecting the second term above and using the LD approximation for the first, one obtains the potential $2\varepsilon_{xc}(n(\vec{r}))$ in the hydrogen atom. This leads to an order-of-magnitude reduction of the error in the eigenvalue as compared to the ordinary LD approximation (from 6.0 eV to 0.6 eV). If we instead apply the WD approximation to the first term and still neglect the second term we obtain the exact result for hydrogen (see the discussion of the WD approximation in Sec. 2.4). This could well be one of the reasons behind the success of Kerker[97] who applied precisely this procedure to bulk silicon and obtained almost the correct band gap. In most cases, however, $\delta\tilde{g}/\delta n$ is not zero and it is not justified to neglect the second term of Eq. 97. In fact, in the electron gas, this term is approximately one third of the first term.

Nevertheless, we believe Eq. 97 to be a good starting point for finding an improved v_{xc}. The first term of Eq. 97 has the intuitively correct physics built into it, i.e., it is the potential from the exchange-correlation hole. Hence this term is of long range and gives, for instance, the correct Coulomb tail $(-1/r)$ outside an atom. It is not unreasonable to assume that this long-range behavior becomes more important in a solid where exchange-correlation tails from many atoms could add up to a significant contribution. We speculate that Langreth-Mehl[53,54] scheme being based on density gradients has difficulties in describing the long-range behavior of v_{xc} and that these difficulties could be responsible for the failure of the method to produce the correct band gap in silicon[71].

The second term of Eq. 97 is however, short ranged. In an atom it decays faster than r^{-4} at large r. Presumably this term is more important in the interior of the atom where the density varies rapidly. This notion is supported by the fact that whereas the LD approximation applied to the first term of Eq. 97 gives an error of 0.6 eV in hydrogen, the same approximation applied to second term introduces an error as large as 5.4 eV. We therefore suggest that our knowledge of the exact exchange-correlation potential for many atoms could be used to find better approximations for this term. Note that there is a sum rule also for the three-particle correlation function $\delta\tilde{g}/\delta n$ which could be exploited in this effort (see Ref. 2).

From the stationary property of the energy with respect to variations in the density, it could perhaps be argued that a better potential would be pointless if one does not succeed in finding also an improved functional for the exchange-correlation energy. In view of the discussion above, a more accurate exchange-correlation potential could, however, considerably improve the ability of the ground-state theory to provide an approximate description of the excitation spectrum.

We have in this section elaborated on the large errors introduced into the ground-state DF eigenvalues by the use of the LD approximation. We will, however, end this section with an example of the application of the ground-state theory in the context of core-hole spectroscopy that illustrates the interpretational virtues of the orbital eigenvalues and their occupation-number dependence. Williams and Lang[80] considered the amount by which the binding energy, or ionization potential I, of an atomic core electron is reduced when the atom is embedded in a metal. This change in the binding energy results from several physical effects : 1) the screening of the core hole by the conduction electrons of the metal, 2) the requirement that an electron photoexcited from an embedded atom traverse the surface dipole-layer potential in order to leave the solid, 3) the electrostatic elevation of the core level by the compression of the valence electron charge that occurs as part of the embedding process, and 4) changes in the electrostatic potential in the core region due to changes in the valence configuration. The binding-energy reduction is typically 5-10 eV and varies considerably from atom to atom. A principal objective of their analysis was to determine how much of this variation should be ascribed to initial-state effects (chemical shifts) and how much to final-state effects (relaxation). The interpretive value of the Taylor-series representation of the ionization potential (Eqs. 85, 86 and 87) is immediately clear for we see that the first term in the series, ε, is manifestly a property of the ground (initial) state, whereas the remaining terms, i.e. the eigenvalue shift Δ, describe the response of the system to the creation of the core hole. If we distinguish quantities associated with the embedded atom with an asterisk, then the total binding-energy shift Δ_{tot} is simply

$$\Delta_{tot} = I - I^{\star} \tag{98}$$

and Eq. 86 immediately provides the desired decomposition into initial- and final-state components, that is

$$\Delta_{tot} = \Delta_{initial} + \Delta_{final} \tag{99}$$

with $\Delta_{initial}$ given by

$$\Delta_{initial} = -(\varepsilon - \varepsilon^{\star}) \tag{100}$$

and Δ_{final} given by

$$\Delta_{final} = \Delta - \Delta^{\star} \tag{101}$$

with the eigenenergy shift Δ given by Eq. 87. This analysis of binding-energy shifts illustrates another practical aspect of the theory. We have shown above that, due to the large self-interaction effects associated with the DF eigenvalues and, to a larger extent with the LD eigenvalues of deep levels, the eigenvalue shift Δ can not be interpreted as a physical relaxation shift. The practical point in question is that, despite the corruption of Δ by the self-interaction the LD approximation can be used to calculate the change in Δ. In other words, the large self-interaction effects in the LD approximation to Δ subtract out of the change in Δ described by the final-state shift Δ_{final}. The reason for this fortunate circumstance is that the major contribution to the change in Δ can be shown[79] to be simply half the Hartree potential from the screening charge induced by the appearance of the core hole. The total and decomposed shift obtained in this way from parameter-free self-consistent calculations are compared with measured values in Fig. 7 taken from Ref. 80. The clear and intuitively appealing interpretation of the measurements provided by the analysis motivates the application of the ground-state formalism far outside its formally justifiable range of applicability.

4.4. Valence-Band Spectra of Transition Metals

The preceding sections have presented the available theoretical information which bears on the connection between the eigenvalues of the ground-state theory and the single-particle excitation spectrum. In this subsection, we consider the relevance of the ground-state eigenvalues to a particularly important class of experiments, spectroscopic studies of transition-metals and their surfaces (e.g., photoemission). These measurements constitute a large and important area of condensed-matter physics ; their special significance in the context of these notes is that many electronic-structure calculations attempt to interpret these measurements on the basis of the eigenvalues of the LD ground-state theory. What can be said about the legitimacy and accuracy of this type of analysis ? Unfortunately, the arguments of the preceding sections indicate only the issues involved (the relative ability of the electron gas to screen electrons having different velocities, the effect of the self-interaction on the DF and the LD eigenvalues, and the lifetime of the excitation) ; these arguments do not indicate the quantitative importance of these effects in the context of transition-metal spectra. It is therefore necessary to rely on empirical evidence.

The work of Andersen[98] and collaborators has greatly clarified the informational content of the single-particle calculations of transition-metal spectra. These spectra reflect primarily[99] the crystal structure, the energy separation of the s and d bands and the width of the d band. Beyond these three aspects, quantities such as the effective mass of the s band provide additional refinement but the d-band position and width are the parameters of greatest importance. (In magnetic transition metals, the energy separation of the spin-up and spin-down bands is of course, also an important parameter). We therefore ask, how well does the

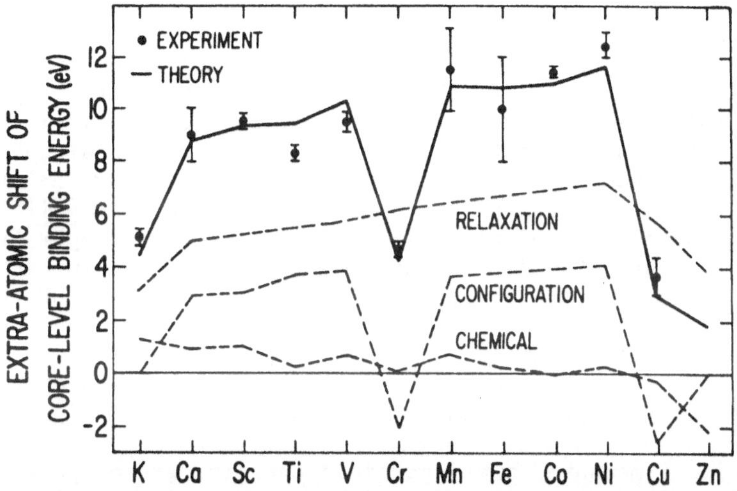

Fig. 7. Comparison of measured and calculated core-level binding-energy shifts for 3d transition-metal atoms. The solid curve shows the total theoretical shift and the dashed curves indicate the contributions of initial- and final-state effects. Relaxation refers to the screening of the core hole by the metal, a final-state effect. Configuration refers to the effect on the core level of changing the electronic configuration from that of the free atom to that of the solid (a particular initial-state effect). The strong variation of the configuration component reflects the fact that only Cr and Cu have a $3d^{n-1}4s^1$ configuration as free atoms; the others have a $3d^{n-2}4s^2$ configuration. The contribution labelled chemical is the remainder of the initial-state shift ; it exhibits the near cancellation of compression and dipole-layer-shift effects. (From Ref. 80).

ground-state theory predict these parameters.* The situation is clearest in the noble metals, where, because the Fermi level does not pass through the d band, the d-band position relative to the s band is experimentally very well determined. In the case of Cu, the d band given by the ground-state theory in the LD approximation is ~ 0.4 eV too high relative to the s band[100]. In Zn[101], where the d band lies deeper in energy (~ 10 eV below E_F) and is narrower than the d band of Cu (1 eV as compared to 3 eV) the LD ground-state d bands are too high by ~ 1.5 eV. It would be quite useful to know whether these discrepancies are due to the LD approximation or to the more fundamental difference between the excitation energies and the DF eigenvalues associated with all but the highest occupied level of the solid. Unfortunately, the available information is insufficient for answering this question. Recalling our discussion of DF and LD eigenvalues in Secs. 4.2 and 4.3 we know that the LD approximation is responsible for most of the discrepancy if the levels can be considered as shallow. In the case of deeper levels approximately half the discrepancy would probably remain even if the exact single-particle potential of the ground-state DF theory were used to calculate the bands. Note, however, that these conclusions were drawn from our experience of localized systems where the exact DF eigenvalues could be determined. In extended systems we do not know the precise energy positions of the DF eigenvalues.

In Fig. 8 measured quasi-particle energies are compared with the results of a calculation in which, as we shall see the self-interaction does not enter. The calculated energy bands are seen to agree quite well with the measured spectra. In particular, the energy position of the narrow d bands relative to the Fermi level does not exhibit the 0.4 eV error seen in the corresponding LD calculations. The effective single-particle potential used in the calculations shown in Fig. 8 is due to Chodorow[102], and was constructed using what has come to be called the "renormalized-atom" procedure[103]. The aspect of that procedure which is relevant to the present discussion is its use of non-local exchange within the atomic cell. In this way the self-interaction effect was avoided. It is generally true for deeper levels that the Hartree-Fock eigenvalues, which do not contain any self-interaction effects, agree better with the corresponding excitation energies than either the LD or the DF eigenvalues which both contain such effects although to a lesser extent in the latter case. The only thing we can be sure of without any further information is that the spectrum of DF eigenvalues will be closer to experiment than the spectrum obtained from the LD approximation.

---

* Because of new findings made after the completion of Ref. 1 the discussion and the conclusions of this subsection differ from those of the corresponding section in Ref. 1.

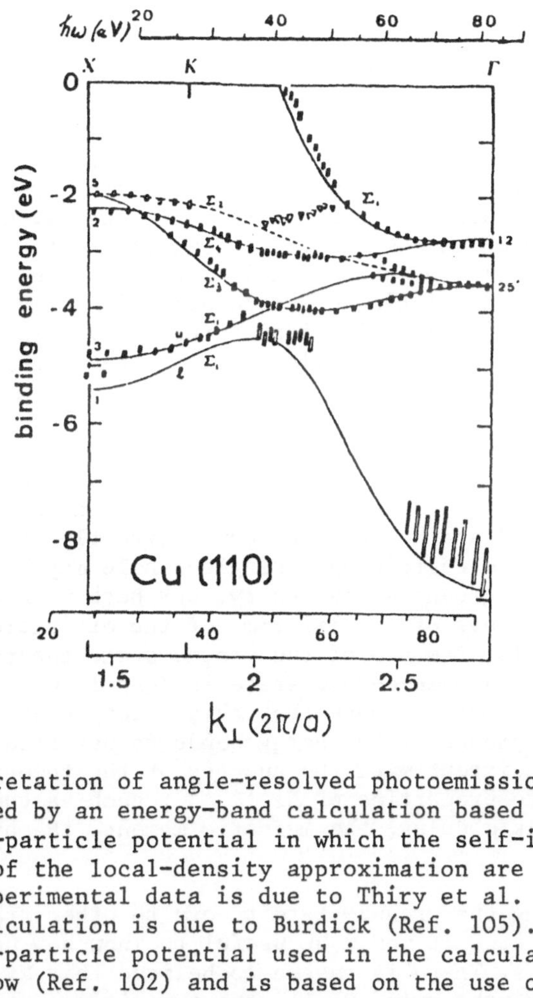

Fig. 8. Interpretation of angle-resolved photoemission data for Cu provided by an energy-band calculation based on an effective single-particle potential in which the self-interaction effects of the local-density approximation are not present. The experimental data is due to Thiry et al. (Ref. 104) and the calculation is due to Burdick (Ref. 105). The effective single-particle potential used in the calculation is due to Chodorow (Ref. 102) and is based on the use of non-local exchange within the atomic cell. The use of non-local exchange avoids the introduction of self-interaction effects which are believed to cause the d bands of similar calculations based on the local-density approximation to lie 0.4 eV closer to the Fermi level than those shown above.

With regard to the width of the d bands experience with semi-empirical adaptations of the ground-state theory suggest that, if the artificial elevation of localized levels by the LD approximation could be eliminated, the eigenvalues of the ground-state theory would provide an interpretation of the excitation spectrum which is adequate for most purposes.

The excitation spectrum of copper's other periodic-table neighbor, Ni, introduces another complication into the connection between ground-state eigenvalues and excitation spectra. Experimental measurements have shown with increasing precision over the past decade that the spectrum is significantly more narrow than the width of the calculated ground-state energy bands. Nickel appears to be a case in which the excitation spectrum reflects a particular type of correlation which, while present in the electron gas on which the LD description of correlation is based, is emphasized in Ni by the spatial localization of the d orbitals and the availability of low-energy excitations within the d shell. Experimental evidence[106] and theoretical analysis[107-111] both indicate that, in Ni, the state of the solid in which a single d hole is propagating in a state below ε_F is strongly coupled to other states in which a second d hole is created by the elevation of a d electron to an empty state above ε_F. The state consisting of two d holes is itself long lived and has been observed directly in several experiments. The state containing a single-d hole may be reasonably described by the ground-state theory, but here we have a case where the discussion above of the lifetime of the excitation is probably relevant. That is, the use of the ground-state theory implicitly assumes that the single-d-hole state is long lived ; in Ni, this assumption is not valid. Note that this interpretation of the failure of the ground-state energy bands to describe the excitation spectrum is consistent with the success of the theory in describing the ground-state properties of Ni[112-114], such as the equilibrium volume, the bulk modulus, the magnetic moment, the hyperfine field etc.

In applying the ground-state theory to other transition metals, their surfaces etc. it would be useful to know how unusual the situation in Ni is. There is reason to believe that Ni is the exception rather than the rule. Of all the transition metals, Ni has the most localized valence d orbitals. As we proceed to the left in the periodic table, the decreasing nuclear charge results in less localization. As we proceed down the Ni column of the periodic table, localization decreases because the valence d orbitals must be orthogonal to the d orbitals of the core. The nuclear charge in Cu is, of course, greater than that of Ni ; we might ask why these two-hole effects are not important in Cu. The d orbitals of Cu are more localized than those of Ni, but Cu lacks the high density of empty states at ε_F which facilitates the creation of the two-hole state in Ni. Thus, it is seems likely that the deviation of the excitation spectrum from the ground-state eigenvalue spectrum in Ni is probably unusual.

We can summarize this section by saying that, in many cases the eigenvalues of the ground-state theory can be a useful guide to the interpretation of single-particle excitation spectra. The errors introduced into the ground-state DF eigenvalues by the use of the LD approximation is often larger than the difference between these and the corresponding excitation energies. But, having said that, we remind the reader that the ground-state eigenvalues are at best only an approximation to the true excitation spectrum. Thus, while the ground-state eigenvalues can be a useful interpretive tool, we must remain alert to the possibility, illustrated by Ni that one or more of the assumptions implicit in their use may be invalid in particular systems.

4.5. The Self-Energy Approach to Excitation Energies

As pointed out above, the orbital eigenvalues of the DF ground-state theory do not, in general, give the exact single-particle excitation energies for many-electron systems. We have also seen, however, that the orbital eigenvalues are often usefully accurate approximations to excitation (quasi-particle) energies, especially when they are not corrupted by self-interaction effects associated with the LD approximation. The most serious inadequacy of the orbital eigenvalues as a description of the excitation spectrum is the fact that they do not represent a well-defined approximation to a description which is correct in principle. There is not, at present, any systematic way in which to improve upon the description they provide. The virtue of the self-energy approach is however that it provides a description of the quasi-particle excitation spectrum which is, in principle, exact. Furthermore, this approach lends itself to systematic and physically understandable levels of approximation.

The basis of the self-energy approach (see e.g. Ref. 83) is the Dyson equation satisfied by the single-particle Green's function, or equivalently, by the quasi-particle amplitudes $f(\vec{r})$

$$\{-\frac{1}{2}\nabla^2 + V_C(\vec{r})\}.f(\vec{r}) + \int \Sigma(\vec{r},\vec{r}';\varepsilon).f(\vec{r}')d^3r' = \varepsilon\, f(\vec{r}) \quad (102)$$

where $\Sigma(\vec{r},\vec{r}';\varepsilon)$ is the self-energy operator and $V_C(\vec{r})$ is the Coulomb potential due to all charges in the system. The fact that the self-energy operator is not hermitian implies that the quasi-particle energies ε are in general complex, as they should be, reflecting the finite lifetime of excitations in an interacting system. The fact that the self-energy depends on the quasi-particle energy means that the amplitudes $f(\vec{r})$ are not orthogonal and cannot be straightforwardly squared and summed to obtain the electron density.

The single-particle equations of the DF ground-state theory[7] differ from the Dyson equation only in the replacement of the self-energy $\Sigma(\vec{r},\vec{r}';\varepsilon)$ by the exchange-correlation potential $v_{xc}(\vec{r})$ defined by Eq. 18. We see in this way that the extent to which the orbital eigenvalues of the ground-state theory approximate excitation energies is determined by the similarity of the expectation values of $\Sigma(\vec{r},\vec{r}';\varepsilon)$ and those of $v_{xc}(\vec{r})$. The rather different mathematicel properties of these two quantities (non-local, non-hermitian and energy-dependent versus local, hermitian and energy-independent), by themselves, provide little hope of a strong similarity. Figure 5 above shows that, for the homogeneous electron gas, the similarity is in fact strong but the discussion of Fig. 5 indicates that the similarity rests on the smallness of the many-body correction to the effective mass[83]. That the similarity persists for slightly inhomogeneous systems is shown by the work of Sham and Kohn[8]. Recently, however, MacDonald[115] has suggested a mechanism by which the similarity might be considerably reduced by the presence of a lattice potential in real solids.

Consider now the calculation of the self-energy and the excitation spectrum for real systems. Most efforts to implement the self-energy approach have been based on many-body perturbation theory[116-122]. (This set of references is representative, not exhaustive). Many-body perturbation theory is really an alternative to, rather than an aspect of, DF theory ; we therefore restrict the present discussion to implementations of the self-energy approach which exploit the ideas of DF theory. The basic idea in this context were given by Sham and Kohn[8] who, in the spirit of the LD approximation, approximated the self-energy by that of a homogeneous electron gas

$$\Sigma(\vec{r},\vec{r}';\varepsilon) = \Sigma_h(\vec{r} - \vec{r}',\varepsilon - \varepsilon_F + \mu_h(n);n) \qquad (103)$$

The physical idea here is that the response of the inhomogeneous electron gas to the passage of a given electron or hole should be similar to that of a homogeneous electron gas. In particular, the two important physical parameters characterizing the response should be i) the energy of the quasi-particle relative to the Fermi level (recall that the imaginary part of the true self-energy vanishes at the Fermi level), and ii) the electron density in the vicinity of the electron or hole considered. The approximation described by Eq. 103 clearly reflects this intuition ; its specification is completed by taking the density parameter n appearing in the right-hand side of Eq. 103 to be the density of the inhomogeneous system at the position midway between $\vec{r}$ and $\vec{r}'$. In addition to its intuitive appeal, this approximation has the virtue of becoming exact for systems in which the density varies slowly. Because, however, the density variations in real systems are not slow, justification for the application of this approximation to real systems remains empirical.

It should be noted that the LD approximation to the self-energy (Eq. 103) retains the non-local and energy-dependent character of the exact self-energy. To eliminate the practical difficulties associated with the non-locality, Sham and Kohn also suggested a second, more approximate, form for the self-energy

$$\Sigma(\vec{r},\vec{r}';\varepsilon) = \Sigma_h\{\vec{p}(\vec{r}),\varepsilon - \varepsilon_F + \mu_h[n(\vec{r})];n(\vec{r})\}.\delta(\vec{r} - \vec{r}') \qquad (104)$$

This approximation is based on arguments similar to those leading to the WKB approximation which are valid for slowly varying electron densities. Here, the self-energy is approximated again by that of a homogeneous electron gas, but in this case for a quasi-particle of momentum $\vec{p}(\vec{r})$. This local momentum is in turn specified by the implicit relationship

$$\varepsilon - \varepsilon_F + \mu_h = \frac{1}{2}|\vec{p}|^2 + \Sigma_h(\vec{p},\varepsilon - \varepsilon_F + \mu_h;n) \qquad (105)$$

with the chemical potential μ_h and the electron density n taking on local values, $n = n(\vec{r})$ and $\mu_h = \mu_h(n)$. (Ref. 16 contains a discussion of the choice of quasi-particle momentum). An important aspect of this approximation (Eqs. 104 and 105 is that, at the Fermi energy ($\varepsilon = \varepsilon_F$), the local momentum $\vec{p}(\vec{r})$ is given by the Fermi momentum of a homogeneous electron gas (possessing the local density). The approximate self-energy given by Eq. 104 then reduces to the exchange-correlation potential of the ground-state theory in the LD approximation. In other words, in the approximation of Eq. 104 the Fermi surface given by the single-particle equations of the ground-state theory is that of the quasi-particle spectrum. In contrast to the ground-state theory, however, the quasi-particle spectrum implied by Eq. 104 contains the many-body correction to both the effective mass and the state density (see Sec. 4.1). The local approximation to $\Sigma(\vec{r},\vec{r}';\varepsilon)$ given by Eq. 104 has been tested in Al where it was found to give a quasi-particle band structure quite similar to that given by the LD ground-state theory[17]. Similar conclusions were reached by Watson et al.[123] for Ni and, more recently, by MacDonald[115] for the alkali metals.

The more elaborate, non-local approximation to the self-energy (Eq. 103) has been applied by Vosko, Rasolt and coworkers. These tests have been carried out for model systems[124,125], for Cu[126], and for the alkali metals[115,127]. These applications indicate, quite interestingly, that the implications of the non-locality can be substantial. In particular, Fermi-surface distortions for the alkali metals were found to be much reduced for the non-local self-energy relative to those given by the non-local theory are much closer to the measured values. In the case of Cu, the non-local theory was shown[126] to lead to corrections of sufficient magnitude to remove the deviation from measured Fermi-surface "neck" radii found by Janak et al.[128] using the LD ground-state theory.

As mentioned above, MacDonald[115] recently pointed out another interesting effect which has its origin in the non-local character of the self-energy. When interactions with the lattice of ions produce a substantial "band mass", the many-body contribution to the effective mass can be enhanced relative to its value without scattering by the lattice. For Li and Cs the corrections to the effective mass were estimated to be 10% and 12%.

The examples considered here suggest that the superior description of the excitation spectrum offered by the non-local self-energy approach provides just the level of refinement required to understand many-body contributions to Fermi surface distortions and effective masses. It should be noted, however, that this approach still employs a "local-density" approximation, and it has been tested primarily on systems in which the electron density is relatively slowly varying at least compared to transition metals and surfaces. We have seen above that there are two principal sources of error in using the LD ground-state theory to describe excitations : i) the use of the LD approximation, and ii) the use of the ground-state theory. The non-local self-energy approximation of Eq. 103 eliminates only the second of these, and we have seen that the first can be the more important source of error in systems containing relatively localized states. Further tests are needed to determine the range of validity of this approximation scheme.

5. ACKNOWLEDGEMENTS

It is a pleasure to acknowledge the long-term collaboration with A.R. Williams which has resulted in, among other things, a jointly written review article on density-functional methods. Some of the material from that article has been used also in the present notes and many of the ideas presented here originate in discussions with A.R. Williams. I am also indebted to R. Car and C.O. Almbladh for fruitful collaboration which has led to some of the new results presented here. I am further delighted to thank D. Langreth for stimulating discussions during the course of this work and G. Grossmann for critical reading of the manuscript. Finally I would like to express my gratitude to the organizers and participants of the NATO Advanced Study Institute on "The Electronic Structure of Complex Materials" for two very rewarding weeks in Gent, July, 1982. These notes are a summary of the authors lectures given at that meeting.

This work was supported in part by The Swedish Natural Science Research Council under contract number F-FU-2163-107.

REFERENCES

1. A.R. Williams and U. von Barth, "Applications of Density-Functional Theory to Atoms, Molecules and Solids" in "Theory of the Inhomogeneous Electron Gas", edited by S. Lundqvist and N.H. March, Physics of Solids and Liquids Series, Plenum (1983).
2. A.K. Rajagopal, Advances in Chemical Physics, edited by I. Prigogine and S.A. Rice, 41, 59, Wiley, New York (1980).
3. D.R. Hartree, Proc.Comb.Phil.Soc. 24, 89 (1928).
4. L.H. Thomas, Proc.Comb.Phil.Soc. 23, 542 (1927).
5. E. Fermi, Rend.Acad.Naz.Lincei 6, 602 (1927).
6. P. Hohenberg and W. Kohn, Phys.Rev. 136, B864 (1964).
7. W. Kohn and L.J. Sham, Phys.Rev. 140, A1133 (1965).
8. L.J. Sham and W. Kohn, Phys.Rev. 145, 561 (1966).
9. M. Levy, Proc.Natl.Acad.Sci. USA, vol.76, 6062 (1979).
10. U. von Barth and L. Hedin, J.Phys.C 5, 1629 (1972).
11. O. Gunnarsson and B.I. Lundqvist, Phys.Rev.B 13, 4274 (1976).
12. N.D. Mermin, Phys.Rev. 137, A1441 (1965).
13. J.F. Janak, Solid State Commun. 25, 53 (1978).
14. A.K. Rajagopal and J. Callaway, Phys.Rev.B 7, 1912 (1973).
15. J.G. Zabolitzky, Phys.Rev. B 22, 2353 (1980).
16. L. Hedin and B.I. Lundqvist, J.Phys.C 4, 2064 (1971).
17. G. Arbman and U. von Barth, J.Phys.F 5, 1155 (1975).
18. M. Gell-Mann and K.A. Brueckner, Phys.Rev. 106, 364 (1957).
19. E.P. Wigner, Trans.Faraday Soc. 34, 678 (1938).
20. K.S. Singwi, A. Sjölander, M.P. Tosi, and R.H. Land, Phys.Rev.B 1, 1044 (1970).
21. D.M. Ceperley, Phys.Rev. B 18, 3126 (1978).
22. D.M. Ceperley and B.J. Alder, Phys.Rev.Lett. 45, 566 (1980).
23. R.F. Bishop and K.H. Luhrmann, Phys.Rev. B 26, 5523 (1982).
24. U. von Barth, Physica Scripta 21, 585 (1980).
25. J. Kübler, J. Magnetism and Magn. Mater. 20, 277 (1980).
26. H.L. Skriver, J.Phys.F 11, 97 (1981).
27. S.H. Vosko, L. Wilk and M. Nusair, Can.J.Phys. 58, 1200 (1980).
28. A.K. Rajagopal, J.Phys.C 11, L943 (1978).
29. A.H. McDonald and S.H. Vosko, J.Phys.C 12, 2977 (1979).
30. M.P. Das, M.V. Ramana and A.K. Rajagopal, Phys.Rev.A 22, 9 (1980).
31. M.V. Ramana and A.K. Rajagopal, Phys.Rev. A 24, 1689 (1981).
32. M.V. Ramana, A.K. Rajagopal and W.R. Johnson, Phys.Rev. A 25, 96 (1982).
33. O. Gunnarsson, B.I. Lundqvist and J.W. Wilkins, Phys.Rev. B 10, 1319 (1974).
34. Uday Gupta and A.K. Rajagopal, Phys.Rev. A 22, 2792 (1980).
35. L.J. Sham, in : Computational Methods in Band Theory, (P.J. Marcus, J.F. Janak, and A.R. Williams eds.), p. 458, Plenum New York (1971).
36. D.J.W. Geldart, M. Rasolt and R. Taylor, Solid State Commun. 10, 279 (1972).
37. L. Kleinman, Phys.Rev. B 10, 2221 (1974).

38. D.J.W. Geldart, M. Rasolt and C.O. Almbladh, Solid State Commun. 16, 243 (1975).
39. A.K. Rajagopal and S. Ray, Phys.Rev.B 12, 3129 (1975).
40. A.K. Gupta and K.S. Singwi, Phys.Rev. B 15, 1801 (1977).
41. A.K. Rajagopal and S.P. Singhal, Phys.Rev. B 16, 601 (1977).
42. D.J.W. Geldart and M. Rasolt, Phys.Rev. B 13, 1477 (1976).
43. M. Rasolt, Phys.Rev. B 16, 3234 (1977).
44. O. Gunnarsson, M. Jonson and B.I. Lundqvist, Phys.Rev. B 20, 3136 (1979).
45. F. Herman, J.P. Van Dyke, and I.B. Ortenburger, Phys.Rev.Lett. 22, 807 (1969).
46. D.J.W. Geldart and R. Taylor, Can.J.Phys. 48, 155 (1970) and 48, 167 (1970).
47. D.C. Langreth and J.P. Perdew, Phys.Rev. B 21, 5469 (1980).
48. D.C. Langreth and J.P. Perdew, Phys.Rev. B 15, 2884 (1977).
49. J.E. Inglesfield and E. Wikborg, Solid State Commun. 16, 335 (1975).
50. J.P. Perdew, D.C. Langreth and V. Sahni, Phys.Rev.Lett. 38 1030 (1977).
51. C.O. Almbladh, Technical Report, University of Lund (1972).
52. B.Y. Tong, Phys.Rev. A 4, 1375 (1971).
53. D.C. Langreth and M.J. Mehl, Phys.Rev.Lett. 47, 446 (1981).
54. D.C. Langreth and M.J. Mehl, Phys.Rev. B 28, (1983).
55. O. Gunnarsson, M. Jonson, B.I. Lundqvist, Phys.Lett. 59A, 177 (1976).
56. O. Gunnarsson, M. Jonson, B.I. Lundqvist, Solid State Commun. 24, 765 (1977).
57. J.A. Alonso and L.A. Girifalco, Phys.Rev. B 17, 3735, (1978).
58. B.Y. Tong and L.J. Sham, Phys.Rev. 144, 1 (1966).
59. O. Gunnarsson and R.O. Jones, Physica Scripta 21, 394 (1980).
60. U. von Barth and C.D. Gelatt, Phys.Rev.B 21, 2222 (1980).
61. V.L. Moruzzi, A.R. Williams and J.F. Janak, Phys.Rev. B 15, 2854 (1977).
62. C.O. Almbladh and U. von Barth, to be published.
63. J.C. Kimball, Phys.Rev.B 14, 2371 (1976).
64. C.O. Almbladh, U. von Barth, Z.D. Popovic and M.J. Stott, Phys. Rev. B 14, 2250 (1976).
65. K. Schwarz, Chem.Phys.Lett. 57, 605 (1978).
66. R.O. Jones, J.Chem.Phys. 71, 1300 (1979).
67. N.D. Lang, Phys.Rev.Lett. 46, 842 (1981).
68. D. Pines, "Elementary Excitations in Solids", p.75, Benjamin, New York (1963).
69. A.K. Rajagopal, J.C. Kimball and M. Banerjee, Phys.Rev. B 18, 2339 (1978).
70. N.D. Lang and W. Kohn, Phys.Rev. B 1, 4555 (1970).
71. U. von Barth and R. Car, to be published.
72. D.R. Hamann, Phys.Rev.Lett. 42, 662 (1979).
73. U. von Barth, Phys.Rev. A 20, 1693 (1979).
74. T. Ziegler, A. Rauk and E.J. Baerends, Theoret.Chim.Acta 43, 261 (1977).

75. J.H. Wood, J.Phys.B 13, 1 (1980).
76. O. Gunnarsson and R.O. Jones, J.Chem.Phys. 72, 5357 (1980).
77. C.E. Moore, "Atomic Energy Levels", Circular of the National Bureau of Standards, Washington, D.C., 1949, vol. I, pp.467.
78. A.K. Theophilou, J.Phys.C 12, 5419 (1979).
79. C.O. Almbladh and U. von Barth, Phys.Rev. B 33, 3307 (1976).
80. A.R. Williams and N.D. Lang, Phys.Rev.Lett. 40, 954 (1978).
81. N.D. Lang and A.R. Williams, Phys.Rev. B 20, 1369 (1979).
82. V. Peuckert, J.Phys.C 11, 4945 (1978).
83. L. Hedin and S. Lundqvist, in Solid State Physics (F. Seitz, D. Turnbull and H. Ehrenreich, eds.), vol. 23, p. 1, Academic Press, New York (1969).
84. A. Zangwill and P. Soven, Phys.Rev. A 21, 1561 (1980).
85. A. Zangwill and P. Soven, Phys.Rev.Lett. 45, 204 (1980).
86. J.C. Slater and J.H. Wood, Int. J. Quantum Chem. Suppl. 4, 3 (1971).
87. J. Janak, Phys.Rev. B 18, 7165 (1978).
88. J.C. Wheatley, Rev.Mod.Phys. 47, 415 (1975).
89. J.E. Demuth, P.M. Marcus and D.W. Jepsen, Phys.Rev.B 11, 1460 (1975).
90. P. Lee and G. Beni, Phys.Rev. B 15, 2862 (1977).
91. U. von Barth and R. Car, to be published.
92. C.O. Almbladh and A.C. Pedroza, to be published.
93. J.D. Talman and W.F. Shadwick, Phys.Rev. A 14, 36 (1976).
94. K. Aashamar, T.M. Luke and J.D. Talman, Phys.Rev. A 19, 6 (1979).
95. H. Siegbahn and L. Karlsson, "Photoelectron Spectroscopy" in "Handbuch der Physik", vol. XXXI, edited by S. Flügge and W. Mehlhorn, pp. 316, Springer Verlag, Berlin (1982).
96. J.C. Slater, Quantum Theory of Molecules and Solids : vol. 4, McGraw-Hill, New York (1974).
97. G.P. Kerker, Phys.Rev. B 24, 3468 (1981).
98. A.R. Mackintosh and O.K. Andersen, in "Electrons at the Fermi Surface", (M. Springford, Ed.) Cambridge Univ. Press, Cambridge (1980).
99. D.G. Pettifor, J.Phys.C 3, 367 (1970).
100. J.F. Janak, A.R. Williams and V.L. Moruzzi, Phys.Rev. B 11, 1522 (1975).
101. F.J. Himpsel, D.E. Eastman, E.E. Koch and A.R. Williams, Phys. Rev. 22, 4604 (1980).
102. M.I. Chodorow, Ph. D. Thesis (MIT 1939), unpublished.
103. C.D. Gelatt, Jr., H. Ehrenreich and R.E. Watson, Phys.Rev. 15, 1613 (1977).
104. P. Thiry, D. Chandesris, J. Leconte, C. Guillot, R. Pinchaux and Y. Petroff, Phys.Rev.Lett. 43, 82 (1979).
105. G.A. Burdick, Phys.Rev. 129, 138 (1963).
106. C. Guillot, Y. Ballu, J. Paigne, J. Lecante, K.P. Join, P. Thiry, R. Pinchaux, Y. Petroff and L.M. Falicov, Phys.Rev.Lett. 19, 1632 (1977).
107. D.R. Penn, Phys.Rev.Lett. 42, 921 (1979) ; J.Appl.Phys. 50, 7480 (1979).

108. A. Liebsch, Phys.Rev.Lett. 43, 1431 (1979).
109. G. Treglia, F. Ducastelle and D. Spanjaard, Phys.Rev. B 21, 3729 (1980).
110. D.M. Edwards, Inst.Phys.Conf.Ser. No. 39, 279 (1978).
111. L.C. Davis and L.A. Feldkamp, J.Appl.Phys. 50, 1944 (1979).
112. J.F. Janak, Phys.Rev. B 16, 255 (1977).
113. J.F. Janak and A.R. Williams, Phys.Rev. B 14, 4199 (1976).
114. J.F. Janak, Phys.Rev. 20, 2206 (1979).
115. A.H. MacDonald, J.Phys.F 10, 1737 (1980).
116. L. Hedin, Arkiv. Fysik 30, 231 (1965).
117. W. Brinkman and B. Goodman, Phys. Rev. 149, 597 (1966).
118. A.B. Kunz, Phys.Rev.B 6, 606 (1972).
119. J.C. Inkson, J.Phys.C 6, L181 (1973).
120. L. Dagens and F. Perrot, Phys.Rev. B 8, 1281 (1973).
121. E.O. Kane, in "Proceedings of the twelfth international conference on the physics of semiconductors" (M.H. Pilkuhn ed.), p. 169, B.G. Teubner, Stuttgart (1974).
122. G. Strinati, H.J. Mattausch and W. Hanke, Phys.Rev.Lett. 45, 290 (1980).
123. R.E. Watson, J.F. Herbst, L. Hodges, B.I. Lundqvist and J.W. Wilkins, Phys.Rev. B 13, 1463 (1976).
124. M. Rasolt and S.H. Vosko, Phys.Rev. B 10, 4195 (1974).
125. S.B. Nickerson and S.H. Vosko, Phys.Rev. B 14, 4399 (1976).
126. J. S-Y. Wang and M. Rasolt, Phys.Rev. B 15, 3714 (1977).
127. M. Rasolt, S.B. Nickerson and S.H. Vosko, Solid State Commun. 16, 827 (1975).
128. J.F. Janak, A.R. Williams and V.L. Moruzzi, Phys.Rev. B 6, 4367 (1972).

DENSITY FUNCTIONAL CALCULATIONS FOR ATOMIC CLUSTERS

J. Harris

IFF der KFA
D-5170 Jülich
BRD

INTRODUCTION

In these lectures we will be concerned with the problem of calculating the adiabatic ground-state energy, $E[\{\vec{R}_i\}]$, for a cluster of atoms. "Adiabatic" means that the motion of electrons is regarded as infinitely rapid compared with that of the nuclei, which are represented by point charges at positions $\vec{R}_i$. The energy $E[\{\vec{R}_i\}]$ then serves as the potential term in the Schrödinger equation governing the nuclear motion. By its nature, the density functional [DF] can make no statement about non-adiabatic processes. For the most part we will consider only the simplest clusters - molecules having energy curves that can be determined via spectroscopic measurement and can usually be calculated accurately by the 'configuration interaction' [CI] method of quantum chemistry. These systems provide a useful testing ground for the density functional, enabling us to establish some simple ideas as to how the method describes chemical bonding, and a "level of credibility" in application to larger systems, for which CI calculations are not feasible and the energy is not directly measurable. It is the potential of the method with respect to these systems that justifies the effort currently being expended in solving the density functional equations, and we should be careful to avoid regarding functional calculations and CI methods as competitors. This is not to say that useful results cannot be obtained for small systems. For example, the first realistic energy curve for the ground state of Be_2 was obtained via a DF calculation[1]. However, we must be aware of the fact that the DF method involves an uncontrolled, and almost certainly uncontrolable approximation for exchange and correlation, and so cannot be regarded as a true ab initio method.

In fact, when confronted with this approximation, the newcomer will certainly discount the possibility that predictions of the scheme could have anything to do with nature! In this, however, we are pleasantly surprised, and much of the following will be devoted to the question of why. There have been several recent reviews of the formalism[2] and I do not intend to compete with them. Instead I concentrate on a central question that is usually assigned a seat in the last row - why "local approximations" work where they should not - and expound and expand on my view that the DF scheme is in essence "molecular orbital chemistry-quantified".

Cluster calculations in solid-state physics

As solid-state physicists we have to do with extended systems and Bloch waves, and the relevance of energy calculations for clusters of atoms is not obvious. There is a strong feeling amongst proponents of such calculations, however, that the energy of, say, an interstitial or chemisorbed atom is mainly influenced by the local environment and only weakly dependent on the extended nature of the host lattice. Though there is a growing amount of evidence in its favour, this 'feeling' is extraordinarily hard to translate into hard mathematics. As an illustration of the basic point, and why it is difficult to quantify, I offer the following consideration.

When an oxygen atom approaches a metal surface, say aluminium metal, its unfilled 2p-shell lies below the Fermi level. This implies, in a one-electron picture, that at some stage metal electrons will hop onto the O-atom, ionizing it and causing it to accelerate sharply and smash into the surface. Something like this probably occurs, and has, indeed, been made the basis of a model for sticking[3]. Once on (or in) the surface, however, the oxygen atom is not ionized but almost neutral, as work function measurements show[4]. This means that a rather drastic charge re-arrangement has taken place during the final stages of chemisorption, and this we would have to describe accurately in calculating the energy. The essential point is that the structure of the "ad-atom" orbitals in the chemisorbed state is quite different from that of the free atom. This means that a simple description e.g. via a Newns-Anderson model cannot be applied at all.

Suppose, now, that we view the adsorption differently. Instead of allowing the free atom to interact directly with the surface we first take out one Al atom, leaving a hole in the surface, and form an AlO molecule. This has a rather similar structure to the CO molecule, but with one electron less in the highest occupied orbital. On making the molecule we gain its binding energy, 4.9 eV, and change the structure of the orbitals in such a way that the "p-shell" of the oxygen is filled and charge is distributed to satisfy rough charge neutrality of both Al and O. If we now put the molecule back into the hole in the solid, a further redistribution of the charge will take place, but this will be less drastic than if the O atom were not pre-bonded to Al. The energy gained on replacing the AlO complex will not be equal to that lost on first creating the hole, but the difference is probably small compared with the chemisorption energy. In fact, the value of 4.9 eV for the chemisorption energy of O on Al is as good a number as is currently available.

The statement that bonding in solids is determined primarily by the local environment implies no more than the convergence of the above "pre-bonding" procedure. As we use more and more host atoms in the "pre-bonding cluster" we expect the energy gained on re-embedding the cluster to approach the energy lost in carving it out of the solid. If this energy difference is small we should be able to treat it via perturbation theory. However, it is rather difficult even to define what the perturbation is, and therefore not easier to generate an expansion in it ! Conventional "embedding" schemes do not attempt to do so, but seek to solve the entire problem, cluster plus host, in a mixed-basis of localized and extended orbitals[5]. This is easier said than done, and we may doubt that a fast, tractable but accurate scheme will be forthcoming in the near future. An alternate route is to rely on the 'feeling' mentioned above and to hope either that one's cluster is big enough for the problem at hand, or that some clever person will figure out how to carry through the "perturbation theory" mentioned above.

Naturally, the 'cluster route' can address only a limited class of problems. Long range lattice relaxation, for example, cannot be described, which highlights a difficulty yet to be overcome - to what extent do we allow host atoms in the pre-bonding cluster to adjust their spacings ? We cannot let them adjust completely without allowing for re-adjustment or embedding in the lattice. Equally, we cannot hold them completely rigid. This problem immediately imposes a limit on the smallest cluster that is likely to give plausible results - two shells of neighbours of the impurity must be included. The first shell can then be allowed to relax with the second shell held fixed. With this in mind, we confront the "minimum problem" of relevance in solid-state applications, i.e. that of evaluating $E[\{\vec{R}_i\}]$ for clusters with $\gtrsim 10$ atoms. Though quantum chemists have made great progress in recent years[6], it is improbable that they will have much to say about such systems in the near future - enter the DF scheme.

The density functional

On the assumption that much has already been said about the density functional I give here only a resumé of a few essential points that have special relevance for localized systems and a few comments that focus on the differences between the DF scheme and "standard" methods of quantum chemistry.

The idea of using the density as a basic variable goes back to the era of Thomas and Fermi[7], whose functional, however, is extremely limited in application because it fails to describe correctly the balance between kinetic- and potential-energy near nuclei. The essential contribution of Hohenberg, Kohn and Sham (HKS)[8] was to show how it is possible to remedy this defect whilst retaining the density as basic variable. This they achieved by an ingenious separation of the functional into "kinetic" electrostatic and exchange correlation terms

$$E[n(\vec{x})] = T_0 + E_{es} + E_{xc} \tag{1}$$

Whereas Thomas-Fermi gave an explicit "local density" form that allows evaluation of T_0 for an arbitrary density, $n(\vec{x})$, HKS propose to generate $n(\vec{x})$ and T_0 pairwise with the aid of an (arbitrary) Slater determinant ψ :

$$n(\vec{x}) = \langle\psi|\hat{n}(\vec{x})|\psi\rangle = \sum_n a_n|\psi_n(\vec{x})|^2 \tag{2}$$

$$T_0[n(\vec{x})] = \langle\psi|\hat{T}|\psi\rangle = \sum_n a_n\int\psi_n^\star(\vec{x}) - \frac{1}{2}\nabla^2\psi_n(\vec{x})d\vec{x} \tag{3}$$

In (2) and (3) the occupation numbers a_n are determined by the choice of determinant, which in turn is governed by the symmetry of the state under study (sometimes several determinants must be included, in which case some a_n may have fractional, though fixed values). As can be seen, T_0 is the kinetic energy of a system of non-interacting electrons which has density $n(\vec{x})$*. Near nuclei the external potential dominates over the electron-electron interaction so (3) is an excellent approximation to the full k.e. associated with density $n(\vec{x})$. The balance of k.e. and p.e. near nuclei is therefore achieved and the main flaw in the Thomas-Fermi functional corrected. In regions where the external field is not dominant, T_0 is only a part of the true kinetic energy**. The remainder contri-

* HKS required this system to be in its ground state, a condition some workers have sought to impose by juggling their occupation numbers in (3). However, the restriction is not necessary, and, though for extended systems this never occurs, it is no contradiction whatsoever if the minimum of the function corresponds to a ψ that is not a ground state[9]. Similarly, a casual remark in HKS's paper has led to a confusion concerning degeneracy. There is no in principle restriction of the theory in this respct, though it is true that an <u>approximate</u> functional will not correctly preserve the exact degeneracies of a high symmetry system.

** Which is why the density functional does not, and should not, obey the "Virial theorem" $E=T_0$, for a system in equilibrium.

butes to the exchange-correlation energy, E_{xc}, about which I will later have quite a lot to say.

The essential feature of the HKS scheme that distinguishes it from CI methods is that no attempt is made to approximate the wave-function of the interacting system. The only wave-functions appearing in the theory, the Slater-determinants ψ mentioned earlier, relate strictly to a system of non-interacting electrons. All effects due to the electron-electron interaction are buried in the potential functionals E_{es} and E_{xc} and not into the ψ, which have no other function than to enable a connection between kinetic energy and density to be established. That we end up minimizing the density functional by solving a Schrödinger-like equation should not be allowed to obscure the fact that its solution relates not to the physical system under study but to a fictitious system of non-interacting electrons whose only connection with the real system is that, in principle, both have the same density*.

In CI, on the other hand, the aim is to approximate the wave-function of the interacting system, ψ_{int}, well enough to take advantage of the variational principle for $<\psi_{int}|\hat{H}|\psi_{int}>$. This is extraordinarily difficult and usually requires the inclusion of many (thousands of) terms in an expansion in Slater determinants. CI is much more ambitious than the DF-scheme in that, since ψ_{int} is approximated, one can presumably determine accurate expectation values for any operator of interest i.e. not only the energy and density. However, CI is not a scheme that gives a simple picture of, say, bonding trends across a series that is understandable to the lay chemist or physicist. In this respect the density functional is clearly superior. It gives a picture of bonding that is simple and intuitive and that chemists will readily recognize as their time-honoured "molecular orbital picture". The difference is that the DF scheme allows us to put "flesh on the bones".

As an illustration of the point of departure of CI and DF methods consider a famous approximation that has a foot in both camps. I refer to Hartree-Fock. Hartree-Fock owes its special place in chemistry to the fact that $E_{HF} = <\psi_{HF}|\hat{H}|\psi_{HF}>$ is the lowest energy that can be attained using a single determinant (or "configuration") for ψ. Thus ψ_{HF} is the best possible "zero-order CI" wave-function. Because ψ_{HF} is a single configuration wave-function it is rather simple. One has well defined symmetry orbitals and orbital energies, which give the theory a one-particle flavour and whose variations can be followed across the periodic table.

---

* On the assumption that a system of independent electrons in some external potential can have the same density as the real system, which can be "proved" only on making the equivalent assumption that the density response function of an arbitrary system of N independent electrons possesses an inverse.

Furthermore, the HF energy $E[\{\vec{R}_i\}]$ is usually fairly accurate and its defects are known and can be accounted for [e.g. poor description of energetics of open-shell states, too small inter-nuclear separations, too large vibration frequencies] . However, precisely because it is a single configuration wave-function, ψ_{HF} is (presumably) an exact eigenfunction of <u>some</u> system of non-interacting electrons. Thus it is an admissable trial function in the DF scheme, and, indeed, would exactly minimize the density functional if, as an approximation for $E_{xc}[n]$, we used the "Fock" expression

$$E_{xc} = -\sum_{\substack{n,n'\\ \sigma}} \int d\vec{x} \int d\vec{x}' \frac{\psi_n^{\sigma\star}(\vec{x})\psi_n^{\sigma}(x')\psi_{n'}^{\sigma}(x)\psi_{n'}^{\sigma\star}(x')}{|\vec{x}-\vec{x}'|} \tag{4}$$

Here σ refers to spin and $\psi_n^{\sigma}(\vec{x})$ is the set of spin-orbitals used in constructing $\psi^{\star}$.

Thus, HF is either "zero-order CI" or a specific approximate density functional, the only difference being one of view. In the former case, ψ_{HF} is regarded as an approximation to ψ_{int}, while in the latter it is the wave-function of the "equivalent non-interacting system" with only E_{HF} and $n_{HF}(\vec{x}) \equiv \langle\psi_{HF}|\hat{n}(x)|\psi_{HF}\rangle$ having physical significance. In going "beyond Hartree-Fock", CI and DF part company. The CI specialist tries to improve the wave-function while retaining the same energy expression, $E = \langle\psi_{int}|\hat{H}|\psi_{int}\rangle$, while the proponent of the DF scheme tries to improve the energy expression while retaining the essential simplicity of a one-determinant wave-function. An important point to bear in mind is that whereas there is a clear, systematic criterion for "improving the wave-function" - i.e. because $\langle\psi_{int}|\hat{H}|\psi_{int}\rangle$ obeys the Rayleigh-Ritz variational principle - the same is not true with respect to "improving the functional". Though one can write down an exact expression for E_{xc} [10], it has not proved possible to work this expression into a form suitable for systematic approximation. As matters stand we are forced to make an uncontrolled approximation that can only be justified a posteriori. This is a serious deficiency and the proponent of the DF-scheme has to stop worrying about rigour and to learn to live with the uncomfortable suspicion of pursuing not a science but a greyish art !

Approximating E_{xc} : the 'local density' approximation

To carry out a DF calculation we need to find an approximation for E_{xc}, which I write, without loss of generality in terms of an exchange-correlation energy density $\varepsilon_{xc}(\vec{x})$

★ To the reader who is upset by the explicit appearance of these orbitals in the <u>density</u> functional E_{xc}, I ask why no objection was raised when the same was done in connection with T_0.

$$E_{xc} = \int d\vec{x}\; n(\vec{x})\; \varepsilon_{xc}(\vec{x}) \tag{5}$$

The matter of approximating $\varepsilon_{xc}(\vec{x})$ has generated an immense literature and the uninitiated bold enough to venture into such muddy water is faced with a bewildering variety of schemes and acronyms, Xα, LD, LSD, HL, GL, HLG, vBH, SIC, VWN, that grows with each fresh batch of publications. The initial impression, quite properly and the final judgement, most probably, is rather clear - we do not know what we are doing !

In fact, most of this activity is peripheral to the central issue in that it refers to attempts to establish the validity of approximations for $\varepsilon_{xc}(\vec{x})$ a posteriori. Since there is no limit to the number of test systems available, some can always be found for which one scheme appears to be more accurate than others. In all of this, the crucial question of why <u>any</u> simple (e.g. local density) approximation should work at all for molecules has received scant attention. For this reason, I focus entirely on this single point and leave the reader free to judge in the light of my subsequent remarks the extent to which it is sensible to hang one's hat on a particular approximation.

In solid-state applications the most common approximation is the original suggestion of HKS that

$$\varepsilon_{xc}(\vec{x}) \simeq \varepsilon^{h}_{xc}(n(\vec{x})) \tag{6}$$

where $\varepsilon^{h}_{xc}(n)$ is the exchange correlation energy density of a homogeneous electron gas with density n. This approximation has the merit that it becomes exact in the limit that gradients of the density are everywhere small. For this reason (6) is referred to as <u>the</u> local density approximation (LDA). HKS did not forsee applications of the LDA in chemistry because for e.g. a molecular density the "slowly varying" condition is manifestly violated virtually everywhere. This obvious fact accounts for the 10 years or so that elapsed between HKS's original work and its first successful application in chemistry - a devastatingly accurate reproduction of the binding energy curve of the H_2 molecule[11]. Since then it has become apparent that the LDA works* for a wide variety of systems, localized and extended, where the density gradients are in no sense small and there is no formal justification for the approximation whatsoever[12-15]. In view of the pain that correlation effects inflict on quantum chemists, this seems like magic. However, it is not and we can go someway towards understanding it by expanding to their logical conclusion some arguments due to Gunnarsson and Lundqvist[16] (who, however, are not to be held accountable for the artistic licence used in their presentation here !).

---

* In the sense that experimentally measured energies are semiquantitatively reproduced.

The energy term E_{xc} should be viewed as a correction to the electrostatic energy of the electron distribution

$$E^{e}_{es} = \frac{1}{2} \int d\vec{x}\; n(\vec{x})\; \phi_e(\vec{x})$$

where $\phi_e(\vec{x})$ is the electronic Coulomb potential. That E^{e}_{es} overcounts the electron-electron repulsion is apparent on noting that $\phi_e(\vec{x})$ is the potential experienced by a test charge that neither "exchanges" with the electrons nor disturbs the electron distribution. To get the full energy, we should replace ϕ_e by $\tilde{\phi}_e$, the potential experienced by an electron of the system. By analogy with ϕ_e, which is calculated using a completely undisturbed electron density

$$\phi_e(\vec{x}) = \int d\vec{x}'\; \frac{n(\vec{x}')}{|\vec{x} - \vec{x}'|} \tag{7}$$

we expect $\tilde{\phi}_e(\vec{x})$ to be given by

$$\tilde{\phi}_e(\vec{x}) = \int d\vec{x}'\; \frac{\tilde{n}(\vec{x},\vec{x}')}{|\vec{x} - \vec{x}'|} \tag{8}$$

where $\tilde{n}(\vec{x},\vec{x}')$ is the density of electrons at $\vec{x}'$ given that there is an electron at $\vec{x}$. The difference density

$$n_h(\vec{x},\vec{x}') = n(\vec{x}') - \tilde{n}(\vec{x},\vec{x}') \tag{9}$$

is called the exchange-correlation hole density, because it describes the 'hole' an electron digs around itself as a result of Pauli principle restrictions and Coulomb correlations. The exchange-correlation energy can then be written

$$E_{xc} = \frac{1}{2} \int d\vec{x}\; n(\vec{x})\; [\tilde{\phi}_e(\vec{x}) - \phi(\vec{x})] = -\frac{1}{2} \int d\vec{x} \int d\vec{x}'\; \frac{n(\vec{x})\, n_h(\vec{x},\vec{x}')}{|\vec{x} - \vec{x}'|} \tag{10}$$

On comparing with (5), we see that the exchange-correlation energy density is

$$\varepsilon_{xc}(\vec{x}) = -\frac{1}{2} \int d\vec{x}'\; \frac{n_h(\vec{x},\vec{x}')}{|\vec{x} - \vec{x}'|} \tag{11}$$

merely half the Coulomb potential due to the exchange-correlation hole*.

---

* This is not quite right because the interaction part of the kinetic energy is missing. However, including this amounts only to a re-definition of n_h [16] (via an "integration over the coupling constant") and the difference need not concern us here.

Though the hole density $n_h(\vec{x},\vec{x}')$ may be a rather complicated function in a bounded system much of its structure has no influence on $\varepsilon_{xc}(\vec{x})$. In fact, the isotropy of the Coulomb interaction allows us to write (11) in a rather simple form. If we define $\vec{y} \equiv \vec{x} - \vec{x}'$, and the spherical average of the hole density about the point $\vec{x}$

$$n_h^{av}(\vec{x},y) \equiv \int d\hat{y}\; n_h(\vec{x},\vec{x}') \tag{12}$$

then $\varepsilon_{xc}(\vec{x})$ may be written

$$\varepsilon_{xc}(\vec{x}) = -\,1/[\,2R(\vec{x})] \tag{13}$$

with

$$R(\vec{x}) \equiv [\int_0^\infty dy\; y\; n_h^{av}(\vec{x},y)]^{-1}$$

This expression makes apparent that all structure in $n_h(\vec{x},\vec{x}')$ that is not spherically symmetric in the variable $\vec{y}$ gives no contribution to ε_{xc}. Furthermore, $\varepsilon_{xc}(\vec{x})$ depends solely on $R(\vec{x})$, which is simply the radius of the spherically averaged hole density n_h^{av}. A very important property, valid for any system having two-body interactions is

$$\int d\vec{x}'\; n_h(\vec{x},\vec{x}') = \int_0^\infty dy\; y^2\; n_h^{av}(\vec{x},y) = 1 \tag{14}$$

which states that the hole contains exactly one electron. This condition enables us to normalize the hole exactly. For example, an exponential approximation to the spherically-averaged hole must have the form

$$n_h^{av}(\vec{x},y) = \frac{1}{2}\,\lambda^3(\vec{x})\; \exp[-\lambda(\vec{x})y] \tag{15}$$

Evaluating (13) we identify the radius of the hole

$$R(\vec{x}) = \frac{2}{\lambda(\vec{x})} \tag{16}$$

In summary, this discussion shows that, however complicated the structure of the exchange-correlation hole surrounding an electron at $\vec{x}$, only its average radius, $R(\vec{x})$, is needed to calculate the energy density $\varepsilon_{xc}(\vec{x})$.

This is as far as we can proceed rigorously, but there is a further step that is very revealing and, at least, highly plausible. Consider the quantity $n_h(\vec{x},\vec{x})$. By definition (9), this is equal to $[n(\vec{x}) - \tilde{n}(\vec{x},\vec{x})]$, with $\tilde{n}(x,x)$ the density of electrons at $\vec{x}$ given that there is already an electron at $\vec{x}$. Now if all electrons had the same spin this would be identically zero by the Pauli principle. Furthermore, if the interaction had a hard core, $\tilde{n}(\vec{x},\vec{x})$ would be zero whatever the spin. The mere singularity of the Coulomb potential does not imply this condition because it is integrable, but it is nevertheless clear that Pauli-principle restrictions and interactions imply $\tilde{n}(\vec{x},\vec{x}) \ll n(\vec{x})$. Thus we have

$$n_h(\vec{x},\vec{x}) = \frac{n_h^{av}(\vec{x},y=0)}{4\pi} \simeq n(\vec{x}) \tag{17}$$

which condition, together with (14), fixes the hole radius precisely for any assumed shape of the hole. For our exponential hole, for instance, we have

$$4\pi n(\vec{x}) = \frac{\lambda^3(\vec{x})}{2}$$

or

$$\varepsilon_{xc}(\vec{x}) = -\frac{1}{2}[\pi n(\vec{x})]^{1/3} \tag{18}$$

The initiated reader will surely recognize the "Xα" approximation !

In fact, the relation $\varepsilon_{xc}(\vec{x}) = \alpha n^{1/3}(\vec{x})$ is valid for a wide class of hole densities, namely those whose spherical average can be represented by

$$n_h^{av}(\vec{x},y) = f(\vec{x})g(\lambda(\vec{x})y) \tag{19}$$

where f and g are arbitrary functions.* The value of α then merely depends on the shape of the function g. If this is a square well, for instance, we get $\alpha = (3\pi^2/4)^{2/3}$.

Now this discussion is not meant to be a plaidoyer for the Xα approximation, but as an illustration of how a local relation between $\varepsilon_{xc}(\vec{x})$ and $n(\vec{x})$ can be obtained even when $n(\vec{x})$ is not "slowly varying". The density enters via the condition (17), which, with (14), determines the radius $R(\vec{x})$ uniquely for any hole density that can be written like (19). The success of "local" formulae for $\varepsilon_{xc}(\vec{x})$ therefore has nothing whatsoever to do with their being local but is perhaps an indication that the shape of the exchange-correlation hole does not vary dramatically from point to point, at least for regions of space that contribute most to the change in the integral

$$E_{xc} = \int d\vec{x}\, n(\vec{x})\, \varepsilon_{xc}(\vec{x}) \tag{20}$$

---

* In case this is not obvious, choose g(0) = 1. Then (17) gives $f(\vec{x}) = 4\pi n(\vec{x})$, while (14) guarantees that $1 = f(\vec{x})\lambda^3(\vec{x})g_2$, with g_n the n-th moment of g. (13) now gives $1/R(\vec{x}) = 4\pi n(\vec{x})\lambda^2(\vec{x})g_1$ so that

$$\varepsilon_{xc}(\vec{x}) = -\frac{1}{2R(\vec{x})} = -\frac{g_1}{2g_2^{2/3}}[4\pi n(\vec{x})]^{1/3}$$

when, say, a molecule forms from two atoms. Different 'local' potentials correspond to different assumptions about hole shape. As we have seen, the "Xα" potential amounts to assuming that the hole shape can be characterized by a single length, the constant multiplying $n^{1/3}$ depending on the precise functional form adopted. The "local density approximation" behaves essentially like $n^{1/3}$, but with deviations that arise because the hole density of an electron gas depend on two lengths, k_F^{-1}, k_{FT}^{-1}, where k_F, k_{FT} are Fermi- and Thomas-Fermi wave-vectors, respectively.* However, neither of these lengths has physical significance for atoms and molecules, so the case for LDA versus Xα cannot be argued on the basis that the former gives a better description of the hole. Similarly, any argument based on exactness of results in the limit of slowly varying $n(\vec{x})$ is irrelevant. The best we can say is that the LDA corresponds to a unique prescription for the shape of the hole and takes into account to some extent changes in this shape on going from regions of high to low density.**

Now, having explained how very reasonable it is that local approximations work even for strongly inhomogeneous systems, I can reveal the other side of the coin ! As those active in the trade are aware (and as the layman surely suspects) the sequence of arguments I have just presented hides a worm. To find this worm, we need merely to apply the above arguments to the only systems where we know the exchange-correlation energy density, $\varepsilon_{xc}(\vec{x})$, exactly i.e. systems having only one electron. If there is only one electron, $\tilde{n}(\vec{x},\vec{x}')$ is exactly zero because if the electron is at $\vec{x}$ it cannot simultaneously be elsewhere. Thus $n_h(\vec{x},\vec{x}') = n(\vec{x}')$ and, by (11),

$$\varepsilon_{xc}(\vec{x}) = -\frac{1}{2}\int d\vec{x}' \frac{n(\vec{x}')}{|\vec{x}-\vec{x}'|} = -\frac{1}{2}\phi_e(\vec{x}) \qquad (21)$$

The exchange-correlation energy then cancels exactly the 'self' electrostatic repulsion of the one electron. If $n(\vec{x})$ is a localized density we can easily determine the value of ε_{xc} where $n(\vec{x})$ is small via a moment expansion

$$\varepsilon_{xc}(\vec{x}) \to -\frac{1}{2|\vec{x}|} \quad , \quad \vec{x} \to \infty \qquad (22)$$

---

* In addition, because it results from an in principle exact theory for a specific system, the LDA automatically takes into account to some extent the fact that (17) is not obeyed exactly.

** Note that $k_F/k_{FT} \propto n^{1/6}$. As a rider to these remarks, it is naturally satisfying to be able to use the same approximation both for rapidly and slowly varying densities.

In fact, this result applies to any three-dimensionally bounded system irrespective of the number of electrons, as one can easily show. Eq. (22) merely expresses the fact that in regions of low density the frequency of electron-electron collisions goes to zero and the only exchange-correlation effect left is the canceling of the 'self' interaction.

According to our 'hole' arguments, however, $\varepsilon_{xc}(\vec{x})$ should vanish as the cube root of the density i.e. should fall off exponentially and not like 1/x. Thus, in spite of the fact that both of the critical relations, (14) and (17), are manifestly satisfied, the argumentation fails completely. What has gone wrong ?

What has gone wrong is illustrated in Fig. I, where I show calculations of the spherical hole density $n_h^{av}(\vec{x},y)$ for the H-atom $[n(x) = e^{-2x}/\pi]$. For completeness, I quote the analytic formula

$$n_h(x,y) = \frac{1}{2xy}\,\{(1+2|y-x|)e^{-2|y-x|} - (1+2(y+x))e^{-2(y+x)}\} \quad (23)$$

which does not look much like the simple form (19), and corresponds to a shape that changes dramatically with x. At x = 0, the nuclear position, the hole is exactly exponential, whereas for $x \gtrsim 1.5$, it becomes extremely extended and shows a weak structure at $y \sim x$. This arises because the actual hole density (i.e. non-spherically averaged) is independent of the position of the electron, $n_h(\vec{x},\vec{x}') = n(\vec{x}')$, and does not follow the electron as it moves away from the nucleus i.e. the curves in Fig. I merely represent spherical averaging of the exponential function about different points.*

In the inset of Fig. I, I show the hole radius function R(x) defined by (13) which is simply x when $n(x) \simeq 0$ in agreement with (22). Also shown are the radii corresponding to exponential and square-well holes of the form (19)

$$R_{exp}(x) = (\pi n(x))^{-1/3} \quad , \quad R_{sq}(x) \simeq 0.6\ (\pi(n(x))^{-1/3}$$

As can be seen, R_{exp} gives the correct radius when the hole is exponential, while R_{sq} is right when the hole is approximately square. However, no simple functional form can be found that will describe the shape variations of the hole over the entire range of x. In fact, for large x the variation in shape dominates the structure of the hole.** Our 'worm' has surfaced. Even a simple

---

* Note that when weighted with the volume element y^2, all these curves integrate to unity.

** In the sense that no function g can be found for which (19) is valid over a range of x. From the footnote on p.16 we see that the ratio of moments $(g_2^{2/3})/g_1$ of the function describing the exact hole at large x must vanish exponentially as $x \to \infty$.

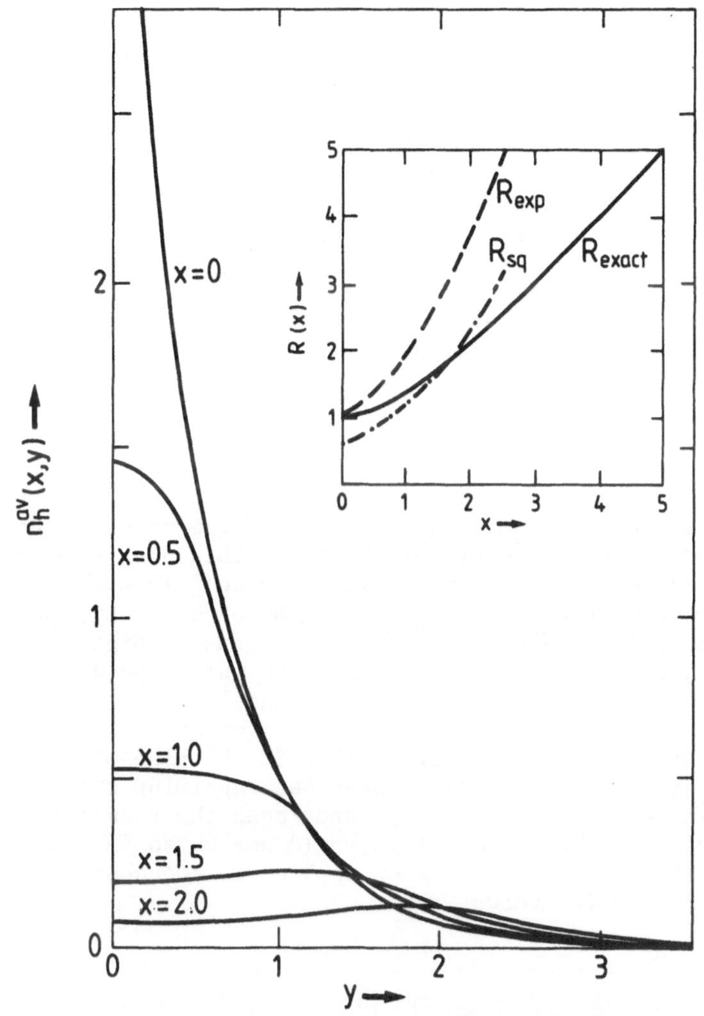

Fig. 1. Spherically averaged exchange-correlation hole for H-atom [Eq. (23)]. Inset : Hole radii (see text).

thing like the radius of a hole can display perversity !

One encouraging feature of Fig. I is that the shape of the hole is reasonably constant between $x \sim 0.5$ and $x \sim 1.5$, where it looks something like a Fermi function. The corresponding 'local' radius, R_F, would therefore be an excellent approximation over a region containing ~ 50% of the electrons. Furthermore, for $x < 0.5$ we would have $R_F < R_{exact}$, while for $x > 1.5$ we would have $R_F > R_{exact}$, leading to some error cancellation when we come to evaluate E_{xc}. As Gunnarsson, Lundqvist and Wilkins[17] have shown, this is precisely the reason why the LDA gives a reasonable energy for the H-atom. The LDA 'hole radius', R_{LDA}, is about 15% below R_{exp}, while the LSDA* radius is about 15% above R_{sq} and crosses the exact radius at $x = 1$, where $x^2n(x)$ is maximum. The corresponding total energies for H are $E_{LDA} = -0.45$ a.u. $E_{LSDA} = -0.49$ a.u.[17]

Though one is rightly wary of extrapolating from so simple a system as the H-atom, where the 'hole' is of very special form, it is at least plausible that the error cancellation mentioned above may be a general phenomenon. I argue in terms of the radius R. Firstly, if $n(x) \simeq 0$, $R_{exact} \to x$ for any bounded system, while R_{LDA} diverges exponentially. Secondly, near nuclei, R_{LDA} refers to the exchange-correlation hole of a homogeneous electron gas having density equal to the maximum density of the system under study. R_{exact}, however, refers to a system which has this density only over a small region of space, and so will 'screen' less effectively. Accordingly, we expect $R_{LDA} < R_{exact}$ near the nuclei, which means that the two must cross somewhere between regions of high and low density, which, if we are fortunate, might be where the density changes most when the nuclear positions change.**

It will not have gone unobserved that in the case of the H atom, where $\varepsilon_{xc}(\vec{x}) = -1/2\,\phi_e(\vec{x})$, I have been applying the 'hole arguments to the electrostatic energy and found the result to be not bad, in spite of the 'worm'. This would not be so in general, however, because condition (14), when applied to the density $n(\vec{x})$ that gives rise to $\phi_e(\vec{x})$ reads

$$\int d\vec{x}\; n(\vec{x}) = N \qquad (24)$$

---

* The spin-unrestricted version of (6) for which E^h_{xc} refers to a spin-polarized electron gas.

** In this, we are assuming that R_{exact} is a smooth function, which is not unreasonable in view of the extensive averaging involved in its definition.

where N is the number of electrons in the system. That is, the 'hole' contains N electrons and any mistake made as a result of approximation would get multiplied by N. In fact, the balance of electrostatic forces in any system of electrons and nuclei is very delicate because of the extreme range of the Coulomb interaction and any attempt to approximate ϕ_e would almost certainly lead to disaster.*

In their separation of the density functional, HKS isolated the only quantity, E_{xc}, that one can even try to approximate simply with some hope of success. It should not be thought, however, that any approximation will do. This point is illustrated in Fig. II, where I show the variation of the three terms in the functional (1) as the internuclear separation, R_e, of the N_2 molecule varies.** The separate contributions are plotted with respect to their values at $R_e = 2.5$. The figure illustrates very dramatically that the kinetic and electrostatic contributions, T_0 and E_{es}, vary on the scale of Hartree, while the total energy, $E_T = T_0 + E_{es} + E_{xc}$, varies by maximally 3 eV. Obviously, the competition between T_0 and E_{es} is crucial. However, the result of this is the dashed curve marked $T_0 + E_{es}$, whose variation with R_e is on the same scale as E_{xc}. The form of the total energy is thus dependent in an essential way on the variation of E_{xc} with R_e. The good correspondence of measured spectroscopic parameters with those calculated using the LDA (or, for that matter, the Xα approximation[13]) can only mean that the approximation gives a reasonable description of this variation. In turn, this suggests that at least over these regions where the density changes most when the nuclei move closer together (in general, the mid-point between the nuclei) the form of the averaged exchange-correlation hole is relatively constant and similar to that of a homogeneous electron gas. Note that an accurate description of the energy near the minimum does not guarantee a good value for the binding energy, E_B, of the molecule. For example, an error in the slope of E_{xc} vs R_e of ~ 10% would have a relatively small

* For example, the core-core repulsion of the N_2 molecule near minimum, ~ 6 a.u., exceeds the binding energy by an order of magnitude. This is cancelled almost exactly by the electron-electron repulsion, which must therefore be evaluated very carefully.

** These results were obtained using an old programme[12] that did not minimize the function adequately in this case. They should therefore be regarded only as a guide to behaviour. This, and appearances notwithstanding, the energy curve E_T gives a reasonable description of the ground state of N_2 i.e.: $R_e = 2.16$ a.u. vibration frequency $\omega_e = 2170\ \mathrm{cm}^{-1}$. Subsequent work[13] has shown that a better minimization leads to results very close to the measured 2.07 a.u., 2358 cm^{-1}.

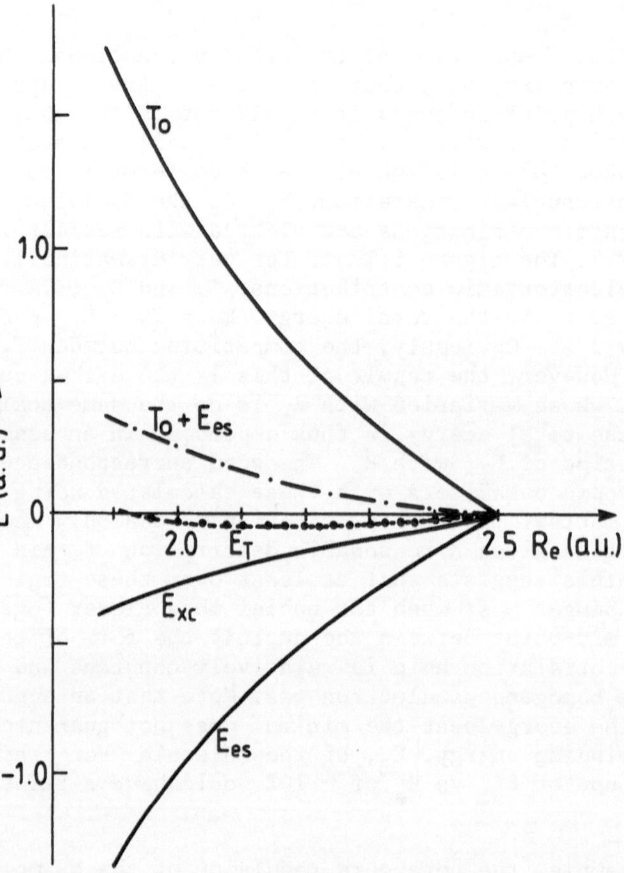

Fig. 2. Behaviour of different terms in the energy functional as function of internuclear separation of N_2 molecule. All quantities plotted relative to their values at $R_e = 2.5$.

influence (~ a few %) on the equilibrium separation and vibration frequency, but would give an error of > 10 % in the energy difference between stable molecule and isolated atoms. This is because the error accumulates as one goes from the molecular minimum R_e^0 to infinite separation, and the total change in E_{xc} when the molecule forms, is substantially greater than E_B.

In this section I have tried to set down what a priori arguments can be presented in favour of local approximations for the exchange-correlation energy density, $\varepsilon_{xc}(\vec{x})$. I have sketched Gunnarsson and Lundqvist's proof that $\bar{\varepsilon}_{xc}(\vec{x})$ depends only on $R(\vec{x})$, the average radius of the exchange-correlation hole surrounding an electron at point $\vec{x}$, and showed that if this has a relatively simple shape that varies little from point to point then $R(\vec{x}) = \alpha n(\vec{x})^{-1/3}$, with α depending on the particular shape involved. The LDA corresponds to a specific choice of hole whose shape varies slightly with density - i.e. that of a homogeneous electron gas having density n(x) - and corresponds to a hole radius $R_{LDA}(\vec{x}) = \alpha(n)n^{-1/3}$, where $\alpha(n)$ is a weak function of n over most of the range. For a specific one-electron system, I showed that the exact hole shape varies dramatically near, and far from the nucleus, but is relatively constant over the region where the electron is most likely to be found. If this result were general, it would go some way towards explaining why the LDA seems to give relatively reliable values for the exchange-correlation energy changes that accompany molecule formation.

I can think of no better way of ending this section than to quote the comment of a colleague who read this section four times, asked twenty five questions, read the section again, and then expostulated in the instant of sudden understanding "Now I get the message ! You mean it's simply a matter of blind good luck !"

Carrying out a DF calculation for a cluster

Since I can do no more than scratch the surface of this topic in a single lecture I will be brief and try to stick to essentials. Though we have to do with a density functional, HKS's way of splitting it up condemns us to introduce orbitals, $\psi_n(\vec{x})$ which we can expand in a basis

$$\psi_n(\vec{x}) = \sum_{i,L} C^n_{iL}\, \chi^i_L(\vec{r}_i) \tag{25}$$

where $\vec{r}_i = \vec{x} - \vec{x}_i^0$, with $\vec{x}_i^0$ the location of nucleus i. Once the basis is defined, the HKS energy can be regarded as a function of the basis parameters, C^n_{iL}, and the extremum condition converted to a matrix eigenvalue problem

$$(\hat{H} - \varepsilon_n \hat{O})(C^n) = 0 \tag{26}$$

where $\hat{H}$ and $\hat{O}$ are Hamiltonian and overlap matrices. In particular,

$$H^{ij}_{LL'} = \langle\chi^i_L|-\frac{1}{2}\nabla^2 + V_{eff}(\vec{x})|\chi^i_{L'}\rangle \tag{27}$$

The energy functional is minimized in the space spanned by the χ^i_L when the one-electron potential $V_{eff}(\vec{x})$ in (27) satisfies the self-consistency condition

$$V_{eff}(\vec{x}) = \phi(\vec{x}) + \mu_{xc}(\vec{x}) \tag{28}$$

Here $\phi(\vec{x})$ is the total electrostatic potential

$$\mu_{xc}(\vec{x}) \equiv \frac{\delta E_{xc}}{\delta n(\vec{x})} \tag{29}$$

and $\phi(\vec{x})$ and $\mu_{xc}(\vec{x})$ are to be calculated with the density corresponding to eigenvectors C^n_{iL}.

Equations (26)-(29) can be solved by iteration. An assumption is made for $V_{eff}(\vec{x})$, e.g. the overlapped potentials of neutral atoms, matrices $\hat{H}$ and $\hat{O}$ calculated, the eigenvalue problem solved, a new potential created using (28) with the density corresponding to eigenvectors C^n_{iL}, and the process repeated with the new potential until convergence is achieved.* In each iteration the value of the functional can be calculated, and this will be found to fall steadily and to saturate long before the eigenvalues in (26) reach their final values. Though this further step is generally not carried out, it is advantageous to perform a variation, or optimization of the exponents of the basis functions χ^i_L. The above procedure can be repeated sequentially for neighbouring exponent values to generate a map of the energy as a function of the exponents. The absolute minimum of the energy, as a function of both the orbital exponents and the linear parameters C^n_{iL}, is then the best estimate of the minimum value of the density function that can be attained with basis functions χ^i_L.

It will be recognized that this procedure is not to be embarked on lightly ! Whatever the basis used in (25), formidable technical problems must be overcome in order to evaluate all the integrals needed with sufficient accuracy. In addition, the basis must be extensive enough to ensure that the sum (25) includes a density that is sufficiently close to $n_0(\vec{x})$, the density corresponding to the exact minimum of the energy functional, to yield a reliable estimate of the minimum energy. While I do not propose to go into detail concerning the above mentioned technical problem, it is perhaps useful to compare two calculational schemes that corres-

---

* In practice part of the old potential must be fed back to avoid instabilities during the iterations.

pond to quite different choices for the basis functions in (25).

The most extensive density functional cluster calculations carried out to date use an LCAO, fixed exponent basis set and rely on a quasi-statistical integration procedure[14,13]. This 'discrete variational method' is rather straightforward and seems in practice to be surprisingly accurate i.e. convergence with respect to orbital basis size and the number of points in the integration mesh seems to be attainable[18]. In an LCAO scheme, the sum over L in (25) involves a sum over partial waves, or spherical harmonics $Y_L(\hat{x}_i)$, each term of which is multiplied by a series of radial functions with different functional forms and exponents. Because the exponents are not optimized many radial functions are needed for each partial wave, each having its own linear variation parameter C. However, this disadvantage is off-set in that the difference energy between cluster and constituent atoms can be evaluated directly. This is achieved by a clever trick - the so-called 'transition state' trick[19] - that is applicable whenever the only quantities that change with the nuclear positions are the coefficients C^n_{iL} in the basis expansion (25). The resulting LCAO discrete-variational/transition-state scheme has been applied successfully to many diatomic molecules and, notably, to the chemisorption of CO on Li, Al, Cu, V, Fe, and Ni surface clusters[20]. In the case of Li it was found possible to carry through the calculation for a 'pre-bonding' cluster containing up to 70 atoms.

As mentioned, the fixed functional form of the radial functions in conventional LCAO methods means that rather large basis sets are required. This is not only computationally expensive, but makes it rather hard to ensure that sufficient functions are included to reproduce reliably the variation of the energy that accompanies changes in the nuclear positions. An alternative kind of basis set is one where the functions $\chi^i_L(\vec{r}_i)$ do not have a fixed dependence on r_i, but adjust to the cluster potential.

The first attempt to incorporate this feature, the so-called "Scattered-Wave" (SW) method[21], was made by straight-forward adaptation of the KKR band-structure scheme to the cluster geometry. The potential in (27) was approximated by its "muffin-tin" part and the cluster boundary condition applied by defining a "concave muffin-tin", the space outside a boundary sphere surrounding the cluster. In spite of initial optimism, the SW scheme fell into disrepute when it was found to yield a very poor description of molecular geometries. For example, the C_2 molecule was found not bound[22], and the water molecule predicted to be straight[23]. This failure was interpreted by some people as implying the inadequacy of the local approximation for exchange and correlation. In fact, it occured because the SW method used a 'muffin-tin construction' to evaluate the energy functional. This is a very bad approximation.

A better procedure is to use the muffin-tin potential to generate the basis functions χ_L^i in (25), but then to evaluate the energy functional exactly within this basis. One then takes full advantage of the variation principle for the energy. A scheme using this approach[12] and taking the χ_L^i to be the 'muffin-tin orbitals' (MTO) that Andersen has used to such good effect in band calculations[24], turned out to give a semi-quantitative description of chemical bonds in a wide variety of diatomic molecules. This scheme retained the partitioning of space used in the SW-method. In particular, the 'concave' muffin-tin was necessary in order to build in the cluster boundary condition. In general applications, this potential construction was found too restrictive, and a new scheme for basis function construction was proposed[25] that retains the essential features of the MTO's, but ensures that each orbital is localized with respect to its own muffin-tin, thus eliminating the need for an outer sphere. I will refer to the χ_L^i resulting from this scheme as LMTO's (where the acronym now stands for 'localized muffin-tin-orbitals' rather than 'linear muffin-tin-orbitals', as Andersen designated his band-functions) and these I now describe.

A given potential, $V_{eff}(\vec{x})$, in (27) is spherically averaged in spheres of radius R_i, centred on the nuclear positions $\vec{x}_i^0$. Using the resulting averaged potentials $V_0^i(r_i)$ in the radial equation, the function $\phi_\ell(\varepsilon_i, r_i)$ and its energy derivative $\dot{\phi}_\ell(\varepsilon_\ell^i, r_i)$ are generated. Here, $\phi_\ell{}^i(\varepsilon, r_i)$ is the solution of the radial equation for angular momentum ℓ, potential $V_0^i(r_i)$ and energy ε. The reference energies ε_ℓ^i are chosen to correspond roughly to the centre-of-gravity of eigenvalues of the cluster that have a large weight in partial waves with angular momentum ℓ on site i. Inside its own sphere, the LMTO χ_L^i is defined by

$$\chi_L^i(x) = [\,\phi_\ell(\varepsilon_\ell^i, r_i) + \omega_\ell^i \dot{\phi}(\varepsilon_\ell^i, r_i)]\, Y_L(\hat{r}_i) : \vec{x} \in \text{sphere } i \tag{30}$$

where ω_ℓ^i is a constant and $Y_L(\hat{r}_i)$ a spherical harmonic. At the boundary of sphere i, χ_L^i is matched differentiably to a linear combination of Hankel functions $K(\kappa, r) \equiv i(i\kappa)^{\ell+1} h_\ell^{(1)}(i\kappa r)$, i.e.

$$\chi_L^i(\vec{x}) = G_\ell^i[\,K(\kappa_\ell^i, r_i) + A_\ell^i\, K(\tilde{\kappa}_\ell^i, r_i)]\, Y_L(\hat{r}_i) : \vec{x} \text{ outside all spheres} \tag{31}$$

The parameters $\kappa_\ell^i, \tilde{\kappa}_\ell^i$, and A_ℓ^i are the variation parameters of the LMTO, and the matching is achieved by adjustment of G_ℓ^i and ω_ℓ^i. Finally, in each other sphere $j \neq i$, the 'tail' of the orbital (31) is re-expressed as a sum of $(\phi, \dot{\phi})$ combinations of the type (30). This is achieved by using expansion theorems obeyed by the Hankel functions to transfer the origin from $\vec{x}_i^0$ to $\vec{x}_j^0$ i.e.

$$K_\ell^i(r_i) Y_L(\hat{r}_i) = \sum_{L',j} B_{LL'}^{ij}\, J_{\ell'}^j(r_j) Y_{L'}(\hat{r}_j) \tag{32}$$

where $B^{ij}_{LL'}$ are structure constants, and then replacing each J^j_ℓ, by a $(\phi,\dot{\phi})$ combination of appropriate amplitude and logarithmic derivative.

If we were to set $A^i_\ell = 0$ in (31) and admit imaginary values of κ (i.e. 'positive' energy, delocalized tails), then χ^i_L would be precisely Andersen's 'band' MTO. The restriction to Hankel functions is necessary because of the cluster boundary condition. The second Hankel function, having a different (larger) exponent, $\tilde{\kappa}^i_\ell$ and negative amplitude, A^i_ℓ, allows the orbital to display 'positive energy' behaviour close to its own sphere and 'negative energy' behaviour further away. The three tail parameters $\kappa^i_\ell, \tilde{\kappa}^i_\ell$, and A^i_ℓ fix the form of the tail and the logarithmic derivatives at all sphere boundaries, and so determine the orbital completely up to a multiplicative constant. In principle, all three could be varied but this gives more variational freedom than is needed. One way of restricting this variation[25] is to link $\tilde{\kappa}^i_\ell$ and A^i_ℓ to the leading exponent κ^i_ℓ using an atom programme. If the orbital concerned is bound in the corresponding isolated atom, a family of contracted or expanded tails can be generated by increasing or decreasing the nuclear charge fractionally from its physical value. For each value of the excess charge, the resulting atomic orbital tails can be least squares fitted to the two Hankel function form (31). This generates families of values of the three tail parameters and enables appropriate values of $\tilde{\kappa}^i_\ell$, A^i_ℓ to be assigned to each value of the leading exponent κ^i_ℓ. With this construction, a set of orbitals χ^i_L depending on just one variation parameter, κ^i_ℓ, is generated. If the orbital concerned is 'atom unbound', the tail parameters cannot be linked this way. For such orbitals, a sensible variation requires two parameters to allow adjustment of both the spatial extent of the orbital and its logarithmic derivative at R_i. In practice, varying the atom unbound orbitals has little influence on the energy, however. In fact, the simple choice $\kappa^i_\ell \simeq (2R_i)^{-1}$ and values of $\tilde{\kappa}^i_\ell$ and A^i_ℓ that give a logarithmic derivative of ℓ at $r_i = R_i$ seems usually to be adequate.

This LMTO basis has several attractive features. Firstly, because the LMTO's adjust to the changing potential of the cluster, a small number of them gives results equivalent to a much larger LCAO basis. Secondly, because each orbital is made up of $(\phi,\dot{\phi})$ combinations in all spheres, all linear combinations of the type (25) are automatically orthogonal to <u>all</u> core orbitals. In conjuction with the 'frozen core' approximation made in calculating the total energy, this feature virtually <u>eliminates the core from the calculation.</u>* Thirdly, again because of the $(\phi,\dot{\phi})$ construction, all

---

* The core density appears in the total energy, but in an essentially trival way that poses no calculational problems. Because of this feature, there is no advantage to be gained by representing the core via a pseudopotential.

large (intra-sphere) contributions to all integrals can be determined quasi-analytically. Finally, because the tails are composed solely of Hankel functions, a partial-wave expansion of the electron density associated with each set of linear combinations (25) can be made in each atomic sphere. This allows a quasi-analytic evaluation of the dominant part of the electrostatic potential and energy.

These advantages notwithstanding, there are serious technical difficulties associated with the scheme that have only recently been overcome. In particular, the evaluation of integrals over the region outside the spheres requires special techniques, and the Coulomb potential due to the electron density outside the spheres can be determined only approximately by least squares fitting this density to a sum of Hankel functions, for which Poisson's equation can be integrated analytically. A consequence is that the energy functional cannot be evaluated exactly.* However, it is possible to arrange the terms so that (i) the fit-error in the energy is quadratic in the difference density $n(\vec{x}) = n(\vec{x}) - n_f(x)$, where n_f is the fit density, and (ii) the only non-trivial integral is rather small (~ 1 a.u.) and has a smoothly varying integrand. For all details I refer to the relevant literature[25 27].

The LMTO scheme has as yet been applied to no system larger than the ammonia molecule. However, the advantages mentioned above and experience with small systems suggest that it should be regarded as a serious competitor to LCAO, especially for clusters of relevance in solid-state and surface physics. For completeness, I quote some results which illustrate the kind of accuracy that can be achieved.

Fig. III shows the calculated binding energy curves of the H_2 and Be_2 dimers[26]. The basis used included s, p and d LMTO's (i.e. 3 independent radial functions) and only a limited variation of the orbital exponents was carried out. Nevertheless, the result shows that the functional is virtually exactly minimized. This can be deduced from the correspondence between energy and the force calculated using a modified Hellmann-Feynman formula[26]. In deriving this formula it is necessary to assume that the exact minimum of the functional has been found. Inadequacy in the orbital basis results in poor correspondence between calculated force and the gradient of the calculated energy. As can be seen in Fig. III, the force lines drawn on the energy curves are in close correspondence with the gradient of the energy, strong evidence that the orbital basis is essentially complete for these systems.

* It should be noted that conventional LCAO schemes are even more strongly reliant on achieving an accurate fit.

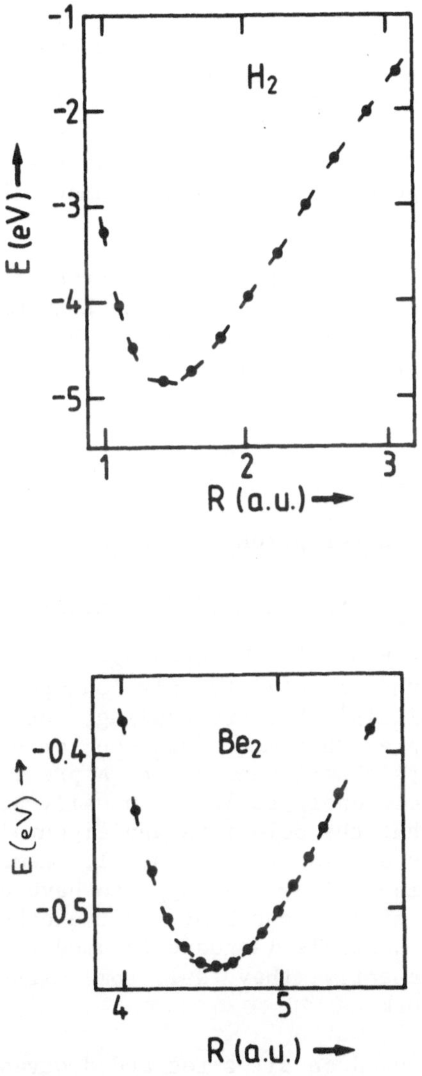

Fig. 3. Density functional binding energy curves of H_2, Be_2 molecules. The bars drawn through the points denote the internuclear force, which is calculated independently (see text).

Figs. IV and V show the variation of the energy of the water molecule as a function of bond angle at fixed OH distance, and bond distance at fixed bond angle[27]. The correspondence between energy and force is less good, though one has to bear in mind that the discrepancies in the figure would result from a mistake of only a few percent in the gradient of the electron Coulomb potential at the nuclei. Thus even the agreement evident in Fig. IV implies that the basis is quite accurate.* These results illustrate not only the adequacy of the LMTO basis, but show also that the local density approximation gives an excellent description of energy variation in the water molecule. A comparison of relevant energy parameters and 'stretch' and 'bend' vibration frequencies with experimental values is included in the figure and shows an agreement that is truly 'better than Hartree-Fock'.

I have described the LMTO scheme for performing density functional calculations in more detail than the LCAO method because it is less well known, has more overlap with band-theory, and, not least, because I use it myself ! This should not be allowed to obscure the fact that LCAO/DVM/TS is the only method that has so far been applied to 'surface' clusters. I do not propose to discuss this work in these lectures, but instead, refer to the very complete summary of results given by Post[20].

The density functional description of bonding in molecules

(i) The 'one-electron' picture : molecular orbitals and eigenvalues

In chemistry text books[28], the bonding of molecules is discussed in a language that is unmistakably 'one-particle'. Electrons occupy 'molecular orbitals' that are bonding, anti-bonding or non-bonding and have energies that move down, up, or stay put when the molecule forms from its constituent atoms. A precise definition of these orbitals and their energies is not usually given, though it is sometimes hinted that the solutions and eigenvalues of the Hartree-Fock equations are meant. In general, chemists do not worry about precise definitions. They are happy to have a few concepts that work. Physicists of rigorous bent, on the other hand, tend to be skeptical when one-particle language is used to describe many-particle systems. Inherently, they feel, some dirty and suspect approximation is at work.

---

* s- and p-waves were used on all sites and d-waves on the O-site giving in all 5 independent radial functions. Adding higher ℓ partial-waves had virtually no effect. The origin of the force discrepancies is probably the absence in the basis of a 3s-like orbital. That is, the $(\phi,\dot{\phi})$ combination describing the oxygen s-function is used over a too wide energy range. This defect can be overcome by adding to the basis a second combination with a different reference energy.

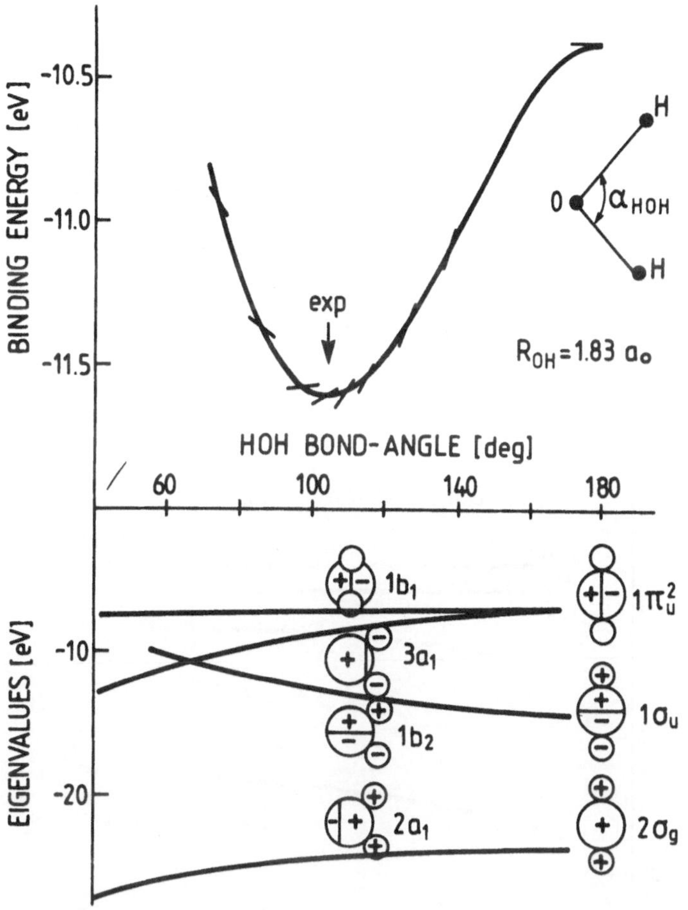

Fig. 4. Density functional energy of H_2O molecule as function of bond angle for fixed OH distance. Equilibrium angle is 104.5^0, 'bend' vibration frequency is 1530 cm^{-1}. Experimental values, 104.52^0, 1595 cm^{-1}. Lower half : Variation of eigenvalues corresponding to molecular orbitals involved in the bond.

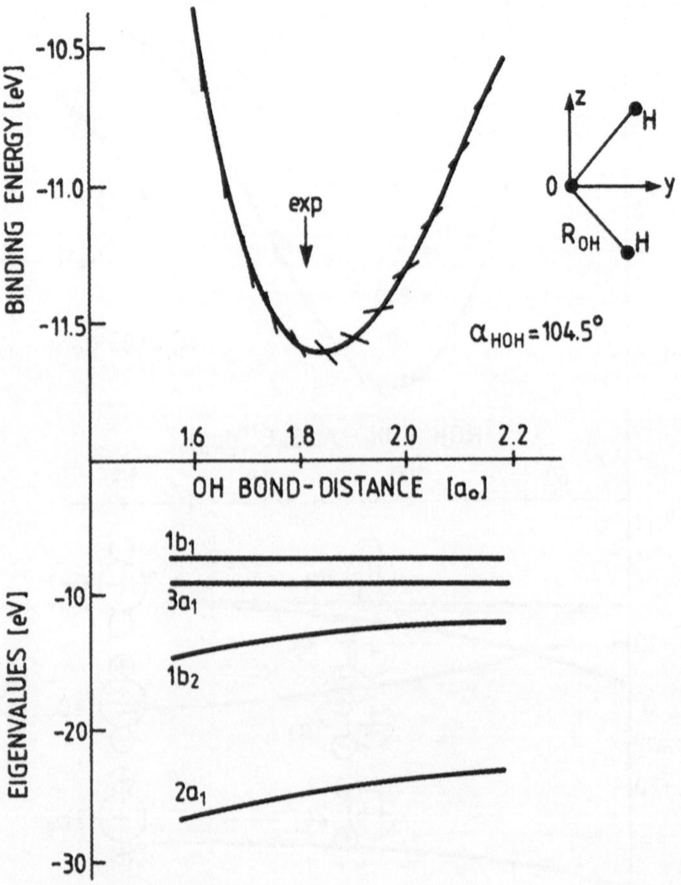

Fig. 5. H_2O energy as function of OH distance for fixed bond angle. Equilibrium value R_{OH} = 1.83 a.u. 'Stretch' vibration frequency is 3420 cm^{-1}. Experimental values 1.81 a.u., 3557 cm^{-1}.

The density functional approach of HKS provides a nice framework for discussing the concepts of molecular orbital chemistry because it implies a formal link between the physical system under investigation and an "equivalent system" of non-interacting electrons. This link can be viewed as follows. Suppose we allow the electron-electron coupling constant, e^2, in our physical system to go smoothly from its physical value of 1 to zero. We do this in infinitesimal steps, and at each step adjust an additional external field so that the electron density remains constant.* When e^2 is zero we have a system of independent electrons in an external field V^0_{eff}, so arranged that the electron density is the same as that of the physical system. From the extremum condition obeyed by the density functional we know that

$$V^0_{eff}(\vec{x}) = \phi_N(\vec{x}) + \phi_e(\vec{x}) + \mu_{xc}(\vec{x}) \tag{33}$$

where $\phi_e(\vec{x})$, $\phi_N(\vec{x})$ are respectively the exact electronic and nucleonic Coulomb potentials of the physical system, and $\mu_{xc}(\vec{x}) \equiv \delta E_{xc}/\delta n(\vec{x})$ is the exchange-correlation potential. The electron density in both physical and equivalent systems is

$$n(\vec{x}) = \sum_n a_n |\psi_n(\vec{x})|^2 \tag{34}$$

where the a_n are occupation numbers, and the $\psi_n(\vec{x})$ obey

$$\{-\frac{1}{2}\nabla^2 + V^0_{eff}(\vec{x}) - \varepsilon_n\}\, \psi_n(\vec{x}) = 0 \tag{35}$$

If we assume $\mu_{xc}(\vec{x})$ to be known exactly, the energetic behaviour of the physical system can now be discussed rigorously in terms of the behaviour of the equivalent system of independent electrons. That is, the energy of the physical system, E_1 - the minimum value of the density functional for external field ϕ_N - can be written in terms of the energy of the equivalent system

$$E_0 = \sum_n a_n\, \varepsilon_n \tag{36}$$

together with correction terms depending on the density (34) i.e.

$$\begin{aligned} E_1 &= T_0 + E_{es} + E_{xc} = E_0 - \int d\vec{x}\, n(\vec{x}) V^0_{eff}(\vec{x}) + E_{es} + E_{xc} \\ &= E_0 + \int d\vec{x}\, n(\vec{x})[\varepsilon_{xc}(\vec{x}) - \mu_{xc}(\vec{x})] + \frac{1}{2}\sum_{i,j} \frac{Z_i Z_j}{R_{ij}} - \frac{1}{2}\int d\vec{x}\, n(\vec{x})\phi_e(\vec{x}) \end{aligned} \tag{37}$$

* A formal procedure for finding this field can be given provided the density-density response function possesses an inverse. Taking a leaf out of Fermi-liquid-theory's book ,we designate as 'abnormal' systems for which an appropriate field cannot be found.

where Z_i, R_{ij} are the nuclear charges and the internuclear separations. Let us examine what happens to E_1 as two atoms combine to form a molecule.

Consider firstly the second term on the r.h.s. of Eq. (37)

$$E_1^{xc} = \int d\vec{x}\; n(\vec{x}) \{\varepsilon_{xc}(\vec{x}) - \mu_{xc}(\vec{x})\} \tag{38}$$

This is positive and increases as the atoms approach each other. To see this, consider two cases. Firstly, for independent electrons we have $\mu_{xc}(\vec{x}) = 2\varepsilon_{xc}(\vec{x}) = -\phi_e(\vec{x})$, so

$$E_1^{xc} = \frac{1}{2} \int d\vec{x}\; n(\vec{x})\phi_e(\vec{x}) \tag{39}$$

which is > 0 and cancels the final term on the r.h.s. of (37) exactly, as it must. Secondly, for a general system, we make the 'constant hole shape' approximation $\varepsilon_{xc}(\vec{x}) = -\alpha n^{1/3}(\vec{x})$. Then $\mu_{xc}(\vec{x}) = -4\alpha/3\; n^{1/3}(\vec{x})$ and so

$$E_1^{xc} = \frac{\alpha}{3} \int d\vec{x}\; n^{4/3}(\vec{x}) = -\frac{1}{3} E_{xc} \tag{40}$$

Again, $E_1^{xc} > 0$ and increases as the atomic densities overlap.

Consider next the final two terms on the r.h.s. of (37)

$$E_1^{es} = \frac{1}{2} \sum_{i \neq j} \frac{Z_i Z_j}{R_{ij}} - \frac{1}{2} \int d\vec{x}\; n(\vec{x})\phi_e(\vec{x}) \tag{41}$$

E_1^{es} is negative for large R_{ij}, positive as R_{ij} goes to zero, and will therefore increase as the atoms approach each other. This merely reflects the de-screening of the nuclei that results when the charge densities of the atoms overlap.

Thus the <u>only</u> term in (37) that can <u>decrease</u> as the atoms approach - i.e. that can lead to a substantial chemical bond energy - is E_0, the energy of the equivalent system of non-interacting electrons. Since E_0 is obtained by simply adding the 'molecular orbital energies', ε_n, of the individual electrons in the equivalent system, we have arrived at the 'molecular orbital picture'. Orbitals whose eigenvalues move down or up as the nuclei approach help or hinder bond formation. Furthermore, the behaviour of a given eigenvalue depends simply on the charge distribution of its orbital. That is, if ΔV_{eff} is the change in the effective potential on displacing the nuclei infinitesimally towards each other, then

$$\Delta\varepsilon_n \simeq \langle\psi_n|\Delta V_{eff}|\psi_n\rangle \tag{42}$$

will be negative or positive depending on whether $\psi_n(\vec{x})$ has most weight in regions where ΔV_{eff} is negative (between the nuclei) or positive (behind the nuclei). This is, of course, in accordance with standard chemistry. The only new feature we have is that the molecular orbitals and their energies relate directly to the 'equivalent system of independent electrons' rather than the physical system under study.

Though this molecular orbital picture allows a pleasing interpretation of results, it gives only limited information. In general, all we can deduce is the likely equilibrium geometry, and a rough estimate of the bond strength. This is because the eigenvalue term E_0 is only one component of E_1, the energy of the physical system. In a 'bonding configuration' - i.e. one in which there are more electrons in bonding orbitals than in anti-bonding orbitals - E_0 falls monotonically with decreasing internuclear separation. The position of the minimum is not determined by the behaviour of E_0 alone but results from a competition between this 'eigenvalue force' which draws the nuclei together, and the contribution $E_1^{xc} + E_1^{es}$ (predominantly E_1^{es}), which is essentially a 'de-screening force' of electrostatic origin that drives the nuclei part. Though the contribution of each orbital to the eigenvalue force can be assigned a unique value, the same is not true for the de-screening force, which is non-linear in the orbital occupation numbers. Roughly, one can see that a 'bonding' orbital will contribute more to the de-screening than an 'anti-bonding' orbital because the latter has a node at the bond centre. This will tend to weaken the contribution that the bonding orbital makes to the overall bond strength by an amount that depends on the detailed structure of the orbital concerned.

In general, there is no universal relation between the values of E_0 and E_1 as a chemical bond forms. However, these do tend to be roughly proportional for internuclear separations close to the equilibrium separation, R_m, and the eigenvalue sum often reproduces reasonably a binding energy trend amongst molecules having similar orbital structures. For example, first row molecules have molecular orbitals composed of atomic 2s and 2p orbitals that contract along the series, but are otherwise rather similar. Empirically, it is found that the binding energies obey[12]

$$E_B \sim \frac{1}{3}\left[E_0^{Atoms} - E_0(R_m)\right] \qquad (43)$$

indicating that the de-screening force at $R \sim R_m$ is roughly proportional to the eigenvalue force across the series. In this case too, the binding energy reflects the net number of bonding electrons, and maximizes at the centre of the series, N_2.* An approximate relation similar to (43) also holds for the series of alkali dimers Li_2, Na_2, Cs_2, but the constant of proportionality is about 1/2 rather than 1/3.

In stressing the importance of the de-screening force, which can only be calculated via a complete solution of Poisson's equation for the full three-dimensional density, I am referring to 'open' systems where the charge density is free to spread throughout space as it likes. For closed-packed bulk metal structures, local charge neutrality confines the electrons to the unit cell. When the metal is squeezed the overall electron density within the unit cell increases and there is increased penetration of the core regions. Both effects lead to an increase in valence electron kinetic energy which results in an upward movement of all the eigenvalues, or band energies contributing to E_0[29]. This confinement of the charge is the reason why interatomic equilibrium separations are larger in bulk solids than in corresponding dimers. The increase in valence k.e. causes the energy to saturate and then to increase before the nuclear 'de-screening' that results from the departure of the unit cell from a spherical shape becomes appreciable. For this reason, it is possible to obtain good values for the cohesive properties of closed-packed metals without the need for a complete solution of Poisson's equation. We should be careful, however, not to translate this experience to surface problems, or to compounds where important changes in the structure within the unit cell might take place.

(ii) An example : the water molecule

The results of a density functional, local-density calculation for H_2O are shown in Figs. IV and V, where the energy is plotted as a function of HOH angle and OH distance. I mentioned these results earlier, and noted that they amount to a quantitative description of the energetics of the molecular ground-state that is substantially "better than Hartree-Fock". Here, I focus on the interpretation in terms of molecular orbitals and eigenvalues.

The orbitals are best viewed in the linear geometry (Fig. IV, far right), where hybridization between the 2s and 2p functions of the oxygen is symmetry forbidden. The OH bonds are due to σ_g and σ_u orbitals and are predominantly combinations of hydrogen

---

* This is not generally so. In the 3d-series the trend is determined by a competition between this effect and a corresponding loss of spin correlation energy that results when bonding orbitals are doubly occupied. The latter effect is so strong that the dimer binding energy actually shows a minimum at the series centre, Cr_2, Mn_2.

1s with oxygen 2s and $2p_z$ functions, respectively.* There are four bonding electrons which give a bond strength of ~ 5 eV per bond. The remaining 4 electrons reside in the oxygen $2p_x$ and $2p_y$ orbitals and are non-bonding because the symmetry forbids interaction with H_{1s}. As the H-atoms move out of line, in say, the x-direction, one of these orbitals, $2p_y \rightarrow 1b_1$ is unaffected, while the other, $2p_x$, hybridizes with O_{2s} to form the $3a_1$ orbital that remains OH non-bonding (see Fig. V), but is HH bonding. Similarly O_{2p_x} mixes into the σ_g orbital $2a_1$, which is both OH and HH bonding. This is reflected in the downward movement of the eigenvalues as the HOH bond angle decreases. The $\sigma_u \rightarrow 1b_2$ orbital, however, is HH anti-bonding, and its eigenvalue moves upwards.** Thus, roughly speaking 4 electrons try to pull the molecule out of line, and 2 tend to restore the linear geometry. The net result is an 'equivalent system' energy, E_0, that decreases steadily with decreasing HOH angle throughout the range shown in Fig. IV. The equilibrium geometry is determined by the competition between this 'eigenvalue force' and the nuclear 'de-screening force', the result of which is a calculated bond angle of 104.5^0 that reproduces the measured value with an accuracy no less than stunning ! (Presumably, this level of accuracy is accidental).

In Figs. IV and V the eigenvalues shown are those of the final self-consistent potential. As remarked earlier these are harder to calculate than the total energy because they are not stationary. However, the general behaviour of the eigenvalues - i.e. whether they move up or down and by how much - is the same for all reasonable potentials and the above arguments could just as well be made using eigenvalues corresponding to overlapped atom potentials.*** In fact, even this is not necessary in that the orbital structure can usually be inferred from a knowledge of the atomic configurations. As an example, it is easy to guess the geometry of CH_3 directly from the orbital structure, which is analogous to the case of H_2O. We find that for fixed CH distance there are, respectively, 3 and 4 electrons in HH bonding and anti-bonding orbitals. This

---

* The pictures in the figure display the symmetry of the orbitals. They have only pedagogic value and obscure the fact that oxygen d- and hydrogen p-orbitals play an important role in the angular variation.

** Lest this seem too obviously 'standard chemistry', I remark that at least one chemistry text has the H_2O $2a_1$-orbital energy <u>increasing</u> as the bond angle decreases. There are actually two effects in play. Hybridization of the higher-lying p-orbital would tend to give an increase, the HH bonding nature a decrease. Note that for $3a_1$, both effects give a decrease.

*** Provided no substantial charge transfer takes place.

implies a planar geometry with a HCH bond angle of 120^0, as is found experimentally. For NH_3, on the other hand, there are equal number of bonding and anti-bonding orbitals so these simple arguments give no information. A preliminary DF-LD-LMTO calculation[27] for this molecule gave the correct out-of-plane geometry (with HNH bond angle within 1^0 of the measured 106.7^0) and an inversion barrier height of 0.18 eV, compared with the measured 0.25 eV. This indicates that the HH 'bonds' are slightly stronger than the anti-bonds by an amount that is given at least semi-quantitatively by the local density formula for E_{xc}.

It is interesting that many 'non-bonding' configurations actually give reasonably large bond energies. NH_3 is a case in point, the Be_2 dimer mentioned earlier is another. When we count bonds and anti-bonds to guess a structure, we must be aware of the fact that they do not all contribute equally. This is particularly true when hybridization plays an important role. For example, the 2σ-shell of the first-row molecules, composed primarily of atomic 2s functions, but with an admixture of $2p_z$, gives a reasonably large net 'bonding' contribution to E_0 throughout the series. That is, $2\sigma_g$ moves down in energy faster than $2\sigma_u$ moves up. The effect is compensated by a relatively weak downward movement of $3\sigma_g$, which, however, is unoccupied in the Be_2 dimer. This hybridization effect is the origin of the substantial binding energy found for the Be_2 dimer in DF-LD calculations[26].* By contrast, hybridization plays no role at all in the He_2 dimer because the 2p resonance is far removed in energy from the 2s energy. As a result, eigenvalues of the $2\sigma_u$ and $2\sigma_g$ molecular orbitals move upwards and downwards by equal amounts as the nuclei approach. Thus the He_2 $^1\Sigma_g$ state is truly non-bonding in the DF-LD scheme as in nature.

(iii) What factors determine the strength of a molecular bond ?

I mentioned earlier that the binding energies of the first row molecules scale roughly with the number of excess bonding electrons. This approximate correspondence holds because the orbitals that make up the bonds are similar for all members of the series. If we compare instead dimers from different series we must take into account the dependence of bond energies on the structure of the orbitals. As an illustration, I show in Fig. VIa) the trends in the spectroscopic parameters for the group 1A dimers Li_2-Cs_2, all of which have a simple σ_g-bond containing two bonding electrons. The DF-LD-LMTO results** for binding energy, E_B, equilibrium separation, R_m, and vibration frequency, ω_0, denoted by the open circles

* ~ 0.5 eV. Currently, no experimental figure is available.

** With the exception of the parameters for Li_2, which have been recalculated using the basis described above[26], these are old numbers[12] and may be in error by some few percent.

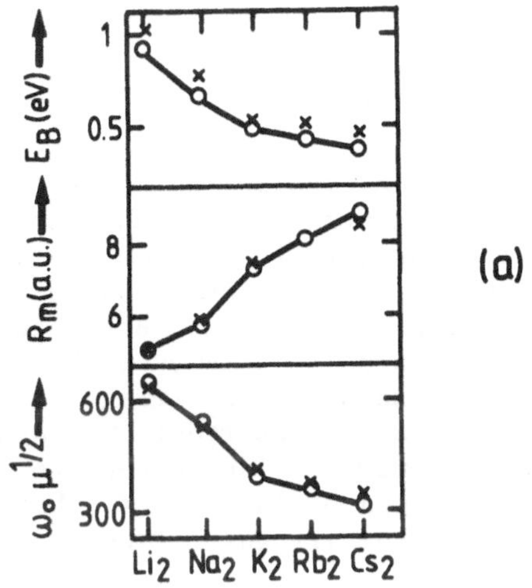

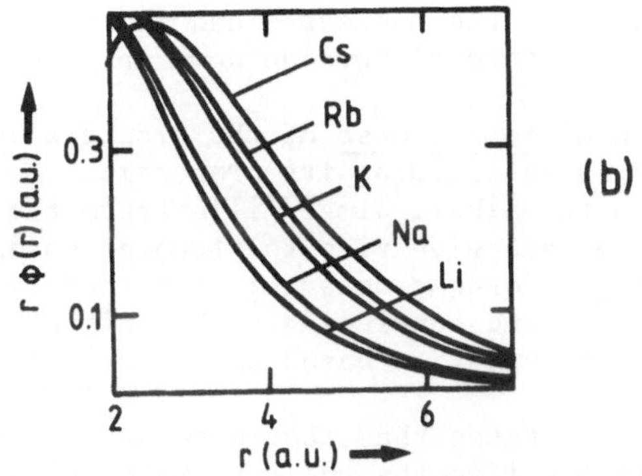

Fig. 6. a) Binding energy curve parameters of alkali dimers. E_B is binding energy, R_m equilibrium separation, W_0 vibration frequency. μ is the reduced mass. Crosses are experimental values. Open circles density functional values.

b) Tails of the valence eigenfunctions for alkali atoms.

joined by full lines, agree well with measured values, and show clearly the trend to smaller E_B, ω_0 and larger R_m down the row. To understand this trend we make use of the fact that the σ_g molecular orbital responsible for the bond is reasonably approximated by the LCAO

$$\psi_{\sigma_g}(\vec{x}) = N_A\{\phi_A(r_1) + \phi_A(r_2)\} \tag{44}$$

where N_A is a normalization constant and $\phi_A(r_1)$, $\phi_A(r_2)$ are appropriate atom s-functions* centred at the nuclear positions. The form of $r\phi_A(r)$ outside the core region for the atoms Li → Cs is shown in Fig. VIb), which illustrates the outward displacement of the atomic orbital tail as the nuclear charge increases. This displacement occurs because a node is added whenever the radial quantum number increases by one unit. The irregularity in the trend between Na and K is due to the intervention of the 2p-shell. The Na 3s-tail is displaced outward with respect to the Li 2s-tail because of imperfect screening of the nuclear charge by the 2p-shell.** This effect accounts for the irregularities in the trends in Fig. VIa).

The displacement of the tails is partially responsible for the increase in the R_m down the row. More interesting, however, is the change in '<u>compactness</u>' of the orbitals - by which is meant, broadly speaking, the decrease in orbital exponents - as the nuclear charge increases. This is due to the orthogonality requirement - the logarithmic derivative of the tails must go to zero further and further out from the nucleus - and is reflected in the upward movement of the valence eigenvalue down the row.

The degree of <u>compactness</u> of the orbitals in a bond is an important factor in determining its properties. Many molecular systems other than the alkali dimers illustrate the general trend : more compact orbitals give rise to stronger bonds, shorter bond lengths and larger force constants. It is not easy, however, to demonstrate simply and convincingly why this should be. I offer the following argumentation based on the Hellmann-Feynman theorem.

This theorem states that the force on a given nucleus i, located at $\vec{R}_i^0$, is equal to the nuclear charge times the gradient of the total electrostatic potential at $\vec{R}_i^0$

---

* In fact, it is better to use somewhat contracted orbitals, but this does not affect the argumentation here.

** Similarly, the 3d-shell draws in the Rb tail somewhat, though the effect is less marked because the 3d-atomic orbital is well localized.

$$\vec{F}_i = - Z_i \vec{\nabla} \{\phi_N(\vec{x}) + \phi_e(\vec{x})\}\Big|_{\vec{x}=\vec{R}_i^0} \tag{45}$$

The theorem follows immediately from the stationarity property for the energy and is valid for the density functional irrespective of the approximation made for E_{xc}.* Consider a homonuclear diatomic molecule with bond axis along the $\hat{z}$-direction, one nucleus being at the origin, the other at $\vec{x} = (0,0,R)$. Then it is easy to see from (45) that the force on the nucleus at the origin is in the $\hat{z}$ direction with magnitude

$$F_0(R) = Z_0 \{-\frac{Z_0}{R^2} + \int d\vec{x} \frac{z}{|\vec{x}|^3} n_R(\vec{x})\} \tag{46}$$

where $n_R(\vec{x})$ is the electron density that minimizes the energy functional with internuclear separation R. The energy curve is given in terms of the force by

$$E_B = E(R) - E(\infty) = - \int_R^\infty dR' \, F_0(R') \tag{47}$$

and has minimum value E_B^0 at $R = R_m$, where

$$F_0(R_m) = 0 \tag{48}$$

Similarly, the vibration frequency ω_0 is given by

$$\omega_0 = (k/\mu)^{1/2} \quad ; \quad k = \frac{d^2E}{dR^2}\Big|_{R=R_m} = \frac{dF_0}{dR}\Big|_{R=R_m} \tag{49}$$

with μ the reduced mass of the molecule.

The force formula (46) is useless for quantitative purposes in all but the smallest systems because it requires an exact minimization of the energy functional, including the polarization of the core[26]. I use the formula here only for illustrative purposes. Suppose that the density which minimizes the functional for a class of similar systems can be characterized by a single length λ i.e.

$$n_R(\vec{x}) = \frac{1}{\lambda^3} \bar{n}(\vec{x}/\lambda) \tag{50}$$

As an example, we might try to model the alkali dimers by making a point core approximation - which amounts to replacing Z_0 by unity - and using for the valence density the one-parameter form

---

* Different functionals, of course, are minimized by different densities and so give different forces. The relation (45) between density at minimum and force is always valid, however.

$$\overline{n}(\vec{x}/\lambda) = \frac{1}{\pi}\frac{1}{(1+S)}[e^{-|\vec{x}|/\lambda} + e^{-|\vec{x}-\vec{R}|/\lambda}]^2 \tag{51}$$

Here, S is the usual overlap integral and depends only on the dimensionless quantity λR. For each dimer we can determine λ by fitting the molecular density outside the core region to the form (51),* and can describe the displacement of the tails due to orthogonality to the core by re-interpreting $R \rightarrow R - \Delta$, where Δ is a shift that depends on radial quantum number.

Using the form (50) in (46), we find

$$F_0(R) = \frac{1}{\lambda^2} G(q) \quad , \quad q = R/\lambda \tag{52}$$

which tells us immediately that the minimum always occurs at $q = q_m$ where $G(q_m) = 0$. Thus

$$(R_m - \Delta) \propto \lambda \tag{53}$$

$$E_b^0 = -\frac{1}{\lambda}\int_{q_m}^{\infty} dq\, G(q) \propto \frac{1}{\lambda} \tag{54}$$

and

$$k = \frac{1}{\lambda^3}\frac{\partial G(q)}{\partial q}\bigg|_{q=q_m} \propto \frac{1}{\lambda^3} \tag{55}$$

We see that the correct trend is predicted for all three quantities. The more compact the orbital tail i.e. the smaller λ, the larger the binding energy and force constant, and the smaller the internuclear separation. Furthermore, combining (53) and (55) we have the relationship between force constant and equilibrium separation

$$k(R_m - \Delta)^3 = \text{constant} \tag{56}$$

This relationship, known in chemistry circles as 'Badger's Rule'[31], was originally established on empirical grounds and holds for a wide variety of molecular states. Here we see how it follows from a simple dimensional argument, which also illustrates the increased bonding energy associated with more compact orbitals.

In addition to the net number of bonding electrons and the compactness of the orbitals, the strength of a given bond depends on hybridization, and on geometric factors that have to do with

---

* Since the orbitals tend to contract as the molecule forms, the compactness parameter will depend on R. However, this dependence will be rather weak compared with changes in λ on going from dimer to dimer.

the distribution of bonding charge. For example, the de-screening of the nucleus is different for σ-bonds, which concentrate charge along the bond axis, and π-bonds (combinations of p_x and p_y orbitals), for which the bonding charge is spread out on either side of the axis.

(iv) Why carbon and silicon chemistries are different

I conclude these lectures by giving an example that illustrates how the simple picture of bonding that emerges from the density functional scheme can help us to understand chemical behaviour. Since this picture is essentially a molecular orbital description with which chemists have worked for many years, one might expect to find answers to all basic questions of chemical behaviour in standard chemistry texts. For the most part this is the case. However, there are some loopholes. For example, though the difference between carbon and silicon chemistries is very striking - an organism dependent on silicon chemistry for its functioning would be a curious one indeed - the reader will be hard pressed to find a satisfactory explanation for it in the chemistry literature. More concerned with practical questions, chemists are content to attribute the difference to the relative abilities of carbon and silicon to form "unsaturated structures" or "pi-bonds", and seem uninterested in probing the matter further. In fact, the above discussion of the relevant factors that determine bond strengths provides a rather simple explanation for the tendency of carbon to form unsaturated structures where silicon does not, as I now demonstrate.*

Fig. VII shows the behaviour of the molecular orbital energies of the C_2 and Si_2 dimers[33]. The orbitals have similar structures and are made up of s-p_z hybrid -orbitals and linear combinations of p_x or p_y atom functions that give rise to degenerate π-orbitals having nodes along the bond axis. 8 electrons must be accommodated, four of which occupy the lower lying σ_g,σ_u orbitals. This leaves four electrons to be shared amongst $2\sigma_g$ and π_u, which can accommodate in all 6 electrons. Both for C_2 and Si_2, the $2\sigma_g$,π_u eigenvalues lie close together. This means that the "equivalent system energies", E_0, of several possible configurations lie close together, and so we may expect the physical energies E_1 to do the same. Consider two cases. Firstly, we fill the π_u shell and leave $2\sigma_g$ empty, obtaining a ${}^1\Sigma_g^+$ molecular state with, roughly speaking, 4 bonding-π electrons. Secondly, we fill the $2\sigma_g$ level and half-fill the π_u, obtaining a triplet ${}^3\Sigma_g^-$ molecule state** having 2

* A somewhat condensed version of this argumentation was buried in Ref. 32.

** The symmetry of the state can be deduced from that of the Slater determinant constructed from the orbitals and is of no concern here.

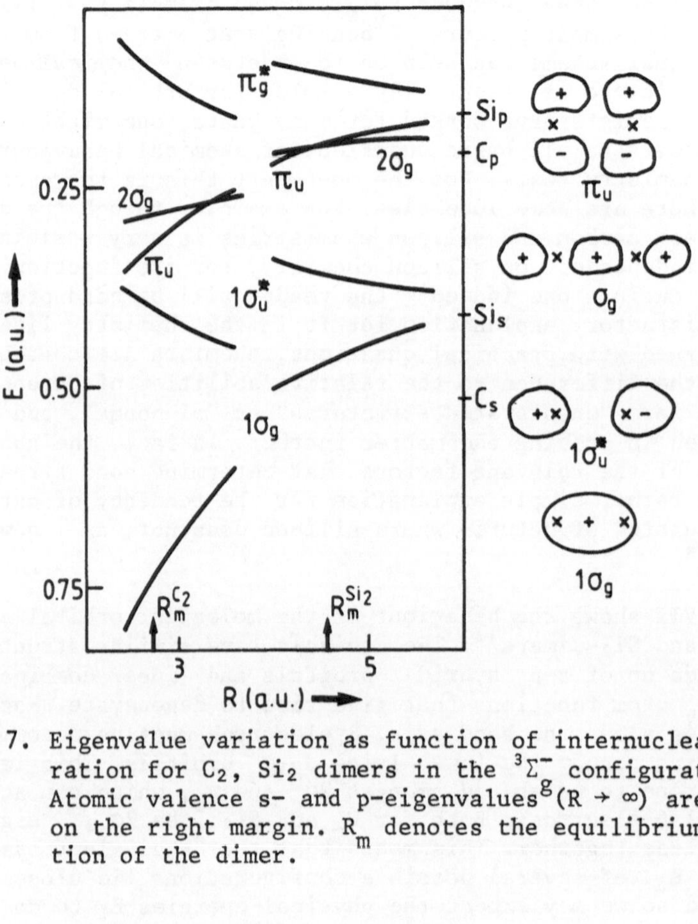

Fig. 7. Eigenvalue variation as function of internuclear separation for C_2, Si_2 dimers in the $^3\Sigma_g^-$ configuration. Atomic valence s- and p-eigenvalues[8] $(R \to \infty)$ are marked on the right margin. R_m denotes the equilibrium separation of the dimer.

bonding-σ and 2 bonding-π electrons. I refer to these configurations as $\pi\pi$ and $\pi\sigma$, respectively. Explicit calculations of the physical energy E_1 are necessary to decide which of these lie lowest,* and showed a striking difference between C_2 and the remaining group-IVA dimers. In C_2, the $\pi\pi$ configuration has a slightly lower energy than $\pi\sigma$, whilst in the remaining dimers $\pi\sigma$ lies considerably lower than $\pi\pi$, in Si_2, by 1.5 eV, which is half the binding energy of the $\pi\sigma$ state. This merely reflects what the chemist knows - that Si π-bonds are weak - and strongly suggests the main feature of silicon chemistry i.e. stable structures tend to be σ-bonded[34].

To see how the relative energy of the $\pi\pi$ and $\pi\sigma$ bonds influences the formation of stable unsaturated molecules consider what happens when we try to make the highly unsaturated molecules C_2H_2, Si_2H_2 by bringing up two hydrogen atoms on either side of the dimer. Only the σ-molecular orbitals are affected. All three combine with H_{1s}, bond the H atoms to the dimer and move downward in energy with the π_u orbital essentially unaffected. However, there are now 2 more electrons, so that the π_u shell must be completed. In C_2, the resulting configuration not only bonds the H-atoms to the C-atoms, but also increases the strength of the C-C bond. To see this, start from the $\pi\pi$ C_2 ground state for which the $2\sigma_g$ orbital is empty. This merely fills when the hydrogen atoms approach. Since the π-bonds remain unaffected and the combination of σ-orbitals is CH and CC bonding, the result must be an increased bonding of the carbon atoms to each other.** What we have uncovered, in fact is the infamous carbon $\pi\pi\sigma$ "triple-bond".

We may view the formation of Si_2H_2 similarly, but must take into account an energy cost of $\sim$ 1.5 eV in 'preparing' the Si_2 dimer in the $\pi\pi$ configuration. Thus, although the $2\sigma_g$ bonding charge that results when the H-atoms are incorporated increases the Si-Si bond strength slightly with respect to its $\pi\pi$ value, the resulting 'triple' bond is still weaker than the $\pi\sigma$ double-bond of the Si_2 ground state.

The reason for the relative weakness of the Si π-bonds is clear from Fig. VIII, where the orbital s- and p-tails, $r\phi(r)$, are plotted for the carbon and silicon atoms. Because the carbon atom has no p-core, the 2p orbital is exceptionally compact. In

---

* In fact, the ordering of the self-consistent eigenvalues depends on the configuration. Those shown in Fig. VII refer to the $\pi\sigma$ configuration.

** This increase, which is rather small because the σ-charge redistributes to maintain rough local charge neutrality, is reflected in the somewhat smaller bond length in C_2H_2 as compared with C_2.

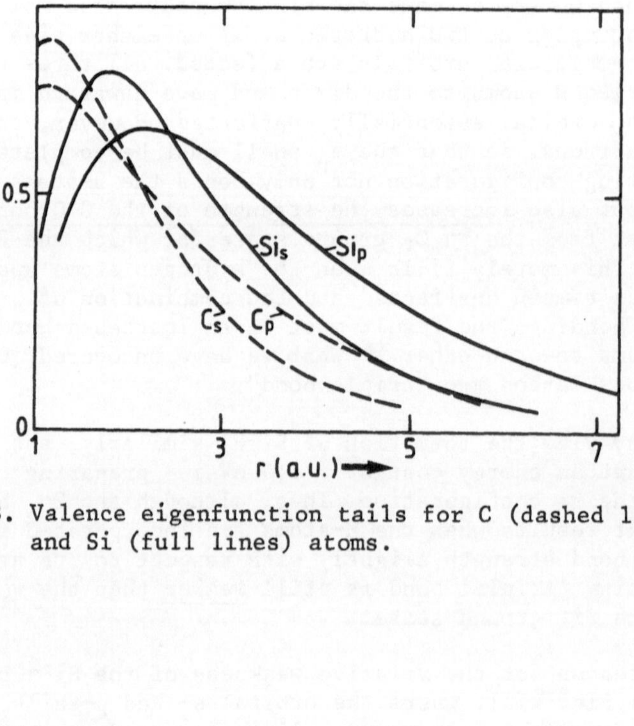

Fig. 8. Valence eigenfunction tails for C (dashed lines) and Si (full lines) atoms.

fact, both 2s- and 2p-carbon orbitals have anti-nodes at the same distance from the nucleus. In the dimer, the π-bonds are constructed from p-orbitals that are almost as compact as the corresponding s-orbitals so that transfer to or from the π-shell does not involve a substantial change in bond strength.* In fact, the dimer $^3\Pi_u$ state, which corresponds to a "$\frac{3}{2}\pi\ \frac{1}{2}\sigma$ bond", has essentially the same energy as ππ or πσ.**

For Si, both the 3s- and 3p-orbitals are less compact than the 2s, 2p carbon functions. This is reflected in the behaviour of the dimer eigenvalues (Fig. VII), and accounts for the lower binding energies of all Si_2 states relative to their C_2 counterparts. More important in the present context is the relative compactness of silicon 3s- and 3p-functions. The 3p-function is substantially more extended than 3s and has its final anti-node further from the nucleus because the 2p-core is more extended than 2s. Invoking the link between compactness of the orbitals and bond strength, we see that those bonds will be strongest in Si-Si linkages that utilize the s-functions maximally. Thus, if we reduce matters to rock bottom, life as we know it is due to the accident that the lowest lying solution of the radial equation for the carbon atom having $\ell = 1$, and the first excited $\ell = 0$ function peak at the same point !

REFERENCES

1. R.O. Jones, J. Chem.Phys. 71, 1300 (1979).
2. See, for example, W. Kohn and P.D. Vashishta, 'General Density Functional Theory', in : "Theory of the Inhomogeneous Electron Gas", S. Lundqvist and N.H. March, eds, Plenum Press, New York (1982).
3. J.K. Nørskov, D.M. Newns and B.I. Lundqvist, Surf.Sci. 80, 179 (1979).
4. P.O. Gartland, Surf.Sci. 62, 193 (1977).
5. See, for example, O. Gunnarsson and H. Hjelmberg, Physica Scripta 11, 97 (1975).
6. For a review, see 'Methods of Electronic Structure Theory', H.F. Schaefer III, ed., Plenum Press, New York (1977).
7. L.H. Thomas, Proc.Com.Phil.Soc. 23, 542 (1927) ; E. Fermi, Z. Physik 48, 73 (1928).

---

* That the p-orbital is somewhat less compact than the s-orbital is compensated for by a reduction in nuclear de-screening for π-orbitals. This reflected in the equilibrium separations of the dimer states. R_m is slightly smaller for ππ than for πσ.

** Experimentally, ππ is the ground state, $\frac{3}{2}\pi\ \frac{1}{2}\sigma$ lies ~ 0.1 eV and πσ ~ 0.7 eV higher[35].

8. P. Hohenberg and W. Kohn, Phys.Rev. 136, B864 (1964) ; W. Kohn and L.J. Sham, Phys.Rev. 140, A1133 (1965).
9. J. Harris, Int.J.Quantum Chem. 13, 189 (1979).
10. J. Harris and R.O. Jones, J.Phys.F 4, 1170 (1974).
11. O. Gunnarsson and P. Johansson, Int.J. Quantum Chem. 10, 307 (1976).
12. O. Gunnarsson, J. Harris and R.O. Jones, J.Chem.Phys. 67, 3970 (1977) ; 68, 1190 (1978) ; 70, 830 (1979).
13. B.I. Dunlap, J.W.D. Connolly and J.R. Sabin, J.Chem.Phys. 71, 3396 (1979).
14. F.J. Baerends and P. Ros, Int.J.Quantum Chem. 125, 169 (1978).
15. V.L. Moruzzi, J.F. Janak and A.R. Williams, 'Calculated Electronic Properties of Metals', Pergamon (1978).
16. O. Gunnarsson and B.I. Lundqvist, Phys.Rev. B13, 4274 (1976).
17. O. Gunnarsson, B.I. Lundqvist and J.W. Wilkins, Phys.Rev. B10, 1319 (1974).
18. For details, see, E.J. Baerends, D.E. Ellis and P. Ros, Chem. Phys. 2, 41 (1973).
19. T. Ziegler and A. Rauk, Theoret. Chim. Acta 46, 1 (1977).
20. D. Post, Doctoral Thesis, University of Amsterdam (1981).
21. K.H. Johnson, Adv. Quantum Chem. 7, 143 (1973).
22. J.B. Danese, J.Chem.Phys. 61, 3071 (1974).
23. J.W.D. Connolly and J.R. Sabin, J.Chem.Phys. 56, 5529 (1972).
24. O.K. Andersen, Phys.Rev. B12, 3060 (1975).
25. J. Harris and G.S. Painter, Phys.Rev. B22, 2614 (1980).
26. J. Harris, R.O. Jones and J.E. Müller, J.Chem.Phys. 75, 3904 (1981).
27. J.E. Müller, R.O. Jones and J. Harris, to be published.
28. e.g. J.C. Slater, "Quantum Theory of Molecules and Solids", Vol. 1, McGraw Hill, New York (1963) ; G. Herzberg, "Electronic Spectra of Polyatomic Molecules", Van Nostrand Reinhold, New York (1966).
29. D.G. Pettifor, J.Phys.F. 7, 613 (1977).
30. J.D. Swalen and J.A. Ibers, J.Chem.Phys. 36, 1914 (1962).
31. R.M. Badger, J.Chem.Phys. 2, 128 (1934).
32. J. Harris and R.O. Jones, Phys.Rev. A19, 1813 (1979).
33. J. Harris and R.O. Jones, Phys.Rev. A18, 2159 (1979).
34. See, e.g. 'Silicon Chemistry I,II', Topics in Current Chemistry Series, Vols. 50, 51, Springer, Berlin (1974).
35. E.A. Ballik and D.A. Ramsay, Astrophys. J. 137, 61 (1963).

THE BAND MODEL FOR d- AND f-METALS*

D.D. Koelling

Argonne National Laboratory
Argonne, Illinois 60439
USA

ABSTRACT

The application of band theory to metallic systems with d- and f-orbitals in the valence and conduction bands will be discussed. Because such an application pushes theory and technique to their limits, several important features will be briefly recapitulated. Within the transition metal systems, the elemental systems will be used to discuss the fundamental formalism being applied and the newer directions into more complex systems will be mentioned. Here we focus more on anisotropic properties and Fermi surface properties. Within the f-orbital systems, the focus will be more on Ce and its compounds because of current interest with a relatively brief discussion of the actinides. The point of view advanced, however, has its orgins in actinide research.

I. INTRODUCTION

Materials involving d- and f-electrons in the conduction bands present a problem of a very different character than the so-called simple (i.e., s-p only) metals because the d- and f-states retain a great deal of site specific atomic character. The remaining s-p like electrons in the system exhibit nearly free electron (NFE) behavior modulated by their orthogonalization and hybridization with the d- and f-states. The most difficult and dominating feature of the d- and f-band systems is to get the relative energies of these different states right-whereas one is usually concerned with more subtle and precise features in the simple metals. Further,

* Work supported by the U.S. Department of Energy.

the physics community feels it is more necessary to reexamine the applicability of a band theory in the case of the d- and f-systems than the simple metals. In actuality one can easily reach the bounds of band theory in both types of systems. They are merely tighter in the case of d- and f-systems. Because this is so, a quick overview of the structure of the theory will be given in Section II to point up some of the features particularly relevant to d- and f-electron systems.

The application of a band theory requires some computational effort. The techniques used are much like the engineering details or equipment and methods of an experimental study. Although they may not be viewed as an essential part of the physics, they do define the credibility of the results and are generally the limiting factor on the complexity of the problems that can be dealt with. In Section III, I will state some of my philosophy and discuss some of the more interesting problems at hand. Again, these issues are more significant for the d- and f-systems because they are quite sensitive to small changes-which is actually one of their most interesting features !

The remaining two sections will deal with specific examples. A number of d- and f-electron systems will be discussed, both in these lectures and those given in companion lectures. The quantity of effort is just too great to hope to give a complete review. Thus I can take the easier task and present a personal point of view. I will choose examples from those with which I am more familiar and choose my subset of topics from those that have most interested and impressed me. No attempt at completeness will be made and omission should definitely not be taken as a value judgement.

II. THEORETICAL OVERTURE

Band theory has some of the character like the definition once given for physics : "Physics is what a physicist does". Similarly band theory is what a band theorist does in the sense that it has several possible formal bases, and being an approximate theory, has different possible boundaries depending on which of those approximations one feels to be an integral part of "band theory", Koelling (1981).

One possible way to come to a band theory is as exercise in density functional theory and that is the view we will take here. A more formal and complete treatment of density functional theory will be presented at this school but it is useful to briefly reiterate and name the various theorem, approximations, and ansatzs of band theory because some of the problems we will be considering push the theory to its limit of applicability (and beyond). No effort will be made to be complete in this exposition of the theoretical structure--it is intended to highlight the features which will later prove to be of concern.

Density functional theory starts from a study of Thomas-Fermi theory by Hohenberg and Kohn (1964) which resulted in two fundamental theorems :

DF1 : The nondegenerate ground state of N particles moving in a local external potential $V_{ext}(\underline{r})$ and their mutual Coulomb interaction is a unique functional of the density $\rho(\underline{r})$.

DF2 : The total energy is a minimal property so long as the number of particles is conserved.

DF2 is merely a restatement of the ground state property but it is extremely useful. The statement of DF1 has truly practical implications as well as formal ones. It means that density mixing schemes to damp the oscillations of a self-consistent calculation will usually work better than those mixing potentials. It also means that the stable way to calculate properties including the total energy is to start with the density, calculate the potential and eigenstates for that density, and use this set of quantities for further calculations. To use the potential and eigenvalues from the previous iteration is equivalent to making the potential the fundamental variable. This will require much tighter tolerances.

The requirements of DF1 that the ground state be studied and that the external potential be local both have serious implications. Most experiments take the system away from a ground state so that one must either assume it is "sufficiently near the ground state" or discuss how the excited state differs. In general, it is necessary to go outside density functional theory to do this. However, some excited state problems can be fit into density functional theory. Mermin (1965) has demonstrated the result for a thermal population and Theophilou (1979) has shown that a similar correspondence can be made between a subspace spanned by the lowest energy eigenstates and their densities. These are restricted but important extensions not yet explored.

This ground state limitation is an especially important point to remember as photoemission continues to rise in popularity. In many cases, useful information can be gotten from comparing to the ground state results but it can also be misleading. A particularly amusing situation exists in the case of Pd and Pt. There angular resolved photoemission compares very well with the results obtained using the overlapping charge density (OCD) model and the full Slater exchange. The use of the stronger exchange is well known to mimic crystal relaxation effects for the Fermi surface so that the Fermi surface is represented quite well. But the d-bands are narrowed compared to the results of an SCF calculation with a more proper exchange-correlation function. This is consistent with the earlier atomic results where full Slater exchange was found to give better transition energies than the improved functionals because it simulated some the relaxation effects. Thus the good agreement is instructive only if one examines it very carefully. The requirement of a local external potential creates difficulties when one wishes to apply a magnetic field which interacts with the velocity. Formal work involving currents has been done but with limited application Rajagopal and Callaway (1973) and Rajagopal (1978). Work involving magnetic fields has circumvented this problem by postulating a fictitious magnetic field which interacts only with the spin. That, of course, requires a spin density functional formalism. Von Barth and Hedin (1972) have demonstrated such a spin density functional theorem.

DF3 : The ground state and its properties are a unique functional of the density matrix $\rho_{\alpha\beta}(\underline{r})$.

The same restrictions apply as for DF1 except that V_{ext} can depend on spin. If one includes spin orbit coupling, there are problems using the fictitious field mentioned above as well as DF3 itself except in special cases.

The above results are the fundamental theorems which define the range of density functional theory. They are, however, existence proofs which do not prescribe the actual form of the functionals. Only their approximate forms are known. One writes the total energy as

$$E[\rho] = T[\rho] + U[\rho] + \int d^3\underline{r}\ \rho(\underline{r})V_{ext}(r) + E_{xc}[\rho] \tag{1}$$

where T is the kinetic energy

$$U[\rho] = \frac{e^2}{2}\int d^3\underline{r}\ d^3\underline{r}'\ \rho(\underline{r})\rho(\underline{r}')/|\underline{r} - \underline{r}'| \tag{2}$$

is the Coulomb repulsion energy, and E_{xc} is the correction for the exchange and correlation effects. V_{ext} is most often the nuclear Coulomb interaction and we have omitted the nuclear-nuclear term. For simplicity, we will not carry a spin index forward except where absolutely necessary.

The major concern in actually applying density functional theory is how to represent $E_{xc}[\rho]$. The most common approximation is the local density approximation.

$$\underline{\text{LDA}} : E_{xc}[\rho] \sim \int d^3\underline{r}\ \rho(\underline{r})\ \varepsilon_{xc}(\rho) \tag{3}$$

where ε_{xc} is the specific exchange energy to be taken from electron gas theories. One can expect LDA to work fairly well for exchange based on Slater's (1951) original arguments for a statistical exchange and the general success of exchange hole type concepts.

The Optimized Effective Potential formalism of Talman and Shadwick

(1976) also demonstrates the strengths (and shortcomings at large and small r) of the LDA. The situation is worse for correlation. (Simple correlation hole based schemes usually encounter problems). One can argue that we have already seen the limitations of LDA and we will do so below. LDA has the considerable advantage that, when the variational solution is sought, it results in a local potential. It is not the only approximation to result in a local potential. An interesting and useful alternative is to introduce the single particle potential as a parametric variable (Talman and Shadwick 1976) to perform the variation. This alternative has some exciting possibilities but is, as yet, largely unexplored except for Hartree-Fock in atoms. Principle efforts in studying LDA have focussed on the best representation of the homogeneous gas results (see for example, Vosko et al. 1980). With the generalization of density functional theory to include excited states one must also consider the modifications of E_{xc} appropriate to the excited states Stoddart and Davis (1982). The situation is slightly complicated for generalizations to a local spin density by the non-existence of the electron gas of arbitrary polarization. The spin density dependence is normally described by interpolation between the paramagnetic and fully polarized ferromagnetic electron gas (Gunnarsson et al. (1974) and Gunnarsson and Lundqvist (1976)). Nonetheless, density functional calculations based on the local spin density approximation (LSDA) correctly yield the result that only Fe, Co, and Ni have ferromagnetic ground states and Cr an antiferromagnetic ground state amongst the elements of the 3d and 4d series. It also produces very good results for the saturation magnetization of those elements. Currently, it seems the spin dependent part of LSDA is very sound and it is only the charge dependent part which exhibits the shortcomings.

Thus far, we have not come to dealing with a single particle theory. That is done by making the single particle ansatz (Kohn and Sham 1965).

SP1 : For an N particle system, $\rho = \sum_{n=1}^{N} |\psi_n|^2$

The content of this ansatz is the truncation of the sum at N. One can always expand in a complete set of functions and take a square root. Note that this is precisely the form of the density in a Hartree or Hartree-Fock approximation. In the case of a natural orbital expansion, one would have more than N terms multiplied by a weight if one insists on ψ_n being normalized. Natural orbital expansions produce expansions where well over 90% of the density is provided by the first N terms but with a very long tail. By making ansatz SP1, one insists on a set of single particle like entities with the correlation effects folded into the effective operators. Based on SP1, one can reorganize Eq. 1 in a very useful

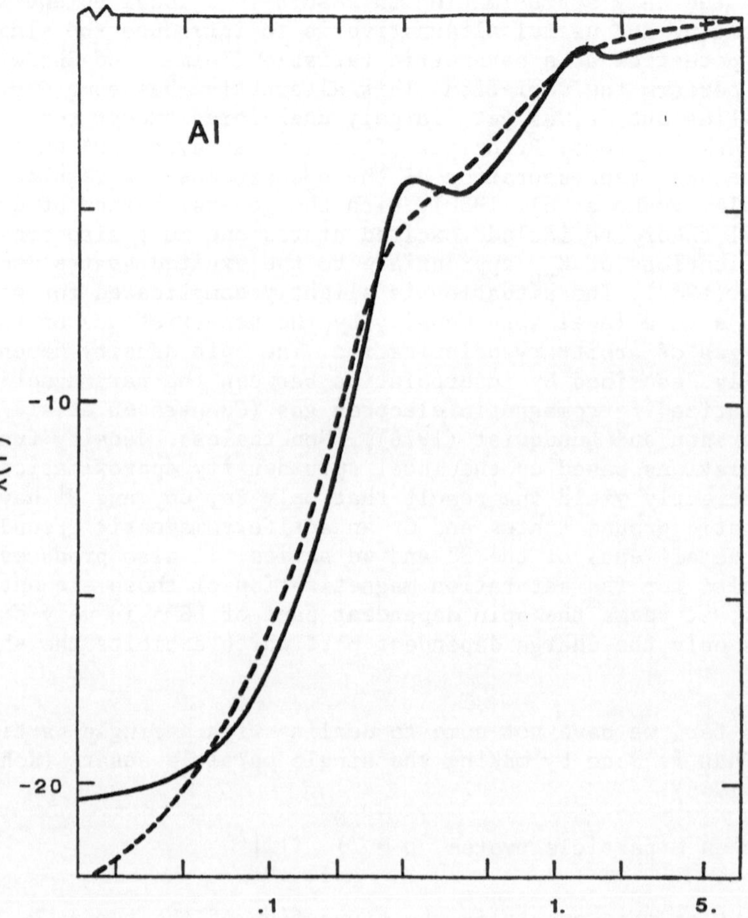

Fig. 1. Comparison of $X_\alpha(\alpha = 0.7)$ potential (dashed line) to Optimized Effective Potential (Talman and Shadwich 1976).

way :

$$\underline{SP2} : T(\rho) \to T_s(\rho) = \sum_{n=1}^{N} \int d^3\underline{r}\psi_n^{\star} t\psi_n$$

$$E_{xc}(\rho) \to E'_{xc}(\rho) = E_{xc}(\rho) + T(\rho) - T_s(\rho)$$

$$t = p^2/2m \qquad \text{(non-relativistic kinematics)}$$

$$= c\underline{\alpha}.\underline{p} \qquad \text{(relativistic kinematics)}$$

That the single particle kinetic energy might not be valid representation of the full kinetic energy can be seen from a Jastrow function representation of the ground state wave function

$$\Psi(\{r_j\}) = J(\{r_j\})\ D(\{\phi\}) \tag{2a}$$

$$J(\{r_j\}) = \exp \sum_{i<j} u(|\underline{r}_i - \underline{r}_j|) \tag{2b}$$

where D represents a Slater determinant. If one assumes that orthonormality of the ϕ's can be recovered with the J^2 kernel, the presence of J will still contribute to T in a way not structured as T_s in SP2. This, however, is a consideration only for correlation-exchange is readily included in the form of SP1 and SP2.

Given SP1 and SP2, the minimization of $E(\rho)$ produces the single particle equations

$$h\ \psi_n = \varepsilon_n\psi_n \tag{3a}$$

$$h \equiv t + v_{eff}(r) \tag{3b}$$

$$v_{eff} = v_{ext} + e^2\int d^3\underline{r}' \frac{\rho(r')}{|r - r'|} + \frac{\delta E_{xc}}{\delta\rho} \tag{3c}$$

and in the case of LDA, one has

$$\frac{\delta E_{xc}(\underline{r})}{\delta\rho} = \varepsilon_{xc}(\underline{r}) + \rho(\underline{r})\ \frac{\delta\varepsilon_{xc}(\underline{r})}{\delta\rho} \tag{4}$$

But now we come to a crucial question. There are more than N solutions of (3). Which of those solutions should be chosen ? The natural (and standard) answer to that question is to select the N lowest eigenvalues. Because one has the relation (Janak 1978)

$$\varepsilon_j = \frac{\delta E(\rho)}{\delta n_j} \tag{5}$$

such a choice will guarantee a local minimum and we can state our third single particle ansatz.

SP3 : The N solutions to be selected for the construction of ρ are those N with the lowest eigenvalues in Eq. 3.

This will guarantee a local minimum where it is successful but not a global minimum. Further, it is an additional and separate ansatz to SP1. SP1 only requires that one choose the N ψ_n which produce a minimum total energy. It is SP3 which requires partially filled d- or f-band to lie at the Fermi surface. To illustrate these ideas, let us consider a fictious system with two ψ_n near the Fermi energy with nearly equal eigenvalue. The first, begin spatially diffuse, has an eigenvalue independent of its occupation. The second, being localized in some region of space, depends roughly linearly on its occupation when the functionals are analytically continued away from the physically accessible values of zero and one. So we have

$$e_1 = \text{const.}$$

$$e_2 = e_2 + e'_{2e} n_2$$

Note that the form of e_2 is appropriate to a very simple Hubbard model. We assume $e_2 < e_1$ and compare the total energies

$$E(n_1 = 1, n_2 = 0) \equiv E_1 \text{ and}$$

$$E_2(n_2)\big|_{n_2=1} = E_1 + \int_0 dn[e_2 + e'_2 n - e_1]$$

$$= E_1 + (e_2 - e_1) + \frac{1}{2} e'_{2\,0}$$

Now $E_2 < E_1$ only if $e'_2 < 2(e_1 - e_2)$. If not, the use of SP3 would yield and infinite oscillation. When ε_2 is empty, it is the lower eigenvalue but when it is filled, it is the higher eigenvalue and should be depopulated. On the other hand, the situation is clear when considering the total energies but would result in an unoccupied eigenvalue below an occupied eigenvalue. We recently looked at such a situation for the Ce atom where the linear variation of the eigenvalues worked out quite well.

Finally, we also include Block's theorem in our list :

BLOCH : $T\psi_n(\underline{r}) \equiv \psi_n(\underline{r} + \underline{T}) = e^{i\underline{k}.\underline{T}}\, \psi_n(\underline{r})$

where T is a primitive translation of the lattice.

This is the direct consequence of v_{eff} and thus h having the symmetry of the lattice in Eq. (3). In LDA, this is guaranteed.

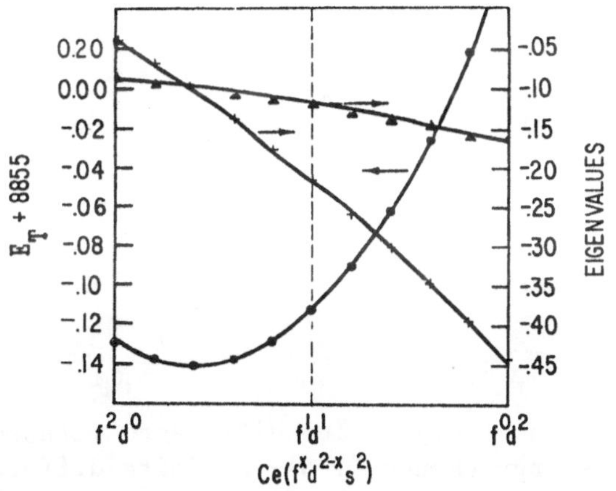

Fig. 2a. Total energy (left hand scale) together with 4f and 5d eigenvalues (righd hand scale) for $Ce(f^{x}d^{2-x}s^{2})$. Note that the minimum is total energy occurs precisely at the point where these eigenvalues cross.

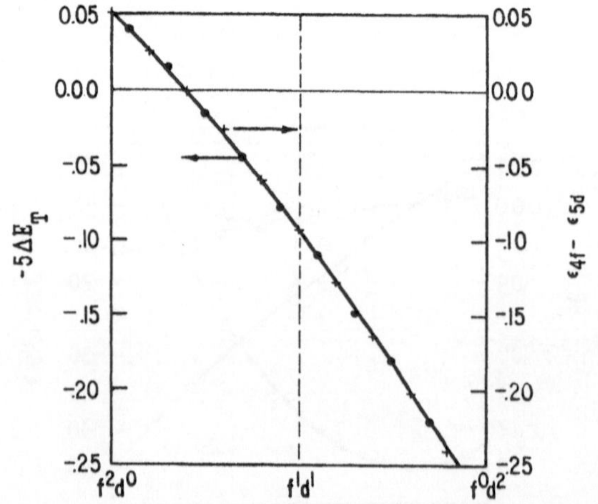

Fig. 2b. Comparison of eigenvalue difference (crosses) to the total energy change (1/5)--a finite difference evaluation of dE/dn plotted at the midpoint between the two configurations. Note that this is very nearly a straight line giving good support to the use of transition state calculations. Note also that the f^2 configuration has the same energy as the f^1 being slightly lower. This is consistent with the assertion that the local density unduly favors the higher angular momentum. But in addition, one sees that screening has essentially eliminated the penalty of double occupation in contradition to many models of the system. After Koelling (1981)--figure is erroneously reversed there.

Interestingly, Hartree-Fock also satisfy this but most formulations of the self interaction correction to not and would not have a BLOCH theorem. I personally am not certain whether it is possible to produce a functional which would break Bloch's theorem for the ground state when correlations are included. It is of course possible for excited states but then one is at the very limit of density functional theory if not beyond.

The Bloch theorem is an extremely useful result. Thus, when faced with a loss of the translational symmetry, it is often reintroduced by supercell techniques. A defect will be replaced by an ordered array of defects or an incommensurate Q-vector replaced by a nearby commensurate value, for example. In that way, the number of atoms per unit cell is greatly increased but one retains the partitioning of states by their translational symmetry index $\underline{k}$. This greatly enhances the reliability of the calculational results so it is a significant consideration whether the artificially introduced symmetry induces unacceptable effects or not. There are, in addition, a large number of systems where the repeating cell becomes very large : (1) complex solids such as the Chevrel phases, ternary tetraborides, low temperature actinides, La_3P_4 structure, UPd_3, Perovskite structures of the $GdFeO_3$ type, etc. to name only a very few. (2) Layered Ultrathin Coherent Structures (LUCS) which are artificially produced composites of immissible materials. These are to be distinguished from the composite modulated structures (CMS) which are actually non-random alloys with a modulated concentration dependence where alloy theories must be applied. (3) Frozen phonon calculations for those selected phonons yielding a simple distortion of the lattice. While this subset of phonons is a relatively small one, it is sufficiently large on the calculational results sufficiently precise to serve as a good basis for further studies on the nature of the electron phonon interaction. They have, for example, been able to lend considerable credence to the importance of phonon induced charge transfer of a disproportionation type as a mechanism for phonon softening. (4) Two dimensional systems such as the thin film or slab geometry used to study surfaces. The best procedure for studying surfaces is to study them two at a time by performing calculations for thin films which are "sufficiently thick that the two surfaces do not interact". Fortunately, for most properties the healing distance is only one or two atomic layers so the film need not be very thick. Of course, one has translational symmetry in the remaining two directions giving rise to a two dimensional Brillouin zone. In all these applications, one must deal with large unit cells but there is still a qualitative difference between having Brillouin zone and not having one. There are other viable alternatives when Bloch symmetry is lost but they are beyond the scope to be considered here.

III. EXPERIMENTAL METHODS

Thus far, we have dealt with the formal tools. But a physicist must deal in brass and beeswax as well. The experimental apparatus that must be dealt with here is the computer with its finite precision and great, but not infinite, computational power. The combination of increasing computer power and improved algorithms has greatly extended the range of feasible problems over the last 25 years and that process has only just begun. The advances being made by computer designers and numerical linear algebraiists can be exploited such that the physicist should focus on the physics and not on the generation of a fast running code. Better algorithms are still an important consideration and we will give a few examples. Criterion like precision and stability are much more significant than speed except when at the very limits of feasibility. Approximations should be made to gain insight more than computational efficiency. Nonetheless, as one deals with even more complex problems and seeks greater precision, one requires an ever increasing bag of computational tools. These are often the result of a clearer understanding of what one is trying to accomplish. I would like to give a couple examples of technological improvements viewed in that light.

A. Linearization of the Traditionally Muffin Tin Methods

The solution of the eigenvalue problem is most frequently attempted using a variational approach. The variational quantity may not always be the same from method to method but it will be similar in nature. This being so, there are basically three elements to any method. Definition of the variational quantity, definition of basis functions, and mechanical details of execution. It is the definition of the basis set which is the driving factor in the process so one can base a qualitative discussion on that feature. That we will do here.

The technique for solving the secular equations of the traditional augmented plane wave (APW) and Green's function (Koringa-Kohn-Rostocker or KKR) methods has been to evaluate the determinant as a grid of points and find its zeroes. The reason for this is that one was dealing with $[H(E)-ES(E)]$ or similar quantity where H and S depend on the trial energy. The origin of this dependence is, of course, that in the traditional methods one always used the radial solutions inside the muffin tin for the trial energy E in the definition of the basis set. Thus one had a non-linear variational parameters E_{rad} in addition to the linear parameters which were the coefficients of the APW or KKR basis functions. Some progress can be made just by recognizing this variational character of E_{rad} and generating a solution scheme which involves alternate diagonalizations to set the linear coefficients and bilinear products to vary E_{rad} (Harmon and Koelling 1974). That technique offers some definite advantages over determinant plotting. But the real advantage of exploring that scheme is the insight one can get from it. First, one sees for the close packed structures that the E_{rad} variation depends only very mildly on the energy used to set up the linear coefficients. This is a very strong support for the single tail energy approximation used in getting the linearized muffin tin orbital (LMTO) (Andersen 1975) and augmented spherical wave (ASW) (Williams et al 1979) methods in their simplest forms. Second, one sees that the E_{rad} dependence of s-p like states is very mild whereas those of d- of f-states is quite pronounced. This is because the radial solution varies dramatically between the principal maximum and the muffin tin sphere radius as one goes from the bottom of the band (where $u_\ell' \sim 0$ at the boundary) to the top (where $u_\ell \sim 0$ at the boundary). Inside the principal maximum, the shape of the radial solution is very much determined by the orthogonalization conditions focussed on by the OPW and pseudopotential methods. In fact, were the muffin tin radius to be shrunk to the principal maximum of the d- or f-orbital, the E_{rad} dependence would be negligible--and the non-muffin tin effects would all be in the plane wave representation. Unfortunately, the basis function convergence would be quite slow.

The variation in radial function, while extreme, is simple in structure. This is why it can be successfully represented in a linear combination of atomic orbital (LCAO) scheme. Thus, one can eliminate the non-linear dependence of the secular equation by utilizing a fixed basis set which has the flexibility to resolve the variation in radial derivative. One of many possible ways to do this is to utilize the energy derivative of the radial solution and insist on matching both function and derivative in the augmentation process. Several aspects of this choice deserve comment as they are instructive. First, note that this is quite a different procedure than letting $u_\ell(\varepsilon) = u_\ell(\varepsilon_0) + (\varepsilon - \varepsilon_0)\dot{u}_\ell$ in the standard formalism. That choice would result in a rational fraction for the secular equation which would be second order in the numerator and first order in the denominator. Second, although it seems quite obvious in hindsight that the radius of convergence of this "linear expansion" should be usefully large, the realization that it would be so is, to my mind, the true insight provided by O.K. Andersen. It is an amusing personal footnote for me that it was precisely on the day that I met O.K. Andersen that I heard about the use of the energy derivative from H. Schosser and P. Marcus (Marcus 1967). Their estimation was that it would be useful over 15-20 mRy. Because of that estimate, the use of energy derivatives was to wait another 5 years for Andersen to point out the large range of convergence. Thereafter, the formulation of a linearized APW method is a trivial exercise in working out integrals. Third, one should observe that it is not necessary to introduce the energy derivative to achieve a linear method. The methods which use principal quantum number expansions inside the muffin tin spheres can all be linearized if one truncates the expansion. The 1953 APW scheme (Slater 1953) truncated at

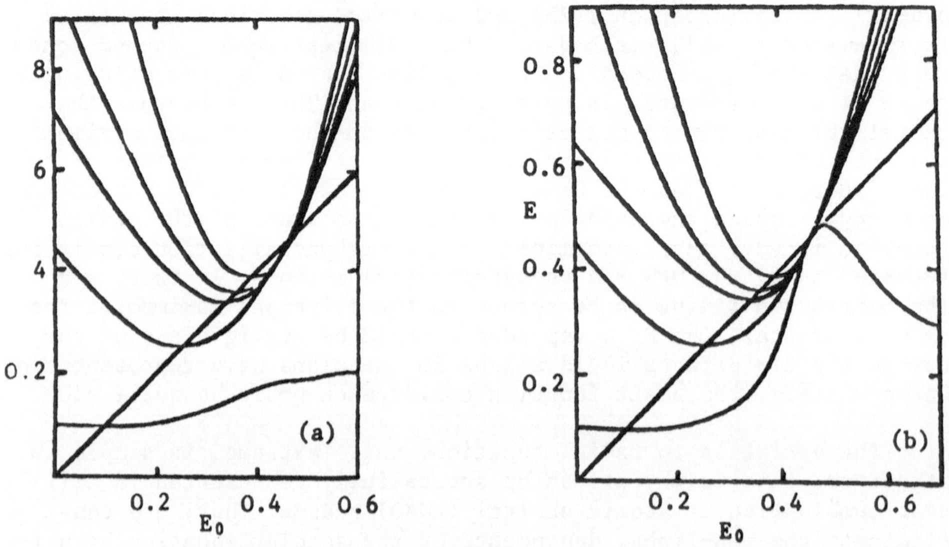

Fig. 3a,b : Analysis of the eigenvalue dependence on energy for the standard APW method (Harmon and Koelling 1974). In all four panels, the resultant eigenvalue is plotted along the ordinate while the energy parameter in question is plotted along the abscissa. In (a) the eigenvalue obtained for a point along the Δ line is plotted as a function of the energy parameter used in setting up the augmenting orbitals. One notes a strong dependence on this energy parameter and that one obtains a variational minimum along the 45° line representing $E = E_0$. In (b) the solutions in the interstitial region (i.e. APW coefficients) were fixed by a single diagonalisation with energy parameter at 0.25 Ryd. This was then kept fixed apart from renormalisation, while the energy parameter determining the augmenting orbitals was again varied. One notes that the upper four curves corresponding to the d bands (Δ_3 is two-fold degenerate) are nearly identical to (a) whereas the bottom nearly-free electron shows a rather strange behaviour.

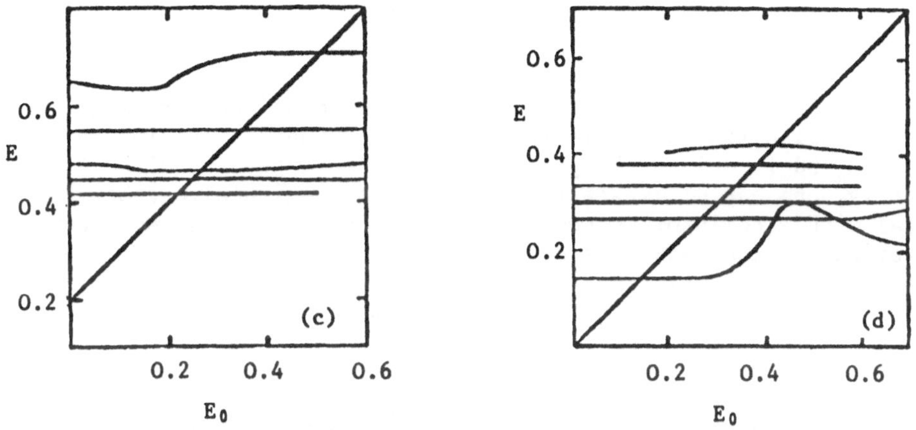

Fig. 3c,d : In (c), the APW expansion coefficients are obtained using augmenting orbitals obtained at energy E_0. However, the eigenvalue plotted is the minimum obtained by the re-variation of the energy parameter as in (b). This, is some sense, shows the sensitivity of the eigenvalue to small errors in the interstitial solution and, as expected, the d states are far less sensitive to this region than the nearly-free-electron-like states. In (d), the same analysis as in (c) is shown for a point of the symmetry line where there is no symmetry selection. Note that there is a greater sensitivity which would become worse in more complex and less close packed systems (the model system here is the Chodorow potential for copper). The places where one sees a maximum are the result of the approximate solutions not maintaining orthogonality to lower solutions - usually the "high" or "soft" or "semi" core states.

one term (which varies for each APW) is essentially the scheme used by Mueller (thesis-Cornell University). The MAPW scheme of Bross (1964) when truncated at two terms yields basically the case of Andersen's Laurent series expansion. This method has been formulated to be linear in the secular equation but have greater variational freedom (Bross et al. 1970). The AAPW scheme of Koelling (1970) linearizes if one chooses one u_ℓ and v_ℓ function and the scheme of Takeda and Kuebler (1979) may be viewed as its natural generalization to using any two functions through the bands. Thus one can achieve a linear secular by a number of possible routes and each of these has its advantages and disadvantages.

To linearize a KKR or spherical wave technique, it is necessary to also fix the variation of the functions in the interstitial region and formulate the problem as an LCAO type expansion. Alternately, one can eliminate the interstitial region altogether and deal with a cellusor method (Fritsche et al. 1979). As pointed out previously, the results are relatively insensitive to this reduction of variational freedom in the interstitial region for close packed structures and can be made even more so by shrinking that region through the use of overlapping spheres.

Linear methods have proven to be very powerful. They offer great efficiencies at only modest loss of precision in most cases. Further, being basis set expansions, they may be used for a full general shaped potential even though they were originally constructed utilizing a muffin tin shaped potential. They are an excellent example of the thesis that the physicist should focus on obtaining the most useful solution in that they arose not from focussing on a faster method but on a more useful and convenient one which more clearly exposed the essential features of the problem. It is worthwhile remembering here that the utility of pseudopotential techniques is not their computational speed but their simplification of the problem to the physicist.

B. Spin Orbit Coupling

The treatment of spin orbit coupling provides another interesting case history. I use it in no small part because I am pleased to announce some recent progress accomplished by a group effort at Argonne National Laboratory (Dongarra et al. 1982 and b). But it is also illustrative of the thesis that progress arises from better understanding of the structure of the problem.

The inclusion of spin orbit coupling has a rather unusual history. A very inconvenient feature of the spin orbit coupling is that it makes the secular equation matrices twice the size and complex. This has the potentiality of increasing the computational effort by well over an order of magnitude. But, when the system has time reversal (no magnetic field or moments) and parity (i.e.

inversion) symmetry, all eigenvalues are doubly degenerate. Relativistic effects were first introduced into the standard determinant plotting methods and a technique developed by Soven (1965) can be used to calculate a factor which is the square root of the determinant. This technique worked so effectively that calculations with spin orbit coupling could be performed with less than a factor of two increase in effort. As a result, of the majority of relativistic calculations performed utilizing determinant plotting techniques include spin orbit coupling directly.

A dilema arose, however, when one wanted to advance to the linearized techniques which required the solution of a generalized eigenvalue problem. In that case, the equivalent of the Soven technique did not exist. The great efficiencies of the linearied methods would then be lost due to the inability to exploit the internal symmetry. (In a simple view, what we have achieved is the equivalent of the Soven determinant technique for the generalized eigenvalue problem). The result has been that most calculations utilizing the linear methods have either completely omitted spin orbit coupling altogether or included it in a second variation. There are a number of formalisms which isolate the spin orbit coupling to be used for this process. Of course, if one merely omits the spin orbit coupling, one has omitted a physical effect. This may be inconsequential in many cases but definitely not all as we shall see in the Ce compounds, for example. For APW schemes, the alternative of merely doubling the matrix size is simply not available because the overlap matrix becomes too nearly singular for a reasonably converged basis set. Thus it is necessary to perform the double variation--first with a spin orbitless like problem (where the nearly singular overlap matrix problem is managable) and then again including spin orbit coupling and utilizing the solutions of the first problem (Koelling 1974). Such a technique is very useful especially when one wants also to include magnetic moment effects or non-spherical effects as well since one is working with a smaller secular equation and the basis functions in an angular momentum representation. But, as the number of basis functions and the number of bands considered rises, it can rapidly become quite expensive. Further, most of the techniques which isolate the spin orbit coupling involve approximations or limitations which are not acceptable in the last row or two of the periodic table. Thus it is desireable to be able to treat the problem with spin orbit coupling included at all stages. The key to doing so lies in developing matrix routines which exploit the internal symmetries in the same way that the Soven technique deals with determinants.

The symmetry we are discussing can be seen from the form of a relativistic APW matrix element

$$H(\underline{k}_1 s_1;\underline{k}_2 s_2) = H_0(\underline{k}_1,\underline{k}_2)\delta_{s_1 s_2} + h_{so}(k_1,k_2,\hat{k}_1.\hat{k}_2)\ i(\hat{k}_1 \times \hat{k}_2).\sigma_{s_1 s_2} \tag{6}$$

In these expressions H_0 and h_{so} are real scalar functions whose exact form will depend on the type of augmentation. The overlap matrix will have precisely the same structure. Considering the 2 X 2 matrices created by considering both spins, one sees that they have the form

$$\underline{\alpha} = \begin{pmatrix} a & b \\ -b^{\star} & a^{\star} \end{pmatrix} \tag{7}$$

It is also possible to organize angular momentum based methods such that they exhibit this structure. The two associated states will then involve the orbital of (jμ) symmetry and its comparison of (j - μ) symmetry on the site related by inversion symmetry.

These 2 X 2 matrices are real multiples of SU2 known as quaternians. They can be used as new elements replacing the scalar elements of the same problem without spin orbit coupling. They add, subtract, multiply, and "divide" with the result also being a quaternian. They do not in general commute but that is not required for the matrix reductions. Such a view is not only computationally expeditious, it is a return to the original development of the spin matrices. Once one knows that one can do the arithmetic with these elements, one finds several things to be obvious. First, the double degeneracy falls out immediately because a diagonal matrix made up of such elements would have to always contain double scalar values. Second, the Soven determinant evaluation technique immediately follows from this organization. Third, because this arithmetic will keep the small roots of the overlap matrix separated, it will be possible to directly manipulate the generalized eigenvalue problem. Fourth, because one need only calculate the upper row for any manipulation, one is guaranteed a factor of two increase in computational speed. Additional efficiency can be obtained by a proper ordering of ones matrix manipulations. By doing so, it can be arranged that in the majority of the required operations, the quaternions involved are scalars times a 2 X 2 unit matrix. The routines produced thus far have achieved an order of magnitude increase in speed over comparable routines which do not utilize knowledge of additional symmetry. (We have some hope that these additional techniques developed by the numerical linear algebraists may be applicable to the case where one does not have inversion symmetry and the matrices are complex multiples of SU2. That, however, is currently only a speculation and must await the results of further research). Fifth and last, one now gains an additional efficiency because the eigenvectors for the two degenerate eigenvalues are directly obtained in quaternian form. That means that one has the eigenvector and its companion related by the product of time reversal and spacial inversion which are trivial operations. Thus one need only store one of the pair and all subsequent manipulations with the eigenvector can include the sum over states before numerical calculation begins.

I am personally very enthused by these developments. Current experimental codes are not yet as fast as the double variational technique (Koelling 1974) but are more stable and improving. I have known that the results should be possible for over a decade and have thus rankled at the tremendous effort expended solving problems with spin orbit. The development is obviously the correct one in that now the physicist can spend his time worrying about what goes into the secular equation and the solution is provided by machinery which he can treat as a black box--only being aware of the fundamental issues it embodies.

C. The Local Density Approximation

The local density approximation is a very useful one in achieving computational simplicity. It is not the only possible route to a local potential in the single particle equation although, in practice, its use far outnumbers any of the alternatives almost to their exclusion. The LDA has been very successful as can be seen from several of the other lectures. But, because many of the d- and f-electron systems are at the limit of applicability for a band approach, one needs to be aware of its limitations as well. That will be the focus of this section. The essential features are :

1. The charge densities calculated in the solid state are insufficiently anisotropic.
2. The momentum densities calculated exhibit too large an anisotropy. By this I do not mean the oscillatory structure (which is more determined by the Fermi surface) but the amplitudes.
3. The LDA unduly favors orbitals of higher angular momentum. This is present when including exchange only and is exacerbated when adding correlation effects.
4. The anisotropy of the Fermi surface is enhanced in the LDA.

These features may all be part of the same picture but it will be useful to arbitrarily distinguish them. The examples chosen in the next two sections will be motivated by these considerations.

The charge density anisotropy is obtained from X-ray scattering factors and the momentum densities from the Compton profiles in the impulse approximation. The X-ray scattering factor experiments must be exceedingly carefully done to get the information but they are very suggestive if not totally proven. Compton profile comparisons look very good when all electrons are included but a sizable part of the contribution is from the core electrons. To make progress, it is necessary to remove the core states from consideration. A very useful technique is to examine the difference in profiles along different symmetry directions. The results seem reasonably sensible but with the calculated anisotropy being too large in magnitude. This may be due to the local density approximation but could equally well be due to the identification of the momentum density as the sum of momentum operator expectation values with the single particle amplitudes (Lam and Platzman 1974).

The undue favoring of the higher ℓ-(or j-) states is a particularly worrisome feature when dealing with transition, rare earth, and actinide systems where the relative positioning of the different states is the critical feature of the system. The phenomena is clearly seen in atomic calculations (Harris and Jones 1978) in the middle of the transition series. It appears not so critical at the top of the series (i.e., nearly filled shell). Although in the atomic calculations one could question whether the controlling factor is the improper treatment of multiplet structure, the same phenomenon is also observed in the solid state calculations where it is less a question. Niobium is the best studied case (Wakoh et al 1975, Boyer et al. 1977, Elyashar and Koelling 1977). When using exchange only, the d- levels are 18 mRy too low relative to the p-levels in the vicinity of the Fermi energy. Further, the inclusion of correlation effects in LDA lowers the d-levels making the situation much worse. The situation is easily characterized. The dominant density dependence of the exchange-correlation functional is the $\rho^{1/3}$ factor. The remaining much milder dependence is then characterizable as a prefactor. To scale it, it is useful to set $\alpha = 2/3$ as the exchange only point in agreement with the scale used in X_α considerations. Correlation effects derived from electron gas considerations then result in this prefactor $\alpha(\rho)$ varying from 2/3 at the high density limit to about 0.85 in the lower density regimes. Because the d- and f-states exist to a greater extent in the interior or the atom where the total density is higer, they will be pulled down when $\alpha(\rho)$ is increased. The result is that the inclusion of correlation in the LDA will inevitably lower the d- and f-states relative to the s/p states where they are already too low using exchange alone - at least in the middle of the series and most likely at the beginning as well. That is why all my cerium and actinide compound work has been done with exchange only. Interestingly, the situation is better in Nb than it is in either V or Ta. This is probably due to two separate causes. In the case of V, the 3d states are more special compact (i.e., local) which increases the problem. In Ta, one would expect the situation to be improved because of the greater extent of the d-states but for the fact that another effect enters. The electrons deep in the cores have not only their kinematics modified by relativistic effects but their interactions as well (Rajagopal 1978, MacDonald and Vosko 1979). These effects are the time retardation and, crudely speaking, magnetic effects. Treated in a local density approximation act, these relativistic interaction effects act to reduce the strength of the functional. In the case of Pt, the relativistic interactions roughly cancelled the effect of the correlations on the Fermi surface (MacDonald 1981). To be fair to the LDA, it should be noted that it works very well

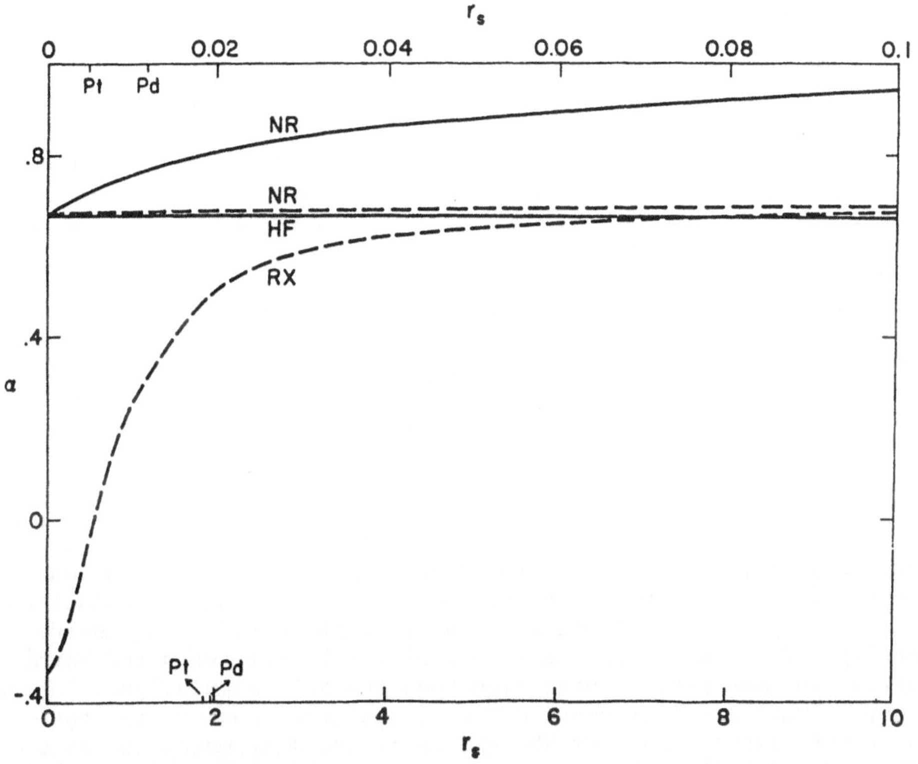

Fig. 4. Exchange-correlation potentials expressed in terms of the parameter α as a function of density in terms of the parameter $r_s = (3/4\ \pi\rho a_0^3)^{1/3}$. Solid line is the exchange-correlation effect and dashed line is the result of including relativistic interaction effects such as time retardation into the functional. The solid curves refer to the upper r_s-scale which expands the high density region. Minimum (upper scale) and maximum (lower scale) r_s values for Pd and Pt are indicated.

for the platinum and noble group metals. In fact, it is seen to work quite well for the famous test case of copper in spite of a few flaps (Janak et al. 1975, Jepsen et al. 1981, MacDonald et al. 1982). That does not mean that theory and experiment are inside their error bars but they do agree quite well.

Much of my interest involves phenomenon at the Fermi surface. To be strictly correct, one must recognize that one is making an additional LDA by identifying the density functional effective potential as a Dyson mass operator. However, that approximation should be as good as the use of Fermi statistics with the single particle eigenvalues. It has been demonstrated by various arguments and model systems that including non-local effects into the mass operator will reduce the anisotropy of the Fermi surface (Nickerson and Vosko 1976, Rosolt et al. 1974a,b, 1975). A most interesting question to then ask is what it would do to relative s/p-d-f- positions. In the LDA, we have just seen that correlations act to pull down the d- and f-levels which are often too low already. Relativistic interaction effects can go either way but often push them up--but they are too small to be essential consideration here. Thus, unless the non-local effects act to raise the d- and f-states, one will have to look to the kinetic energy effects of correlation or the single particle ansatz (SP1) itself. This is unlikely, however. The argument is as follows. A considerable body of experience shows that the Fermi surface can be represented very well using the solutions of a local potential. A practised user of the overlapping charge density model can generate a potential which represents the Fermi surface to much greater precision than the SCF calculations. As one example, Mattheiss' original OCD potential does much better the later SCF calculations for Nb. And there are many more. The existence of model (local) potentials suggests that a formalism must exist-perhaps not based on LDA-which does as well. One believes this to be so because analyses of the eigenvalue problem suggest that there are very few critical features to be determined which, once determined, fix the remainder. Now consider the "experimental" situation. In the case of Cr, a Hartree SCF calculation gives the Fermi surface with about 10% precision. In Nb, another system from the middle of the series, the exchange only calculation yields errors in the 5-6% range being an overshoot. If one were to do an X_α calculation with α reduced from the exchange only value of 2/3 to just below 0.65, the Fermi surface would be very well reproduced. But this is in the opposite direction as the correlation effects included in the LDA ! It seems improbable that if the roughest approximation (LDA) does so well one cannot find an improvement which does better and that would bring one inside the experimental error bars.

Thus far, I have advanced my opinion that for effects near the Fermi surface, one is better off not including correlations in the LDA until the d- or f-shell is very nearly filled. This creates some difficulties, however. One sees that the total energy properties are greatly improved by the inclusion of correlation in the LDA (Boyer et al. 1977, Moruzzi et al. 1978). This is consistent with arguments used by Gunnarsson (Gunnarsson et al. 1979) when discussing his approach to moving away from the LDA. These deal with the LDA getting only the spherical average properties of the exchange-correlation hole correct. Comforting as it is that the discrepancy should be understandable, it is still disturbing since the Fermi surface must have its effect on the total energy--its distortions are a critical feature of the frozen phonon calculations for example. However, the 5-10% errors we are discussing are usually smaller that the grid fo points used to sample the Brillouin zone. The problem is real and present but will only be seen as one seeks greater precision for the total energy dependent phenomenon. Of course, one can dismiss this discussion with the simple observation that it is the total energy which is the variational quantity. But then one is merely refusing to look further into the problem.

The much more dramatic need for correlation effects is seen when one looks at the spin-polarized calculations. In that case, the omission of correlation has disastrous effects where the calculated moment of Fe, for example, explodes ! (DeCicco and Kitz 1967). When the correlation effects are included, the calculated moments agree very well with experiment for all elements below the rare earths. Of course, the magnetic systems are nearer the top of the series where the underlying LDA proves out much better. To provide even more support for the success of LSDA with correlations, one can observe that for Pd both a finite magnetic field calculation (Jarlborg and Freeman 1981) and an enhanced susceptibility calculation (Liu et al. 1979) yield the susceptibility very well ($6.5 + 5 \times 10^{-6}$ emu/g and 7.7×10^{-6} emu/g calculated versus $6.91 \overline{\times} 10^{-6}$ emu/g experimental. The situation is less clear in the rare earths. One very natural system to examine is Gd as it is an S state with a half filled f-shell (Harmon et al. 1978). In that case, SCF band calculations find the spin-down f-states creeping into the calculation. This is very small (order o.03 electron or so). But it should be remembered that the Coulomb interaction is very strong and acts to prevent this occupation. This small occupation actually represents a sizable energetic effect. The most probable explanation is that the underlying LDA unduly favors the f-states. In the case of $CeAl_2$, one correctly finds that the antiferromagnetic state is the stable one (Jarlborg et al. 1982) just as in Cr (Kuebler 1980). This may be more driven by the non- f states, however, and is thus not as relevant to discussions of quality of the LSDA. The model exchange driven disproportionation discussed by Williams et al. (1982) is probably the appropriate one. The moment is properly determined in these calculations for both Cr and $CeAl_2$. To me, further probes of the quality of the spin dependent properties of LSDA will come from elicidating the magnetic properties of $TiBe_2$, $ZrZr_2$, and some of the magnetic Ce compounds as well as the enhanced susceptibilities of the Sc, Ti, Y, Zr, La, and non-moment forming Ce systems. In these systems, one is on less sure ground for the LDA and this may influence the quality of the spin dependent representation. It is thus of interest that a calculation for $TiBe_2$ incorrectly predicts a moment (Jarlborg et al. 1981). As a very sensitive system, this result bears verification but it may be an interesting exception to the success of LSDA in predicting magnetic order. When compared to the configuration self consisted OCD APW results (de Groot et al. 1980) which have had reasonable success in relating to the limited de Haus van Alphen data, the SCF-LMTO results appear to reverse the order of states in the critical L-W region. The result is only suggestive but reasonable if slightly large results are obtained for $ZrZn_2$ where the reversal is not observed.

IV. TRANSITION METAL SYSTEMS

The gross electronic structure of the basic transition metal systems has been pretty well illucidated. These systems are now in a more mature phase of study where it is appropriate to perform more careful calculations examining improvements in theory and extend considerations to more and more of their properties. The alternate avenue of advancement is to examine the more complex systems. This is more visible route. Work is being done on binary and ternary systems with more complicated crystal structures are being studied : A15's, C15's, $L1_2$ (Au_3Cu), the dichalcogenide layered structures, Perovikites and even the building block systems such as hexaborides, ternary tetraborides, Chevrel phases, and so on. Further, one finds surfaces being dealt with by calculations on two dimensionally infinite thin slabs ; phonons by calculations on distorted systems represeting the phonon frozen in place; random alloys amorphous solids, and defects by calculations d on very large supercells ; and the artifical layered systems (LUCS) by calculations on specific examples. Admittedly, in many cases these calculations must be done for idealized approximations but they can be viewed both as an approximation to the real system and as calibration points for the more approximate theories developed for the less idealized problem. Both the careful examination of detailed results in the simpler systems and the advanced into new frontiers are reasonable efforts. They simply offer progress in different ways.

Here, let us look first at a few results from the structurally simple systems to illustrate comments made earlier. Then we can pick a few interesting examples from the structurally complex systems to consider some general ideas. In discussing LDA, a distinction was made between the beginning, center, and end of the series. Since I have been involved in work with La (of the beginning Sc, Y, La/Lu), the group IVB elements from the middle, and the Pt and noble metal elements at the end, the choices to be made are obvious. One needs to consider the cohesive energies, the charge and spin densities, the Compton profiles, the Fermi surface parameters and masses, and the susceptibilities. The cohesive energies will be exhaustively covered at this school. The Compton scattering results are really only useful if one looks at the figures comparing theory and experiments. They are too many to include here but one can summarize that the anisotropy calculated is much larger than what is observed. (See Rath et al. (1973), Wakoh et al. (1976)). To relate the Compton measurements to a Compton profile, it is necessary to make the impulse approximation and neglect relativistic effects. This limits the acuracy of the comparison which is better for low atomic number and lower photon energy. On the theoretical side, the density functional purest can question the identification of the momentum density with the momentum operator acting on the amplitudes (Lam and Platzman 1974). But none of these can overshadow the effect seen.

Transition metal x-ray form factors are only available for the 3d elements, V, Cr, Ni, and Cu because of the extreme difficulties encountered with increasing atomic number. It was this fact that caused V to be suggested as the ideal testing ground-being the first transition metal with a simple (body centered) cubic structure. Throughout the transition metals, one must look at small differences in numbers with large error bars. The situation has been recently summarized for V by Weiss (1978) who concludes that the charge anisotropy is larger than that calculated. This we show in Table I. He observes that this could be realized using superposed atomic wavefunctions if the triply degenerated t_{2g} functions were to be contracted, the doubly degenerate ε_g functions were to be expanded, and their relative occupation were to be 70-30 instead of the spherical value of 60-40. The theoretical calculations, which are consistent amongst themselves, yield less anisotropy. General shaped potentials have been used so this not a muffin tin approximation effect and adequate flexibility has been provided in the calculations so that radial functions can differ. Clearly, there are the two naturally separated classes of d-electrons. The ε_g orbitals are the (100) and the t_{2g} orbitals are the (110) or (111) directed d orbitals. In bcc V, the ε_g states point in the direction of the interstices while the t_{2g} states are directed towards the nearest neighbors at (111). One can find a correlation between the effectiveness of the electron-phonon interaction with the fractional ε_g character at the Fermi energy (Mueller and Myron 1978) and this is interpreted as the greater freedom to respond of the ε_g orbitals. In addition, the t_{2g} orbitals occur in the occupied states near the center of the Brillouin zone whereas the occupied ε_g orbitals occur nearer the surface of the Brillouin zone. i.e. the t_{2g} orbitals are associated with orbitals with less rapidly varying structure than the ε_g orbitals. So it is clear that one should be able to interpret the anisotropies in terms of differences in these orbitals. The actual effective mechanism is a question worthy of speculation.

Table I. Charge form factors for V

k	Laurent[a] et al.	Koelling[b]	Wakoh[c] & Yamashita	elec.sc.[d]	X-ray[e]	X-Ray
(110)	15.75	15.72	15.84	15.89 ± .98	15.94 ± .08	15.90 ± .18[f]
(200)	13.11	13.06	13.15	13.11 ± .38	13.16 ± .04	13.22 ± .17[f]
(211)	11.43	11.40	11.36		11.30 ± .04	
(22)	10.23	10.20	10.12			10.41 ± .05
(330/511)	1.004	1.004	1.003			1.024 ± .001[g] 1.0085[h]

[a] D.G. Laurent, C.S. Wang, and J. Callaway, Phys.Rev.B 17, 455 (1978) (Includes a Hartree-Fock core).

[b] Unpublished.

[c] Extracted from discussion of form factor by M.V. Linkaoho, Physica Scripta 5, 271 (1972).

[d] From S. Wakoh and J. Yamashita, J.Phys.Soc.Jpn. 35, 1394 (1973).

[e] U. Korhonen, E. Rantovuori, M. Linkaoho, Ann Acadm. Sc. : Fenn Ser A VI 361, 1 (1971) (performed at 80°K. A correction for core motion by Linkaoho ref. c and reported by S. Wakoh and Y. Kubo, J.Phys.F. 10, 2707 (1980)).

[f] O. Terasaki, Y. Uchida, and D. Watanabe, J.Phys.Soc.Jpn. 39, 1277 (1975).

[g] R.J. Weiss and J.J. Demarko, Phys.Rev. 140, A1223 (1965).

[h] M. Diana and G. Mazzone, Philos.Mag. 32, 1227 (1975).

The effect is less pronounced but clearly present in Cr as can be seen from Table II.

The picture is even less clear when considering the charge anisotropies at the top of the series as can be seen from Tables III and IV. I believe the contraction of the orbitals near the top of the band to be very well given but that is better seen in the spin densities. I have argued before that there is a reduction in the experimental data for along the 100 direction not seen in the band calculations. This would correspond to an expansion of the ε_g orbitals consistent with the observation of Weiss for V. That anisotropy does not extend to Cu as should be expected with the d-band filled. In Table IV, we consider the results for Cu. In Table IV, the first three columns are experimental results. The remaining five are theroretical results which can be used to examine the effects of various procedures. The calculations by Arlinghaus used the Chodorow potential to calculate the valence electron density and a Hartree-Fock atomic calculation to determine the core density. The Chodorow potential is determined from a Hartree-Fock atomic calculation as $\varepsilon_{3d} + (\nabla^2\psi_{3d})/\psi_{3d}$ which is a potential representation of the Fock operator. If one argues that the 4s-p band should be relatively insensitive to the potential using pseudopotential orthogonalization arguments, then the Chodorow potential results should be very much like the first iteration of a Hartree-Fock calculation. Because SCF procedures tend to overshoot, the solid state effects should be overemphasized. But clearly the technique has worked very well. Next note that the two exchange only LDA calculations (XO and S now $\alpha = 2/3$) do not differ greatly from it. The major differences between the two exchange only calculations are three : 1) slightly different lattice constant which has an effect less that 0.05 ; 2) XO used relativistic kinematics - which should not affect these short wave vectors greatly although that has not been tested. One would expect a relativistic contraction and is consistent with the XO results being slightly higher ; and 3) the Snow calculation was performed using muffin tin shaped approximations and XO was not. This is the most probable cause for the difference at (200). The remaining values at (111) and (222) are essentially identical. The Snow $\alpha = 1$ calculation clearly shows the effect of increasing the strength of the attractive exchange-correlation functional. The charge density is contacted and the form factors increased. The use of a local density correlation potential behaves in the same way only to a lesser degree. It is tempting, although not entirely justified, to conclude that with exchange alone the calculations are already too well bound and that correlation effects in the local approximation make it worse.

The magnetization densities of Co, Fe and Ni have been extensively studies using neutron scattering. In Table V, a comparison of the spin density for Ni is shown. This comparison involves the neglect of orbital contributions which is probably the major source

Table II. Charge form factors for Cr

k	Laurent et al[a]	Rath & Callaway[b]	Wakoh & Yamashita[c]	Diana & Mazzone[d]	Cooper[e]	Fuijimote et al[f]
(110)	16.29	16.27	16.32	15.74 ± .2	15.88	16.30 ± .12
(200)	13.39	13.31	13.46	13.06 ± .17	13.14 ± .34	
(211)	11.66	11.60	11.56	11.37 ± .15	11.23 ± .34	
(220)	10.39	10.33	10.37	10.10 ± .14	9.97 ± .5	
(330/411)	1.008	1.008	1.003	1.013 ± .007		

[a] D.G. Laurent, J. Callaway, J.L. Fry, and N.E. Brener, Phys.Rev.B 23, 4977 (1981).
[b] J. Rath and J. Callaway, Phys.Rev.B 8, 5398 (1973).
[c] S. Wakoh and J. Yamashita, J.Fhys.Soc.Jpn. 35, 1394 (1973).
[d] M. Diana and G. Mazzone, Phys.Rev.B 5, 3832 (1972).
[e] M.J. Cooper, Philos.Mag. 7, 2059 (1962) ; 10, 177 (1964).
[f] M. Fijimoto, O. Terasaki, and D. Watanabe, Phys.Lett. 41A, 159 (1972).

Table III. Charge form factor for Ni

k	WC[a] KSG	WC[a] vBH	WY[b]	Diana[c] et al.	Hoyowa & Fukomachi[b]
(111)	20.39	20.43	20.28	20.10 ± .16	20.78
(200)	19.05	19.08	19.05	18.55 ± .16	19.29
(220)	15.39	15.40	15.35	15.34 ± .12	15.6
(400)	11.47	11.47	11.35	11.18 ± .11	11.49
(511/333)	1.044	1.055		.989 ± ?	

[a] C.S. Wang and J. Callaway, Phys.Rev.B 15, 298 (1977).

[b] S. Wakoh and J. Yamashita, J.Phys.Soc.Jpn. 30, 422 (1971).

[c] M. Diana, G. Mazzone and J.J. De Marko, Phys.Rev. 187, 495 (1966).

of discrepancy in the overall magnetude although the neglect of relativistic effects will make some contribution. Again, the anisotropy is too small. Because the d-states producing the orbital contribution are so contracted, the oribtal contribution can be evaluated using an atomic approximation. (It even works for Cr). Thus one can assume that there will be little anisotropy in the orbital contribution. Thus this observation of insufficient anisotropy is very probably real and quite consistent.

For the paramagnetic systems, the spin density can be studied by applying a magnetic field if the susceptibility is large enough. The first question, however, is how well can be done for the susceptibility itself. If spin-orbit coupling can be neglected, it appears that one can do quite well. The significance of spin orbit coupling is primarily through the g-factor since one still has a double "spin" degeneracy at zero field even after spin orbit coupling has been included. A formal theory does not exist in the case of spin orbit coupling. A variational treatment (Vosko Perdew 1975) for the (enhanced) response of the system in its simplest form

Table IV. Comparison of Experimental and Theoretical Scattering From Factors for Cu

(hkl)	Jennings et al.[a] (1964)	Freund[a]	Schneider et al.[c] (1981)	XO (a=6.809)	NX (a=6.831)	Snow[b] (a = 1) (a=6.831)	Snow[b] (a = $\frac{2}{3}$) (a=6.831)	Arlinghaus[a] (Chodorow)
(111)	21.52(10)	22.63	21.51(5)	21.65	21.73	22.33	21.63	21.54
(200)			20.22	20.32	20.39	21.04	20.04	20.25
(222)	14.01(10)	14.64(10)	14.07(5)	14.19	14.25	14.53	14.16	13.90
(333)	9.41(10)	9.54(10)	9.49(6)	9.63	9.66			
(511)			9.53(6)	9.64	9.67			9.51

[a] J.R. Schneider, N.K. Hansen, and H. Kretschmer, Acta.Cryst. A37, 711 (1981).
[b] E.C. Snow, Phys.Rev. 171, 785 (1968).
[c] A.H. MacDonald, J.H. Daams, S.H. Vosko, and D.D. Koelling, Phys.Rev.B 25, 713 (1982).

Table V. Spin density for Ni

	WC[A] KSG	WC[A] vBH	WY[b]	Exp[c]
(111)	.790	.762	.766	.793 ± .009
(200)	.697	.669	.665	.703 ± .008
(220)	.440	.423	.419	.447 ± .005
(311)	.309	.297	.296	.321 ± .005
(222)	.295	.285	.287	.311 ± .004
(400)	.160	.152	.154	.157 ± .003
(331)	.153	.149	.151	.168 ± .003
(333)	.086	.085	.089	.109 ± .003
(511)	.042	.039	.045	.036 ± .004

[a] C.S. Wang and J. Callaway, Phys.Rev. B15, 298 (1977).
[b] S. Wakoh and J. Yamashita, J.Phys.Soc.Jpn. 30, 422 (1971).
[c] H.A. Mook, Phys.Rev. 148, 495 (1966).

yields a Stoner theory like form

$$\chi_p = \frac{\chi_s}{1 - In(\varepsilon_F)} \tag{8a}$$

$$\chi_s = \mu_B^2 n(\varepsilon_F) \tag{8b}$$

$$I = \mu_B^2 \int d^3\underline{r}\gamma^2(\underline{r},\varepsilon_F)\,[\chi_0^{-1} - \chi_h^{-1}] \tag{8c}$$

$$\gamma(\underline{r},\varepsilon) = \sum_i \delta(\varepsilon - \varepsilon_i)|\phi_i(\underline{r})|^2/n_{(\varepsilon_F)} \tag{8d}$$

In these expressions, μ_B is the Bohr magneton ; $n(\varepsilon_F)$ is the density of states at the Fermi energy ; γ is the charge density at the Fermi energy which is taken as an approximation of the spin density. The quantity I has been evaluated through a local approximation consistent with LSDA. Strictly speaking, this expression is only a lower bound involving a restriction in variational freedom that the exchange enhanced field be a constant times the applied field. That limitation will be of interest when discussing the f-electron systems. Janak (1977) has applied this formalism to the metals in the first half of th periodic table. We have already mentioned that both the finite field calculation and this formalism do quite well for Pd. Using the finite temperature generalization, the temperature dependence of the susceptibility can be reasonably described for Pd but not for Pt. In Pd and Pt, the enhancement I appears to be sensitive to anisotropy in the calculation but not to the exchange-correlation functional and is probably temperature dependent. $n(\varepsilon_F)$ is the more sensitive component.

The total susceptibility is helped by having a variational formulation and being an integral property. Anisotropies are much more sensitive but offer more information. Much of the initial work has been done with the OCD model potentials with reasonable success. It appears as through enhancement is roughly uniform in space such that the induced spin density approximated by the charge density at the Fermi energy is the dominant factor. This, of course, will need further investigation but it gives a good starting point. In the case of Sc, one is able to see the expansion of the d-orbitals at the bottom of the d-band. Further, the charge density at the Fermi energy is a strong p-d hybrid generating a strong L = 3 component. In light of the strong anisotropy dependencies observed for I in Pd and Pt, it is interesting to consider what would happen in the case of Sc which is known to be strongly enhanced and yet described reasonably well in the muffin tin approximation (Mac Donald et al. 1977).

Returning to the top of the transition series, we return to Pd. Again the gross properties are gotten correctly in the OCD model and the self consistent calculation. That simply involves the contraction of the d-orbital at the top of the d-band. Using the measured g-factors, one finds that orbital effects are about 10 % of the total for Pd and $23\frac{1}{2}$ % for Pt (Waston-Yang et al. 1976 and 1977). In the case of Pd, it would appear that the anisotropy is again underestimated although no by much and one is really inside the experimental error bars. In the case of Pt, one can only make estimates based the crystal field model of Weiss and Freeman applied to the data of Maglic et al. (1978). In so doing, it is estimated that the fractional ε_g character is $.30 \pm .17$. The OCD model-which represents the Fermi surface quite well-yields a number of 0.07 (Watson-Yang et al. 1977). The self consistent model-which also represents the Fermi surface quite well-yields 0.30 (MacDonald et al. 1981). Clearly, the wavefunctions can vary considerably and yet exhibit very similar energetics. With the observation that calculations with very different wavefunctions properties can yield very similar Fermi surfaces, we turn to the question of Fermi surface properties.

In the Pt and noble group metals, a rather careful examination of the effect of correlation and relativistic interactions in the LDA has been performed. For the noble metals, the Fermi results were successfully analyzed in terms of relative s/p-d shifts by considering the wavefunction character of the different states on the Fermi surface. As pointed out by Janak et al. (1975), the neck area proves a very sensitive indicator. In their non-relativistic muffin tin calculation, they found that they needed to increase from the value of 0.7 which represented their exchange-correlation calculation quite well to a value of 0.77 in order to adjust the neck radius. This was not at all satisfying as the value 0.7 very nicely yields the proper bulk property behavior (Snow 1973) and corresponds to the average of the electron gas results. Similar discrepancies of about 10% were obtained in calculations which included the non spin-orbit relativistic effects but retained the muffin tin shape approximation (Jepsen et al. 1981).

Table VI

Some Fermi surface calipers for the noble metals. k_F^{100} and k_F^{110} are the Fermi radii in the <100> and <110> directions while k_F is the free-electron Fermi radius. k_F^N is the neck radius on the <111> face. The theoretical results obtained using exchange and correlation (XC), plus relativistic interactions (SCR), and exchange only (XO) potentials are listed.

		Expt[a]	SC[b]	SCR[b]	XO[b]
	Cu	1.058	1.079	1.078	
$k_F^{100}/{}^o_F$	Ag	1.048	1.059	1.057	
	Au	1.123	1.146	1.142	1.156
	Cu	0.951	0.949	0.950	
k_F^{110}/k_F^o	Ag	0.963	0.968	0.969	
	Au	0.942	0.948	0.950	0.944
	Cu	0.189	0.193	0.193	
k_{RAD}^N/k_F^o	Ag	0.136	0.134	0.133	
	Au	0.179	0.178	0.174	0.186

[a] P.B. Allen, Phys.Rev.Lett. 37, 1638 (1976) ; P.T. Coleridge and I.M. Templeton, J.Phys.F 2, 643 (1972).

[b] A.H. MacDonald, J.M. Daams, S.H. Vosko, and D.D. Koelling, Phys. Rev. B 25, 713 (1982).

Table VII. Pd Fermi Surface Areas (au^{-2})

H	A (theor.)[a]	A (exp.)[b]
	Γ-centered Electron Surface	
<100>	0.739	0.731
<110>	0.839	0.827
<111>	0.653	0.648
	Open surface holes	
<100> α	0.058	0.072
<100> ε	2.038	1.969[8]
<110> β	0.314	0.308★
84° β	.238	.285
63° β	.197	.218
	L-centered holes	
<100>	0.0063	0.0061
<110> LKΓ	0.0090	0.0088
<110>	0.0063	0.0058
<111> LKW	0.0053	0.0051
<211> LWΓ	0.0087	0.0087
	X-centered holes	
<100> XWΓ	.021	0.024
<110>	.021	0.024
<111>	.081	.020

[a] A.H. MacDonald, J.M. Daams, S.H. Vosko and D.D. Koelling, Phys.Rev. B23, 6377 (1981)

[b] D.H. Dye, S.A. Campbell, G.W. Crabtree, J.B. Ketterson, N.B. Sandersara and J.J. Vuillemin, Phys.Rev. B23, 462 (1981).

★ Inferred from KKR fit.

Table VII (cont.). Pt Fermi Surface Areas

Fermi surface areas are compared for the exchange only (HF), exchange-correlation (XC), and exchange-correlation plus relativistic interaction (XCR) calculations with the experimental results. Units are a.u.$^{-2}$.

Γ-centered Electron Surface

H	A (HF)[a]	A (XC)[a]	A (XCR)[a]	A (exp)[c]
<100>	0.765	0.743	0.761	.770
<110>	0.851	0.824	0.846	.857
<111>	0.684	0.665	0.678	.678
Open hole surface				
<100> ε	1.931	1.883	1.922	1.890
<100> α	0.069	0.078	0.069	0.074

[c] D.H. Dye, J.B. Ketterson and G.W. Crabtree, J.Low Temp.Phys. 30, 813 (1978).

When the muffin tin approximation was removed, the neck areas was found to be within about 3% of the experimental value (MacDonald et al. 1982). Clearly, one must be careful that approximations in technique do not mask the information sought. One important feature to note is that the inclusion of the correlation in the LDA improves the results in all the noble metals by lowering the d-states relative to the NFE states. The story is basically the same for Pd but Pt requires a more careful consideration together with Au. In the case of Au, adding correlations in the LDA dramatically improves the results. Including relativistic interaction effects move the results in the same direction as including correlations and produces an overshoot. The exchange only calculation is pretty good for Pt and the inclusion of correlations creates considerable mischief on the electron surface but improves the hole surface. Including relativistic interaction effects acts in the opposite direction and yields generally improved results. In this case the relativistic interaction effects have acted to move the d-states upward. Several aspects of this puzzle can be pointed out : (1) the inclusion of relativistic interactions in the local approximation is a very approximate scheme. (2) The use of the scalar relativistic approximation in the SCF calculation may have its effect since eigenvalues are affected by as much as 4 milli Rydberg in going from SO-LAPW to the RAPW calculation and spin orbit couplings themselves are much larger. (3) The size of the s-d shift due to relativistic interaction effects is probably about right i.e. ~ 3 mRy in the 4d series and ~ 8 mRy in the 5d series.

Thorium may be viewed as a representative of the beginning of the series and is a prime factor in my original interest in the effects of relativistic interactions. Thorium occurs at the beginning of the actinide series and should have very little f character. However, when using the OCD model one must use the correct exchange only functional ($\alpha = 2/3$) rather than increasing to full Slater to make up solid state effects. (Koelling and Freeman 1975). The crucial feature is the placement of the f-resonance as it critically influenced the size of the so called superegg at the center of the Brillouin zone where the f-character first appears. This model was adequate to identify the fact that the principle effect of pressure was to move towards opening a hybridization gap so that all areas decrease with pressure (Schirber et al. 1977). This was verified by SCF-LMTO calculations (Skriver and Jan 1980). The SCF-LMTO calculation also resulted in a superegg eith areas in very good agreement with the dHvA data and good agreement with the equation of state under extreme pressure. It would seem Th should be considered a well understood case except that I have also performed an SCF-APW calculation (Koelling 1979) and find the superegg to be too large. These are not equivalent calculations. The SCF-LMTO was performed for an exchange-correlation functional and the SCF-APW for exchange only which should imply that the LMTO calculation should have the larger superegg. However, the LMTO calculation made potential shape and frozen core approximations not APW calculation. Further, one would expect the results obtained from the APW calculation in that the SCF process should mimic an increase in α. Thus I think the APW results are a closer representation of the model and one of the LMTO approximations has produced a cancelling effect. Certainly, it is not a clear cut case. Nonetheless, I would claim that exchange only already unduly favors the f-character in Th consistent with what is seen in the VB elements, La, and Ce. The possibility of a weakened exchange correlation functional in this very heavy would then be attractive.

In the noble metals, the electron-phonon enhancement is quite small. Thus one gets a good idea of the electron-electron enhancement effects by comparing the base density of states to the electronic specific heat corrected for the (small!) electron phonon enhancement. One finds electron-electron enhancement of 14% for Cu, 7% for Ag, and 15% for Au. (As a parenthetical comment, note that they correlate nicely with the location of the d-bands).

The enhancement factor for Cu agrees reasonably well with the value of 16% calculated by Sacchetti (1982). It is also consistent with the 8% factor used earlier (Jansk et al. 1975). It should be noted that these are all considerably larger than the 3% to be obtained for the electron gas at the same density. Such is the prediction of MacDonald (1980) based on his work in the alkali metals. We will return to a discussion of masses after considering the group VB metals.

One would like to start a discussion of the VB metals, V, Nb, and Ta with the data for vanadium. Unfortunately, the data is really incomplete and it is wise to begin with Nb and simply include V by comparison as Nb is the best characterized system of the three. The Fermi surfaces of all three have the same structure. There are three pieces of surface : the ellipses centered at the point N (110 face center) of the Brillouin zone ; a childs "jungle gym" like structure with tubes running in the (100) directions and intersecting at the points Γ (center) and H (100 corner) of the Brillouin zone ; and an "octahedron" structure centered in the zone. The octahedron structure at Γ which is made up of nearly pure d states and the ellipsoids at N which are made up of d- and p-orbital admixtures will be the subjects of our considerations. The more structurally complex jungle gym is comparatively insensitive and thus can be neglected in this discussion. In Nb, all pieces of the Fermi surface are seen. In V, only the ellipses are seen. This can be due to three reasons : (1) Materials problems are more severe in V and the ellipses have the lightest masses making them the easiest to see. (2) The frequency of the octahedron is masked by harmonics of the ellipsoids, and (3) Electron-electron (paramagnon) effects will be greatest on the octahedron and jungle gym. (Clearly, it would be most interesting to point to the paramagnon effect but that is hardly justified at this time. On the other hand, this would be consistent with the fact that de Haas van Alphen (DVHA) signals have not been seen in Sc whereas they have been seen in Y and the Sc has been prepared to higher purit).

The results for Nb can be characterized in a simple way : the calculated areas are too large for the ellipses and are too small for the octahedron. This observation comes from three separate studies and differs in detail according to the quality of the calculation performed but never in its presence. I have spent a great deal of effort insuring that no detail of the calculation should give rise to the effect other than the physical model. It is so. There are only d and p orbitals at the Fermi energy so these results can be interpreted as a relative d-p shift. In my most careful calculation, (an improvement which more carefully sampled the Brillouin zone) of the original Elyashar and Koelling (1977) work I found that excellent results could be obtained if the d-orbitals were shifted upward by 18 milli Ryd. This is quite consistent with the results found in atomic calculations that the LDA unduly favored the d-states in the center of the transition series. Further, that 18 milli Ryd shift is for an exchange only functional. The inclusion of correlations in the LDA functional would lower the d-orbitals and make matters worse. Wakoh et al. (1975) used $\alpha = 0.8$ and then shifted their $dq(t_{2g}$ up by 20 milli Ry and the $d\varepsilon(e_g)$ up by 40 milli Ry. The results of Boyer et al. (1977) can also be interpreted in the same way. Returning to V, one sees that the ellipses are again too large by an even larger amount. This is again found in more than one study and is for exchange alone. The effect is also larger in Ta but, in that case, one must consider whether the relativistic interaction effects might not enter.

Using the empirically adjusted bands, the mass enhancements can be examined (Crabtree et al. 1979). The experimental DHVA masses are fit using the energy derivatives of the phase shifts in a KKR scheme. Thus the "mass" or velocity is available from the fit at all points on the Fermi surface. That fit, when integreted over the entire Fermi surface, yielded the measured electronic specific heat to 1.5%. Comparing to the bare band velocities, it was found that the rigid muffin tin approximation to the electron-phonon interaction gave the wrong anisotropy while the tight binding based methods gave a more nearly correct variation. Further, an enhancement factor of $\lambda = 1.33$ was found. Analyses of the superconductivity properties would place the electron-phonon coupling $\lambda_{e.ph.} = 0.95$ so we again see evidence for an appreciable electron-electron enhancement of the mass. This electron-electron enhancement is formally within the range of application of density functional theory as it can be obtained from a thermal equilibrium ensemble.

Now let us consider some of the more complex systems. Calculations can be made to very high precision on these systems. Especially illustrative is the APW based film code currently in use by the Northwestern University group. That package has carefully eliminated all potential and charge density shape approximations such that the calculational results truly reflect the assumed model of a density functional theory applied to a infinite slab. This is not accomplished with cost, of course. The typical environment of that code is a CRAY macine although it is also running (and running and running ...) on Digitial Equipment VAX 11/780, Prime 750 and IBM 370.

Of particular interest to me is the experiment with an ordered array of almost non-interacting O_2 molecules (Wimaner et al. 1981). To get good results for that test system, the technique must have a good representation of the non-muffin tin like variations of the potential. But, in addition, it must have adequate flexibility in the basis set. This has proven to be the case. The results were compared to those of an LCAO calculation on the same system. It was found that the LCAO calculation could be brought into agreement with the APW based calculation by introducing orbitals which were not centered at the atomic site. Later analysis has found that basis sets using orbitals from various degrees of ionization can do as well. Note that the interesting point here is not efficiency. The APW based code used almost an order of magnitude more basis functions than the LCAO calculation. After all, one can always use a minimal basis set if one already knows the answer. The truly interesting point is that the additional flexibility revealed an unexpected feature that could then be interpreted simply. This Full potential Linearized Augmented Plane Wave (FLAPW) package is gaining considerable credibility through a number of successful application (Posternak et al. 1982, Wimmer et al. 1981b, Wang et al. 1981). Clearly, the ability to study the end point of an oxidation system (O_2 itself) is a necessary part of the whole problem. With the techniques used, it is the hardest part. Surface studies are now able to go beyond just looking for surface states and surface resonances to consider questions of surface relaxations, reconstruction, and even reactions. Self consistency is essential in such studies as some of these surface specific states are only bound through the relaxation of the electronic structure occuring at the surface. Further, correlation effects are very important at the surface where one is dealing with densities dropping to zero-a very bad case for the LDA ! To more precisely describe the situation, the Wigner interpolation formula for low densities is often used.

A great deal of coordinated effort has gone into the FLAPW package. It is now somewhat over 10,000 records long so it ranks with a number of the large scale chemistry codes. A physicist's taste and resouce limitations often do not permit such an extensive solution. This is indeed unfortunate as the systems of interest become more complex. It means that one must be ever more sensitive to whether an effect is real or is a computational artifact. Results for the more complex systems will usually be less precise. To illustrate, consider a simple fcc or bcc transition element system. The controlling parameter is the relative population of d and s/p states and they are limited by a simple sum rule so that one has a single parameter system. If one has a simple compound, the number of parameters is immediately increased to three as now one has possible charge transfer and the relative adjustment between s/p and d-states. As one has also introduced a new bonding mechanism--ionic bonding. It is a well known observation that precision drops rapidly with the number of parameters in the problem.

The calculations on frozen phonons are illustrative of the charge transfer phenomenon and their complications. Further, the charge transfer considerations in this system are significant for understanding the underlying physics of soft phonons and phase transformations as well as for the computational problem. (See Sinha 1980 and ref. therein). Soft phonons are those for which there is a local (in reciprocal space) dip in the frequency spectrum. In some cases a given phonon frequency going to zero (requiring no energy to create) is seen as a precurser of a phase transition. For a phonon to be well screened, there must be many electron states available with favorable energetics. Matrix element effects are important but still a secondary effect at least in locating the proper region of q-space. Thus, it is of interest to study the generalized susceptibility

$$\chi_0(q) = \sum_{\underline{k},n,n'} \frac{|M^{\underline{k},\underline{k+q}}_{nn'}|^2 [f_{n,\underline{k}} - f_{n',\underline{k+q}}]}{E_{n',\underline{k+q}} - E_{n\underline{k}}} \tag{9}$$

$M \rightarrow 1$ (approximation)

This quantity can develop peaks either from Fermi surface nesting or from "volume effects" based on parallel electron-hole bands near the Fermi energy. Of course, this is a bare susceptibility and as the atoms are moved there will be a screening charge developed resulting in a screened susceptibility with a structure similar to a Stoner enhancement :

$$\chi(q) = \chi_0(q)/(1 - V(\underline{q})\ \chi_0(\underline{q})) \tag{10}$$

There is, of course, the more free electron like screening of the s-p electrons but an atomistic view proves quite interesting for the d- (and f-) orbitals. (Sinha and Harmon 1975). In that case, as the distortion occurs, one discusses the energetics of depopulating the d- orbitals on one site and increasing the population of the d- orbitals on another--conceptually very similar to disproportionation as discussed by Goodenough. The critical feature then is the relative electro static potential gain from forming this array of "ions" and kinetic energy penalty paid to put the extra charge on the populated ion. This phenomena shows up in structure in V(q). Thus, it is possible that structure in V(q) can produce a singular response where there is no peak in $\chi_0(q)$ but one might expect the two effects to occur together more often than not. The analysis of calculations on frozen phonons can be quite useful in sorting out the relative significance of the two phenomenon. Such calculations are in progress. However, they must be done with great care. Again, I take an example because of the considerable impression it made on me. It occured in a study of a L(2/3, 2/3, 2/3) phonon in Mo and Nb which distorts the bcc lattice to one where the unit cell contains three atoms (Harmon and Ho 1980). Considerable difficulty was experienced getting the calculation to converge as charge kept bouncing back and forth between the three atoms from iteration to iteration. That is not an uncommon phenomenon in SCF calculations involving one or more set of the mode local orbitals. It is no doubt exacerbated by the fact that the orbitals are for the same atom. But a particularly interesting experiment was performed : the displacement was taken to zero. In that case, the calculation is for a bcc crystal only not utilizing fully all the translational symmetry. One still had the same charge hopping phenomenon from iteration to iteration--which of course eventually went to zero. One can view this in one of a least two ways. That the SCF process has considerable difficulty in adjusting charges properly in response to the Coulomb interactions unless a response function is used--which is true ! Or that these charge fluctuations are not hard to drive--which seems also to be true. Certainly, it is impressive that describing a truly bcc system as one with three atoms/cell should create so much additional effort.

As mentioned before, band calculations have been performed for transition metals in many of the more complicated structures (A15, C15, $LaNi_5$, LaB_6, $ErRh_4B_4$, and idealized Chevrel phases). Of these, the cubic Laves (C15) phase systems are an excellent choice to use as an example. They are adequately complex, occur frequently, are of interest for both superconductivity and magnetism, and have a crucial role in our understanding of f-electron systems. Probably the most famous C15 systems are $TiBe_2$ and $ZrZn_2$. These are particularly interesting systems as itinerate magnets and, possibly, as test cases for triplet superconductors (under pressure). Ce doped $LaAl_2$ is the classic reentrant superconductor, and $CeRu_2$ has been the host for many studies looking for coexistence of magnetism and superconductivity.

The C15 structure is a cubic structure with the formular unit AB_2. There are two formula units or six atoms per unit cell. The A atoms are arranged on a diamond lattice. The diamond lattice itself can be viewed as a bcc lattice with alternate sites empty. The B atoms of the Laves system form tetrahedra which fill those sites. The formula unit is not AB_4 because all vertices are shared. In fact, the B atoms form intersecting chains somewhat like a geodesic network. Atomic sizes are governed by A-A and B-B contacts rather than A-B contacts so the definition of muffin-tin and/or Wigner-Seitz radii is relatively straightforward. Because the holes in the diamond sublattice have been filled, the cubic Laves phase is a close packed structure (71% filling). Thus it is actually very favorable for techniques originally designed for close packed metals such as the LMTO. However, the site symmetry is tetrahedral rather than cubic so an L = 3 component enters rather than the first term being L = 4. This happens in hcp crystals as well and the A15 linear chain atoms and the $L1_2$ (Au_3Cu) face centered atoms actually have L = 2 components. The significance of the lower L potential components can be appreciated by considering the possible couplings that can occur. For a cubic site with L = 4, one finds d-d (t_{2g}-e_g), (f-p), and (g-s) couplings. The tetrahedral site symmetry permits additional (d-p) and (f-s) couplings. Thus spherical approximations can be worrisome as the symmetry is lowered. Certainly, the shape approximations entail some error but surprisingly not as much as might be expected. In the A15's, the major changes are found locally about the M point (Jarlborg 1979, Klein et al. 1978, Pickett et al. 1979, van Kessel et al. 1978) which is also sensitive to Brillouin zone sampling. In the C15 structure, the comparisons are not as complete but sensitivity has been observed in the region around X for $TiBe_2$ and $ZrZn_2$. However, in a number of these systems, the dominant feature in the vicinity of the Fermi energy is bond forming properties of the diamond lattice A atom d- or f-orbitals. A great deal can be said about $ZrZn_2$ for example, by performing a model calculation for a Zr diamond lattice (Koelling et al. 1971). The formation of those bonds is kinetic energy driven and is much less sensitive to potential anisotropy. (The wave functions, of course, are highly directional which is critical to the success of the model calculation).

Because the atomic sizes are individually controlled by the sublattices, a very convenient model can be used. Given the resultant Wigner-Seitz radii for each site, a potential, and the assumption of metallic electronic structure, it is easy to determine the energy range of the bands arising from a particular orbital species using Wigner-Seitz type rules. If one is willing to construct a model partial density of states and sum to get a total density of states, one can determine a Fermi energy and a set of occupation numbers for each atom. One can then use those occupation numbers in the atomic calculations used to construct an OCD potential. This can be carried to configurational consistency. This model, while quite crude, does reasonably well in adjusting the potential. In the case of YAl_2, the potential constructed in that fashion was used to start the LMTO-SCF calculations (Jarlborg and Freeman 1980). It was found that within the occupied bands the results of the initial potential very nicely coincided with those of the final potential. I personally have been involved in two efforts dealing with the cubic Laves phases. In one, we used such partially or configuration self-consisted potentials and then studied the systems with fully relativistic (spin orbit included) APW techniques including the interstitial potential variations directly (warped muffin tin (WMT) approximation) and the non-spherical potential variations inside the spheres perturbatively (de Groot et al. 1980, van Duersen et al. 1981). Consistent with experience in elemental transition element systems, the Slater exchange was used to stimulate relaxation effects. In the other, we used the SCF-LMTO method with semi-relativistic (no spin orbit coupling) kinematics (Jarlborg et al. 1981). The proper LDA and LSDA exchange correlation functionals were used. These calculations were performed with the standard spherical potentials and the interstitial tail function energy equal to zero. Both these approximations may be expected to have some effect at the Fermi energy. One of my objectives in working with both efforts was to bring about a detailed comparison of various effects. Unfortunately, the realities of having the two efforts separated by the Atlantic ocean have severely limited the accomplishment of that objective. Nonetheless, it has proven useful.

In the case of $ZrZn_2$, and $TiBe_2$, the two calculational efforts yield similar results with the OCD model potentials giving better Fermi surface results. One would be tempted to conclude that this is consistent with experience in the elements. This can not be not a clear cut conclusion, however, because of the various other approximations made.

The SCF calculations for spin polarized $ZrZn_2$ yield a very extended spin density which agrees with the strongly itinerant character observed using neutron scattering. The calculations do show an additional large spin density deep within the core (< 0.5 au) but this is too compact to be seen by the neutron experiment as it did not extend far enough in reciprocal space. The calculated moment is somewhat smaller (0.14 versus 0.18 μ_B observed) and the spherical component of the form factor appears somewhat less extended than the experimental results. The significance of this discrepancy is difficult to assign because $ZrZn_2$ normally forms with Zr defects. The radial spin density shows a strong delocalization typical of the bottom of a band or of the formation of a bond.

$TiBe_2$ is a special puzzle. The ground state of the pure material is probably a very strongly enhanced bat paramagnetic state. When alloyed with copper, the copper substitutes the magnetically inactive Be. With 10% Cu (i.e., $TiBe_{1.8}Cu_{0.2}$), the system is ferromagnetic. Both by considering the Stoner enhanced paramagnetic state (Jarlborg and Freeman 1980) and the spin polarized state (Jarlborg et al. 1981), the SCF-LMTO calculations predict a ferromagnetic state with 0.19 μ_B moment for $TiBe_2$. When compared to the form factor measured on the alloy, it is found that the calculations properly reflect a tendancy for the spin density to be more local moment like centered on the Ti. Even in the Cu alloy, the moment is only 0.11 μ_B. This is very interesting as it thus is a good candidate for the exception to the rule that the LSDA always correctly predicts the existence or non-existence of a moment. However, the question of whether it is the LSDA model or approximations in its application is still open. $TiBe_2$ is strongly enhanced and has a very high density of states at or near the Fermi energy. Therefore, very small errors in calculating the enhancement factor in the paramagnetic calculations will have a precipitous effect. In the spin-polarized calculation, the question of metamagnetism arises because the calculation is started by first applying a driving field. The results are somewhat sensitive to Brillouin sampling. One must also consider the limitations of the method applied in that the system is critically sensitive to the Fermi surface. If such questions are resolved favorably and the discrepancy persists, then one suspects the spin independent part of the functional will prove the difficulty.

$TiBe_2$ and $ZrZn_2$ have a great deal in common with Pd and Pt in that the critical Fermi surface region is near the hexagonal face around L and between L and W. There are some very flat bands there giving rise to the high density of states. In the configuration consisted model potential calculation for $TiBe_2$, the bands are below the Fermi energy at W whereas in the SCF calculation they are at the Fermi energy. Thus although the model potential calculation places the d- bands lower (≈ 20 mRy) and makes them roughly 10% narrower, it yields as lower density of states at the Fermi energy. This could be due to neglected non-spherical potential terms as both at L and at W, the states in question are d-p admixtures. As mentioned previously the L = 3 term in this structure can induce coupling of those components. If it is not this approximation which is the cause, then the misplacement of the relative position of the d orbitals by the LDA is the next natural candidate.

V. SYSTEMS WITH f-ORBITALS

The systems with f-orbitals in the conduction band region strain the limitations of density functional theory and its band theory realization. The reason lies in the compact nature of the orbitals as can be seen from atomic considerations. The principal quantum number n is the most significant parameter determining the radial extent of an orbital. The angular quantum number P actually has a secondary effect--the orbitals having more extended tails the higher the value of ℓ. Now one notes that the partially filled shells are the 4f, 5d, and 6 s-p for the rare earth elements and the 5f, 6d and 7 s-p for the actinides. This means that the principal maximum of the f orbital is just inside that of the core p states (5p or 6p) in most cases. The exceptions are Ce and the lighter actinides. The Coulomb interactions are very important and the atomic multiplet structure is often dominant with very weak solid state interactions. As a result, a very useful treatment of the rare earth systems is to view the 4f shell as an atomic shell having only weak interactions (other than the direct Coulomb field) with the conduction electrons and then perform band calculations for the remaining conduction states alone. This model relates well to the magnetic structures in the rare earths through Fermi surface nesting features describing the peaks in the response function. It also does reasonably well in describing the Fermi surface of Gd where de Haas-van Alphen data is available (Young et al. 1973, Schirber et al. 1976). In such a model, the rare earths may be viewed as having a definite valence : all trivalent (3+) except Eu and Yb which are divalent (2+). A great deal of systematics has been built up with that view in terms of size of ions and known behavior. In some cases, some of the rare earths behave as though they had a different but definite valence Eu^{3+} or Yb^{3+}. This behavior, though most extreme for the 4f's, is not unique. There are a number of 3d systems which can be understood in the same model. These systems in this model are not the systems of interest for this section as the conduction bands are then those of a transition metal and the f-orbitals are the perview of those who would discuss multiplet structure.

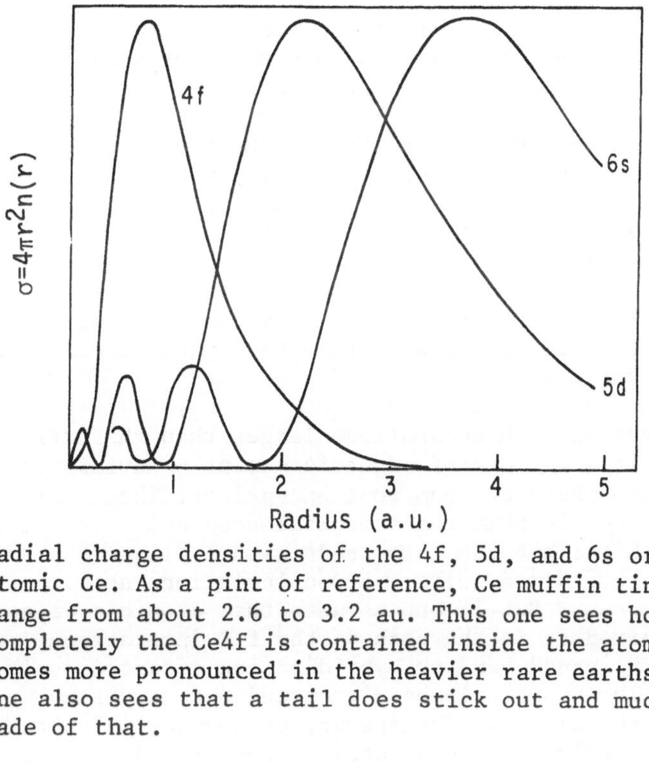

Fig. 5. Radial charge densities of the 4f, 5d, and 6s orbitals in atomic Ce. As a pint of reference, Ce muffin tin radii range from about 2.6 to 3.2 au. Thus one sees how nearly completely the Ce4f is contained inside the atom. This becomes more pronounced in the heavier rare earths. However, one also sees that a tail does stick out and much is to be made of that.

Many of the light actinide systems and a number of the rare earth (especially Ce) systems do not behave as if they had a particular valence. The f-orbitals in these systems not only have their principal maximum outside that of the p-core orbitals, but have much more extended tails. This raises the possibility of an f-band and led the late H.H. Hill to make a series of plots in which he compared the actinide (or Ce) separation (adjusted to 12 fold coördination) to the occurence of a magnetic moment or of superconductivity (Hill 1970). These so called Hill plots are quite informative. There appears a critical separation below which no magnetic moment appears unless driven by Co, Fe, or Ni in the system. The interpretation of this is that an f-band is formed and has enough width that the energy balance is not favorable to magnetism. There are almost no exception to this rule for small separations. Using the Hill plots, we can justify our previous assertion that the cubic Laves phases hold a key position in the understanding of f-orbitals systems. One finds that in the Laves phases, the Ce or actinide sites are quite close together. In Np, the Laves phases span the critical separation and this has provided the opportunity to investigate the effect by alloying (Haldned $CePt_2$ particularly interesting as it also occurs below the critical separation and yet exhibits a moment. These Hill empirical correlation plots are not limited to f-electron systems. They work very well for V and Cr as well. The 3d transition metals have much in common with the f-electron systems.

Most systems with separations larger than this critical separation exhibit some form of magnetic moment behavior. In this regime, however, there are numerous exceptions. The interpretation of those systems is that the f-orbital need not interact only with another f-orbital but can form bonds with a "ligand" p- or d-orbital. We have performed quite a few calculations showing that that does indeed occur. But if one assumes that atomic overlap is all that is required as in the case of the f-f interaction, one must ask why the "ligand" bond or hybridization effect doesn't occur more often. The f-orbital doesn't extend out very far but the atomic ligand orbital does. The answer, of course, is that the f-orbital resides mostly inside the atom so the ligand orbital must be orthogonalized to the underlying core states in that region. And, as we know from pseudopotential theory, that process acts as a repulsive potential. Thus, the formation of bonded or hydridized states in these large separation systems is more a property of the alloyed atom. It is revealing that UsI_3, UGe_3 and USn_3 are described as itinerant f systems while $CeSn_3$ is a mixed valent system and URh_3 is an itinerant f system while $CeRh_3$ is a "tetravalent" system--with one f electron. UPd_3 is not a Au_3Cu structure so its relation to mixed valent $CePd_3$ is less clear.

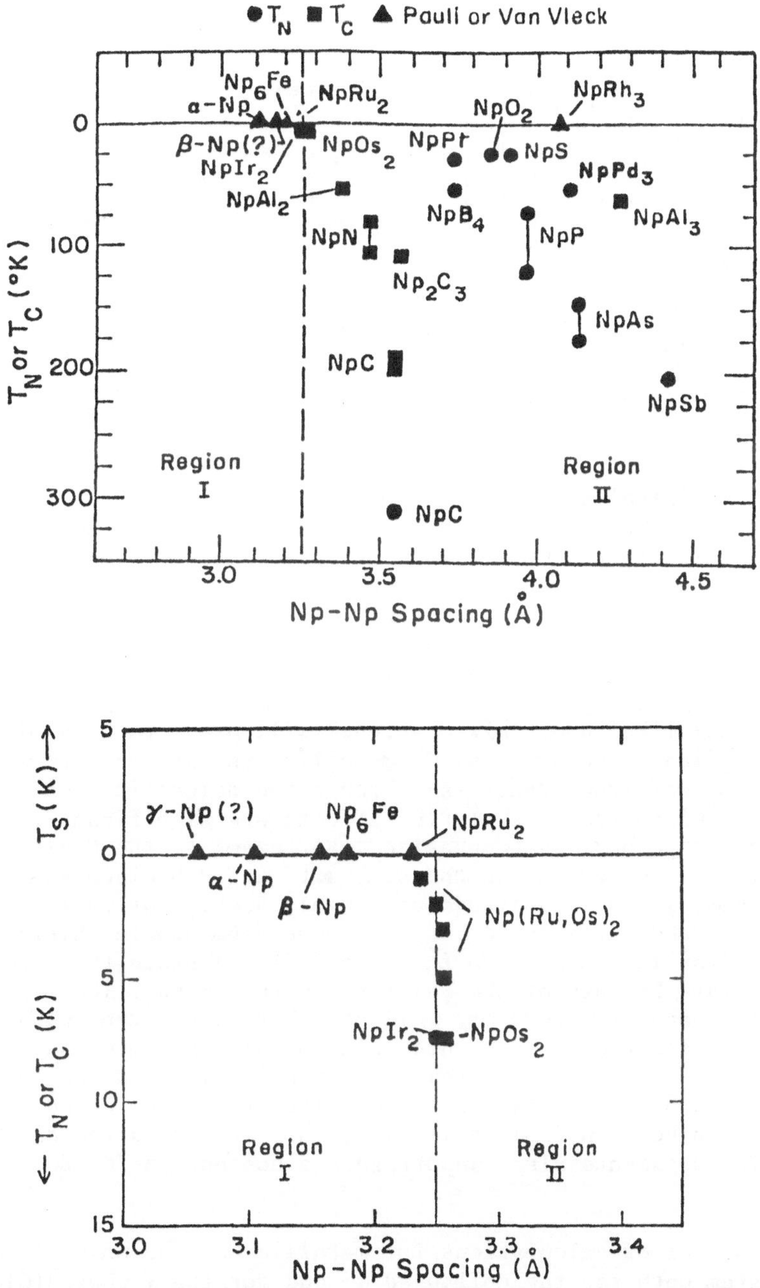

Fig. 6. Empirical correlation plots made by Hill (1970) for Np. Note the significant role played by the cubic Laves phases. Undated and improved versions were kindly supplied by J.L. Smith.

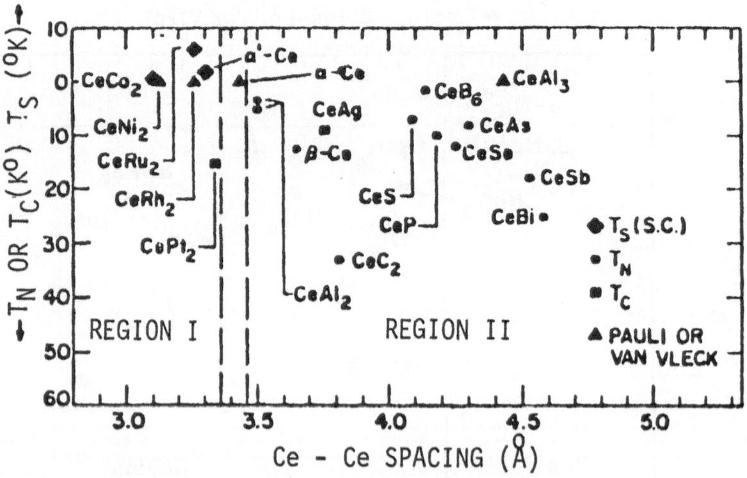

Fig. 7. Empirical correlation plots made by Hill (1970) for Ce. $CeSn_3$, $CePd_3$, and $CeRh_3$ would appear far to the right in this figure. $CeMg_2$, which acts as a true local moment rare earth compound occurs to the right of the dashed lines indicating the critical region. Thus one can see that the cubic Laves phases play a pivotal role in understanding Ce f-orbital structure.

In all likelihood, mixed valence will prove to be a misnomer where applied to Ce systems. Ce metallic systems are intermediate valent the way iron itself is--through the delocalization of the orbitals. Of course, correlation effects are significant in these systems but perhaps overemphasized. For example, the Coulomb repulsion of two f electrons on the same site is much discussed. In most Ce systems, it is established as roughly 5 eV. That Coulomb integral is indeed quite large but one knows from atomic physics that it takes far less energy to form the $Ce(f^2s^2)$ state than the Ce (d^2s^2) state because of the relaxations that occur. The implication is that a band structure may well be a better starting point for discussing electronic structure than normally assumed and that a tetravalent metallic Ce system is energetically quite unfavorable. We have performed a number of band calculations for Ce compounds and, as can be seen from Table VIII, this is indeed the case. Alloy modelling arguments also support such a contention (de Boer et al. 1979).

A number of calculations for metallic Ce find an f-electron in the system both for the collapsed α- and for the γ-phase (Glötzel 1978, Pickett et al. 1981). A detailed examination of the properties of the system shows the band results including an f-electron to be consistent with the low temperature properties. Of particular interest is the very approximate result that the specific heat would require an enhancement such that $\lambda_{tot} \approx 0.9$ and an estimation

Table VIII. Some results of SCF calculations on a series of La and Ge compounds. (Self consistency is achieved omitting the spin orbit coupling but is included for final analysis of the system. Except for the density of states with spin orbit coupling, these results omit its effect as it is believed to be small for these parameters). The partial charges give the charge coming from wavefunctions with that character within the muffin tin spheres. These are single site quantities that should not be confused sith LCAO-Mulliken analyses. Density of states units are Ry^{-1} $spin^{-1}$ and lengths are in atomic units.

	Material	$LaSn_3$	$CeSn_3$	$LaPd_3$	$CePd_3$	$CeRh_3$	$LaRu_2$	$CeRu_2$	$CeRh_2$
	a	9.000	8.863	8.000	7.765	7.600	14.586	14.240	14.24
	q_s	0.16	0.17	0.14	0.13	0.12	0.23	0.25	0.15
	q_p	0.18	0.14	0.18	0.20	0.24	0.24	0.24	0.22
R.E.	q_d	1.15	1.20	0.95	0.90	1.11	1.18	1.38	1.31
	q_f	0.20	1.16	0.37	1.34	1.25	0.20	0.97	1.01
	Q_{TOT}	1.80	2.79	1.67	2.59	2.78	1.88	2.91	2.79
	R_{MT}	3.18	3.13	2.82	2.765	2.687	3.158	3.082	3.082
	q_s	1.55	1.52	0.54	0.50	0.48	0.43	0.45	0.46
	q_p	1.71	1.71	0.34	0.33	0.34	0.27	0.33	0.34
X	q_d	0.17	0.17	8.75	8.57	7.65	5.82	5.95	7.07
	Q_{TOT}	3.49	3.47	9.69	9.47	8.55	6.59	6.80	7.95
	R_{MT}	3.18	3.13	2.82	2.765	2.687	2.517	2.517	2.517
	$n(\varepsilon_F)$-noSO	n.av.	32	7.8	103	24	62	51	81
	$n(\varepsilon_F)$-wSO	17	30	n.av.	86.7	42	n.av.	n.av.	n.av.

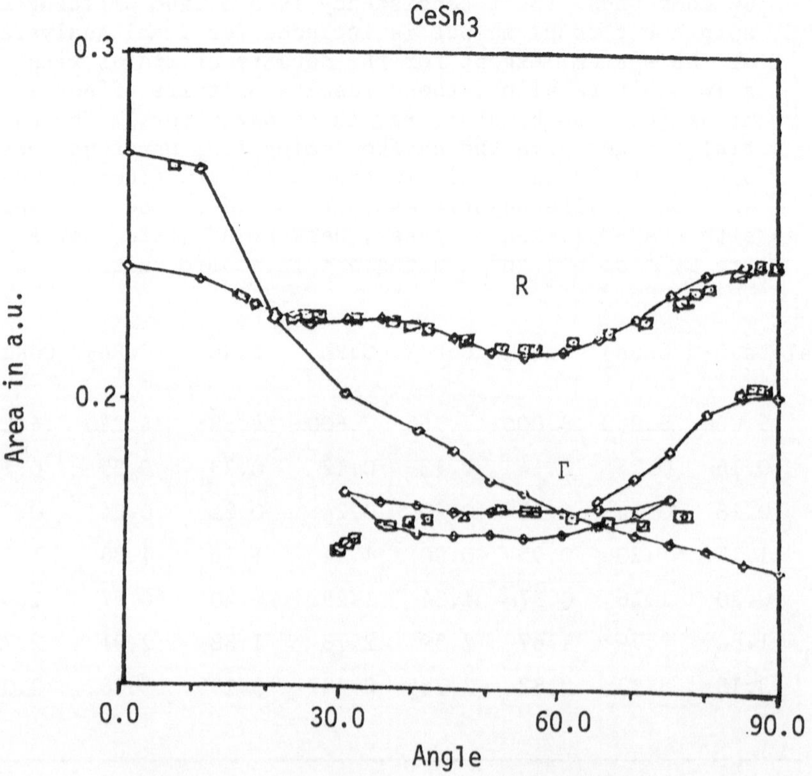

Fig. 8. Comparison of calculated extremal areas (diamonds) and dHvA frequencies (squares) for large orbits in $CeSn_3$. The branch marked R is for the octahedron centered at R in the Brillouin zone. This surface contains a relatively mild admixture of f-character and may be viewed as the simplication of a similar structure in LaSn . The branches marked Γ are for a Γ-centered surface which has supplanted those found there in $LaSn_3$. This the wavefunctions for this furface are predominantly f- in character. The centro symmetric orbits which reach to 0.27 au^{-2} at <100> (0 deg) have an extremely small amplitude factor which, when combined with the large mass, would make them almost impossible to see.

of the electron phonon enhancement would account for about half. The remaining electron-electron enhancement would then be expected to be responsible for the other half. Considerations of the superconducting properties would require spin-fluctions to destroy the superconductivity consistent with this sizable electron electron enhancement. In the paramagnetic calculations for α- and γ-Ce, there is an obvious localization of the 4f density. When spin polarization is also considered (allowing only the ferromagnetic state), it was found that γ-Ce did polarize. In LSDA, polarization makes for a more attractive potential and would enhance the localization in γ-Ce. This can, in fact, produce a phase transition. If spin-orbit effects were to be incorporated in the polarized model, it is probable that the density functional-band theory approach would properly describe the onset of the transition. However, the use of Fermi statistics in the γ-phase (SP3) might not prove adequate. That would not take one outside my definition of the boundaries of band theory. Not everyone would agree with that definition. If, however, SP2 is also not adequate, there is no argument.

The material preparation problem is servere for α-Ce which is unfortunate. An experimental measurement of the α-Ce Fermi surface and associated masses would be a very useful piece of information. The calculated Ce Fermi surface looks a great deal like that of Th except that the Γ centered superegg is not present. This would be the natural consequence of the f-character entering as discussed before for Th.

The Fermi surface has been examined in the case of $CeSn_3$ (Johansen et al. 1981). The mere observation of dHvA signals is already useful information in that it requires coherence over hundreds of Angstroms. Thus we know that the ground state must be described in terms of quantities with coherence over that distance. A hundred Angstroms is an infinity for our considerations. The experiment does not tell us that these entities can be single particle orbitals but they are a natural first choice. When the calculations are performed using an exchange only functional (Koelling 1982), the agreement of extermal areas with observed dHvA frequencies is very good. From this agreement, one can make several assertions : 1) Density Functional theory in its band theory form is still applicable to a system with correlation as strong as $CeSn_3$. 2) More particularly, the use of Fermi statistics (SP3) to occupy states has not broken down. 3) The inclusion of correlation in the LDA is inadequate. In those cases where correlations have been included in the LDA for Ce compounds and even La compounds, its attractive nature has pulled the f-character down and resulted in too much f character much in the same way as one found too much d character in the group VB elements. The reader is specifically pointed to the works of Hasegawa and Yanase to examine this effect.

It is interesting to compare the $LaSn_3$ and $CeSn_3$ Fermi surfaces.

In $LaSn_3$, there exists a large piece of surface of a very complex structure which could be viewed as centered at R (the <111> corner of the cubic Brillouin zone) and two Gamma centered pieces which actually contain some small f-character admixture. In $CeSn_3$, the R-centered surface remains but has been shrunk to a simple octahedron which is very well described by the band structure calculations. The two Gamma centered pieces in $LaSn_3$ have been replaced by a single convolated sheet of very high f-orbital content states. Note that here, as in the comparison of α-Ce and Th, the f-occupany occurs near the center of the zone.

The consideration of mass enhancements reveal a new feature. In $LaSn_3$, the comparison of the bare band density of states to the electronic specific heat yields an enhancement consistent with the electron phonon enhancement of a good superconductor. $LaSn_3$ is a six degree superconductor. On the other hand, $CeSn_3$ would require an enhancement factor of seven ! ($\lambda = 6$). If we assume the electron-phonon enhancement is the same as $LaSn_3$($\lambda = 6$)--it is probably somewhat larger, then one has $\lambda_{e-e} = 5.4$! Much of this may well be due to spin fluctuations since an applied field of 10 tesla reduces λ by 2. However, since within density functional theory the electron mass enhancement can be much larger than the electron gas results, it will be important to explore just what value might result before seeking the more exotic explanations.

Another experiment that reveals the delicate nature of the system is the induced neutron form factor (Stassis et al. 1979). That experiment can be explained by assuming a Ce d(eg)-Ce f hybridization. This is the most likely Ce-Ce interaction as the $(3z^2-r^2)$ e_g orbital and the $z(5z^2 - 3r^2)$ f orbital are both (001) directed which is the Ce near neighbor direction in $CeSn_3$. In the case of the not unrelated UGe_3, we (Arko and Koelling 1978) had looked explicitly for such hybrids and found none. In $CeSn_3$ there appears to be some admixture mediated by Sn-p character. But they are all a half volt below the Fermi energy. These would not be relevant to a model which assumes that the induced spin density is a simple splitting and repopulation at the Fermi energy because that would reveal only the charge density of the Fermi energy states. That experiment will require a more sophisticated examination of the response.

The essence may well be similar to an argument I have used for an aspect of the UGe_3, induced magnetization experiment (Lander et al. 1980). In that experiment, one found a reasonable U form factor but also evidence of a patch of magnetization very close to the Ge atom (about 1 au). That data could be fit assuming something like a Rh d orbital on the Ge site which of course had significance only in the compactness of the magnetization. The following simple model might explain that result. One must assume that the response of the f-orbitals to the applied magnetic field is large (it is) and that the changes in those orbitals has a greater effect on the remaining states than the applied field. Then one can get an effect like a transferred core polarization. If that were to be represented as a simple constant potential shift between the two "spin-states", the orbital change (not a population change !) would yield a density like the energy derivative. The energy derivative of the Ge p orbital has a large maximum in precisely that region ! Further, in the later stages of the SCF process where the f-occupation is still mildly oscillating, one observes that the maximum Ge deviations occur precisely in that region. This model, while unproven, is rather plausible and does have some suggestive support.

If this were to prove the resolution of the experimental data, it could still be compatible with the good agreement of zero field areas with 10-15 Tesla dHvA experimental data as one knows that wavefunction properties can change first. (Note the case of Pt anisotropy). An example is that energies are usually stable in the SCF process quite a bit before the charge density is. Nonetheless, the response of $CeSn_3$ to even the mild excitation and perturbation is still an open problem.

It is dangerous to pin too much on the analysis of a single material. Thus, $CePd_3$ would make a nice companion to $CeSn_3$ since the Pd d-states are buried by 2 eV and one expected Pd to offer mostly p states at the Fermi energy. It turns out that through f-d hybridization there is some Pd-d character at the Fermi energy. But another interesting feature appears. In $CeSn_3$, the f-states are not at all pure j states but are more appropriate to 1 = 3 with a spin orbital. In $CePd_3$, the f states (which are all at the Fermi energy) are nearly pure $f_{5/2}$ states, the spin orbit coupled orbital. Further, the Pd-d states which admix are very predominantly $d_{5/2}$ states. $CeRh_3$ gives some insight here. In $CeRh_3$, much of the Ce f characteris hydridized down is the Rh d-bands. That part is strongly j mixed. The f character at the Fermi energy just above the Rh-d bands is much more predominantly j = 5/2 although less so than in $CePd_3$. It would seem that one could formulate a rule of thumb that the better the hydridization, the more 1-s like the orbitals and the weaker the hydridization, the more j like the orbitals. Clearly spin orbit coupling is an important feature in these systems. Note the difference in the density of states in Table VIII.

$CeAl_2$ presents another interesting paradox which would like to mention briefly. $CeAl_2$ is a cubic Laves phase system with a Ce-Ce separation larger than the critical separation. One therefore expects minimal direct f-f interaction except by interacting through the Al-p orbitals. If one includes the f-orbitals in the band calculation, one correctly finds that $CeAl_2$ should be antiferromagnetic with a moment of 0.89 μ_B. In this simplified antiferromagnetic structure, the Al's can carry no net moment. Then, on the other hand, much can be understood using the original rare earth model where the f's are atomic like and the remaining conduction electrons are treated separately. Most interestingly, the Q-vector of $CeAl_2$ would be a nesting vector for the $LaAl_2$ calculation (Jarlborg et al. 1983).

Before departing the rare earths, it would be remis not to mention the situation for Gd. A spin polarized calculation for Gd (Harmon et al. 1978) including the 4f's would get a reasonable moment except that some very small 4f spin down character becomes occupied (0.03 el). This one might expect to trace to the undue favoring of f-states by the charge density part of the LSDA. Of course, spin orbit and neglected non-spherical effects could enter but I obviously have my prejudgices.

Because I was first involved in actinides, they have greatly influenced my views of the Ce systems. I believe rightly so. Let us consider a few aspects of calculations in actinide systems.

The actinide elements present something of a problem because they not only exhibit f-orbitals in the conduction bands but have reasonably complex crystal structures. It is not that there are too many atoms per unit cell ; α-U has only two, for example ; but that they have very low site symmetry. α-U may be viewed as stacked up sheets of corrugated cardboard. One is, in fact, suspicious that these complicated crystal structures are another manifestation of the presence of the f-orbitals. These systems also require relativistic kinematics and are even near the point where the single particle Dirac equation breaks down for the point nucleus approximation. That occurs at $Z = 137$. Oddly enought the early non-relativistic self consistent calculations did better than one would expect (Hill and Kmetko 1975). This is probably due to the way in which the radial solutions were started at small r. Nonetheless, those results were more similar to the early non-self consistent but fully relativistic calculations than were predicted by setting $c \to \infty$ except that they far underestimated the s-p character as should be expected. The SCF-LMTO calculations performed to examine the cohesive energy properties were performed for idealized fcc structured systems (Skriver et al. 1978). Those calculations very effectively demonstrated the critical role of the f-orbitals play in the bonding. They also raise a number of questions as well which point to new questions which should be the basis of future research. I have already expounded on what I think those questions are so I will leave the discussion of the actinide metal cohesive energy studies to Prof. Andersen.

The actinide compounds are often simpler conceptually than the elements because, although there may be more atoms per unit cell, they may be more symmetrically arranged. Actinide metallurgy is sufficiently difficult that the ability to prepare samples is a strongly controlling factor in the development of the field. It is thus very fortunate that the rock salt structures of pnictides and chalcogenides can be prepared as single crystals. They have been studied extensively theoretically by several theoretical groups (Brooks and Gloetzel 1980, Erbudak and Keller 1979, Weinberger et al. 1979, 1980, Podloucky and Weinberger 1981). However, they exhibit magnetic moments and in many cases may be more amenable to a definite valence rare earth type approach. UN and UC are definite exceptions and there is evidence one should not abandon the band approach too soon for the remainder. In these systems the actinide is ionized such that it retains little or no s-p character but still retains d- and f-orbital character which are strongly admixed into the nominally pnictide of chalcogenide p bands. The admixed d's tend to be found more towards the bottom of the p bands and the f's toward the top. Their is an Fano antiresonance effect whereby the d character is forced out of the energy regime of the f-orbitals. This is een as a negative polarization of the d-states in photoemission. It is nicely shown by the cluster Green's function calculations of Keller (Erbudak et al. 1979).

The inclusion of spin orbit coupling is also critical especially when dealing with the heavier pnictides and chalcogenides (Weinberger, priv. comm.). In that case, the oft used approximation that the large component of the radial function is the same for $j = 1 \pm 1/2$ is just not adequate as was found for lead.

URh_3 (Arko et al. 1975), UIr_3 (Arko 1976) and UGe_3 (Arko and Koelling 1978) have all been grown as single crystals and dHvA data is available. Considerable success has been achieved using OCD model potentials for these systems. We have now performed self consistent calculations for these systems. In the case of UIr_3 the SCF calculation appears to improve the Fermi surface properties of the calculation. UIr_3 has also been studied by angular resolved photoemission. Somewhat surprisingly, this excitation spectrum appears to relate well to the ground state band structure calculation without relaxation effects.

REFERENCES

Aldred, A.T., Dunlap, B.D., Lam, D.J. and Nowik, I., (1974), Phys. Rev.B 10, 1011-1019.
Aldred, A.T., Lam, D.J., Harvey, A.R. and Dunlap, B.D., (1975), Phys.Rev.B 11, 1169-1175.
Andersen, O.K., (1975), Phys.Rev.B 12, 3060-3083.
Arko, A.J., (1976), Proc. 2nd Int. Conf. on the Electronic Structure of the Actinides, Ossolinuem, p. 309-314.
Arko, A.J. and Koelling, D.D., (1978), Phys.Rev.B 17, 3104-3114.
Arko, A.J., Brodsky, M.B., Crabtree, G.W., Karim, D., Koelling, D.D. and Windmiller, L.R., (1975), Phys.Rev.B 12, 4102-4112.
Boyer, L.L., Papaconstantopoulos, D.A. and Klein, B.M., (1977), Phys.Rev.B 15, 3695-3693.
Brooks, M.S.S. and Gloetzel, D., (1980), Physica 102B, 51-58.
Bross, H., (1964), Physik Kondens. Mat. 3, 119-38.
Bross, H., Bohn, G., Meister, G., Schubö, W., Stöhr, H., (1970), Phys.Rev.B 2, 3098-103.
Crabtree, G.W., Dye, D.H., Karim, D.P. and Koelling, D.D., (1979) Phys.Rev.Lett. 42, 390-393.
De Cicco, P.D. and Kitz, A. (1967), Phys.Rev. 162, 486-491.
De Boer, F.R., Dijkman, W.H., Mattens, W.C.M. and Miedema, A.R., (1979), J.Less-Common Met. 64, 241-253.
De Groot, R.A., Koelling, D.D. and Mueller, F.M., (1980), J.Phys.F 10, L235-240.
Dongarra, J., Gabriel, J.R. and Wilkensen, J.H., (1982a), to be published.
Dongarra, J., Gabriel, J.R. and Koelling, D.D., (1982b), to be published.
Elyashar, N. and Koelling, D.D., (1977), Phys.Rev.B 15, 3620-3632.
Erbudak, M. and Keller, J.,(1979), Z.Physik B 32, 281-286.
Erbudak, M. and Keller, J., (1979), Phys.Rev.Lett. 42, 115-118.
Erbudak, M., Greuter, F., Meier, F., Reihl, B., Vogt, O. and Keller, J. (1979), J.Appl.Phys. 50, 2099-2101.
Erbudak, M., Greuter, F., Meier, F., Reihl, B. and Keller, J. (1979), Sol.St.Commun. 30, 439-441.
Freeman, A.J. and Jarlborg, T., (1979), J.Appl.Phys. 50, 1876-1879.
Fritsche, L., Rafat mehr, M., Glocker, R. and Noffke, J., (1979), Z.Physik B 33, 1-12.
Glötzel, D., (1978), J.Phys.F 8, L163-L168 : correction 8, L205.
Gunnarsson, O. and Lundqvist, B.I., (1976), Phys.Rev.B 13, 4274-4298.
Gunnarsson, O., Lundqvist, B.I. and Wilkins, J.W., (1974), Phys. Rev.B 4, 1319-1327.
Gunnarsson, O., Jonson, M. and Lundqvist, B.I., (1979), Phys.Rev.B 20, 3136-3164.
Harmon, B.N. and Ho, K.M., (1980), in Superconductivity in d- and f-band metals, H. Suhl and M.B. Maple (eds.), Academic Press, London, p. 173-180.
Harmon, B.N. and Koelling, D.D., (1974), J.Phys.C 7, L210-L215.

Harmon, B.N., Schirber, J.E. and Koelling, D.D., (1978), Inst.Phys. Conf. Ser. 39, 47-51.
Harris, J. and Jones, R.O., (1978), J.Chem.Phys. 68, 3316-3317.
Hasegawa, A., (1981), J.Phys.Soc.Jpn. 50, 3313-3320.
Hasegawa, A. and Yanase, A., (1977), J.Phys.F 7, 1245-1260.
Hasegawa, A. and Yanase, A., (1980), J.Phys.F 10, 847-58.
Hill, H., (1970), Plutonium 1970 and other Actinides, W.N. Miner (ed.) Nuclear Metallurgy 17, 2-19.
Hill, H.H. and Kmetko, E.A., (1975), in Heavy Element Properties, W. Mueller and H. Blank (eds.), North-Holland, p. 17-27.
Hohenberg, P. and Kohn, W., (1964), Phys.Rev. 136B, 864-871.
Janak, J.F., (1977), Phys.Rev.B 16, 255-262.
Janak, J.F., (1978), Phys.Rev.B 18, 7165-7168.
Janak, J.F., Williams, A.R. and Moruzzi, Y.L., (1975), Phys.Rev.B 11, 1522-1536.
Jarlborg, T., (1979), J.Phys.F 9, 283-305.
Jarlborg, T. and Freeman, A.J., (1980), Phys.Rev. 22, 2332-2342.
Jarlborg, T. and Freeman, A.J., (1981), Phys.Rev.B 23, 3577-2579.
Jarlborg, T., Freeman, A.J. and Watson-Yang, T.J., (1977), Phys. Rev.Lett. 39, 1032-1034.
Jarlborg, T., Freeman, A.J. and Koelling, D.D., (1981), J.Magn. and Magn. Mat. 23, 291-298.
Jarlborg, T., Freeman, A.J. and Koelling, D.D., (1982), J.Appl.Phys. 53, 2140-2141.
Jarlborg, T., Freeman, A.J. and Koelling, D.D., (1983) (to be published).
Jepsen, O., Glotzel, D. and Mackintosh, A.R., (1982), Phys.Rev.B 23, 2684-2696.
Johanson, W.R., Crabtree, G.W., Edelstein, A.S. and McMasters, O.D., (1981), Phys.Rev.Lett. 46, 504.
Klein, B.M., Boyer, L.L., Papaconstantopoulos, D.A. and Mattheiss, L.F., (1978), Phys.Rev.B 18, 6411-6438.
Koelling, D.D., (1970), Phys.Rev.B 2, 290-298.
Koelling, D.D., (1974), Int.J.Quant.Chem. 85, 473-81.
Koelling, D.D., (1979), J. de Physique C4, 117-123.
Koelling, D.D., (1981), Rep.Prog.Phys. 44, 139-212.
Koelling, D.D., (1982), Sol.St.Commun. (in press).
Koelling, D.D. and Freeman, A.J., (1975), Phys.Rev.B 12, 5622.
Koelling, D.D., Johnson, D.L., Kirkpatrick, S. and Mueller, F.M., (1971), Solid State Commun. 9, 2039-2043.
Kohn, W. and Sham, L.J., (1965), Phys.Rev. 140A, 1133-1138.
Kuebler, J., (1980), J.Mag. and Magn.Mat. 20, 277-284.
Lam, L. and Platzman, P.M., (1974), Phys.Rev.B 9, 5122-7.
Lander, G.H., Reddy, J.F., Delapalme, A. and Brown, P.J., (1980), Phys.Rev.Lett. 44, 603-606.
Liu, K.L., MacDonald, A.H., Daams, J.M., Vosko, S.H. and Koelling, D.D., (1979), J.Mag. and Magn.Mat. 12, 43-57.
MacDonald, A.H., (1980), J.Phys.F. 10, 1737-51.
MacDonald, A.H., Liu, K.L. and Vosko, S.H., (1977), Phys.Rev.B 16, 777-784.

MacDonald, A.H., (1979), J.Phys.F 9, L99-106.
MacDonald, A.H. and Vosko, S.H., (1979), J.Phys.C 12, 2977-2990.
MacDonald, A.H., Daams, J.M., Vosko, S.H. and Koelling, D.D., (1981), Phys.Rev.B 23, 6377-6398.
MacDonald, A.H., Daams, J.M., Vosko, S.H. and Koelling, D.D., (1982), Phys.Rev.B 25, 713-725.
Maglic, R., Brun, T.O., Felcher, G.P. and Chang, Y.K., (1978), J.Mag.Magn.Mater. 9, 318.
Marcus, P.M., (1967), Int.J. of Quant.Chem. 15, 567-588.
Malik, S.K., Arlinghaus, F.J. and Wallace, W.E., (1982), Phys.Rev.B 25, 6488-6491.
Mermin, N.D., (1965), Phys.Rev. 137A, 1441-1443.
Moruzzi, V.L., Janak, J.F. and Williams, A.R., (1978), Calculated Electronic Properties of Metals, Pergamon, New York.
Mueller, F.M. and Myron, H.W., (1978), Int.J.Quant.Chem. 11, 1035-41.
Nickerson, S.B. and Vosko, S.H., (1976), Phys.Rev.B 14, 4399-4406.
Pickett, W.E., Ho, K.M. and Cohen, M.L., (1979), Phys.Rev.B 19, 1734-1750.
Pickett, W.E., Freeman, A.J. and Koelling, D.D., (1981), Phys.Rev.B 23, 1266-1291.
Podloucky, R. and Weinberger, P., (1981), Z.Phys.B 42, 107-11.
Posternak, M., Krakauer, H. and Freeman, A.J., (1982), Phys.Rev.B 25, 755-761.
Rajagopal, A.K., (1978a), Phys.Rev.B 17, 2980-2988.
Rajagopal, A.K., (1978b), J.Phys.C 11, 1943-948.
Rajagopal, A.K. and Callaway, J., (1973), Phys.Rev.B 7, 1912-1919.
Rasolt, M. and Vosko, S.H., (1974a), Phys.Rev.Lett. 32, 297-301.
Rasolt, M. and Vosko, S.H., (1974b), Phys.Rev.B 10, 4195-4204.
Rasolt, M., Nickerson, S.B. and Vosko, S.H., (1975), Sol.St.Commun. 16, 827-830.
Rath, J., Wang, C.S., Tawil, R.A. and Callaway, J., (1973), Phys. Rev.B 8, 5139-42.
Sacchetti, F., (1982), J.Phys.F 12, 281-92.
Schirber, J.E., Schmidt, F.A., Harmon, B.N. and Koelling, D.D., (1976), Phys.Rev.Lett. 36, 448-450.
Schirber, J.E., Schmidt, F.A. and Koelling, D.D., (1977), Phys.Rev. B 16, 4235-4238.
Sinha, S.K., (1980), Ch I, Phonons in Transition Metals, in Dynamical Properties of Solids, G.K. Horton and A.A. Maradudin (eds.) North-Holland.
Sinha, S.K. and Harmon, B.N., (1975), Phys.Rev.Lett. 35, 1515.
Skriver, H.L., Andersen, O.K. and Johansson, B., (1978), Phys.Rev. Lett. 41, 42-45.
Skriver, H.L. and Jan, J-P, (1980), Phys.Rev.B 21, 1489-1496.
Slater, J.C., (1951), Phys.Rev. 81, 385-390.
Slater, J.C., (1953), Phys.Rev. 92, 603-608.
Snow, E.C., (1973), Phys.Rev.B 8, 5391-5397.
Soven, P., (1965), Phys.Rev. 137, A1706-1717.
Stoddart, J.C. and Davis, K., (1982), Sol.St.Commun. 42, 147-148.

Stassis, C., Loong, C.K., McMasters, O.D. and Moon, R.M., (1979), J.Appl.Phys. 50, 2091-2103.
Takeda, T. and Kuebler, J., (1979), J.Phys.F 9, 661-672.
Talman, J.D. and Shadwick, W.F., (1976), Phys.Rev. A14, 36-40.
Theophilou, A.K., (1979), J.Phys.C 12, 5419-5430.
Van Deursen, A.J.P., van Ruitenbeek, J.M., Verhoef, W.A., de Vroomen, A.R., Smith, J.L., de Groot, R.A., Koelling, D.D. and Mueller, F.M., (1982), Proc.Low Temperature Conf. 16.
Van Kessel, A.T., Myron, H.W. and Mueller, F.M., (1978), Phys.Rev. Lett. 41, 181-184.
von Barth, U. and Hedin, L., (1972), J.Phys.C 5, 1629-1642.
Vosko, S.H. and Perdew, J.P., (1975), Can.J.Phys. 53, 1385-1397.
Vosko, S.H., Wilk, L. and Nusair, M., (1980), Can.J. of Physics 58, 1200-1211.
Wakoh, S. and Yamashita, J., (1973), J.Phys.Soc.Japan 35, 1394-1401.
Wakoh, S., Fukamachi, T., Hosoya, S. and Yamashita, J., (1975), J.Phys.Soc. Japan 38, 1601-1606.
Wakoh, S., Kubo, Y. and Yamashita, J., (1975), J.Phys.Soc.Japan 38, 416-430.
Wakoh, S., Kubo, Y. and Yamashita, J., (1976), J.Phys.Soc.Japan 40, 1043-1047.
Wakoh, S. and Kubo, Y., (1980), J.Phys.F 10, 2707-15.
Wang, D.S., Freeman, A.J. and Krakauer, H., (1981), Phys.Rev.B 24, 3092-3143 ; Phys.Rev.B 24, 3104-3107.
Watson-Yang, T.J., Harmon, B.N. and Freeman, A.J., (1976), J.Mag. Magn.Mater. 2, 334.
Watson-Yang, T.J., Freeman, A.J. and Koelling, D.D., (1977), J.Mag. Magn.Mater. 5, 277.
Weinberger, P., Podloucky, R., Mallett, C.P. and Neckel, A., (1979), J.Phys.C 12, 801-817.
Weinberger, P., Mallett, C.P., Podloucky, R. and Neckel, A., (1980), J.Phys.C., 173-187.
Weiss, R.J. and Freeman, A.J., (1959), J.Phys.Chem.Solids 10, 147.
Weiss, R.J., (1978), Phil.Mag. B 38, 645-648.
Williams, A.R., Kübler, J. and Gelatt, C.D., Jr., (1979), Phys.Rev. B 19, 6094-6122.
Williams, A.R., Moruzzi, V.L., Gelatt, C.D., Jr., Kuebler, J. and Schwartz, K., (1982), J.Appl.Physics 53, 2019-2023.
Wimmer, E., Krakauer, H., Weinert, M. and Freeman, A.J., (1981), Phys.Rev.B 24, 864-875.
Wimmer, E., Weinert, M., Freeman, A.J. and Krakauer, H., (1981b), Phys.Rev.B 24, 2292-2294.
Young, R.C., Jordon, R.G. and Jones, D.W., (1973), Phys.Rev.Lett. 31, 1473.

ELECTRONIC STRUCTURE OF HYDROGEN IN METALS

Michèle Gupta

Le Centre de Mécanique Ondulatoire Appliquée du CNRS
23 rue du Maroc
75019 Paris, France

and

L'Université Paris-Sud Bâtimet 506
91405 Orsay, France

I. INTRODUCTION

In these lectures, I shall give an overview of the electronic structure of hydrogen in metals focussing mostly on transition metals. I will try to show that using the different techniques to study the electronic structure which have been developed at length during this Summer Institute, one can gain insight on the important problem of the electronic properties of hydrogen in metals. I would like to point out at the outset that this paper is not meant to be an exhaustive and complete review and that I will rather be giving my personal views on the problem and develop the points which I have found particularly interesting.

The hydride forming metals exist over a wide range of hydrogen concentrations which extend from the very dilute limit where hydrogen forms a solid solution to the concentrated stoichiometric phases[1] ; in between these two extremes we find the continuous domain of non-stoichiometric compounds MH_x. The phase diagrams of some metal hydrides, such as the group V bcc transition metals, can be quite complex. The methods used to study the metal-hydrogen systems will thus cover a broad spectrum from the problem of H as an impurity, to the case of the ordered periodic stoichiometric compounds which can be studied using the standard techniques of band structure calculations (for example the augmented plane wave (APW), the Korringa-Kohn-Rostoker (KKR) Green's function methods, or the linear muffin tin orbitals[2] (LMTO) and the augmented spherical wave[3] (ASW) techniques which have been presented at this Institute). The non-stoichiometric compounds are best treated with the presently available methods within the coherent potential approximation[4] (CPA) also presented at this Institute. The choice of the methods used to study the electronic structure will depend obviously upon the nature of the metallic matrix ; this will appear clearly in solving the impurity problem where we will treat on a different footing the screening of a proton in a simple metal and in a transition metal.

The study of the electronic structure of hydrogen in metals is particularly instructive for a theorist since it gives an example of interaction between a light interstitial element hydrogen, the simplest of all with no core structure ; and a metal. As we go along, we will focuss mostly on the problems of the screening of the proton, the chemical nature of the metal-hydrogen and hydrogen-hydrogen interactions, the modifications of the electronic properties of the metal upon hydrogenation which will be used to interpret several experimental results ; we will also study the factors which control the stability of the hydrides and try to explain why some metal hydrides are stable while others do not form. Another problem which has attracted the attention of the theorists and of the experimentalists alike is the occurrence of superconductivity in some metal-hydrogen systems with, in some cases, fairly high values of the superconducting critical temperature T_c. On the other hand, a high concentration of hydrogen has a negative effect on the superconducting properties of some high T_c metals (Nb, La etc.) ; we will therefore also discuss the electron-phonon interaction of some of the metal-hydrogen systems.

It should be also pointed out that besides the theoretical interest of the study of the metal-hydrogen systems, a large amount of experimental work is currently being devoted to some aspects of technological applications. The study of very low concentrations of H in metals and alloys is particularly interesting for the metallurgists since the presence of traces of H in a metallic matrix can modify its mechancial properties and lead to the problem of hydrogen embrittlement. For such studies, the knowledge of the interaction of H with the lattice defects (vacancies, grain boundaries, etc.) and also the interaction of H with other impurities in the metal matrix (such as the H-H or the H- other metal impurity) is essential. The role of the diffusion of H in the metallic matrix is of course crucial in the understanding of these problems. The migration of hydrogen atoms can lead to the formation of gas bubbles in metals which may result in blistering when these bubbles are formed close to the metal surface. The technological applications are also numerous for the highly concentrated metal-hydrogen systems[5]. Some metallic matrices act as sponges which can absorb and desorb hydrogen under suitable experimental conditions, the density of hydrogen being for some of these matrices larger than that of liquid hydrogen ; such metallic matrices are thus considered as possible hydrogen storage materials. Since hydrogen could become an important element as a non-polluting energy source, the problem of its storage should thus be studied. In the case of production of large amounts of hydrogen, say by electrolysis of water using the excess of energy provided by nuclear plants, the storage of H in metallic matrices would be unrealistic ; only the storage in large natural cavities of the underground can be envisaged in such a situation. However, the storage of much smaller quantities of hydrogen in metallic matrices for some quasi stationary applications, for example solar energy applications, is currently under experimental investigation. The metal matrices provide a safer storage than liquified hydrogen and do not require the use of high pressure and low temperatures technologies. Other applications of hydrogen in metallic matrices such as hydrogen burning engines for motor vehicles, chemical heat pumps, electrodes in fuel cells etc. are being tested. In view of these interesting potential applications, it appears important to broaden the fundamental knowledge of the properties of hydrogen in metals.

The present lectures will be limited to some aspects of the electronic structure of these systems and are divided as follows. Section II is devoted to the electronic properties of H as an impurity in simple and in transition metals ; in section III we will focuss on the electronic properties of stoichiometric hydrides of transition metals and lanthanides ; we shall take examples amongst monohydrides (PdH, NiH), dihydrides (TiH_2, ZrH_2, NbH_2, LaH_2, ErH_2, TbH_2) and trihydrides (LaH_3) ; a large portion of this section will be devoted to the interpretation of the experimental data in the light of the results of the electronic structure studies. We will also discuss our present understanding on the trends in the heats of formation of metal hydrides in a series of transition metals. In section IV we present very recent results on the electronic structure of the hydrogen storage materials which have a fairly complex crystal structure of the hydrogen storage materials which have a fairly complex crystal structure such as FeTiH, $FeTiH_2$, Mg_2NiH_4. In section V we will discuss briefly the electronic properties of non-stoichiometric compounds with an emphasis on the evolution of such properties as a function of hydrogen concentration ; this shall in particular bridge the gap between the electronic properties of the impurity studied in section II and the electronic properties of stoichiometric hydrides discussed in sections III and IV. In section VI we shall discuss the problem of the occurrence of superconductivity in some metal hydrides ; the discussion will be based on the theoretical estimates of the electron-phonon coupling in these materials and on a brief review of the phonon properties available from experimental work. The role of the H atoms in the metallic matrix which leads to the electron optical phonon coupling mechanism will be particularly emphasized ; we will also discuss the modifications in the electron acoustic phonon coupling due to the presence of the H atoms in the metallic matrix. A summary and concluding remarks are given in section VII.

II. ELECTRONIC STRUCTURE OF H IMPURITY IN METALS

As already mentioned in the introduction, the study of the electronic structure of a hydrogen impurity in metals has, besides its intrinsic fundamental theoretical importance, a great relevance for its applications in the field of metallurgy. In this section we shall be concerned mostly by the study of the screening of the proton, the existence of a bound state etc. We shall naturally separate the problems of the simple and of the transition metals. The proper treatment of the screening of the proton together with an adequate description of the lattice ions play an important role in the study of the heat of solution of H in metals, in the determination of the preferred site occupancy of the H atoms (why does H occupy preferentially octahedral or tetrahedral interstices in the different lattices ?), in the characterization of the diffusion paths and in the estimation of the activation energies for the over barrier jump processes involved in the classical diffusion mechanisms, in the calculation of the frequencies of the local mode vibrations etc. These questions, although very important, will be only very briefly addressed in the present section. The problem of the strain field around the H impurity and the subsequent lattice relaxation around H will just be mentioned ; this constitutes, especially in transition metals, a difficult problem for the theorists and ab-initio studies of this effect are still lacking. We wish also to mention that the study of H as an impurity can be extended to other isotopes of H such as deuterium and tritium and also to the problem of the positive muons in solids whose mass is approximately one ninth of the proton mass.

A. <u>Screening of a H impurity in simple metals</u>

(a) <u>Jellium model</u>

The Coulomb potential of the proton creates a strong perturbation in the metal matrix and the screening of the proton cannot be properly treated in the linear response theory. This point was emphasized already in the pioneering work of Friedel[6] in which he studied the heats of solution of H in noble metals. The starting point of his calculation was the self-consistent solution of a set of single particle Schrödinger equations to study the screening of a proton with one bound electron interacting with the conduction electrons via the direct Coulomb and the Slater exchange interactions ; the correlation corrections and the valence electrons exchange were not included.

The development of the density functional formalism by Hohenberg and Kohn[7] and Kohn and Sham[8], together with the availability of computers made possible self-consistent calculations of the non-linear screening of a proton in an electron gas including exchange and correlation through the local density approximation. Such calculations were first performed by Popovic et al[9], Almbladh et al[10] and later by a large number of groups[11-14]. In these calculations, the simple metals are described within the jellium model in which the ion cores of the metal form an infinite uniform positive background ; the positive charge introduced by the proton is screened by the interacting electron gas corresponding to the average conduction electron density $\rho_0 = (3/4\pi)(r_s a_0)^{-3}$, the electron density parameter r_s ranges from 2. to 5. for metallic densities. Since the density functional formalism has been discussed in detail by Von Barth[15] and Harris[16] at this Summer Institute, I shall briefly summarize the main results. The total energy E of the ground state of an interacting N particles system in an external potential V_{ext} can be expressed in terms of a unique functional of the particle density $\rho(\vec{r})$

$$E[\rho(\vec{r})] = T_0[\rho(\vec{r})] + e\int d\vec{r}\rho(\vec{r})V_{ext}(\vec{r}) + \frac{e^2}{2}\int\frac{\rho(\vec{r})\rho(\vec{r}\,')d\vec{r}d\vec{r}\,'}{|\vec{r}-\vec{r}\,'|} + E_{xc}[\rho(\vec{r})] \qquad (1)$$

where T_0 and E_{xc} are respectively the kinetic and the exchange and correlation energies. The ground state density is the one which minimizes the energy functional $E[\rho(\vec{r})]$.

The application of the variational principle leads to the self-consistent solution of a set of N one particle Schrödinger equations :

$$\{-\nabla^2 + V_{eff}[\rho(\vec{r}),\vec{r}]\}\,\psi_i(\vec{r}) = \varepsilon_i\psi_i(\vec{r}) \qquad (2)$$

from which the wavefunctions ψ_i and energy eigenvalues ε_i are generated. In the independent particle model, the total electron density is the sum over all the occupied states of the one electron densities

$$\rho(\vec{r}) = \sum_i^{occ} |\psi_i(\vec{r})|^2 \qquad (3)$$

The effective one electron potential is the sum of the external potential, the average electrostatic potential and the exchange and correlation potential V_{xc}

$$V_{eff}(\vec{r}) = V_{ext}(\vec{r}) + \int\frac{2\rho(\vec{r}\,')d\vec{r}\,'}{|\vec{r}-\vec{r}\,'|} + V_{xc}[\rho(\vec{r})] \qquad (4)$$

V_{xc} is the functional derivative of the energy functional $E_{xc}[\rho(\vec{r})]$

$$V_{xc}[\rho(\vec{r})] = \frac{\delta E_{xc}[\rho(\vec{r})]}{\delta\rho(\vec{r})} \qquad (5)$$

Using the local density approximation (LDA) E_{xc} can be expressed in terms of $\varepsilon_{xc}(r_s)$, the exchange and correlation energy per particle for an homogeneous electron gas

$$E_{xc}[\rho(\vec{r})] \simeq \int \rho(\vec{r})\varepsilon_{xc}[\rho(\vec{r})]\, d\vec{r} \tag{6}$$

The success and limitations of the LDA have been discussed already at this Summer Institute and the reader is referred to the corresponding papers[15,16]. In the case of the proton, the external potential is spherically symmetric

$$V_{ext}(\vec{r}) = -\frac{e^2}{r} \tag{7}$$

The single particle wavefunctions can be written in terms of partial waves $U_{\ell k}(r)$ through

$$\psi_{k\ell}(\vec{r}) = [U_{\ell k}(r)/r]\, Y_{\ell m}(\hat{r}) \tag{8}$$

The partial waves satisfy the radial Schrödinger equation:

$$\left(-\frac{d^2}{dr^2} + V_{eff}(r) + \frac{\ell(\ell+1)}{r^2} - k^2\right)U_{\ell k}(r) = 0 \tag{9}$$

where k is the electron wave vector labelling the state of energy $\varepsilon_k = k^2$. At large distances, the impurity is completely screened and the partial wave has the well-known asymptotic form:

$$U_{\ell k}(r) \underset{r\to\infty}{=} \cos\delta_\ell(\varepsilon_k)\, j_\ell(kr) - \sin\delta_\ell(\varepsilon_k)\, n_\ell(kr) \tag{10}$$

where δ_ℓ's are the scattering phase shifts for ℓ th partial wave and j_ℓ and n_ℓ are the spherical Bessel and Neuman funcations of order ℓ, respectively. The phase shifts at the Fermi energy must satisfy the Friedel sum rule :

$$Z = \frac{2}{\pi}\sum_\ell (2\ell + 1)\delta_\ell(\varepsilon_F) \tag{11}$$

where Z is the charge of the proton. The condition of occurrence of bound states corresponding to the ℓth partial wave is

$$\delta_\ell(\varepsilon_k = 0) = n\pi \tag{12}$$

where $2n(2\ell+1)$ is the number of bound states. The bound states correspond to negative energy eigenvalues for Eq. 9. In the present problem only s-wave bound states occur. Their wavefunction must satisfy the boundary condition

$$U_{bound} \sim e^{-k_b r} \tag{13}$$

with $|\varepsilon_{bound}| = k_b^2$ and

$$\psi_{bound} = \frac{1}{(4\pi)^{1/2}}\, U_{bound}(r) \tag{14}$$

The results of Popovic et al[9] are shown in Fig. 1 ; the displaced electron density $\Delta\rho(r) = \rho(r) - \rho_0$ around the proton is plotted as a function of the distance for a value of $r_s = 2.064$ a.u. appropriate to Al. We can see that the proton in an electron gas is well screened, better than in the free H atom since we observe a larger pile up of the electron cloud around the positive charge of the nucleus.

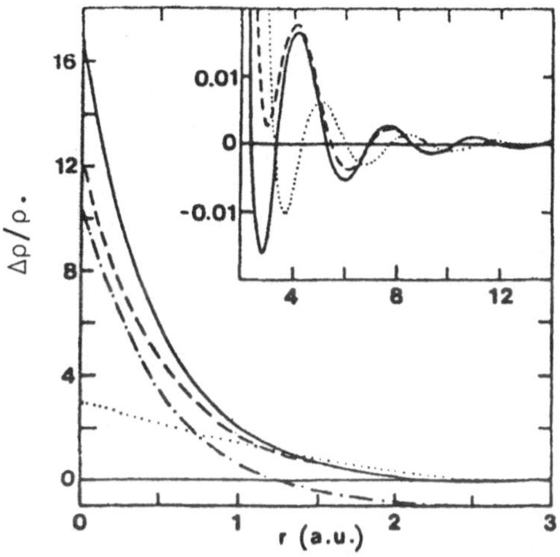

Fig. 1 : Theoretical results of Popovic et al. from Ref. 9. Displaced electron density $\Delta\rho(r) = \rho(r) - \rho_0$ in units of ρ_0 plotted against r (a.u.) for a proton in Al ($r_s = 2.064$ a.u.). The non linear theory with exchange and correlation corrections (solid line), the nonlinear Hartree theory (dash line), the linear response theory (dotted line), and the hydrogen atom (dash-dot line). Friedel's oscillations are shown in a magnified scale in the inset.

At distances greater than 2 a.u., the inset of Fig. 1 shows in a magnified scale the well known Friedel oscillations. In Fig. 1 a comparison is also made with the results of the linear response theory. It appears very clearly that the linear response theory grossly underestimates the magnitude of the screening of the proton and moreover that the phase of the Friedel oscillations is shifted by about 90° from the results of the non-linear response. This behaviour is expected to lead to erroneous answers in the calculation of the heats of solution of H when the interaction between the electron screening charge and the metal ions is calculated. It has also been shown that the correct behaviour of the phase of the Friedel's oscillations plays an important role in the study of the nuclear quadrupole interactions between the impurity nucleus and the nucleus of the matrix atoms located up to several atomic shell distances away from the impurity[17]. These results[9] clearly demonstrate that the proton-electron interaction is strong and cannot be treated properly in linear response theory.

Another feature found in the studies of the non-linear screening of a point charge in an electron gas is the existence of a very shalow bound state which is found for all values of r_s in the metallic density range ; this bound state is very close to the bottom of the conduction band. The binding energies reported vary somewhat but are always very small. Jena and Singwi[13] obtained values of the binding energy of the order of 16mRy for r_s = 5 and 1mRy for r_s = 2. In Fig. 2 the radial charge density for H in Al obtained by Norskov[12] is plotted and shows separately the contribution of the bound state density which is extremely extended in real space. Thus, the picture which emerges from the study of the self consistent non linear screening of the proton is that of a very extended H^- ion compensated by an equally extended hole in the electron gas. One should notice however that since the binding energy E_b is extremely small, the bound state may disappear when for example, the potentials of the metal ions are introduced in the calculation or even by a broadening of the level due to lifetime effects. The bound state has thus been sometimes characterized as an incipient bound state.

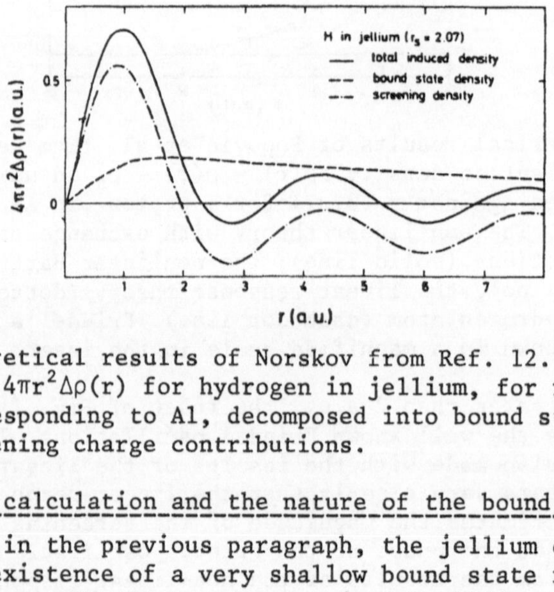

Fig. 2 : Theoretical results of Norskov from Ref. 12. Induced density $4\pi r^2\Delta\rho(r)$ for hydrogen in jellium, for r_s = 2.07 a.u. corresponding to Al, decomposed into bound state and screening charge contributions.

(b) Supercell calculation and the nature of the bound state

As shown in the previous paragraph, the jellium calculations indicate the existence of a very shallow bound state for the metallic density range. If the bound state is destabilized and merges in the bottom of the conduction band, the screening of the proton characterized by the displaced electron density profile around the positive charge is not sensibly affected ; however, the discussion of the existence of a bound state becomes very important when one wants to understand how the electronic structure of H in metals evolves as a function of hydrogen concentration. Indeed, the existence of a bound state and thus the picture of an extended H^- ion suggests that, when the H concentration increases, the bound state will evolve as an impurity band which will accommodate, when the full stoichiometry MH is achieved, two electrons of opposite spin ; since each H atom brings only one electron per unit cell, this picture suggests in turn that the impurity band should be filled at the expense of the conduction band of the metal. This would imply a depopulation of the metal conduction band in favour of the impurity band and would thus support the anionic model. We will show in section III that this picture is indeed erroneous for stoichiometric hydrides. A more precise understanding of the existence and of the nature of the bound state is thus needed. Such an understanding can be provided by a theoretical study of the impurity problem using the supercell method ; this method has been used for simple[18] and also for transition metal matrices[19]. We shall discuss here the results obtained by Gupta[18] concerning the problem of H in Al. The calculation was carried out for a supercell having a lattice parameter three times larger than that of the primitive fcc cell ; the band structure results without and with an H atom located at an octahedral site are plotted respectively in Figs. 3 and 4. The size of the supercell is such that the H-H distances of 8.6 Å are large enough that the H-H interactions can be neglected and thus, although the periodicity has been artificially introduced, this calculation can accurately represent the problem of the H impurity in an Al matrix. A comparison of the energy bands plotted in Figs. 3 and 4 shows that the presence of hydrogen in the lattice leads to the splitting of the lowest occupied energy band from the rest of the conduction band by about 0.72 eV. The second band in $Al_{27}H$ originates from the splitting of the lowest fourfold degenerate band shown in Fig. 3, which gives rise to a triply and a singly degenerate band ; in fact a comparison of Figs. 3 and 4 shows that even some bands located at much higher energies are affected by the metal-hydrogen interaction. The important result of this calculation is that the split-off band at low energy was previously occupied in pure Al and thus, the 'bound state' is not an additional state. The extra electron brought by the H atom is necessarily accomodated at the Fermi energy E_F of the metal ; thus E_F is displaced towards higher energies by the addition of hydrogen contrary to what could be expected from an extrapolation of the jellium results. The metal states which form the split-off band in $Al_{27}H$ have a strong s character at the H site. They can be truly considered as resulting from the metal-hydrogen bonding. The small dispersion of the split-off band cannot be ascribed to the H-H interaction since the H-H distances are large ; it is rather due to the different metal atoms environment for the various $\vec{k}$ points of the Brillouin Zone (BZ) due to the finite unit cell size ; as this size increases, this level should become dispersionless. The electron density-profile obtained from the supercell $Al_{27}H$ calculation is plotted in Fig. 5. It shows that the total screening is not dominated by the split-off state contribution which provides only 0.29 electrons inside the muffin-tin H sphere, compared to a total value of 0.95 electrons. As in the jellium calculations, the proton is better screened than in the case of a free H atom ; however, the jellium calculations

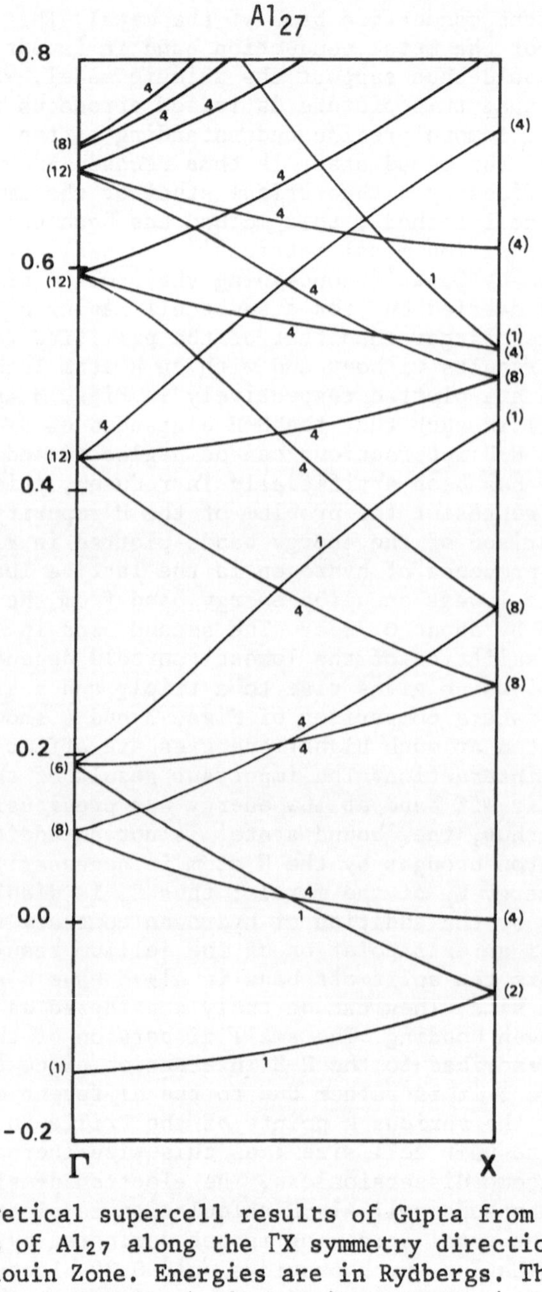

Fig. 3 : Theoretical supercell results of Gupta from Ref. 18. Energy bands of Al_{27} along the ΓX symmetry direction of the fcc Brillouin Zone. Energies are in Rydbergs. The degeneracies of the states are indicated in parenthesis at the high symmetry points.

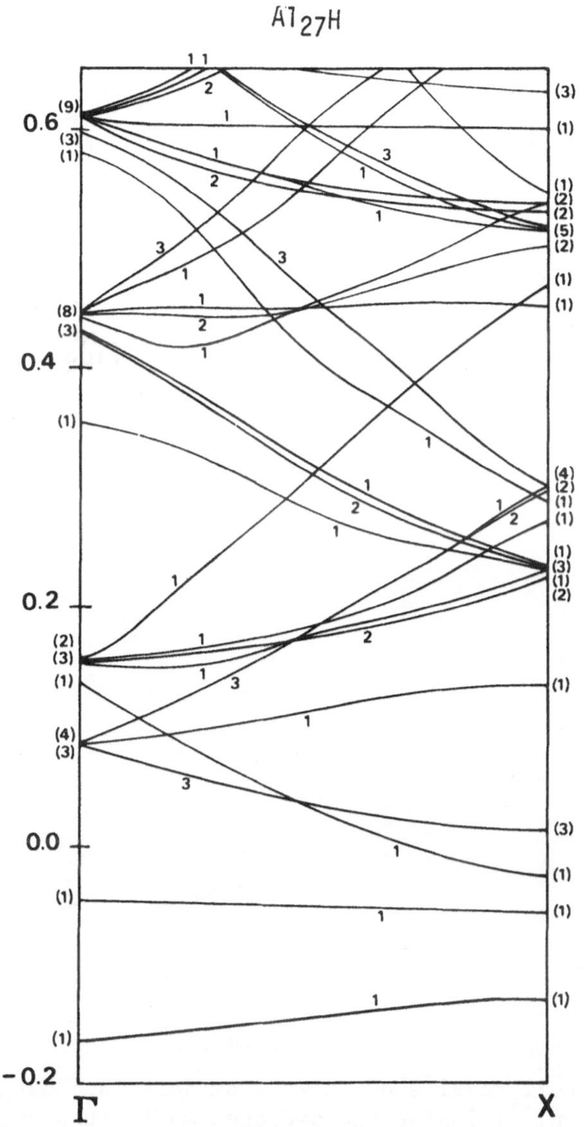

Fig. 4 : Supercell results of Gupta from Ref. 18. Energy bands of $Al_{27}H$ (H at octahedral position) plotted along the ΓX symmetry direction of the fcc Brillouin Zone. Energies are in Rydbergs. The degeneracies of the states are indicated in parenthesis at the high symmetry points.

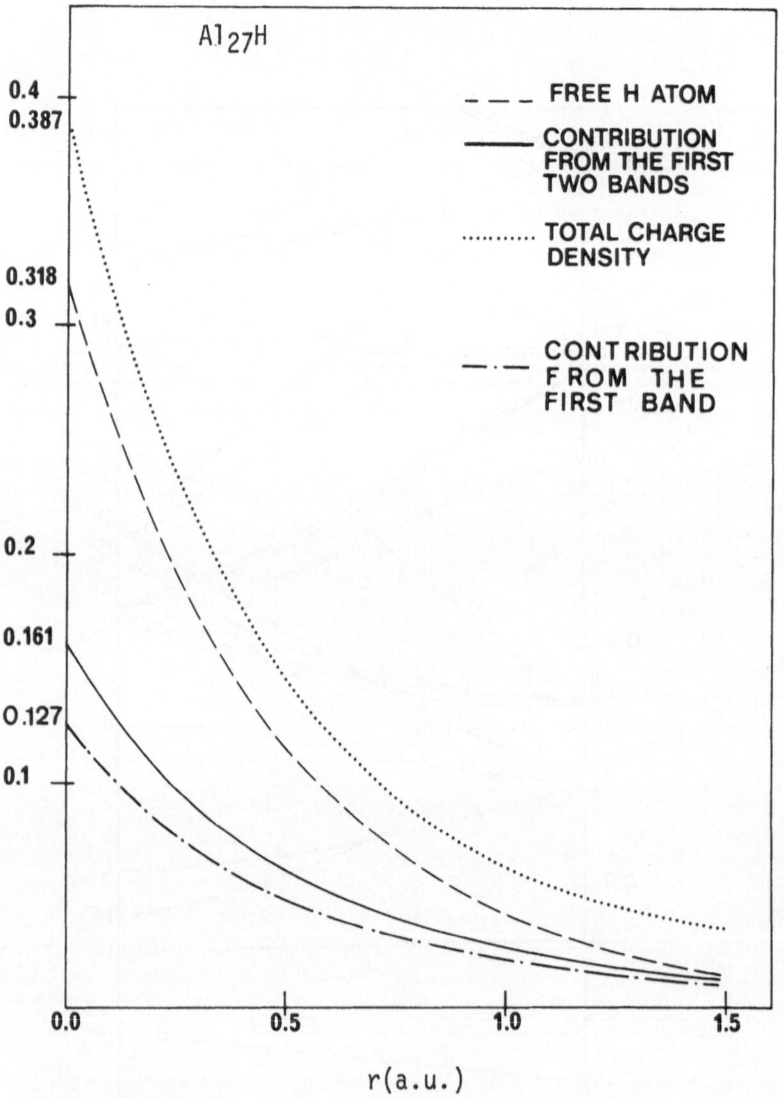

Fig. 5 : Supercell results of Gupta from Ref. 18. Charge densities (in a.u.$^{-3}$) inside the hydrogen muffin tin sphere for a $Al_{27}H$ supercell ; the total contribution (dotted line) ; the contribution from the first band (dot dashed line) and the contridution from the first two bands (full line). The charge density for a free H atom is plotted for comparison (dashed line).

appear to overestimate the screening since, at the proton site the electron density is[13] 0.519 a_0^{-3} compared to only 0.387 a_0^{-3} in the case of $Al_{27}H$. The reduction of the screening charge in the latter case is due to the attractive potentials of the neighboring Al ion cores which are of course ignored in the jellium calculations. The important point which emerges from the supercell calculation[18] is that the split-off state below the conduction band of the metal, which is characteristic of the metal hydrogen interaction, is not an additional state and thus, in the very low concentration limit, the additional electrons brought by the H atoms are accomodated necessarily at the Fermi level of the metal. Similar conclusions hold also for H in transition metals[19].

B. Screening of a H impurity in transition metals

Due to the presence of the metal d electrons, the study of H in transition metals (TM) cannot be treated as in the case of simple metals using the jellium model. A variety of other methods have been used to treat this complex problem, ranging from the parametrized tight binding Hamiltonians[20] to ab-initio calculations such as the cluster models[21-24], the supercell method[19] and the generalization to interstitial defects of the KKR Green's function approach[26-28].

(a) Cluster methods

These methods can be divided essentially into two categories, the non-embedded clusters and the embedded clusters.

The non-embedded cluster models[21-24] suffer from several problems : (i) the limited cluster size, (ii) the boundary conditions which lead to spurious surface states. The methods used to solve the eigenvalue eigenvector problem are those employed by the theoretical chemists for molecules, at best configuration interaction or Hartree-Fock linear combination of atomic orbitals (LCAO) methods, for small size clusters. Further approximations are often made to treat large size clusters such as the local exchange approximation. Muffin-tin (MT) potentials have also been used, for example, in the multiple scattering X-α method of Messmer et al[21]. Other simplifications to evaluate multicenter integrals in the LCAO methods have been introduced such as the discrete variational method[22]. The quality of the results obtained depends on one hand on the accuracy of the method used to solve the molecular problem and on the other hand on the size of the cluster. It has been shown in this connection that even before introducing the H impurity, a large number of metal atoms are necessary to reproduce the bulk density of states[24] (DOS). In the case of Ni clusters, after removing spurious surface states, the DOS of a 79 atoms Ni cluster reproduces approximately the bulk DOS[24], in contrast to the results of a 6 Ni atoms cluster[21]. A careful analysis of the variation of the electronic properties as a function of the size of the metallic cluster[23] shows that the metal conduction s states are very poorly represented in small metallic clusters since the corresponding atomic level falls above the d states unlike in the bulk. The renormalization effects on the metal s states are strongly dependent on the cluster size. The introduction of an H atom in the metallic cluster leads to the presence, below the conduction band of the metal, of a state which is characteristic of the metal-hydrogen bonding. The position of this state is however sensitive to the method used to solve the molecular cluster problem and it also depends upon the position of the H atom at the center or at the periphery of the cluster. Thus, the non-embedded cluster models give indications about the metal-hydrogen bonding ; however, the results should be examined critically.

The second type of cluster treatment, namely the embedded clusters, is more satisfactory. Without entering into the details of the techniques we want to refer the reader to the recent work of Inglesfield[25] who developed an embedding theory of a self-consistent cluster into the bulk metal. It can be shown that one has then to solve the cluster Hamiltonian with an additional surface term derived from the Green's function of the bulk substrate. In his first numerical application, Inglesfield[25] limited the cluster to one hydrogen atom surrounded by four Cu atoms embedded in bulk Cu ; the wavefunctions and Green's functions were expanded into linear muffin-tin orbitals although the method is not restricted to such functions and any other convenient basis could be used.

(b) KKR scattering theory

The Green's function method based on the KKR formalism proposed by Beeby[29] to treat substitutional impurities has been generalized to interstitial defects by Deutz et al[26], Podloucky et al[27] and applied to the problem of H in Pd, Ni, Cu, Ag. Although, in principle, the perturbation created by the H impurity on the neighboring metal atoms can be included within the framework of this method, the numerical applications have been carried out with self-consistency only in the impurity cell. The corresponding results[26] which are plotted in Fig. 6 show the presence of an s-type bound state below the conduction band of the metal host. However, as we shall see later in the discussion of stoichiometric and non-stoichiometric hydrides, the bound state found in these impurity calculations appears to be much too low in energy ; this effect indicates the need of including changes in the potentials of the neighboring host sites.

Similar but non-self-consistent technique has been applied by Katayama et al[28] to treat the screening of μ^+ in ferromagnetic Ni. We wish to mention also here the recent generalization of the impurity problem developed by Temmerman[30] who used the multiple scat-

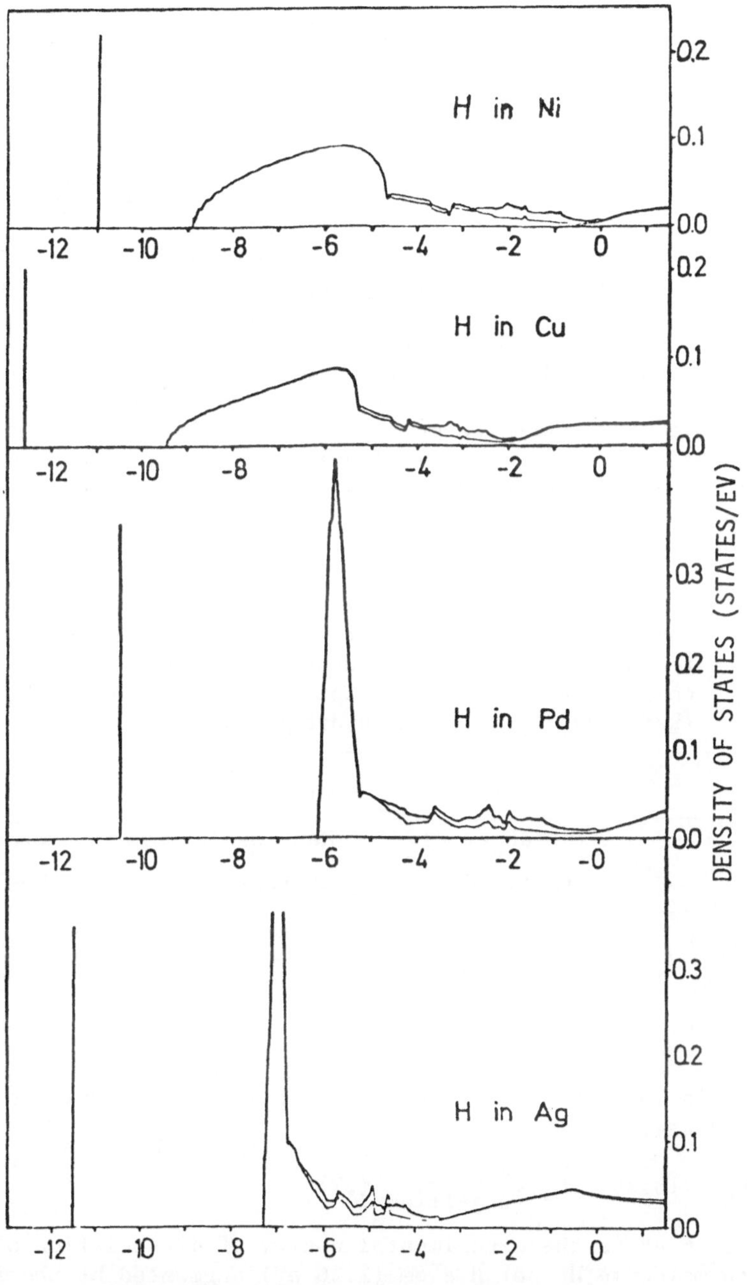

Fig. 6 : Theoretical results of Deutz et al. from Ref. 26 showing the presence of a bound state below the conduction band of the host metal.

tering formalism to derive the perturbed Green's function in the whole crystal on the introduction of an impurity atom. No numerical application has yet been reported using this recently developed formalism.

C. Energetics of hydrogen in metals

After the first attempts of Friedel[6] for the evaluation of the heat of solution of H in noble metals, most of the theoretical work has up to date been performed on simple metals such as Al, Mg, etc. Using the local density functional formalism summarized in paragraph (A) and ignoring for a while the interaction with the lattice ionic cores and the relaxation effects, the part of the total energy due to the hydrogen-conduction electron interactions can be calculated. The results of the jellium energy E_H defined as the energy of hydrogen in jellium minus the energy of the jellium substrate is given as a function of r_s in Table I. The lowest values of E_H are obtained for the largest values of r_s ; this trend is due to the fact that as r_s decreases, the electron density becomes increasingly contracted around the proton site and this results in an increase of the kinetic energy contribution to the total energy.

Table I : Comparison of energies of hydrogen in jellium E_H at different r_s by different authors. The energies should be compared with the local spin-density value of 13.38 eV (Ref. 32) for the free hydrogen atom.

r_s	E_H (eV)		
	Almbladh et al. (Ref. 10)	Manninen et al. (Ref. 11)	Worskov (Ref. 12)
2.07 (Al)	- 12.3	- 12.7	- 12.1
2.65 (Mg)	- 14.2	- 14.3	- 14.4
3.93 (Na)	- 15.0	-	- 15.0

Within this approximation, the heat of solution of hydrogen can be estimated as :

$$\Delta H_H = 15.86 + E_{H \text{ in jellium}} \text{ (eV)} \tag{15}$$

where 15.86 eV is the experimental energy of dissociation of the hydrogen molecule H_2 per H atom (2.26 eV) augmented by the ionization energy of the hydrogen atom (13.6 eV). The corresponding theoretical values obtained in the local density approximation are 2.40 eV per H atom to dissociate the H_2 molecule[31] and 13.38 eV to ionize the H atom[32]. The experimental values of the heats of solution obtained from the slope of the logarithmic plot of the H solubility versus 1/T have been determined to be 0.66 eV for H in Al[33] and 0.25 eV for H in Mg[34]. The larger solubility of H in Mg than in Al is thus reproduced already by the jellium calculations; however, the absolute values are rather different. This fact is not surprising since the jellium results do not introduce the lattice ions. Such a contribution is expected to play and important role and should be taken into account.

Popovic and Stott[9] and Popovic et al[9] have proposed an approximate treatment of this effect using the local pseudopotential theory; this calculation of the electronic contribution to the heat of solution is based on linear screening; however, as explained in paragraph (A), non linear effects are included in the study of the screening of the proton. The local pseudopotential $W(\vec{r})$ can be taken to be of the Heine-Abarenkov type

$$W(\vec{r}) = -Z/DR_m \qquad r < R_m$$
$$= -Z/r \qquad r > R_m \tag{16}$$

where Z is the valence of the host ions; the core radius R_m and the parameter D can be adjusted to give good values of the equilibrium lattice constant and of the crystal binding energy, to second order in perturbation theory. Neglecting relaxation effects, the energies of H at several sites of octahedral and tetrahedral symmetries are given in Table II according to several authors. The differences between Popovic's[9] and Norskov's[12] results have been ascribed by Norskov to differences in the charge density of the electron screening cloud around the proton due to the approximate self-consistent treatment of Popovic et al[9]. The results of Manninen and Nieminen[11] and of Kahn et al[14] have been obtained within the approximation of the spherical solid model (SSM) proposed by Almbladh and von Barth[35]. In this model, the ion pseudopotentials are spherically averaged around the impurity site. This approximation is of course most questionable when the packing of metal atoms around the H interstitial site is not very dense. It is thus better to use this approximation for H at an octahedral site in an fcc metal such as Al than for H at an octahedral site in a bcc metal since in the latter case H is surrounded only by two nearest neighbors and four next nearest neighbors metal atoms. For H at octahedral sites in Al, the non-spherical corrections to the SSM have been estimated to be small[14], about 2.7% of the heat of solution. Table II shows a dispersion of the theoretical results which clearly points out to the numerical difficulties of the total energy calculations. It is to be noted that the experimental heats of solution of H in simple metals are small and are not too far from the limit of accuracy of the calculations.

Another effect which could play a non negligible role on such small values of ΔH is the lattice relaxation around the impurity site. This effect has been estimated[14] within the framework of the

Table II : The heat of solution E_H of a proton at octahedral and tetrahedral positions in an Al host according to several theoretical calculations

	E_H (eV) octahedral	E_H (eV) tetrahedral	ΔE_H (eV) (octahedral - tetrahedral)
Popovic et al. (Ref. 9)	- 15.41	- 15.28	- 0.13
Norskov (Ref. 12)	- 14.4	- 14.6	+ 0.20
Manninen and Niemienen (Ref. 11)	- 15.59	- 14.84	- 0.75
Khan et al. (Ref. 14)	- 15.267	- 14.985	- 0.282

SSM assuming that the relaxed positions of the lattice sites $\underline{R}'_l$ correspond to radial displacements of the form

$$\underline{R}'_l = (1 + \lambda_l)\underline{R}_l \tag{17}$$

where $\underline{R}_l$ are the unrelaxed lattice positions. The total energy $E\{\underline{R}'_l\}$ expressed as a function of $\underline{R}'_l$ is then expanded up to λ^2 using second order perturbation theory and the values of λ_l are thus determined by minimizing E. The values of the radial displacements obtained for the first and second shell of Al atoms around an octahedral H site are respectively[14] $2.41\ 10^{-2}$ a.u. and $-\ 0.25\ 10^{-2}$ a.u. compared to the values of $\underline{R}_l$ of 3.819 a.u. for the first shell and 6.615 a.u. for the second shell. The linear relaxation effect is thus smaller than 1% for the first shell of metal atoms and two orders of magnitude smaller for the second shell. The positive sign stands for an outward relaxation effect and the heat of solution of H in Al amounts to about 10% of the total value[14].

The total energy calculation at several levels of approximation[9,11,14] have also been carried out as a function of the position of the proton in the lattice. From these results, the activation energies of H in simple metals for overbarrier classical diffusion as well as the diffusion paths can be estimated. An order of magnitude of the value of the local mode frequencies of H in simple metals can be also obtained from the calculations described above by making the approximation of a harmonic potential well around the occupied lattice sites[9,11,14].

The problem of possible vacancy trapping of the H atoms in simple metals has also been considered. In the jellium model, a vacancy can be represented by removing from the positive background the charge contained in a Wigner-Seitz cell. The sensitivity of the electron density profile around a proton in an electron gas to the values of r_s indicates that the charge pile up around a proton in a vacancy is expected to be different from that of a proton in an electron gas. The density profile needs thus to be recalculated in the presence of a vacancy, unlike in the work of Popovic et al[9] in which it was assumed to be unchanged. Indeed, for H in a vacancy, a full non linear calculation leads to a decrease of the charge pile up around the positive charge[11,12,14]. This will have obviously an important consequence on the energetics. As already mentioned above, it is difficult to obtain accurate calculations of total energies and a general consensus has not yet been reached amongst theorists. The main results can be however summarized as follows. If only the jellium part of the energy is calculated[12], one finds that the vacancy trapping of H is possible in Al (with a binding energy to the vacancy of the order of 3 eV) but not in Na. However, if the lattice ions are included using for example the spherical solid model[11,14], the energy of H at substitutional positions is always found to be higher than at interstitial positions indicating on the contrary that the vacancy trapping of H does not occur in simple metals. This result appears to be in agreement with the experimental data concerning H in Al[36].

The electronic structure of H in transition metals including an estimation of the dipole force tensor and the size effect has been studied using the tight binding method[20].

III. ELECTRONIC STRUCTURE OF BINARY STOICHIOMETRIC HYDRIDES

Leaving aside the ionic hydrides of alkali and alkaline earths we will focuss here on the electronic properties of hydrides of transition metals (TM) and lanthanides. We shall essentially discuss (i) the modifications of the electronic properties of the metal upon hydrogonation these are characterized by the deformation of the metal conduction band as observed by several spectroscopic techniques (photoemission, X-ray emission) and by the drastic change in the properties at the Fermi energy E_F such as the electronic specific heat coefficient γ, the Pauli susceptibility χ_p, the geometry of the Fermi surface, the conductivity etc. (ii) the nature of the metal-hydrogen (M-H) and hydrogen-hydrogen (H-H) interactions, and (iii) the factors which control the stability of the hydrides in particular, the role of the attractive H potential which lowers the metal d bands, the role of the M-H bonding and of the H-H interactions, the charge transfer, the shift in the position of the Fermi level etc.

Prior to the results from the band structure calculations, most of the experimental data were interpreted either in terms of the protonic or of the anionic rigid band models. The protonic model was essentially used for the hydrides of the end of the TM series, such as PdH. In Fig. 7(a), we sketched the density of states (DOS) of a TM such as Pd which is characterized by the narrow metal d bands overlapped and hybridized by a wider metal s-p band. The Fermi level of the pure metal Pd falls below the top of the metal d bands. According to the protonic model, when the metal hydride is formed, the electron brought by the H atom in each unit cell is accommodated at the Fermi level of the pure metal ; the Fermi level of the hydride is thus shifted towards higher energies as indicated by the dashed line in Fig. 7(a). Thus, according to the protonic rigid band model, PdH should have filled d bands and the Fermi level (E_F) should fall in the metal s-p band as in the isoelectronic metal Ag. This interpretation was supported by the experimental observation that while Pd has a large DOS at E_F which leads to a large value of the electronic specific heat coefficient γ and a large Pauli susceptibility χ_p ; the DOS at E_F decreases drastically upon hydrogen uptake[37] as it does upon alloying Pd with Ag. However, there are already some experimental indications of the quantitative deficiency of the protonic model : while there are only about 0.36 holes in the d bands of Pd[38], one needs to reach a hydrogen to metal ratio of 0.6 to fill the d bands and obtain a diamagnetic compound[39]. We will show in what follows the limitations of the protonic model. The anionic model has been mostly used to explain the electronic properties of the hydrides of the beginning of the TM series and of the lanthanides. In this model, as shown schematically in Fig. 7(b), an impurity band is formed below the metal conduction bands ; since this hydrogen induced band can accommodate two electrons while each H atom brings only one additional electron per unit cell, the low-lying band has to be filled at the expense of the metal conduction band. Thus, contrary to the case of the protonic model, the metal conduction band is depopulated and the Fermi level of the hydride is shifted towards lower energies. The experimental evidence for the existence of a structure in the DOS at low energy is given by the spectroscopic techniques such as photoemission and X-ray emission. However, we shall show later that the anionic model, as well as the protonic model, is oversimplified. The limitations of these crude models show clearly that realistic studies of the electronic structure of the hydrides using band structure methods are needed.

We shall not discuss here the electronic properties of the monohydrides of the bcc metals. In what follows, we shall give three examples : the monohydrides (PdH, NiH), the dihydrides (TiH_2, ZrH_2, NbH_2, LaH_2, TbH_2, ErH_2) and the trihydrides (LaH_3). I will be presenting here mostly the results of my own calculations[40] which have been performed using the APW-method ; I took into account the departure of the potential from a constant value outside the muffin-tin (MT) spheres, the so-called warped MT corrections. The energy eigenvalues have been calculated ab-initio at 89 points in

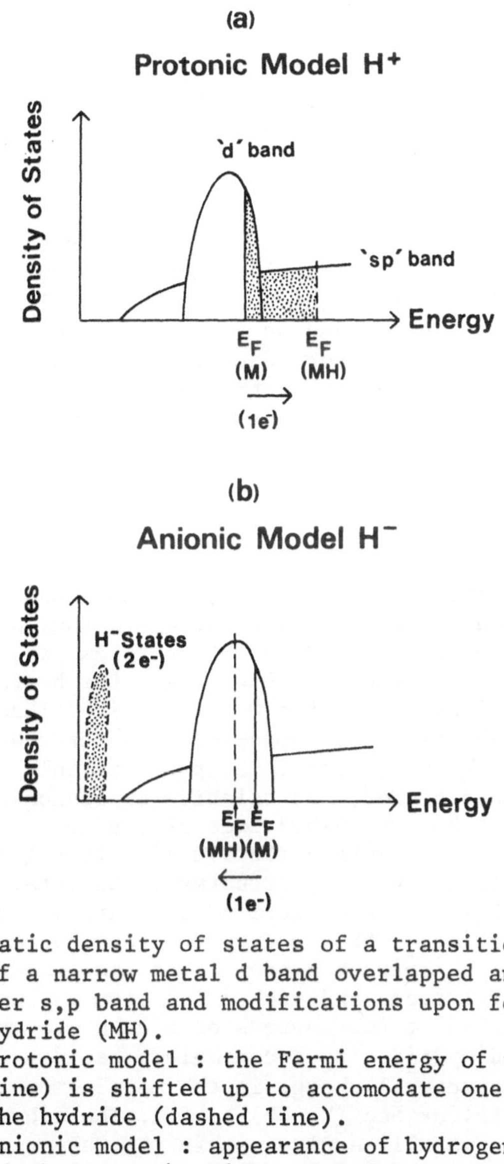

Fig. 7 : Schematic density of states of a transition metal (M) formed of a narrow metal d band overlapped and hybridized with a wider s,p band and modifications upon formation of a metal hydride (MH).

(a) Protonic model : the Fermi energy of the metal (full line) is shifted up to accomodate one more electron in the hydride (dashed line).

(b) Anionic model : appearance of hydrogen induced states (dashed curve) which accomodate 2 electrons at low energy ; the Fermi energy of the metal (full line) is shifted down due to the depopulation of the d band in the hydride (dashed line).

the 1/48th irreducible wedge of the fcc Brillouin zone (BZ) ; the densities of states and partial wave analysis of the DOS have been obtained accurately using the linear energy tetrahedron method with 6048 tetrahedra in the 1/48th wedge of the BZ. I will refer also to the work of Switendick[41] who first started studying the electronic structure of metal hydrides in the early 70's ; since then, other groups[42-46] have also been working on some of these compounds, the references to their work will be also included.

A. Structures

As shown in Table III concentrated hydrides form on one hand with the TM of the end of the series such as Ni and Pd which form monohydrides (notice that Pt which is a good catalyst does not absorb hydrogen in the bulk), and on the other hand with early members of the TM series which form dihydrides[1]. Dihydrides exist for the TM on the left of column VI with the exception of Ta. The TM on the right of column V do not absorb easily hydrogen ; we shall explain later in this section the reasons why such a behaviour is observed. The lanthanides form also dihydrides, the trivalent rare earth and yttrium can absorb even larger amounts of hydrogen and form trihydrides.

Unlike the hydrides of intermetallic compounds, the binary hydrides under study in this section have a rather simple crystal structure. The monohydrides PdH and NiH crystallize within the rocksalt structure ; as shown in Fig. 8(a), the H atoms occupy the octahedral interstices of the fcc metal lattice. Thus, in this case hydrogen uptake is not accompanied by a structural change in the metal lattice which remains fcc ; there is only a lattice expansion. In the case of PdH, the relative expansion of the lattice parameter is $\Delta a/a \sim 5\%$. The dihydrides of the early TM are also cubic ; they crystallize within the fluorite (CaF_2) structure. As shown in Fig. 8(b), in the CaF_2 structure, the metal atoms form an fcc lattice; the H atoms which form a cubic array are located at the 1/4 (111) positions ; they occupy the tetrahedral interstices of the metal lattice, so each H atom is at the center of a regular tetrahedron of metal atoms and each metal atom is surrounded by eight H atoms located at the corners of a cube. For the early TM, the formation of dihydrides is accompanied, besides the lattice expansion, by a structural change in the metal atom lattice from hcp (for Sc, Ti etc) or bcc (V, Nb) to fcc. The dihydrides of the rare earths (RE) crystallize also within the CaF_2 structure. The trihydrides of the early trivalent RE up to Nd have the BiF_3 structure which is analogous to the fluorite structure shown in Fig. 8(b) in which the third H atom of the unit cell occupies the octahedral interstice which is empty in the corresponding dihydride. The late members of the RE trihydride series and Y have the hexagonal HoD_3 structure ; this crystallographic phase transformation leads to an increase in the H-H distances.

Table III : Occurrence of binary transition and rare earth metal hydrides. From Ref. 41.

ScH_2	TiH_2	VH VH_2	CrH	(Mn) --	(Fe) --	(Co) --	NiH
YH_2 YH_3	ZrH_2	NbH NbH_2	(Mo) --	(Tc) --	(Ru) --	(Rh) --	PdH
See Rare Earth Series	HfH_2	TaH	(W) -	(Re) --	(Os) --	(Ir) --	(Pt) --

LaH_{2-3}	CeH_{2-3}	PrH_{2-3}	NdH_{2-3}	Pm ?	SmH_2 SmH_3	EuH_2	GdH_2 GdH_3	TbH_2 TbH_3	DyH_2 DyH_3	HoH_2 HoH_3	ErH_2 ErH_3	TmH_2 TmH_3	YbH_2 YbH_3(?)	LuH_2 LuH_3

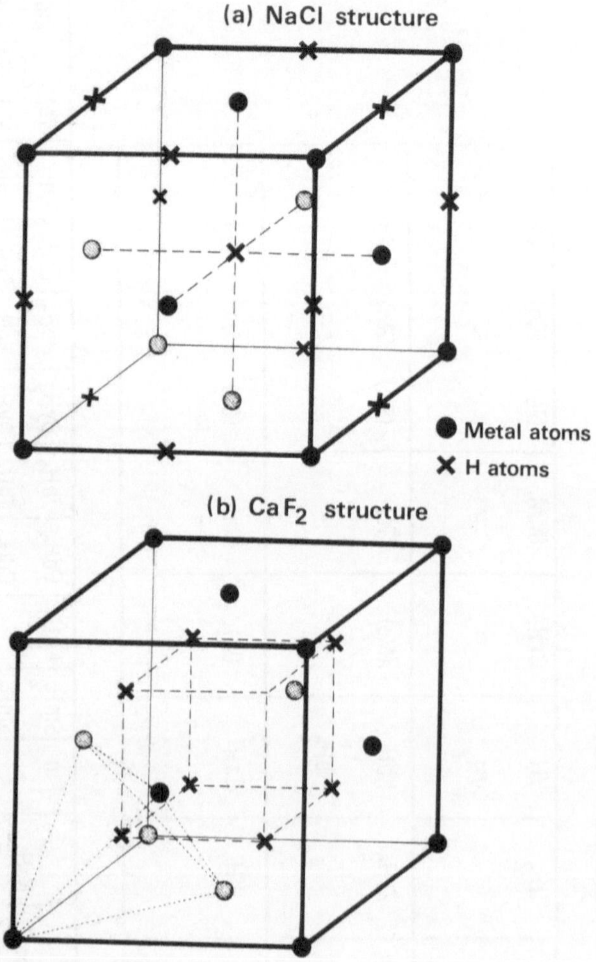

Fig. 8 : Crystal structure of fcc binary hydrides.
(a) Rocksalt structure. The H atoms have an octahedral environment of metal atoms.
(b) Fluorite structure. The H atoms are located at $\frac{1}{4}[111]$ and have a tetrahedral environment of metal atoms.
• black circles represent the metal atom positions.
x crosses represent hydrogen atom positions.

B. Monohydrides PdH and NiH

If we compare, as done in Fig. 9 the energy bands of Pd and PdH we notice essentially two differences (i) the presence in PdH of a low-lying band, below the metal d bands, and (ii) the difference in the Fermi level position which falls below the top of the metal d bands (marked by the X_5' point) in pure Pd, while in PdH the d bands are filled and the Fermi level cuts only one band, the metal s-p band as in Ag.

The low-lying band in PdH is mostly formed out of metal states such as Γ_1, X_1, W_2' which were already filled in pure Pd and which have been lowevered drastically (by about 2.5 eV) by the metal-hydrogen interaction. However, we notice that a full branch of metal states around L_2', located above E_F in pure Pd, are lowered by about 5 eV by the metal-hydrogen interaction and falls below E_F in PdH. This feature is particularly important since the lowering of this branch of metal 5p states brings about 0.25 additional electrons below E_F in PdH. We can now understand from the electron count that since most of the states forming the low-lying band in PdH were previously filled in pure Pd, the Fermi level has to go up in the hydride ; however since some additional states appear at low energy in PdH, less than one electron has to be acommodated at the Fermi level of Pd. Only about 0,76 of the one electron brought by the H atom is accommodated at E_F so, although the Fermi level goes up with hydrogen uptake ; the limitations of the rigid band model appear clearly from the analysis made here[19,40.b]. Moreover, Fig. 9 shows clearly the deformation of the metal d bands due to the metal-hydrogen interaction. This appears also in the DOS plotted in Fig. 10 where we observe for PdH a structure at low energy and a drastic reduction of the DOS from that of pure Pd. The DOS at E_F in PdH is very low being 6.81 states/Ry unit cell compared to 28.9 states/ Ry unit cell in pure Pd[38] for which the Fermi level falls close to a peak in the DOS just below the top of the metal d bands. This explains the drastic reduction of the electronic specific heat coefficient from $\gamma = 9.48$ m J.mole^{-1}K^{-2} in pure Pd[47] to $\gamma = 1.54 \pm 0.13$ m J. mole^{-1}K^{-2} in PdH_x[37] for x = 0.876 as well as the decrease of the Pauli susceptibility χ_p[39] .

The lowering of metal states by the metal-hydrogen interaction is an important factor for the stability of the compound ; on the contrary, the rising of the Fermi level costs energy. It is to be noted in connection with the filling of the metal d bands that taking into account the 0.36 holes in the d bands of Pd and the 0.24 additional electrons at low energy in the hydrogenated compound, the filling of the metal 4d bands should occur for a hydrogen to Pd ratio of about 0.6 ; it is remarkable that experimentally, it is easy to load H in Pd up to ratios of about 0.6 but that hydrogen has the tendency of escaping the simple for larger H concentrations. Indeed, for loading H/Pd ratios larger than 0.6, the Fermi level

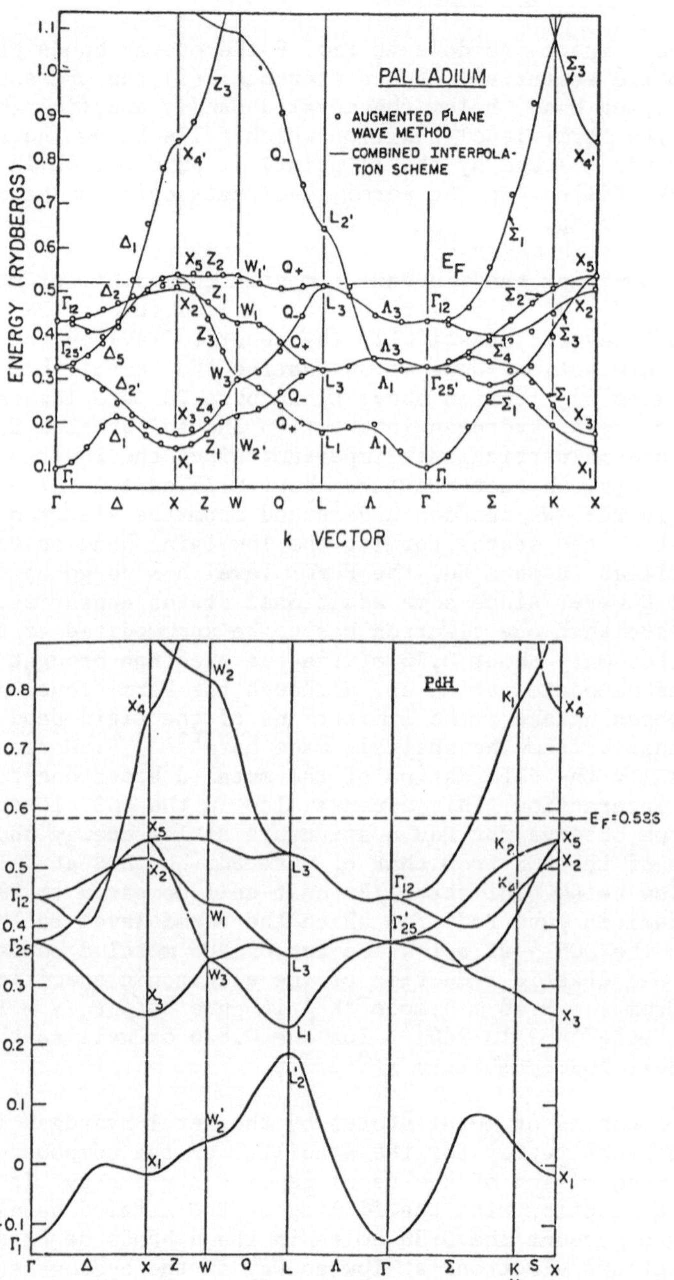

Fig. 9 : APW energy bands of (a) Pd from Ref. 38 (b) PdH from Ref. 40-b. The bands are plotted along several high symmetry directions of the fcc Brillouin Zone.

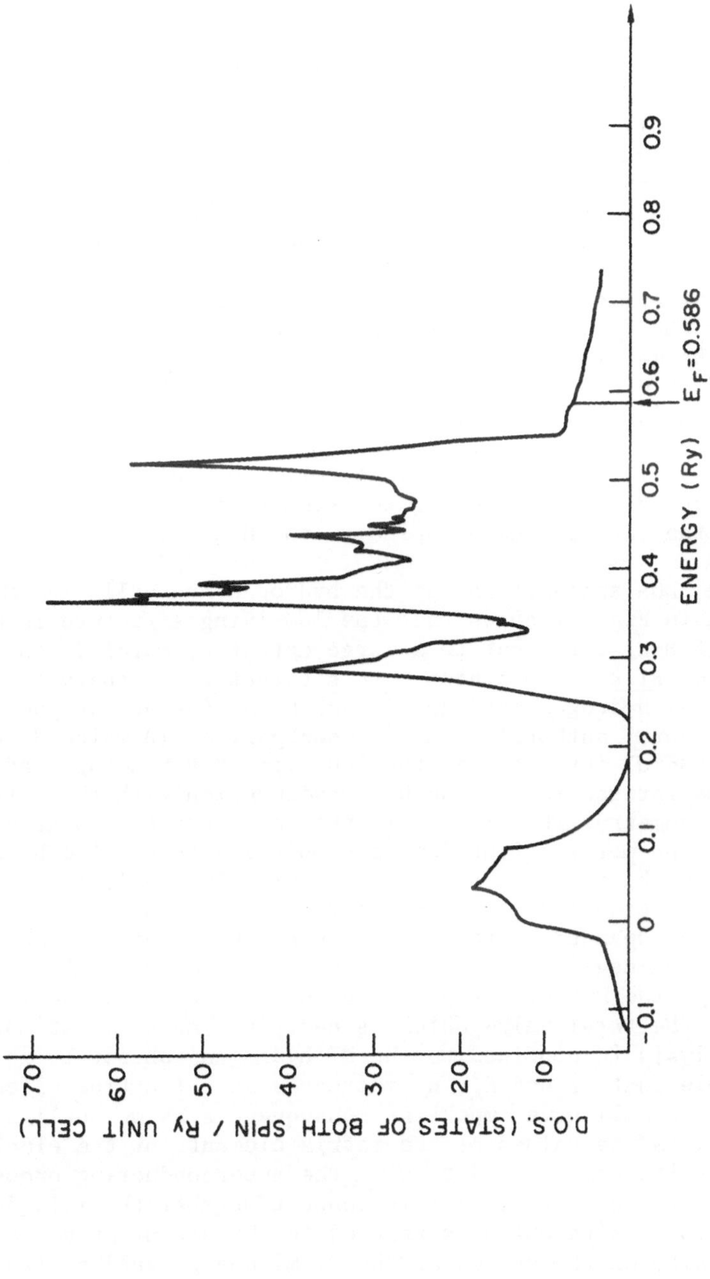

Fig. 10 : Total DOS of PdH from Ref. 40-b.

which is then in the metal s-p band needs to be raised substantially; this situation is energetically less favorable than when E_F lies in the narrow high DOS metal d bands.

In order to understand why some metal states such as Γ_1, X_1, W_2', L_2' etc are drastically affected by the metal-hydrogen interaction while other states such as $\Gamma_{25'}$, Γ_{12}, L_3, W_3 etc shown in Fig. 9 are practically not affected, it is interesting to study the behaviour of the corresponding wavefunctions[43] as done in Fig.11. We can see that all the metal wavefunctions which have a finite density at the hydrogen interstitial site are strongly modified by the metal-hydrogen (M-H) interaction and take a strong H-s character at the hydrogen site while the metal wavefunctions having a node at the hydrogen site are almost unperturbed by the hydrogen potential. This interesting feature can be explained easily by the fact that the hydrogen potential is characterized by large s phase shifts and scatters only the s waves while the p,d waves are not effected. We notice that the metal states affected by the H potential can have either a s, p or d symmetry at the metal site ; what matters for the M-H interaction is the symmetry of the metal wavefunction at the H interstitial site. With such a feature, one should be able to predict for any hydride which metal states will be modified by the M-H interaction by simply studying the symmetries of the metal wavefunctions at the interstices occupied by the H atoms.

A partial DOS analysis inside the hydrogen and palladium MT spheres given in Fig. 12 shows that the low-lying structure is made up not only of H-s states but to a large extent of metal d, and also of some metal s and p states ; this structure is characteristic of the metal hydrogen bonding. It has been observed by photoemission by several authors[48]. In the monohydrides in which H occupies the octahedral interstices, the low-lying M-H bonding band results from the interaction of the H-s wavefunction with the metal states of e_g symmetry. This is due to the fact that the lobes of the corresponding metal d functions are pointing toward the H interstitial site in the fcc structure. The partial DOS analysis shows that the contribution inside the H sphere is vanishingly small in the energy range spanned by the metal d bands but that its importance becomes non-negligible when the metal d bands are filled. For PdH, at E_F, the DOS of s-type at the H site, n_s^H, represents about 7.5% of the total value which is certainly an underestimate since the analysis is confined to the MT hydrogen sphere. It is also remarkable that around E_F, n_s^H increases as a function of energy. This feature will have important consequences as we shall see in section VI, on the values of the matrix elements of the electron-optical phonon interaction and thus on the superconducting properties of PdH_x. One can notice in this connection that the spin density at the proton site which is related to the amount of s character at the H site of the states at the Fermi energy varies also in the same way as n_s^H, namely it increases with energy. An experimental

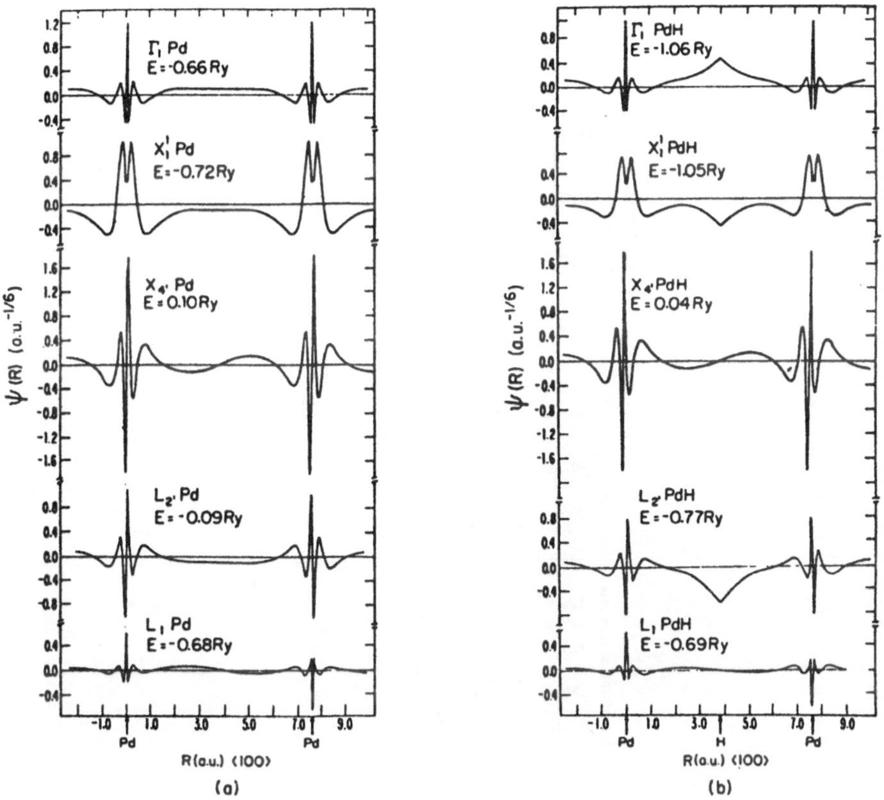

Fig. 11 : (a) Pd and (b) PdH wavefunctions from Ref. 43 are plotted along the [100] direction for the symmetry states Γ_1, X_1, X_4', L_2' and L_1 (see Fig. 9).

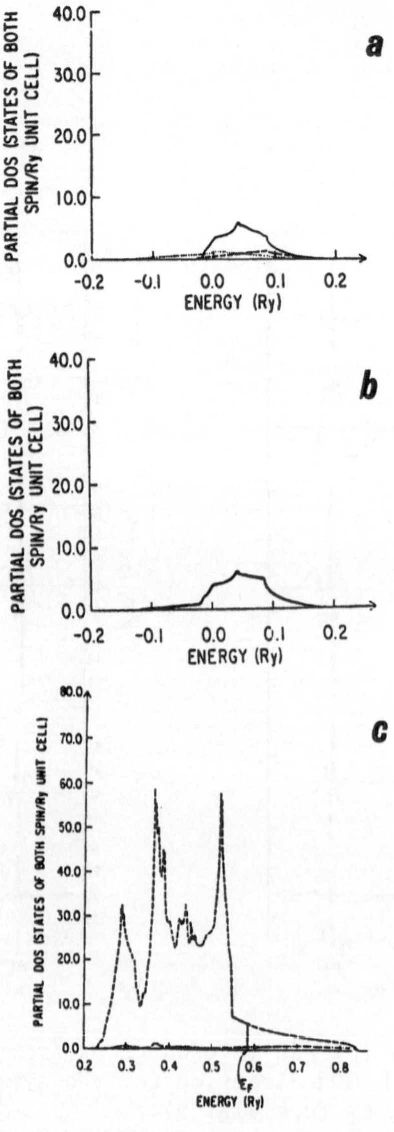

Fig. 12 : Partial DOS analysis of PdH from Ref. 40-b. From top to bottom we have plotted the contributions from (a) The low lying band inside the Pd sphere ; $\ell = 0$ (dotted line) ; $\ell = 1$ (dashed line) ; $\ell = 2$ (full line). (b) The low-lying band inside the H sphere ; $\ell = 0$ (full line) other components are almost not visible on the same scale. (c) The conduction bands inside the MT Pd sphere (upper dashed curve) ; inside the H sphere (lower dashed curve).

evidence of this variation is given by the measurements of the spin-lattice relaxation rate T_1^{-1} of the proton in PdH_x obtained from nuclear magnetic resonance (NMR) experiments[49]. At low temperatures (between 3 and 77 K) where the diffusion mechanism of the proton does not contribute to the nuclear spin relaxation process, Wiley et al[49] have obtained for H in stoichiometric PdH relaxation times which obey the Korringa law with $T_1T = 48 \pm 2$ sec.K. Assuming that the hyperfine interaction between the conduction electrons and the nuclear spins is dominated by the Fermi contact term, the spin-lattice relaxation rate T_1^{-1} can be shown to be proportional to the temperature and to the square of the spin density at the H site arising from the electronic states at the Fermi surface (FS).

$$T_1^{-1} = \frac{\Omega^2}{9\pi^3} \hbar^3 \gamma_N^2 \gamma_e^2 k_B T \left| \int_{FS} |\psi^s_{\vec{k}} \text{ (H site)}|^2 \frac{dS_{\vec{k}}}{|\nabla_{\vec{k}} E_{\vec{k}}|} \right|^2 \tag{18}$$

where Ω is the unit cell volume, k_B the Boltzmann constant, γ_e and γ_N are respectively the gyromagnetic ratios of the electron and of the nucleus (here the proton). The integral in Eq. (18) is performed on the FS ; at the H site, only the s-wave part of the wavefunction of the FS electrons ψ^s has a nonzero contribution. This contribution has been accurately calculated[40b] for PdH by generating the wavefunctions at 15000 $\vec{k}$ points at the FS. Equation 18 is often written for convenience under the factorized form

$$T_1^{-1} = 4\pi\gamma_N^2 k_B T (N_s \; H_{hfs})^2 \tag{19}$$

where N_s is the partial DOS of s type at the Fermi energy and H_{hfs} is the hyperfine field. The theoretical value $(N_s H_{hfs})_{theory}$ = $4{,}4 \times 10^{15}$ $G.erg^{-1}$ has been found to be in satisfactory agreement with the experimental result of Wiley et al[49] $(N_s H_{hfs})_{expt.}$ = $(4.0 \pm 0.1) \times 10^{15} G.erg^{-1}$. Wiley et al[49] have also studied the spin-lattice relaxation rate for non stoichiometric compounds. For an $x = H/Pd$ ratio of $0.7 \leqslant x \leqslant 0.8$, $T_1T = 70 \pm 3$ sec.K this increase over the value of stoichiometric PdH corresponds, if other factors are equal, to a decrease of 21% in the value of the partial s DOS at E_F for $x < 1$. This result is in agreement with the trend found in the calculated values of the partial DOS of s type using, in the close vicinity of the stoichiometry PdH, the rigid band model. We shall see later that this approximation is not too bad if applied with care, only in the close vicinity of the full stoichiometry, to the band structure of PdH.

Nickel hydride NiH is isoelectronic to PdH and its DOS plotted in Fig. 13 shows essentially the same features as that of PdH plotted in Fig. 10. We observe for NiH, at low energy, a structure which results from the M-H bonding, the Ni-d bands are filled and the Fermi level of NiH falls in the Ni s-p band. The main differences with PdH are that in NiH the d bands are narrower and that the M-H binding structure is located at lower energies; this is due to the difference in the relative position of the metal d states relative

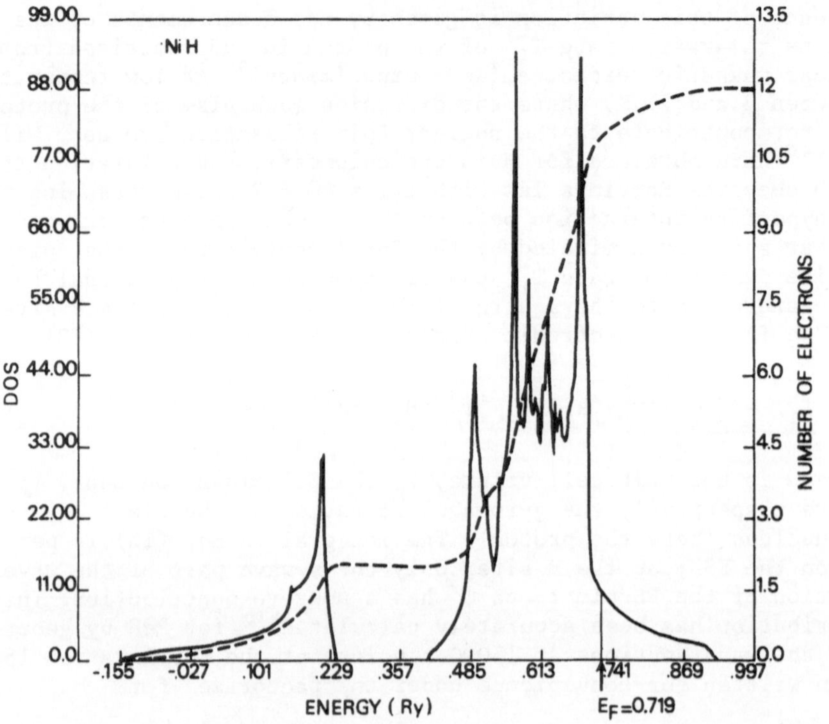

Fig. 13 : The total DOS of NiH (full line curve, left hand side scale) units are states of both spin per Rydberg unit cell ; the number of electrons (dashed line and right hand side scale). From Ref. 40-d.

to the bottom of the s-p band in Ni metal compared to Pd pure metal. In NiH, the Fermi level is closer to the top of the d bands than in PdH ; this is mainly due to the fact that there are more holes in the d bands of paramagnetic Ni (of the order of 0.6 holes) than in Pd (about 0.36 holes) ; consequently, in NiH, the DOS at E_F is larger than in PdH, the amount of metal d character is larger and the value of n_s^H is smaller. We shall discuss in section VI the consequences of this fact on the superconducting properties.

Since for both PdH and NiH the Fermi level cuts only the metal s-p band, the geometry of the FS is expected, as sketched in Fig. 14, to be similar to that of the noble metals Ag, Cu ; it is characterized by an electron surface centered around Γ with necks along the (111) ΓL directions. Nevertheless, since the Fermi level of the hydrides is closer to the top of the d bands than in a noble metal, the asphericity of the FS is expected to be larger ; this appears clearly in Fig. 15 where a cross section of the belly of the FS of NiH in the (100) direction shows substantial departure from a perfect circle, and in the results listed in Table IV in

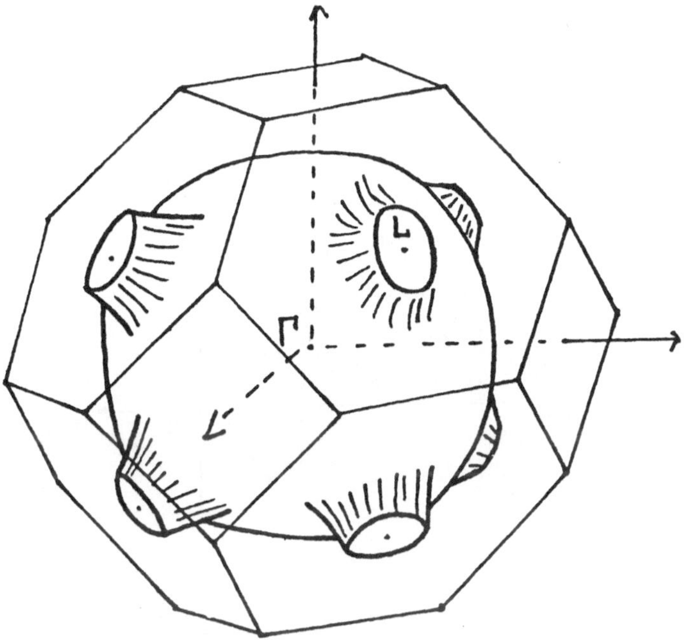

Fig. 14 : Sketch of the Fermi surface of PdH (or NiH) in the first Brillouin Zone of the fcc structure. It is a deformed sphere centered around Γ with necks along the ΓL [111] directions.

Table IV : Comparison of the belly radii R (in units of the free electron sphere radii) of NiH with PdH and noble metals (a) From Ref. 40-d ; (b) From Ref. 40-b ; (c) From Ref. 123 ; (d) From Ref. 124.

	$R_{[100]}$	$R_{[110]}$	$R_{[100]}/R_{[110]}$
Ni H (a)	1.104	0.8780	1.25
Pd H (b)	1.024	0.889	1.15
Cu (c)	0.979	0.958	1.02
(d)	1.049 ± 0.021	0.965 ± 0.0021	1.09
Ag (c)	1.066	0.975	1.03

which the asphericity of the FS is characterized by the ratio of the belly radii in two directions of the fcc BZ.

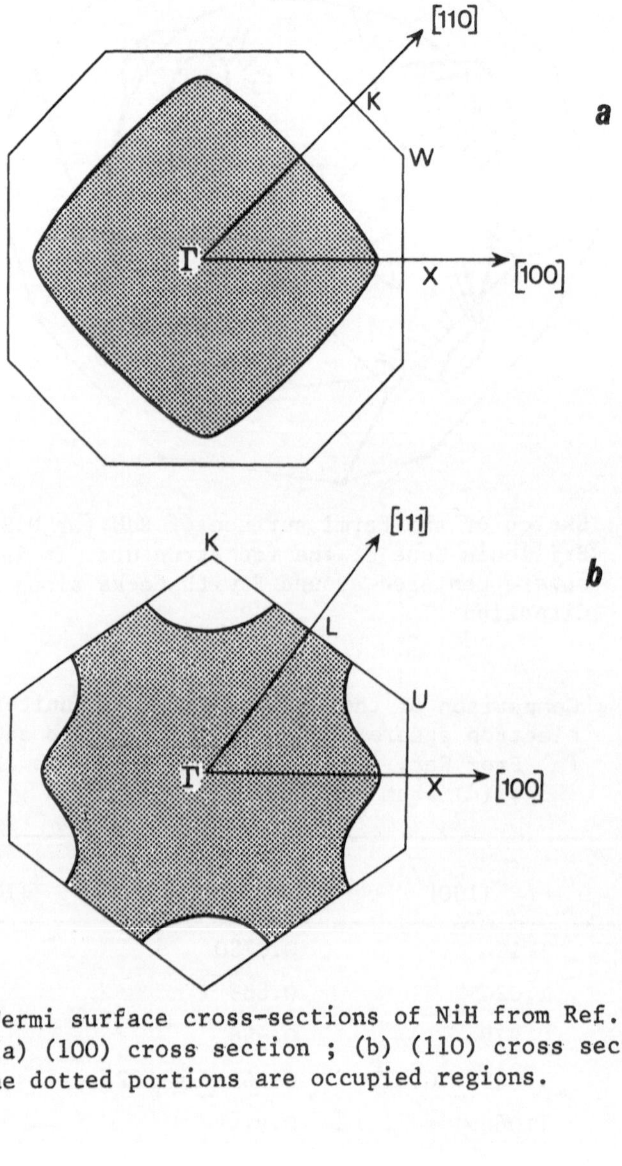

Fig. 15 : Fermi surface cross-sections of NiH from Ref. 40-d ; (a) (100) cross section ; (b) (110) cross section. The dotted portions are occupied regions.

Returning to the question of the stability of the hydrides, the lowering of some metal states by the M-H interaction discussed previously contributes favorably to the stability, in particular, the lowering below E_F of states which are empty in the pure metal is of great importance since, by bringing additional electrons below E_F this will also reduce the shift of E_F towards higher energies, a factor which on the contrary affects negatively the stability of the hydride. The fact that PtH does not exist could be understood in this picture ; in pure Pt, due to relativistic effects, the L_2' branch is already below E_F[50] thus, in that case the totality of the extra electron brought by the H atom would have to be accommodated at E_F leading to a greater shift in E_F than in the case of palladium hydride. It is particularly favorable for the hydride formation to accommodate some electrons at the Fermi level of the metal when E_F falls in a high DOS region as for Pd and Ni ; we can surmize that for the noble metals which have filled d bands, the situation is not very good since the DOS at E_F is low and it will cost a significantly larger amount of energy to raise the Fermi level. Indeed, noble metals do not absorb easily hydrogen.

The analysis of the electronic structure of monohydrides presented here shows clearly that, although the Fermi levels of the pure metals Pd and Ni are shifted towards higher energies upon formation of the corresponding hydrides, the protonic model is not quantitatively correct. We have emphasized in this connection the importance of the M-H bonding and the deformation of some portions of the metal d bands in the bonding process.

C. Dihydrides of early TM and RE

(a) General features

As shown in Figs. 16 and 17, the electronic structure of the dihydrides is characterized by the presence of two low-lying bands below the metal d states ; in addition to the lowest band which is also observed in the case of the monohydrides, a second band appears. At the BZ center, a wavefunction analysis reveals that the lowest Γ_1 state is a bonding combination of the metal-s and of the two H atoms-s wavefunctions while the Γ_2' state is an antibonding combination of the two H s wavefunctions with a slight admixture of metal f states. The presence of this second additional state at low energy in the dihydride is an important factor for the stability of the compound. Since Γ_2' is an antibonding combination of the two hydrogen 1s wavefunctions, its position in energy depends sensitively upon the H-H distance. Thus, as seen in Figs. 16 and 17, for the TM dihydrides, there is an overlap between the antibonding H-H band and the metal d bands located at higher energies whereas in LaH_2 which has a much larger lattice constant the antibonding band is more stable and is separated by a gap from the metal d bands. This feature is certainly connected to the larger stability of the

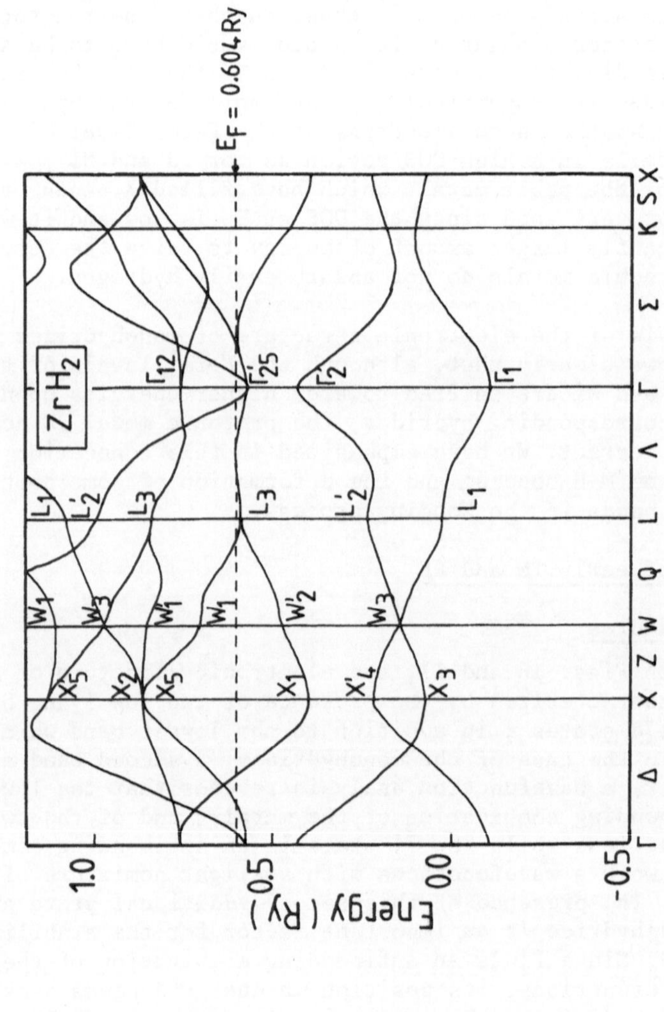

Fig. 16 : The energy bands of cubic ZrH_2 (a = 9.014 a.u.) along several high-symmetry directions from Ref. 40-h. Energies are in Rydberg.

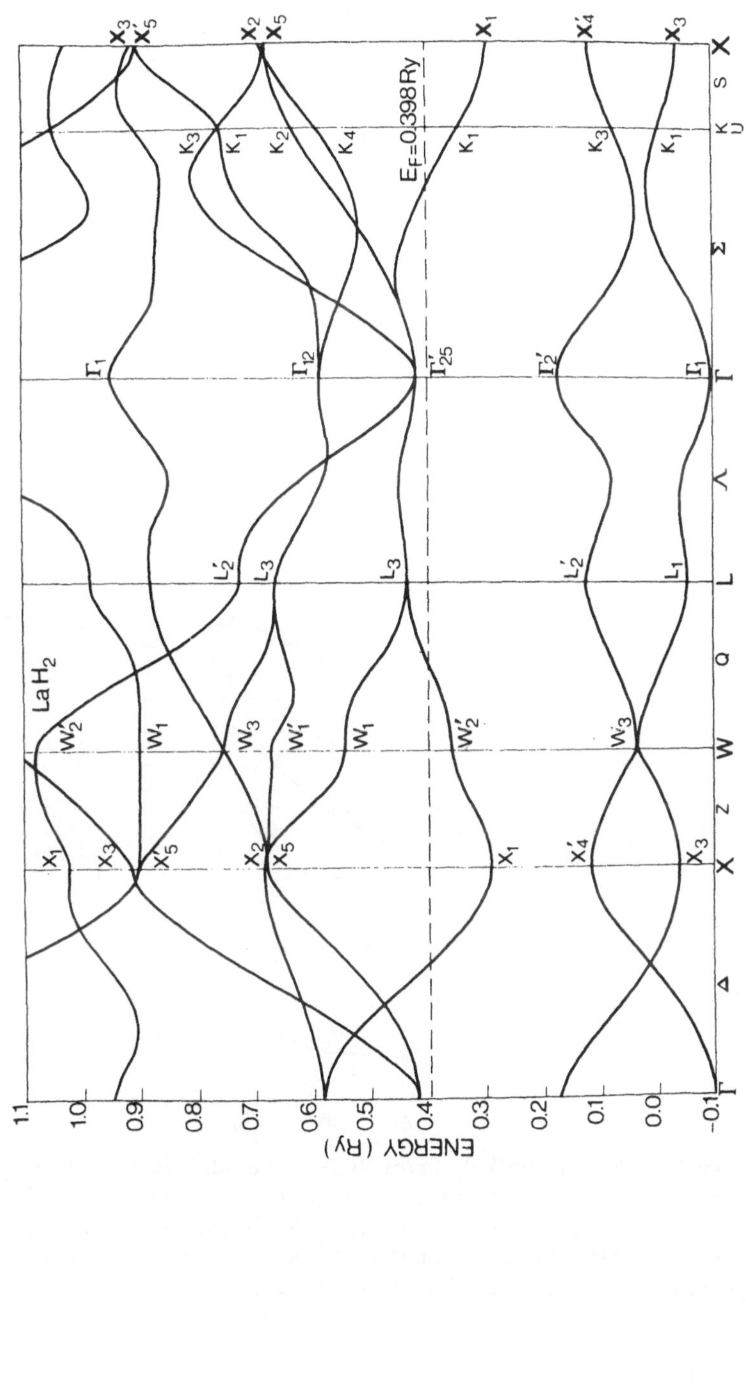

Fig. 17 : Energy bands of LaH_2 (a = 10.7128 a.u.) along several high symmetry directions of the fcc Brillouin zone, from Ref. 40-e. Energies are in Rydberg.

RE dihydrides compared to the TM dihydrides and can also explain why the dihydrides do not form with the TM on the right of column V. As we go from the left to the right of the TM series in the Periodic Table, the lattice parameter of the pure metals contracts as shown in Fig. 18, consequently, the H-H distances decrease, the antibonding band is destabilized[41.e] as illustrated in Fig. 18 by the variation in the position of the Γ_2' state relative to the position of the Fermi level of the dihydride ; when Γ_2' is located above E_F as it occurs on the right of column V, the formation of the dihydride becomes energetically unfavorable.

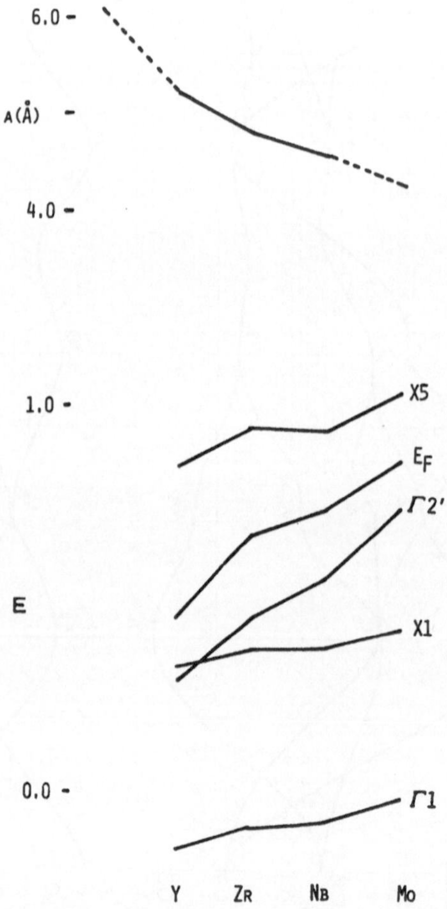

Fig. 18 : Results of Switendick from Ref. 41-e showing in the 4d series the contraction of the lattice parameter A (Å) and the energy variation of several high symmetry levels, particularly the antibonding Γ_2' state, and the Fermi energy E_F. The energy scale is in Rydberg.

A calculation performed for the fictitious fluorite structure compound PdH_2, using the lattice constant of PdH shows that the antibonding H-H band is well above the metal d bands and consequently, the dihydride of palladium does not form[41-e]. We want to point out however that the position of the antibonding H-H band is related not only to the H-H distance which appears to be in the hydrides always larger than about 2Å, but also to the relative position of the metal s-p band and the d bands in the pure metal which is related to the position of the metal atom in the periodic table. Thus, for example, a calculation of fictitious PdH_2 even when assuming a large lattice constant does not lead to a stabilization of the antibonding H-H band below the metal d states[51].

As illustrated in Figs. 19-21, the DOS of the dihydrides is characterized at low energy by a double peak structure which, as explained previously, is due to the first two low-lying bands and thus originates from the M-H and H-H interactions. We can see that this structure is more important in energy width and in magnitude than in the case of monohydrides since it accommodates four electrons compared to only two in the case of monohydrides. A partial DOS analysis plotted in Figs. 22 (a) and (b) in the case of ZrH_2 shows clearly the composition of the low-lying bands which are not formed only out of H-s states but also, out of metal d states with a smaller metal s and p contribution. In the case of the fluorite structure dihydrides, the H-s states interact with the metal d states of t_{2g} symmetry unlike in the case of the monohydrides ; this is due to the fact that the H atoms are now located at tetrahedral interstices. Since the low-lying bands of the dihydrides accommodate four electrons, there are Z-2 electrons (where Z is the valency of the metal atom) filling the bottom of the portion of the metal d bands which extends at higher energies. It thus appears, at first sight, that the metal d bands have been depopulated in favour of the low-lying bands. This statement has however to be taken with caution since, as shown by the partial DOS analysis of Fig. 22(a), a substantial portion of the metal d bands has been deformed and shifted to lower energies by the M-H interaction and forms an important part of the low-lying bands. In view of this important deformation of the metal d bands in the hydride, the appearent depopulation of the metal d states should not thus be understood in a rigid band model sense and the anionic model often used to interpret some experimental data on the dihydrides is thus not correct. In the dihydrides of the group III TM and of the trivalent RE, only one electron is accommodated at the bottom of the metal d bands, while there are two and three electrons respectively in the case of the dihydrides of the group IV and V TM. The divalent RE dihydrides are semiconductors since there is a gap between the filled two low-lying bands and the empty metal d states. In fact, a comparison of the structure of the metal d bands of the dihydrides plotted in Figs. 19-21 shows a remarkable similarity which is due to the fact that all the dihydrides considered crystallize within the same structure.

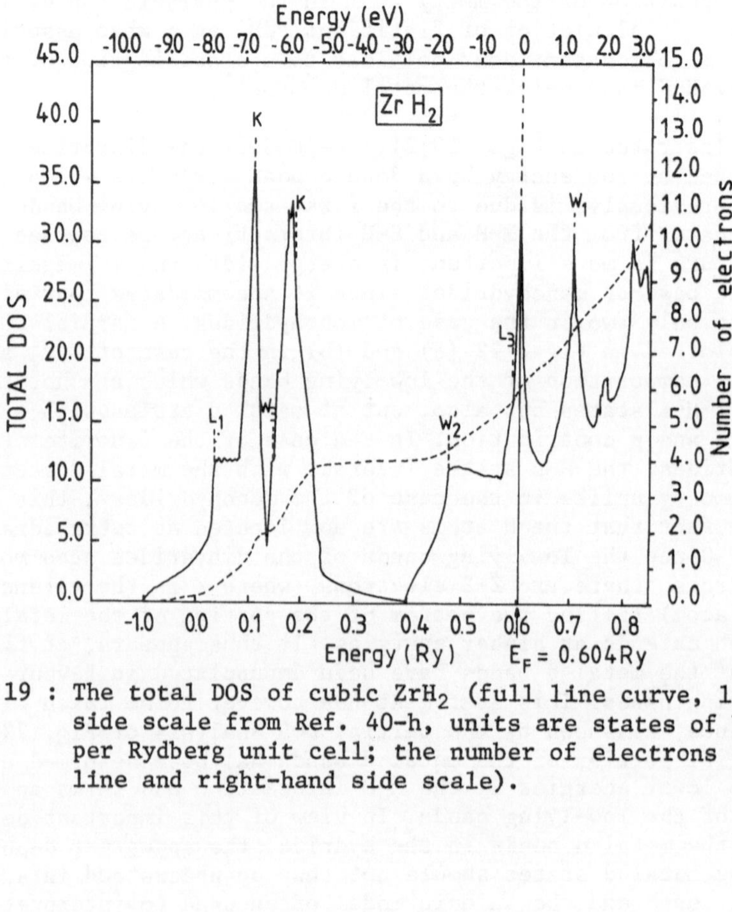

Fig. 19 : The total DOS of cubic ZrH_2 (full line curve, left-hand-side scale from Ref. 40-h, units are states of both spin per Rydberg unit cell; the number of electrons (dashed line and right-hand side scale).

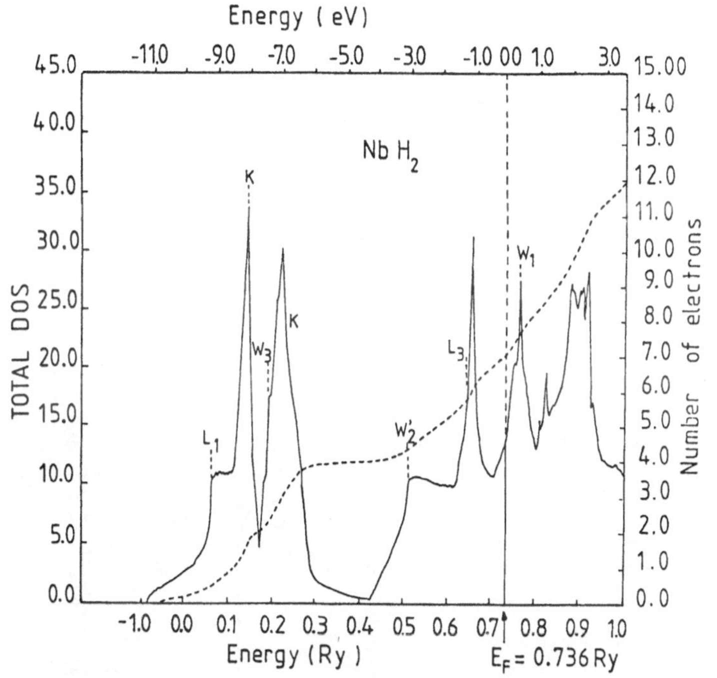

Fig. 20 : The total DOS of NbH_2 (full line curve, left-hand-side scale) from Ref. 40-h; units are states of both spin per Rydberg unit cell; The number of electrons (dashed line and right-hand side scale).

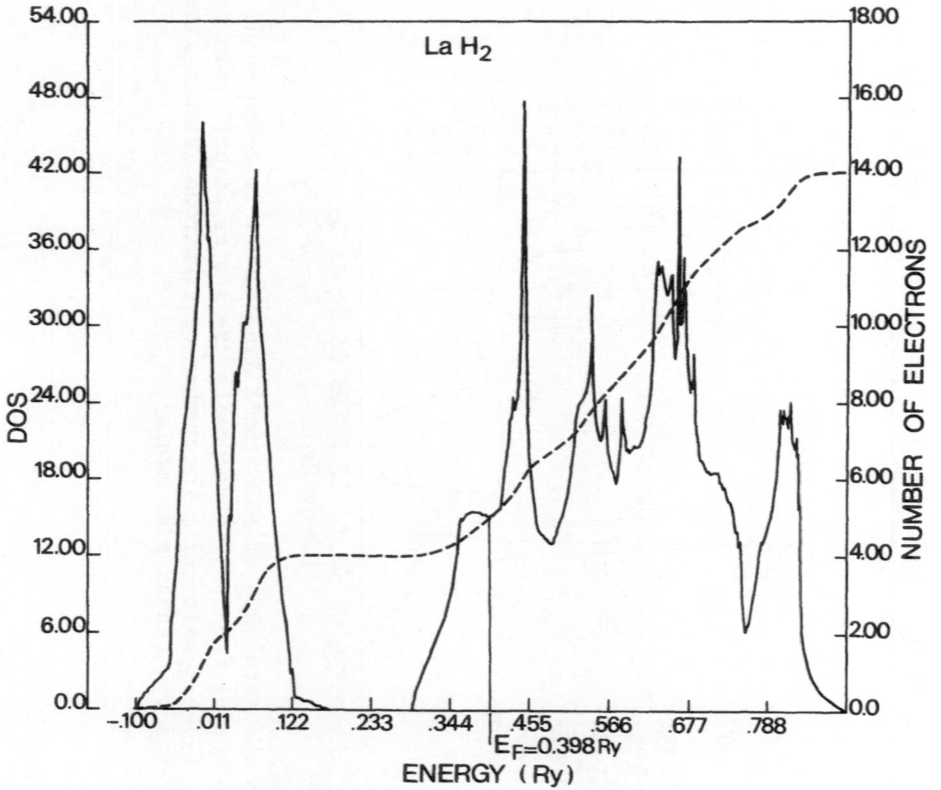

Fig. 21 : The total DOS of LaH_2 (full line curve, left-hand-side scale) from Ref. 40-e ; units are states of both spin per Rydberg unit cell ; the number of electrons (dashed line and right-hand-side scale).

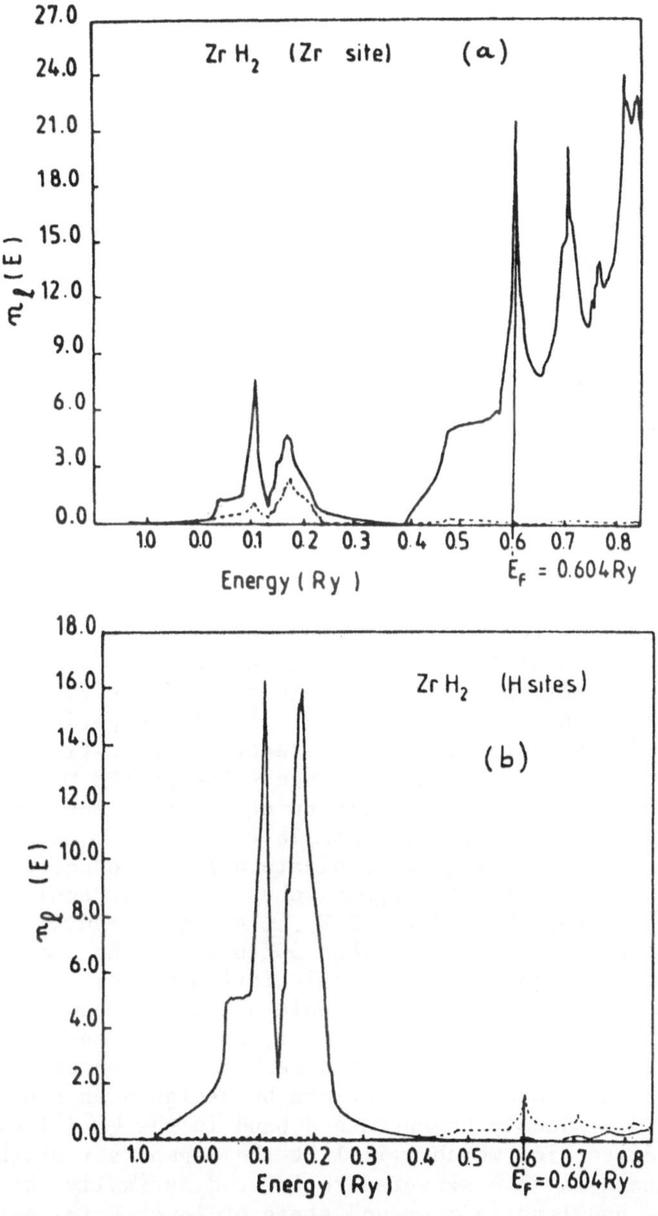

Fig. 22 : The angular momentum DOS analysis, n_ℓ, of cubic ZrH_2 from Ref. 40-h. (a) Inside the Zr muffin-tin sphere $\ell = 0$ (dotted curve), $\ell = 1$ (dashed curve), and $\ell = 2$ (full line). (b) Inside the two H muffin-tin spheres $\ell = 0$ (full line), $\ell = 1$ (dashed line) and $\ell = 2$ (dotted line). Units are states of both spin per Rydberg unit cell.

With the filling of the d bands by one electron, the Fermi level falls in a shoulder of the DOS, as shown in Fig. 21 for LaH_2 ; adding one more electron, the Fermi level falls in a very high peak of the DOS, as for ZrH_2 or TiH_2 (see Fig. 19) ; adding two electrons, E_F falls as for NbH_2 in between two large peaks of the DOS (see Fig. 20). Thus, as one goes from one dihydride to the next in the periodic table, the metal d bands which present the same structure get progressively filled and one can obtain qualitative information concerning the Fermi level position by applying the rigid band model from one dihydride to another dihydride of the same series ; the reader should however be cautioned that this model would be very erroneous if applied to the bands of the pure metal to obtain information about the electronic structure of a hydride.

(b) Properties of the group IV dihydrides

All the dihydrides of the group IV metals Ti, Zr, Hf undergo a cubic to tetragonal distortion[52] below a critical temperature, for hydrogen concentrations in the vicinity of the stoichiometry $x = H/M \sim 2.0$. For example T_iH_x is cubic at all temperatures for $1.6 \leqslant x \leqslant 1.8$ but for $1.8 \leqslant x \leqslant 2.0$ it is cubic for $T > 294$ K and tetragonal below $T < 294$K, the c/a ratio being 0.976. This second order tetragonal distortion is accompanied by anomalies in the temperature dependence of the magnetic susceptibility[53], the proton spin-lattice relaxation rate[54], the electronic specific heat[55], the Hall coefficient and the resistivity[56]. The experimental data on the magnetic susceptibility, the electronic specific heat and the thermoelectric power reveal a large value of the DOS at E_F in the cubic phase and a strong variation of the electronic properties with the crystal structure. For example the magnetic susceptibility data on TiH_x plotted in Fig. 23 indicate a large reduction of the DOS from the cubic to the tetragonal phase ; an extrapolation of Stalinki's results to $T = 0°K$ for $TiH_{1.98}$ gives a reduction of the magnetic susceptibility by more than 30% between the two phases. The results of the band structure calculation on the cubic phase[40-c,41] point out to an electronic origin of the cubic to tetragonal distortion. We have shown that the Fermi level of the cubic group IV dihydrides falls in a peak of the DOS as seen in Fig. 19. This structure in the DOS can be assigned as shown in Fig. 16 to a flat doubly degenerate d band in the ΓL (Λ) direction. The lattice distortion which will lift the degeneracy of this dispersionless band is thus expected to lower drastically the value of the DOS at E_F and lead to a ground state of lower total energy. The cubic to tetragonal distortion observed in the group IV dihydrides can thus be attributed to a Jahn-Teller effect which by lifting the degeneracy of the states at E_F, lowers the total energy of the system. This effect is similar to the theoretical model proposed by Labbé and Friedel[57] to account for the tetragonal distortion of several superconducting A-15 compounds. With this interpretation

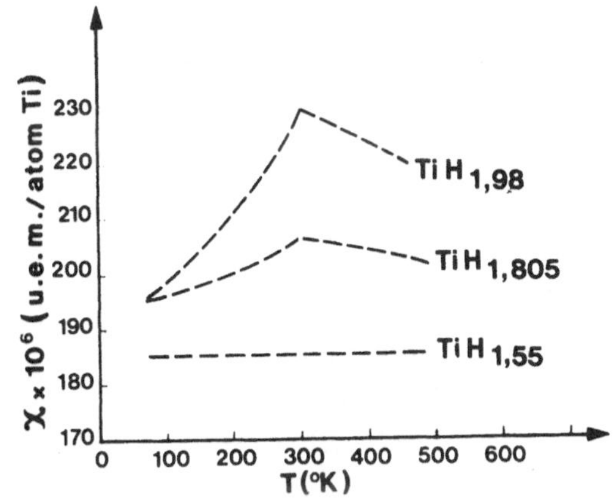

Fig. 23. Experimental variation of the magnetic susceptibility χ of TiH_2 as a function of temperature and hydrogen concentration from Stalinski et al. (Ref. 53).

we can qualitatively understand that above a critical temperature, when the states giving rise to the peak structure in the DOS are thermally populated, the system cannot lower its energy by undergoing a structural transformation and the cubic phase remains stable. The same remark can be used to explain some of the features of the $Ti_{1-x}V_xH_y$ phase diagram[41-d], starting from the tetragonal phase, an increase in vanadium concentration (which has one more electron than Ti) stabilizes the cubic phase. The electronic origin of the tetragonal distortion and the sensitivity of the shape of the FS to the lattice distortion is revealed also by the Hall effect experiments[56] which show a change in the Hall coefficient from negative in the tetragonal phase to positive in the cubic one. The FS of the cubic phase is determined by three bands which, as shown in Fig. 16 belong to the lowest portion of the d band complex. The FS cross sections of cubic TiH_2 are given in Fig. 24 ; the bands have been labelled by increasing order of energy. The symmetry lowering will lift, as indicated above, the degeneracy of band 3 in the Λ direction and the holes of this band are thus expected to be filled in the tetragonal phase, a result compatible with the sign change of the Hall coefficient.

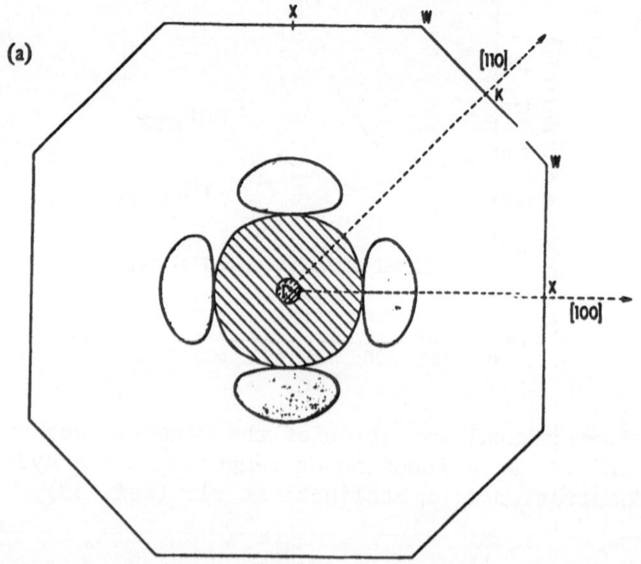

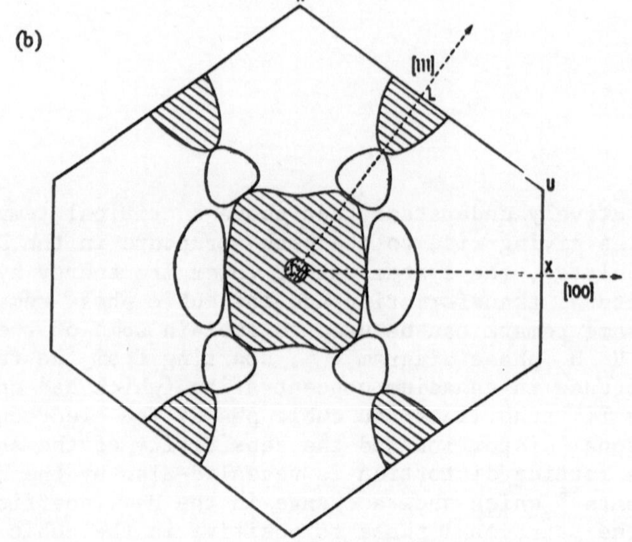

Fig. 24 : Fermi surface cross-sections of three bands of TiH_2 in two high symmetry planes from Ref. 40-c.
(a) (100) cross section ; (b) (110) cross section.
Matched portions are occupied regions due to band 4 (\\\ ) and band 5 (///) ; the shaded portions are band 3 hole regions.

(c) Properties of the dihydrides of lanthanides

As mentioned previously in the general discussion of the electronic properties of the dihydrides, besides the two low-lying metal-hydrogen bonding and H-H antibonding bands which are filled, one electron is left at the bottom of the metal d bands of the dihydrides of the trivalent RE. The Fermi level falls in a shoulder of the DOS as shown in Figs. 21, 25, 26 ; the DOS at E_F is consequently much lower in the dihydride than in the corresponding pure RE. For example we obtain[40-e] $N(E_F)$ = 15.05 states of both spin per Ry-cell for LaH_2 compared to $N(E_F)$ = 27.48 in the same units for pure fcc La[58]. This drastic decrease in the value of $N(E_F)$ explains why both the electronic specific heat[59] and the Pauli susceptibility[60] of La decrease upon hydrogenation. It is to be noted in connection with the decrease of γ, however, that the electron-phonon mass enhancement decreases also from pure La to LaH_2 as we shall see in section VI. A comparison of Figs. 21, 25, 26 shows the remarkable similarity between the electronic structure of the dihydrides of all trivalent RE. The main differences are due to the lattice contraction of the lanthanides and of their hydrides, as Z increases.

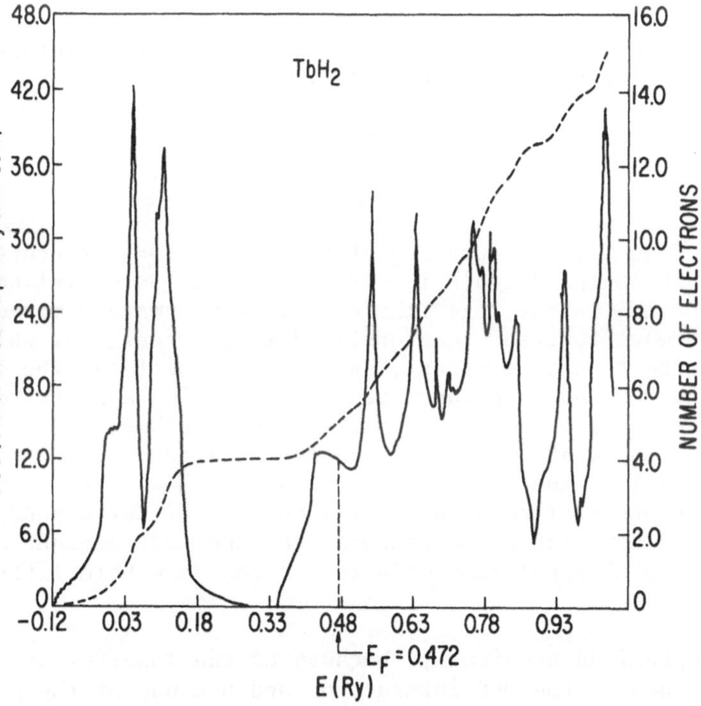

Fig. 25 : The total DOS of TbH_2 (full line and left hand side scale) the number of electrons (dashed line and right hand side scale).

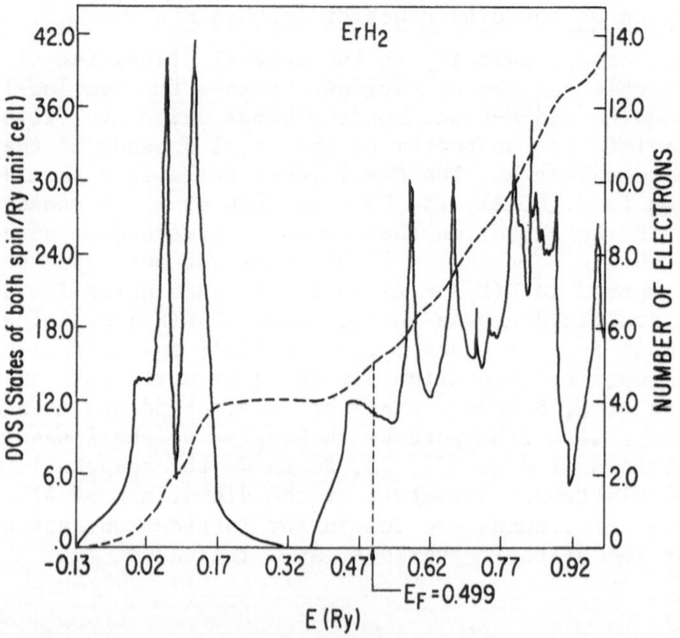

Fig. 26 : The total DOS of ErH_2 from Ref. 40-a (full line and left hand side scale) ; the number of electrons (dashed line and right hand side scale).

In the present work, we treated the RE hydrides on the same footing as TM hydrides ; the occupation of the f shell was of course taken into account in the calculation of the crystal potential however, the f resonance in the logarithmic derivatives were removed since we did not want to treat the localized f electrons, for which correlation effects are important, using a band picture. The structure of the DOS found in the present calculation appears to be in good agreement with the photo-emission results[60] since, as shown in Fig. 27, the width of the La d bands of about 1.5 eV is reproduced, as well as the two peaks structure at low energy. Notice that the theoretical results plotted in Fig. 27 correspond to a bare DOS and have not been broadened to take into account neither the instrumental resolution effects nor the core hole lifetime broadening.

As emphasized previously, because of the lowering of some metal states due to the M-H interaction and because of the presence of additional H-H antibonding states at low energy, the net result appears as a depopulation of the d bands which in the trivalent RE dihydrides accommodate only one electron. This decrease in the number of conduction electrons and also the large expansion of the

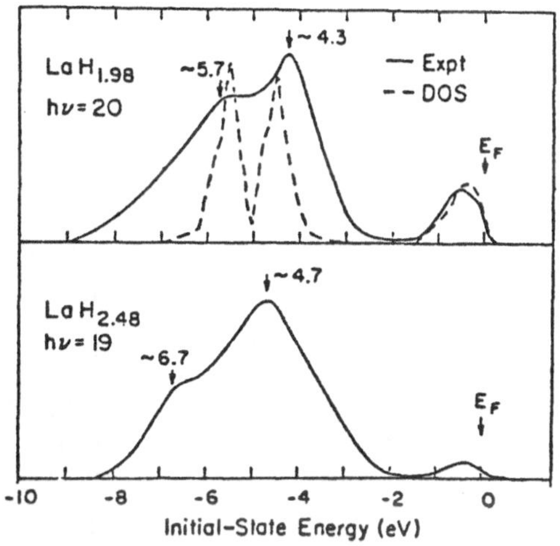

Fig. 27 : Photoemission spectra of Weaver et al. from Ref. 61 for two hydrogen concentrations in La. The dashed curve is the theoretical DOS of Ref. 40-e.

lattice upon hydrogenation should lead to a net decrease in the strength of the Ruderman-Kittel-Kasuya-Yoshida (RKKY) indirect exchange interaction of the RE local moments mediated by the conduction electrons. This can explain why the magnetic transition temperatures are far lower for the dihydrides than for the pure RE metals. For example, the Neel temperature is 229K for pure Tb while TbH_2 shows an antiferromagnetic order below 17K only[62].

For all the RE dihydrides, as shown in Fig. 17 in the case of LaH_2, only one band cuts the Fermi level so, unlike for the RE metals, the FS of the dihydrides are very simple. They can be described as a hole surface which is a deformed cube centered around Γ, with necks along the ΓL directions, leading to a multiply connected hole surface. The cross sections of the FS of ErH_2 are given in Fig. 28. The geometry of the hole surface of the RE dihydrides bears some similarities with the electron surface of the noble metals but with a warped cube rather than a warped sphere around Γ. The FS of the RE dihydrides possesses important nesting features. In the case of ErH_2, the nesting vectors are $\vec{q} = (0.8, 0.0, 0.0)$ $2\pi/a$. The importance of the nesting properties of the hole surface should lead to maxima in the generalized susceptibility function $\chi(\vec{q})$ of the conduction electrons for the values of $\vec{q}$ equal to the nesting vectors. The nesting properties have not yet been correlated to the magnetic ordering of these hydrides.

Another interesting property of the trivalent RE dihydrides is that they are all much better conductors than the corresponding RE (the dihydrides of divalent RE, as explained previously, are

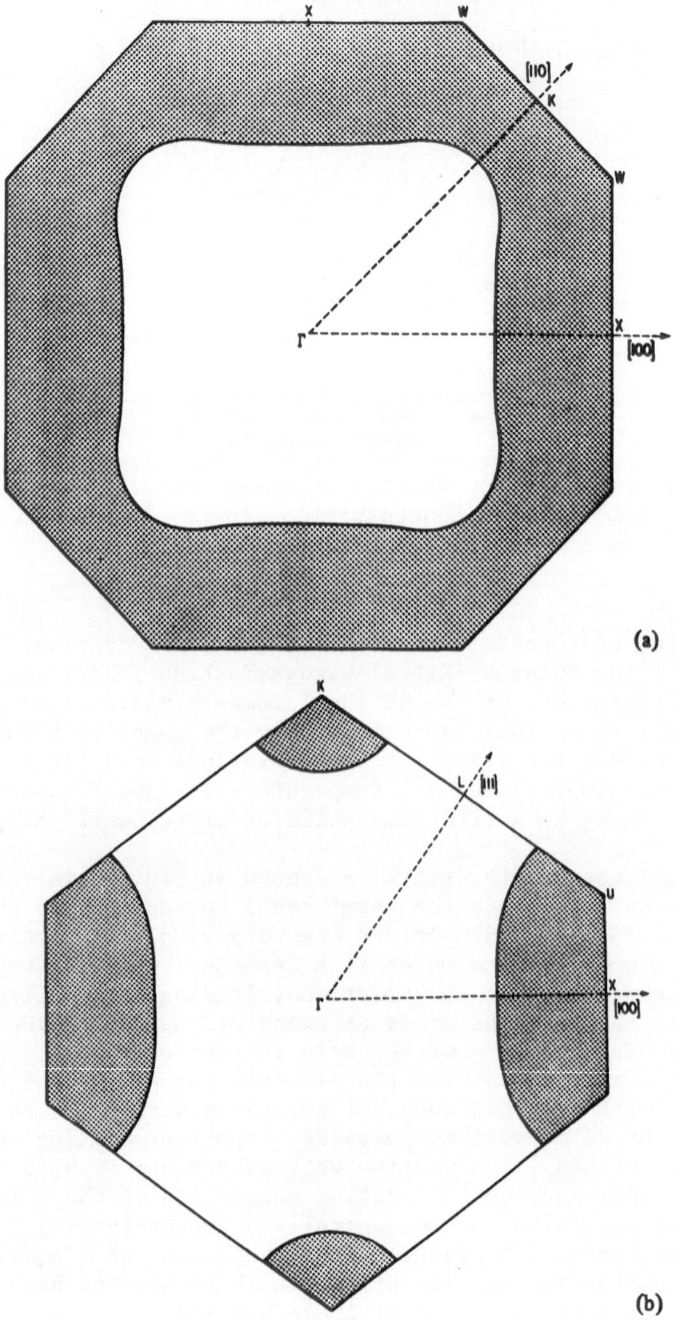

Fig. 28 : The Fermi surface cross-sections of ErH_2 by two high symmetry planes from Ref. 40-a (a) (100) plane ; (b) (110) plane. The dotted protions are occupied regions.

semiconductors) ; the resistivity of the dihydrides[63] is reduced from that of the corresponding RE by a factor which varies from 2 to 7. The change in the electronic properties of the RE upon hydrogenation which is characterized by a depopulation and deformation of the metal d bands and by a reduction of the DOS at E_F leads certainly to a different value of the s → d scattering mechanism of the electrons at the Fermi surface and also, as we shall see in Section VI to a change in the electronic term of the matrix elements of the electron-phonon coupling. Moreover, the hydrogenation process which is accompanied by a large lattice expansion and a change in the screening of the phonon field, results in drastic changes in the phonon properties. This is indicated by the variation of the value of the Debye temperature which is Θ_D = 140K for pure La[64] while it becomes Θ_D = 243K for $LaH_{2.03}$[65]. This increase of Θ_D leads also to the reduction of the phonon resistivity from the pure RE to the dihydrides.

(d) Comments on the anionic model used for the dihydrides of TM and RE

Indeed, numerous experimental data on these dihydrides have been often interpreted in terms of the anionic H^- model. Using the model of Lea Leask and Wolf[66] (LLW), the splitting pattern of the rare earth ions in the crystal field of neighboring H^- or H^+ ions can be determined. In Fig. 29 we give as an example the splitting pattern of Pr^{3+} in PrH_2 determined by Wallace[67]. This author has shown from the analysis of the magnetic properties of the RE dihydrides that the experimental data are compatible with the H^- model ; similarly, the interpreation of the polarized neutrons experiments[62] as well as the Mössbauer effect[68] on RE dihydrides is compatible only with the H^- model. This is also compatible with the electropositivity of the RE. Other studies on dihydrides such as the Compton profile experiments[69] also point out to the adequacy of the anionic rather than to the protonic model. The existence of a charge

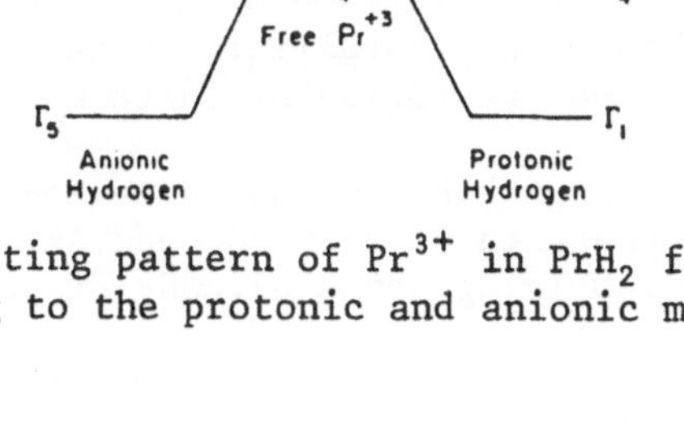

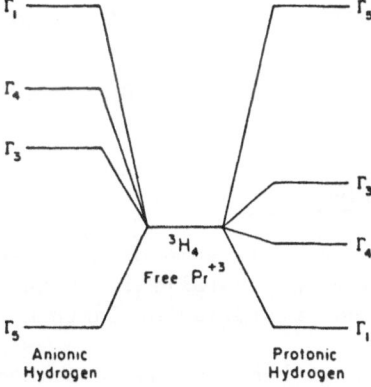

Fig. 29 : The splitting pattern of Pr^{3+} in PrH_2 from Wallace Ref. 67, according to the protonic and anionic models.

transfer from the metal to the hydrogen atoms is also indicated by the study of the core level shifts obtained from photoemission experiments[61,70]. The results indicate a shift of the metal core levels towards higher binding energies, upon hydrogenation. This in turn points out, at least qualitatively, to the existence of some charge transfer from the metal to the H atoms. For example, in the case of $ZrH_{1.86}$ it has been found that the 4p levels of Zr are shifted by about 1 eV towards higher binding energies[70].

The detailed theoretical results of the electronic structure indicate, as emphasized previously, that inspite of an apparent depopulation of the metal d bands in the dihydride, the study of the bonding effects and the deformation of the d bands upon formation of a compound show that the picture of H^- ions is oversimplified. The proton is well screened in the dihydrides however, part of the charge found in the MT hydrogen spheres is due to a charge overlap rather than to a charge transfer. Moreover, the metal-hydrogen interaction cannot be considered as fully ionic ; covalent and metallic interactions are also important. An example of charge analysis[44a] inside the MT spheres is given in Table V, it shows clearly that the sign of the charge transfer from the metal to the hydrogen MT sphere which is estimated to be 0.14 electrons is in agreement with the trend observed experimentally ; however, its magnitude is too small to justify quantitatively the anionic model.

Table V : Charge transfers for ScH and YH from the theoretical work of Ref. 44-a.

Charge inside muffin-tin sphere	ScH_2, a = 4.783 Å Sc	H	Intersitial	YH_2, a = 5.204 Å Y	H	Interstitial
Atomic charge	19.019	0.536	...	36.793	0.578	...
Overlapping atoms	19.531	0.707	2.055	37.283	0.756	2.205
Charge after final iteration	19.323	0.844	1.989	37.116	0.896	2.093

D. Rare earth trihydrides

We shall discuss here only some of the electronic properties of the cubic RE trihydrides crystallizing within the BiF_3 structure, taking as an example the trihydride of lanthanum.

The band structure of LaH_3 is characterized[40-e,41-b] as shown in Fig. 30 by the appearance of a third band at low energy, in addition to the two low-lying bands already present in the dihydride.

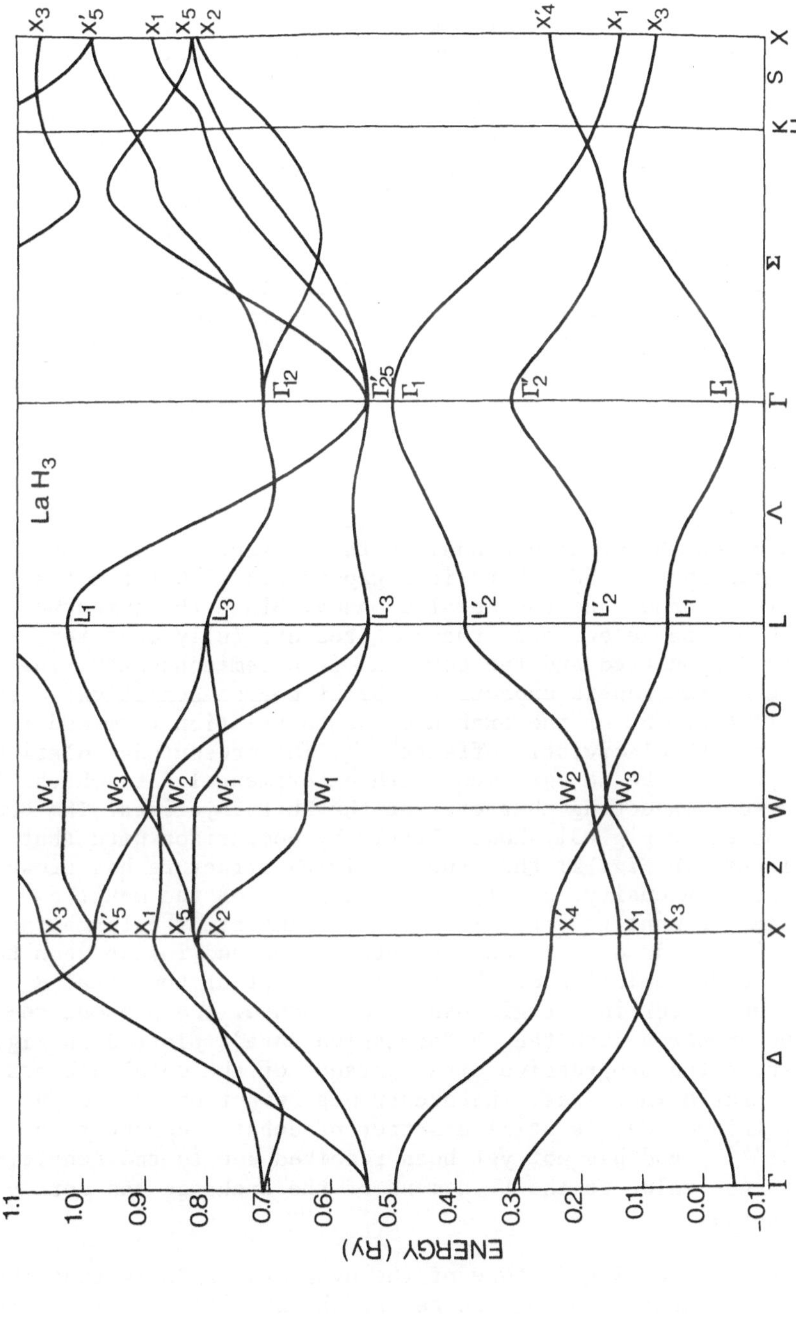

Fig. 30 : Energy bands of LaH_3 (a = 10.5864 a.u.) from Ref. 40-e, along several high symmetry directions. Energies are in Rydberg.

Thus, at the BZ center Γ we find by increasing order of energy the Γ_1 state which is a bonding combination of the metal 6s wavefunction with the 1s wavefunctions of the two H atoms in tetrahedral and one H atom in the octahedral positions ; Γ_2' is an antibonding combination of the 1s wavefunctions of the tetrahedral H atoms and Γ_1 is an antibonding combination of the metal 6s and the two tetrahedral H 1s wavefunctions with the 1s wavefunction of the octahedral H atom. The presence of a third band at low energy is an important factor for the stability of the trihydride. Since the corresponding states, as discussed above at Γ, are an antibonding combination of the metal and the tetrahedral hydrogen wavefunctions with the octahedral hydrogen wavefunction, the position of the third band is sensitive to the distances between tetrahedral and octahedral H atoms and thus to the size of the unit cell. The larger the unit cell, the more stable are the second and third low-lying bands and thus the more stable is the corresponding trihydride. It is interesting to notice that Sc which has a smaller lattice constant than Y does not form a trihydride. Also, in the lanthanides series, with the lattice contraction observed as Z increases, the third band is destabilized and it can be surmized that the lattice distortion which the metal lattice undergoes for the late members of the series and which is accompanied by an increase of the H-H distances takes place to allow for the stabilization of the third low-lying band below the metal d bands. In the present calculation of LaH_3, as shown in Fig. 30, there is a gap of 0.4 eV between the three low-lying bands and the metal d bands. Since the compound possesses 6 valence electrons, these states are fully occupied, the d bands are depopulated and the compound is a semiconductor. The exact value of the gap is expected to be of course sensitive to the theoretical treatment of the exchange and correlation term and to the inclusion of relativistic effects[44-b]. The present calculation however appears to be in agreement with experimental data which indicate a semiconducting character of the RE trihydrides. The DOS of LaH_3 plotted in Fig. 31 shows clearly by comparison with that of LaH_2 plotted in Fig. 21 that the low-lying structure has grown up in width and intensity, partly at the expense of the metal d bands. Again, we want to emphasize here the important deformation of the metal d bands due to the M-H interaction and insist upon the fact that the depopulation of the remaining part of the metal d bands does not occur in a rigid band model sense. The present result can be compared with the photoemission data[61] plotted in Fig. 27 which shows the progressive disappearance of the metal d bands as the H/La ratio increases. The occurrence or not of an energy gap in the trihydrides is still a matter of debate amongst theorists[40,41,44-46] and has not yet been resolved due to the sensitivity of the gap value to the treatment of the exchange and correlation potential.

Another interesting feature of the hydrides of RE is that the metal-insulator transition occurs before the full stoichiometry is

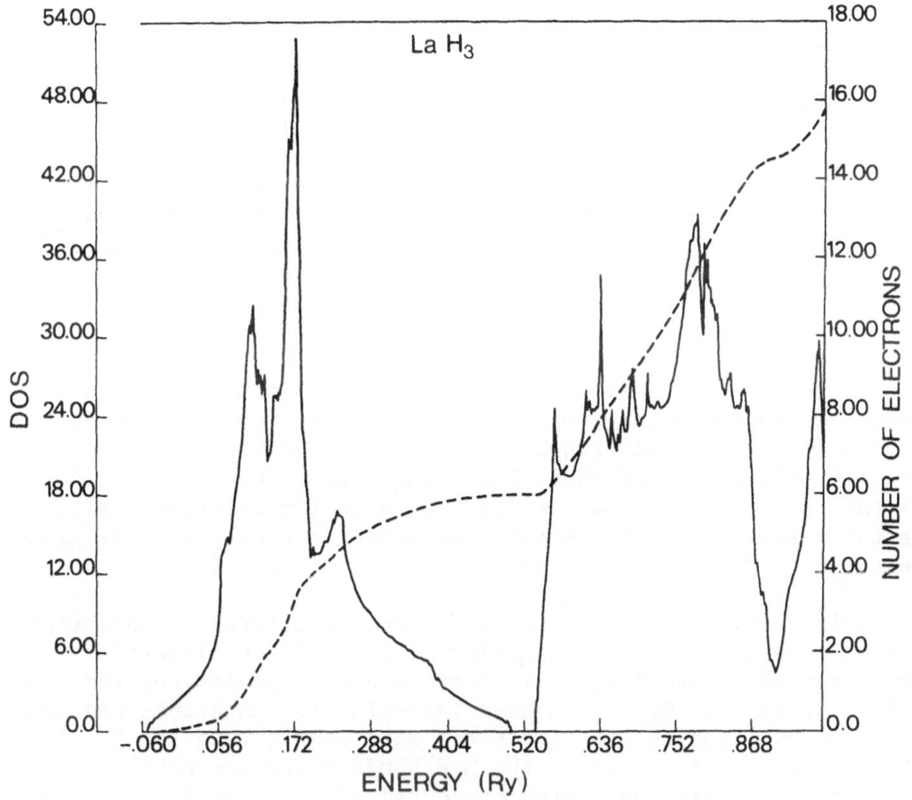

Fig. 31 : Total DOS of LaH_3 from Ref. 40-e (full line curve and left hand side scale) units are states of both spin per Rydberg unit cell ; the number of electrons (dashed line curve and right hand side scale).

achieved. Several possible interpretations of this phenomenon have been proposed[72].

E. Stability of binary metal hydrides

We will discuss here only some aspects of the stability of the TM and RE hydrides whose electronic properties are presented above; we will omit the case of the ionic hydrides of alkali and alkaline earths which are very stable and for which the Madelung contribution to the total energy plays certainly an essential role. In the discussion of the electronic properties of TM metal hydrides we have shown that the stability results from the interplay of several competiting factors. The factors which positively affect the stability are (i) the general lowering of the metal d bands by the attractive H potential ; (ii) the lowering of the metal states which have an s symmetry at the H sites, by the M-H interaction. This effect is particularly important if the metal states lowered were previously empty in the pure metal as discussed for example in the case of the metal 5p branch in Pd. In this connection, we have pointed out an important difference between Pd and Pt which does not form a hydride probably, in part, because due to relativistic effects, no empty metal states can be lowered below E_F by the M-H interaction (iii) the H-H interactions which lead to the presence of additional states below the metal d bands such as the antibonding states observed in the di- and trihydrides of TM and RE. We shave shown that the destabilization of these low-lying states due to the increase of the H-H distances and the difference in the relative positions of the metal d and s bands as Z increases in a series of TM, leads to the non-existence of dihydrides beyond column VI. We have seen also why, for similar reasons, the trihydrides form only when the metal lattice parameter is large (Sc does not form a trihydride) and we gave a possible explanation of the lattice distortion of the heavy RE trihydrides. This feature also explains why the dihydrides of lanthanides which have large lattice constants and thus have low antibonding H-H states are more stable than the TM dihydrides (iiii) the nature of the M-H bonds which are partially covalent and partially ionic with of course also a metallic contribution affect also favorably the total energy of the metal hydrides.

We have also indicated the factors which adversely affect the stability of the hydrides in particular the upward shift of the Fermi level of the pure metal. In fact, as shown previously in this section, there is never an exact balance between the number of additional empty states brought below E by the M-H interaction and the number of additional electrons due to the presence of H atoms in the unit cell. Besides the effects of the deformation of the metal d bands, the Fermi level of the metal necessarily shifts upon formation of a hydride. The most unfavorable case for the hydride formation corresponds to a large upward shift of E_F ; this occurs when the DOS of the pure metal is low, as in the noble metals, and when few empty metal states are lowered below E_F by the M-H interaction. The experimental data of the enthalpies of formation of hydrides plotted in Fig. 32 show a general upward trend of the curve as we go from left to right in the periodic table, the most stable hydrides form on the left. In addition to this general trend in the curve of the heats of formation, an anomalous dip is observed for Pd and Ni. The factors which control the stability of the hydrides have been studied in a more quantitative way for the TM of the end of the series Co to Cu and Ru to Ag by means of a total energy calculation by Williams et al[73]. These authors have used the ASW method to evaluate total energies and could reproduce the dip in the enthalpy of formation curve which occurs for Pd and Ni. This calculation however reproduces only the trends but not the exact magnitudes since the enthalpies of formation of the hydrides studied are small and of the same order of magnitude as the accuracy of the total energy calculations. The analysis of the different contributions to the total energies shows that as Z increases in a series, the M-H bonding states are destabilized, this explains the general upward trend in the enthalpy ΔH curve. It is the variation of the chemical potential of the metal which explains the dip in ΔH observed for Ni and Pd. In fact, the Fermi energy of the pure TM decreases from the middle to the end of the TM series then increases drastically when the d bands are filled, for the noble metals and beyond. This interpretation of the results of the total energy calculation corroborates a previous semi-quantitative study of Gelatt et al[43] and are also in agreement with the qualitative arguments drawn in this section from the band structure results.

IV. ELECTRONIC STRUCTURE OF INTERMETALLIC HYDRIDES

The hydrides of intermetallic compounds, in particular $FeTiH_x$, Mg_2NiH_x, $LaNi_5H_x$ etc are technologically very promising as materials[5] for hydrogen storage, chemical heat pumps, electrodes in fuel cells etc ; these applications have stimulated an extensive experimental effort. The choice of these materials is guided by several factors such as the maximum hydrogen capacity per volume and also per weight unit, the easyness of the activation process, the kinetics of absorption and desorption processes, the small sensitivity of the matrix to impurities in order to be used for a large number of absorption and desorption cycles, the temperature and pressure conditions of absorption and desorption of hydrogen and of course the cost of the material used for the storage. Inspite of the technological importance of these materials, no theoretical study has yet been reported till recently[40-k,l] due to the complexity of the crystal structures of these compounds. I will discuss here the essential features of the electronic structure of FeTiH, $FeTiH_2$, Mg_2NiH_4 and use the theoretical results to interpret several experimental data.

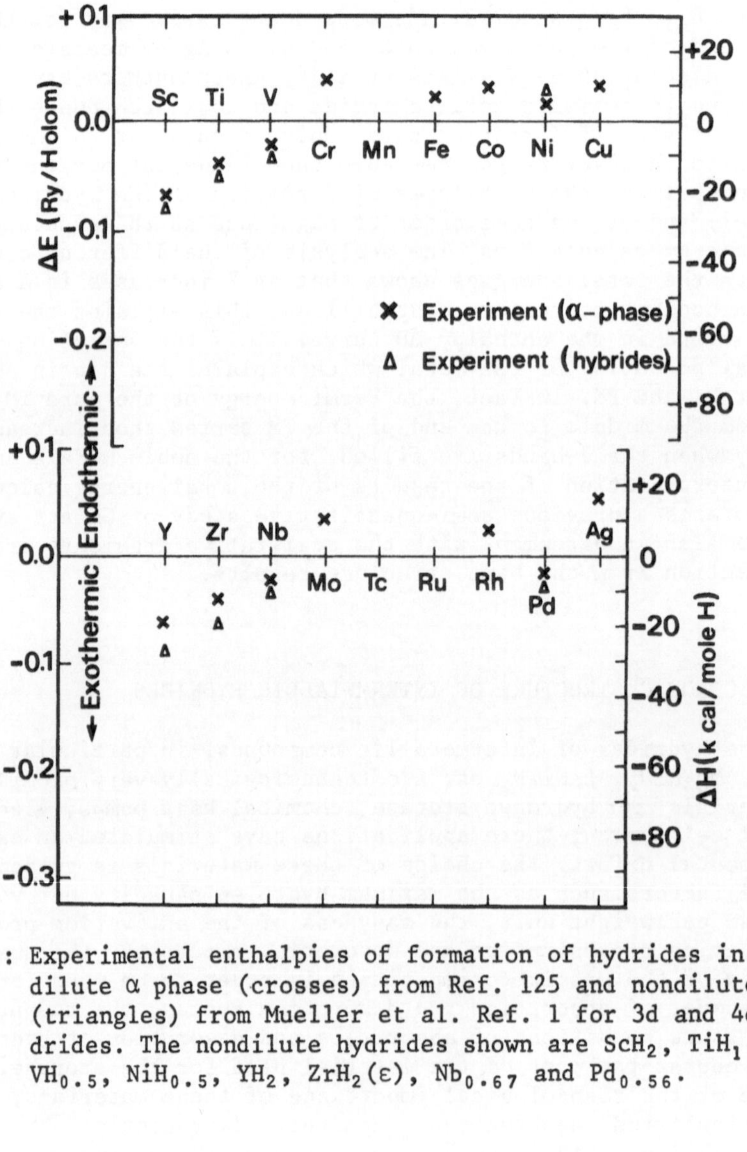

Fig. 32 : Experimental enthalpies of formation of hydrides in the dilute α phase (crosses) from Ref. 125 and nondilute (triangles) from Mueller et al. Ref. 1 for 3d and 4d hydrides. The nondilute hydrides shown are ScH_2, $TiH_{1.61}$, $VH_{0.5}$, $NiH_{0.5}$, YH_2, $ZrH_2(\varepsilon)$, $Nb_{0.67}$ and $Pd_{0.56}$.

A. FeTiH and $FeTiH_2$

As first shown by Reilly and Wiswall[74], FeTi absorbs hydrogen reversibly and forms, besides the dilute α phase observed up to $x \sim 0.1$, two concentrated β and γ phases[75] based respectively upon the composition $x \sim 1.0$ and $x \sim 2.0$. Recent neutron scattering data[76] indicate that the β phase is formed out of two phases, the β_1 phase under study here and the β_2 phase corresponding to $x \sim 1.37$. The unit cell of β_1 - FeTiH is shown in Fig. 33. Elastic neutron scattering data[75] on this phase have been interpreted in terms of an orthorhombic structure which can be viewed as resulting essentially from the doubling of the cesium chloride unit cell of pure FeTi, followed by a large tetragonal distortion due to an expansion along the (110) directions of the cubic CsCl lattice, and a further orthorhombic distortion. The H atoms occupy distorted octahedral interstices with two Fe atoms as first nearest neighbors and four Ti atoms as second nearest neighbors ; the H atoms form strings along the original (110) direction of the CsCl cell of FeTi. It is not clear why the hydrogen atoms occupy octahedral positions instead of the more usual tetrahedral ones characteristic of the bcc metals, why do they form chains and why they are much closer to the two Fe

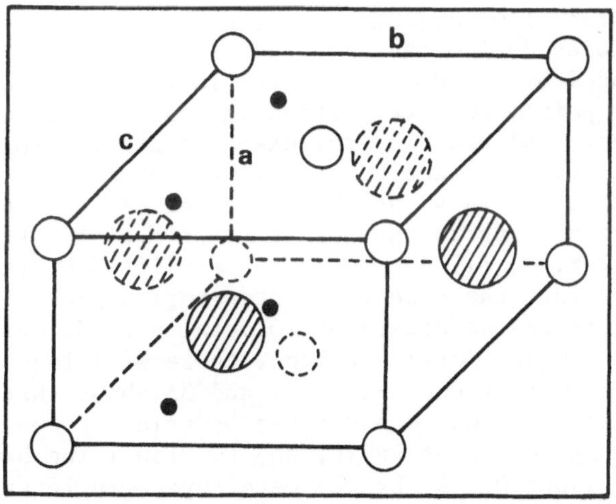

Fig. 33 : The orthorhombic unit cell of β_1 - Fe TiH (see Ref. 75); $a = 2.956$ Å, $b = 4.543$ Å ; $c = 4.388$ Å.
● H atoms ; ○ Fe atoms ; ⦶ Ti atoms.

atoms (d_{Fe-U} = 1.579 Å) than to the four Ti atoms (d_{Ti-H} = 2.163 Å) inspite of the greater affinity of H for Ti than for Fe. A similar situation is encountered in $LaNi_5H_6$ where very short Ni-H distances are observed. The rule of reverse stability proposed by Miedema[77] to explain the stability of the hydrides relative to that of the pure intermetallic is not of much help here since this rule is based on bond breaking arguments of the two components of the intermetallic while there is no bond breaking between Fe and Ti in FeTiH due to the particular positions occupied by the H atoms. The unit cell of β_1-FeTiH as well as that of γ-$FeTiH_2$ contain two formula units. γ-$FeTiH_2$ has a monoclinic structure[75] which can be regarded as resulting from a distortion of a tetragonal cell ; the H-atoms are located at octahedral interstices. Among the four H octahedral interstices in the monoclinic cell, three are located in the basal plane at positions similar to those observed for β_1-FeTiH while the fourth octahedral H atom is located in the a/2 plane where it is surrounded by two Fe and four Ti atoms. Since for both hydrides, a large fraction of the unit cell volume is located outside the touching MT spheres (about 50% in the case of β_1-FeTiH), the departure of the crystal potential from a constant value in the interstitial region - the so-called warping corrections - have been included. (All our calculations presented in this paper include indeed warping corrections). In the present work, the monoclinic distortion with an angle of about 97° in $FeTiH_2$ was ignored for simplicity and the latter was treated as tetragonal.

The energy bands[40-k] of β_1-FeTiH are plotted in Fig. 34. Keeping in mind the remarks made above concerning the crystal structure, our results can be understood (i) by a folding of the bands of FeTi due to the doubling of the size of the CsCl cell (ii) the expansion of the cell in the (110) direction of the CsCl structure and lifting of the degeneracies due to the orthorhombic distortion and (iii) by the role played by the H atoms in the lattice. In order to illustrate the first point, we have plotted in Fig. 35 the bands of pure FeTi, we have indicated by the dashed lines the folding of the bands in the (110) direction of the CsCl BZ due to the doubling of the unit cell size. Thus, at the BZ center of β-FeTiH we obtain the states corresponding to the points Γ and M of the BZ of pure FeTi ; in the hydride, all degeneracies are lifted by the orthorhombic distortion. Since the electronic structure of FeTi in the CsCl simple cubic BZ is characterized by 10 metal d bands overlapped and hybridized with a wider metal s-p band, these will lead to 22 bands in β-FeTiH. The comparison of Figs. 34 and 35 shows that the presence of the H atoms leads to a drastic lowering of two bands ; for example, at the BZ center of FeTiH, the two low-lying bands have been lowered by about 0.3 Ry by the metal-hydrogen interaction. The lowering of metal states having an s symmetry at the H interstitial site by the H potential which strongly scatters s waves, is a feature common to all the metal hydrides previously studied. The two low-lying bands give rise to the structure of the DOS observed in

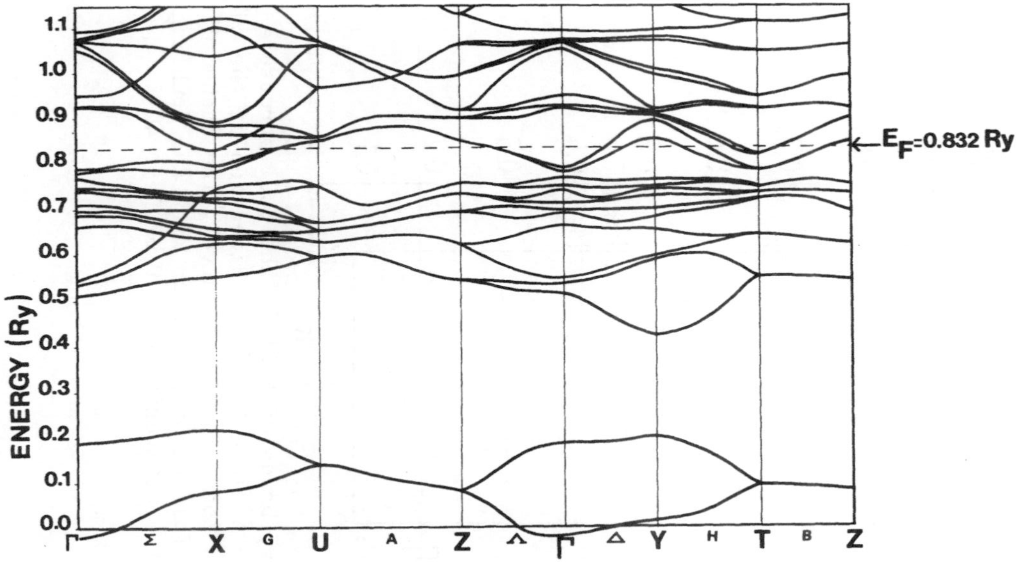

Fig. 34 : The enrgy band of β_1-FeTiH from Ref. 40-j along several high symmetry directions of the orthorhombic Brillouin Zone. Energies are in Rydbergs. The dashed lines indicates the Fermi level position.

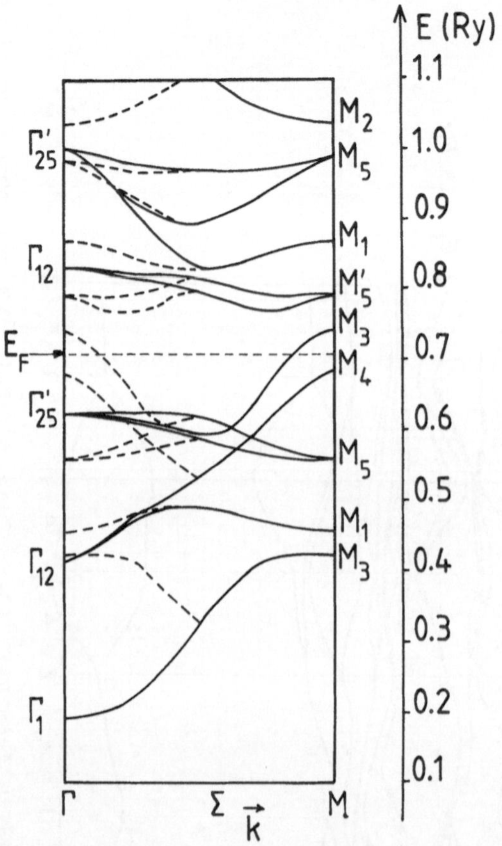

Fig. 35 : The energy bands of FeTi in the CsCl structure ; a = 2.976 Å along the (110) direction of the cubic Brillouin Zone (full line curve). The dashed lines indicate the folding of the bands with the doubling of the size of the CsCl unit cell. From Ref. 40-ℓ.

Fig. 36 ; the second band in Fig. 34 should be understood as resulting from the folding of the low-lying metal-hydrogen band due to the doubling of the size of the unit cell. The lowering of metal states by the H potential is an important factor for the stability of the hydride. The orthorhombic distortion leads to important splittings in the d bands, for example, at the BZ center, the Γ_{12} and Γ'_{25} states of the cubic BZ are split by as much as 25 mRy in the hydride. The metal-hydrogen low-lying bands are formed out of states already filled in the pure intermetallic ; however, a full branch of metal p states around the M'_5 point of the BZ of pure FeTi which appears in Fig. 35 at about 1,5 eV above E_F is lowered by the M-H interaction by about 3.4 eV and brought inside the metal d bands below the Fermi level of β-FeTiH. These additional states do not participate in the low-lying bands however, they bring some additional electrons, about 0.15 electrons, below E_F. As a consequence, less than one electron brought by the H atom has to be accommodated at the top of the d bands and thus, although the Fermi level of the pure intermetallic is shifted towards higher energies in the hydride, the protonic rigid band model is not quantitatively correct. Further, the metal d bands are deformed in going from the intermetallic to its hydride. The DOS of β-FeTiH plotted in Fig. 36 can be characterized (i) by a structure centered at 9 eV below E_F which is due to the M-H interaction, and (ii) the metal d bands at higher energies in which we identify, as in pure FeTi[78], the two peaks structure characteristic of the bonding and antibonding metal states in the bcc metals. The width of the lowest metal d states measured up to the energy of the valley of the DOS is slightly smaller in FeTiH than in FeTi due to the lowering of the lowest portion of the Fe-d states by the H potential ; the increase in the lattice parameter plays also a role in the slight narrowing of the d bands. The increase by 2,3% of the Fe-Ti bond length from FeTi to β_1-FeTiH results also in a weakening of the bond and a larger overlap between the bonding and antibonding metal d states in FeTiH, the valley of the DOS being narrower in the hydride. Pure FeTi is isoelectronic to Cr and the Fermi energy falls in the valley of the DOS[78], the Fermi level of FeTiH is located at higher energies in the antibonding states but, as discussed above less than one electron brought by the H atom is added at the top of the d bands. The number of states at E_F increases substantially from FeTi to FeTiH. We find $N(E_F)$ = 23.93 states of both spin/Ry-FeTiH which corresponds to an electronic specific heat coefficient of 2.02 $mJ.mole^{-1}K^{-2}$ in the noninteracting electron model and without electron-phonon enhancement factor. This is to be compared with the values of $1mJ.mole^{-1}K^{-2}$ and 0.53 $mJ.mole^{-1}K^{-2}$ given respectively by the authors of Ref. 78 for pure FeTi.

A partial DOS analysis into its angular momentum components inside the H and metal MT spheres is plotted in Figs. 37-39 and reveals that the low-lying energy states have essentially a H-s

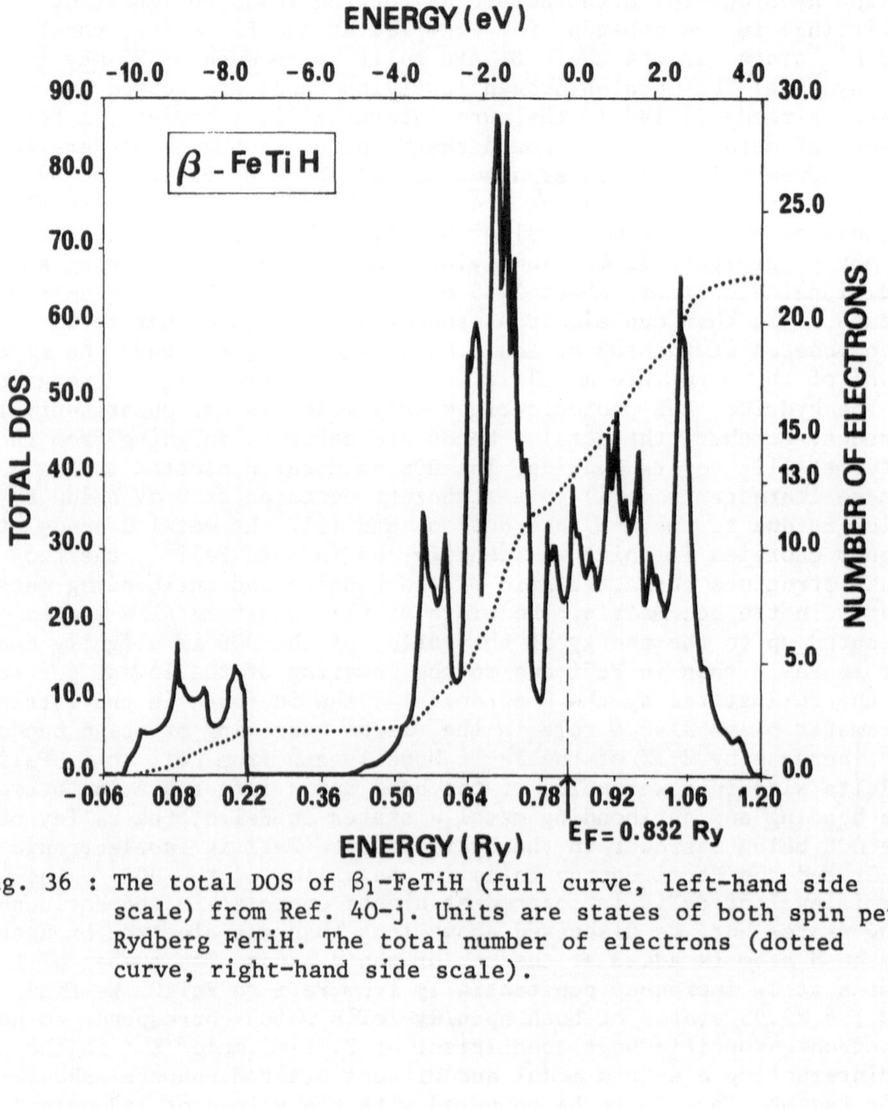

Fig. 36 : The total DOS of β_1-FeTiH (full curve, left-hand side scale) from Ref. 40-j. Units are states of both spin per Rydberg FeTiH. The total number of electrons (dotted curve, right-hand side scale).

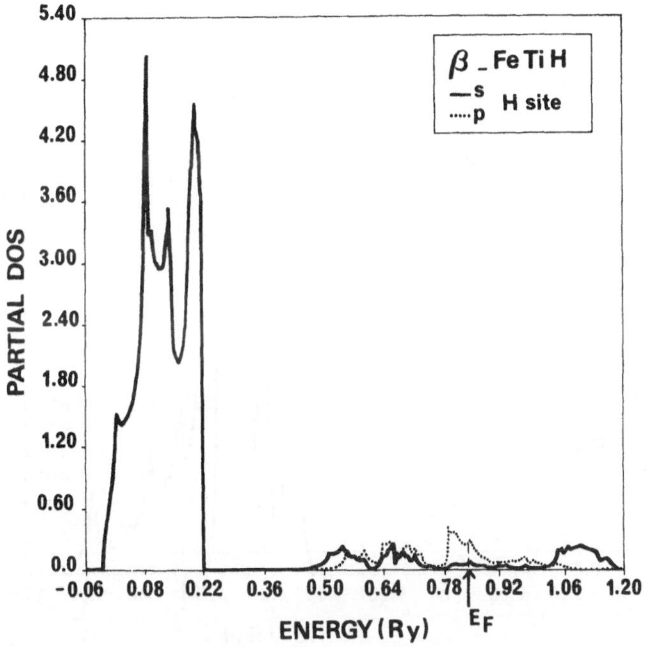

Fig. 37 : The angular momentum DOS analysis $n_\ell(\ell = 0,\ \ell = 1)$ of β_1-FeTiH inside the hydrogen MT sphere, from Ref. 40-j. Units are states of both spin per Rydberg FeTiH.

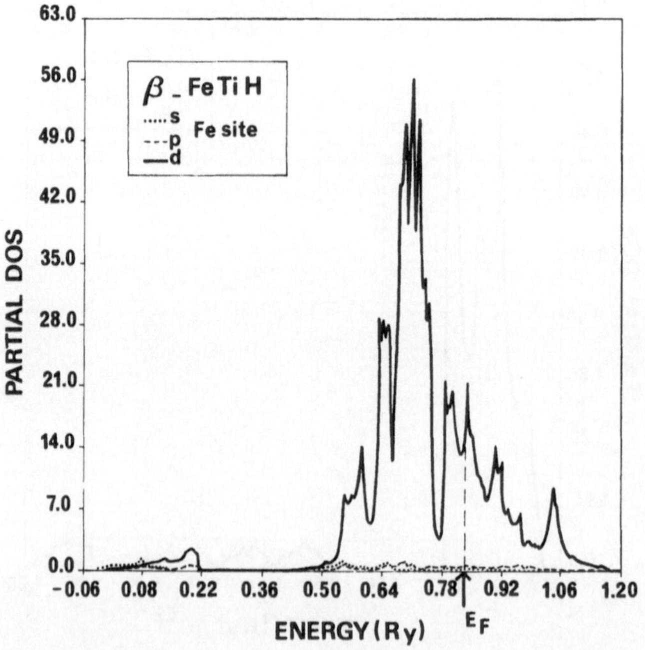

Fig. 38 : The angular momentum DOS analysis $n_\ell(\ell = 0, \ell = 1, \ell = 2)$ of β_1-FeTiH inside the Fe MT sphere, from Ref. 40-ℓ. Units are states of both spin per Rydberg FeTiH.

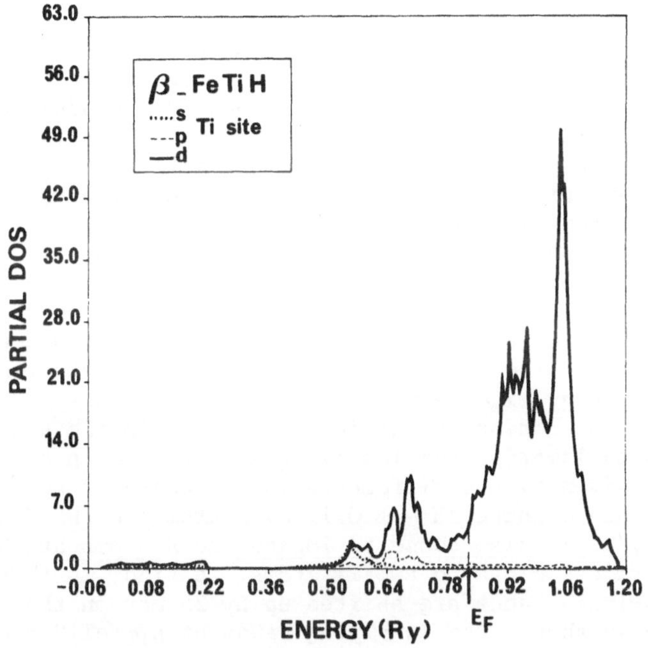

Fig. 39 : The angular momentum DOS analysis n_ℓ (ℓ = 0, ℓ = 1, ℓ = 2) of β_1-FeTiH inside the Ti MT sphere, from Ref. 40-ℓ. Units are states of both spin per Rydberg FeTiH.

and also a Fe-d character ; the Ti-d contribution and the metal s and p components are substantially smaller. Similar partial DOS analysis shows that the lowest portion of the metal d bands is essentially dominated by Fe-d states while Ti-d states have their most important contribution for energies larger than that of the valley in the DOS ; this feature reminiscent of the relative position of the atomic d levels, is also observed in the pure intermetallic FeTi. It would be very instructive to compare the features of the band structure described above with photoemission and X-ray emission spectra.

With the aim of assessing the role of the H-H interactions along the chains in β_1-FeTiH, we chose to break these chains by locating the two H atoms at (0 1/4 1/4) and (0 3/4 3/4) instead of (0 1/4 1/4) and (0 1/4 3/4) as in β_1-FeTiH. The breaking of the chains results in a shift of the low-lying metal hydrogen bands towards higher energies by 15 mRy on the average. With this new location of the H atoms in addition to the plane URTZ of the orthorhombic BZ which is doubly degenerate in β-FeTiH, we obtain another plane TYRS of doubly degenerate eigenvalues. The metal d states and the position of E_F are essentially unchanged except for the states at the point Y which are shifted up in the new structure and become doubly degenerate like the point Z. Thus, from the study of the one electron energy eigenvalues, we can conclude that the H-H interactions along the chains appear to play a role in the stability of the compound by lowering the low-lying M-H states. An evidence of this role is given by the difference in the heats of solution between the dilute α phase $\Delta H = +\ 0.11$ ev/H atom and the β phase[74] $\Delta H = -0.16$ eV/H. We also chose to locate the H atoms at the octahedral interstices close to two Ti atoms. We find, in this case, that the low-lying bands are shifted up by 25 mRy on the average. Since, as shown above, the low-lying bands of β_1-FeTiH are mostly formed of H and Fe states, the rising of these states in the new structure can be interpreted by the decrease in the strength of the Fe-H interaction due to the new location of the H atoms close to Ti. In addition, we observe sensible modifications of the metal d bands thus, the final conclusion concerning the stability of this intermetallic hydride needs to await for total energy calculations. Besides the role of the electronic interactions evidenced here by the change in the position of the low-lying M-H states, and also by the modification of the d bands, the size of the interstices[80] available for H may also be an important factor for the preferred site occupancy of H close to the Fe atoms.

A comparison of the DOS of $FeTiH_2$[40-ℓ] in Fig. 40 to the DOS of FeTiH reveals that upon hydrogen uptake, the M-H related structure at low energy grows in width and in intensity. Indeed, this structure which corresponds to two bands in FeTiH is formed from four bands in $FeTiH_2$ and thus accommodates twice as many electrons. The additional two bands found for $FeTiH_2$ result from the presence

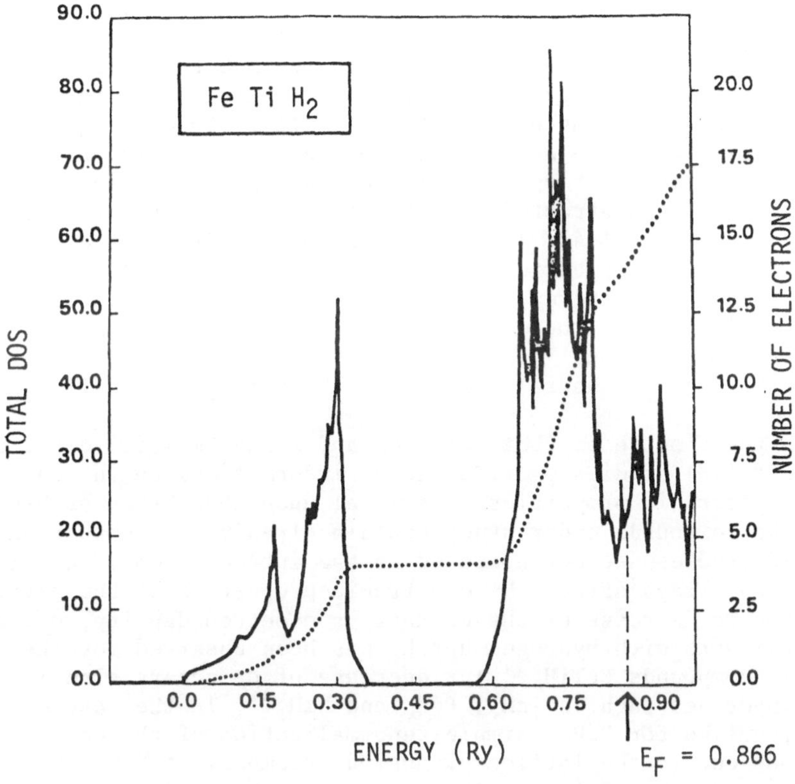

Fig. 40 : The total DOS of $FeTiH_2$ from Ref. 40-ℓ (full line curve, left hand side scale). Units are states of both spin per Rydberg $FeTiH_2$. The total number of electrons (dotted line and right hand side scale).

of two additional H atoms in the unit cell ; the H states forming these low-lying bands are also hybridized with the metal-d and to a lesser extend with the metal s and p states. The Fermi energy of $FeTiH_2$ falls, as for FeTiH, above the valley of the DOS however, it is closer to the bottom of the valley of the DOS in the case of $FeTiH_2$ due to the presence of a large number of additional states in the low-lying bands. For $FeTiH_2$ we obtain $N(E_F) = 22.82$ states of both spin/Ry - $FeTiH_2$ which corresponds to an unenhanced value of $\gamma = 1.93$ mJ.mole^{-1}K^{-2}. The results of γ obtained for FeTiH and $FeTiH_2$ are in agreement with the large increase observed experimentally in the electronic specific heat coefficient[81] after hydrogenation of FeTi[82]. The large increase in the DOS from FeTi to its hydrides results, in agreement with experimental data, in an increase in the Pauli susceptibility χ_p [83]. It is to be noted that magnetization data are not trivial to analyse since in order to obtain χ_p, the important contribution of superparamagnetic Fe particles has to be subtracted. The partial DOS analysis of $FeTiH_2$ is given in Figs. 41-43. It is interesting to note that as for FeTiH, the H-s contribution is very important in the low-lying bands ; the main difference is that in the case of $FeTiH_2$, the Ti-d contribution to the metal-hydrogen bonding is as important as the Fe-d contribution. This is due to the peculiar environment of one of the four octahedral H sites in the $FeTiH_2$ unit cell.

In the absence of photoemission and X-ray emission data, isomer shift (IS) studies provide useful informations on the change in the electronic properties of a metal upon absorbtion of hydrogen. Since the compounds under study contain already a Mössbauer nucleus, iron, IS studies are not hindered by the introduction of an impurity atom which always affect the electronic properties of the matrix. A systematic decrease of the contact -s electron density, ρ_s, at the metal site with hydrogen uptake has been observed for the intermetallic compounds $FeTiH_x$ [84] ; a decrease of ρ_s occurs also in binary hydrides such as TaH_x, PdH_x and NiH_x [85]. In the case of $FeTiH_x$, unlike for PdH_x, simple renormalization of the 4s wavefunctions due to the lattice expansion accompanying H uptake did not appear to be sufficient to explain the observed decrease of ρ_s. We thus performed a microscopic calculation of ρ_s at the Fe nucleus in FeTiH and $FeTiH_2$ to interpret the data of Scwartzendruber et al[84]. The valence electron contributions are given separately in Table VI and compared with the theoretical value obtained for pure Fe metal by Callaway et al[86] and with the experimental results of Shinohara and Fujioka[87]. For β_1-FeTiH we obtain a value ρ_s (Fe) which is nearly the same as for pure Fe. This result is in very good agreement with the data of Swartzendruber et al[84] since for the sample $FeTiH_{0.9}$ these authors did not observe any shift of the Mössbauer line as compared to pure Fe metal ; indeed a negative IS = -0.14 mm.s^{-1} is observed from pure Fe metal to pure FeTi and this shift disappears with hydrogenation when β_1-$FeTiH_{0.9}$ is formed. As shown in Table VI we obtained a 10,4% decrease of the valence contribution

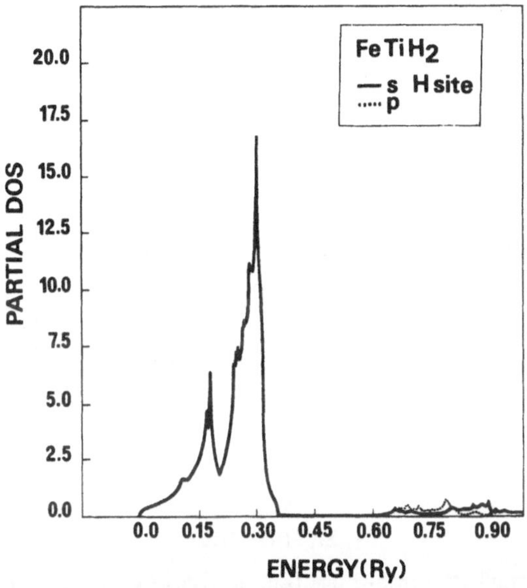

Fig. 41 : The angular momentum DOS analysis n_ℓ (ℓ = 0, ℓ = 1) of $FeTiH_2$ inside the hydrogen MT spheres, from Ref. 40-ℓ. Units are states of both spin per Rydberg $FeTiH_2$.

to ρ_s (Fe) from β_1-FeTiH to $FeTiH_2$. We can calculate the corresponding IS difference

$$\Delta\varepsilon = \alpha(\rho_{FeTiH} - \rho_{FeTiH_2}) \tag{20}$$

where the calibration constant α which is related to the change in nuclear radius during the transition has the value given by Ingalls[88], $\alpha = -0.37\ a_0^3.mm.s^{-1}$. We find $\Delta\varepsilon = 0.23\ mm.s^{-1}$, a result which is in good agreement with the experimental value

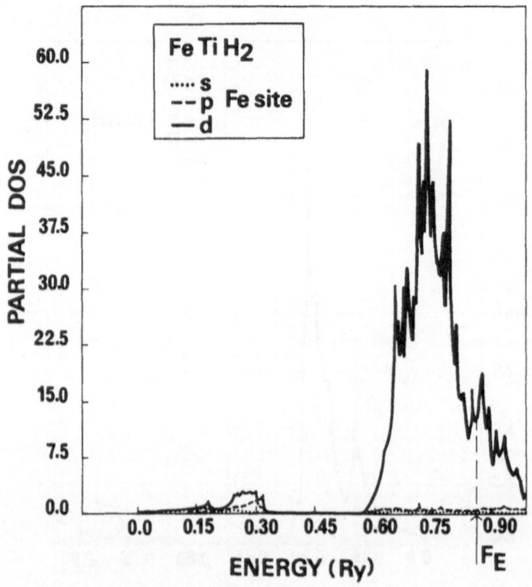

Fig. 42 : The angular momentum DOS analysis $n_\ell(\ell = 0, \ell = 1, \ell = 2)$ of $FeTiH_2$ inside the Fe MT sphere from Ref. 40-ℓ. Units are states of both spin per Rydberg $FeTiH_2$.

$$\Delta\varepsilon = \alpha(\rho_{FeTiH_{0.9}} - \rho_{FeTiH_{1.7}}) = 0.27 \text{ mm.s}^{-1} \tag{21}$$

The decrease of ρ_s (Fe) with H uptake from FeTiH to $FeTiH_2$ cannot be ascribed to a pure effect of the 4s wavefunction renormalization due to volume expansion. Assuming that this effect on the IS is proportional to - Δvolume/volume, it has been estimated that the volume expansion accounts for only half of the IS observed between the dilute α phase of $FeTiH_x$ and γ-$FeTiH_{1.7}$. In order to understand

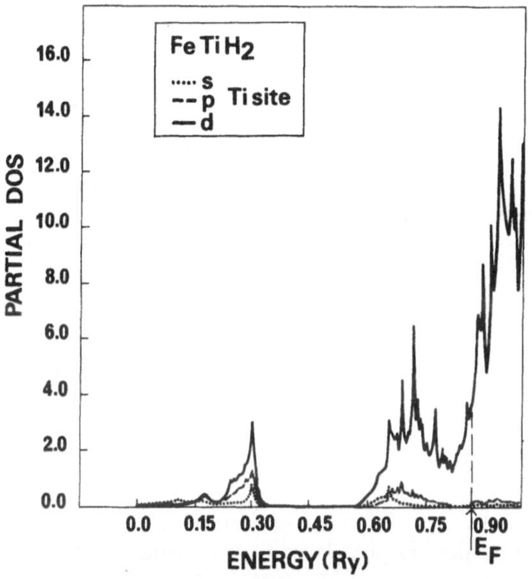

Fig. 43 : The angular momentum DOS analysis n_ℓ (ℓ = 0, ℓ = 1, ℓ = 2) of $FeTiH_2$ inside the Ti MT sphere from Ref. 40-ℓ. Units are states of both spin per Rydberg $FeTiH_2$.

the origin of the further depletion of the 4s electron density at the Fe site, we calculated separately the contribution to ρ_s (Fe) due to the low-lying metal-hydrogen bonding bands. From the results listed in Table VI, it appears that 93,6 % of the decrease in ρ_s (Fe) from FeTiH to $FeTiH_2$ is due to the decrease in the contribution of the low-lying bands. Since, as discussed in the previous paragraph, the metal states which contribute to this band are very much affected by the M-H bonding, we can conclude from our theoretical

Table VI : Contact charge density ρ_s of band electrons and Isomer Shifts (IS) at the Fe site in metallic Fe, in FeTiH and $FeTiH_2$; (a) From Ref. 40-k ; (b) The experimental data of Ref. 84 correspond to $FeTiH_{0.9}$ and $FeTiH_{1.7}$; (c) Theoretical results for metallic Fe from Ref. 86 ; (d) Experimental result for metallic Fe from Ref. 87.

	Present results				Experiment (b)	
	ρ_s (Fe) low-lying bands	ρ_s (Fe) metal-d bands	ρ_s(Fe) total	I.S. $mm.s^{-1}$	I.S. $mm.s^{-1}$	ρ_s (Fe)
Fe					0.0	5.823 (c) 5.53±0.46 (d)
FeTiH (a)	2.735	3.123	5.858	+ 0.01	0.0	
$FeTiH_2$ (a)	2.163	3.084	5.247	+ 0.23	+ 0.27	

study that the depletion of the 4s metal states at the Fe site observed upon H uptake is in large part due to the M-H bonding.

B. Mg_2NiH_4

We present here the result of the first attempt to the theoretical investigation of the electronic properties of Mg_2NiH_4 which is a material potentially important for technological applications.

Mg_2Ni absorbs hydrogen reversibly and leads to high hydrogen capacity compounds since the Mg_2Ni-H system forms an α phase up to H/M = 0.3 and a β phase for higher hydrogen concentrations, which is based upon the stoichiometric Mg_2NiH_4 composition. The hydrogen uptake is accompanied by a structural transformation since the pure intermetallic has an hexagonal crystal structure of the C-16 $CuAl_2$ type[89] while the β-phase X-ray diffraction data of Reilly and Wiswall[90] were interpreted in terms of a tetragonal crystal structure with a = 6.464 Å and c = 7.033 Å. There is also a substantial volume expansion ΔV = 4.80 $Å^3$ per H atom. Recently Gavra et al[91] reported the existence of a high temperature cubic phase with lattice constant a = 6.490 Å occuring between 220°C and 240°C. The existence of a high temperature cubic phase has also been confirmed independently by the X-ray diffraction data of Genossar and Rudman[92] (a = 6.525 Å above 250°C), of Darnaudery et al[93] (a = 6.49 Å above 240°C) and by the neutron diffraction experiment of Yvon et al[94] (a = 6.507 Å at 280°C). In this phase, the metal atoms have an antifluorite type arrangement with the Ni atoms forming an fcc lattice, the Mg atoms occupy the (1/4, 1/4, 1/4) positions in the cubic cell which are those of the cations in the fluorite CaF_2. A general consensus has not yet been achieved for the structural properties of the low temperature phase. The exact location of the H atoms in the high temperature cubic phase has not yet been fully determined ; neutron diffraction data on the family of compounds M_2RuD_5 and M_2IrD_5 (M = Ca, Sr, Eu)[95] indicate that the coordinate of D could be (1/4, 1/4, 0) or (~ 1/4, 0, 0). Since a comparative study of the proton NMR in MgH_2 and Mg_2NiH_4 shows that the nature of the Mg-H bonding does not seem to be perturbed by the presence of the Ni atoms[96], I did not locate the H atoms at the (1/4, 0, 0) positions since these would lead to rather short Ni-H distances, I thus located the H atoms at the (1/4, 1/4, 0) positions which are compatible with the neutron diffraction data. Since in the cubic cell only 1/3 of such sites are occupied, the following choice has been made for the H occupancy : (1/4, 1/4, 0), (0, 1/4, 1/4), (1/4, 0, 1/4), (-1/4, 1/4, 0). These positions for the H atoms lead to H-H distances which are larger than the minimum distance of 2.1 Å always observed from a compilation of the data on metal hydrides ; we obtain H-H distances of about 2.3 Å and 3.25 Å respectively for first and second neighbors. This is in general agreement with the NMR data analysis[96] which indicates proton distances 2.44 Å $< d_{H-H} <$ 2.83 Å for Mg_2NiH_4 at 480K where the sites are randomly occupied.

Since all the structural data concerning the cubic phase agree on the metal atom positions, I first studied the band structure of the fictitious Mg_2NiH_0 compound corresponding to the antifluorite cubic structure observed for Mg_2NiH_4 with lattice constant a = 6.507 Å, where the H atoms have been removed. This study has been performed to show clearly the role played by the H atoms in the electronic structure. The corresponding bands along several high symmetry directions of the fcc BZ are plotted in Fig. 44. The electronic structure of the intermetallic lattice is characterized by a very narrow Ni-d band complex overlapped and hybridized with a wider metal s-p band. The width of the Ni-d bands from X_3 to X_2 (see Fig. 44) is only about 0.42 eV compared to the value of about 4 eV found from similar calculation for paramagnetic Ni[40-d]. This is due to the large increase in the value of the Ni-Ni distances in Mg_2NiH_0 (d_{Ni-Ni} = 4.6 Å) compared to d_{Ni-Ni} = 2.49 Å in pure Ni, which results in a drastic reduction of the metal-metal interaction. A partial wave analysis of the wavefunctions inside the MT spheres at some high symmetry points reveals the strong hybridization between the Ni s-p and the Mg s-p states. Because of the large value of the Mg_2NiH_0 unit cell volume, the Ni 4s wavefunctions are not as strongly compressed and renormalized as in pure Ni ; consequently in pure paramagnetic Ni, the lowest Ni-4s Γ_1 state lies at about 6 eV below the Ni-d Γ'_{25} state, while in Mg_2NiH_0, the corresponding Ni-4s Γ_1 state, which is also strongly hybridized with an Mg-s state, is much closer to the Ni d bands since it lies at only

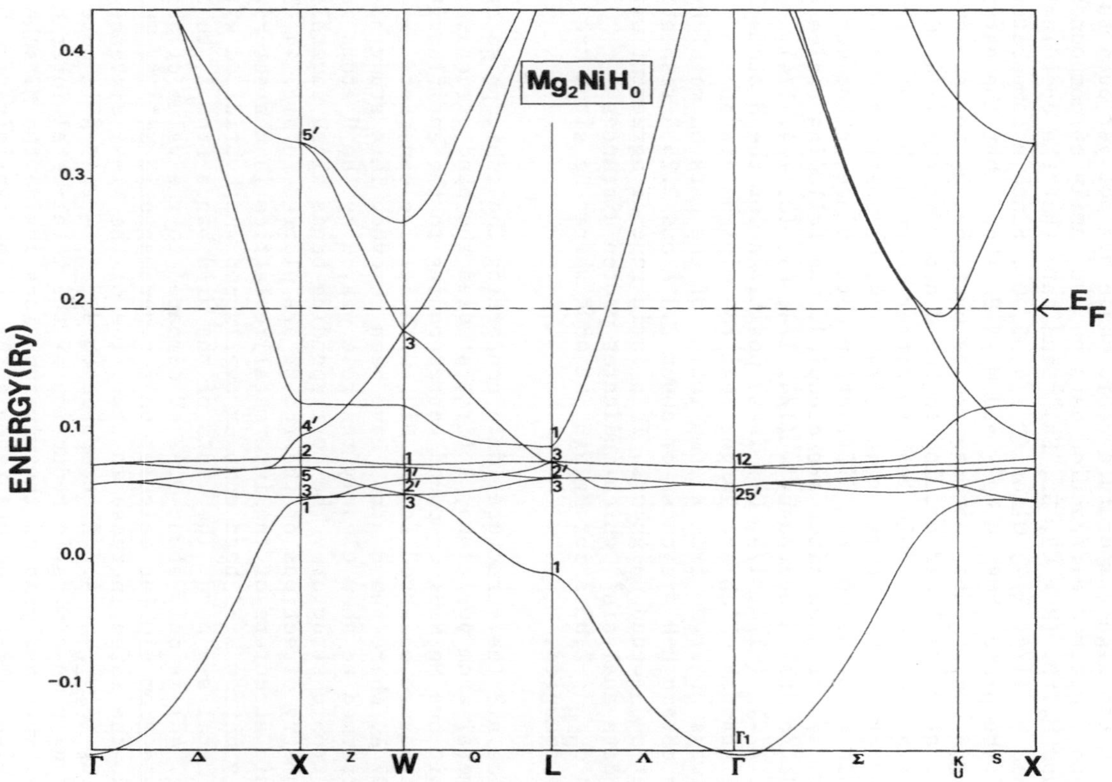

Fig. 44 : The energy bands of cubic Mg_2NiH_0 (antifluorite structure) along several high symmetry directions of the fcc Brillouin Zone. Energies are in Rydberg. The dashed line indicates the position of the Fermi energy E_F.

2.85 eV below Γ'_{25}. The lowest metal s-p band overlaps the Ni-d bands. Above the narrow Ni-d bands, the metal states have a rather free-electron like character. A wavefunction analysis indicates the large delocalization of the corresponding states which have mostly a Mg s-p, Ni s-p character ; above the metal-d bands, with increasing value of the energy, the metal p components become very important. From the DOS curve plotted in Fig. 45, we can observe clearly the Ni-Mg s-p bands extending down to 2.7 eV below the narrow Ni d bands ; the d bands give rise to very large values of the DOS. Since the fictitious intermetallic Mg_2NiH_0 has 14 valence electrons, its Fermi level falls as shown in Fig. 44 and 45 at 1.6 eV above the top of the metal d bands, in a region where the states have a metal p-s character. The analysis of the results obtained in the present calculation indicates that a good model of the electronic structure of the intermetallic Mg_2NiH_0 could be obtained by studying the Ni s-p and Mg s-p interaction, using for example the orthogonalized plane wave method, while treating the Ni-d states as weakly interacting atomic like states.

A comparison between Fig. 45 and Fig. 46 where the DOS of the hydride Mg_2NiH_4 is plotted, shows clearly the role of the hydrogen atoms on the electronic structure of the intermetallic compound under study. The band structure of the hydride is characterized, below the Ni-d bands which are centered around 0.35 Ry, by the presence of two separate structures in the DOS which result from the presence of four metal-hydrogen induced bands. By increasing value of the energy the two low-lying structures in the DOS of Mg_2NiH_4 observed in Fig. 46 are due respectively to one and to three metal-hydrogen bands. A wavefunction analysis indicates a strong H-s character taken by the states forming the low-lying first band in the hydride as well as a large depletion of the metal s-p charge in favor of the hydrogen muffin-tin sphere. This effect is particularly important for the Mg s-p states. We can thus conclude from this analysis that the first low-lying M-H bonding band in Mg_2NiH_4 results from the interaction between the H-s and the delocalized metal s-p charge. The Mg s-p, H-s interaction is particularly strong because of the vicinity of the two atoms and results undoubtly in a charge transfer from the metal to the H sites. This character is different from what we previously observed for transition metal hydrides in which the metal-hydrogen low-lying band was mostly due to a H-s, metal-d interaction. From the present result, we predict a large metal s charge depletion upon formation of the hydride, which could be observed by Mössbauer IS experiments ; such data, previously discussed in the case of $FeTiH_x$ are not yet available for Mg_2NiH_4. The next three low-lying bands which give rise to the second structure of the DOS centered just below the narrow Ni-d states result mostly from the H-s interaction with the delocalized metal states of mostly p character, and from the H-H interaction. We observe a very small Ni-d contribution in these metal-hydrogen induced bands. This contribution is much smaller than in the case

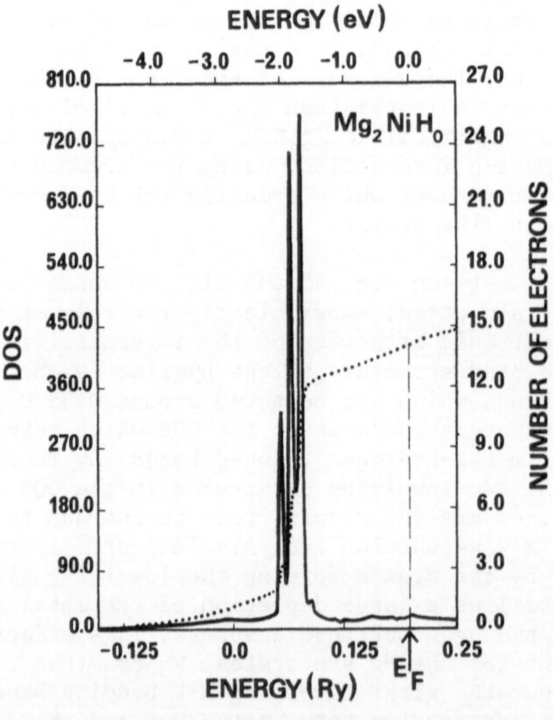

Fig. 45 : The DOS of cubic Mg_2NiH_0 full line curve and left hand side scale units are states of both spin per Rydberg unit cell ; the number of electrons, dotted line and right hand side scale. The arrow indicates the position of the Fermi energy E_F.

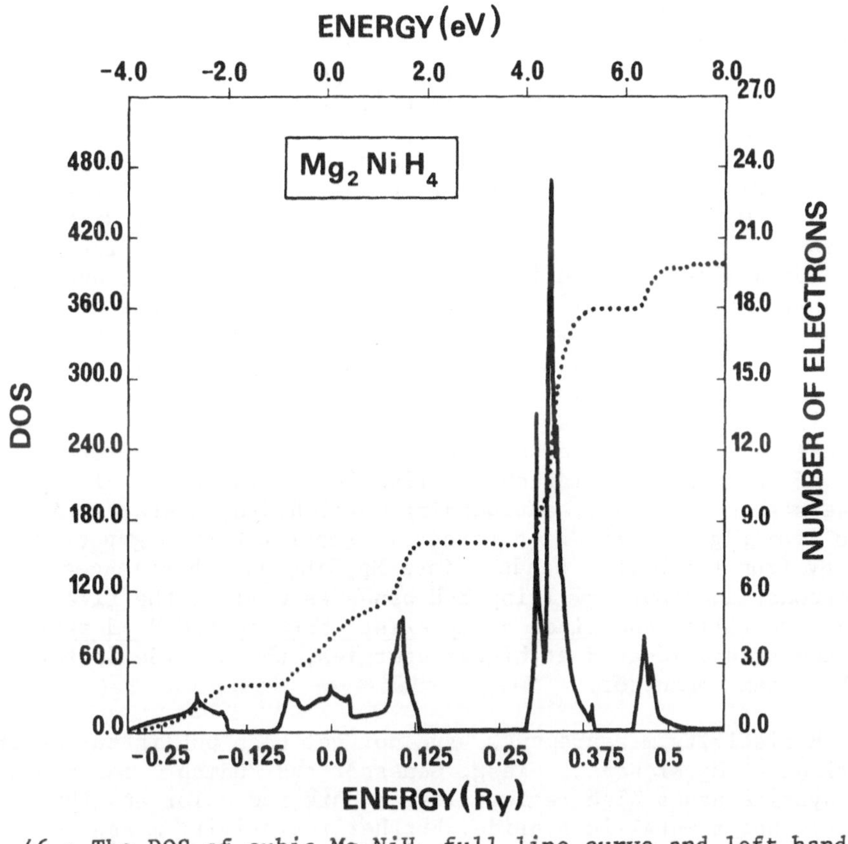

Fig. 46 : The DOS of cubic Mg_2NiH_4 full line curve and left hand side scale units are states of both spin per Rydberg unit cell ; the number of electrons, dotted line and right hand side scale.

of transition metal hydrides ; this is due to the relatively large Ni-H distances ~ 2.3 Å in Mg_2NiH_4 compared to ~ 1.86 Å in NiH. An important difference between the lowest lying M-H band and the next three bands is that a large fraction of the latter are formed out of metal states previously empty in pure Mg_2NiH_0. The lowering of empty metal states by the M-H interaction in Mg_2NiH_4 appears to be much stronger than in previously studied metal hydrides ; this is due to the fact that the metal states which participate in the bonding in Mg_2NiH_4 are delocalized and give rise to an ionic contribution. This is at variance with the TM hydrides in which the localized d states were contributing largely to the M-H bonding. The four low-lying M-H induced bands are separated by a gap of ~ 2.1 eV from the five Ni-d bands. From a comparison of Fig. 45 and 46 we notice that the width of the d bands has increased substantially from 0.4 eV in Mg_2NiH_0 to 1.3 eV in the hydrides. Since the Ni-Ni distances have not changed between the two compounds, the increase of the d band width has a different origin ; it is in fact due to the large crystal field splitting of the d bands due to the presence of the H atoms in the lattice. Let us recall that the H atoms in this structure are located half-way between the nearest neighbors Ni atoms and thus the lobes of the Ni-d t_{2g} functions cannot interact directly. The states located above the Ni-d bands are largely delocalized since a wavefunction analysis shows that a large fraction of the corresponding charge lies in the interstitial region ; these states are mostly antibonding metal hydrogen states. As seen from Fig. 46, the Ni-d bands are separated by a gap of about 0.8 eV from the higher bands. Since Mg_2NiH_4 has 18 valences electrons, the four low-lying M-H bands as well as the five Ni-d bands are filled and since a gap exists between the Ni-d states and the states located at higher energies, the hydride is formed to be a semiconductor.

Resistivity measurements have not yet been published for the hydrides of Mg_2Ni. Nevertheless, Genessar and Rudman[92] mentioned that the hydride has a high resistivity, a dark red color and thus it is clearly not a metallic hydride. Further resistivity measurements are underway in this group and their preliminary yet unpublished results indicate that for both the low and the high temperature phases, the conductivity increases exponentially with temperature[97], a behaviour which is in agreement with the present theoretical results. It is worth to mention here that the electrical resistivities of other ternary hydrides of the same family are extremely high[95] ; this is the case especially of Ca-Ir-H, Sr-Ir-H, Ca-Ru-H, Sr-Ru-H systems which have conductivities ranging from 6.4×10^{-8} to $2.6 \times 10^{-7} \Omega^{-1} cm^{-1}$; the conductivities of Ca-Rh-H and Sr-Rh-H are also small but much higher than those of the other systems since there are of the order of $10^{-1} \Omega^{-1} cm^{-1}$.

If we consider the family of compounds $M_2M'H_x$ where M is a divalent metal (M = Mg, Ca, Sr, Eu, Yb) and M' a transition metal

(M' = Ni, Rh, Ir, Ru), it is remarkable that the maximum hydrogen content which can be achieved corresponds to a number of valence electrons of 18. As an example, the hydrides of highest capacity observed are: Sr_2RuD_6, Sr_2IrD_5 and Mg_2NiD_4. This suggests that the capacity of absorption of H in this family of intermetallic compounds could be governed by a factor of electronic origin. Indeed, our results of the electronic structure study of Mg_2NiH_4 shows that an energy gap exists between the Ni-d bands and the higher states. The filling of the Ni-d bands is achieved for a number of valence electrons of 18, a feature which is due to the presence at low energies, of four M-H induced bands. It can be surmized, in view of the present results, that the filling of the high energy bands above the Ni-d states is energetically unfavorable due to the existence of an energy gap. Further work in other hydrides of the family which can accommodate up to 5 and 6 H atoms is now underway to check this prediction.

Finally our theoretical results point out to an important ionic component in the Mg-H bonding. This result is in agreement with the experimental observation of Senegas et al[96] who found, from a comparative study of the NMR line of H in MgH_2 and Mg_2NiH_4 that the Mg-H bonding seems to have the same characteristics in the two compounds.

In view of the technological interest of Mg_2NiH_4 it would be very desirable to obtain additional experimental information on several fundamental physical properties which have not yet been investigated we have already mentioned for example the lack of resistivity measurements.

V. ELECTRONIC STRUCTURE OF NON-STOICHIOMETRIC METAL HYDRIDES

The electronic properties of some of the non-stoichiometric TM hydrides have been studied theoretically. Several methods have been used such as the supercell method[18,19] which is valid essentially in the very low concentration limit but requires significant computational efforts since it is dealing with large unit cells containing numerous TM atoms ; for the higher hydrogen concentration range, this method necessarily assumes an order in the hydrogen sublattice, which is certainly arbitrary. Other methods have thus been proposed. The virtual crystal approximation (VCA) suffers from the fact that the energy eigenvalues are real, there is no damping of the states due to the disorder ; moreover, the weight of the split-off metal-hydrogen bonding level is too large since it accommodates two electrons independently of the hydrogen concentration. The average t matrix approximation (ATA), which leads to complex energy bands has also been used[98,43]. Up to date however, the coherent potential approximation (CPA) remains the best method for studying the electronic properties due to the disorder in the hydrogen sublattice. Since the details of the CPA method are developed at length in other lectures of this Institute, particularly by Stockes[4] and also by Gyorffy[99], we refer the reader to the corresponding papers for the general features of the method. We shall indicate only here some of the aspects of the results obtained for non stoichiometric TM hydrides. The CPA has been first applied to the study of PdH_x in its tight binding form[100]. In that case of course, the accuracy of the results depends on the tight binding parameters which can be determined from a Slater-Koster fit to the energy bands of the stoichiometric hydride. More recently, the KKR form of the CPA has been applied to the problems of hydrides of Pd-Ag alloys. A detailed account of these results are discussed in detail in a seminar given by Pindor and Temmerman[101] at this summer institute. We shall here briefly recall the main conclusions obtained from the studies of non-stoichiometric metal hydrides and establish a link between the electronic structure of the impurity and of the stoichiometric compounds. A study of PdH_x for several hydrogen concentrations shows, as it appears in Fig. 47 that the main modifications are (i) the low-energy structure which grows in width and intensity as a function of increasing H concentration. The calculation[100] does not indicate any visible shift in the position of the low-energy structure as a function of H concentration, and (ii) the shift in the position of the Fermi level with the progressive filling of the metal d bands as the ratio H/Pd increases.

In the intermediate concentration range, to our knowledge, a detailed study of the count of the electron states appearing at low energy and those that are accommodated at the Fermi level, has not yet been performed. The situation appears to be better understood in the very dilute limite or for very small departure from the fully stoichiometric composition. As we discussed in Section II, in the dilute limit, the supercell calculations[18,19] performed for simple and for transition metals have the advantage over other treatments of the impurity problem in that they show unambiguously that the electrons brought by the H atoms are accommodated at the Fermi level of the metal since no previously empty metal states are lowered below E_F. In the opposite high hydrogen concentration range, we have seen in section III and IV that for stoichiometric hydrides, additional empty states of the metal are lowered below E_F by the metal-hydrogen interaction. Although detailed studies of the electron count for non stoichiometric hydrides are not available, it can be surmized that there exist a critical H concentration above which the M-H interaction is strong enough to lower some additional metal states below E_F. Once this hydrogen concentration is reached, the supplementary electrons brought by the H atoms upon increasing H/M, are accommodated at the Fermi level. This interpretation is consistent with the observation made from the CPA calculations performed on PdH_x for H/Pd ratios close to the value of 1.0, that if we consider integrated properties such as the DOS at E_F, the rigid band model applied to the stoichiometric hydride is not too bad[100].

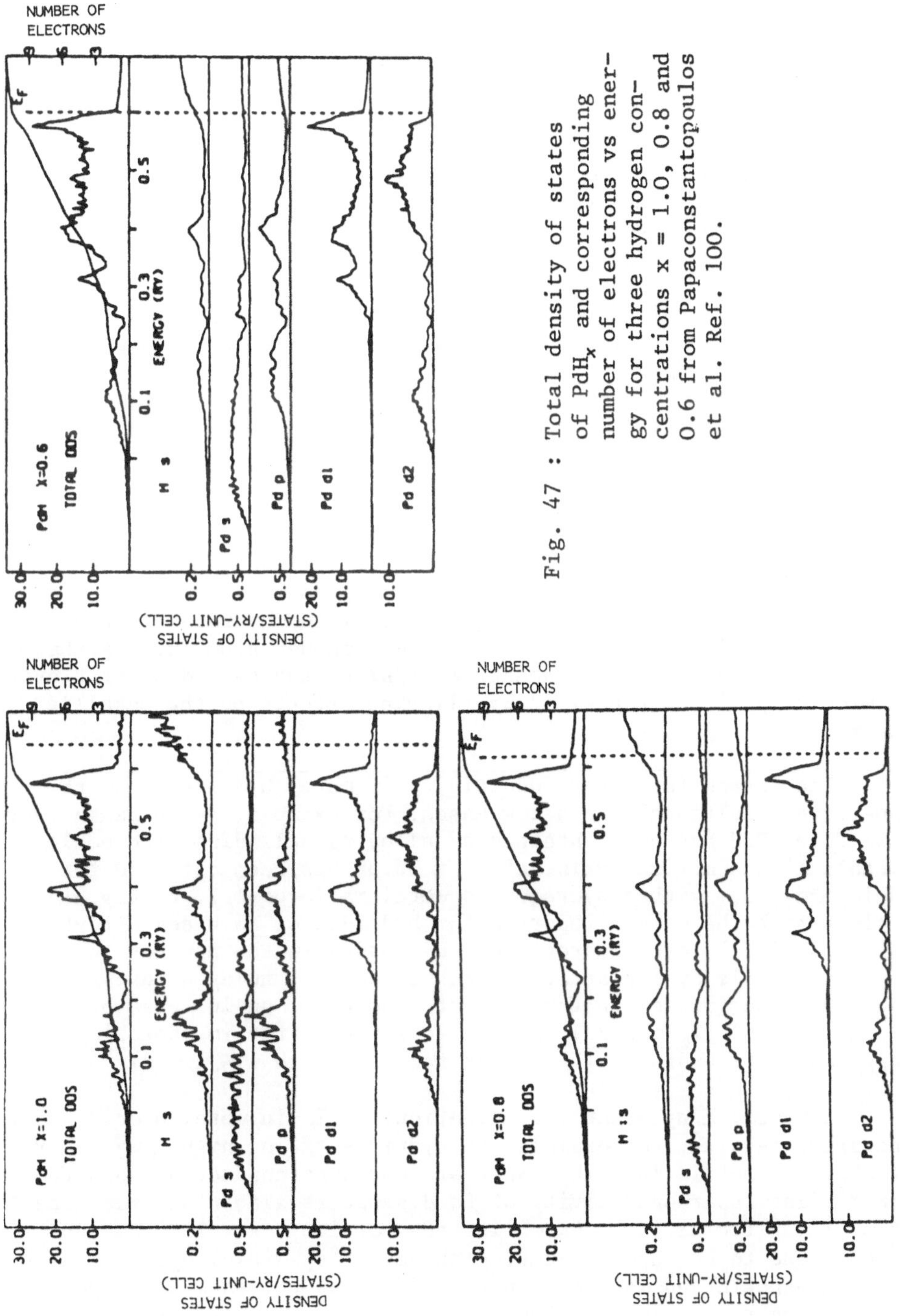

Fig. 47 : Total density of states of PdH_x and corresponding number of electrons vs energy for three hydrogen concentrations x = 1.0, 0.8 and 0.6 from Papaconstantopoulos et al. Ref. 100.

This model should however always be used with caution and only in the very low hydrogen concentration range or in the vicinity of the stoichometric composition. We caution the reader again that this model either applied to the bands of the pure metal or to the bands of the stoichiometric hydride is very erroneous in the intermediate H concentration range. We should also mention that even in the two extreme limits where this model is quite satisfactory for providing informations about integrated quantities such as the DOS, it does not give reliable data concerning the detailed deformation of the bands as it has been shown by Fermi surface measurements made on PdH_x in the very dilute limit[102].

The CPA calculations will hopefully shed some light on the detailed electronic structure of the hydrides in the intermediate concentration range.

VI. SUPERCONDUCTIVITY IN METAL HYDRIDES

After the discovery of superconductivity with fairly high values of the superconducting critical temperature T_c in Th_4H_{15} ($T_c \sim 8K$)[103] and in PdH ($T_c \sim 9K$)[104] with the occurrence of a puzzling reverse isotope effect ($T_c \sim 11K$ for PdD), many other metal hydrides have been investigated as possible high T_c superconductors. Another interesting feature of the hydrides is the possibility of varying the average electron per atom ratio either by varying the hydrogen concentration since the hydrides exist over a wide composition range, or by alloying effects on the metallic matrix.

The search for superconductivity in metal hydrides lead to positive results only in a few cases. For example, an increase over the T_c of PdH has been obtained by alloying palladium with noble metals[105] the maximum value of $T_c = 16.6K$ has been obtained for the alloy $Pd_{55}Cu_{45}$ with a hydrogen to metal ratio of 0.7. Fairly high values of T_c have been reported for hydrides of Nb based alloys with Pd, Ru[106], etc. Hydrogen can be loaded in the non-forming-hydride metals by implantation-techniques[105] ; numerous experiments have been performed. In the case of aluminium, an increase over the T_c of pure Al ($T_c \sim 2.2K$) has been observed after implantation of high doses of hydrogen ($T_c \sim 6.8K$ for AlH_x with $1 \leqslant x \leqslant 2$)[107].

However, many other investigations of T_c in metal hydrides gave negative results. For example the hydrides of the high T_c group V transition metals TaH, VH_2, NbH_2 are not superconducting down to 1K[108]. The superconductivity of La disappears also upon hydrogenation ; for fcc La, $T_c \sim 6K$ while LaH_x for $1.8 < x < 2.36$ has not been found to be superconducting above 1K[109]. Nickel hydride which is diamagnetic and isoelectronic to PdH does not show any superconducting transition[110] above 1K. Other examples of the experimental search for superconductivity in metal hydrides can be found in the recent review of Strizker and Wühl[105].

The metal hydrides pose thus several interesting questions ; while some non-superconducting metals like Pd become superconductors after formation of a hydride, hydrogen on the other hand destroys superconductivity in many high T_c metals. In section III and IV we have shown that the formation of hydrides is accompanied by drastic modifications of the electronic properties of the corresponding pure metals ; in addition, the lattice expansion observed upon hydrogenation, which is accompanied in some cases by structural changes in the metallic matrix as well as the modifications of the electronic properties which will obviously affect the screening of the phonon field are expected to lead to drastic changes in the electron-phonon coupling. Moreover, the presence of a light interstitial in the matrix leads to an additional electron-optical phonon coupling mechanism.

We shall present here theoretical estimates of the electron-phonon coupling constant λ, using the McMillan formalism[111], the electronic contribution η is calculated within the rigid ion approximation by means of our band structure results while experimental data are used to give estimates of the phonon contribution. We will try to explain the existing data on stoichometric hydrides such as PdH, NiH, TiH_2, ZrH_2, NbH_2, LaH_2, AlH and AlH_2 and from the trends derived from our calculations[40-g] we will try to make some speculations about the possible occurrence of high T_c superconductors amongst metal hydrides.

A. Method

Following the work of McMillan[111] for transition metals, an approximate expression of λ has been proposed for compounds with a large mass ratio between the constituent atoms

$$\lambda \sim \lambda_{metal} + \lambda_H = \frac{\eta_{metal}}{M_{metal} \langle\omega^2\rangle_{acoustic}} + \frac{\eta_H}{M_H \langle\omega^2\rangle_{optic}} \quad (22)$$

where η is the 'electronic contribution', M the atomic mass, and $\langle\omega^2\rangle$ the second moment of the renormalized phonon frequencies defined by McMillan. Using the rigid ion approximation in which the renormalization effects accompanying the lattice vibrations are ignored, Gaspari and Gyorffy have shown that the mean square of the electron phonon matrix element can be conveniently expressed in terms of quantities obtained from ab-initio band structure calculation. They have shown that

$$\eta \sim \sum_K \frac{E_F}{N_\uparrow(E_F)\pi^2} \sum_\ell 2(\ell+1)\sin^2(\delta^K_{\ell+1} - \delta^K_\ell) \frac{n^K_\ell(E_F)n^K_{\ell+1}(E_F)}{n^{K(1)}_\ell(E_F)n^{K(1)}_{\ell+1}(E_F)} \quad (23)$$

where the summation on K runs on all the atoms in the unit cell, δ_ℓ^K is the single-site scatterer phase shift at the Fermi energy E, n_ℓ^K the partial DOS of angular character ℓ at site K, $n_\ell^{K(1)}$ is the corresponding partial DOS of a free scatterer.

B. Electron-optical phonon coupling

We shall first examine the trends in the variation of λ_H in the metal hydrides under study. For the PdH_x system which has been thoroughly investigated both from the experimental[113] and the theoretical[42,40-i] point of view, it is now well established that, in contrast to most of the superconducting transition metal compounds, the electron-optical phonon coupling plays a dominant role. The importance of the low energy optical phonons ($\hbar\omega \sim 56$ meV for PdH)[114] in the electron phonon coupling mechanism has been revealed by superconducting tunneling experiments[113]. This is illustrated in Fig. 48 which shows from an analysis of the tunneling characteristics that the maxima of the Eliashberg function $\alpha^2(\omega)F(\omega)$ which give the dominant contribution to λ occur for values of the phonon frequencies which, as indicated by the neutron scattering data, correspond to the optic mode range. Theoretical estimates of the electron-phonon coupling constant[42,40-i] have shown, in agreement with the prediction of Ganguly[115], that the largest contribution to λ is provided by the electron-optical phonon coupling. In spite of a general lowering of the acoustic phonon frequencies by 20 to 30% from the pure metal to $PdD_{0.63}$, the electron-acoustic phonon coupling is not for this system the essential mechanism which leads to a high value of T_c. It is thus interesting to find out wether the strength of the electron-optical phonon coupling (or its weakness) is responsible for the high value of T_c (or the lack of superconductivity) in other metal hydrides. In the case of Th_4H_{15} for which $T_c \sim 8K$, the role of the electron-optical phonon coupling has

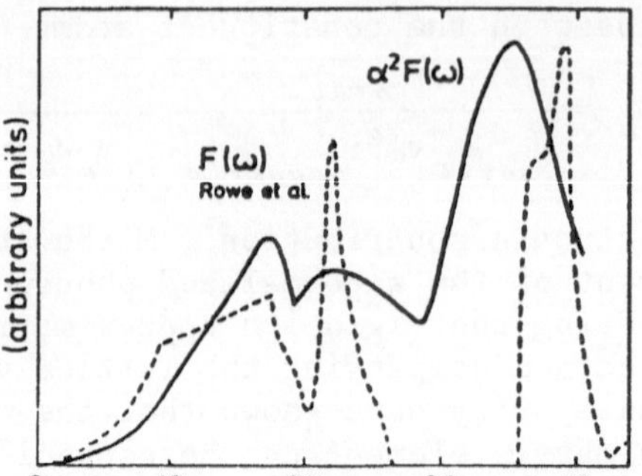

Fig. 48 : Tunneling spectrum $\alpha^2F(\omega)$ of a D/Pd ≅ 0.9 alloy from Ref. 113 (full line curve) compared with the phonon spectrum $F(\omega)$ of a D/Pd = 0.63 alloy from neutron scattering experiment of Ref. 114 (dashed curve).

not yet been clearly assessed ; nevertheless, the presence of optic modes at low energy[116] extending down to 50 meV, which are absent in the phonon DOS of the non superconducting ThH_2 seem to indicate the possibility of a mechanism similar to that evidenced for PdH. This is corroborated by the fact that the acoustic modes of Th_4H_{15} are harder than those of pure Th, in spite of a lattice expansion, and thus are less favorable for superconductivity. From Eq. 22 we can see that large values of λ_H are expected when substantial values of the electronic contribution η_H are associated with small values of the phonon contribution. We shall first analyze the variation of η_H in the series of stoichiometric metal hydrides under study here.

(a) Trends in the values of η_H

The results listed in Table VIII show that the hydrogen potential scatters strongly the s waves ; the s(ℓ = 0) scattering phase shift δ_0^H is large at the Fermi energy for all the metal hydrides, δ_0^H is always close to a resonance $\delta_0^H \sim \pi/2$. Fig. 49 shows a typical plot of δ_0^H in NiH as a function of energy. In the energy range spanned by the metal d bands, the δ_0^H phase shift remains large and is a very slowly varying function of energy ; the same behaviour is observed for dihydrides ; this indicates that all metal hydrides of the beginning as well as of the middle and of the end of the TM series will have large values of δ_0, no matter where the Fermi energy lies in the metal d bands. The results listed in Table VIII

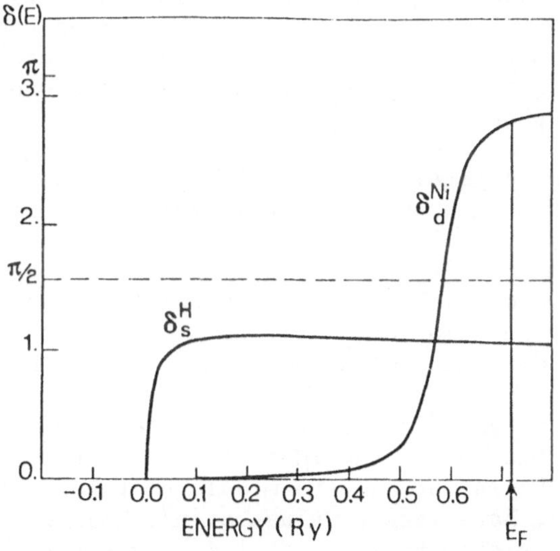

Fig. 49 : The d phase shift of the Ni potential and the s phase shift of the H potential in NiH as a function of energy. Phase shifts are in radians. E_F indicates the position of the Fermi level in NiH.

show that for all the hydrides, the phase shifts of higher angular momentum components at the H site are very small ; consequently the value of η_H is dominated by the s-p scattering mechanism and its variation across a series of metal hydrides is controlled by the magnitude of the partial s and p DOS at E_F, relative to the value of the total DOS. As shown in Table VII, the value of n_s at the H site is found to be sizeable for the TM hydrides at the end of the series, especially for PdH, and for the simple metal hydrides. From the analysis of the band structure results made in Sections III and IV, we have seen that the metal states having an s symmetry around the H site are considerably lowered in energy by the H potential and hybridize strongly with the H-s state ; they form a low-lying metal-hydrogen bonding band characteristic of the electronic structure of metal monohydrides. In the case of dihydrides, a second low-lying band is also formed, mostly from an antibonding combination of the two H-s states. Since for the early transition and rare earth metal hydrides the Fermi level falls at the bottom of the metal d bands, it is not surprising to find small values of n_s^H in this energy range. For PdH and NiH, the d bands are filled and the values of n_s^H at E_F become sizeable ; the Fermi level is not high enough to fall in the antibonding metal hydrogen band ; nevertheless, in this energy range some metal states of s symmetry at the H site have been lowered by the H potential, but not enough to fall below the d bands. The difference in the values of n_s^H between PdH and NiH is due to the fact that for NiH, the Fermi energy is closer to the top of the d bands than in PdH. The increase of n_s^H as a function of energy, if the rigid band model is applied to the Fermi level of PdH has been invoked[42] to explain (in part) the increase of T_c in the Pd-noble-metal-H_x systems. We remind the reader that this is not the only factor which explains the increase of T_c upon alloying since there is also experimental evidence of a softening of the acoustic phonons.

For the simple metal hydrides[40g,i] in contrast to the early transition and reare earth (RE) metal hydrides, the value of n_s^H as E_F is rather large. This is due to a larger hybridization of the s-p metal states with the H-s states ; in the case of AlH[40-f] the presence of a metal-hydrogen antibonding band at E_F is responsible for the large value of n_s^H ; this will lead to an enhancement of the electron-optical phonon coupling in simple metals.

The DOS of p type ($\ell = 1$) at the H site is vanishingly small for hydrides of the end of the TM series such as PdH and NiH while hydrides of the beginning and the middle of the series, for which n_s^H is very small, have larger values of n_s^H. This contribution to the partial DOS arises from the metal d states having a p symmetry at the H site, which have not been perturbed by the H potential and remain almost unperturbed from the pure metal to the hydride. Thus the s-p scattering mechanism which essentially determines the electronic contribution η_H in the angular momentum representation pre-

Table VII : The partial wave analysis n_ℓ of the DOS inside the MT metal and hydrogen spheres at the Fermi energy E_F for several stoichiometric metal hydrides (from Ref. 40-i). $N_\uparrow(E_F)$ is the total DOS per spin at E_F.

		n_s	n_p	n_d	n_f	$N_\uparrow(E_F)$
AlH	Al	0.621	0.757	0.217	0.037	3.203
	H	0.325	0.168	0.011	0.000_5	
AlH_2	Al	0.684	0.520	0.124	0.016	3.155
	1xH	0.073	0.107	0.005	0.000_3	
LaH_2	La	0.017	0.168	3.250	0.014	7.550
	1xH	0.017	0.154	0.0168	0.000_7	
TiH_2	Ti	0.0015	0.0590	18.7405	0.035	23.519
	1xH	0.015	0.8755	0.0535	0.0015	
ZrH_2	Zr	0.004	0.073	10.805	0.035	16.460
	1xH	0.008	0.432	0.021	0.000_8	
NbH_2	Nb	0.004	0.051	4.369	0.034	6.440
	1xH	0.016	0.119	0.010	0.001_2	
NbH	Nb	0.0925	0.3555	9.3800	0.066	12.425
	H	0.0202	0.3049	0.0283	0.004	
NbH_0	Nb	0.337	1.464	11.405	0.1652	15.41
PdH	Pd	0.058	0.163	2.618	0.012	3.405
	1xH	0.255	0.029	0.001_6	0.001	
NiH	Ni	0.048	0.133	4.741	0.009	5.390
	H	0.161	0.035	0.002_5	0.001	

sently used should not be viewed as an intra atomic effect since the partial DOS of p type at the H site arises from the tails of the metal d states. We can conclude from the results listed in Table VIII that, except for the simple metal hydrides which have large values of η_H and for the hydrides of the end of the TM series such as PdH, the values of η_H per H site remain small for most of the metal hydrides studied here.

(b) Optical phonons in metal hydrides

To date, ab-initio prediction of the position of the optic modes in metal hydrides have not yet been made ; nevertheless some neutron scattering data are available for these compounds[116]. A compilation of the experimental results shows that the occupation of the octahedral sites by the H atoms such as in the fcc Pd metal or in the fcc β $VH_{0.4}$ seems to lead to lower optic modes ($\hbar\omega_{opt}$ ~ 50 meV) than the occupation of tetrahedral sites. In the

Table VIII : Values of the various parameters entering the calculation of λ from Ref. 40-i. Symbols are defined in Eqs. (22) and (23). The angular-momentum-dependent phase shifts δ_ℓ are given in radians. (a) Theoretical values obtained by Butler (Ref. 121) for the corresponding transition metals

		δ_0	δ_1	δ_2	δ_3	η (eV/Å^2)	$M\langle\omega^2\rangle$ (eV/Å^2)	λ
AlH	Al	0.3365	0.4014	0.0528	0.0015	0.294		
	H	1.1850	0.0404	0.0012	0.0	2.292		
AlH_2	Al	0.2536	0.3714	0.0556	0.0018	0.224		
	1xH	1.1738	0.0379	0.0010	0.0	0.744		
LaH_2	La	− 1.1006	− 0.4556	0.5326	0.0006	0.753	7.35	0.103
	1xH	1.5290	0.0400	0.009	0.0	0.043	3.35	0.013
TiH_2	Ti	− 0.7913	− 0.2262	0.3156	0.0016	3.898		
	1xH	1.2502	0.0374	0.0009	0.0	0.067		
ZrH_2	Zr	− 0.9504	− 0.3662	0.9042	0.0049	2.352 (3.87)[a]	6.24	0.377
	1xH	1.2462	0.0373	0.0009	0.0	0.088	4.92	0.018
NbH_2	Nb	− 1.0219	− 0.4075	1.3749	0.0067	2.975 (7.39)[a]		
	1xH	1.1593	0.0408	0.0011	0.0	0.102		
NbH	Nb	− 0.8077	− 0.2852	1.3732	0.0054	2.855		
	H	1.2413	0.0405	0.0012	0.0	0.0898		
NbH_0	Nb	− 0.5105	− 0.1402	1.3396	0.0032	3.6848		
PdH	Pd	− 0.5115	− 0.1094	2.8066	0.0030	0.886	5.971	0.15
	H	1.1931	0.0280	0.0006	0.0	0.641	1.062	0.60
NiH	Ni	− 0.3857	− 0.0235	2.7916	0.0025	0.810	10.0	0.08
	H	1.0604	0.0318	0.0008	0.0	0.275	3.44	0.08

cubic CaF_2 structure dihydrides or in the bcc metals, the energies of the optic modes is of the order of $\hbar\omega_{opt} \sim 120$ meV. The difference in the position of the optic phonons is due to the size of the interstices (which is larger for the octahedral than for the tetrahedral holes) and to the differences in the electron-phonon interaction of electronic origin. Only detailed microscopic calculations can give an answer for the trends in the variation of the second factor. For compounds having essentially the same electronic structure, like all the RE dihydrides, it is the distance metal-hydrogen which appears to control the position of the optical phonon since for dihydrides of La through Ho for which experimental data exist, the average frequency of the optic phonons increase as the power 2/3 of the inverse metal-hydrogen distance[117]. In the case of NiH, although H occupies the octahedral interstices, there is some experimental evidence that the optic modes have a much higher energy than in PdH. A study of the temperature dependence of the electrical resistivity of NiH[110] shows that the optical phonon contribution occurs at higher temperature than in PdH. The Einstein temperature of the optic modes of NiH has been estimated to be at least a factor of 1.8 larger than that of PdH. This feature is probably due in part to the fact that the metal-hydrogen distance is smaller in NiH than in PdH, and certainly to differences in the electron-phonon interaction of electronic origin between two compounds of different rows of the periodic table. For the hydrides of the intermetallic compound FeTi, neutron scattering data[118] indicate that the optic modes of β-$FeTiH_{1.4}$ are centered around 100 meV; for the γ phase we used the estimate of Wenzl and Pietz[119] from their specific heat data ($\hbar\omega_{opt} \sim 77$ meV).

The values given in Table VIII for the phonon contribution $M_H \langle\omega^2\rangle_{optic}$ are taken from realistic neutron scattering data in the case of PdH. In view of the lack of data for other dihydrides, we used a scaling factor derived from the average position of the optic modes when the corresponding data are available[116].

(c) Trends in the values of λ_H

From the results obtained in the present work, we find that the values of the electronic parameter η_H are sizeable for the TM hydrides with filled d bands, especially when E_F is not too close to the top of the d bands. The value of λ_H in NiH is smaller than in PdH for two reasons : because η_H is smaller and also because the phonon contribution is larger. These differences can explain why the two isoelectronic compounds PdH and NiH present differences in their superconducting properties. We have obtained very large values of η_H in the case of simple metal hydrides. These compounds appear to be particularly favorable for a large electron-optical-phonon coupling, provided that the energies of the optic phonons, which are not known experimentally, are not too high. For the dihydrides of the beginning of the TM series, for LaH_2 and the hydrides of FeTi, we obtained small values of η_H. Moreover, for these compounds, the energies of the optic phonons are high and this leads to small values of the electron-optical phonon coupling. From our study of the electronic properties of the TM dihydrides we could however expect to obtain large values of η_H for the unstable dihydrides of the middle of the TM series, since in this case the antibonding H-H band cuts E_F and gives a large density of H-s states at the Fermi level. For these compounds, which could eventually be obtained by the ion implantation technique, one could expect sizeable values of T_c due to the electron-optical phonon coupling provided that the optic-mode frequencies are not too high.

C. Electron-acoustic phonon coupling

We shall first summarize here the changes in the values of η_{Metal} occurring from the pure metal to the corresponding stoichiometric hydrides under study.

(a) Electronic contribution η_{Metal}

From the results listed in Table VII we can see that for the TM hydrides, the s and p phase shifts at the metal site are negative ; this indicates a repulsive character of the metal potential for the s and p waves, due to the orthogonalization conditions to the corresponding core states. The d wave phase shifts at the Fermi energy are positive and increase with the filling of the metal d bands thus δ_2^{Metal} is small for hydrides whose Fermi energy falls at the bottom of the metal d bands like LaH_2 ; it increases and reaches a resonance $\delta_2^{Metal} \sim \pi/2$ in the middle of the d bands, the sharpness of the resonance being related to the width of the bands. When the d bands are filled like for PdH of NiH, the d wave phase shift is large and close to the value of π. A typical energy dependence of δ_2^{Metal} for a TM hydride is illustrated by the plot of the d wave phase shift of Ni in NiH given in Fig. 49. A study of the partial wave decomposition of the DOS at E_F shows that for all the TM hydrides, LaH_2 and the hydrides of FeTi, the 'd' character is dominant at the metal site as it can be seen from the results listed in Table VII. From the trend obtained in the values of the phase shifts and of the partial DOS at the metal site we found that the value of η_{Metal} for the TM hydrides of the middle and of the end of the series is dominated by the d-f scattering mechanism while the p-d mechanism is important also for the early TM and LaH_2. As an example, 60% of the contribution of η_{La} in LaH_2 arises from the p-d scattering term and 37% from the d-f contribution while for TM hydrides of the end of the series such as NiH and PdH more than 80% of the value of η_{Metal} is provided by the d-f scattering mechanism. We wish at this point to remind the reader that in the angular momentum representation used here, the d-f scattering should not be considered as an intra-atomic effect since the partial DOS of f type at one metal site is provided by the tails of the metal d functions of the neighboring sites ; thus the physical origin of this term should rather be understood in terms of a metal d-d interaction, in a tight-binding picture[120].

For the simple metal hydrides AlH and AlH_2, ~ 80% or more of the total value of η_{Al} is provided by the p-d scattering term ; the d-f mechanism is negligible since the d phase shifts and the partial DOS of d and f type are small for simple metals. The s-p scattering term is small in spite of the importance of the partial DOS of s and p type at E_F because the s and p phase shifts are nearly equal and lead to a cancelation of the phase shift dependent term in Eq. 23. A comparison of the values of η_{Metal} in the hydrides with the values of the corresponding pure metals[121] shows, in most of the cases under study here, an important decrease of η_{Metal} upon formation of the hydride ; the theoretical values for the pure TM obtained using similar method[121] are listed in parenthesis in Table VIII. It is interesting to focuss on the change of η_{Metal} observed for the high T_c metals such as Nb and fcc La, upon formation of a dihydride.

In order to understand the origin of the disappearance of superconductivity between bcc Nb and the dihydride of Nb which has an fcc metal lattice, we calculated η_{Nb} for pure Nb assuming an fcc phase which we will call here after NbH_0 since it corresponds to the NbH_2 lattice where the H atoms have been removed. We obtained a strong reduction of η_{Nb} from 7.39 eV/Å^2 in bcc Nb[121] to η_{Nb} = 3.68 eV/Å^2 in fcc NbH_0[40-h] in spite of the fact that the total DOS at E_F increases from 11.41 states/Ry spin unit cell in bcc Nb to 15.41 states/Ry spin unit cell in NbH_0. This reduction in η_{Nb} can be ascribed to the lattice expansion effect (we assumed the same lattice constant for fcc NbH_0 and NbH_2) which leads to an increase of the Nb-Nb nearest neighbor distance from 5.369 a.u. in the bcc phase to 6.088 a.u. in the fcc phase and thus reduces the strength of the d-d interaction.

As shown in Table VIII, the value of η_{La} decreases from 2.62 eV/Å^2 in pure fcc La[58] to 0.753 eV/Å^2 in LaH_2. It is also smaller than η_{Metal} in trivalent pure metals like Y and Sc[121]. This decrease is due (i) to the 6.7% increase in the lattice constant which leads to a decrease of the width of the d bands and thus to a change in the metal d-d interaction. A similar effect is observed in the pure metal since η_{La} decreases under presssure (ii) to the reduction of the DOS at E_F from 13.74 states/Ry spin unit cell[58] to 7.55 states/Ry spin unit cell[40-e]. We have previously emphasized the electronic differences between pure fcc La and LaH_2 which are characterized by the deformation of the lower portion of the metal d band and its apparent depopulation in favor of the low-lying states in the dihydride.

(b) Acoustic phonons in metal hydrides

Of course, even when the electronic contribution η_{Metal} decreases upon hydrogenation, an enhancement of λ_{Metal} could still be obtained if the acoustic modes becomes soft enough in the hydrides. Unfortunately, experimental data on the acoustic phonon frequencies of all the hydrides investigated here are not always available[116]. For PdH, the value of $M_{Pd}\langle\omega^2\rangle_{acoustic}$ listed in Table VIII has been obtained from the experimental phonon DOS of Rowe et al[114]. Since similar data are not available for other hydrides, we have thus used the approximation

$$M_{Metal}\langle\omega^2\rangle_{acoustic} \sim \frac{1}{2} M_{Metal} \Theta_D^2$$

for the hydrides whose Debye temperature Θ_D is known.

For LaH_2 for example, the increase of Θ_D from the pure metal (Θ_D = 140K)[64] to the dihydride (Θ_D = 243K for $LaH_{2.03}$)[65] leads to a reduction of λ_{Metal} ; since we have seen in the study of λ_H that the electron-optical phonon coupling provides a negligible contribution in LaH_2, we find a drastic decrease in the total value of λ upon formation of the dihydride, which leads to a vanishingly small value of T_c. Thus, according to our theoretical estimate, LaH_2 should not be a superconductor ; this result is in agreement with the experimental investigation of Merriam and Schreiber[109].

For the bcc metals of group V such as Nb, there exists some experimental evidence of a hardening of the acoustic phonons with hydrogenation[122] ; it has been observed in this connection that the anomalous dips in the acoustic branches of bcc Nb disappear upon absorption of hydrogen ; this factor will further reduce the decrease of λ_{Metal} caused by the reduction of η_{Metal}.

For NiH we have used the value[110] of Θ_D = 366K which is larger than that of PdH and leads to a reduction of λ_{Metal}. Our theoretical values of the electronic contribution η together with the experimental estimates of the phonon contribution lead to values of the electron-phonon coupling λ which explain the experimental trends. Thus we found that substantial values of λ are obtained for the TM with filled d bands such as PdH. We have explained why NiH is not superconducting above 1K, unlike PdH. The large theoretical values obtained for η_H in the hydrides of Al can explain the observed increase of T_c upon implantation of H in Al. Our results explain also why the dihydrides of the early TM and LaH_2 are not superconducting down to 1K, in agreement with the experimental observations[109]. From our calculations we can speculate that large values of the electron-optical phonon coupling could be obtained for the TM dihydrides of the middle of the series if they could be stabilized and prepared by techniques such as the ion implantation. Moreover, our simulations of the effect of alloying on the metallic matrix show that sizeable values of η_{Metal} could be reached ; thus, nothing prevents in principle the occurrence of large values of the electron acoustic phonon mechanism in metal hydrides as the Fermi energy sweeps through the metal d bands. These speculations will hopefully stimulate further experimental investigations.

VII. CONCLUSIONS

Although the essential features of the electronic structure of Metal Hydrides have already been clarified, the theoretical studies would still benefit from improvements on the crystal potential (better treatment of the exchange and correlation term, introduction of non-spherical corrections to the MT potential, self-consistency effects etc.), and from the development of accurate total energy calculations which could help in understanding the important problem of the stability of the metal hydrides.

In our view, the most promising theoretical studies of the problem of hydrogen in metals currently underway are (i) the electronic structure investigations of intermetallic hydrides which have great potential technological applications (ii) the development of CPA calculations for non stoichiometric compounds (iii) the search for superconductivity in metal hydrides (iv) the theoretical treatment of the impurity problem especially in transition metals and its extension to the understanding of the diffusion properties of hydrogen in metals. Up to date, theoretical studies on the phonons of metal hydrides are totally lacking.

The presently available band structure calculations on stoichiometric metal hydrides have however already successfully explained the modifications of the electronic structure of a metal upon formation of a compound ; numerous experimental data have been interpreted and the limitations of the schematic protonic and anionic models used previously have been clearly assessed.

The present summary of some of the theoretical achievements already obtained and the evocation of the numerous yet unsolved problems will hopefully stimulate further theoretical investigations on the rich and important field of hydrogen in metals.

REFERENCES

1. G.G. Libowitz,"Solid State Chemistry of Binary Metal Hydrides", ed. W.A. Benjamin, Inc., New York (1965) ; W.M. Mueller, J.P. Blackledge and G.G. Libowitz eds., "Metal Hydrides", Academic Press, New York (1968).
2. O.K. Andersen, in this volume.
3. A.R. Williams, in this volume.
4. M.C. Stocks , in this volume.
5. R. Wiswall in "Topics in Applied Physics, vol.29, Hydrogen in Metals II", ed. by G. Alefeld and J. Völkl, Springer-Verlag Berlin,Heidelberg, New-York 1978, chap. 5.
6. J. Friedel, Phil.Mag. 43, 153 (1952).
7. H. Hohenberg and W. Kohn, Phys.Rev. 1313, B864 (1964).
8. W. Kohn and L.J. Sham, Phys.Rev. 140, A1133 (1965).
9. Z.D. Popovic and M.J. Stott, Phys.Rev.Lett. 33, 1164 (1974) ; Z.D. Popovic, M.J. Stott, J.P. Earlotte and G.R. Piercy, Phys.Rev. B 13, 590 (1976).
10. C.O. Almbladh, U. von Barth, Z.D. Popovic and M.J. Stott, Phys. Rev.B 14, 2250 (1976).
11. M. Manninen, P. Hautojärvi and M. Nieminen, Solid State Commun. 23, 795 (1977) ; M. Manninen and N. Nieminen, J.Phys.F. 9, 1333 (1979).
12. J.K. Norskov, Solid State Commun. 24, 691 (1977) ; ibid. Phys. Rev.B 20, 446 (1979).
13. P. Jena and K.S. Singwi, Phys.Rev.B 17, 3518 (1978).
14. L.M. Kahn, F. Perrot and M. Rasolt, Phys.Rev.B 21, 5594 (1980) ; F. Perrot and M. Rasolt, Phys.Rev.B 23, 6534 (1981).
15. U. von Barth, in this volume.
16. J. Harris, in this volume.
17. W. Kohn and S.H. Vosko, Phys.Rev. 119, 912 (1960).
18. R.P. Gupta, J. of the Less.Common.Met. 88, 299 (1982).
19. A.C. Switendick, Ber. Bunsenges.Phys.Chem. 76, 535 (1972).
20. C. Demangeat, M.A. Kahn, G. Moraitis and J.C. Parlebas, J. de Physique 41, 1001 (1980).
21. R.P. Messmer, S.K. Knudson, K.H. Johnson, J.B. Diamond and C.Y. Yang, Phys.Rev.B 13, 1396 (1976).
22. D.E. Ellis and G.S. Painter, Phys.Rev.B 2, 2887 (1970).
23. P.A. Cox, M. Bénard and A. Veillard, Chem.Phys.Lett. 87, 159 (1982).
24. R.W. Simpson, N.F. Lane and R.C. Chaney, J. of N. Mat. 69 and 70, 581 (1978) ; P.G. Rudolf and R.C. Chaney, to be published.
25. J.E. Inglesfield, to be published.
26. J. Deutz, P.H. Dederichs and R. Zeller, J.Phys.F. 11, 1787 (1981) ; P.H. Dederichs and R. Zeller, Festkörperprobleme (Advances in Solid State Physics) vol. 21, 243, ed. J. Treusch, Vieweg, Braunschweig 1981.
27. R. Podloucky, R. Zeller and P. Dederichs, Phys.Rev.B 22, 5777 (1981).

28. H. Katayama, K. Terkura and J. Kanamori, Solid State Commun. 29, 431 (1979).
29. J.L. Beeby, Proc.Roy.Soc. London, A302, 113 (1967).
30. W.M. Temmerman, to be published.
31. O. Gunnarsson and P. Johansson, Int.J.Quantum Chem. 10, 307 (1976).
32. O. Gunnarsson, B. Lundqvist and J.W. Williams, Phys.Rev.B 10, 1319 (1974).
33. W. Eichenauer, Z. Metallkd. 59, 613 (1968).
34. Z.D. Popovic and G.R. Piercy (unpublished) quoted in Ref.9.
35. C.O. Almbladh and U. von Barth, Phys.Rev.B 13, 3307 (1976).
36. J.P. Bugeat and E. Ligeon, Phys.Lett. 71A, 93 (1979).
37. C.A. Mackliet, D.J. Gillespie and A.I. Schindler, Solid State Commun. 15, 207 (1974).
38. F.M. Mueller, A.J. Freeman, J.O. Dimmock and A.M. Furdyna, Phys.Rev.B 1, 4617 (1970).
39. F.A. Lewis, "The Palladium Hydrogen System" Academic Press, New York (1967).
40. a) Michèle Gupta, Solid State Commun. 27, 1355 (1978).
b) M. Gupta and A.J. Freeman, Phys.Rev.B 17, 3029 (1978).
c) M. Gupta, Solid State Commun. 29, 47 (1979).
d) ibid, J.Phys.F. 10, 2649 (1980).
e) ibid, Phys.Rev.B 22, 6074 (1980).
f) ibid, J.of the Less-Common Met. 73, 321 (1980).
g) ibid, in Metal Hydrides ed. by G. Bambakidis, Plenum Press (1981) p.255.
h) ibid, Phys.Rev.B 24, 7099 (1982).
i) ibid, Phys.Rev.B 25, 1027 (1982).
j) ibid, J.Phys.F. 12, L57 (1982).
k) ibid, Solid State Commun. .42, 501 (1982).
l) ibid, J. of the Less-Common Met. 88, 221 (1982).
41. a) A.C. Switendick, Solid State Commun. 8, 1463 (1970).
b) ibid, Int.J.Quantum Chem. 5, 459 (1971).
c) ibid, "Hydrogen in Metals - A New Theoretical Model" in Hydrogen Energy, Part B, ed. by T.N. Veziroglu (Plenum Press, New York, 1975) p.1029.
d) ibid, J. of the Less-Common Met. 49, 283 (1976).
e) ibid, in "Transition Metal Hydrides, Advances in Chemistry Series" N° 167 ed. Robert Bau, (The Americal Physical Society 1978) p.264.
f) ibid, in "Topics in Applied Physics", vol.28, Hydrogen in Metals, ed. by G. Alfefeld and J. Völkl, Springer Verlag 1978, p. 101.
42. a) D.A. Papaconstantopoulos and B.M. Klein, Phys.Rev.Lett. 35, 110 (1975).
b) B.M. Klein, E.N. Economou and D.A. Papaconstantopoulos, Phys. Rev.Lett. 39, 574 (1977).
c) D.A. Papaconstantopoulos, B.M. Klein, E.N. Economou and L.L. Boyer, Phys.Rev.B 17, 141 (1978).

d) D.A. Papaconstantopoulos in Metal Hydrides, ed. by G. Bambakidis, Plenum Press (1981).

43. C.D. Gelatt,Jr., J.A. Weiss and H. Ehrenreich, Phys.Rev.B 17, 1940 (1978).

44. a) D.J. Peterman, B.N. Harmon, J. Marchiando and J.H. Weaver, Phys.Rev.B 19, 4867 (1979).
b) D.J. Peterman and B.N. Harmon, unpublished.

45. N.I. Kulikov, V.N. Borzunov and D.A. Zvonkov, Phys.Stat.Sol.(b) 86, 83 (1978) ; N.I. Kulikov, Phys.Stat.Sol.(b) 91, 753 (1979).

46. A. Fujimori, F. Minami and N. Tsuda, Phys.Rev.B 22, 3573 (1980); A. Fujimori and N. Tsuda, Solid State Commun. 41, 491 (1982).

47. U. Mizutani, T.B. Massalski and J. Bevk, J.Phys.F : Metal Phys. 6, 1 (1976).

48. D.E. Eastman, J.K. Cashian and A.C. Switendick, Phys.Rev.Lett. 27, 35 (1971) ; L. Schlapback and J.P. Buiger, J.de Physique 43, L273 (1982) ; P. Bennett and J.C. Fuggle, Phys.Rev.B, to be published.

49. C.L. Wiley, G. Cinader and F.Y. Fradin, Bull.Am.Phys.Soc. 21, 404 (1976).

50. F.Y. Fradin, D.D. Koelling, A.J. Freeman and T.J. Watson-Yang, Phys.Rev.B 12, 5570 (1975) ; O.K. Anderson, Phys.Rev.B 2, 883 (1970).

51. A. Traverse, N. Bernas, L. Dumoulin and Michèle Gupta, Solid State Commun. 40, 725 (1981).

52. A.D. McQuillan, Proc.Roy.Soc. London 204, 309 (1950) ; S.S. Sidhu, N.S. Satya Murthy, F.P. Campos and D.D. Zauberis, Non Stoichiometric Compounds Advances in Chemistry Series 39, p.87, American Chemical Society Washington 1963 ; R.K. Edwards and E. Veleckis, J.Phys.Chem. 66, 1657 (1962).

53. W. Trebiatowski and B. Stalinski, Bull.Acad.Polon.Sci. 1, 131 (1953) ; B. Stalinski, C.K. Coogan and H.S. Gutowsky, J.Chem. Phys. 34, 1191 (1961).

54. C. Korn, Phys.Rev.B 17, 1707 (1978).

55. B. Stalinski and Z. Bieganski, Bull.Acad.Polon.Sci., Serie Sci. Chim. Geol. Geogr. 8, 243 (1960).

56. K. Gesi, Y. Takagi and T. Takeuchi, J.Phys.Soc. Japan 18, 306 (1963).

57. J. Labbé and J. Friedel, J.Physique 27, 153 (1966).

58. W.E. Pickett, A.J. Freeman and D.D. Koelling, Phys.Rev.B 22, 2695 (1980).

59. Z. Bieganski and B. Stalinski, Phys.Status Solidi A2, K161 (1970) ; Z. Bieganski and M. Drulis, Phys.Status Solidi A44, 91 (1977).

60. W.E. Wallace and K.H. Mader, J.Chem.Phys. 48, 84 (1968) ; B. Stalinski, Bull.Acad.Pol.Sci.Cl. 35, 997 (1957).

61. D.J. Peterman, J.H. Weaver and D.T. Peterson, Phys.Rev.B 23, 3906 (1981) ; L. Schlapbach, J. Osterwalder and H.C. Siegmann, J. of the Less-Common Met. 88, 291 (1982).

62. H. Shaked, J. Faber Jr., M.H. Mueller and D.G. Westlake, Phys. Rev.B 16, 340 (1977).

63. B. Stalinski, Bull.Acad.Pol.Sc. Cl 35, 1001 (1957) ; J.N. Daou, C.R. Acad.Sci. 250, 3165 (1960) ; J.N. Daou and J. Bonnet, C.R. Acad.Sci. 261, 1675 (1965).
64. D.L. Johnson and D.K. Finnemore, Phys.Rev. 158, 376 (1967).
65. Z. Bieganski, D. Gonzalez-Alvarez and F.W. Klaaysen, Physica 37, 153 (1967).
66. K.R. Lea, M.J.M. Leask and W.P. Wolf, J.Phys.Chem.Solids 23, 1381 (1962).
67. W.E. Wallace, Ber.Bunsenges Phys.Chem. 76, 832 (1972).
68. G.K. Shenoy, B.D. Dunlap, D.G. Westlake and A.E. Dwight, Phys. Rev.B 14, 41 (1976).
69. R. Lässer and B. Lengeler, Phys.Rev.B 20, 1390 (1979).
70. B.W. Veal, D.J. Lam and D.G. Westlake, Phys.Rev.B 19, 2856 (1979).
71. G.G. Libowitz, Ber. Bunsenges. Phys.Chem. 76, 837 (1972) and references therein.
72. W.G. Bos, J.Chem.Phys. 53, 855 (1970).
73. A.R. Williams, J. Kübler and C.D. Gelatt Jr., Phys.Rev.B 19, 6094 (1979).
74. J.J. Reilly and R.H. Wiswall Jr., Inorg.Chem. 13, 218 (1974).
75. P. Thompson, M.A. Pick, F. Reidinger, L.M. Corliss, J.M. Hastings and J.J. Reilly, J.Phys.F. 8, (1978) L75 ; P. Fisscher, W. Hälg, L. Schlapbach, F. Stucki and A.F. Andresen, Mat.Res. Bull. 13 ; (1978) 931 ; D. Fruchart, M. Commandré, D. Sauvage, A. Rouault and R. Tellgren, Journal of the Less-Comm. Met. 74 (1980) 55 ; W. Schäfer, G. Will and T. Schober, Mat.Res.Bull. 15 (1980) 627 ; P. Thompson, J.J. Reilly, F. Reidinger, J.M. Hastings and L.M. Corliss, J.Phys.F. Metal Phys. 9 (1979) L61.
76. F. Reidinger, J.F. Lynck and J.J. Reilly, J.Phys.F 12, L49 (1982).
77. A.R. Miedema, K.H.J. Buschow and H.N. van Mal, J.Less-Commun. Met. 49, 463 (1976).
78. Y. Yamashita and S. Asano, Progr.Theor.Phys. 48, 2119 (1972) ; D.A. Papaconstantopoulos, Phys.Rev.B 11, 4801 (1975).
79. G. Arnold and J.M. Welter, in Bericht der Deutschen Gesellschaft für Metallkünde über Gase in Metallen (Deutsche Gesellschaft für Metallkünde, Darmstadt, 1979).
80. D.G. Westlake, J.Less-Commun. Met. 75, 177 (1980).
81. R. Hempelmann, O. Ohlendorf and E. Wicke, Proc.Int.Symp. on Hydrides for Energy Storage, Geilo Norway (1977), (London-Pergamon Press).
82. E.A. Starke, C.H. Chen and P.A. Beck, Phys.Rev. 126, 1746 (1962).
83. F. Stucki, Ph.D. Dissertation number 6835, E.T.H. Zurich (1981).
84. L.J. Swartzendruker, L.H. Bennett and R.E. Watson, J.Phys.F 6, L331 (1976).
85. A. Heideman, G. Kaindl, D. Salomon, H. Wipf and G. Wortmann, Phys.Rev.Lett. 36, 213 (1976) ; J.S. Carlow and R.E. Meads, J.Phys.F 2, 982 (1972) ; G.K. Wertheim and D.N.E. Buchanan, J.Phys.Chem.Solids 28, 225 (1967).
86. J. Callaway, R.A. Tawil and C.S. Wang, Phys.Lett. 46A, 161 (1973).
87. T. Shinohara and M. Fujioka, Phys.Rev.B 7, 37 (1973).

88. R. Ingulls, Phys.Rev. 155, 157 (1967).
89. K. Schubert and K. Anderko, Z. Metallkunde 42, 321 (1951).
90. J.J. Reilly and R.H. Wiswall, Inorg.Chem. 7, 2254 (1968).
91. Z. Gavra, M.H. Mintz, G. Kimmel and Z. Hadri, Inorg.Chem. 18, 3595 (1979).
92. J. Genossar and P.S. Rudman, J.Phys.Chem.Sol. 42, 611 (1981).
93. J.P. Darnaudery, M. Pezat, B. Darriet and P. Hagenmuller, Mat. Res.Bull. 16, 1237 (1981).
94. K. Yoon, J.Schefer and F. Stucki, Inorg.Chem. 20, 2776 (1981).
95. R.O. Moyer, C. Stanistski, J. Tanaka, M.I. Kay and R. Kleinberg J. of Solid State Chem. 3, 541 (1971).
96. J. Senegas, M. Pezat, J.P. Darnaudery and B. Darriet, J.Phys. Chem.Sol. 42, 29 (1981).
97. J. Genossar, private communication.
98. A. Bansil, R. Prasad, S. Bessendorf, L. Schwartz, W.J. Venema, R. Feenstra, F. Blom and R. Griessen, Solid State Commun. 32, 1115 (1979).
99. B.L. Gyorffy, in this volume.
100. J.S. Faulkner, Phys.Rev.B 13, 2391 (1976) ; D.A. Papaconstantopoulos, B.M. Klein, J.S. Faulkner and L.L. Boyer, Phys.Rev. B 18, 2784 (1978).
101. A.J. Pindor and W.M. Temmerman, to be published.
102. W.J. Venema, thesis (Vrije Universiteit, Amsterdam) 1980 ; H.L.M. Bakker, R. Feenstra, R. Griessen, L. Huisman and W.J. Venema, Phys.Rev.B 26, 5321 (1982).
103. C.B. Satterthwaite and I.L. Toepke, Phys.Rev.Lett.25, 741 (1970).
104. T. Skoskiewicz, Phys.Stat.Sol. (a) 11, K123 (1972) ; B. Stritzker and W. Bückel, J.Phys. 257, 1 (1972).
105. B. Stritzker and H. Wühl, in "Topics in Applied Physics, vol.28, Hyrdogen in Metals", ed. G. Alefeld and J. Völkl, Springer Verlag, Berlin, Heidelberg, New York 1978, p.243.
106. C.G. Robbins and J. Muller, J.Less Common.Met. 42, 19 (1975) ; C.G. Robbins, M. Ishikawa, A. Trey and J. Muller, Solid State Commun. 17, 903 (1975).
107. A.M. Lamoise, J. Chaumont, F. Meunier and H. Bernas, J.Phys. Lett. 36, L271 (1975).
108. C.B. Satterthwaite and D.T. Peterson, J.Less Common. Met. 26, 361 (1972).
109. M.F. Merriam and D.S. Schreiber, J.Phys.Chem.Solids 24, 1375 (1963).
110. D.S. McLachlan, I. Papadopoulos and T.B. Doyle, J.de Phys. Paris C6, 430 (1978).
111. W.L. McMillan, Phys.Rev. 167, 331 (1968).
112. G.D. Gaspari and B.L. Gyorffy, Phys.Rev.Lett. 29, 801 (1972) ; I.R. Gommersall and B.L. Gyorffy, J.Phys.F 4, 1204 (1974).
113. A. Eichler, H. Wühl and B. Stritzker, Solid State Commun. 17, 213 (1975).
114. J.M. Rowe, J.J. Rush, H.G. Smith, M. Mostoller and H.E. Flotow, Phys.Rev.Lett. 33, 1297 (1974).

115. B.N. Ganguly, Z.Phys. 265, 433 (1975).
116. T. Springer, in "Topics in Applied Physics, vol.28, Hydrogen in Metals", Ed. G. Alefeld and J. Völkl, Springer Verlag, Berlin, New York (1978) p. 75 and references therein.
117. D.G. Hunt and D.K. Ross, J. Less Common Met. 49, 169 (1976).
118. J. Eckert, J.A. Goldstone and D. Richter, J.Phys.F 11, L101 (1981).
119. H. Wenzl and S. Pietz, Solid State Commun. 33, 1163 (1980).
120. S. Barisic, J. Labbé and J. Friedel, Phys.Rev.Lett. 25, 919 (1970).
121. W.H. Butler, Phys.Rev.B 15, 5267 (1977).
122. J.M. Rowe, N. Vagelatos, J.J. Rush and H.E. Flotow, Phys.Rev. B 12, 2959 (1975).
123. R.W. Morse, "The Fermi Surface of Metals" ed. by W.A. Harrison and M.B. Webb (New York : Wiley 1960) p. 214.
124. G.A. Burdick, Phys.Rev.Lett. 7, 156 (1961).
125. R.B. McLellan and W.A. Oates, Acta Metall. 21, 181 (1973).

FIRST PRINCIPLES LATTICE DYNAMICS OF TRANSITION METALS

Werner Weber

Kernforschungszentrum Karlsruhe
Institut für Nukleare Festkörperphysik
Postfach 3640, D-7500 Karlsruhe
Federal Republic of Germany

1. INTRODUCTION

1.1. Model Theories as a Guide to First Principles' Methods

For a very long time, lattice dynamics (LD) has essentially been a theory of model assumptions on the forces in a solid. Its starting point has been the work of M. Born and Th. v. Kármán in 1912, and in the following four decades the field of LD was mainly left to M. Born and his collaborators (see Born and Huang 1954). Until the late 1950's there was only modest interest in LD among the solid state theorists, since the phonon dispersion relations away from the long wavelength limit were practically not accessible by experiment. Then, however, inelastic neutron scattering technique appeared, and LD drew enormous interest, as the phonon dispersion curves of an ever-growing number of the materials became available.

The great time of model theories began. For most crystals, Born-and-von-Kármán models assuming a small number of spring constants between a few near neighbor atoms turned out to be insufficient to describe the phonon dispersion curves. It soon became obvious that in the various classes of materials, the electronic medium is screening and thus modifying the bare internuclear Coulomb forces in rather different ways. Assumptions about the respective electronic medium were made and led to a variety of models. However, the knowledge about the electrons as provided by the energy band theory - which started to flourish at about the same time - was used only in a global manner. For most classes of materials there was no direct link between LD and band structure theory. The only exception has been the LD of simple metals, derived from the limit of free electrons (for a review see Brovman and Kagan 1974).

As model theories have been a guide to the development of useful first principles theories of LD, I will now present three different model theories, each applied to a different class of materials. The choice of these models is not aimed to be a representative summary on model theories, instead this selection reflects the author's experience in this field. When going through this article, it may be amusing for the reader to realize, how much the various model assumptions have indeed been verified or have been disproved by microscopic theory. We should point out here that comprehensive surveys of the various model theories in LD as well as the developments in first principles theories are found in the series "Dynamical Properties of Solids", especially Vol.s 1 and 3 (Horton and Maradudin 1974, 1980). Model theories also have been reviewed by papers of Cochran (1972) and Sinha (1973).

The first example is the shell model, applied mainly to the LD of alkali halides. Its starting point was the necessity to correctly describe the splitting of the longitudinal optic (LO) and transverse optic (TO) phonons at the center of the Brillouin zone (BZ). These modes are split only by the presence of the macroscopic electric field induced by the component of the displacement dipoles parallel to the phonon wavevector. This electric field acts as an additional restoring force for the LO phonons. If the displacement dipoles were made up only of rigidly moving ions of charge $Z = \pm 1e$, the Lyddane-Sachs-Teller relation

$$\omega_{LO}^2/\omega_{TO}^2 = \varepsilon_0/\varepsilon_\infty$$

would only be fulfilled under the unphysical assumption $\varepsilon_\infty = 1$ for the electronic part of the dielectric constant. It was therefore concluded that the valence electrons also contribute to the displacement dipoles. An intuitive way was to assume (Dick and Overhauser, 1958) that the valence electrons form a shell around the ion core displaceable under the acting forces of neighbor atoms (see Fig. 1). The finite core-shell force constant k and the charge Y of the shell lead to an electronic polarizability and thus may fix ε_∞. As a consequence, the optic phonons are in much better agreement with experiment than those of a rigid ion model (see Fig. 2). The additional forces between atoms include, however, not only the long range Coulombic interactions involving the induced dipoles, there is also another type of potentially long range forces. When an atom is displaced, the shell of a neighbor atom moves instantaneously - we assume the adiabatic approximation for the electronic shells - to a new equilibrium and thus transmits a force to a further distant atom, although the springs are only acting between nearest neighbors. This "mechanical" effect leads to a flattening of the dispersion curves near the BZ boundary.

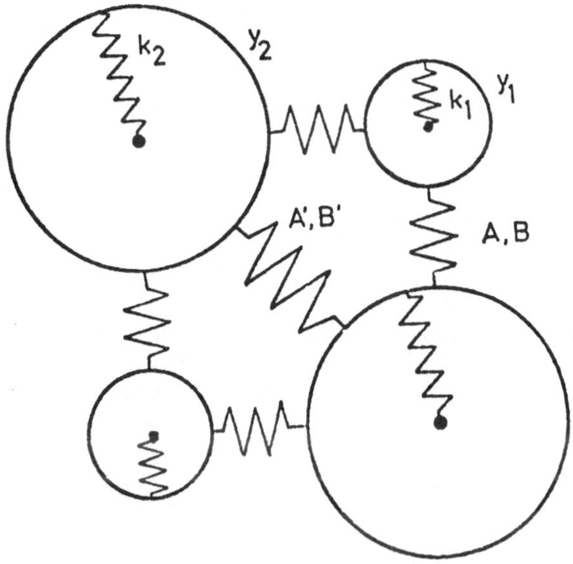

Fig. 1. Simple shell model for ionic crystals. The ions are represented by cores plus electronic shells with charges Y_1 and Y_2, coupled to the cores by coupling constants k_1 and k_2. Nearest and next nearest neighbor shells are interacting via force constants A,B and A',B'.

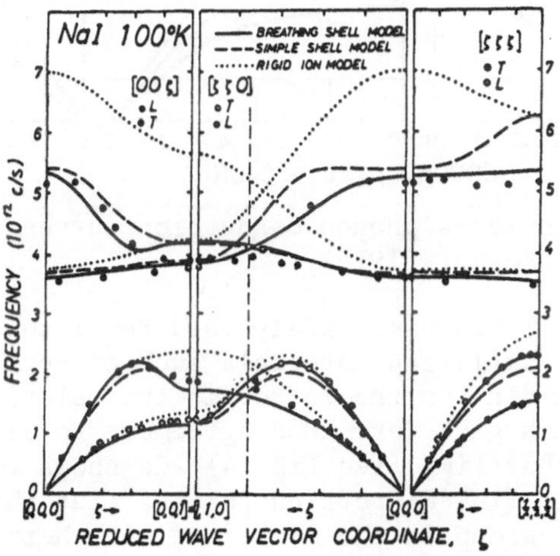

Fig. 2. Comparison of various model calculations for the phonon dispersion curves of NaI. The simple shell model yields the essential improvement for the LO branches as compared to the rigid ion model. The "breathing shell model" of Schröder (1966) gives almost perfect agreement with experiment, removing a remaining discrepancy of the shell model at the L point of the BZ (from Schröder 1966).

There have been extensions of the shell model by allowing not only a dipolar degree of freedom for the shell, but also a "breathing" deformability (Schröder, 1966) or a quadrupolar deformability in the case of Ag halides (Fischer et al., 1972).

It is noteworthy that Zeyher (1975) has presented a first principles theory of LD for ionic crystals, developed in terms of highly localized nonorthogonal wave functions. He has shown that his approach is in close correspondence to the shell model and leads to a microscopic understanding of the model parameters. His theory is limited to large gap ionic materials with small wavefunction overlap. However, his approach is similar to the first principles theory of LD presented in the bulk of this article. Since Zeyher has used Hartree-Fock theory for the electrons, his calculations are restricted to systems of light atoms such as LiD (see Fig. 3).

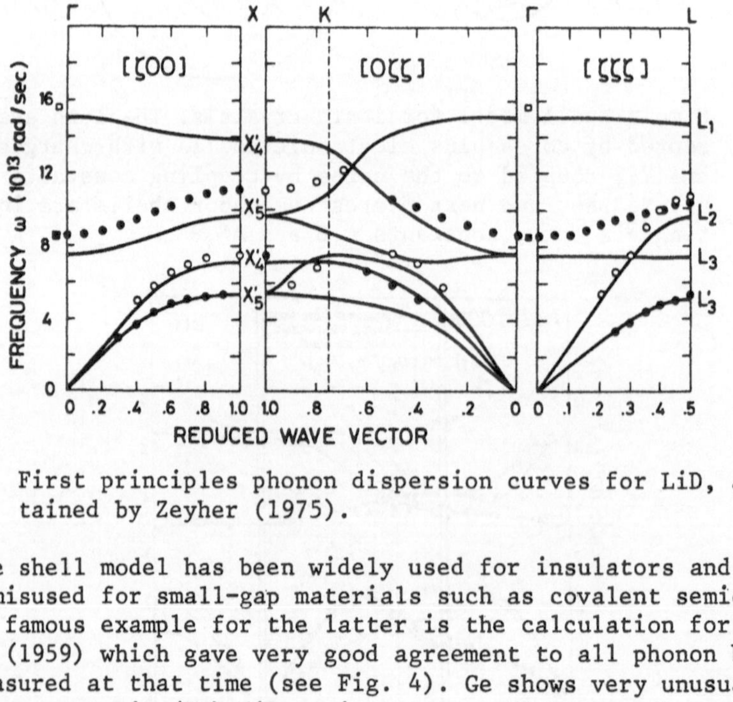

Fig. 3. First principles phonon dispersion curves for LiD, as obtained by Zeyher (1975).

The shell model has been widely used for insulators and also widely misused for small-gap materials such as covalent semiconductors. A famous example for the latter is the calculation for Ge by Cochran (1959) which gave very good agreement to all phonon branches measured at that time (see Fig. 4). Ge shows very unusual transverse acoustic (TA) dispersion curves, flat through most of the BZ, but rather steep in the elastic limit. Herman (1959) had shown before that a correct description of this feature required spring constants to rather distant neighbors, while Cochran used a few nearest neighbor springs. However, his shell model was criticized that the electrons in a bond should not be divided between two atoms when they are shared among them (Phillips, 1968a). Based on work of Phillips (1968b) and Martin (1969), I have proposed a bond charge model for the LD of covalent semiconductors, (Weber,

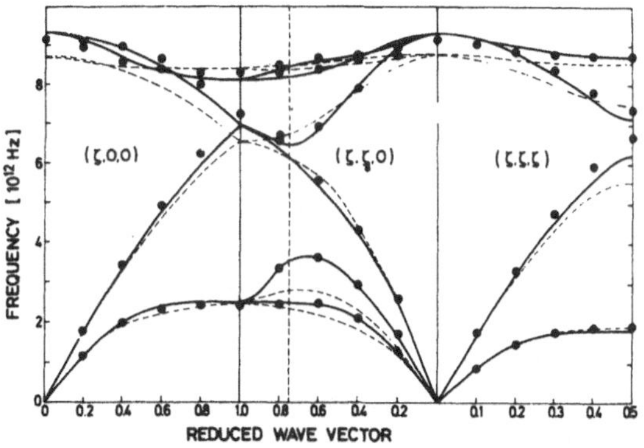

Fig. 4. Phonon dispersion curves for Ge. Solid lines represent the results of the adiabatic bond-charge model of Weber (1974). Dashed lines show the results of Cochran's shell model (1959).

1974, and 1977) where the bond charges move adiabatically like the shells. This model, again using very few parameters, could reproduce the flattening of the TA branches and gave even better agreement with experiment than the shell model (see Fig. 4). The basic idea was (see Fig. 5) that the forces between ions and bond charges are weak as compared to the angular stiffness. It is easy to see that at long waves the bond charges have to move in phase with the atoms which leads to large shear elastic constants because of the large angular stiffness. At short waves, however, the bond charges may move away from their midway position. These displacements are especially large at the BZ boundary. In Sect. 4 results of first

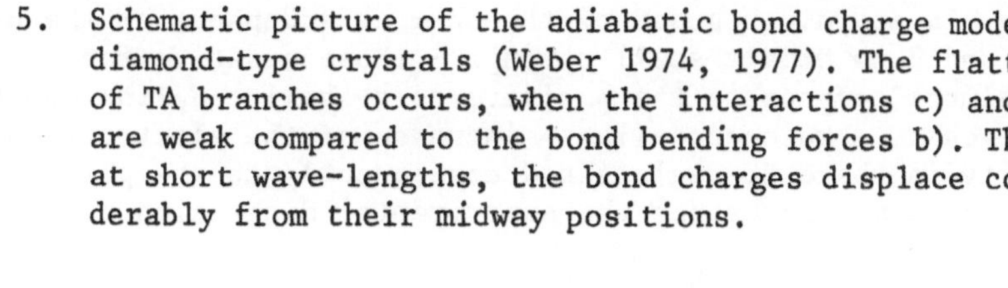

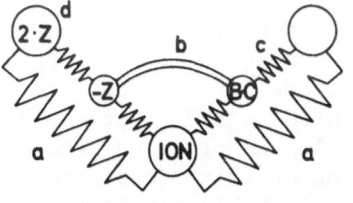

Fig. 5. Schematic picture of the adiabatic bond charge model for diamond-type crystals (Weber 1974, 1977). The flattening of TA branches occurs, when the interactions c) and d) are weak compared to the bond bending forces b). Then, at short wave-lengths, the bond charges displace considerably from their midway positions.

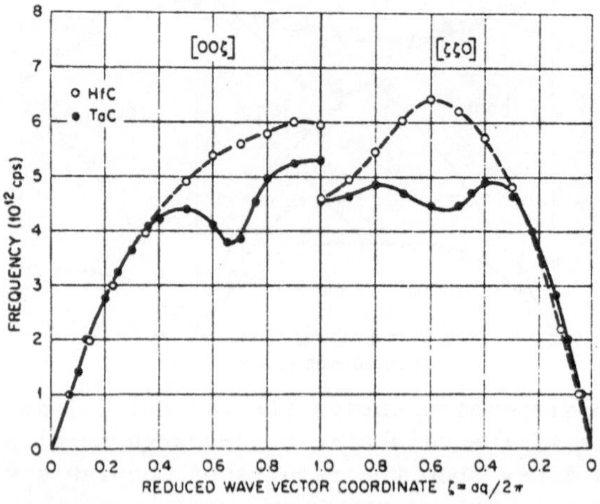

Fig. 6. Longitudinal acoustic phonons for HfC (superconducting temperature T_c < 1 K) and TaC (T_c = 10 K) (Smith and Gläser 1970).

principles calculations are presented which indeed confirm this motion of the bond charges.

My third example of a model theory deals with the LD of transition metals and compounds (TMC). In 1970, Smith and Gläser presented a comparison of the phonon dispersion curves of TaC (superconducting transition temperature $T_c \approx 10$ K) and HfC ($T_c < 1$ K), where in TaC remarkable dips were found in some specific regions of the BZ, which were completely absent in HfC (see Fig. 6).

Quite obviously, these phonon anomalies and the increase of T_c seem to be correlated - a conclusing which could also have been drawn from the earlier neutron study of Powell et al. (1968) on the phonons of Nb-Mo alloys. In the electronic structure the essential difference between TaC and HfC, or between the related compounds NbC and ZrC, is a different band filling. In TaC or NbC, with one more valence electron than ZrC of HfC, a new complex of d-bands is being filled. We therefore proposed a model (Weber et al. 1972, Weber 1973) where these additional d-electrons are represented by a second shell at the TM atoms - the first shell should describe the polarizability of all the other valence electrons (see Fig. 7). The basic assumption of this "double shell model" is that the springs between second shells are taken to be negative. In certain regions of the BZ - this depends on the blend of 1 nn and 2 nn double shell springs - these negative force constants almost compensate the positive core-shell coupling constant, which leads to a resonance-like increase of the electronic polarizability in these BZ regions and consequently to the phonon softening. The model describes the experimental dispersion curves

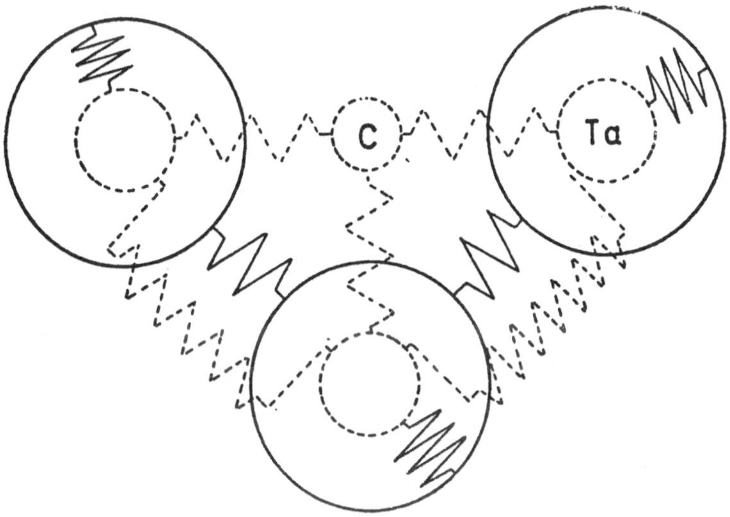

Fig. 7. Double shell model to describe the phonon anomalies of TaC (or NbC). The second shells are defined only at the metal sites. Their shell-shell coupling constants are negative to produce a resonance-like phonon instability (Weber 1973).

very well (see Fig. 8), using only few fitting parameters, even a further dip in a (110) branch was correctly predicted (Smith 1972). From the planar symmetry of the force constants between the second shells it was concluded that the new d-bands in TaC or NbC should consists of orbitals with T_{2g} ($\Gamma_{25'}$) symmetry. This conclusion has been confirmed by a tight binding analysis of augmented-plane-wave (APW) energy band calculations for NbC (Schwarz 1975).

In spite of all this apparent success, today we think that the double shell model is misleading with respect to the physical origin of the phonon anomalies. The model suggests a real space explanation for the anomalies. As we see it now, they are casued by a specific type of Fermi surface nesting which means a k-space origin. We will come to this problem of the microscopic origin of the phonon anomalies in transition metals and compounds in Sect. 3 of this paper.

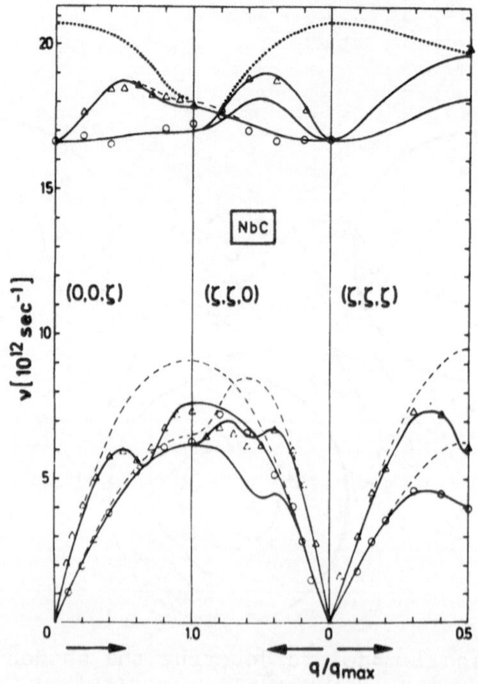

Fig. 8. Phonon dispersion curves of NbC, as calculated with the double shell model. Dashed lines show the same model without the second-shell contribution (Weber 1973).

1.2. Survey of First Principles Methods

The conventional theory of LD has been derived a while ago (Baym 1961, Sham 1969, Pick et al. 1970). Its formulation is leaning on the LD for simple metals which is based on the nearly free electron approximation. We may see this from the expression of the force constant matrix between two atoms ℓ and ℓ' (sublattice indices κ, κ' are omitted for simplicity ; α,β denote Cartesian coordinates and R_ℓ is a lattice vector)

$$\begin{aligned}\phi_{\alpha\beta}(\ell\ell') &= \partial^2/\partial R_{\ell\alpha}\,\partial R_{\ell'\beta}\; Z^2e^2 \,/\, |\vec{R}_\ell - \vec{R}_{\ell'}| \;- \\ &\int d^3r \int d^3r' \,(\partial/R_{\ell\alpha}\, V_b(\vec{r}-\vec{R}_\ell))\;\chi(\vec{r},\vec{r}\,') \\ &\qquad (\partial/R_{\ell'\beta}\, V_b(\vec{r}\,'-\vec{R}_{\ell'})) \\ &= \nabla\nabla V_{cc} - \nabla\nabla V_{i-e-i} \end{aligned} \tag{1.1}$$

Here the first term is the direct Coulomb interaction between the ion cores at ℓ and ℓ', while the second term represents the indirect interaction via the electronic medium. V_b is the bare ion core potential and

$$\chi(\vec{r},\vec{r}') = \int d^3r'' \ \varepsilon^{-1}(\vec{r},\vec{r}'') \ \chi_0(\vec{r}'',\vec{r}') \qquad (1.2)$$

is the full response function (full susceptibility) of the electrons, while ε^{-1} is the inverse dielectric function and χ_0 is the response function of the non-interacting electrons (bare susceptibility). In the nearly-free-electron approximation, as generally assumed for simple metals, the calculation of χ is rather easy as $\chi(\vec{r},\vec{r}') \rightarrow \chi(|r - r'|)$ which is imple to treat in reciprocal space (see Brovman and Kagan 1974). For transition metals and compounds, as well as for semiconductors or ionic crystals, the calculation of χ is a big problem. As χ and ε^{-1} are nonlocal, there is the question how to invert ε. Procedures for doing the inversion have been described (Sinha et al. 1971, Hanke 1973), they base on a factorization scheme which employs localized orbitals defined on the ideal lattice sites.

To exemplify the numerical problems of the inversion method we note that the term $\nabla\nabla V_{cc}$ may be represented by the so-called ion plasma frequency

$$m\omega_{pl}^2 = 4\pi Z^2 \ . \ e^2/v_a$$

the q = 0 contribution of $\nabla\nabla V_{cc}$ to the dynamical matrix (e is the unit charge and v_a the atomic volume). In the case of Nb with Z = 5 we obtain $m\omega_{pl}^2 \approx 300$, while the average force constant $\langle m\omega^2\rangle_{exp} \approx 20$, in units of e^2/v_a. This means that also the term $\nabla\nabla V_{i-e-i}$ of Eq. (1.1) is one order of magnitude larger than $\langle m\omega^2\rangle_{exp}$ and thus has to be computed with extreme care.

Using the inversion method, some calculations for the LD of Nb and NbC have been performed where rather simplifying model assumptions on the electronic structure of TMC have been made (Sinha and Harmon 1975, Hanke et al. 1976, see also reviews by Sinha 1980 and Allen 1980). These authors have proposed general and physically simple real space arguments for the existence of a screening resonance at specific phonon wave vectors - an interpretation of the phonon anomalies in the spirit of the "double shell model". The only inversion method calculation for TMC with realistic energy bands and wave functions, has so far been carried out by J. Cooke (1978, 1980) for Nb and Mo. He uses a somewhat different inversion procedure based on the Kohn-Korringa-Rostoker (KKR) method. Because of enormous computational problems he could present numerically stable results only for the $\Delta(\xi,0,0)$ direction of the Brillouin zone (see Fig. 9). There he obtains a correct description of the phonon anomaly in Nb near (0.7,0,0) and in Mo at (1.0,0,0). He points out, however, that ther is no indication of a screening resonance at the anomalous wave vectors. This finding

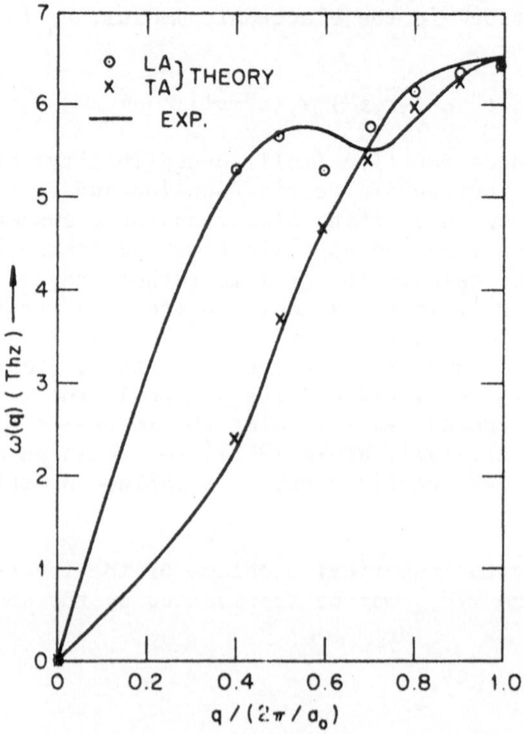

Fig. 9. First principles calculations of the phonons of Nb along the (ξ,0,0) direction, using the inversion method (Cooke, 1978).

is in agreement with our interpretation of the microscopic origin of the phonon anomalies in TMC (see Sect. 3.1).

An alternate method for the LD of TMC has been proposed by Pickett and Gyorffy (1976), and by Varma and Dynes (1976) to overcome the problems of the inversion method. The basic idea is to regroup the terms of Eq. (1.1) in order to directly evaluate the "band structure" contribution $\nabla\nabla E_{bs}$ to $\phi_{\alpha\beta}$. As in $\nabla\nabla E_{bs}$ the electron-electron interaction $\nabla\nabla V_{ee}$ has been counted twice, a correction term $-\nabla\nabla V_{ee}$ appears. Thus $\phi_{\alpha\beta}$ reads as

$$\phi_{\alpha\beta} = (\nabla\nabla V_{cc} - \nabla\nabla V_{ee}) + \nabla\nabla E_{bs} \tag{1.3}$$

This regrouping has the advantage that the two very big and long range terms $\nabla\nabla V_{cc}$ and $\nabla\nabla V_{ee}$ almost cancel each other so that only forces between "neutral objects" are left, which are essentially of short range.

Similar regrouping ideas have been proposed by various authors (Sinha 1969, Pick 1973, Eschrig 1973) for the LD of ionic crystals, semiconductors and noble metals. Further, Zeyher's (1975) first principles method for the LD of ionic crystals implies the regrouping by starting out from the rigid ion limit. Altogether we think that this alternate approach is more in the spirit of the model theories than the inversion method as it allows to divide the dynamical matrix rather easily into a rigid atom (or ion) part and into a polarization part.

Various different bandstructure schemes has been used to formulate the LD in the alternate approach. Pickett and Gyorffy (1976) used the KKR method and applied the rigid-muffin-tin approximation. So far, no actual computations have been carried out using this way.

Ashkenazi and Dacorogna (1981) are employing the linearized-muffin-tin-orbitals (LMTO) method, again with the rigid-muffin-tin approximation, and, in addition, some further simplifying assumptions. They have reported results for shear elastic constants of various transition metal elements which are in good agreement with experiment. Calculations of the full phonon dispersion curves of transition metals are in progress.

Winter (1981) has proposed a real space representation which allows for corrections beyong the rigid-muffin-tin approximation. He uses a cluster method for the calculation of the electronic structure and thus calculates force constants between near neighbor atoms. Results for Al are very satisfactory.

The most extensive calculations of the LD of TMC have been carried out with the tight binding method (Varma and Weber 1977, 1979 ; Ashkenazi et al., 1978 ; Weber et al., 1979, Weber 1980a, 1980b, 1982). In this approach the idea is used (Fröhlich 1966, Friedel 1969) that the tight-binding orbitals move with the displaced ion cores. As a consequence, a very economic way of computing first principles phonon dispersion curves becomes possible. In addition, this method provides a simple and intuitive understanding of the phonon anomalies and their relation to the occurrence of superconductivity.

The bulk of this article will deal with the non-orthogonal tight binding (NTB) method and its applications to various TMC systems (Sect. 2 and 3, and Appendix). We should mention that the NTB method is not restricted to TMC. For instance, also the LD of insulating silver halides has been studied using the NTB scheme (Kleppmann and Weber 1979).

In particular, the NTB method is outlined in Sect. 2. There the expressions of the dynamical matrix and of the electron-phonon coupling parameters are given. Also, some NTB technicalities are described. Furthermore, the NTB method is compared to the conventional inversion approach, and finally the method of calculation is sketched. Details of these subjects are left to the Appendix. There, not only various terms of the dynamical matrix in the NTB method are derived ; also, a very general program package is outlined, applicable to a great variety of crystal structures. Finally tables of NTB matrix elements and their derivatives are given in the Appendix.

In Sect. 3, we present results of calculations for three TMC systems : i) $Nb_{1-x}Mo_x$ crystals, ii) the refractory materials like NbC etc., and iii) the A15 compound Nb_3Sn. Each subsection starts with a discussion of the energy bands of the respective system. Then the theoretical phonon dispersion curves are presented, and the origin of the phonon anomalies is discussed. Finally calculations of the electron-phonon coupling parameters are reported. In particular, results for the Eliashberg function $\alpha^2F(\omega)$ are given and compared to tunneling data.

At present, certain terms of the tight binding formalism, which give rise to short range forces, are treated as parameters. They are obtained using experimental information. One way to complement the tight binding approach are the "frozen phonon" calculations. Modern energy band methods allow to compute very accurate total energy curves for certain lattice configurations such as displacement patterns corresponding to "frozen" phonons of high symmetry. In Sect. 4, a brief survey of some of the "frozen phonon" calculations is presented, particularly in view of their complementary role for the NTB method. A summary of this article is given in Sect. 5.

2. OUTLINE OF THE NON-ORTHOGONAL TIGHT BINDING APPROACH FOR THE THEORY OF LATTICE DYNAMICS

2.1. Dynamical Matrix

The Hamiltonian for a system of electrons and ions is

$$H = T_e + T_c + V_{cc}(\{R\}) + V_{ee}(\{r\}) + V_{ec}(\{r\}, \{R\}) \tag{2.1}$$

In Eq. (2.1), T is the kinetic-energy operator, and V is the potential energy ; the subscript c refers to the ion cores and e to the valence electrons ; $\{r\}$ denotes the electronic coordinates and $\{R\}$ the ion coordinates. In the adiabatic approximation for the electron-ion system, the electrons are assumed to respond instantaneously to the ionic motion. Thus we may write for the total energy of the crystal (at T = 0, neglecting zero point motion)

$$E_{tot}\{R\} = V_{cc}(\{R\}) + E_e(\{R\}) \tag{2.2}$$

where E_e is the total electronic energy which depends parametrically on the ionic configuration $\{R\}$. When we expand E_{tot} up to second order around the equilibrium configuration of the solid, we arrive at the usual harmonic approximation for the lattice dynamics, with the force constant matrices

$$\phi(\ell\kappa\alpha,\ell'\kappa'\beta) = \frac{\partial^2}{\partial u_\alpha(\ell\kappa)\partial u_\beta(\ell'\kappa')} E_{tot}\{\vec{R}(\ell\kappa),\ldots,\vec{R}(\ell'\kappa')\} \quad (2.3)$$

Here ℓ,ℓ' denote the unit cells and κ,κ' the sublattice atoms. The vectors $\vec{R}(\ell,\kappa)$ are now considered to be the equilibrium positions of the ions and the vectors $\vec{u}(\ell,\kappa)$ are the (small) displacements from equilibrium. The force constant matrices $\phi(\ell\kappa\alpha,\ell'\kappa'\beta)$ are the basis for the formal theory of LD (see Appendix A1), leading to the dynamical matrix

$$D(\kappa\alpha,\kappa'\beta|q) = \sum_{\ell-\ell'} \phi(\ell\kappa\alpha,\ell'\kappa'\beta) \exp[-i\vec{q}\vec{R}(\ell\kappa,\ell'\kappa')] \quad (2.4)$$

by transformation into reciprocal space. Here, we use the short hand notation $\vec{R}(\ell\kappa,\ell'\kappa') = \vec{R}(\ell\kappa) - \vec{R}(\ell'\kappa')$.

The main problem of first principles lattice dynamics is the proper treatment of the total electronic energy E_e. Our approach is to express $E_e(\{R\})$ in terms of the eigenvalues of the self-consistent one-electron Hamiltonian. In the adiabatic approximation, the wave functions and the eigenvalues of the electrons are determined using the instantaneous configuration of the ions, $\{R\}$ in V_{ec}, as a parameter. In the one electron approximation, the two-body term $V_{ee}(\{r\})$ is approximated in some self-consistent fashion to a one-body form. Because the wave functions for the electrons depend parametrically on the instantaneous ionic configuration, so does the self-consistent electron-electron interaction ; we denote the latter by $\overline{V}_{ee\{R\}}(\{r\})$. This interaction, $\overline{V}_{ee\{R\}}(\{r\})$, can be written as the sum of the Hartree potential and a one-body approximation to the exchange and correlation potential

$$\overline{V}_{ee\{R\}}(\{r\}) = V^H_{\{R\}}(\{r\}) + V^{xc}_{\{R\}}(\{r\}) \quad (2.5)$$

$$V^H_{\{R\}}(\{r\}) = e^2\int d^3r' \frac{\rho_{\{R\}}(\{r'\})}{|\vec{r}-\vec{r}'|} \quad (2.6)$$

In Eq. (2.6), $\rho_{\{R\}}(\{r\})$ is the electronic density in a prescribed configuration of ions. There are various approximate schemes for V^{xc}, the most popular among these is to express it as a local functional of the density $\rho(\{r\})$ (Kohn and Sham, 1965).

For the configuration in which the ions are at the equilibrium lattice sites, one can introduce the wave-vector index k in the first Brillouin zone and the band index μ as quantum numbers for the electronic wave function so that the one-electron Schrödinger equation for the electrons is

$$(T_e + V_{ec} + V_{ee})\psi_{\vec{k}\mu} = \varepsilon_{\vec{k}\mu}\,\psi_{\vec{k}\mu} \tag{2.7}$$

where we have omitted the labels {r},{R} for simplicity.

We wish to express E_e in terms of the self-consistent eigenvalues $\varepsilon_{\vec{k}\mu}$. Towards this end, we first define

$$E_{bs} = \sum_{\vec{k}\mu} \varepsilon_{\vec{k}\mu}\, f_{\vec{k}\mu}$$

where $f_{\vec{k}\mu}$ is the Fermi function. E_{bs}, the "band structure" energy is the sum of the energies of the occupied electronic states. In obtaining $\varepsilon_{\vec{k}\mu}$, the two-body electron-electron interaction terms have been approximated by one-body terms. Therefore E_{bs} includes the interaction energy twice. The total electronic energy E_e is

$$E_e = E_{bs} - \langle\overline{V}_{ee}\rangle \; ; \; \langle\overline{V}_{ee}\rangle = \langle V^H\rangle + \langle V^{xc}\rangle \tag{2.8}$$

where < > denotes the expectation value. Thus

$$\langle V^H\{R\}\rangle = \frac{1}{2}\, e^2 \int d^3r \int d^3r' \,\frac{\rho_{\{R\}}(r)\rho_{\{R\}}(r')}{|\vec{r} - \vec{r}\,'|} \tag{2.9}$$

We define

$$V_0(\{R\}) = V_{cc}(\{R\}) - \langle\overline{V}_{ee}(\{R\})\rangle \tag{2.10}$$

In Eq. (2.8), $\langle V^{xc}\{R\}\rangle$ is a short-range function of the distance between the ions, and the infinite range Coulomb term in V_{cc} is exactly cancelled by V^H for a metal with one atom per unit cell. V_{cc}.$-$ V^H is also much smaller than either V_{cc} of V^H. This cancellation is the better the more tightly bound the atomic orbitals are.

Now, the total crystal energy is re-written as

$$E_{tot}\{R\} = V_0\{R\} + E_{bs}\{R\} \tag{2.2a}$$

and the dynamical matrix D derived from Eq. (2.2a) has the form

$$D = D_0 + D_1 + D_2 \tag{2.11}$$

Here, D_0 arises from the force constants $\nabla\nabla V_0$, while D_1 and D_2 stem from the "band structure" contributions $\nabla\nabla E_{bs}$.

To evaluate $\nabla\nabla E_{bs}$, we must consider the shift in the eigenvalues $\varepsilon_{\vec{k}\mu}$ when the ions are displaced from the equilibrium lattice sites. D_1 arises from 1st order correction to $\varepsilon_{\vec{k}\mu}$ due to second order displacements, and D_2 results from 2nd order corrections to $\varepsilon_{\vec{k}\mu}$ due to first order displacements. The two terms are represented by these diagrams

D_1

D_2

(Note that for D_1 we deal with a second order vertex).

2.2. The Tight-Binding Formalism

The tight-binding representation for the electrons in TMC is a very convenient scheme to derive expressions for D_1 and D_2. In the Appendix A2-A6 a detailed derivation is given, including the corrections due to the non-orthogonality of adjacent atomic orbitals. In the following we give the most important formulae of the non-orthogonal-tight-binding (NTB) method.

In NTB a Bloch function is given by

$$\psi_{k\mu} = N^{-1/2} \sum_{\ell\kappa m} A(\kappa m|k\mu)\ \phi(\ell\kappa m)\exp[\, i\vec{k}\vec{R}(\ell\kappa)] \tag{2.12}$$

Here $\phi(\ell\kappa m) = \phi_m[\vec{r} - \vec{R}(\ell\kappa)]$ is the localized orbital with quantum number m at the site (ℓ,κ).

The matrix of eigenvectors A satisfies the eigenvalue equation of Hamiltonian (H) and overlap matrices (S)

$$HA = SAE$$

or

$$S^{1/2}\, H\, S^{-1/2}\, U = UE\ , \text{ with } A = S^{-1/2}\, U \tag{2.13}$$

Here $E(k)$ is the diagonal matrix of the energy eigenvalues $\varepsilon_{k\mu}$. We note that Eq.s (2.13) are regarded as completely equivalent to Eq. (2.6), thus the elements of the Hamiltonian matrix H are self-consistent quantities including the effect of the electron-electron interactions.

Then we obtain

$$D_2(\kappa\alpha,\kappa'\beta|q) = -\sum_{\substack{k\mu\mu' \\ k'=k+q}} \frac{f_{k'\mu'} - f_{k\mu}}{\varepsilon_{k\mu} - \varepsilon_{k'\mu'}}\, g^{\kappa\alpha}_{k\mu,k'\mu'}\, g^{\kappa'\beta}_{k'\mu',k\mu} \tag{2.14}$$

with

$$g^{\overline{\kappa}\alpha}_{k\mu,k'\mu'} = \sum_{\substack{\kappa m \\ \kappa' m'}} A^{\star}(\kappa m|k\mu)[\gamma_\alpha(\kappa m,\kappa' m'|k')\delta_{\kappa\overline{\kappa}} - \gamma_\alpha(\kappa m,\kappa' m'|k)\ \delta_{\kappa'\overline{\kappa}}]\ A(\kappa' m'|k'\mu') \tag{2.15}$$

being the electron-phonon matrix or electron-ion form factor. The quantities γ_α are defined as

$$\gamma_\alpha(\kappa m,\kappa' m'|k) = \sum_{\ell-\ell'} [\nabla_\alpha H(\ell\kappa m,\ell'\kappa' m') - \varepsilon\ \nabla_\alpha S(\ell\kappa m,\ell'\kappa' m')]\ \exp[-i\vec{k}\vec{R}(\ell\kappa,\ell'\kappa')] \tag{2.16}$$

with

$$\varepsilon = 1/2\ (\varepsilon_{k\mu} + \varepsilon_{k'\mu'})$$

We note that Eq. (2.16) implies the two-center approximation in the NTB scheme, furthermore any "crystal field" terms have been ignored. How to include the latter terms is given by Eqs. A35-A37.

The contribution to the phonon linewidth $\gamma(qj)$ originating from electron-phonon coupling is given by

$$\gamma(qj) = \pi\hbar/2 \sum_{\substack{\kappa\kappa' \\ \alpha\beta}} (m_\kappa m_{\kappa'})^{-1/2}\ e(\kappa\alpha|qj)\ e(\kappa'\beta|-qj) \times \Gamma(\kappa\alpha,\kappa'\beta|q) \tag{2.17}$$

where $e(\kappa\alpha|qj)$ is the phonon polarization vector associated with the phonon frequency $\omega(qj)$ (see Eq. A5), and m_κ is the atomic mass. The matrix Γ is the imaginary part of D_2 and is given by a Fermi surface integral

$$\Gamma(\kappa\alpha,\kappa'\beta|q) = \sum_{\substack{\mu\mu' \\ k'=k+q}} \iint_{FS} d\sigma_{k\mu}\ d\sigma_{k'\mu'} \times g^{\kappa\alpha}_{k\mu,k'\mu'}\ g^{\kappa'\beta}_{k'\mu' k\mu}\ \delta(k-k'-q) \tag{2.18}$$

with

$$d\sigma_{k\mu} = dS_{k\mu}\ /\ |v_{k\mu}| \text{ and } v^{\alpha}_{k\mu} = \partial\varepsilon_{k\mu}/\partial k_\alpha$$

The phonon linewidths γ_{qj} are related to the so-called Eliashberg function $\alpha^2F(\omega)$ of the strong-coupling theory of superconductivity (see, e.g. McMillan 1968) by

$$\alpha^2F(\omega) = [\,2\pi\hbar N(E_F)]^{-1} \sum_{qj} \gamma_{qj}/\omega_{qj}\ \delta(\omega - \omega_{qj}) \tag{2.19}$$

and to the electron-phonon coupling parameter λ by

$$\lambda = [\,\pi\hbar N(E_F)]^{-1} \sum_{qj} \gamma_{qj}/\omega^2_{qj} \tag{2.20}$$

where $N(E_F)$ is the density of states of one spin at the Fermi energy. In the case of one atom per unit cell, McMillan (1968) has introduced a factorization

$$\lambda = N(E_F)\ \langle I^2\rangle/m\langle\omega^2\rangle$$

with $\langle\omega^2\rangle$ being an appropriate second moment of the phonon density of states. The quantity $\langle I^2\rangle$ is then given by

$$\langle I^2\rangle = N(E_F)^{-2} \sum_{q,\alpha} \Gamma(\alpha,\alpha|q) \tag{2.21}$$

D_1 consists of five different contributions (see Appendix A5). The first two yield a force constant matrix

$$\phi_1^{(1+2)}(\ell\kappa\alpha,\ell'\kappa'\beta) =$$

$$2N^{-1} \sum_{k\mu} f_{k\mu} \sum_{m,m'} A^{\star}(\kappa m,k\mu)\gamma_{\alpha\beta}(\ell\kappa m,\ell'\kappa'm')\ A(\kappa'm',k\mu) \tag{2.22}$$

with

$$\gamma_{\alpha\beta}(\ell\kappa m,\ell'\kappa'm') = \nabla_\alpha\nabla_\beta H(\ell\kappa m,\ell'\kappa'm') - \varepsilon_{k\mu}\ \nabla_\alpha\nabla_\beta S(\ell\kappa m,\ell'\kappa'm') \tag{2.23}$$

Note that the range of $\phi_1^{(1+2)}$ is the same as the range of energy transfer or overlap matrix elements, i.e., in TMC these forces are of short range.

The expression $D_1^{(1+2)}$ is obtained from $\phi_1^{(1+2)}$ using Eq. (2.4). The terms $D_1^{(3-5)}$ give a somewhat lengthy expression (Eq. A42). The forces involved in $D_1^{(3-5)}$ extend at most up to twice the longest overlap distance. $D_1^{(3-5)}$ can be included into D_2 by slightly changing the expression for γ_α in Eq. (2.16), (see comment in Appendix A6).

When the ions displace from equilibrium, the wave functions $\psi_{k\mu}$ of Eq. (2.12) and thus the charge density change in two ways :

(i) The orbitals $\phi(\ell\kappa m)$ follow the ions, thus the charge around is moving with the ion.
(ii) The coupling of states $k\mu$ and $k'\mu'$ due to a periodic wave of vector q leads to a modification of the $A(\kappa m|k\mu)$ in $\psi_{k\mu}$. This

changes the orbital character of the occupied states and thus also alters the charge density.

These modifications of the charge density lead to variations in $\overline{V}_{ee}$ and have to be incorporated both in ∇H and $\nabla\nabla H$ in a self-consistent (or renormalized) manner. In principle, these gradients are a function of the wave vector q of the displacement field, as for each q the induced charge density is different. However, for practical purposes we assume that these gradients are constants independent of q. If this approximation was not fulfilled within ≈ 10-20%, our approach would loose a lot of its advantages. The validity of the constant gradient approximation is discussed at the end of the following section.

2.3. Determination of the Gradients

In two-center approximation, the derivatives $\nabla_\alpha H(\ell\kappa m,\ell'\kappa' m')$ have two terms, the so-called radial and angular parts. This can be seen by looking at a typical TB matrix element, as given in Table I of Slater and Koster (1954). Taking two d_{xy} orbitals at adjacent lattice sites, the matrix element is

$$E_{xy,xy} = 3\hat{x}^2\hat{y}^2(dd\sigma) + (\hat{x}^2 + \hat{y}^2 - 4\hat{x}^2\hat{y}^2)(dd\pi) + (\hat{z}^2 + \hat{x}^2\hat{y}^2)(dd\delta)$$

Here $\hat{x},\hat{y},\hat{z}$ are the direction cosinus of the distance vector $R(\ell\kappa,\ell'\kappa')$ of length r. The integrals $(dd\sigma)$, $(dd\pi)$ and $(dd\delta)$ are assumed to depend only on the distance between the atoms, in the spirit of the two-center approximation.

Then, e.g.,

$$(\partial/\partial x)E_{xy,xy} = R^x_{xy,xy} + A^x_{xy,xy} \tag{2.24}$$

with the radial part

$$R^x_{xy,xy} = 3\hat{x}^3\hat{y}^2(dd\sigma)' + \hat{x}(\hat{x}^2 + \hat{y}^2 - 4\hat{x}^2\hat{y}^2)(dd\pi) + \hat{x}(\hat{z}^2 + \hat{x}^2\hat{y}^2)(dd\delta)'$$

and the angular part

$$A^x_{xy,xy} = 6\hat{x}\hat{y}^2(1 - 2\hat{x}^2)(dd\sigma/r) + (16\hat{x}^3\hat{y}^2 - 8\hat{x}\hat{y}^2 + 2\hat{x}\hat{z}^2)(dd\pi/r) + (2\hat{x}\hat{y}^2 - 2\hat{x}\hat{z}^2 - 4\hat{x}^3\hat{y}^2)(dd\delta/r)$$

Similarly the second derivatives $\nabla\nabla H$ terms can be found. Of the two parts of ∇H or ∇S, the angular part A is completely given by the NTB parameters $(dd\sigma)$ etc. at equilibrium. The gradients $(dd\sigma)'$ etc. enter only the radial terms R. We now put

$$f = f_0 e^{\alpha r}$$

where f stand for any of the overlap or energy transfer integrals. Then

$$f' = \alpha f$$

and we need to know the exponentials

$$\alpha = \frac{\partial f}{\partial r} f^{-1} \approx \frac{\Delta f}{\Delta r} f^{-1} \tag{2.25}$$

The derivatives ∇S, ∇H can be determined from fits of NTB matrix elements to self-consistent energy band calculations at different lattice configurations. Then all renormalization effects are included. The simplest way of varying the lattice configuration is to change the lattice constant. Another way is to analyze frozen phonon calculations, where the total crystal energy (including the band energy) is calculated for static displacement patterns corresponding to high symmetry phonons (see Sect. 4). There are, however, numerical problems when the derivatives are determined by NTB fits to energy bands at slightly different lattice constants. Each NTB parameter has a certain inaccuracy, and setting up difference quotients means increasing this inaccuracy considerably. As a consequence, only the difference quotients of the largest matrix elements can be determined sufficiently reliable (≈ 10-20 percent). There is another practical problems in this procedure. The magnitude of the NTB parameters depends on the energy zero (see Appendix A8). When we vary the lattice configuration, the energy bands may shift with respect to the energy zero, in other words, the work function may change. Thus, for a correct determination of the gradients we would also need to know, where the vacuum energy is situated with respect to E_F. This problem can be avoided by using an orthogonal TB scheme for the gradient determination.

We have also chosen a different way to obtain the gradients. Using atomic wave functions and potentials, we have calculated the overlap and energy integrals as a function of distance. For the large NTB matrix elements the results for the exponentials α agreed within 10-20% with those obtain from energy band fits.

It turns out that only the gradients of these largest matrix elements are important for the D_2 or electron-phonon coupling calculations ; i.e., the magnitude of the form factors $g^{\kappa\alpha}$ is dominated by the biggest ∇H, ∇S terms. The small H and S matrix elements are merely necessary to achieve the correct shape of the energy bands - this is crucial for the energy denominator in D_2 or for the correct Fermi surface in the calculations of the electron-phonon-coupling.

The "radial" and "angular" derivatives contribute approximately by the same amount to $g^{\kappa\alpha}$. Thus a 10% change in the "radial" derivatives modifies the $g^{\kappa\alpha}$ values by ≈ 5% and D_2 or Γ vary by ≈ 10%. Its effect on $D_2(q)$ of $\Gamma(q)$ is predominantly a scaling of the magnitude, but is hardly at all of any influence on the q-dependence. Thus, prominent structures in D_2 or Γ at certain anomalous wavevectors q, do not change their positions in q-space, when the magnitude of the gradients is varying.

As a consequence, for phonon calculations, the magnitude of the gradients need not be known with extreme accuracy, as the overall magnitude of D_2 can be compensated to a large extend by a different choice of the short range force constants for $(D_0 + D_1)$. Only when we attempt to calculate correctly the phonon linewidths or the electron-phonon coupling constant λ, we need to know the gradients as accurately as possible.

2.4. Comparison with the Inversion Approach

The dielectric theory of LD calculates the dynamical matrix directly from V_{cc} and E_e. If electron-electron interactions are ignored, the second-order-perturbation change in E_e with displacements of ions is given by

$$\delta E_e = g_0 \, \chi_0 \, g_0 \tag{2.26}$$

where g_0 is the bare electron-ion interaction and χ_0 is the electronic susceptibility (see Fig. 10a). To include the effect of the electron-electron interactions, χ_0 in Eq. (2.26) must be replaced by a renormalized susceptibility ; for instance in random-phase approximation (RPA), δE_e is given by

$$\delta E_e^{RPA} = g_0 \, \chi \, g_0 = g \, \chi_0 \, g_0$$

$$\chi = \varepsilon^{-1} \chi_0 \, , \; g = \varepsilon^{-1} g_0 \tag{2.27}$$

where ε is the dielectric function, δE^{RPA} is show in Fig. 10(b). As indicated above, the inversion of ε in Eq. (2.27) is a very formidable problem for materials with rapidly varying charge density.

In D_2 of Eq. (2.14), we calculate $g \, \chi_0 \, g$ which is represented in Fig. 10(c). As shown there, $g \, \chi_0 \, g_0$ is equal in RPA to $D_2 - \langle \nabla\nabla V_H \rangle$. We have explicitly included $\langle \nabla\nabla H_H \rangle$ in D_0, that part of the dynamical matrix derived from V_0 of Eq. (2.10). Thus, the two methods are formally equivalent in RPA. This equivalence can be shown more generally (Gyorffy 1978).

By expressing the dynamical matrix in terms derived from V_0 and E_{bs} (Eq. 2.11), the cancellation problem of the inversion method has been overcome. Furthermore, the NTB method has several computational advantages.

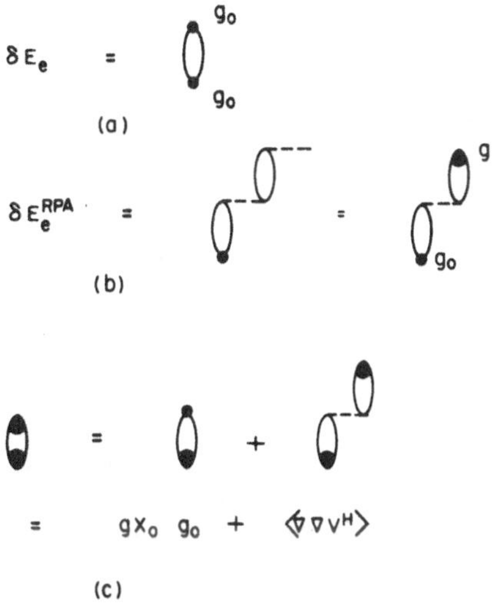

Fig. 10. Equivalence of the present method of calculating the dynamical matrix to the inversion method is shown within RPA. In (b), the sum of ring diagrams is symbolized by a single diagram (from Varma and Weber 1979).

(i) The changes in the Hamiltonian due to ion motion are expressed in terms of derivatives of matrix elements taken at discrete lattice points ($\ell\kappa$) and ($\ell'\kappa'$) etc. Thus we do not explicitly require the knowledge of the wavefunctions and potential gradients at all points in space and yet include local field effects.

(ii) The effect of variation in screening, exchange and correlations is included in the terms ∇H, ∇S etc., as long as these quantities are determined from self-consistent energy band calculations at different lattice configurations.

(iii) The significant ∇H and ∇S terms are confined to a few nearest neighbors.

The conceptual advantage of working in the tight binding representation is that we are able to express the electron-phonon coupling in rather simple pictures. This will become evident in the discussion of our results (see Sect. 3).

2.5. Method of Calculation

We have mostly been interested in understanding the anomalous features in the phonon dispersion of TMC as, for instance, discussed in Section 1.1. These anomalies are indications of long range force fields which arise from the term D_2 of Eq. (2.14) as we will see in Section 3. On the other hand, D_1 yields only short range forces (see Sect. 2.2) so does essentially D_0, too. There are some long-range effects in D_0, due to the interaction of the induced charge density. However, these long-range effects are rather small, as we will discuss in the Appendix A9. Any calculation of $D_0 + D_1$ thus leads to short range forces, extending not much further than 2nd nearest neighbors. Yet the numerical effort appears to be considerable : For D_0, explicit knowledge of orbital shape is necessary to calculate V^H and V^{xc} arising from the interpenetrating charge clouds. To compute D_1 or ϕ_1 of Eq. (2.22) it appears to be insufficient to include only valence electrons. The largest forces arise from the repulsion of valence orbitals when penetrating into neighbor core orbitals. This means that core states have to be included in a calculation of D_1. So far we have not attempted to carry out these tedious computations of $D_0 + D_1$. Instead we have parametrized these terms by a few near neighbor force constants and have focussed our efforts on the first principles calculation of D_2.

We have developed a rather generally applicable program package for computing the phonon dispersion curves in the NTB method. This package is described in some detail in the Appendix A7. The most important parts of the package are

(i) A NTB energy band code in two-center approximation to calculate $\varepsilon_{k\mu}$ and $A(\kappa m|k\mu)$ for arbitrary k-points. A modified version is used to determine the NTB matrix elements by fitting to first principles energy band calculations. The latter version also allows to determine the ∇H, ∇S terms. The NTB energy band code has been developed in collaboration with L.F. Mattheiss (Mattheiss and Weber 1982).

(ii) The D_2 program, including calculation of the linewidth matrix Γ (Eq. 2.18). This is the biggest part of the package, and also the one which requires the most computing time. For instance, calculations of materials like A15 compounds with 8 atoms per unit cell approach the limit of present day computing facilities.

(iii) A phonon program to calculate ω_{qj} and γ_{qj}. This program includes Born-and-von-Kármán force constants to parametrize $D_0 + D_1$. A modified version is used to determine these short range force constants by fitting to experimental phonon frequencies.

It is possible to completely avoid any experimental input in a phonon calculation when this program package is linked to frozen phonon calculations (see Sect. 4). The latter, carried out for high symmetry phonon modes, yield enough information to determine the short range force constants $D_0 + D_1$. Then, the $D_2(q)$ calculations can be seen as an interpolation scheme to determine ω_{qj} throughout the whole Brillouin zone. This tandem scheme seems to be the most economical way to carry out fully first principles calculations of the phonon dispersion in TMC.

3. NTB METHOD : RESULTS AND DISCUSSION

Three systems of transition metals and compounds have been studied using the NTB method. These systems are i) Zr-Nb-Mo (Varma and Weber 1977, 1979, Simons and Varma 1980, Gompf et al. (1981), ii) the refractory compounds like NbC, NbN or VN (Weber et al. 1979, Weber 1980a, 1980b) and iii) the A15 compounds Nb_3Sn, V_3Si and Nb_3Sb (Mattheiss and Weber 1982, Weber and Mattheiss 1982, Weber 1982). The most important results for these three groups of materials are discussed in Sections 3.1 - 3.3. We should mention that calculations of elastic constants in Nb-Zr alloys have been carried out by Ashkenazi et al.(1978) using an orthogonal tight binding method. This scheme, developed in a paper by Peter et al. (1974), is similar to ours, with the main difference that overlap corrections are neglected. It has further been used for a study of anisotropy effects of the electron-phonon coupling on T_c (Peter et al. 1977, see Sect. 3.1.3).

3.1. The Nb-Mo System

3.1.1. Energy Bands

The energy bands for the Nb-Mo system have been obtained by fitting the matrix elements of a NTB Hamiltonian with a nine-orbital s-p-d basis to the non-selfconsistent augmented-plane-wave (APW) calculation for Nb of Mattheiss (1970). These NTB matrix elements (see Table A10.1) have been used both for the calculations of $\langle I^2 \rangle$ (Eq. 2.21) (Varma et al. 1978 ; Varma et al. 1979) as well as for the calculation of the phonon dispersion curves of the bcc Zr-Nb-Mo system (Varma and Weber 1977, 1979, Simons and Varma 1980). The root mean square deviation of this fit was ≈ 15 mRy. Most importantly, the shape of the Nb Fermi surface did not completely agree with that given by Mattheiss (1970). In particular, the ellipsoids around N were connected to the Γ-H jungle gym by a small pipe of diameter ≈ 0.1 k_H around the Σ line (k_H is the wave-vector of point H of the BZ).

As we later worried that these deficiencies would affect some of the results in the above-mentioned papers, we have established a much better fit (r.m.s ≈ 3 mRy), where those discrepancies have been removed (Mattheiss and Weber 1979). These NTB parameters are also given in Table A10.1. We want to mention that a recalculation of the phonons as well as of $\langle I^2 \rangle$ in Nb-Mo did not give any substantial changes as compared to the old results (Richter and Weber, 1983).

A plot of the (old) NTB energy bands of Nb is given in Fig. 11. The electronic density of states (DOS) is shown in Fig. 12.

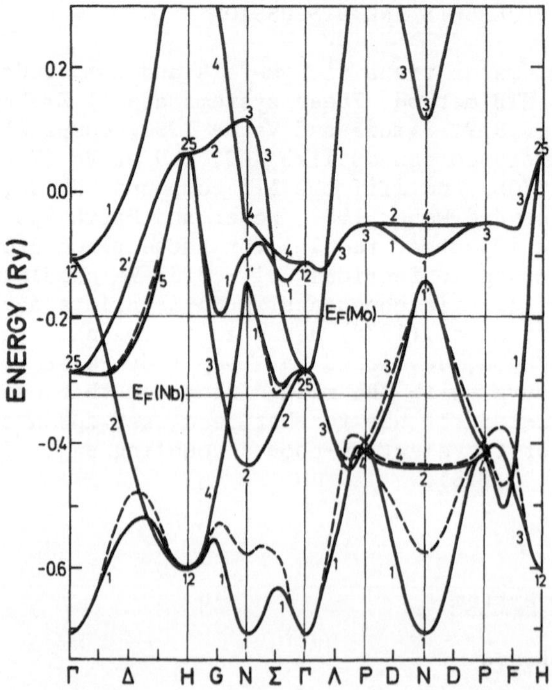

Fig. 11. NTB band structure (fitted to that of Mattheiss, 1970) used for Nb, Mo, and their alloys. Where the first principles band structure visibly deviates from the NTB band structure, is shown as a dashed line (VW 1979).

The energy bands of Mo are very similar to those of Nb (see, e.g., Pickett and Allen 1974), so that we have used a rigid band model for the Nb-Mo system. We note that this "transium" model is well justified by various studies, e.g., by specific heat data (see McMillan 1967), or by coherent-potential-approximation calculations for Nb-Mo alloys (Guiliano et al. 1978).

In the "transium" model the only difference in the electronic structure of the $Nb_{1-x}Mo_x$ alloys is the number of available valence electrons and thus the position of E_F. As a function of Mo concentration, E_F is moving from a high DOS region at x = 0 (E_F = - 4.2 eV) through the deep minimum near x = 0.75 towards a $N(E_F)$ value slightly beyond this minimum at x = 1 (E_F = - 2.7 eV). We note that the values of the superconducting transition temperature T_c follow the $N(E_F)$ values rather closely, from T_c = 9.2 K for Nb down to T_c < 0.5 K at $Nb_{0.75}Mo_{0.25}$ and up again to T_c = 0.9 K for Mo.

3.1.2. Phonons: Analysis of Phonon Anomalies

In the "transium" model for the $Nb_{1-x}Mo_x$ alloys, D_2 is calcu-

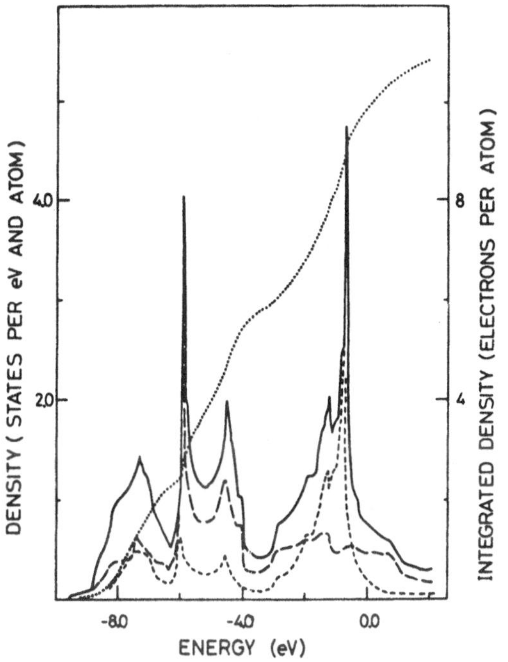

Fig. 12. Density of states and integrated density of states for Nb, Mo, and their alloys. The e_g density of states (short dashes) and the t_{2g} density of states (long dashes) are also shown. E_F = - 4.2 eV for Nb, and E_F = - 2.8 eV for Mo (VW 1979).

lated as a function of x. In these calculations we have included all band-to-band transitions possible in the nine valence bands of our basis. The NTB gradients have been obtained both from atomic calculations and from fits to self-consistent APW calculations of Nb at two different lattice constants (Anderson et al. 1973). Typical values of the gradients are given in Table A10.1. (D_0 + D_1) are represented by 1nn and 2nn force constants and are fitted so that (D_0 + D_1 + D_2) gives best agreement with the experimental dispersion curves of the $Nb_{1-x}Mo_x$ alloys (Powell et al. 1968). These 1nn and 2nn force constants vary rather smoothly and, in general increase with x.

The phonon dispersion curves for Nb, $Nb_{0.25}Mo_{0.75}$, and Mo are shown in Figs. 13-15. We note that all anomalous features in these curves are correctly reproduced. These are, for Nb, along (ξ,0,0), the dip in the longitudinal (L) branch near ξ = 0.7, along (ξ,ξ,ξ), the very deep minimum in L near ξ = 0.7, and the kink in L near 0.4. Also reproduced are the negative curvatures in the transverse (T) branches in (ξ,0,0) and (ξ,ξ,0) including the very small value of c_{44} and the crossing of the two T branches in (ξ,ξ,0).

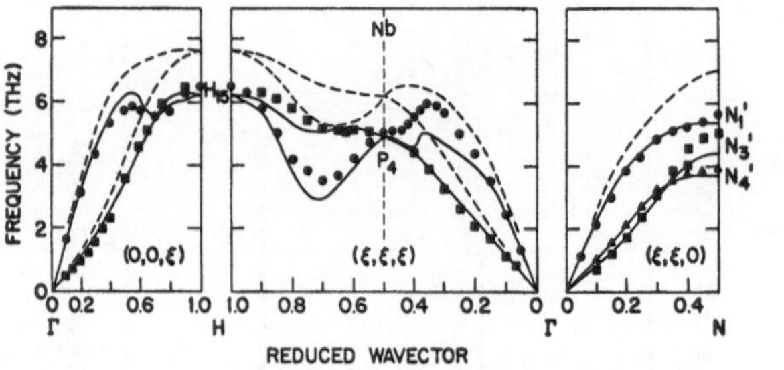

Fig. 13. Calculated phonon dispersion curves for Nb compared with experiments (Powell et al. 1968). Dashed curves are the dispersion curves obtained if $D_2^<$ is omitted (VW 1979).

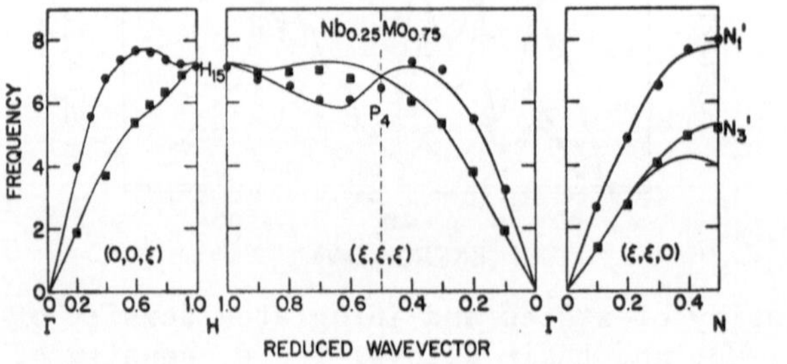

Fig. 14. Calculated phonon dispersion curves for $Nb_{0.25}Mo_{0.75}$ compared with experiment (Powell et al.) (VW 1979).

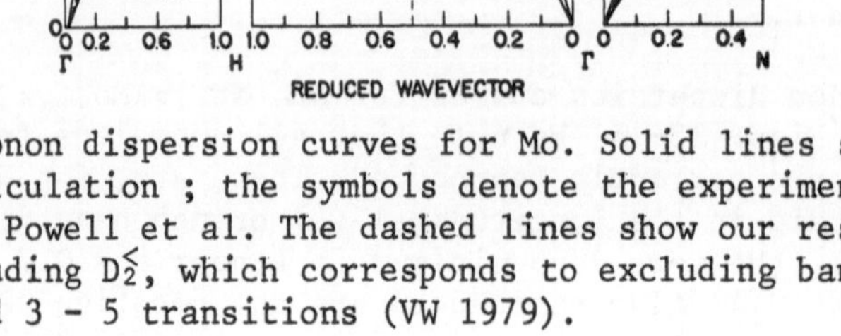

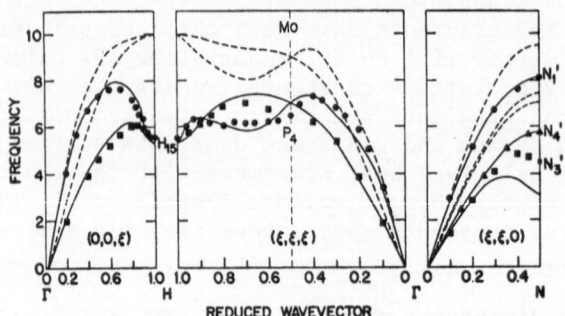

Fig. 15. Phonon dispersion curves for Mo. Solid lines show our calculation ; the symbols denote the experimental values of Powell et al. The dashed lines show our results excluding $D_2^<$, which corresponds to excluding band 3 - 4 and 3 - 5 transitions (VW 1979).

When going toward Mo-rich alloys all the anomalous features in the dispersion curves fade away and disappear around x = 0.75

(see Fig. 14). Near Mo (see Fig. 15) completely new features appear. The most dramatic is the strong softening of the phonons around the H point, and the crossing of L and T branches near $\xi = 0.9$ in (ξ,ξ,ξ). Similarly, one T branch in $(\xi,\xi,0)$ near N, especially the $N_{3'}$ mode starts to decrease rapidly.

We have carefully analyzed our calculations to trace the source of the anomalies and the nature of the overall spectrum. We find that almost all anomalous features in the dispersion curves of Nb, Mo, and the Nb-Mo alloys arise from $D_2^<$; these are scattering processes in D_2 with energies $\varepsilon_{k\mu}$ and $\varepsilon_{k'\mu'}$ close to E_F, approximately in the range $\pm$ 0.5 eV. The concave curvature in the T branches of Nb is the only exception to this finding (see discussion below).

The importance of the small-energy scattering $D_2^<$ is demonstrated in Fig. 13 (Nb) and Fig. 15 (Mo). When $D_2^<$ is omitted, all the anomalous structure is absent. It should be noted that in Nb and Nb-rich alloys $D_2^<$ is equivalent to intraband scattering of band 3 (counted from the bottom, see Fig. 11). For Nb this band yields $\approx$ 90% of $N(E_F)$. In Mo, $D_2^<$ stems from interband scattering between bands 3 and 4. The various contributions to D_2 for Nb are also illustrated in Fig. 16. In a first step, all band 3 intraband scattering, i.e., $D_2^<$ is omitted. Here the negative curvature of the long wavelength T branches is still present.

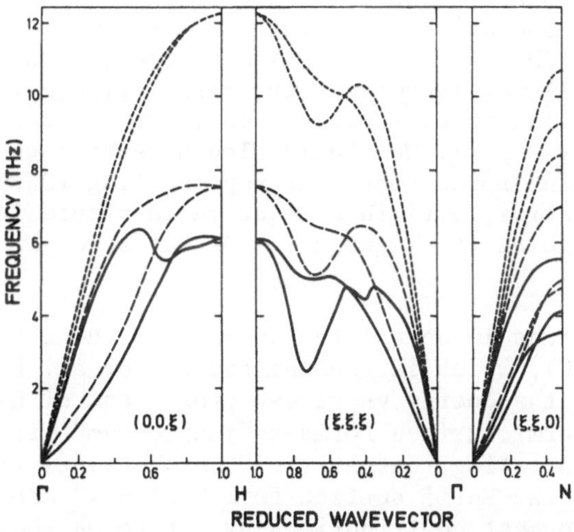

Fig. 16. Phonon dispersion curves for Nb calculated using (i) complete D (full lines), (ii) omitting band 3 intraband scattering in D (long-dashed lines) and (iii) omitting all scattering within bands 1 - 6, which are mainly d-like bands (short-dashes lines) (VW 1979).

In a second step we also omit all scattering within the lowest six bands which are mainly d-like. Now the negative curvature has also disappeared. The rest of D_2 scattering from the d-like bottom to the p-like top of the band complex with an average gap of $\approx$ 10 eV is completely structureless, hence the phonon curves show very smooth behavior indicative of very short-range force constants.

Thus all structure in the dispersion curves originates from the scattering within the mainly d-like bands. In these bands, the wave functions have only 10% - 15% s- or p-like character. We have found that s-d or p-d scattering is unimportant mainly because of the small admixture of s and p orbitals, and to a lesser extent because they have smaller gradients ∇H (see Table A10.1). An analysis of the relative importance of the various coupling gradients ∇H shows that (ddσ) (1nn and 2nn) and (ddπ) (1nn) dominate, all other matrix elements contribute less than 10%. The uncertainties in determining the less important coupling gradients (for instance those involving the s orbitals) are therefore of little consequence for our results. A variation of the magnitude of the d-d gradients causes a change in the magnitude of D_2, but the q dependence of D_2 is not significantly altered. Similarly, the omission of the ∇S term in g^{α}, only leads to an increase of D_2 but does not change its q dependence.

The negative curvature of the T branches is caused by large D_2 values at long wavelengths due mainly to scattering from band 2 to band 4, around the N point. Here the average $\Delta E_{kk'} \approx 4$ eV. As compared to the other structures, the T anomaly extends rather far in q space, approximately between $0 \leqslant \xi \leqslant 0.6$ along Γ-H. Also, the result $D_2(T)/D_2(L) \approx 2$ near $\xi \approx 0$ is found both for Nb and Mo and the Nb-Mo alloys. Therefore, the ratio $c_{11}/c_{44} \gg 1$ in all materials (in a 1nn force constant model for the bcc structure we would obtain $c_{11} = c_{44}$). We should also note that though any fit of the 1nn and 2nn force constants representing $(D_0 + D_1)$ yields large c_{11}/c_{44} ratios, a visible negative curvature in Nb is only achieved by weighing the T phonons somewhat stronger than the others.

We now discuss in detail the anomaly in the L branch of Nb at $q_a \approx (0.7,0,0)$, which is representative for all L anomalies in Nb. Consider the energy vs wave-vector plane of band 3 (see Fig. 17). The Fermi surface forms a "jungle-gym" with the junctions of the interpenetrating rods at the Γ and H points (see Fig. 18). Other pieces of the Fermi surface form "islands" around the N points. The strongest contributions by far to $D_2^{<}$ come from the region $k \approx (0.15, 0.45 \pm \delta, 0.1 \pm \varepsilon)$ with $\delta \approx 0.15$ and $\varepsilon \approx 0.05$, and the corresponding $k + q_a$ region, as indicated in Fig. 17. The two areas have almost parallel pieces of the Fermi surface with nesting vector q_a. This means that for this scattering, a relatively large phase space of states k and $k + q_a$ around E_F is available.

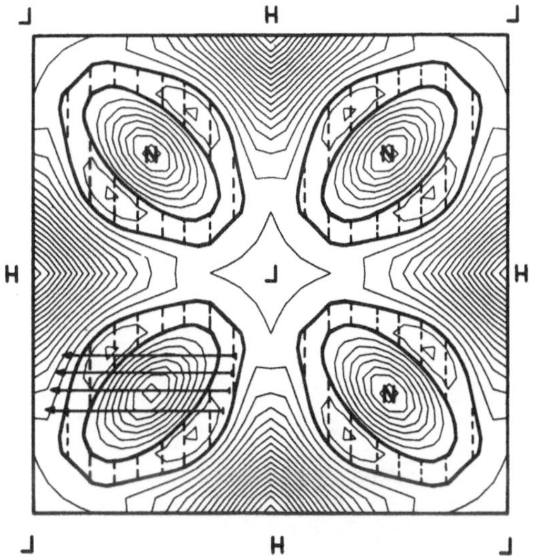

Fig. 17. Contour plot of the band 3 energy of wave-vector plane. The coordinates are the k_x and k_y directions ; $k_z = 0.06$. The E_F contour lines are specially marked. All areas below E_F are indicated by hatching. The arrows indicate the scattering area around $k_x = 0.15$, $k_y = 0.45 \pm 0.15$ and $\vec{k} + \vec{q}_a$ ($q_a \approx 0.7,0,0$) which yields the largest contributions of $D_2^<$ (VW 1979).

In Mo, the dominant contributions to the anomalous q-vector at H arise from interband scattering between bands 3 and 4 (see Fig. 19). Here, band 3 forms a hole-like octahedron around the H point while band 4 generates an electron-like octahedron around Γ of slightly smaller dimensions. (Because of the interaction of bands 4 and 5, the top pieces of the octahedron are extended and the Fermi surface actually is the "electron jack").

It is, however, misleading to look just for nesting of states E_k and E_{k+q} with favorably small energy differences. This is seen from a calculation of the "bare" intraband susceptibility

$$\chi_{33}(\vec{q}) = \sum_{\substack{k,k'=k+q \\ \mu=\mu'=3}} (f_{k\mu} - f_{k'\mu'})/(\varepsilon_{k\mu} - \varepsilon_{k'\mu'}) \tag{3.1}$$

which corresponds to $g_{kk'}$ = const. From an expansion of the denominator in Eq. (3.1)

$$(\varepsilon_{\vec{k}} - \varepsilon_{\vec{k}'}) = \Delta E_0 + (v^\alpha_{k_0} - v^\alpha_{k_0'})(k^\alpha - k^\alpha_0) \tag{3.2}$$

we would expect a maximum in $\chi(\vec{q})$ for $\Delta E_0 \approx 0$ and $v^\beta_{\vec{k}} \approx v^\beta_{\vec{k}'}$; i.e. for parallel or dispersionless bands. $\chi_{33}(\vec{q})$ along $(0,0,\xi)$ shows

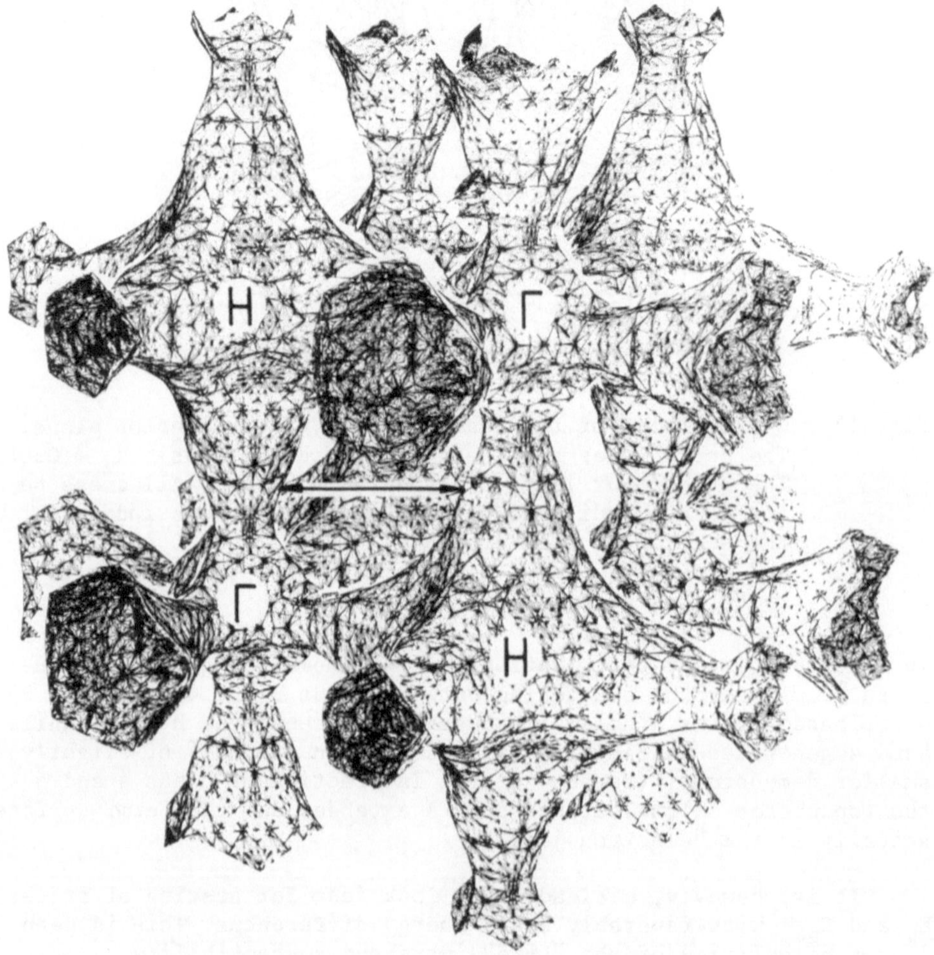

Fig. 18. "Jungle Gym" Fermi surface of Nb (band 3) shown in an extended zone scheme. The cone-shaped rods intersect at points Γ and H of the BZ. Also shown is the vector q_a = (0.7,0,0) of the longitudinal anomaly, indicating the nesting areas. For clarity the N point ellipsoids have been omitted. This plot has been generated using the improved NTB bands.

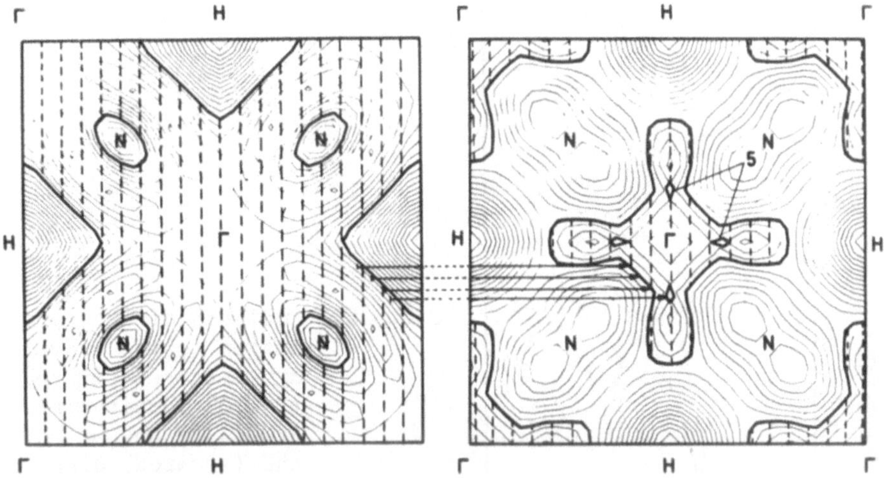

Fig. 19. Contour plots of band 3 and 4 energy vs wave vector planes for Mo. The coordinates are the k_x and k_y directions ; $k_z = 0$. The E_F contour lines are specially marked all areas below E_F are indicated by hatching. The arrows indicate dominant scattering areas for the H point anomaly (VW 1979).

a maximum near $q \approx 0.5$, but actually has a minimum for $q_a = 0.7$ (see Fig. 20). Also, along (ξ,ξ,ξ) the structure in $\chi_{33}(\vec{q})$ does not coincide with peaks in $D_2(L)$. On the other hand, if the q dependence of the energy denominators is ignored in the calculation of D_2 by replacing it with $N(\varepsilon_F)$, and

$$N(\varepsilon_F) \sum_k |g^{\alpha}_{k,k+q}|^2 (f_k - f_{k+\vec{q}}) \tag{3.3}$$

is plotted (see Fig. 20), the anomalous features of the true $D_2(q)$ are qualitatively reproduced.

It is therefore clear that the phonon anomalies have to do with the q dependence of the electron-ion form factors g^{α}. The physical origin of the anomalies is best illustrated if we use the approximation for

$$g^{\alpha}_{k\mu,k'\mu'} \propto (v^{\alpha}_k - v^{\alpha}_{k'\mu'}), \tag{3.4}$$

where $v^{\alpha}_{k\mu}$ is the velocity $\partial\varepsilon_{k\mu}/\partial k_{\alpha}$. This approximation is exact in an energy band model with one s-like orbital and 1nn energy transfer integrals. In a qualitative way, it holds rather well even for transition metals. In the one-orbital model, the velocity v is

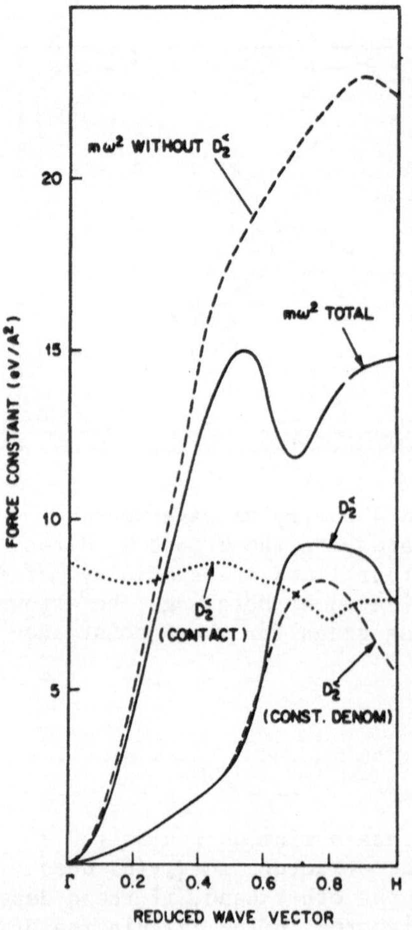

Fig. 20. Analysis of the contribution to the calculated dynamical matrix for the $(\xi,0,0)$ longitudinal branch in Nb. Without $D_2^<$ the (squared) dispersion curves do not exhibit the minimum near $\xi = 0.7$. The dominant contribution to $D_2^<$ does not arise from the $\chi_0(q)$ or $D_2^<$ (contact) term, which actually has a slight minimum near $\xi \approx 0.7$. Instead it arises from the electron-phonon factor $g^{\alpha}_{k,k'}$, as is evident from the curve D_2 (const. denom.) (VW 1977).

directly related to the transfer integral and thus to the bandwidth.

The phonon anomalies arise because lattice distortions at certain wave vectors cause a relatively large change in the electronic energy, and it is precisely this variation that we are calculating in D_2. The reduction in energy can obviously be most significant if local gaps in the electronic structure open up at the Fermi surface for certain lattice distortions. Inserting (3.4) in Eq. (2.14) for D_2, we conclude that a lattice distortion of wave vector $q = k - k'$ will reduce the energy of a state $k\mu$ near the Fermi surface significantly if a state $k'\mu'$ near the Fermi surface exists such that $|v^{\alpha}_{k\mu} - v^{\alpha}_{k'\mu'}|$ is large. Note in Figs. 17 and 19 the many energy contour lines for the k and k' states connected by the vector q_a of the anomalous phonons. This indicates large and opposite band velocities v^{α}_{k} and $v^{\alpha}_{k'}$. The anomaly will appear in the longitudinal mode for α parallel to q and in the transverse modes for α

orthogonal to q. This effect is proportional to $|v^{\alpha}_{k\mu} - v^{\alpha}_{k'\mu'}|^2$, while the number of states $k\mu$ near the Fermi surface is $\sim|v_{k\mu}|^{-1}$. Only the latter effect is present in the usual Kohn-Overhauser theory (Kohn 1959, Overhauser 1960), so that for nesting of bands near the Fermi surface, it predicts phonon anomalies for small $|v^{\alpha}_{k\mu}|$, whereas we predict phonon anomalies for large $|v^{\alpha}_{k\mu}|$. In the above discussion α is the nesting direction. In D_2 one must integrate over the other two components of the wave vector also. To get a large effect from these, only the energy denominator is relevant. For the energy denominator to be as small as possible, the initial and final bands in the scattering process must be nearly parallel. This is generally likely if the bands are flat in these directions. Very anisotropic energy bands with large dispersion in one k-direction and small dispersion in the other k-directions will occur much more likely in TMC than in simple metals, because the d-orbitals have much more direction dependence than the s-p orbitals. We therefore conclude that materials with very anisotropic bands near the Fermi surface give rise to the largest phonon anomalies. This anisotropy is also evident in the shape of the Fermi surface, e.g., the jungle-gym of Nb (see Figs. 17, 18) or the two octahedra around Γ and H in Mo (Fig. 19).

3.1.3. Electron-Phonon Coupling

For the $Nb_{1-x}Mo_x$ alloys, the computations of $\langle I^2 \rangle$, the Fermi surface average of the squared electron-phonon matrix elements (Eq. 2.21) yield good agreement with the empirical values of $\langle I^2 \rangle$ (Varma et al. 1978, 1979). Furthermore, the NTB values of $\langle I^2 \rangle$ are very close to those obtained from calculations (Butler, 1977) using the rigid muffin-tin approximation (RMTA). The increase of $\langle I^2 \rangle$ by ≈ 100% between Nb and Mo (see Fig. 21) is related to the increase of the energy band velocities, when moving from the high density of states region in Nb to the low DOS region in Mo (see Figs. 11 and 12). Similarly as relation (3.4) is obtained, it is possible to show that to a good approximation

$$\langle I^2 \rangle \propto \langle v_F^2 \rangle \, a_d^2 \tag{3.5}$$

holds, where $\langle v_F^2 \rangle$ is the Fermi surface average of the squared band velocities, and a_d^2 is the squared average of the d-orbital components of the Bloch states at E_F.

The quantitative agreement of $\langle I^2 \rangle$ was achieved by incorporating the non-orthogonality corrections ∇S. If these are omitted like in the work of Peter et al. (1977), the $\langle I^2 \rangle$ values are too large by ≈ 100%. In fact, these authors have rescaled their matrix elements by a factor ≈ 0.7 to get close to the experimental T_c value of Nb. This approximation seems to be well justified in view of our experience (see Sect. 3.1.2) that the ∇S corrections only rescale the magnitude of the matrix elements $g^{\alpha}_{kk'}$.

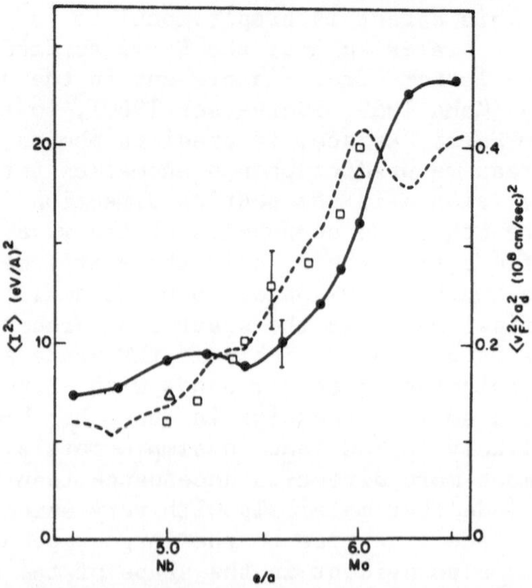

Fig. 21. Theoretical values for $<I^2>$ in the Nb-Mo system (circles). Also shown are empirical values (open squares and triangles). The dashed line depicts the quantity $<v_F^2>a_d^2$ (Varma et al. 1978).

The anisotropy of the electron-phonon coupling in Nb has been investigated quite intensely. Peter et al. (1977) have solved the Eliashberg equations for calculating T_c, by incorporating the full k,k' dependence of the matrix elements $g^{\alpha}_{kk'}$ over the Fermi surface. As compared to the "dirty limit" calculations, where the anisotropy is ignored, they found an increase of T_c by 0.25 K.

The variation of the electron-phonon coupling on the various Nb Fermi surface sheets has been studied by Crabtee et al. (1979). They find that the three sheets "jungle gym" (see Fig. 18), N-point ellipsoid (see Fig. 17) and Γ point octahedron exhibit rather different coupling strength. Peter et al. (1977) and Simons et al. (1981) have calculated these anisotropies using the tight binding method, other groups have used the rigid-muffin-approximation (RMTA) (Butler et al. 1979, Harmon and Sinha 1977). It appears that the tight binding results agree better - certainly not extremely well - with the Crabtree et al. data than the RMTA results.

The anisotropy of the electron-phonon coupling also becomes evident in the "intrinsic" phonon linewidth γ_{qj} (Eq. 2.17). This quantity indicates how much each phonon couples to the electrons at the Fermi energy. For longitudinal phonons along (ξ,0,0) and (ξ,ξ,0), γ values from our NTB calculation (Weber 1980b) are shown in Fig. 22. It becomes evident that γ(q) follows very closely to

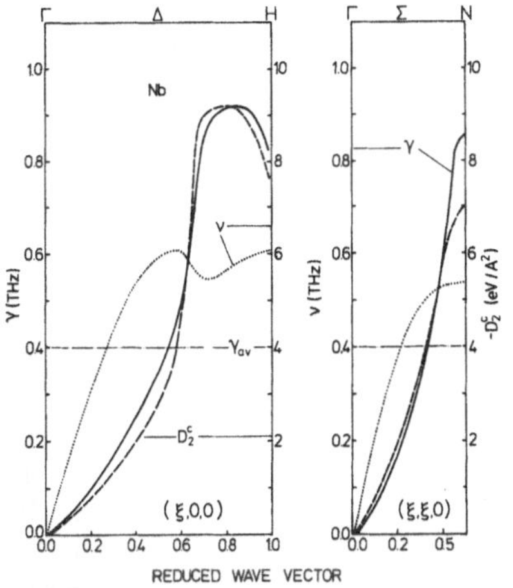

Fig. 22. Linewidths γ (solid lines) along (ξ,0,0) and (ξ,ξ,0) for longitudinal phonons in Nb. Dashed lines show the quantity $D_2^<$ (scattering processes ≈ ± 0.5 eV around E_F). The corresponding phonon frequencies γ are given as dotted lines. γ_{av} denotes the value of the average linewidth of all Nb phonons (Weber 1980a).

$D_2^<(q)$, that is, all anomalous phonons also show very strong electron-phonon coupling. Our results agree fairly well with the RMTA calculations of Butler et al.(1977). The q-dependence along (ξ,0,0) and (ξ,ξ,0) is similar, the magnitude of the (ξ,ξ,0) as compared to the (ξ,0,0) linewidths is however rather different. In addition, the RMTA results at long wavelengths appear to be consistently larger than the NTB results.

In Fig. 23(a) we show NTB results for a calculation of the Eliashberg function $\alpha^2F(\omega)$ (Eq. 2.19). In this calculation both the phonon frequencies ω(qj) and the linewidths γ(qj) are obtained from the same NTB matrix elements. The electron-phonon coupling constant λ (Eq. 2.20) is λ = 0.9. The function

$$\alpha^2(\omega) = \alpha^2F(\omega)/F(\omega) \qquad (3.6)$$

is smoothly increasing by about a factor 2 from the low frequency to the high frequency regions. This means that high frequency phonons couple stronger to the electrons than the low frequency modes. This has also become evident from the γ_{qj} calculations, where the high frequency modes consistently showed larger γ/ω values than the low frequency phonons. We can also locate these strong coupling phonons in k-space : They arise predominantly from the Δ(0.5,0,0)-N-P-H tetrahedron of the bcc BZ.

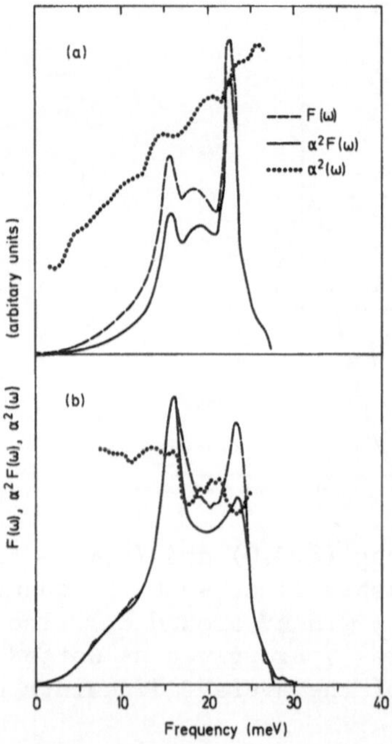

Fig. 23. Comparison of phonon densities of states $F(\omega)$ (dashed lines) and Eliashberg functions $\alpha^2F(\omega)$ (solid lines) for Nb. Also shown are the $\alpha^2(\omega)$ curves (dotted lines). Part (a) shows our NTB results, in (b) the experimental $F(\omega)$ curve (Gompf 1984) and the tunneling data of Geerk et al. (1982) are displayed.

Our results are not in agreement with the RMTA studies, where $\alpha^2(\omega)$ is found to be rather constant (Butler et al. 1977, Glötzel et al. 1979). They disagree even more with the tunneling data (Geerk et al. 1982) as shown in Fig. 23(b). In the experimental $\alpha^2(\omega)$ curve, obtained with the help of neutron data (Gompf 1984), a decrease of $\alpha^2(\omega)$ with increasing ω is observed.

In conclusion, the various theoretical and experimental results on the electron-phonon coupling in Nb are not terribly consistent. NTB and RMTA agree on average quantities like $\langle I^2 \rangle$, but diagree to some extend for spectral quantities like γ_{qj} or α^2F. The Crabtee et al. data are quite well explained by tight binding but not by RMTA. On the other hand the NTB $\alpha^2F(\omega)$ curve disagrees more with the tunneling data than the $\alpha^2F(\omega)$ results of RMTA. We point out here that a detailed discussion on the limitations of the RMTA has been given by Butler (1981).

3.2. The Refractory Compounds

3.2.1. Energy Bands, Bonding Properties

The electronic structure of the refractory compounds with rocksalt lattice like NbC, ZrC, TaC, HfC, TiC, TiN, NbN, etc. has been subject of many energy band calculations, starting with the paper of Bilz (1958), which already provided the correct qualitative description of the bonding properties. The most extensive studies have been carried out by the Vienna group (Schwarz 1971, 1975, 1977, Neckel et al. 1976). Other recent energy band computations have been performed by Gupta and Freeman (1976a,b) and by Klein et al. (1976a,b).

We have carried out NTB fits to the APW calculations of Schwarz (1971, 1977) for NbN and NbC, respectively, and to those of Neckel et al. (1976) for VN. Nine valence orbitals have been used, nonmetal s and p and TM d-orbitals. The NTB parameters for the energy bands of these compounds are given in Table A10.2. We have also obtained NTB fits for the NbC calculations of Gupta and Freeman (1976 a,b) and Klein et al. (1976 a,b). Although there exist some discrepancies between the three calculations for NbC - especially in the ordering of the Γ point states -, these differences are hardly visible in the NTB matrix elements, and, as a consequence, in the phonon calculations.

The energy bands of the refractory compounds are very similar to each other. A typical band-structure is that of NbC as shown in Fig. 24 (Schwarz 1977). In Fig. 25, the electronic density of states of NbC is displayed, and Fig. 26 shows the relevant energy transfer integrals. Non-metal (NM) s orbitals form the lowest valence band of NbC. They hybridize to some extend with TM $d(e_g)$ orbitals (see Fig. 25). In addition, they interact with the TM s orbitals and cause the shift of the TM s band to energies far above E_F. With increasing NM electronegativity; i.e., when going from the carbides via the nitrides to the oxides, the NM s-band splits off more and more from the other valence bands and becomes increasingly core-like.

The key to the understanding of the unusual stability of the refractory compounds - very high melting points, enormous hardness etc. - is the strong hybridization between NM p- and TM d-orbitals. In NbC, the three p bands above the NM s band actually have only 55 % p-orbital character, the $d(t_{2g})$ component is 25% and the $d(e_g)$ component is another 20%. The strong covalent bonding between p and $d(e_g)$ orbitals via the energy transfer integral $(pd\sigma) \approx 2.5$ eV and p and $d(t_{2g})$ orbitals via the integral $(pd\pi) \approx 1.2$ eV (see Fig. 26) is the source of the enormous stability of these materials. In NbC, the TM orbital energy E_d is only ≈ 1.5 eV above E_p of the NM (see Table A10.2), which explains the almost 1 : 1 admixture of p- and d-orbitals.

With increasing electronegativity difference, the separation of the orbital energies E_d and E_p becomes larger (e.g. ≈ 2.5 eV for NbN) and the p-d hybridization is less effective. As the bonding looses its covalent character, the nitrides and oxides are less stable than the carbides.

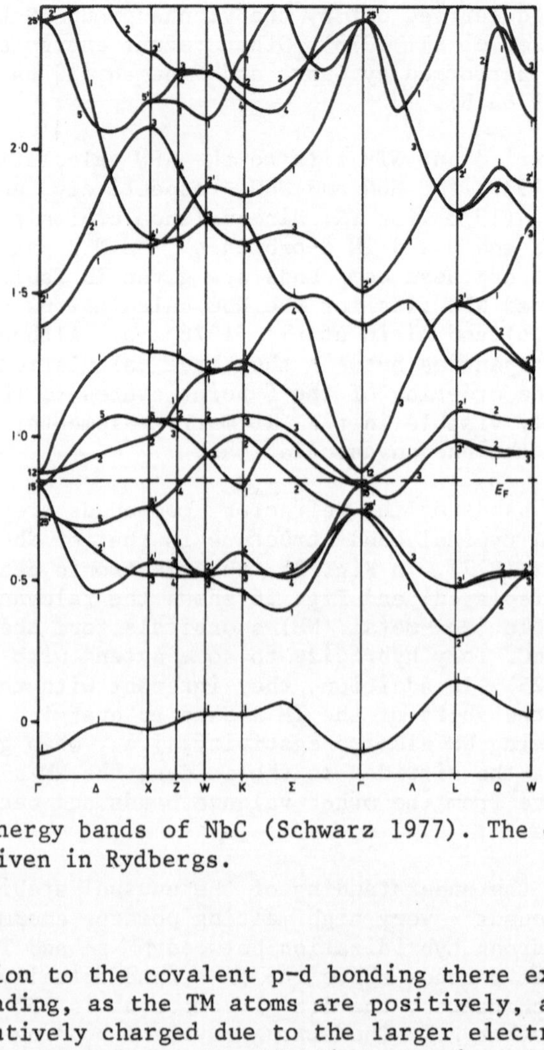

Fig. 24. Energy bands of NbC (Schwarz 1977). The energies are given in Rydbergs.

In addition to the covalent p-d bonding there exists also some ionic bonding, as the TM atoms are positively, and the NM atoms are negatively charged due to the larger electronegativity of the NM atoms. The charge transfer is approximately 1 electron/atom.

The TM d bands are well separated from the NM p bands. In the carbides and nitrides, there is an intermediate region of very low electronic density of states, as the p- and the d-bands overlap to some extend, especially near Γ. Materials with 8 valence electrons (VE) like ZrC have their Fermi energy in this low density region. In the case of the oxides, the p- and d-bands are completely separated, so that a gap exists.

In contrast to the TM elements, the e_g and t_{2g} components of the d-bands are located in different energy regions. The highest d-bands are made up predominantly of $d(e_g)$ orbitals which

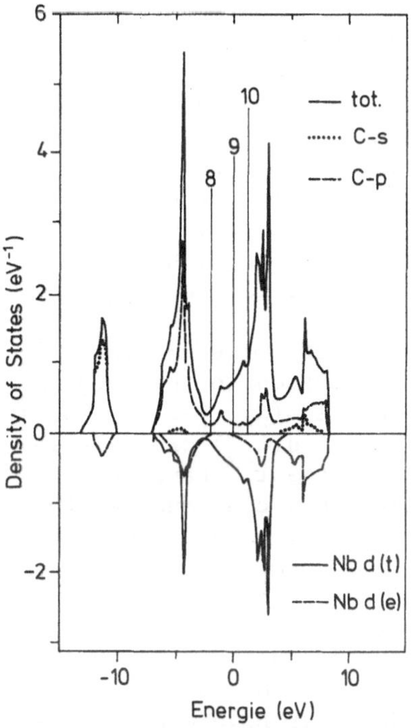

Fig. 25. Electronic density of states for NbC, as obtained from the nine orbital NTB fit to the APW calculations of Schwarz (1977). Also shown are the partial density of states of C s and p and of Nb d(t_{2g}) and d(e_g) orbitals. The positions of E_F for 8,9, and 10 valence electrons, is indicated (corresponding to ZrC, NbC and NbN respectively).

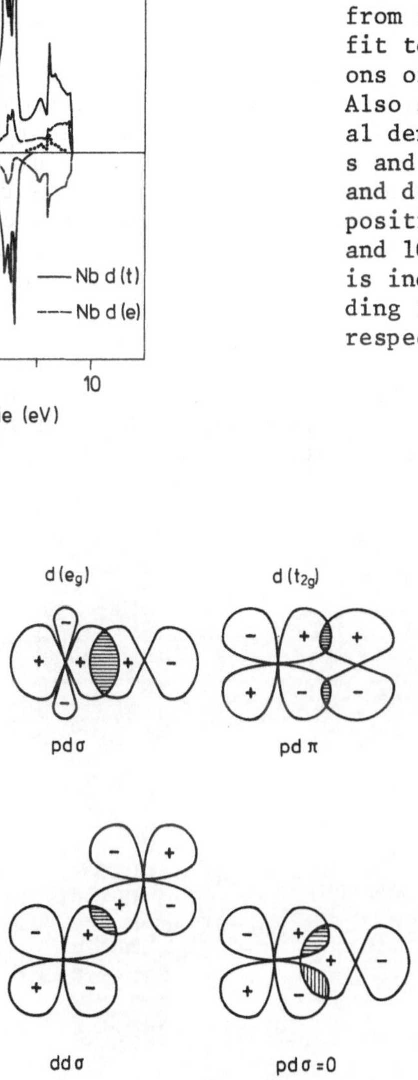

Fig. 26. Relevant energy transfer integrals for refractory compounds. Note that, because of symmetry, there is no σ-type p - d(t_{2g}) coupling.

form anti-bonding states both with the C-p and C-s orbitals (see Fig. 25). Because of these large bonding-antibonding band splitting - due to the big integrals (pdσ) and (sdσ) ≈ 2.0 eV - there is hardly any $d(e_g)$ orbital character in the lower d-bands where E_F of NbC with 9 VE and NbN (10 VE) is situated. These d-bands near E_F consist mainly of $d(t_{2g})$ orbitals with some C-p admixture. The $d(t_{2g})$ bands may be called non-bonding bands, since the relatively weak anti-bonding (pdπ) interaction and the bonding (ddσ) coupling of adjacent $d(t_{2g})$ orbitals are of very similar magnitude. We note that there is no σ-type $p\text{-}d(t_{2g})$ coupling possible (see Fig. 26).

As the t_{2g} orbitals have a planar symmetry, the coupling of the orbitals in one plane is very strong compared to the coupling between planes. Each of the three t_{2g} orbitals (xy, yz, zx) spans a plane orthogonal to each other. Coupling of $d(t_{2g})$ orbitals in orthogonal planes is also very weak so that the t_{2g}-bands have a very pronounced two-dimensional character. As a consequence, the Fermi-surface of NbC consists of three mutually perpendicular cylinders with junctions at the Γ points thus forming a "jungle gym" (see Fig. 27). Actually, there exist two interpenetrating "jungle gym" networks, one intersecting at even integer reciprocal lattice vectors and the other at odd integer vectors. When we compare the shapes of the Nb and NbC Fermi surfaces (Figs. 18 and 27), it becomes evident that the energy bands of NbC near E_F are much more anisotropic than those of Nb.

Fig. 27. "Jungle Gym" Fermi surface of NbC (band 5) in an extended BZ scheme. The cylindrical rods intersect at even integer (hkl) Γ points. A second "jungle gym" shifted by (1,1,1) is not shown.

A remarkable feature of all refractory compounds is their tendency to form vacancies. For instance, NbC is stable in the rocksalt structure over a range $NbC_{0.75}$ to $NbC_{0.99}$. While in the 8 and 9 VE materials only NM vacancies are formed, there appear also TM vacancies in the 10 VE and even more so in the 11 VE compounds. For instance, TiO is found in the nominal composition of $Ti_{0.85}O_{0.85}$, and NbO even forms superlattices of both types of vacancies in the composition $Nb_{0.75}O_{0.75}$. The chemical origin for this tendency to form vacancies is to date not well understood. It has been shown that vacancy states or bands appear in the gap region between p and d bands (Ries and Winter, 1980, Bohnen et al. 1980, Wimmer et al. 1982). Another aspect of this puzzle is that it is impossible to suppress the vacancy concentration below a certain limit (1% in NbC, a few percents in NbN).

When the $d(t_{2g})$ bands become partially occupied as in the case of NbC (9 VE) or NbN and VN with 10 VE, the refractory compounds exhibit high superconducting transition temperatures (T_c = 11 K for NbC, T_c = 16 K for NbN, and $T_c \approx$ 18 K for $NbC_{0.3}N_{0.7}$). In addition, the phonon dispersion curves show remarkable dips in some regions of the BZ (see Figs. 6 and 8). For comparison 8 VE materials like ZrC or HfC are not superconductors (T_c < 0.4 K), and phonon anomalies are absent.

3.2.2. Phonons : the "Dormant Interaction"

The phonon dispersion curves for NbC, NbN and VN (Weber 1980 a,b, Weber et al. 1979) have been calculated in a way very similar to that for the Nb-Mo system (see Sect. 3.1.2). A nine orbital basis has been used (s and p orbitals at NM atoms, d orbitals at TM atoms). The derivatives of the NTB matrix elements have again been determined from atomic overlap calculations, using both neutral and ionic configurations of adjacent atoms. In addition, we have carried out NTB fits to the APW calculations of Schwarz (1971) for NbN, where the energy bands for two different lattice constants have been given. The gradients, as obtained from these fits, agree within 10 - 20% with the atomic values from both neutral and ionic configurations. It is noteworthy that the latter values are very similar. In Table A10.3, the values of the NTB gradients are given for NbC, NbN, and VN. We note that we have put E_F = 0, then it is a good approximation to neglect all S terms in Eq. (2.16). To represent the terms (D_0 + D_1), we have used five 1nn and 2nn force constants ; i.e. two NM-TM and three TM-TM force constants, those between 2nn NM atoms have been neglected.

The acoustic phonon dispersion curves for NbC are shown in Fig. 28. As for the Nb-Mo system, all anomalies are correctly reproduced, both in the longitudinal and the transverse branches. The positions of the anomalies are near (0.65,0,0) along the Δ line, near (0.55,0.55,0) along Σ, and at (0.5,0.5,0.5), the end point L

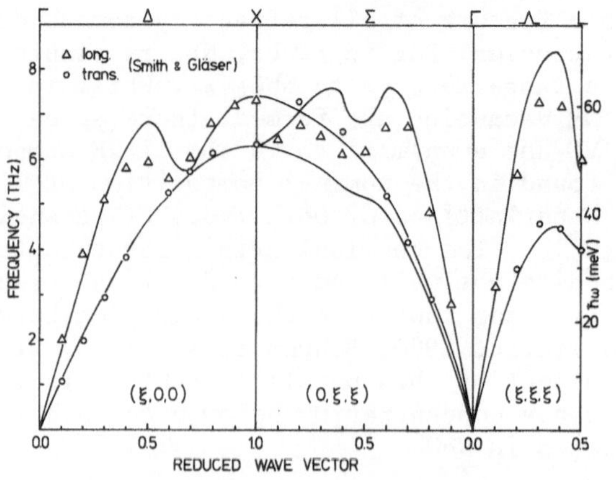

Fig. 28. Acoustic phonon dispersion curves for NbC (Weber 1980a). Experimental data are taken from Smith and Glaser (1971).

of the Λ line. These positions can be traced back to spanning vectors across the cylindrical rods of the NbC Fermi surface (see Fig. 27) in the vicinity of the X point. We note that, in contrast to Nb (or NbN), the bare susceptibility $\chi_0(q)$ (Eq. 3.1) also shows some structure at these positions, as has been pointed out by Gupta and Freeman (1976 a,b), and Klein et al. (1976 a,b). However, the dominant contributions to D_2 arise again from the electron ion form factors $g^{\alpha}_{kk'}$.

Using the velocity difference approximation for $g^{\alpha}_{kk'}$, (Eq. 3.4) it is easy to see why all branches except the transverse acoustic modes with Δ_5 and Σ_4 symmetry exhibit the anomalies. Let us look at a spanning vector $q_a \approx (0.65,0,0)$ across a cylindrical rod (parallel to the X-W line as indicated in the energy contour plot of the "jungle gym" band in Fig. 29). Only those components of the band velocities v^{α}_k, $v^{\alpha}_{k'}$ which are parallel to q_a, are large and of opposite sign. The components v^{β}_k, $v^{\beta}_{k'}$ perpendicular to q_a (and parallel to the Γ-X line) are very small. As a consequence, only the longitudinal Δ_1 modes but not the transverse Δ_5 modes show anomalous features. Similar arguments hold for the other symmetry lines Σ and Λ. For Σ_1 the spanning vectors $q_a \approx (0.55,0.55,0)$ also lie perpendicular to the Γ-X line, but no longer parallel to the X-W line (instead they have a 45° angle to X-W). For Λ_1 the spanning vectors $q_a = (0.5,0.5,0.5)$ also have a component along the Γ-X line. In these cases not only longitudinal but also transverse phonon polarizations may couple large and opposite band velocity components. Only the transverse Σ_4 modes, these are (0,0,1) displacements for $q = (\xi,\xi,0)$ couple to the same small velocity components v^{β}_k, $v^{\beta}_{k'}$ parallel to the Γ-X line, as do the Δ_5 modes (both the Δ_5 and Σ_4 modes end at the same elastic constant c_{44} at long waves). From this discussion it may also become evident that the minima in the dispersion curves are not confined to the symmetry directions Δ, Σ, and Λ, but occur also in intermediate regions. We finally note that the anomalous contributions to D_2 involve states k,k' in the energy region of $\approx \pm 1$ eV around E_F. The coupling of states $\epsilon : E_F$ leads, for the same reasons, to large phonon linewidths (see Sect. 3.2.3).

So far, our analysis has been very similar to that of Sect. 3.1.2 in the case of Nb. It is remarkable however, that the anomalies are much more pronounced in NbC than in Nb. On the other hand the electronic density of states at E_F is only half that of Nb, yet T_c (NbC) = 11 K is even larger than T_c (Nb) = 9 K. All these features lead to the conclusion that the electron-phonon coupling in NbC is much stronger than in Nb. In the following we will show that this enhancement is caused by a new component in the electron-phonon interaction which is absent in Nb.

For this we focus again on the LA anomaly near $\vec{q}_a = (0.65,0,0)$, as displayed in the contour map of Fig. 29. The many energy contour lines parallel to the Γ-X direction indicate large and opposite band velocity components v_k^α, $v_{k'}^\alpha$, for all spanning vectors $\vec{q}_a$ across the cylindrical rod of the Fermi surface, not just near the X point.

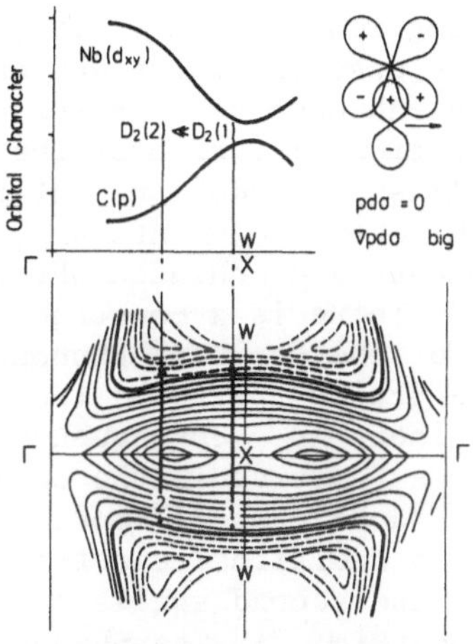

Fig. 29. Contour plot in the Γ-X-W plane of that energy band of NbC which causes the phonon anomalies. Spanning vectors for the LA anomaly at (0.65,0,0) are indicated for areas 1 and 2. Also shown is the variation of the orbital character for states along the Fermi surface rod between Γ and X. The contributions to D_2 from areas 1 and 2 vary strongly because of the influence of the "dormant" pdσ interaction, for which the pdσ integral is zero, but the gradient is very big (Weber 1980b).

Therefore we have expected to find equally large scattering contributions to D_2 both from the area 1 near the W-X-W line and from regions closer to Γ such as area 2. Not only are the band velocity differences similar, also the phase space of area 1 and that of area 2 are comparable, as the velocity components perpendicular to $\vec{q}_a$ are similarly flat. However, we find a strong decrease - about a factor 4 - in the contributions to D_2 when we move the spanning vector $\vec{q}_a$ along the rod from the X point area towards Γ. This result cannot be understood from the velocity difference approximation for $g_{kk'}$ which we have found to be a good qualitative description for Nb. In the upper part of Fig. 29, we have plotted the orbital character of the states on the Fermi line of the contour map, moving parallel to the Γ-X direction. Near Γ the $d(t_{2g})$ or d_{xy} orbitals dominate, however towards X, the NM p orbitals become almost equally important. Thus the scattering between states with large p-d hybridization seem to cause the enhancement in the D_2 contributions. As we have discussed in Sect. 3.2.1, the dominant energy transfer integrals for the bandstructure are pdσ, pdπ and ddσ, while there is no σ-type coupling of p and $d(t_{2g})$ orbitals. Although the latter orbitals penetrate strongly into each other, the overlap integral is zero at equilibrium positions, because of the symmetry of the lattice. This "dormant" interaction therefore does not at all influence the energy bands, and thus does not affect the band velocities. However, when the atoms are displaced from their equilibrium positions, in the way indicated at the top of Fig. 29, the pdσ coupling leads to very big gradients ∇H. Its influence on D_2 is quite dramatic : for NbC and similarly for NbN and VN, the total D_2 is enhanced by a factor of 3 to 4 over the (ddσ) and (pdπ) contributions ; i.e. over the "velocity difference" part. The latter as well as the "dormant interaction" part contribute about evenly to the matrix elements $g_{kk'}$ of area 1, so that the squared matrix elements are enhanced there by the factor 3 to 4 as compared to area 2. There is a constructive interference of the two contributions for the acoustic phonons ; for the optic modes the "dormant interaction" part and the (pdπ), (ppσ) parts appear to cancel each other. As a consequence, there do not occur any significant anomalous features in the optic modes. Here we want to recall that the "double shell" model (Weber 1973) employed a resonance mechanism via additional electronic shells to produce the phonon anomalies. These second shells have been placed only at the TM atom sites (see Fig. 7). The phonon dispersion curves of this model agree very well with experiment, yet the model predicts negligible NM components of the phonon polarization vectors at anomalous wavevectors. On the other hand, our tight binding calculations yield considerable NM components in the polarization vectors of the anomalous modes caused by the effects of the "dormant interactions". The neutron measurements do not only provide phonon frequencies but also data on the dynamical structure factors, where the phonon polarization vectors enter. It appears that the tight-binding results are in better agreement with experimental structure factors than the double shell model calculations (Pintschovius, 1979).

As we mentioned above, the optic modes do not show anomalous structures. These phonon branches look in fact rather like the optic phonons of alkali halides, i.e. they exhibit a large splitting of longitudinal and transverse branches which is caused by the fact that the TM and NM constituents are not neutral, but carry a charge $Z \approx \pm 1$. However, this Lyddane-Sachs-Teller-like splitting of the optic modes (see Sect. 1.1) cannot persist up to the Γ point, as we deal with a metal. There, the long wavelength macroscopic electric field is completely screened by the conduction electrons. Thus the LO and TO modes have to be degenerate at Γ. In the shell model study of the refractory compounds (Weber 1973), the dynamical charges have been screened in an approximate, Thomas-Fermi-like, manner. These screened Coulomb interactions provide a good description of the splitting of the optic modes. On the other hand, the acoustic modes are negligibly affected by these forces.

In tight-binding theory, the dynamical charges, and as a consequence the LO-TO splitting at Γ should originate from the crystal field terms (Eqs. A36 and A37) in D_2 and from corresponding terms in D_1 and D_0. When the two constituents are charged, these crystal field terms have long-range Coulombic components, which lead to the LO-TO splitting. In principle, these crystal field terms can be included in tight-binding calculations. There is however the additional problem that the dynamical charges are screened by the conduction electrons. This screening has so far not been incorporated in the tight-binding formalism. (In elemental crystals, there is no such screening problem as the atoms are neutral). An approximate way to include this effect would be a procedure similar to that used in the old shell model calculation.

At present, we have put zero all crystal field terms, and consequently we have avoided to deal with this problem. However, we could not expect that the optic dispersion curves are in equally good quantitative agreement with experiment as the acoustic modes. Thus we have only reproduced the center-of-mass values of the optic phonons by choosing appropriate TM-NM 1nn force constants representing $(D_0 + D_1)$, and have neglected the NM-NM 2nn force constants.

A challenging problem to both theory and experiment had been the form of the phonon dispersion curves of NbN ($T_c \approx 16$ K) and VN ($T_c \approx 9$ K). These materials have nominally ten valence electrons per unit cell, one more than NbC or TiN. Theory had been challenged because different explanations for the phonon anomalies had led to different predictions for NbN or VN. On the one hand, if the general idea of a screening resonance (Sinha and Harmon 1975, Hanke et al. 1976) had been correct, the positions of the anomalies in q-space should have remained constant. On the other hand, if the properties of the energy bands near E_F had caused the anomalies, one would have expected - because of band filling - a change in the anomaly positions. Indeed, calculations of D_2 as a function of valence electron concentration resulted in big changes between NbC and NbN or between TiN and VN (see Fig. 30). The maximum in $-D_2$ along $(\xi,0,0)$ shifts from $(\approx 0.65,0,0)$ to the X point at the zone boundary. In addition, this maximum becomes stronger and very sharp. In the inset of Fig. 30, it is indicated, how the change in band filling causes the shift in the anomalous phonon vector. As a consequence, a pronounced minimum at X should appear in the phonon dispersion curves (see Fig. 31). A further prediction of the tight-binding calculation had been that the L point anomalies in NbC should disappear when going to NbN.

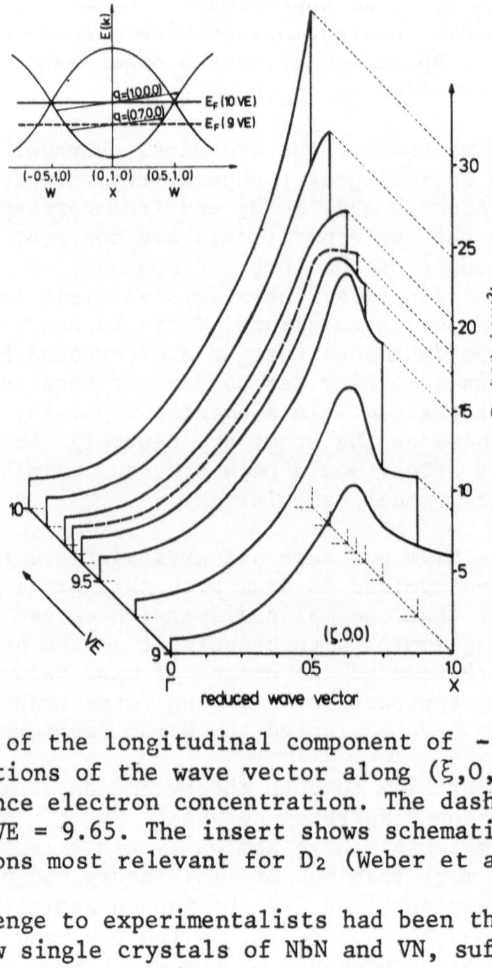

Fig. 30. Plot of the longitudinal component of $-D_2$ in VN as functions of the wave vector along $(\xi,0,0)$ and of the valence electron concentration. The dashed curve is for VE = 9.65. The insert shows schematically the transitions most relevant for D_2 (Weber et al. 1979).

The challenge to experimentalists had been the enormous difficulty to grow single crystals of NbN and VN, sufficiently large for neutron scattering studies. For VN there was the additional problem, that the scattering cross section of V is completely incoherent. When the nitride single crystal problem has been overcome, neutron scattering studies both for VN_x (Weber et al. 1979, x = 0.86) and for NbN_x (Reichardt and Scheerer 1979, x = 0.84, and Christensen et al. 1979, x = 0.93) have been carried out to measure the phonon dispersion curves of these materials. All the experimental results confirmed the predictions of the tight-binding theory, i.e. shift of the anomaly from (0.65,0,0) to X and the disappearance of the L point minimum (see Fig. 31 and 32).

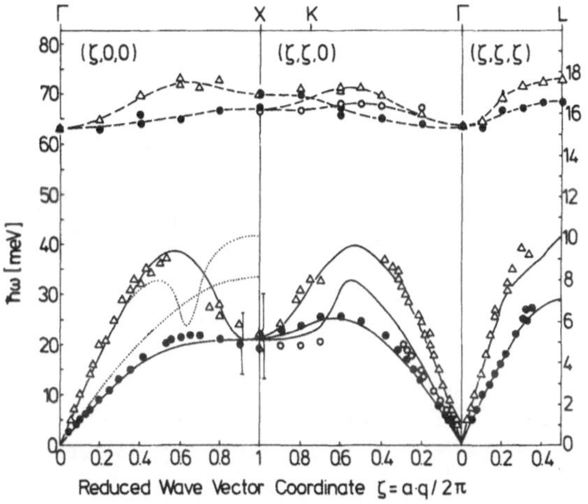

Fig. 31. Phonon dispersion curves for $VN_{0.86}$ (from Weber et al. 1979). Open triangles denote longitudinal phonons ; open circles show tranverse phonons. Bars indicate the full width at half maximum of phonons near point X. Full lines are calculated from the NTB theory ; broken lines are a guide to the eye. The dotted lines represent the dispersion curves of $TiN_{0.98}$ (Kress et al. 1978).

For NbN and VN, the theoretical minimum at X is too sharp. The source of this discrepancy lies probably in the non-stoichiometry of the $NbN_{0.84}$ and $VN_{0.86}$ samples. In order to estimate the influence of N vacancies on D_2, one may look at the change of D_2 as a function of valence electron concentration in a rigid-band-scheme (see Fig. 30). The numer of valence electrons in $NbN_{0.84}$ or $VN_{0.86}$ is ≈ 9.3. There, the peak in $-D_2$ is still far away from X. Best agreement with experiment is found for ≈ 9.65 valence electrons (see Figs. 31 and 32). Here the maximum in $-D_2$ has shifted to the X point but is still much broader than for 10 valence electrons.

Perhaps we can best understand this result by considering the effects of vacancies on band filling in two extreme limits : (i) in the free-electron limit N vacancies would lower E_F in strict accordance with a rigid-band model ; (ii) in the case of non-

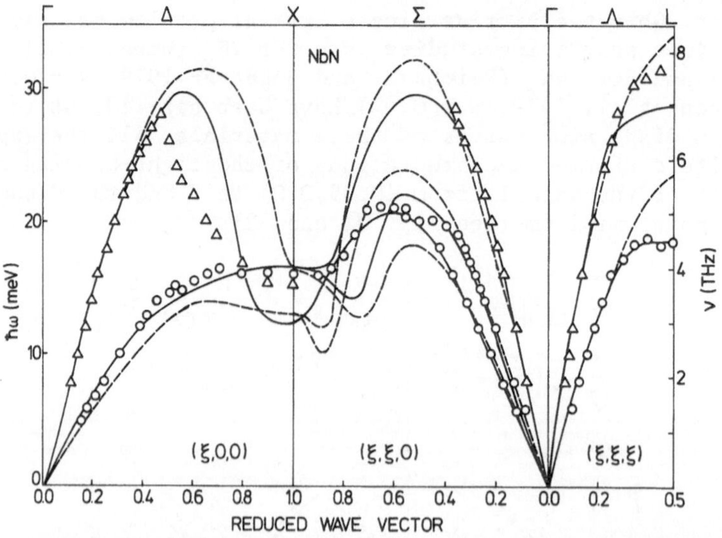

Fig. 32. Acoustic phonon dispersion curves of NbN. Experimental values for $NbN_{0.84}$ from Reichardt and Scheerer (1979). Dashed lines show predicted curves for fully stoichiometric NbN. Solid lines represent calculations which take into account the non-stoichiometry (Weber 1980a).

interacting NM s-p and TM d-bands, N vacancies would remove states from the s-p bands but would not affect E_F. In NbN_x and VN_x the situation is probably intermediate, as E_F lies within the T_{2g} bands which consist mainly of TM d_{xy}-type orbitals with some admixture of NM p orbitals.

A different explanation for the too sharp theoretical X point minimum is that the NTB gradients might exhibit some q-dependence. If they decreased considerably near anomalous phonons, we would also obtain a smoother phonon curve near X.

Another effect of the non-stoichiometry is probably the extremely large and often anisotropic phonon lineshapes observed in the anomalous phonon regions of VN (Weber et al. 1979, see Fig. 31) and NbN (Reichardt and Scheerer 1979). These lineshapes are much larger than the "intrinsic" ones due to electron-phonon coupling (see below) and often exhibit double peak structures. In non-stoichiometric NbC_x, a similar observation has been made, while in stoichiometric NbC, i.e., in $NbC_{0.98}$, the phonon lineshapes do not show this feature (Smith and Gläser 1971). Also, non-stoichiometric ZrC_x or other 8 valence electron compounds do not exhibit big linewidths. In a model calculation using a simplified version of the double shell model, Splettstößer (1977) has shown that large linewidths occur when the resonance-like coupling of the electronic shells is perturbed by vacancies. A microscopic theory of this phenomenon has not been presented up to now.

3.2.3. Electron-Phonon Coupling

Calculations of the "intrinsic" phonon linewidths for NbC are shown in Fig. 33. Similarly as for Nb (see Fig. 22) the values of the linewidths γ follow very closely the values of $D_2^<$, the contributions to D_2 from scattering processes $\approx 0.5 - 1.0$ eV around E_F. In contrast to Nb, large values of γ occur only in anomalous phonon regions. As a consequence, the average γ_{av} of all phonon linewidths is much smaller than in Nb. This very strong anisotropy in the electron-phonon coupling is actually twofold : not only a few selected phonon modes exhibit a strong coupling to the electrons, in addition only a few k-states at E_F (only a fraction of the already rather low Fermi density) are strongly coupled wia the anomalous phonons. In NbC, only the states on the cylinders near the X points have large values of $g^{\alpha}_{kk'}$, coupling values of all other Fermi sur-

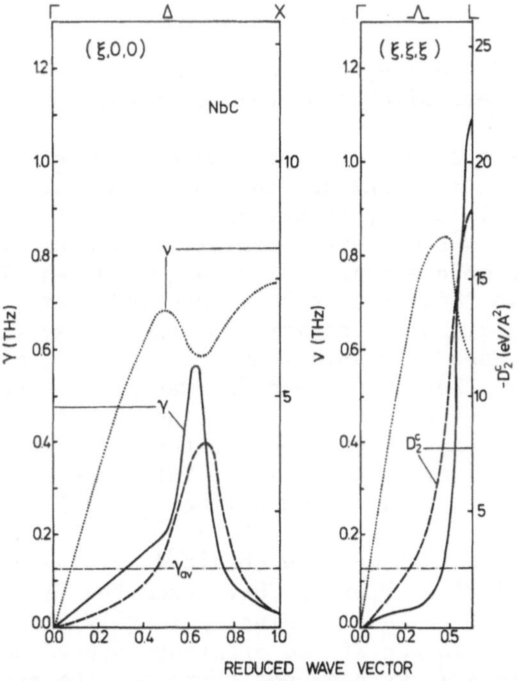

Fig. 33. Longitudinal acoustic phonons ν, $-D_2^<$(Nb-Nb) and phonon linewidths γ along two symmetry directions in NbC. $D_2^<$ includes scattering processes $\lesssim \pm 0.5$ eV around E_F. Also shown in γ_{av}, the average value of all linewidths γ_{qj} in the Brillouin zone (Weber 1980a).

face states are very small. Similar results are obtained for NbN or VN.
Altogether, the anisotropy of the electron-phonon coupling is much larger in the refractory compounds than in Nb. This may have an interesting consequence of the superconducting temperature T_c. In Nb, anisotropy correction in the Eliashberg equations have been found to increase T_c by only ≈ 0.25 K (Peter et al. 1977). Possibly, the increase of T_c in the refractory compounds is much larger than in Nb, if the number of vacancies could be lowered to such an extend that the dirty limit approximation for the Eliashberg equations is nog longer valid. Estimates are that this would occur for 0.1 % vacancy concentration or less (Schopohl, 1982), which means a formidable problem to experimentalists. Until now, in NbC only the composition $NbC_{0.99}$ has been accomplished by ion implantation and consequent annealing. This improved stoichiometry has led to a 0.5 K increase of T_c over the 11.2 K value of $NbC_{0.98}$ (Geerk and Langguth, 1977).

Calculations of the Eliashberg function $\alpha^2F(\omega)$, and of the phonon density $F(\omega)$ for NbC are shown in Fig. 34(a). These results are obtained from the same NTB calculations which yielded the phonon dispersion curves of Fig. 28. The theoretical coupling function $\alpha^2(\omega)$ shows two pronounced peaks at those frequency regions where the transverse and the longitudinal anomalies of NbC occur. Like in Nb, $\alpha^2(\omega)$ also exhibits an overall increase towards larger ω.

Fig. 34(b) shows the corresponding experimental data, both $F(\omega)$ from neutron scattering and the tunneling data (Geerk et al. 1975). For the $F(\omega)$ curves, theory and experiment agree fairly well. However, the $\alpha^2F(\omega)$ curves exhibit severe discrepancies, which become even more evident in the $\alpha^2(\omega)$ curves. In the experimental $\alpha^2(\omega)$ curve, only one peak is observed, yet it does not occur at the (experimental) position of the transverse anomalies. The longitudinal peak is missing, and, in general, $\alpha^2(\omega)$ decreases with increasing ω. These discrepancies may in part be caused by a possibly poor quality of the NbC tunneling junctions. On the other hand, the Nb junctions, which produced the tunneling data of Fig. 23(b), have been of very good quality. Yet in Nb, similar discrepancies in the theoretical and experimental $\alpha^2(\omega)$ are found.

We finally note that we have obtained a value of $\lambda = 0.45$ in this $\alpha^2F(\omega)$ calculation. There we have used the NTB gradients of Table A10.3, which have been obtained primarily from atomic calculations. In NTB the values of the gradients depend on the energy zero (see Appendix A9). We have not corrected the gradients when choosing $E_F = 0$ in our phonon calculations. These corrections should lead to $\lambda \approx 0.55$, a value somewhat smaller than the result $\lambda = 0.64$ obtained from RMTA calculations (Winter et al. 1978), and also smaller than the empirical value of 0.7.

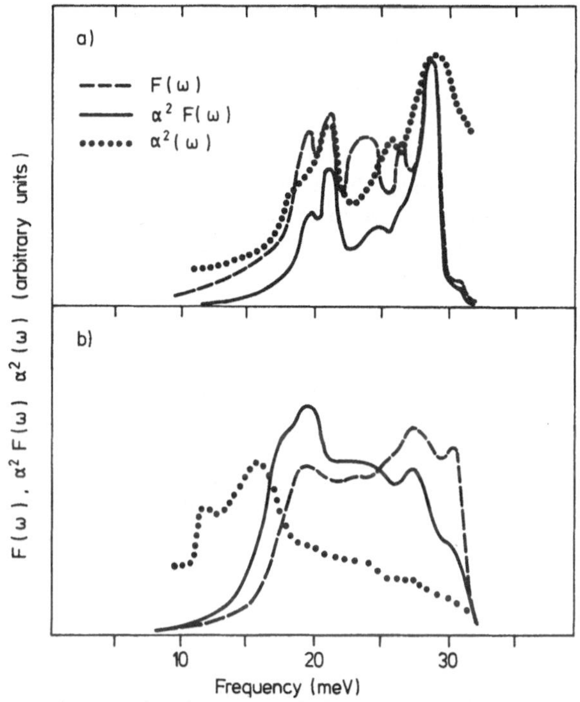

Fig. 34. Comparison of phonon densities of states F(ω) and Eliashberg functions $\alpha^2F(\omega)$ for NbC. Part (a) shows NTB results, in part (b) the neutron and tunneling data of Geerk et al. (1976) are displayed.

3.3. The A15 Compound Nb_3Sn

Various A_3B compounds with the A15 structure, such as Nb_3Sn, Nb_3Ge, Nb_3Ga, Nb_3Al, V_3Si, V_3Ga and others exhibit a number of very unusual physical properties (for reviews see Testardi 1973 and 1975). These compounds not only have the highest superconducting temperatures so far observed in any class of materials, some of them like Nb_3Sn and V_3Si also undergo a cubic-to-tetragonal phase transition at low temperatures, and show mode softenings as precursor effects of this transition. In addition, a strong temperature dependence of the magnetic susceptibility is observed in the cubic phase. The most prominent compound of this family is Nb_3Sn, which will be dealt with in the following.

3.3.1. Energy Bands

A. The Flat Γ_{12} Bands near E_F

The electronic structure of the A15 compounds has provided a continuing challenge to band theorists ever since 1961, when

Clogston and Jaccarino suggested that the anomalous physical properties of A15 materials are due to extremely fine structure in the electronic density of states N(E) near the Fermi energy E_F. Most early speculations (Weger 1964 ; Labbé and Friedel 1966) concerning the origin of this fine structure in N(E) focussed on a geometrical feature of the A15 structure, namely, the existence of nonintersecting A-atom chains along the three coordinate axes (see Fig. 35). Model calculations based on a nearly one-dimensional band structure enjoyed considerable success in explaining the anomalous properties of V_3Si and Nb_3Sn.

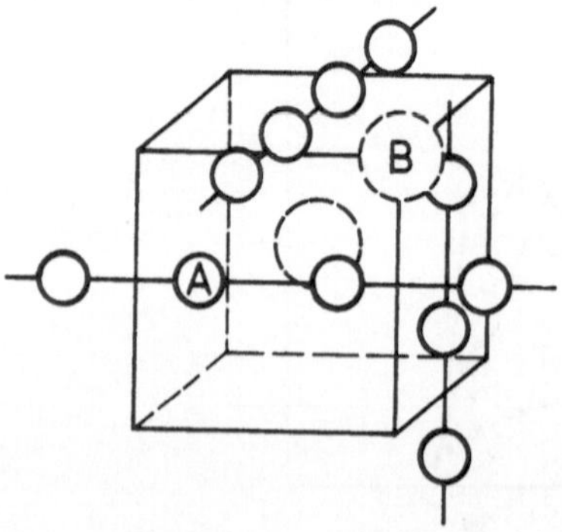

Fig. 35. Structure of the simple cubic unit cell of A15 compouns. The A atoms, shown as solid circles, form three linear, nonintersecting chains. The B atoms (dashed circles) form a bcc lattice.

On the other hand, accurate first-principles calculations of the electronic structure of these A15 compounds have been hindered by the complexity of the crystal structure, with its two A_3B molecules per unit cell. For example, the earliest APW calculations by Mattheiss (1965) showed that approximately 400 APW's were required in order to obtain eigenvalues which were converged to the 1-mRy range. This precluded calculations at general wave vectors $\vec{k}$ in the Brillouin zone (BZ), though calculations were feasible at symmetry points and along symmetry lines where the corresponding secular equation could be block-diagonalized.

However, recent improvements in band-structure methods and computer capabilities have produced a number of rather accurate first-principles band-structure and N(E) calculations for a variety of A15 compounds (Mattheiss 1975 ; Jarlborg and Arbmann, 1977 ; van Kessel et al., 1978 ; Klein et al., 1978 ; Pickett et al., 1979 ; Ho et al., 1979). These have involved varied combinations of band-structure methods and interpolation techniques. By far the most extensive study is that of Klein et al., who have carried out selfconsistent APW calculations for a total of ten V_3X and Nb_3X compounds (X = Al, Ga, Si, Ge, and Sn). The Klein et al. results for Nb_3Sn, Nb_3Ge, and Nb_3Al are quite similar to those of van Kessel et al. (Nb_3Sn) and Pickett et al. (Nb_3Ge and Nb_3Al).

A feature common to all first-principles energy-band results for these A15 compounds is the presence of two relatively flat bands near E_F which originate from a doubly degenerate Γ_{12} state at the BZ center (see Fig. 36). Near Γ, these bands are well isolated from other bands, yet at the BZ boundary, especially near M and R, a number of other bands with rather large dispersion are present and intersect the Γ_{12} bands. The energy band calculations of the various groups differ mostly in the ordering of the various M and R point states, probably because the different band-structure methods produced slightly different atomic potentials. Also, in the Klein et al. calculations for the various A15 compounds, the variation of A or B atom components results in a re-ordering of the M and R point states near E_F, but leaves the isolated Γ_{12} bands rather unchanged. It should be noted that already the first APW calculation of Mattheiss (1965) predicted the occurrence of a Γ_{12} state near E_F in conflict with the one-dimensional models (Weger 1964, Labbé and Friedel 1966), which predicted triply degenerate states at Γ, due to the existence of the three non-intersecting chains.

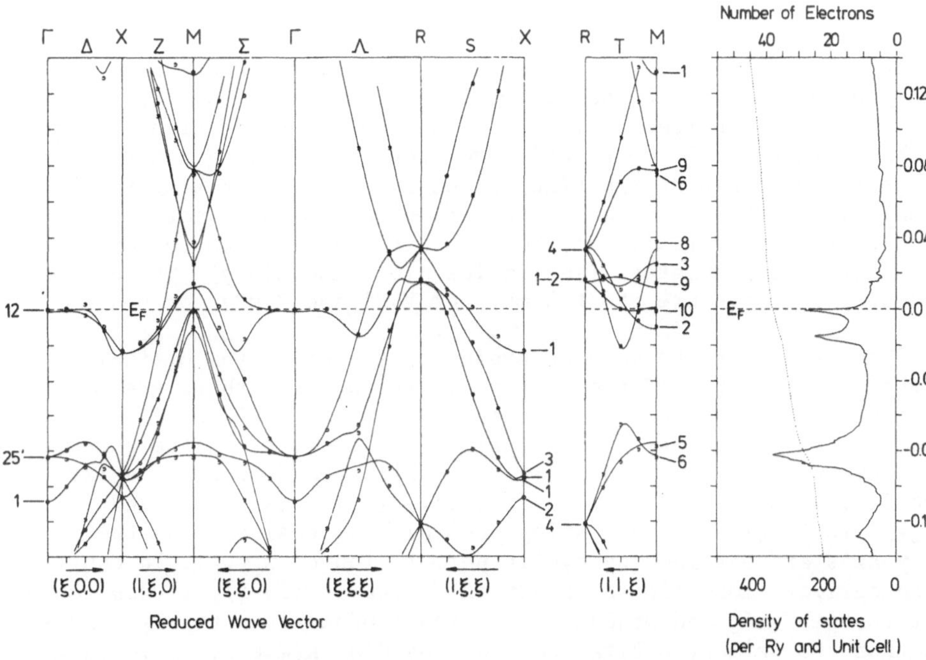

Fig. 36. Energy bands and electronic density of states for Nb_3Sn near E_F. Full lines show the NTB results of Mattheiss and Weber (1982), the small circles depict the APW results of Klein et al. (1978). Horizontal lines indicate the Fermi energies for Nb_3Sn and Nb_3Sb (in a rigid band model).

For most compounds, the Γ_{12} bands produce a pair of sharp peaks in N(E) (see Fig. 36). The peak at higher energy arises from states in the region near Γ, where the Γ_{12} bands are particularly flat over a considerable fraction of the BZ volume, irrespective of wave-vector direction. In the case of the compounds such as V_3Si and Nb_3Sn which undergo a martensitic transformation to a tetragonal phase at low temperatures, this flat portion of the Γ_{12} bands lies within a few mRy of E_F. As a result, the higher-energy N(E) peak very nearly coincides with E_F in both V_3Si and Nb_3Sn.

The lower peak arises from the saddle point area near (0.6, 0.6,0.6) in the Λ direction. In materials like Nb_3Ga or Nb_3Al, E_F is situated in the vicinity of the lower peak. This indicates that for a certain range of valence electron (VE) concentration (4.5 to 4.75 per atom) E_F is pinned in the region of the flat Γ_{12} bands. In this range of 4.5 to 4.75 VE's all high T_c A15 compounds are found. In Nb_3Sb ($T_c \approx 0.2$ K) with 5 VE's per atom, E_F has been pushed above the Γ_{12} bands and lies in a low N(E) region (see Fig. 36).

Obviously the Γ_{12} bands and their orbital character are a key to the understanding of many of the unusual properties of the A15 compounds. A tight-binding analysis of the first principles energy band results by Mattheiss and Weber (1982) provided new insight regarding the nature of the Γ_{12} bands. This analysis was based on highly accurate fits to the energy bands of Klein et al. (1978) for V_3Si and Nb_3Sn (see Fig. 36). Altogether 40 bands at 35 points in the BZ (about 1400 energies) have been fitted using a NTB scheme with s,p,d orbitals at the V(Nb) sites and s,p orbitals at the Si(Sn) sites. This resulted in 62 dimensional H and S matrices, which contained approximately 80 - 100 independent two-center energy and overlap parameters (see Tables A10.4 and A10.5). The corresponding rms errors are about 3 mRy for V_3Si and 2 mRy for Nb_3Sn.

The Γ_{12} bands are made up predominantly of A atom orbitals. As the site symmetry of the A atoms is tetragonal, there are more different orbital components present than for a cubic site symmetry. With respect to an A-atom in a chain along the z-axis, the p components are $p(\pi) \equiv (x,y)$ and $p(\delta_1) \equiv z$, while the d orbitals split into $d(\sigma) \equiv (3z^2 - r^2)$, $d(\pi) \equiv (xz,yz)$, $d(\delta_1) \equiv (x^2 - y^2)$ and $d(\delta_2) \equiv (xy)$. (The corresponding orbitals for chains along the x and y axes are obtained by cyclic permutation of x,y,z). In contrast to an earlier, more limited study of Mattheiss (1975), it was found that in both V_3Si and Nb_3Sn, the dominant orbital in the Γ_{12} bands and thus the primary orbital near E_F has $d(\sigma)$ symmetry, involving $(3z^2 - r^2)$ orbitals directed <u>along</u> the chains (see Fig. 37). Second in importance are the $d(\pi)$ or (xz,yz) orbitals, while the $d(\delta_1)$ orbitals - which in the study of Mattheiss (1975) have been proposed to be principal component of the Γ bands - actually contribute to a lesser extent. In addition, a 10 % A-atom p contribution is found, in agreement with the results of Klein et al. (1978).

Orbital	Fraction in % at Γ_{12}	E_F
$d(\sigma) \equiv (3z^2-r^2)$	60	38
$d(\pi) \equiv (xz, yz)$	-	22
$d(\delta_1) \equiv (x^2-y^2)$	22	16
$p(\delta_1) \equiv z$	15	8
others	3	16

Fig. 37. Orbital character of the Γ_{12} state and of the bands at E_F. Due to symmetry the Γ_{12} state does not have any $d(\pi)$ component (Mattheiss and Weber 1982).

This p contribution consists mainly of $p(\delta_1)$ orbitals which again are directed along the A atom chains. The strongest admixture of $p(\delta_1)$ orbitals occurs at the Γ_{12} state and in the very flat band region around Γ, this is the region of the upper of the two peaks in N(E) produced by the Γ_{12} bands. A similar observation is made for the $d(\delta_1)$ orbitals, these are also more strongly present in the region near the Γ_{12} state (see Fig. 37).

The dominant orbitals in the Γ_{12} bands point along the chain directions ; i.e. there is a very large overlap between nearest neighbor orbitals ($dd\sigma$, $dd\pi$, $pd\sigma$). Thus one would expect large band dispersion, but, surprisingly, the Γ_{12} bands are very flat.

The principal source of the Γ_{12} band flattening is the interaction of the $d(\sigma)$ and $d(\pi)$ subbands (see Fig. 38). The bands, made up from either $d(\sigma)$ or $d(\pi)$ orbitals alone, are plotted along the Λ and Δ lines in Fig. 38(a) and (c), respectively. In Fig. 38(a), the overall $d(\sigma)$ bandwidth is primarily determined by the nearest-neighbor energy and overlap parameters ($dd\sigma$) and [$dd\sigma$] along the axes of the A-atom chains. The $\Gamma_{12} - \Gamma_1$ separation reflects the interchain coupling of the $d(\sigma)$ orbitals. Due to the large overlap all σ-bands have strong dispersion, including the bands which arise from the Γ_{12} state. A similar situation is found for the $d(\pi)$ subbands of Fig. 38(c). Again, there is a large bandwidth due to the $(dd\pi)_1$ energy integral. In addition, the interchain coupling is also very large, as is manifest in the splitting of the Γ_{15} and Γ_{25} states. Due to symmetry, all $d(\pi)$ states are triply degenerate at Γ, so that no Γ_{12} state occurs.

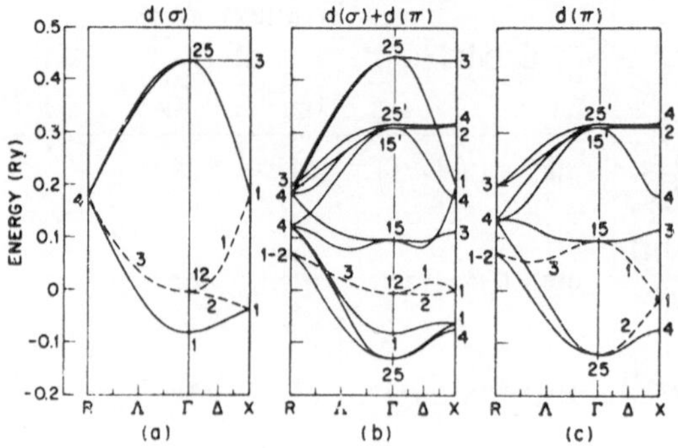

Fig. 38. Effect of d(σ) — d(π) hybridization on reducing the Γ_{12} subband width. Panels (a) and (c) show the independent d(σ) and d(π) bands, respectively, while (b) includes the σ-π interaction. The Γ_{12} subbands are identified by the dashed curves (Mattheiss and Weber 1982).

The interaction of the d(σ) and d(π) orbitals produced a significant flattening of the Γ_{12} bands (see Fig. 38(b)). Away from the zone center, these bands now become strongly hybridized, involving comparable d(σ) and d(π) components. This flattening is further enhanced when the other orbitals are included.

We note that the σ-π hybridization occurs only between orbitals on different chains. The intrachain coupling of d(σ) and d(π) orbitals is zero because of symmetry (see, e.g. Fig. 37). This zero σ-π intrachain coupling will present another case of a "dormant" interaction for electron-phonon coupling (see Sects. 3.3.2 and 3.3.3). As the σ-π hybridization is an interchain effect, the sharp structure in the N(E) curves arising from the flat Γ_{12} bands is not at all an effect of one-dimensional energy bands.

B. The Energy Bands of Tetragonal Nb_3Sn

Based on NTB results for cubic Nb_3Sn, an energy band calculation of tetragonal Nb_3Sn (see Fig. 39) has been carried out by Weber and Mattheiss (1982). In this calculation, the NTB matrix elements for tetragonal Nb_3Sn have been obtained from the corresponding cubic values with some provisions to account for the slight changes in bond-lengths. A principal feature of the tetragonal results in the 6-8 mRy splitting of the flat Γ_{12} bands near E_F. This causes the sharp Γ_{12} peak in N(E) to be split symmetrically into two subpeaks, one of which is raised above E_F (see Fig. 40). The splitting of the Γ_{12} state near E_F due to the tetragonal distortion has also been confirmed by non-selfconsistent APW calculations. The energy gain due to the tetragonal distortion is ≈ 40 K, of the order of the structural transition temperature, but much smaller than the ≈ 1000 K splitting of the Γ_{12} state. The

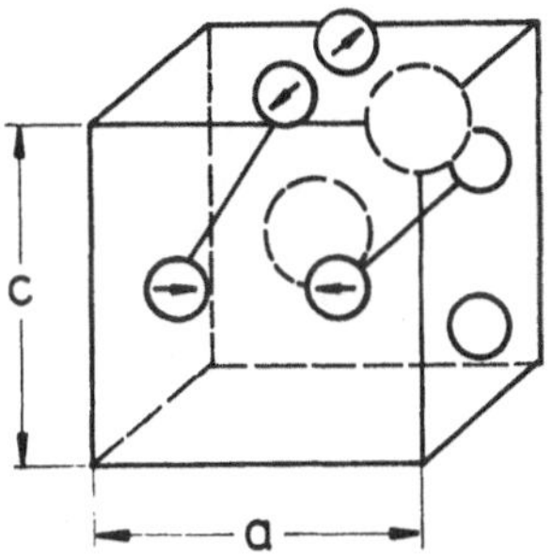

Fig. 39. Tetragonal distortion of the A15 structure. Apart from the tetragonality $\varepsilon = c/a-1$, there is a second new lattice parameter δ, the dimerization of the two A atom chains parallel to the a-axes. Note that the pairing is out of phase, corresponding to condensed Γ_{12} optic phonon.

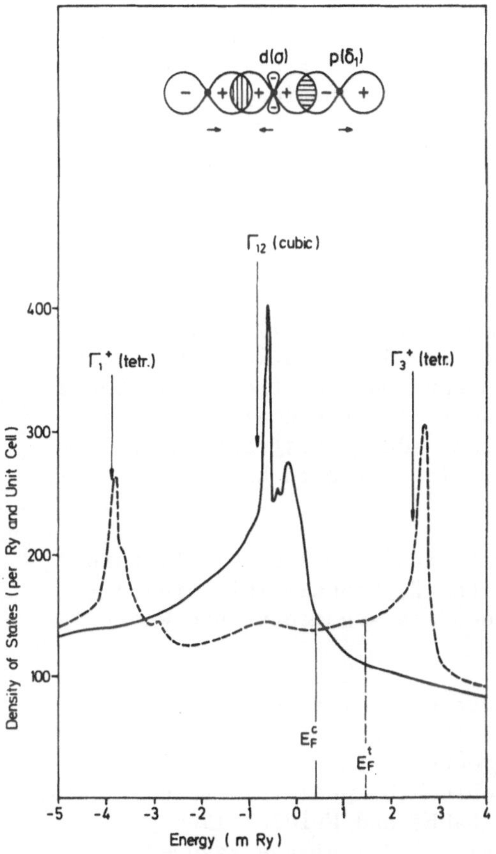

Fig. 40. N(E) curves for cubic and tetragonal Nb_3Sn very close to E_F. Indicated at the top are a pair of p-d overlap integrals, which do no longer cancel each other when the Nb atoms dimerize. This is one reason for the big splitting of the Γ_{12} state (Weber and Mattheiss 1981).

reason is that the areas under each subpeak correspond to only 1/8 electron (when the rather constant background density is subtracted).

The primary cause of the big Γ_{12} splitting is not the tetragonality $\varepsilon = c/a-1$ but the pairing or dimerization parameter δ of the Nb atoms in the chains (see Fig. 39). When assuming only a dimerization δ without any tetragonality ε, about 80 % of the splitting is achieved, in the reversed case (only $\varepsilon \neq 0$), the splitting is ≈ 20 % of its full value. There are two reasons for the big splitting caused by the dimerization δ. One is an additional hybridization between the $d(\sigma)$ and the $p(\delta_1)$ orbitals as indicated at the top of Fig. 40. With respect to the central $d(\sigma)$ orbitals the l.h.s. and r.h.s. $(pd\sigma)_1$ integrals add up to zero in the cubic case, due to opposite signs. When the atoms displace in the dimerization pattern δ, the two $(pd\sigma)_1$ integrals do no longer have the same magnitude.

The second contribution to the splitting Δ of Γ_{12} state arises from an additional interchain coupling of $d(\sigma)$ and $d(\delta_1)$ states. The importance of this effect has not been realized in the paper of Weber and Mattheiss (1982) but has become obvious only later in the course of my further phonon work on Nb_3Sn. While intrachain coupling integrals of type $\langle 3z^2 - r^2|H|x^2 - y^2\rangle$ are zero due to symmetry (and remain zero with tetragonal distortion), the values of interchain integrals of type $\langle 3z^2 - r^2|H|y^2 - x^2\rangle$ are also very small (≈ 0.005 Ry), mainly because the lobes of the wavefunctions hit each others nodal planes. Dimerization leads to a big gradient of this matrix element, another example for a "dormant" interaction as described in Sect. 3.2.2.

The two effects i) intrachain $(pd\sigma)$ asymmetry in the coupling of $d(\sigma)$ and $p(\delta_1)$ states and interchain $d(\sigma) - d(\delta_1)$ "dormant" interaction add up constructively for the dimerization δ which corresponds to a condensed Γ_{12} optic phonon. On the other hand, for a displacement pattern like the Γ_2 mode (all three chains dimerize in the same way), the two contributions add up destructively. As a consequence one cannot expect a Γ_2 distortion to be energetically favorable.

Our results further predict that sign(δ) = sign(ε). In this case we obtain bigger splittings of the Γ_{12} bands. This finding can also be understood from simple bonding arguments (see Weber and Mattheiss, 1982).

Another question is, what effects determine the sign of the tetragonality ε. As two bands split, the two N(E) subpeaks have the intensity ratio 1 : 1. As a consequence the position of E_F with respect to the center of the cubic N(E) peak is irrelevant to sign(ε). Earlier model theories (Labbé and Friedel 1966 ; Gor'kov 1973 ; Bhatt 1977 ; Lee et al., 1977) always assumed 2 : 1 or 1 : 2 ratios of the split peaks as they all were guided by the idea that the splitting occurs for triply degenerate states, which again were linked in one way of the other to the existence of the three A-atom chains.

We concluded that the sign(ε) is due to secondary effects such as the alignment or splitting of the remaining bands near E_F in the vicinity of the M and R points. There, always 2 : 1 splittings occur which may be sufficient to "tip the balance" that determines sign(ε).

Our results are thus not only consistent with the fact that sign(ε) is different for V_3Si and Nb_3Sn, but also with the recent observation that sign(ε) changes as a function of temperature in the $Nb_3Sn_{1-x}Sb_x$ system for $x \approx 0.1$ (Fujii et al. 1982). In addition this work has confirmed our prediction that sign(δ) = sign(ε).

3.3.2. The Phonons : a Lot to Predict

A. The Experimental Situation

Up to very recently, the phonon dispersion curves of A15 compounds have been a rather unknown area. There are two reasons for this situation i) the shortage of single crystals, sufficiently large for inelastic neutron scattering and ii) the big difficulty in evaluating the raw scattering data, as 24 phonon branches have to be dealt with.

A Nb_3Sn single crystal of appreaciable size (originally 50 mm^3) has been grown by Hanak in 1967. With this "RCA" crystal, various unusual properties of Nb_3Sn have been studied, such as softening of the shear elastic constant $c_{11} - c_{12}$ and the temperature dependence of the magnetic susceptibility (Rehwald et al. 1972). In addition, Shirane and Axe (1973) have investigated by neutron scattering, how far the softening of the Σ_4 acoustic mode extends into the BZ (see Fig. 41). These authors have also published some further data on other acoustic modes along $\Delta(\xi,0,0)$, $\Sigma(\xi,\xi,0)$ and $\Lambda(\xi,\xi,\xi)$ (Axe and Shirane 1978). The phonon density of states for Nb_3Sn has been measured by Schweiß et al. (1976). Recently, Raman data for the optic modes Γ_{12} and $\Gamma_{25'}$ at the BZ center have been reported (Schicktanz et al. 1980).

Altogether, the experimental information on the phonons of Nb_3Sn is very sparse. Only about two of the 24 phonon branches have been measured along the symmetry lines Δ, Σ, Λ, and only two of the eight Γ point optic modes are known, not to speak about the dispersion of the optic branches away from Γ, which is completely unknown.

The situation is even worse for V_3Si. For this material, the pioneering ultrasonic measurements of Testardi et al. (1965) had shown that the softening of the shear elastic constant $c_{11}-c_{12}$ in the cubic phase is the precursor of the tetragonal distortion. But only a few neutron data for some acoustic modes exist (Shirane et al. 1971 ; Rumyantsev et al. 1977), in spite of the fact that big single crystals of V_3Si (1000 mm^3) are available. The reason is that V is a very strong incoherent scatterer and therefore only phonon modes with big Si amplitude can be detected. (We will

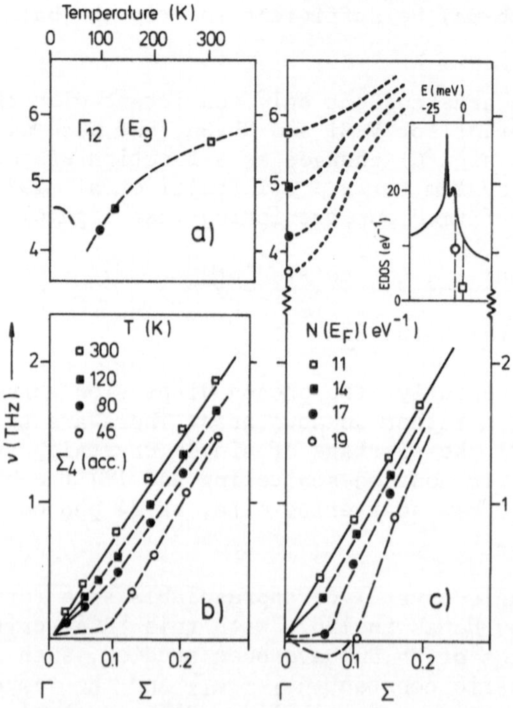

Fig. 41. Softening of Γ_{12} and Σ_4 (c_{11}-c_{12}) modes. Part (a) shows experimental values of the Γ_{12} softening with temperature (Schicktanz et al. (1980), (b) depicts the neutron data of Shirane and Axe (1973). (c) shows theoretical results as a function of $N(E_F)$. Dotted lines in the upper part of (c) indicate the approximate dispersion of the Σ_4 branch originating from Γ_{12}. The inset in (c) indicates the very small changes in E_F when $N(E_F)$ is varied.

later see that the most interesting modes should be those where only V atoms vibrate). The phonon density of states has been measured by Schweiß et al. (1976). Recently, also Raman data on the Γ_{12} mode have been presented (Schicktanz et al. 1978 ; Wipf et al. 1978), but the $\Gamma_{25'}$ mode has not yet been observed by Raman scattering.

Since 1981, strong experimental efforts are under way to improve the knowledge on the phonon dispersion of A15 compounds. The interest first focussed on compounds like Nb_3Sb ($T_c \approx 0.2$ K) and Cr_3Si (no T_c found so far), which can be used as "reference" materials for studying the electron-phonon coupling of high T_c A15 compounds in a way very similar to the refractory materials, where 8 VE compounds like TiC or ZrC are the "reference" systems for NbC, NbN etc.

For Nb_3Sb and Cr_3Si, large single crystals have become available. But it soon turned out that it was extremely difficult to disentangle the many optic modes, unless good lattice dynamical models existed to provide reliable structure factors. As a consequence, the neutron scattering measurements and model calculations were closely intertwined. First the experimental information on low frequency modes provided the starting input for a force constant model. Then the model predicted those areas in reciprocal space where to find further phonon peaks. In consequent steps, the experimental information was used to refine the model, which in turn provided more reliable identification of optic modes. This self-consistent process was carried out most thoroughly for Nb_3Sb where along Δ, Σ, and Λ, all 24 phonon branches have been measured and identified for their symmetry (Pintschovius et al. 1982 and 1983a). The study of the Cr_3Si phonons has been less complete, only phonons ranging up to medium optic frequencies have been measured (Weiss and Rumyantsev, 1981, Jørgensen et al. 1982). In addition, the phonon density of states curve for Cr_3Si has been determined. These data allowed further improvement of the lattice dynamical model for Cr_3Si (Kobbelt et al. 1982). The basic result of these investigations is that the force constants in the low T_c A15 materials Nb_3Sb and Cr_3Si are of short range, not extending beyond one lattice constant.

Some high T_c materials have also been investigated by neutron scattering. These are $Nb_{3.2}Ge_{0.8}$ (T_c = 6 K) and $Nb_{3.1}Ga_{0.9}$ (T_c = 12 K). The Nb_3Ge data, taken from a 100 mm^3 single crystal, are only slightly less complete than the Nb_3Sb results (Pintschovius et al. 1982 ; Smith et al. 1983). The data indicate the existence of longer ranging forces than those of Nb_3Sb, but pronounced phonon anomalies like in Nb or NbC have not become obvious. Similar results are found for Nb_3Ga, where the data have been taken from five aligned 10 mm^3 samples (Pintschovius et al. 1982, Pintschovius et al. 1983b).

The calculation of the phonon dispersion curves of Nb_3Sn is particularly attractive due to the lack of experimental information, and may serve as a stimulus for neutron work, as a lot of theoretical predictions can be examined.

B. Computational Procedure

Our NTB calculation (Weber 1982, Weber 1983) proceeds along the lines described in Sect. 2. We have utilized the general program package described in the Appendix A7. For the energy bands and wavefunctions we have used the highly accurate NTB bands of Mattheiss and Weber (1982) (see Sect. 3.3.1). The gradients of the NTB matrix elements have been chosen in accordance with those used in our earlier work (see Table A10.6). To some extent we have adjusted them to be consistent with the decrease of 1 nn and 2 nn Nb-Nb matrix elements (see Table A10.5).

In spite of all program optimization (see Appendix A7), the computing time for $D_2(q)$ was prohibitively large if all band-to-band transitions of Eq. 2.14 had been included in the calculation. The dimension of the NTB Hamiltonian and overlap matrices is 62, thus a total of approx. 62 x 62/2 ≈ 2000 band-to-band transitions had to be dealt with. In each of these transitions, the 24 dimensional electron-phonon vectors $g^{\kappa\alpha}$ are constructed by matrix multiplications involving 62 dimensional vectors $A(\kappa m|k\mu)$ and $B_\alpha(\kappa m|k'\mu')$ (Eq. A49). We therefore have limited the D_2 calculations by considering only 10 bands around E_F, thus taking into account 55 μ,μ' transitions for D_2. These bands are situated in an energy range $\approx \pm$ 2.5 eV around E_F. We think that this approximation is well justified from the experience made in the studies of Nb-Mo and the refractory compounds (see Sects. 3.1 and 3.2). For simplicity, we have also put zero all gradients of Nb-Sn matrix elements, as in the 10 bands around E_F the fraction of Sn s or p orbitals is very small. Furthermore we set $E_F \approx 0$, and thus all ∇S terms in Eq. (2.16) have been neglected. In the same D_2 calculation we have also computed the linewidth matrices Γ (Eq. 2.18) to be used for the calculation of phonon linewidths γ (Eq. 2.17) and the Eliashberg function $\alpha^2F(\omega)$ (Eq. 2.19). These results are discussed in Sect. 3.3.3.

For the computation of D_2 mesh factors of N = 8,10 and 12 were used (see Appendix A7). We found that N = 10 provided sufficient numerical convergence for D_2. This corresponds to 285 k-points in the even mesh and 220 k-points in the odd mesh of the irreducible BZ. For the linewidth matrices Γ, the numerical convergence of N = 10 was not very good. The reason is that only rather few scattering events of the Fermi surface sheets contribute to Γ while much more scatterings occur in D_2. We think however that integral quantities like $\alpha^2F(\omega)$ or λ are not much affected by this convergence problem.

Another problem of the Nb_3Sn phonon calculation was how to determine the short range forces (D_0 + D_1). In previous work, the (D_0 + D_1) forces have mostly been determined by fit to a few experimental phonon modes. This was not possible in the case of Nb_3Sn, as the experimental information was too sparse. We thus have chosen a different way, utilizing the neutron data on Nb_3Sb (Pintschovius et al. 1983). A second D_2 calculation for Nb_3Sb has been performed, assuming a rigid band model and shifting E_F by approximately 0.45 eV. This corresponds to an increase in band filling by 1/4 electron per atom from 4.75 to 5.0 electrons per atom. E_F of Nb_3Sb is now situated in a region of very low electronic density of states (see Fig. 35). We note that van Kessel et al. (1980) and Klein and Boyer (1980) have presented APW calculations for Nb_3Sb, carried out using the respective procedures as for their Nb_3Sn energy band calculations (van Kessel et al. 1978, Klein et al. 1978). Both groups point out that, to a large extend, a rigid band behavior is observed between Nb_3Sn and Nb_3Sb.

Using the D_2 calculation for Nb_3Sb, the short-range forces (D_0 + D_1) have been determined by fitting to the neutron data (r.m.s. error ≈ 0.18 THz). In (D_0 + D_1), we have assumed general forces for the 1 nn Nb-Nb (intrachain) springs (3 force constants, see Appendix A7) and axially symmetric forces for 1nn Nb-Sb and 2nn Nb-Nb (interchain) springs, altogether 7 force constants. Little indication was found for the existence of longer ranging forces in D_2 of Nb_3Sb ; i.e. forces extending much beyond one lattice constant. This result is in agreement with the Born-and-von Kármán model analysis of Pintschovius et al. (1983) and also with our previous experience on the lattice dynamics of low $N(E_F)$, low T_c transition metals and compounds such as $Nb_{0.75}Mo_{0.25}$ or ZrC.

The (D_0 + D_1) force constants of Nb_3Sb have then been taken over in the Nb_3Sn phonon calculation. Thus our NTB theory does not contain any adjustable parameter.

The use of the Nb_3Sb (D_0 + D_1) forces is probably well justified for various reasons. First, there is very little change of the lattice constant (≈ 0.5 %) between Nb_3Sn and Nb_3Sb which indicates very similar ionic radii of Sn and Sb. Then, the change in band-filling corresponds to only 1/4 electron per atom. The quantities D_0 and D_1 depend on all valence electrons, there is no special weight to the states near E_F, as there is in D_2.

C. Prediction of Anomalies

We have calculated the phonon dispersion curves $\omega(qj)$ for Nb_3Sn at the symmetry points Γ,X,M and R of the BZ and along all symmetry lines. In addition, we have computed $\omega(qj)$ (as well as $\gamma(qj)$) on an evenly spaced grid of 56 q-points (corresponding to N = 5), in order to generate phonon density-of-states curves $F(\omega)$ and the Eliashberg function $\alpha^2F(\omega)$ (see Sect. 3.3.3).

The dispersion curves of Nb_3Sn are displayed in Figs. 42 - 44. They agree well with the limited experimental information, both the neutron and the Raman data.

Compared to Nb_3Sb, we find a general 10 - 15 % phonon softening, caused by the shift of E_F from a low to a high electronic-density-of-states region. Especially for the Δ and Σ lines, this impression of a general softening prevails (see Figs. 42 and 43).

Some specific modes are lowered much more than 10 %, especially some branches along the Λ direction (see Fig. 44). The lowest Λ_1 (longitudinal acoustic) branch now ends at (the doubly degenerate) R_3 instead of (the sixfold degenerate) R_4 in Nb_3Sb. This

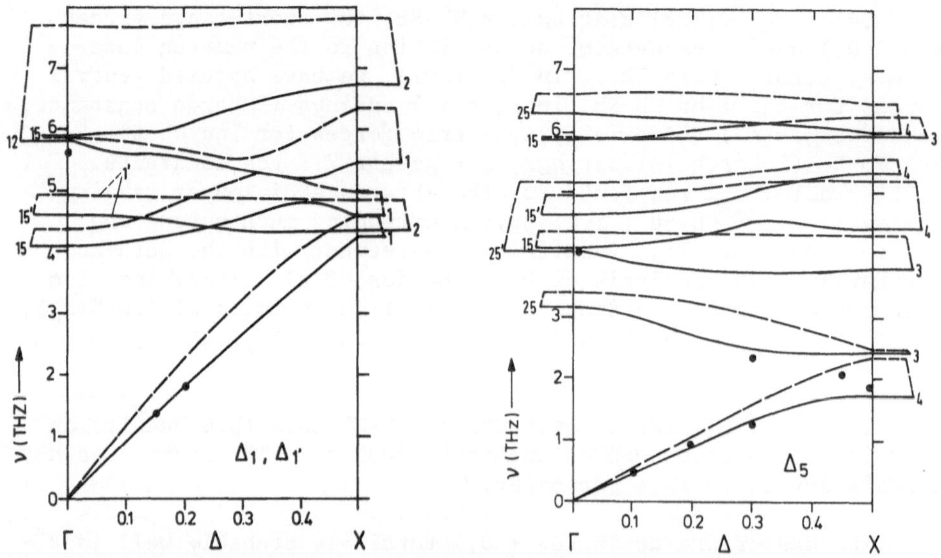

Fig. 42. Phonon dispersion curves of Nb_3Sn along Δ (solid lines) (Weber 1983). Also shown are the experimental dispersion curves for Nb_3Sb (dashed lines) of Pintschovius et al. (1983a). Not shown are the Δ_2 and $\Delta_{2'}$ representations. For comparison, the neutron data of Shirane and Axe (1973) and Axe and Shirane (1978), and the Raman data of Schicktanz et al. (1980) are presented.

Λ_1 branch exhibits anomalies near q = 0.25 and at R. The same behavior is found for the lowest Λ_2 branch, also ending at R_3 (not shown here). Similar minima at R are seen in the dispersion of the Λ_3 (transverse) branches ending at the fourfold degenerate R_{1-2}. It is not quite accidental that only the lowest lying branches of the Λ_1, Λ_2 or Λ_3 representations exhibit these typical features of "phonon anomalies" familiar from Nb and Mo or from the refractory compounds. In the higher frequency regions the appearence of deep local minima in phonon dispersion curves of a certain representation is disguised by the interactions with the other branches of the same representation which are not affected from the anomalous mechanism. For instance, the Γ_{12} modes are drastically lowered as compared to Nb_3Sb, yet when we look at the dispersion curves, e.g., along Δ_1, there is little indication of a deep local minimum. The Δ_1 and Δ_2 branches originating from Γ_{12} do not seem to show much dispersion. However, an analysis of the polarization vectors of these branches indicates that they are quickly changing with q, and close to 0.2, the highest lying Δ_1 or

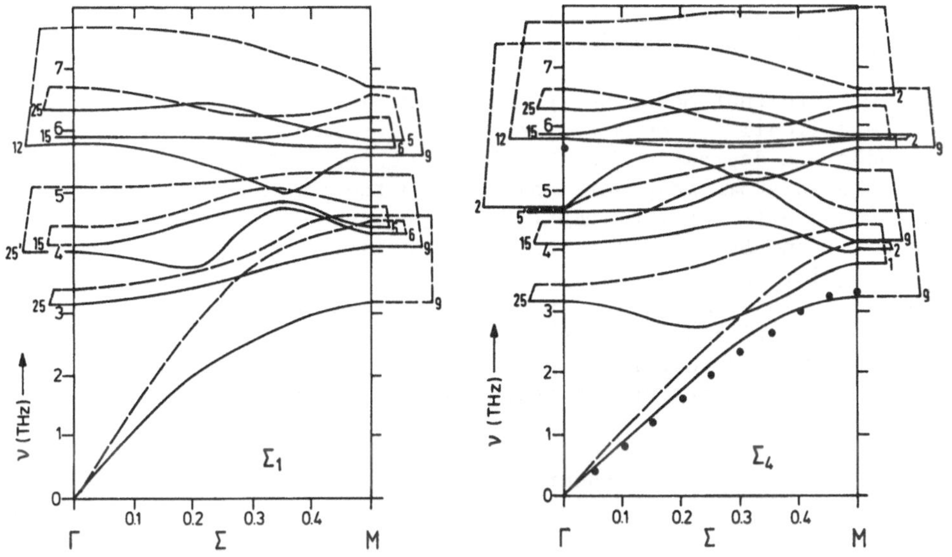

Fig. 43. Phonon dispersion curves for Nb_3Sn and Nb_3Sb along Σ. Not shown are the Σ_2 and Σ_3 representations (Weber 1983).

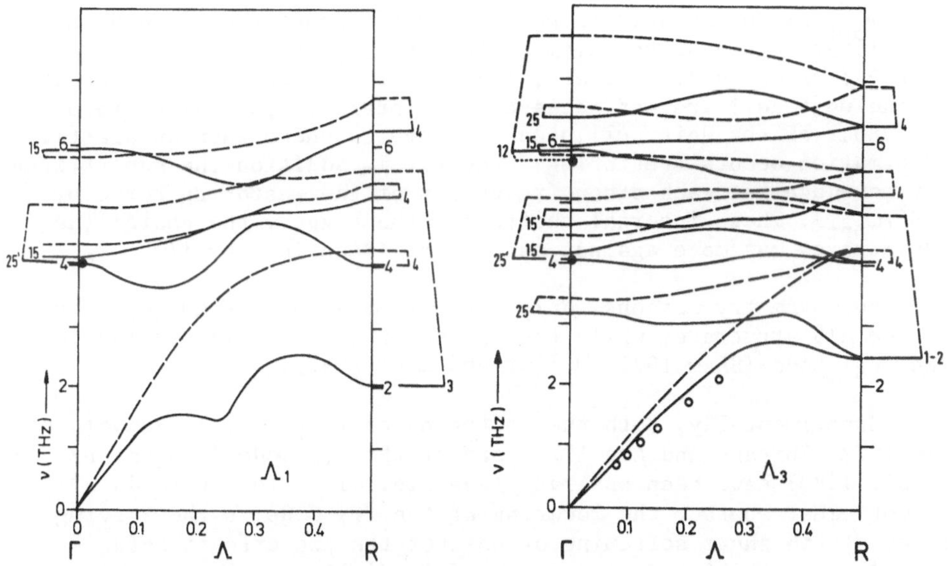

Fig. 44. Phonon dispersion curves for Nb_3Sn and Nb_3Sb along Λ (Λ_2 is not shown). Note the strong depression of the Nb_3Sn R_{1-2} and R_3 states (Weber 1982).

Δ_2 branches have taken over the displacement pattern typical for the Γ_{12} mode. This behavior is indicated in Fig. 41c.

We note that modes where only Nb atoms vibrate, are lowered most drastically as compared to Nb_3Sb. For instance, the R_{1-2} and R_3 modes are depressed from $\approx$ 5 THz in Nb_3Sb to $\approx$ 2-3 THz in Nb_3Sn. Here the Nb atoms move perpendicular to the chain directions, the Sn atoms are at rest. Of the Γ point phonons, the longitudinal modes Γ_{12} and Γ_2 as well as the transverse modes $\Gamma_{25'}$ are lowered most. Again, in all these modes only the Nb atoms vibrate, with the two chain atoms moving in opposite directions. The lowering of the phonon frequencies is also indicative of a strong electron-phonon coupling, therefore we will discuss the nature and electronic origin of strong and weak coupling modes in more detail in Sect. 3.3.3.

D. The Softening of the Shear Mode

Another important result of our study concerns the behavior of the Γ_{12} optic phonon and its role in the mode softening of the shear elastic constant $(c_{11} - c_{12})$, which is the precursor of the cubic-to-tetragonal transition. The Γ_{12} mode is coupled to the shear by a relation of the form

$$c_{11} - c_{12} = A - B^2/\omega^2(\Gamma_{12}) \quad (3.7)$$

where A and B are certain force constants. Terms of the form $- B^2/\omega^2_{opt}$ may occur only in non-symmorphic space groups such as the A15 structure or the diamond lattice (see Born and Huang, 1954, p. 229 ; or Maraduding et al., 1971, p.21). There, the atomic sites in the unit cell are not necessarily centers of inversion. Then the atoms of the unit cell need not move in phase when an elastic deformation occurs (restoring force A). In addition the sublattices may move against each other (reduction of the restoring force by $- B^2/\omega^2_{opt}$). These internal shifts are the bigger, the easier the sublattices may move against each other (described by $1/\omega^2_{opt}$).

For symmetry reasons, only the Γ_{12} mode couples to $(c_{11} - c_{12})$ in the A15 structure, while the c_{44} shear constant is coupled to the $\Gamma_{25'}$ mode (Sham 1971, Miller and Axe, 1967).

Experimentally, both the softening of $(c_{11} - c_{12})$ (Rehwald et al. 1972, Shirane and Axe 1971) and of the Γ_{12} mode (Schicktanz et al. 1980) have been observed (see Fig. 41). From these data it is not clear whether the lowering of the Γ_{12} mode is the driving force of the shear softening or whether the two effects occur parallel, as the force constant A of Eq.(3.7) may also exhibit a strong temperature dependence. Earlier model theories which have included effects of the Γ_{12} optic mode (Sham 1971, Bhatt and McMillan 1976) have given little weight to the optic mode softening.

As mentioned above, in our calculation the frequency of the Γ_{12} mode of Nb_3Sn is considerably lowered compared to that of Nb_3Sb. Its value is close to the 300 K Raman result ; similarly, the transverse acoustic Σ_4 branch agrees well with the 300 K neutron data (see Figs. 41c and 43). We have simulated the T dependence of these modes by very slightly shifting E_F towards the peak in the electronic density of states (see inset of Fig. 41) that is, towards the electronic Γ_{12} state. This shift strongly increases $N(E_F)$, the Fermi density. We have found that $\omega(\Gamma_{12})$ decreases very much with increasing $N(E_F)$, thereby driving $(c_{11} - c_{12})$ to zero. The extension of the acoustic Σ_4 softening in q-space is very similar to the neutron data ; in addition $(c_{11} - c_{12}) \approx 0$ is found for a value of $\omega(\Gamma_{12})$ close to the low temperature Raman results. In our calculation, the Γ_{12} phonon appears to be the only driving force of the tetragonal transition, we do not find any evidence that the quantities A and B of Eq. (3.7) depend much on $N(E_F)$.

These results are consistent with those of the energy band study for tetragonal Nb_3Sn of Weber and Mattheiss (1982), where it was found that the splitting of the Γ_{12} peak in the electronic density of states is primarily caused by the Γ_{12}-like distortion.

We note that none of the other strong coupling modes depend very sensitively on $N(E_F)$. Even less dependence on $N(E_F)$ is observed for the "normal" modes such as Γ_{15}, Γ_{25} or R_4.

The main contributions to $D_2(\Gamma_{12})$ arise from very flat portions of the Γ_{12} bands around the Γ point. There, the band velocities near E_F are comparable to the sound velocity. As a consequence, the adiabatic approximation used throughout this study is not expected to be valid for the Γ_{12} mode. We think that for an improved theory of the Γ_{12} and $(c_{11} - c_{12})$ mode softening, a "vibronic" approach is needed which treats the electrons around the Γ_{12} state and the Γ_{12} phonon on equal footing. For this reason, we did not attempt to link the $N(E_F)$ dependence of these modes in our calculation to a temperature effect caused by Fermi factor smearing.

3.3.3. Electron-Phonon Coupling

A. "Internal" and "External" Vibrations

The evaluation of the phonon linewidths $\gamma(qj)$ for Nb_3Sn confirms the observations already made in similar calculations for Nb or NbC : modes, strongly depressed as compared to the reference material, also have big $\gamma(qj)$ values. As was mentioned above, the "anomalous" modes in Nb_3Sn are those where the atoms within one chain move against each other. These modes are most easily visualized at the Γ point (see Fig. 45). There are two kinds of internal motion of chain atoms. The first is longitudinal, the atoms move along the chains. There exist two modes of this kind, Γ_2 and Γ_{12}. In Nb_3Sb, they have the highest frequencies of all modes, because of the stretching of the very strong Nb-Nb intrachain bonds. In Nb_3Sn, they are considerably lower, because of the strong electron-phonon coupling ; but they still lie in the high frequency region of the phonon spectrum. The second kind of internal modes is transverse, the atoms move perpendicular to the chain directions, like the $\Gamma_{25'}$ and $\Gamma_{15'}$ modes of Fig. 45 and the modes R_{1-2} and R_3 at R.

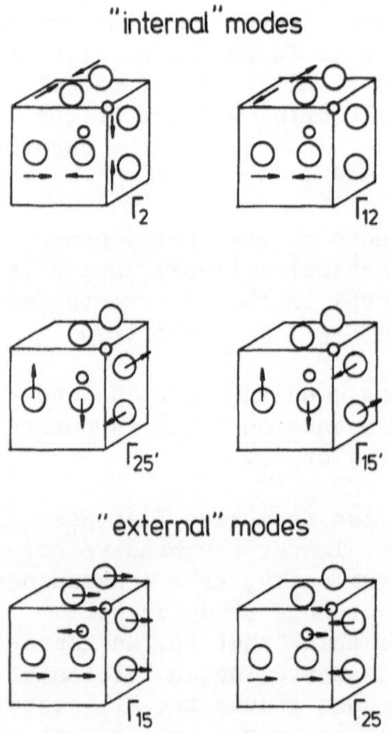

Fig. 45. Γ point phonon displacements for A15 compounds. The "internal" modes, where only A atoms move, the displacement patterns are purely given by the respective symmetries. For the "external" modes Γ_{15} and Γ_{25}, also the forces between the atoms influence the eigenvectors.

These modes are found in the low frequency regions of the Nb_3Sn spectrum. In "external" modes such as the Γ_{15}, Γ_{25}, and R_4 phonons, the two atoms of a chain move in phase. Compared to Nb_3Sb these modes have changed little and they have small linewidths. They are found in the intermediate and high frequency region of the spectrum. We note that a clean distinction between "internal" and "external" modes is only possible at the symmetry points Γ and R, and to some extent also at X and M and along the Δ line. For all the other q-points, the phonons have both "internal" and "external" components ; but of course, this may vary rather strongly for the individual modes.

How is it possible that, quite obviously, the chain structure of Nb_3Sn plays the important role in distinguishing strongly and weakly coupling phonons? Furthermore, why are k-space nesting arguments apparently inoperative ? The answer lies in the special nature of the electronic Γ_{12} bands near E_F. As was discussed in Sect. 3.3.1, the Γ_{12} bands are flat not because of small orbital overlap, but because of a hybridization effect involving strongly overlapping $d(\sigma)$ and $d(\pi)$ orbitals. The big overlap of the orbitals is also the source of the large electron-phonon coupling of "internal" modes. In the systems Nb-Mo or NbC-NbN there is no special hybridization effect leading to flat bands, thus the large d-orbital overlap leads to wide bands, and, therefore, to large electron-phonon matrix elements, when states of large and opposite band velocities are coupled. The strong coupling does not occur at long wave phonons when neighbor atoms move in phase, but at short wave phonons, when neighbor atoms move against each other. The band topology near E_F then provides the most favorable situation for an "anomaly" : a nesting vector connecting many states k,k' with large matrix elements.

However in Nb_3Sn, big electron-phonon matrix elements may occur, even if the bands are flat. They occur for instance, when neighbor $d(\sigma)$ orbitals, the principal orbital component of the Γ_{12} bands, move against each other. This motion creates a large change of the large $(dd\sigma)_1$ intrachain energy integrals, which appears to be the dominant source of the large coupling of the longitudinal "internal" modes like Γ_2 and Γ_{12}. The main source for the large coupling of the transverse "internal" modes like $\Gamma_{25'}$, $\Gamma_{15'}$, R_{1-2} and R_3 is the "dormant" $d(\sigma)$ - $d(\pi)$ intrachain coupling (see : e.g., Fig. 37). Interchain contributions are not really negligible, if they were Γ_2 and Γ_{12} or $\Gamma_{25'}$ and $\Gamma_{15'}$ should behave identically. But the dominant contribution is intrachain. If this is missing, as for the "external" modes Γ_{15} or $\Gamma_{25'}$ the coupling is much smaller.

In Nb_3Sn k-space nesting is not of the same importance as in the other systems, as the flat Γ_{12} bands provide many states k,k' with small energy difference. For Γ_{12} and Γ_2, intraband and interband scattering of the Γ_{12} bands throughout large areas of k-space seem to participate. In addition, there is a particularly strong contribution to Γ_{12} from k states near Γ. For the R_{1-2} and R_3 modes it appears that scattering from states in the Γ_{12} bands near Γ to states in the many bands near R gives the dominant contributions. This might be interpreted as some sort of nesting. We note that we so far could not analyze those areas in k-space which yield the main contributions to the Λ_1 and Λ_2 anomaly near q = (0.25, 0.25, 0.25).

For other high T_c A15 compounds we would expect results quite similar to those of Nb_3Sn, since E_F is pinned to the flat Γ_{12} bands. The only exception is the Γ_{12} mode and, as a consequence the

(c_{11} - c_{12}) softening. Only when E_F lies very close (≈ ± 5 mRy) to the Γ_{12} peak of the electronic density of states, we expect a drastic lowering of the Γ_{12} phonon and the possibility of a tetragonal distortion. According to the energy band calculations of Klein et al. (1978), this is the case for V_3Si and V_3Ga. In the other high T_c materials, E_F lies far enough away from the Γ_{12} state, so that the Γ_{12} phonon is not expected to become the driving force for a tetragonal transition.

B. The Eliashberg Function

Finally, we may compare theoretical and experimental results for the phonon density of states $F(\omega)$ and the Eliashberg function $\alpha^2F(\omega)$ (see Fig. 46). The phonon density shows a three peak structure very similar to the neutron data of Schweiss et al. (1976). The $F(\omega)$ curves mainly differ in the shape of the lowest peak which is less isolated in experiment. Reasonable agreement is also found between the calculated $\alpha^2F(\omega)$ curve and the tunneling data of Schneider and Geerk (1983). We note that we have obtained a theoretical value of λ = 1.5.

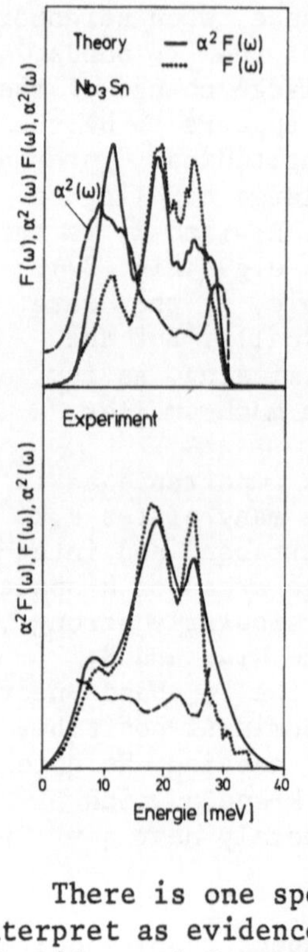

Fig. 46. Comparison of phonon densities of states $F(\omega)$ and Eliashberg functions $\alpha^2F(\omega)$ for Nb_3Sn. Upper part shows our NTB results, lower part exhibits the neutron data of Schweiß et al. (1976) and the tunneling data of Schneider and Geerk (1983).

There is one special feature in the tunneling data which we interpret as evidence for the existence of the Λ_1 anomaly. Schneider and Geerk find a shoulder in $\alpha^2F(\omega)$ near 5.5 meV. In this very low frequency region a structureless ω^2-like behavior of $\alpha^2F(\omega)$ or $F(\omega)$ is expected, since the phonon dispersion should still be linear. This shoulder becomes much more pronounced in the d^2I/dV^2 characteristic and is also seen in Nb_3Ge at somewhat larger frequencies. (The neutron data did not extend down to this frequency range). A very similar shoulder near 6 meV is found in the theoretical curves, arising from the Λ_1 minimum near q = 0.25. These phonons couple strongly to the electrons, in contrast to the transverse acoustic Δ_5 phonons near X which may also cause a singularity in $F(\omega)$ in this frequency range. However, these are "external" modes and contribute very little to $\alpha^2F(\omega)$.

The theoretical coupling function $\alpha^2(\omega) = \alpha^2F(\omega)/F(\omega)$ exhibits three regions of strongly and weakly coupling phonons. $\alpha^2(\omega)$ is particularly large in the area of the low frequency peak between about 5 and 16 meV. There, the strongly coupling transverse "internal" modes are found. In the region of the center and the high frequency peak, the weakly coupling "external" modes dominate, while near the upper end of the spectrum the longitudinal "internal" modes cause the second maximum of $\alpha^2(\omega)$.

A somewhat similar behavior is found in the experimental $\alpha^2(\omega)$ curve. There is a region of relatively strong coupling at low frequencies, followed by an intermediate area of less strongly coupling phonons. Towards the upper end of the spectrum there are indications that $\alpha^2(\omega)$ increases again. We note that the tunneling curve of Kwo and Geballe (1982) exhibits a stronger low frequency peak and thus is in better agreement with our theory than the results of Schneider and Geerk. This discrepancy between the two groups arises from different sample preparation. The samples of Schneider and Geerk exhibit hardly any proximity effect and are probably of better quality than those of the other group.

4. "FROZEN PHONON" CALCULATIONS

Using modern band-structure methods, combined with the local density approximation for exchange and correlation, the total energy of solids at given lattice configurations can be calculated with high accuracy. The results are, of course, valid only for T = 0, and the zero point motion of the atoms is neglected. The only input in these calculations are the atomic numbers of the constituents and a prejudice on the crystal structure. The computations of total energy allow to distinguish among competing crystal structures and to determine the lattice constant(s). In addition, the curvature of the total energy function at the minimum is related to the bulk modulus, and deviations from a quadratic curve indicate the pressure or volume dependence of the bulk modulus. When a stable lattice configuration has been found, the unit cell may be deformed in certain ways, for instance by homogeneous deformations or by displacement patterns of atoms corresponding to high symmetry phonons. Then an analysis of the total energy curves will provide the values of the elastic constants, of phonon frequencies and also of anharmonic force constants.

For each point of a total energy curve, self-consistent energy band calculations have to be carried out. This requirement shows the limitations of "frozen phonon" calculations : it is a procedure which may consume a lot of computer time. As a consequence, only rather highly symmetric crystal structures have been studied so far. The phonon dispersion curves cannot be calculated directly ; it is only possible to study phonons at very few selected high symmetry q-points, such as zone boundary points or at Γ, if optic phonon branches exist for the lattice.

One of the earliest investigations of this kind has been the work of Moruzzi, Janak and Williams (1978), who calculated ground state properties for as many as 32 elemental metals with atomic number $Z < 50$. These authors used a fast version of the KKR method, employing the muffin-tin approximation, and assumed cubic lattices for all cases considered. Their results for the lattice constants are within 1 - 2 % deviation from experimental values ; for the bulk moduli they report agreement within $\approx$ 10 %.

Calculations of phonons frequencies have first been reported for semiconductors (Wendel and Martin 1978, 1979). There, the computations are simpler than for metals, because of the energy gap between valence and conduction bands. This allows to utilize a few "special" k-points to calculate energy eigenvalues and wavefunctions and, from the latter, the charge density and crystal potential to be incorporated for the next step in the self-consistency procedure.

One of the most interesting results of Wendel and Martin are the charge density redistributions when the atoms are displaced from equilibrium positions. In particular they found for transverse acoustic modes at the BZ boundary that the centers of the bond charges move away from the midway positions between the atoms (see Fig. 47). This result thus confirms the basic idea of the adiabatic bond charge model (Weber 1974 and 1977), where it was assumed that the bond charges move adiabatically to adjust to the forces acting on them (see Sect. 1.1). We note that Maschke and Baldereschi (1979) have obtained very similar results for the motion of the bond charges in zone boundary transverse acoustic phonons.

Wendel and Martin's results have been somewhat preliminary, as they used a pseudopotential which did not yield the correct lattice constant. This problem has been overcome by the introduction of norm-conserving pseudopotentials, extracted from ab initio atomic calculations (Hamann et al. 1979). Using these norm-conserving pseudopotentials, Yin and Cohen (1980) presented a very successful investigation of Si, where they studied various competing lattice structures and also calculated some zone boundary phonon frequencies and the optic zone center mode. The deviations from experiment are < 1 % for the lattice constant and < 5 % for phonon frequencies. Cohen and his group as well as Martin and collaborators have continued studies on semiconductors using the pseudopotential method ; a review of this work is given by Kunc (1983).

Using the LMTO method with the atomic-sphere approximation (ASA), Glötzel et al. (1980) report very satisfactory results for the ground state properties of Si. Their work shows that the very efficient LMTO-ASA scheme (Andersen, 1984) is applicable even for semiconductors with open crystal structures.

"Frozen phonon" calculations and the tight binding theory of lattice dynamics could be combined to provide a first principles computation of phonon dispersion, thereby avoiding any experimental input. The "frozen phonon" calculations could supply not only the energy bands for the equilibrium lattice and for certain displaced lattice configurations, but also the values of elastic constants and the frequencies of some high symmetry phonons. Thus, not only the NTB matrix elements and their gradients, but also the short range forces in (D_0 + D_1) could be determined. Then a $D_2(q)$ computation would provide the full phonon dispersion relations.

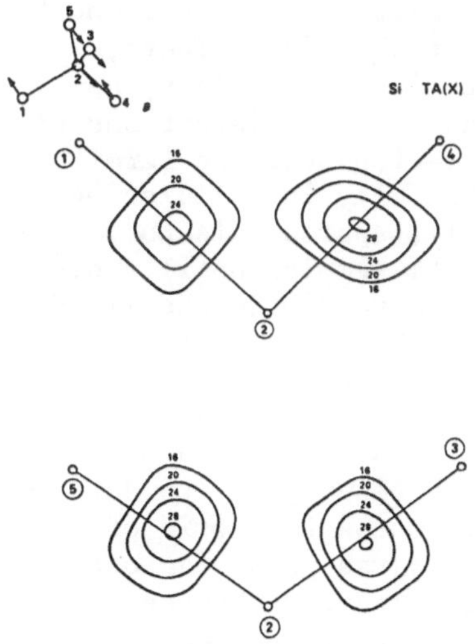

Fig. 47. Charge density for Si with a TA(X) displacement pattern. The inset defines the labeling of the bonds. A center of symmetry is maintained for bonds 1 - 2 and 2 - 4, but is lost for bonds 2 - 3 and 2 - 5. The "rotation" of the bonds is apparent in the 3 - 2 - 5 plane in the lower figure (Wendel and Martin 1979).

Among the various energy band methods for "frozen phonon" calculations we have chosen a first principles linear-combination-of-atomic-orbitals (LCAO) method (Appelbaum and Hamann, 1978). There have been two reasons for this choice. First, the computer code of this specific LCAO scheme is very easily adjustable to any given lattice structure. Then, the LCAO energy band results may provide a firm basis for the physical interpretation of the tight binding results. For instance, the claim that certain orbitals dominate a specific energy band is more convincing, if it arises not only from the (fitted) NTB Hamiltonian, but also from the basic LCAO results.

The applicability of this LCAO method for total energy calculations has been investigated for Si (Harmon et al. 1981 and 1982). It has been demonstrated that this method yields as good results as other schemes. For Si, the theoretical numbers deviate from experimental values by ∿ 1 % for structural properties like the lattice constant and ∿ 10 % for vibrational properties like the bulk modulus or phonon force constants (this corresponds to ∿ 5 % deviation in phonon frequencies). Third order anharmonicities may deviate from experiment by up to 30 %.

In addition, it has been shown in these calculations that the "frozen core" approximation is well justified in total energy calculations for Si (see Fig. 48). In the "frozen core" approximation it is assumed that the core orbitals do not vary their shapes when the lattice configuration is changed ; the core states may however change their energies. In total energy calculations based on norm-conserving pseudopotentials, the validity of this assumption is implied.

The LCAO method has also been used to perform "frozen phonon" calculations in transition metals (Harmon et al. 1981). There, the trick of using a few "special" k-points is not possible, as some of the bands are partially filled. Thus, small distortions of the lattice cause changes in the band occupation near the Fermi energy which must be accounted for very carefully in calculating changes in total energy. One, however costly, approach is to simply keep increasing the number of $\vec{k}$-points sampled until convergence is reached. A more economic approach is to divide the irreduzible BZ into a number of tetrahedra and take their center k vectors as a sample grid. At each iteration in the self-consisting procedure, a NTB fit, as described in Appendix A7, is made to the energies of these k points. Then a density of states calculation is carried out, using NTB energies in a much finer grid, to determine an accurate Fermi energy and surface. As the fine grid of tetrahedra is obtained by sub-dividing the coarse grid, the occupied volume of each energy band inside the large tetrahedron can be determined very accurately. These volumes are then used to provide the weight factors for the states of the coarse grid. As a side result, the NTB matrix elements for equilibrium and distorted lattices are obtained.

Using this weighting procedure, the LCAO results for lattice constants, bulk moduli and H point phonon frequencies of Nb and Mo

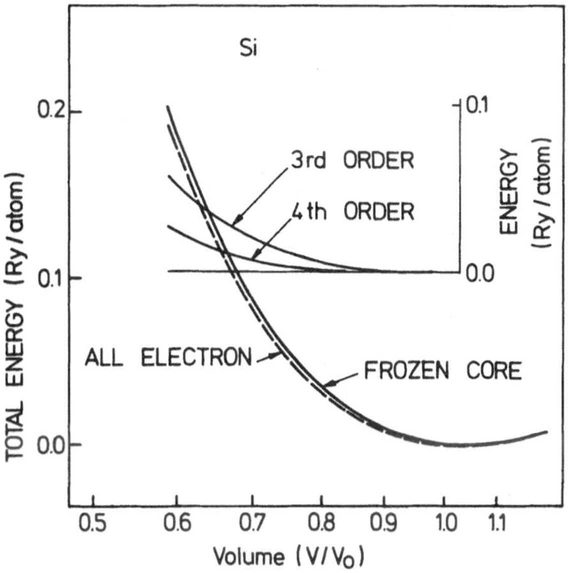

Fig. 48. Fourth-order polynomial fits to the total energy versus volume for the frozen core and all electron calculations in Si (V_0 is the experimental volume). The third- and fourth-order contributions to the all-electron results are also shown (Harmon et al. 1982).

are of the same quality as the results for Si (Harmon et al. 1981, Ho et al. 1982). In Mo, the H point phonon is anomalously depressed. In Sect. 3.1.2., it has been pointed out that the strong coupling of states close to the two Fermi surface octahedra around Γ and H (see Fig. 19) is causing the phonon softening. This observation has been verified by the "frozen" phonon calculations. The energy bands near E_F which produce the two Fermi surface octahedra in the undistorted case, interact very strongly when an H point phonon displacement is applied. This strong coupling leads to rather big shifts of the involved states away from E_F. This effect decreases considerably the energy cost due to the phonon displacement and thus leads to a rather small restoring force.

A similar, even more spectacular case of phonon softening is the frequency decrease of the (2/3, 2/3, 2/3) longitudinal phonon, when going from Mo via Nb to bcc Zr (see Fig. 49). For Mo and Nb, the curvatures of the total-energy-versus-displacement curves correspond very well to the experimental phonon frequencies. For Zr, the calculated total energy curve is highly anharmonic, showing two minima, a very shallow one at the bcc configuration, and a considerably deeper minimum for a situation when two of the three (111) planes of the superlattice unit cell have collapsed. This

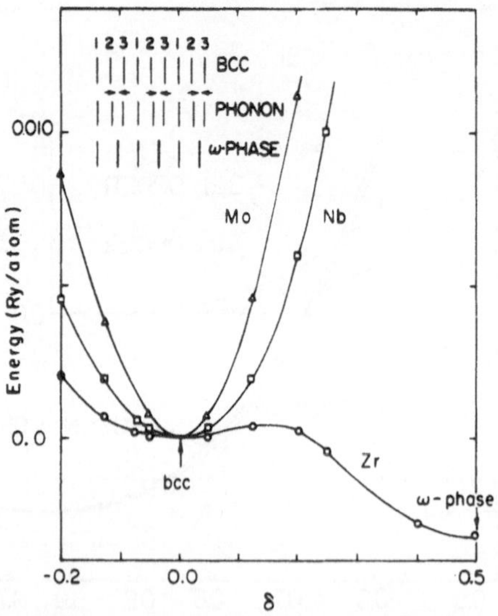

Fig. 49. The calculated energy vs displacement curves for the L(2/3, 2/3, 2/3) mode in Nb, Mo, and bcc Zr. Inset : the displacements of the (111) planes. Plane 1 is fixed, plane 2 is displaced by $\delta(a/6, a/6, a/6)$ and plane 3 by $-\delta(a/6, a/6, a/6)$ (Ho et al. 1982).

structure is known as the ω-phase, found at low temperature in a wide variety of Ti, Zr and Hf alloys and also in pure Zr and Ti under pressure.

We note that in the paper of Ho et al. (1982) the (2/3, 2/3, 2/3) phonon results have been obtained from a pseudopotential method. The latter has also been used to calculate the structural properties, as well as H point phonon frequencies. It has been shown that this method yields results of the same quality as the LCAO scheme. Furthermore, the pseudopotential method turned out to be more economical, allowing at comparable computer times the calculations of many more k-points than the LCAO scheme. As a consequence it was not necessary to implement very elaborate weighting procedures for states near E_F.

5. SUMMARY

Over the past decade, the field of lattice dynamics has developed from a mainly phenomenological state to a situation where ab initio calculations of phonon dispersion curves have become possible. Various schemes of treating lattice dynamics from first principles have been put forward ; often these schemes have been generated using ideas of earlier model theories (see Sect. 1). In this article we have mainly dealt with the lattice dynamics of transition metals and their compounds. In these materials valence electrons with energies close to the Fermi energy E_F may influence drastically the shapes of the phonon dispersion curves, especially in crystals with high superconducting transition temperatures T_c.

The lattice dynamics of transition metals has been investigated most extensively using the tight binding theory. This method allows a very economic way of computing phonon dispersion curves as well as the electron-phonon coupling. It also provides a simple and intuitive understanding of phonon anomalies and their relation to superconductivity. A very promising complementary way are the "frozen" phonon calculations, where total crystal energy curves for displaced lattice configurations yield accurate values of high symmetry phonon frequencies.

The tight binding theory is based on two ideas (see Sect. 2 and Appendix) : i) the changes of the electronic charge caused by ionic displacements is best described by orbitals moving with the ions. ii) The terms of the phonon dynamical matrix D are rearranged such that the ion-ion and electron-electron Coulomb interactions cancel each other as much as possible. Altogether, the dynamical matrix is divided in three parts $D = D_0 + D_1 + D_2$. D_1 and D_2 are the first and second order perturbation contributions due to changes of the single particle energies $\varepsilon_{k\mu}$ with ionic displacements. In addition to these "band-structure" terms, there is D_0, the above mentioned difference between the ion-ion and electron-electron interactions (the latter has been doubly counted in the single particle terms).

It has been shown that for elemental crystal, $D_0 + D_1$ give rise to essentially short range forces, which do not much extend beyong the range of direct orbital overlap. In the phonon calculations, these terms are at present parametrized by a few near neighbor force constants. The remaining (second order) term

$$D_2(\kappa\alpha,\kappa'\beta|q) = -\sum_{\substack{k,\mu,\mu' \\ k'=k+q}} \frac{f_{k'\mu} - f_{k\mu}}{\varepsilon_{k\mu} - \varepsilon_{k'\mu'}} g^{\kappa\alpha}_{k\mu,k'\mu'} g^{\kappa'\beta}_{k'\mu',k\mu} \tag{2.14}$$

includes the long range forces which are the source of the phonon anomalies. This term is computed using a non-orthogonal tight binding Hamiltonian with an appropriately chosen valence orbital basis. Its matrix elements are found from fits to first principles energy band calculations. The electron-ion form factors $g^{\kappa\alpha}$ contain the first order changes of the tight binding matrix elements with ionic displacements. These gradients are understood to include all effects of screening, exchange and correlation. In general, they are also obtained from first principles energy band calculations, computed for displaced lattice configurations. The values of these gradients are assumed to be independent of the phonon wavevector q. All experience to date shows that this approximation is correct within 10-20 % for transition metals.

In Sect. 3, results of phonon calculations for three different transition metal systems are presented and discussed. First the $Nb_{1-x}Mo_x$ alloys are dealt with. The tight binding calculations reproduce all observed phonon anomalies. It is shown that the anomalies arise indeed from D_2, more precisely from the strong coupling of certain electronic states (k,μ) and $(k + q,\mu')$ close to E_F ($\approx \pm$ 0.5 eV). However, these anomalies are not Fermi surface anomalies in the usual Kohn-Overhauser sense ; i.e., they are <u>not</u> caused by singularities in the "bare" electronic susceptibility (the energy denominator term of Eq.(2.14)). Instead, they arise mainly from the strong q-dependence of the electron-ion form factors, roughly approximated as

$$g^{\alpha}_{k\mu,k'\mu'} \propto (v^{\alpha}_{k\mu} - v^{\alpha}_{k'\mu'})$$

the difference of the electron velocities of the coupled states. The form factors g^{α} are large when states of large and opposite band dispersion along the α-th component of $(k\mu)$ and $(k'\mu')$ are involved. In addition, the other two components of the band velocities should be small to provide a large phase space in the sum of Eq. (2.14). Therefore very anisotropic band topologies near E_F and as a consequence, very anisotropic Fermi surfaces are favourable for the occurence of the phonon anomalies. This is the case much more in transition metals than in simple metals. In transition metals, large two-center energy integrals between near neighbor d-orbitals may cause large band dispersion in one direction in k-space, while due to the strong directionality of the d-orbitals the bands are rather flat in other k-directions.

Similar results as for Nb-Mo are obtained for the refractory compounds like NbC or NbN (see Sect. 3.2). There appears however an additional feature concerning the magnitude of the form factors $g^{\kappa\alpha}$, which is unimportant in the Nb-Mo study. Because of symmetry, certain tight-binding matrix elements are identically zero at equilibrium, although the orbitals strongly interpenetrate each other ; for instance a σ-type interaction of Nb d_{xy} and C p orbitals. Consequently these integrals are not affecting the energy bands nor the band dispersion.

When the atoms are displaced, large gradients of these matrix elements appear which may enhance the form factors g^{α} far over the normal velocity difference part. We have called this type of coupling "dormant" interactions. They are more important in compounds where in general more orbitals with comparable orbital energies are present than in elemental systems.

A comparison of the phonon dispersion of NbC and NbN shows very clearly that the anomalies are indeed a Fermi energy related k-space effect : If the energy bands are futher filled, when going from NbC to NbN, the anomalous wave vectors move towards the Brillouin zone boundary.

The third system presented here (see Sect. 3.3) is Nb_3Sn, one of the various A_3B compounds with A15 structure, which exhibit rather unusual physical properties such as very high values of T_c. The A15 structure has the peculiar geometrical feature of three mutually perpendicular, nonintersecting chains of A atoms. Nb_3Sn undergoes a cubic-to-tetragonal phase transition at low temperatures and shows mode softenings as precursor effects of this transition.

In Nb_3Sn as well as in the other high T_c A15 materials, a pair of very flat bands, originating from a Γ_{12} state at the zone center, dominates the region near E_F. These Γ_{12} bands appear to be the source of all unusual properties of the high T_c A15 materials. They give rise to a pair of sharp peaks in the electronic density of states near E_F. The principal orbital component of the flat Γ_{12} bands consists of $d_{3z^2-r^2}$ Nb orbitals pointing <u>along</u> the chain directions. Thus the small band dispersion <u>cannot</u> arise from small orbital overlap, but is caused essentially <u>by</u> (interchain) hybridization with Nb d_{xz}, d_{yz} orbitals, the second most important orbital component in these bands.

A tight binding energy band calculation for tetragonally distorted Nb_3Sn has shown that a pairing of Nb atoms in two of the three chains is the primary source of the distortion. This pairing corresponds to a displacement pattern of a Γ_{12} phonon in the cubic phase. The tetragonal distortion has been found to split the peak in the density of states related to the Γ_{12} state. The distortion is energetically favorable when E_F lies very close ($\approx \pm$ 50 meV) to the electronic Γ_{12} state. According to various energy band calculations, this is the case for Nb_3Sn and V_3Si, but need not be true for other high T_c A15 materials.

Complementary to this static study are the results of a tight binding calculation of the Nb_3Sn phonon dispersion curves. Since hardly any experimental data of Nb_3Sn phonons have been available, the ample neutron data of the low T_c reference material Nb_3Sb have been used to determine the short range forces of (D_0 + D_1). The Nb_3Sn calculations agree well with the very limited experimental information. In particular, it has been found that the Γ_{12} phonon frequency is considerably depressed in Nb_3Sn as compared to Nb_3Sb. The Γ_{12} mode is further lowered when E_F is slightly shifted closer to the electronic Γ_{12} state. The Γ_{12} mode softening is driving the shear elastic constant (c_{11} - c_{12}) towards zero, in a form very similar as observed in experiment. Consistent with the static results, the Γ_{12} distortion appears to be the driving force of the tetragonal transition.

It should be pointed out however that the Γ_{12} - Γ_{12} electron-phonon coupling is a situation where the adiabatic approximation, used throughout these calculations, is not longer valid.

Apart from the specific Γ_{12} mode softening, the Nb_3Sn phonon calculations predict that other specific modes are strongly depressed as compared to Nb_3Sb. These are the "internal" modes, where the two Nb atoms within one chain vibrate out of phase, either along or perpendicular to the chain directions. "External" modes, where the two chain atoms move in phase are not much lowered compared to Nb_3Sb. As a consequence, the "internal" R_{1-2} and R_3 phonon modes at the R point of the Brillouin zone have become the lowest R point modes, and phonon anomalies appear along the Λ direction, especially in branches ending at R_{1-2} and R_3.

Because of the peculiar Γ_{12} bands in Nb_3Sn, arguments of k-space nesting between states with large and opposite band dispersion cannot be used to trace the origin of phonon anomalies. A more appropriate (and also more fundamental) criterion is to look for vibration patterns, where large gradients of those tight binding integrals are generated which involve the dominant orbitals of the flat Γ_{12} bands. The biggest gradients occur when neighbor atoms in one chain move against each other ; i.e. for "internal" modes. Then either a longitudinal coupling of $d_{3z^2-r^2}$ orbitals is induced (affecting Γ_{12} and Γ_2), or a transverse coupling involving a "dormant" interaction of $d_{3z^2-r^2}$ and d_{xz}, d_{yz} orbitals. These big gradients are inoperative for the "external" modes (Γ_{15}, Γ_{25}, R_4), thus these modes are lowered very little compared to Nb_3Sb.

In addition to the phonon dispersion curves, also calculations of electron-phonon coupling quantities have been presented. The coupling strength of a phonon to the electrons at E_F is described by the "intrinsic" phonon linewidth. For all three transition metal systems, it has been found that the "anomalous" phonons also exhibit very large linewidths γ_{qj}.

Compared to a nonsuperconducting reference material without phonon anomalies, the electron-phonon coupling parameter

$$\lambda \propto \sum_{qj} \gamma_{qj}/\omega_{qj}^2 \tag{2.20}$$

is increased both by the γ_{qj} values of the anomalous phonons, and by their frequency depression. In general, the linewidth effect is considerably larger than that of the frequencies. Since λ essentially determines the superconducting temperature

$$T_c \approx \langle\omega\rangle \exp[-(1+\lambda)/\lambda]$$

it is the strong coupling of the anomalous phonons to the electrons near and at E_F, which causes high values of T_c. Thus the two phenomena, high transition temperatures and phonon anomalies are indeed closely related.

The anisotropy of phonon linewidths has been found to be much more pronounced in the refractory compounds and the A15 materials than in elemental Nb. There, the largest γ_{qj} values exceed the average by ≈ 2, while for NbC and Nb_3Sn even factors 10 have been obtained. In the refractory compounds, a detailed analysis has shown that not only a few anomalous phonons in very limited q-space regions couple very strongly to the electrons, but also only a few electronic states in very limited k-space regions at E_F are very strongly coupled via the anomalous phonons. For Nb_3Sn, a different picture arises : Here, the mode character can be used for the classification of the coupling : only "internal" modes, but not the "external" modes couple strongly to the electrons at E_F.

Compared to other theoretical investigations of the electron-phonon interaction, such as the rigid-muffin-tin-approximation treatment, the tight-binding results agree fairly well. However, there are some discrepancies, for instance, the tight binding results give considerably weaker coupling at small phonon frequencies than the RMTA method.

The experimental results for the anisotropy of the electron-phonon coupling on the various Nb Fermi surface sheets, found from de Haas-van Alphen data, agree quite well with the tight-binding calculations, while they agree less with RMTA results. On the other hand, tight binding calculations of the Eliashberg function $\alpha^2F(\omega)$ for Nb deviate more from tunneling data than do the RMTA results. For NbC the tight-binding $\alpha^2F(\omega)$ curve agrees even less with experiment, but this may be due largely to experimental problems. The relatively best agreement between theoretical $\alpha^2F(\omega)$ curves and tunneling data is found for Nb_3Sn.

Altogether, the various experimental and theoretical results on the electron-phonon coupling are not very consistent. It appears that on the theoretical side, a detailed comparison of different approaches is necessary to find the source(s) of discrepancy. It is possible that the approximation of constant gradients in the tight binding method leads to an overestimation of the anisotropy. On the other hand, corrections beyond the rigid-muffin-tin approximation need to be carefully investigated. What the experiments concerns, it is not easy to see how for Nb the tunneling data and the results from de Haas-van Alphen measurements are compatible. To resolve all these problems, it would be very helpful if high precision measurements of phonon linewidths in Nb could be performed, for instance using the neutron spin echo method.

In the past few years, various modern band-structure methods have been utilized to obtain accurate total crystal energy curves as functions of lattice configuration coordinates. These calculations not only allow to determine structural properties within ≈ 1 % accuracy, they also yield values of elastic constants and high symmetry phonon frequencies within 5 - 10 % accuracy. In Sect. 4 a brief survey of these "frozen" phonon calculations has been given, thereby focussing on results of a first principles linear-combination-of-atomic-orbitals method. For the test material Si it has been shown that the LCAO method yields as good results as other methods. For transition metals, the LCAO method has been supplemented by a tight binding interpolation scheme which provided a very precise weighting of states near E_F. Results for Nb and Mo are as accurate as for Si. It appears however that the pseudopotential method is a more economical scheme for total energy calculations.

"Frozen" phonon calculations may complement the tight binding method not only to provide the equilibrium matrix elements and their gradients, but also to give the information to determine the short range force constants for $D_0 + D_1$. In this way it appears to be possible to avoid any experimental input and yet to calculate the full phonon dispersion curves and the electron-phonon coupling of transition metals and their compounds within a few percent of accuracy.

APPENDIX

A1. Formal Theory of Lattice Dynamics

In harmonic approximation, (Maradudin et al. 1971, p. 6ff) the Hamiltonian for the lattice vibrations can be written as

$$H_L = \frac{1}{2} \sum_{\ell\kappa\alpha} m_\kappa [\dot{u}(\ell\kappa\alpha)]^2 + \frac{1}{2} \sum_{\substack{\ell\kappa\alpha \\ \ell'\kappa'\beta}} \phi(\ell\kappa\alpha,\ell'\kappa'\beta) u(\ell\kappa\alpha) u(\ell'\kappa'\beta) \quad \text{(A1)}$$

Here $u(\ell\kappa\alpha)$ is the displacement in direction α of the κ-th atom of mass m_κ in the ℓ-th unit cell from an equilibrium position $\vec{R}(\ell\kappa)$. $\phi(\ell\kappa\alpha,\ell'\kappa'\beta)$ is the force constant matrix. Because of lattice periodicity we can expand

$$u(\ell\kappa\alpha) = N^{-1/2} \sum_q u_q(\kappa\alpha) \exp[i\vec{q}\,\vec{R}(\ell\kappa)] \quad \text{(A2)}$$

with wavevectors q taken from the first BZ. Inserting (A2) in (A1) we get

$$H_L = \frac{1}{2} \sum_q \{ \sum_{\kappa\alpha} m_\kappa \dot{u}_q(\kappa\alpha) \dot{u}_{-q}(\kappa\alpha) + \sum_{\substack{\kappa\alpha \\ \kappa'\beta}} D(\kappa\alpha,\kappa'\beta|q) u_q(\kappa\alpha) u_{-q}(\kappa'\beta) \} \quad \text{(A3)}$$

with the dynamical matrix

$$D(\kappa\alpha,\kappa'\beta|q) = \sum_{\ell-\ell'} \phi(\ell\kappa\alpha,\ell'\kappa'\beta) \exp[-i\vec{q}\,\vec{R}(\ell\kappa,\ell'\kappa')] \quad \text{(A4)}$$

(We use the short hand notation $R(\ell\kappa,\ell'\kappa') = R(\ell\kappa) - R(\ell'\kappa')$ and indicate, if necessary, 3-dimensional vectors by arrows).

The dynamical matrix satisfies the eigenvalue equation

$$\omega^2(qj) e(\kappa\alpha|qj) = \sum_{\kappa'\beta} (m_\kappa m_{\kappa'})^{-1/2} D(\kappa\alpha,\kappa'\beta|q) e(\kappa'\beta|qj) \quad \text{(A5)}$$

where $\omega(qj)$ are the phonon frequencies of wave vector q and branch index j, and $e(\kappa\alpha\, qj)$ are the corresponding polarization.

As a consequence of the fact that the displacements $u(\ell\kappa\alpha)$ are real, we can put

$$e(\kappa\alpha|-qj) = e^\star(\kappa\alpha|qj)$$

or (A6)

$$u_{-q}(\kappa\alpha) = u_q^\star(\kappa\alpha)$$

This is the Born-Huang convention (Maradudin et al., Eq. 2.1.57).

In the following derivation we will assume that a displacement field $u(\ell\kappa\alpha)$ is associated with one specific wavevector q. Then Eq. (A2) reads

$$u(\ell\kappa\alpha) = u_q(\kappa\alpha)\exp[i\vec{q}\,\vec{R}(\ell\kappa)] + u_{-q}(\kappa\alpha)\exp[-i\vec{q}\,\vec{R}(\ell\kappa)]$$

$$= u_q(\kappa\alpha)\exp[i\vec{q}\,\vec{R}(\ell\kappa)] + \text{c.c.} \qquad \text{(A2a)}$$

Inserting (A2a) in (A1) leads to

$$H_L = \frac{1}{2}\sum_{\kappa\alpha} m_\kappa\{\dot{u}_q(\kappa\alpha)\,\dot{u}_q^\star(\kappa\alpha) + \dot{u}_q^\star(\kappa\alpha)\dot{u}_q(\kappa\alpha)\}$$

$$+ \frac{1}{2}\sum_{\substack{\kappa\alpha\\ \kappa'\beta}} \{D(\kappa\alpha,\kappa'\beta|q)u_q(\kappa\alpha)u_q^\star(\kappa'\beta) + D(\kappa\alpha,\kappa'\beta|-q)u_{-q}^\star(\kappa\alpha)u_{-q}(\kappa'\beta)\}$$

$$= H_L(q) + H_L(-q) = 2H_L(q) \qquad \text{(A7)}$$

A2. The Non-Orthogonal Tight Binding Scheme

We now define

$$\phi(\ell\kappa m) = \phi_m[\vec{r} - \vec{R}(\ell\kappa) - \vec{u}(\ell\kappa)] \qquad \text{(A8)}$$

to be the orbital of character m situated at the κ-th atom of the ℓ-th unit cell. We assume that the orbital is moving with the displaced atom. The corresponding "Bloch orbital" is then

$$\Phi_k(\kappa m) = N^{-1/2}\sum_\ell \phi(\ell\kappa m)\exp[i\vec{k}\,\vec{R}(\ell\kappa)] \qquad \text{(A9)}$$

Note that the electron wavevectors are always labelled by k or k', and the phonon wavevectors by q.

At lattice equilibrium, i.e., if all $u(\ell\kappa\alpha) \equiv 0$, we obtain a Hermitian Hamiltonian matrix

$$H(\kappa m,\kappa' m'|k) = \langle\Phi_k(\kappa m)|H|\Phi_k(\kappa' m')\rangle$$

$$= \sum_{\ell-\ell'} \exp[-i\vec{k}\,\vec{R}(\ell\kappa,\ell'\kappa')]\,.\int d^3r\,\phi^\star(\ell\kappa m)H\phi(\ell'\kappa' m')$$

$$= \sum_{\ell-\ell'} \exp[-i\vec{k}\,\vec{R}(\ell\kappa,\ell'\kappa')]\;H(\ell\kappa m,\ell'\kappa' m') \qquad \text{(A10)}$$

Here, H is the electronic part of the total Hamiltonian of Eq. (2.1) ; i.e.

$$H = T_e + V_{ec}(\{r\},\{\ell\}) + V_{ee}(\{r\},\{\ell\})$$

with $\{\ell\}$ being the equilibrium configuration of the ions. Replacing the operator H by 1 leads to the overlap matrix $S(\kappa m,\kappa' m'|k)$. In matrix notation, the non-Hermitian eigenvalue equation for the one-electron states reads as

$$HA = SAE \tag{A11}$$

Here, E(k) is the diagonal matrix of the energy eigenvalues $\varepsilon_{k\mu}$, and A is the matrix of non-unitary (right-hand-side) eigenvectors. When we chose a certain matrix L such that

$$L^+ L = S^{-1}$$

we may transform Eq. (A11) to a Hermitian eigenvalue equation. In particular we use

$$L = L^+ = S^{-1/2}$$

and obtain

$$S^{-1/2}\, H\, S^{-1/2}\, U = UE \tag{A12}$$

Then a Bloch state $\psi_{k\mu}$ of energy $\varepsilon_{k\mu}$ (μ is band index) is given by

$$\psi_{k\mu} = \sum_{\kappa m} A(\kappa m|k\mu)\Phi_k(\kappa m) = N^{-1/2} \sum_{\ell\kappa m} A(\kappa m|k\mu)\phi(\ell\kappa m)\exp[i\vec{k}\,\vec{R}(\ell\kappa)] \tag{A13}$$

Further, there are the relations

$$A = S^{-1/2}U \quad \text{and} \quad A^+ = U^+ S^{-1/2} \tag{A14}$$

A3. Lattice Dynamics in NTB

To derive the changes $\Delta\varepsilon_{k\mu}$ in the one-electron energies up to second order in lattice displacements $u(\ell\kappa\alpha)$, we assume a phonon field of wavevector q such that

$$u(\ell\kappa\alpha) = u_q(\kappa\alpha)\, \exp[i\vec{q}\,\vec{R}(\ell\kappa)] + u_q(\kappa\alpha)\, \exp[-i\vec{q}\,\vec{R}(\ell\kappa)] \tag{A2a}$$

Now, Bloch orbitals Eq. (A9) of wavevectors k and k' will be coupled, where $k' = k \pm q$. We thus construct a generalized Hamiltonian matrix $\tilde{H}$ with the elements

$$\tilde{H}_{ij} = H(\kappa m,k|\kappa' m',k') = \langle\Phi_k(\kappa m)|H|\Phi_{k'}(\kappa' m')\rangle \tag{A15}$$

The indices i,j and j' are short hand notations for the groups of indices (κm,k), (κ'm',k + q), and (κ"m",k−q), respectively. Note that the index i represents one fixed value of k, but (κ,m) may run through all index values ; i.e., (ii) = (κm,k|κ'm',k). The overlap matrix $\tilde{S}$ is found correspondingly to H. We keep in mind that the Bloch functions $\psi_k(\kappa m)$ are constructed from displaced orbitals $\phi(\ell\kappa m)$ of Eq. (A8) and expand now the matrices $\tilde{H}_{ij}$ and $\tilde{S}_{ij}$ up to $O(u^2)$. This leads to

$$\tilde{H} = \begin{pmatrix} H_j + h_{jj} & h_{ji} & h_{jj'} \\ h_{ij} & H_i + h_{ii} & h_{ij'} \\ h_{j'j} & h_{j'i} & H_{j'} + h_{j'j'} \end{pmatrix} \tag{A16}$$

$$= H_0 + h$$

Similarly

$$\tilde{S} = S_0 + \Delta \tag{A17}$$

The matrix H_i is the equilibrium matrix $H(\kappa m,\kappa' m'|k)$ of Eq. (A10), and analogously defined are H_j, $H_{j'}$, S_i etc. As we will see below, the matrices h_{ii} and Δ_{ii} etc. in the diagonal are of order $O(u^2)$ and higher, while the off-diagonal matrices h_{ij}, $h_{ij'}$, Δ_{ij}, $\Delta_{ij'}$ are of order $O(u)$.

Again we arrive at a non-Hermitian eigenvalue equation

$$\tilde{H}\,\tilde{A} = \tilde{S}\,\tilde{A}\,E$$

Chosing a matrix

$$\tilde{L} = S_0^{-1/2}\,(1 - 1/2\,\Delta\,S_0^{-1} + 3/8\,\Delta\,S_0^{-1}\,\Delta\,S_0^{-1}) \tag{A18}$$

such that

$$\tilde{L}^{+}\tilde{L} = (S_0 + \Delta)^{-1} \quad \text{up to} \quad O(u^2) \tag{A19}$$

we deal with the Hermitian problem

$$\tilde{L}\,\tilde{H}\,\tilde{L}^{+}\,\tilde{U} = \overline{H}\,\tilde{U} = \tilde{U}\,E \tag{A20}$$

The elements of the matrix $\overline{H}$ are

$$\overline{H}_{ii} = S_i^{-1/2} H_i\, S_i^{-1/2} + S_i^{-1/2} h_{ii}\, S_i^{-1/2}$$
$$- 1/2\{S_i^{-1/2}\Delta_{ii}\, S_i^{-1} H_i S_i^{-1/2} + \text{c.c.}\}$$

$$
\begin{aligned}
&- 1/2\ \{S_i^{-1/2}\ \Delta_{ij}\ S_j^{-1}\ h_{ji}S_i^{-1/2} + \text{c.c.}\} \\
&+ 3/8\ \{S_i^{-1/2}\ \Delta_{ij}\ S_j^{-1}\Delta_{ji}\ S_i^{-1}H_iS_i^{-1/2} + \text{c.c.}\} \\
&+ 1/4\ \{S_i^{-1/2}\ \Delta_{ij}\ S_j^{-1}\ H_jS_j^{-1}\Delta_{ji}\ S_i^{-1/2}\} \qquad \text{(A21)} \\
&\qquad + \text{terms with } j \to j'
\end{aligned}
$$

$$
\begin{aligned}
\overline{H}_{ij} &= S_i^{-1/2}h_{ij}S_j^{-1/2} - 1/2\{S_i^{-1/2}\Delta_{ij}S_j^{-1}H_jS_j^{-1/2} + \text{c.c.}\} \\
&\qquad + \text{terms with } j \to j' \qquad \text{(A22)}
\end{aligned}
$$

We perform now a unitary transformation to band space

$$\overline{H}' = \overline{U}^{+}\ \overline{H}\ \overline{U}$$

with

$$\overline{U} = \begin{pmatrix} U_j & 0 & 0 \\ 0 & U_i & 0 \\ 0 & 0 & U_{j'} \end{pmatrix}$$

where U_i is the eigenvector matrix of Eq. (A12).
Then we obtain the energy of the (occupied) state (kμ) up to $O(u^2)$ in first and second order perturbation theory

$$
\begin{aligned}
\varepsilon'_{k\mu} &= \overline{H}'_{k\mu,k\mu} \\
&+ \sum_{\substack{\mu' \\ k'=k\pm q}} \frac{f_{k\mu}(1 - f_{k'\mu'})}{\varepsilon_{k\mu} - \varepsilon_{k'\mu'}}\ \overline{H}'_{k\mu,k'\mu'}\ \overline{H}'_{k'\mu',k\mu} \qquad \text{(A23)}
\end{aligned}
$$

We have introduced the Fermi factors $f_{k\mu}$ to guarantee that the occupied states can couple only to unoccupied ones.

When we sum over all occupied states (kμ) we obtain the contribution of the sum of single particle energies to the potential energy V of the crystal up to $O(u^2)$. There are two contributions
i) from first order perturbation - involving the terms in $\overline{H}'_{k\mu,k\mu}$ second order in displacements. They lead to D_1.
ii) from second order perturbation - involving terms $\overline{H}'_{k\mu,k'\mu'}$ first order in displacements. They lead to D_2.
We now focus on ii) and will deal with i) later.

We will show in Sect. A4 that we can write

$$\overline{H}'_{k\mu,k\pm q\mu'} = \sum_{\kappa\alpha} g^{\kappa\alpha}_{k\mu,k\pm q\mu'}\ u_{\pm q}(\kappa\alpha) \qquad \text{(A24)}$$

Then we get two sums

$$V = \sum_{\substack{\kappa\alpha \\ \kappa'\beta}} \sum_{\substack{k\mu\mu' \\ k'=k+q}} \frac{f_{k\mu}(1 - f_{k'\mu'})}{\varepsilon_{k\mu} - \varepsilon_{k'\mu'}} g^{\kappa\alpha}_{k\mu,k'\mu'} g^{\kappa'\beta}_{k'\mu',k\mu} u_{-q}(\kappa\alpha) u_q(\kappa'\beta)$$

$$+ \sum_{\substack{\kappa\alpha \\ \kappa'\beta}} \sum_{\substack{k\mu\mu' \\ k'=k-q}} \frac{f_{k\mu}(1 - f_{k'\mu'})}{\varepsilon_{k\mu} - \varepsilon_{k'\mu'}} g^{\kappa\alpha}_{k\mu,k'\mu'} g^{\kappa'\beta}_{k'\mu',k\mu} u_q(\kappa\alpha) u_{-q}(\kappa'\beta)$$

By relabelling the indices in the first term such that $(\kappa'\beta) \to (\kappa\alpha)$ and $(\kappa\alpha) \to (\kappa'\beta)$, and in the second term that $(k\mu) \to (k'\mu')$ and $(k'\mu') \to (k\mu)$, we can combine the two sums to

$$V = - \sum_{\substack{\kappa\alpha \\ \kappa'\beta}} \sum_{\substack{k\mu\mu' \\ k'=k+q}} \frac{f_{k'\mu'} - f_{k\mu}}{\varepsilon_{k\mu} - \varepsilon_{k'\mu'}} g^{\kappa\alpha}_{k\mu,k'\mu'} g^{\kappa'\beta}_{k'\mu',k\mu} u_q(\kappa\alpha) u_{-q}(\kappa'\beta) \quad \text{(A25)}$$

A comparison with Eq. (A7) gives

$$D_2(\kappa\alpha,\kappa'\beta|q) = - \sum_{\substack{k\mu\mu' \\ k'=k+q}} \frac{f_{k'\mu'} - f_{k\mu}}{\varepsilon_{k\mu} - \varepsilon_{k'\mu'}} g^{\kappa\alpha}_{k\mu,k'\mu'} g^{\kappa'\beta}_{k'\mu',k\mu} \quad \text{(A26)}$$

Note i) Eq. (A7) includes D(q) and D(-q), but has a factor 1/2 in front of the sum, which Eq. (A25) does not. ii) the minus sign in front of Eqs. (A25) or (A26) indicates that the contributions of D_2 to the dynamical matrix always lower all eigenvalues. iii) the sum over k should actually read as sum over k and spins.

A4. The Detailed Expression for $g^{\kappa\alpha}_{k\mu,k'\mu'}$

From Eq. (A22) we get for (remember that $i = (\kappa m,k)$ and $j = (\kappa' m',k')$)

$$\begin{aligned} \overline{H}'_{k\mu,k'\mu'} &= U^{\star}_{i\mu} \overline{H}_{ij} U_{j\mu'} \\ &= U^{\star}_{i\mu} S_i^{-1/2} \{h_{ij} - 1/2\, (\Delta_{ij} S_j^{-1} H_j + \\ &\quad H_i S_i^{-1} \Delta_{ij}\} S_j^{-1/2} U_{j\mu'} \end{aligned} \quad \text{(A27)}$$

Using

$$S_i^{-1/2} H_i S_i^{-1/2} U_{i\mu} = U_{i\mu} \varepsilon_{k\mu}$$

and inserting Eqs. (A14) we get

$$H'_{k\mu,k'\mu'} = \sum_{\substack{\kappa m \\ \kappa' m'}} A^{\star}(\kappa m|k\mu)\ [h(\kappa m,k|\kappa'm'k') - 1/2\ (\varepsilon_{k\mu} + \varepsilon_{k'\mu'})\ .\ \Delta(\kappa m,k|\kappa'm'k')] A(\kappa'm'|k'\mu') \qquad (A28)$$

We now focus on the term h_{ij}, the quantity Δ_{ij} can be obtained analogously. Using Eqs. (A15) and (A9) we get

$$h_{ij} = h(\kappa m,k|\kappa'm',k') = \langle\Phi_k(\kappa m)|H|\Phi_{k'}(\kappa'm')\rangle$$
$$= N^{-1} \sum_{\ell,\ell'} H(\ell\kappa m,\ell'\kappa'm') \exp[-\, i\vec{k}\,\vec{R}(\ell\kappa) + i\vec{k}'\vec{R}(\ell'\kappa')] \qquad (A29)$$

There are various ways to change $H(\ell\kappa m,\ell'\kappa'm')$ because of lattice displacements $u(\ell\kappa\alpha)$. We may think of the one-electron Hamiltonian as

$$H = T + \sum_{\ell,\kappa} V(\ell,\kappa)$$

i.e., as a sum of potentials situated at the lattice sites (ℓ,κ) plus the kinetic energy operator. Then we get three different types of terms, apart from trivial on-site terms.

i) two)center terms involving orbitals at sites $(\ell\kappa) \neq (\ell'\kappa')$ with the potential situated on one or the other side.

ii) "crystal field" terms with the two orbitals at the same site $(\ell\kappa)$ and the potential at a neighbor site $(\ell'\kappa')$.

iii) three center terms with orbitals and the potential at three different sites.

In the following, we will mainly deal with terms of type i), but we will also discuss terms of type ii). Three center terms iii) are so far not included in the formalism. We note that the overlap terms in Δ_{ij} are of type i).

$$H(\ell\kappa m,\ell'\kappa'm') = H^0(\ell\kappa m,\ell'\kappa'm') + \sum_{\alpha}[\partial/\partial u_\alpha(\ell\kappa,\ell'\kappa')\ H(\ell\kappa m,\ell'\kappa'm')]\,u_\alpha(\ell\kappa,\ell'\kappa')$$
$$+\ 1/2 \sum_{\alpha\beta} [\partial^2/\partial u_\alpha(\ell\kappa,\ell'\kappa')\partial u_\beta(\ell\kappa,\ell'\kappa')H(\ell\kappa m,\ell'\kappa'm')]\ .\ u_\alpha(\ell\kappa,\ell'\kappa')\ u_\beta(\ell\kappa,\ell'\kappa') \qquad (A30)$$

For h_{ij} and Δ_{ij}, it is easy to see that only the term linear in u is needed in harmonic approximation. We denote

$$\nabla_\alpha H(\ell\kappa m,\ell'\kappa'm') = \partial/\partial u_\alpha(\ell\kappa,\ell'\kappa')H(\ell\kappa m,\ell'\kappa'm') \qquad (A31)$$

and define similarly $\nabla_\alpha S$, $\nabla_\alpha\nabla_\beta H$ and $\nabla_\alpha\nabla_\beta S$. Then

$$\gamma_\alpha(\kappa m,\kappa' m'|k) = \sum_{\ell-\ell'} [\nabla_\alpha H(\ell\kappa m,\ell'\kappa' m') - \varepsilon\nabla_\alpha S(\ell\kappa m,\ell'\kappa' m')] \exp[-i\vec{k}\,\vec{R}(\ell\kappa,\ell'\kappa')] \quad (A32)$$

with

$$\varepsilon = 1/2\,(\varepsilon_{k\mu} + \varepsilon_{k'\mu'})$$

Thus we can write for $g^{\kappa\alpha}$ as defined in Eq. (A24)

$$g^{\overline{\kappa}\alpha}_{k\mu,k'\mu'} = \sum_{\substack{\kappa m\\ \kappa' m'}} A^\star(\kappa m|k\mu)[\gamma_\alpha(\kappa m,\kappa' m'|k+q)\delta_{\kappa\overline{\kappa}} - \gamma_\alpha(\kappa m,\kappa' m'|k)\,\delta_{\kappa'\overline{\kappa}}]\,A(\kappa' m'|k'\mu') \quad (A33)$$

The derivation of Eq. (A33) requires a somewhat tedious book-keeping of the individual displacements $u_q(\kappa\alpha)$. We thus give an example for $g^{\kappa\alpha}$ in the case of two atoms $(\kappa = 1,2)$ per unit cell.

$$g^{1\alpha}_{k\mu,k'\mu'} = \sum_{\substack{m_1,m_2\\ m_1',m_2'}} [A^\star(1\,m_1|k\mu),A^\star(2m_2|k\mu)] \left\{\begin{pmatrix}\gamma_\alpha(1\,m_1,\,1\,m_1'|k') & \gamma_\alpha(1\,m_1,\,2m_2'|k')\\ 0 & 0\end{pmatrix} - \begin{pmatrix}\gamma^\alpha_\alpha(1\,m_1,\,1\,m_1'|k) & 0\\ \gamma^\alpha(2\,m_2,\,1\,m_1'|k) & 0\end{pmatrix}\right\}\begin{bmatrix}A(1\,m_1'|k'\mu')\\ A(2\,m_2'|k'\mu')\end{bmatrix} \quad (A34)$$

Here the sums over the orbital indices m and m' are explicitly written as sums over m_1 and m_2, and m_1' and m_2', respectively. This is to indicate that a subset m_1 of the set m of orbital is associated with particle 1 etc.

We now give an expression for the "crystal field" terms ii) in the electron-phonon matrix $g^{\kappa\alpha}$. Starting from Eq. (A29) we have

$$h^{cf}_{ij} = N^{-1}\sum_\ell \exp[-i\,(\vec{k} - \vec{k} - \vec{q})\,\vec{R}(\ell\kappa)]\,F(\ell\kappa m,\ell\kappa m'|\ell'\kappa') \quad (A35)$$

where the labels $(\ell'\kappa')$ denote the potential site. Taking a term linear in displacements, we obtain

$$h^{cf}_{ij} = \sum_{\ell',\alpha} \nabla_\alpha F(\kappa m,\kappa m\,|\ell'\kappa')\,(u^\star(\kappa\alpha) - u^\star(\kappa'\alpha)\,\exp[i\vec{q}\,\vec{R}(0\kappa,\ell'\kappa')])$$

We put

$$C(\kappa m,\kappa m'|q) = \sum_{\ell'} \nabla_\alpha F(\kappa m,\kappa m'|\ell'\kappa') \, . \, \exp[\, i\vec{q}\, \vec{R}(0\kappa,\ell'\kappa')]$$

and obtain

i) in the case $\kappa' = \kappa$ (same sort of atom but in a neighbor cell)

$$g^{\kappa\alpha}_{k\mu,k'\mu'} = \sum_{\kappa mm'} A^\star(\kappa m,k\mu)\, [\, C(\kappa m,\kappa m'|0) - C(\kappa m,\kappa m'|q)]\, A(\kappa m',k'\mu') \qquad (A36)$$

ii) in the case $\kappa' \neq \kappa$ we have

$$g^{\kappa\alpha}_{k\mu,k'\mu'} = \sum_{\kappa mm'} A^\star(\kappa m,k\mu)\, C(\kappa m,\kappa m'|0)\, A(\kappa m',k'\mu')$$

$$g^{\kappa'\alpha}_{k\mu,k'\mu'} = \sum_{\kappa mm'} A^\star(\kappa m,k\mu)\, C(\kappa m,\kappa m'|q)\, A(\kappa m',k'\mu') \qquad (A37)$$

A5. <u>The First Order Perturbation Term D_1</u>

Eq. (A21) gives five different terms to $\overline{H}_{ii}$ of order $O(u^2)$. We do not present the detailed algebra how to arrive at D_1, instead we sketch some important steps. The first two terms of Eq. (A21) are combined and we arrive at a force constant matrix

$$\phi(\ell\kappa\alpha,\ell'\kappa'\beta) = 2\, N^{-1} \sum_{k\mu} f_{k\mu} \sum_{mm'} A^\star(\kappa m,k\mu)\gamma_{\alpha\beta}(\ell\kappa m,\ell'\kappa'm')A(\kappa'm',k\mu) \qquad (A38)$$

with the matrix

$$\gamma_{\alpha\beta}(\ell\kappa m,\ell'\kappa'm') = \nabla_\alpha\nabla_\beta\, H(\ell\kappa m,\ell'\kappa'm') - \varepsilon_{k\mu}\, \nabla_\alpha\nabla_\beta\, S(\ell\kappa m,\ell'\kappa'm') \qquad (A39)$$

defined in analogy to Eqs. (A29,A30). Note that the factor 2 is <u>not</u> a spin factor. The derivation of ϕ is again based on the two-center approximation, leading to $\gamma_{\alpha\beta} = \gamma_{\beta\alpha}$. Crystal field terms have to be treated separately.

Proceeding along the lines given by Eqs. A1 - A4, we then obtain $D_1^{(1+2)}$, where the superscript (1 + 2) should denote the first two terms of Eq. (A21).

There is one important point to note from Eq. (A38). The range of force constants $\phi_{\alpha\beta}(\ell\kappa\alpha,\ell'\kappa'\beta)$ in $D_1^{(1+2)}$ is identical with the range of energy transfer or overlap matrix elements in $\gamma_{\alpha\beta}(\ell\kappa m,\ell'\kappa'm')$. This means, $D_1^{(1+2)}$ is of short range, when the orbital overlap is short range.

We now consider terms 3 - 5 of Eq. (A21). These are mixed terms involving intermediate states j. It is appropriate to use the identity

$$S_j^{-1} = S_j^{-1/2}\, U\, U^+\, S_j^{-1/2} \tag{A40}$$

We define in analogy to Eq. (A33)

$$h^{\overline{\kappa}\alpha}_{k\mu,k'\mu'} = \sum_{\substack{\kappa m\\ \kappa' m'}} A^{\star}(\kappa m|k\mu)\,[H_\alpha(\kappa m,\kappa' m'|k')\delta_{\kappa\overline{\kappa}} - H_\alpha(\kappa m,\kappa' m'|k)\delta_{\kappa'\overline{\kappa}}]\;A(\kappa' m'|k'\mu') \tag{A41}$$

and the quantity $d^{\overline{\kappa}\alpha}_{k\mu,k'\mu'}$ for the corresponding overlap derivative matrices. Transformation of Eq. (A21) to band space, summation over all occupied states, insertion of Eq. (A40) and use of Eqs. (A41) gives

$$D_1^{(3-5)}(\kappa\alpha,\kappa'\beta|q) = -\frac{1}{2}\sum_{\substack{k\mu\mu'\\ k'=k\pm q}} f_{k\mu}(d^{\kappa\alpha}_{k\mu,k'\mu'}\,h^{\kappa'\beta}_{k'\mu',k\mu} + \text{c.c.}) + \sum_{\substack{k\mu\mu'\\ k'=k\pm q}} f_{k\mu}\,d^{\kappa\alpha}_{k\mu,k'\mu'}\,d^{\kappa'\beta}_{k'\mu',k\mu}(3/4\,\varepsilon_{k\mu}+1/4\varepsilon_{k'\mu'}) \tag{A42}$$

When looking at the range of forces in $D_1^{(3-5)}$ we find they essentially do not extend beyond twice the longest overlap distance. This can be seen from a real space analysis of terms $\Delta_{ij}\, h_{ji}$ etc. of Eq. (A21).

A6. Differences to the Previous Derivation

In the derivation given by Varma and Weber (1979) the matrix γ_α of Eq. (A32) differed by the prefactor $\varepsilon_{k\mu}$ instead of $1/2\,(\varepsilon_{k\mu} + \varepsilon_{k'\mu'})$ in front of the $\nabla_\alpha S$ term. The prefactor $\varepsilon_{k\mu}$ leads to an asymmetry in $g^{\alpha}_{k\mu,k'\mu'} \neq g^{\star\alpha}_{k'\mu',k\mu}$ so that $g_{k'\mu',k\mu}$ had to be taken with the $\varepsilon_{k\mu}$ factor too.

This caused some problems with the individual translational invariance of D_2 in the numerical calculations. It can be shown, however, that the difference between the two expressions for D_2 lead to the additional terms (3-5) in D_1, which are absent in the previous derivation of D_1. In addition, Eq. (21) of VW exhibits a prefactor 1/2 instead of 2.

A7. Description of the General NTB Program Package

As was mentioned in Sect. 2.4, the NTB phonon programs consists of three main parts, the energy band code, the D_2 program and the $\omega(qj)$ part. These programs are now described in some detail.

A7.1. The NTB Energy Band Programs

These computer programs have been developed in collaboration with L.F. Mattheiss and have so far been applied mainly to analyze the energy bands of the A15 compounds Nb_3Sn and V_3Si (Mattheiss and Weber 1982, Weber and Mattheiss 1982). The programs are applicable to materials with arbitrary crystal structures. They allow us to change quite readily from one crystal structure to another simply by altering a few program statements and the input data. In particular, it is possible to choose NTB basis functions consisting of Bloch sums formed from any combination of s, p, or d orbitals at each inequivalent site in the primitive unit cell. Furthermore, one can easily vary the number of shells of neighbors for which energy, overlap, and crystal-field-type matrix elements are included in a given calculation. The principal limitations of the program are : i) it is not equipped to handle orbitals with f or higher angular-momentum quantum numbers ; ii) it applies the two-center approximation, thereby neglecting all three-center Hamiltonian integrals.

In the case of a simple cubic crystal with an s, p, d orbital basis and three shells of neighbors, the NTB program automatically generates, for example, the Hamiltonian matrix elements H that are listed in Table III of the Slater-Koster (1954) paper. In addition, however, it also generates the corresponding overlap matrix S. This has the same analytic form as H, except that the orbital energies

$$E_{\kappa m} = \int d^3r\ \phi^{\star}(\ell\kappa m)\ (T + V(\ell\kappa))\ \phi(\ell\kappa m) \qquad (A43)$$

are replaced by unity and the two-center energy integrals $H(\ell\kappa m,\ell'\kappa'm')$ of Eq. (A10) are replaced by the corresponding overlap integrals $S(\ell\kappa m,\ell'\kappa'm')$. The basic constituents of $H(\ell\kappa m,\ell'\kappa'm')$ are the integrals $(mm'\mu)$ (see Table I of Slater-Koster 1954), e.g. $(sp\sigma)$, $(dd\pi)$ etc. We will denote the corresponding overlap integrals by $[mm'\mu]$. The crystal field terms

$$F(\ell\kappa m,\ell\kappa m'|\ell'\kappa') = \int d^3r\ \phi^{\star}(\ell\kappa m)\ V(\ell'\kappa')\phi(\ell\kappa m') \qquad (A44)$$

can be divided into two parts. The first is

$$\varepsilon_{\kappa m}(\ell\kappa,\ell'\kappa') = \int d^3r\ \overline{\rho}_m(|\vec{r} - \vec{R}(\ell\kappa)|)\ \ V(\vec{r} - \vec{R}(\ell'\kappa')) \qquad (A45)$$

with

$$\overline{\rho}_m(r) = P_0\{\phi^{\star}(\ell\kappa m)\phi(\ell\kappa m)\}$$

the monopole component of the charge density associated with the orbital $\phi(\ell\kappa m)$. $\varepsilon_{\kappa m}$ is independent of the orientations of the potential sites $(\ell'\kappa')$ with respect to the orbital site $\ell\kappa$. This term can be added to the orbital energy E_{im}, and is therefore not treated explicitly.

The second contribution to the crystal field matrix elements F depends on the orientation of $V(\ell'\kappa')$ relative to $(\ell\kappa)$ in a way which is identical the σ-type two-center integrals which are tabulated in Table I of Slater and Koster (1954). For example, if $\hat{x},\hat{y},\hat{z}$ are the direction cosinus of $\vec{R}(\ell\kappa,\ell'\kappa') = R(\ell\kappa) - R(\ell'\kappa')$ and $\phi(\ell\kappa m)$ is chosen as p_x orbital, then

$$<\phi(\ell\kappa x)|V(\ell'\kappa')|\phi(\ell\kappa x)> = \varepsilon_{\kappa p}(i) + \hat{x}^2(pp)_i$$

where i stands for $(\ell\kappa,\ell'\kappa')$.
Here we omit the redundant σ in the bracket (pp) to distinguish these crystals field terms from the two center energy integrals. We note that orientally dependent crystal field terms also allow an off-diagonal coupling of orbitals $m \neq m'$. In cubic symmetry for $(\ell\kappa)$, such terms vanish, when the summation over a neighbor shell is carried out. There, only the well-known $e_g - t_{2g}$ splitting of the d orbital energies survives. At lower site symmetries of $(\ell\kappa)$, however, e_g, t_{2g} and p orbitals may further split and also off-diagonal terms may appear. In our program, sufficiently many terms $(mm')_i$ can be included to provide the most general splitting and off-diagonal coupling allowed by symmetry. We should note that all these $(mm')_i$ crystal field terms turn out to be much smaller than the corresponding energy integrals $(mm'\sigma)_i$.

While one version of our NTB energy band program is used to calculate the $\varepsilon_{k\mu}$ and $A(\kappa m|k\mu)$ necessary for the D_2 calculation, another version is used for determining the NTB matrix elements $E_{\kappa m}$, $(mm'\mu)_i$, $[mm'\mu]_i$ and $(mm')_i$ by means of a non-linear-least-squares fit to first principles energy band calculations.

In carrying out this fit, it is essential to identify the symmetry of the NTB eigenstates at each stage of the iteration procedure in order to ensure that each NTB eigenvalue is compared to the proper first principles result. This has been achieved in the following manner. First, we solve the NTB eigenvalue equation for each of the wave vectors $\vec{k}_n$ that is involved in the fit, using an approximate set of NTB parameters. The symmetry of the individual eigenstates for each $\vec{k}_n$ is identified by standard group-theoretical techniques. Then, the eigenvectors for each $\vec{k}_n$ are rearranged according to symmetry to form a unitary matrix $V(\vec{k}_n)$. It is readily shown that by applying the unitary transformations $V(\vec{k}_n)^{\dagger}H(\vec{k}_n)V(\vec{k}_n)$ and $V(\vec{k}_n)^{\dagger}S(\vec{k}_n)V(\vec{k}_n)$, one reduces $H(\vec{k}_n)$ and $S(\vec{k}_n)$ to block-diagonal form at any subsequent stage of the parameter iteration procedure.

This approach has an added benefit in that one is required to evaluate only the reduced eigenvalue equations during the iteration process. In the case of ν nondegenerate states of a given symmetry, reduced matrices of dimension ν are treated. However, some difficulties can occur when degenerate states are involved, and we have not attempted to block-diagonalize completely the $H(\vec{k}_n)$ and $S(\vec{k}_n)$ matrices in these cases. If there are ν states with degeneracy d at a given wave vector $\vec{k}_n$, then $H(\vec{k}_n)$ and $S(\vec{k}_n)$ are reduced only to dimension νd.

During the nonlinear-least-squares fitting procedure, it is necessary to repeatedly find incremental changes of all the energies due to successive variations of each of the NTB parameters. We found that it was sufficiently accurate to apply first-order perturbation theory to calculate these energy shifts, rather than solving the reduced eigenvalue esuations directly. We note that the first order expansion of the overlap terms $S^{-1/2}$ in such a perturbation treatment is very similar to the expansion given by Eq. (A18).

It is important to exercise some caution in the early stages of a fitting procedure. There is considerable danger of becoming trapped in some local minimum of parameter hyperspace, the result being that the NTB parameters will assume rather unphysical values.

We have avoided these difficulties by varying only a restricted subset of NTB parameters during the early iterations. In general, these have included the large "diagonal" matrix elements such as (ssσ), (ppσ), (ppπ), (ddσ) and (ddπ). In addition, we have found it useful to relate the "off-diagonal" parameters (mm'μ) to "diagonal" ones using the Pandey-Phillips approximation (1975)

$$(mm'\mu)^2 = |(mm\mu)\ (m'm'\mu)|$$

choosing the sign of (mm'μ) in accordance with simple physical arguments. Another possibility to restrict the number of NTB parameters is to assume that

$$[mm\mu]_i = \nu_{i,m}(mm\mu)_i$$

Then, only the smaller number of $\nu_{i,m}$ parameters need to be varied.

Each of these restrictions may be gradually lifted until, in the final stages of iteration, all of the NTB parameters are allowed to vary independently. In all cases, care should be taken to keep the number of fitted band energies at least a factor of 4 larger than the number of varied parameters.

We note that the plain NTB scheme for calculating the $\varepsilon_{k\mu}$, A(kμ) consists of about 2000 FORTRAN statements, while the fit version has about 1000 statements more.

A7.2. The D_2 Program

The computation of

$$D_2 = - \sum_{\substack{k\mu\mu' \\ k'=k+q}} \frac{f_{k'\mu'} - f_{k\mu}}{\varepsilon_{k\mu} - \varepsilon_{k'\mu'}} g^{\kappa\alpha}_{k\mu,k'\mu'} g^{\kappa'\beta}_{k'\mu',k\mu} \qquad (2.14 \text{ or } A26)$$

is a summation over the full BZ. It requires i) the evaluation of principal value integrals of type (we skip the band indices μ for simplicity)

$$\chi(k,k+q) = \int d^3k \frac{f_k - f_{k+q}}{\varepsilon_k - \varepsilon_{k+q}} \qquad (A46)$$

as well as ii) the determination of the matrix elements $g^{\kappa\alpha}$. The summation over the BZ is carried out by dividing the BZ into a grid of appropriate micro-units. These are cubes in the case of cubic crystals and may be trigonal prisms in a hexagonal lattice. The principal value integral $\chi(k,k+q)$ is fully calculated within each micro-unit, while the matrix elements $g_{k,k'}$ are calculated for values k,k' taken at the centers of the micro-volumes. The $g_{k,k'}$ are assumed to be constant within each micro-unit. The density of the grid is chosen such that the numerical convergence is good enough. A division by 2N = 20 of the $\Delta(\xi,0,0)$ line in a cubic crystal usually yields sufficient numerical stability.

In our computation we deal with two different k-point grids, one for the center and the other for the corner points of the micro-units. The first grid we call the "odd" mesh, as in cubic crystals we construct it from triples of odd integers. The second, the "even" mesh is set up from triples of even integers. For example, in the case of a simple cubic lattice, the symmetry points X and R are given by (20,0,0) and (20,20,20), respectively. The k-points of the odd mesh are (1,1,1), (3,1,1), (3,3,1), etc. As the difference of two odd numbers is an even number, the grid of phonon wave vectors q, for which D_2 can be evaluated, is the even mesh. In our example N = 10 phonons (2,0,0), (4,0,0) etc. can be computed along the $\Delta(\xi,0,0)$ line.

For the computation of $\chi(k,k+q)$ there exist two numerical methods. Both assume that the energies ε_k vary linearly with k within the micro-unit. The method of Diamond (1971) is developed in the spirit of the Gilat-Raubenheimer (1966) technique for density-of-states calculations. Diamond's method is based on cubes as micro-volumes. Therefore it is hard to employ it for ; e.g., hexagonal lattices. Input are the energies $\varepsilon_k, \varepsilon_{k'}$ and the band velocities $\vec{v}_k$, $\vec{v}_{k'}$ taken at the cube centers ($\vec{v}_k = \partial\varepsilon_k/\partial\vec{k}$) ; i.e.

at the odd mesh. This program has been used in our earlier calculations (Varma and Weber 1977, 1979 ; Weber et al. 1979).

The other numerical scheme is the tetrahedral method (Lehmann et al. 1970, Lehmann and Taut 1972, Jepsen and Andersen 1971). A detailed description of this method for $\chi(q)$ calculations is given by Rath and Freeman (1975), and our $\chi(q)$ code is programmed according to this paper. The input are the $\varepsilon_k, \varepsilon_{k'}$ values taken at the corner points of the tetrahedra. We think that this method is more flexible than that of Diamond as any micro-unit can be easily divided into tetrahedra. On the other hand we found that Diamond's program is almost a factor of 10 faster than our tetrahedral $\chi_0(q)$ program. However, since the $g^{\kappa\alpha}$ calculation is the most time-consuming part of the program, this is not a severe problem. We finally note that our $\chi_0(q)$ program consists of ≈ 1000 FORTRAN statements.

A rather involved problem is to compute the matrix elements

$$g^{\overline{\kappa}\alpha}_{k\mu,k'\mu'} = \sum_{\substack{\kappa m \\ \kappa' m'}} A^{\star}(\kappa m|k\mu)[\gamma_\alpha(\kappa m,\kappa' m'|k')\delta_{\kappa\overline{\kappa}} - \gamma_\alpha(\kappa m,\kappa' m'|k)\delta_{\kappa'\overline{\kappa}}]\, A(\kappa' m'|k'\mu') \tag{2.15 or A33}$$

with

$$\gamma_\alpha(\kappa m,\kappa' m'|k) = \sum_{\ell-\ell'} [\nabla_\alpha H(\ell\kappa m,\ell'\kappa' m') - \varepsilon\nabla_\alpha S(\ell\kappa m,\ell'\kappa' m')] \exp[-i\vec{k}\,\vec{R}(\ell\kappa,\ell'\kappa')] \tag{2.16 or A32}$$

The quantities $\nabla_\alpha H$ and $\nabla_\alpha S$ are the derivatives of the real space tight binding matrix elements (Eq. A31). Because of the factor $\varepsilon = 1/2\ (\varepsilon_{k\mu} + \varepsilon_{k'\mu'})$ in front of $\nabla_\alpha S$, it is advantageous to separately construct matrices $\nabla_\alpha H(k)$ and $\nabla_\alpha S(k)$. Our program has the same flexibility in computing the $\gamma_\alpha(k)$ matrices, as the NTB energy band code has in computing the matrices H(k) or S(k). This means, it is very easy to vary the crystal structure, number of atomic sites per cell, the orbital character at a given site etc. The subroutines to generate the $\gamma_\alpha(k)$ matrices consist of ≈ 2000 FORTRAN statements.

A conceptually very simple way to obtain $g^{\kappa\alpha}$ would be to get the vectors $A(k\mu)$, $A(k'\mu')$ by solving the NTB eigenvalue equation HA = SAE and by computing the matrices γ_α each time needed for a pair of vectors k,k'. This way is however extremely time-consuming, especially due to the eigenvalue problem. Instead we generate the vectors A(k), A(k') from vectors $A(k^\star)$, where $k^\star$ is lying in the wedge of the irreducible BZ. This is done by applying the necessary space group operations. In the general case of a crystal with a non-symmorphic space group, these operations not only imply rotations of orbital components, such as the x,y,z of p-orbitals components, but also an interchange of sublattice particles κ. When k' lies outside the first BZ, multiplications by phase factors

$$p = \exp[-i\vec{G}(\vec{R}(\kappa) + \vec{\tau})] \tag{A47}$$

are also required. In this case $\vec{G}$ is the reciprocal lattice vector which transfers k back to the first BZ and τ is the non-primitive translation associated with the corresponding point group operation.

We have further accelerated the computation by introducing the matrices

$$B_\alpha(\kappa m|k\mu) = \sum_{\kappa' m'} \gamma_\alpha(\kappa m,\kappa' m'|k)\ A(\kappa' m'|k\mu) \tag{A48}$$

Then

$$g^{\bar{\kappa}\alpha}_{k\mu,k'\mu'} = \sum_{\kappa m} A^\star(\kappa m|k\mu)\ B_\alpha(\kappa m|k'\mu')\delta_{\kappa\bar{\kappa}} - \sum_{\kappa' m'} B^\star_\alpha(\kappa' m'|k\mu)\ A(\kappa' m'|k'\mu')\delta_{\kappa'\bar{\kappa}}$$

or

$$g^{\kappa\alpha}_{k\mu,k'\mu'} = \sum_m A^\star(\kappa m|k\mu)\ B_\alpha(\kappa m|k'\mu') - B^\star_\alpha(\kappa m|k\mu)A(\kappa m|k'\mu') \tag{A49}$$

The matrices $B_\alpha(k)$ are generated from the matrices $B_\alpha(k^\star)$ in a very similar way as described above for the vector A(k).

To make full use of this acceleration, the vector $A(k^\star\mu)$ and the matrices $B_\alpha(k^\star\mu)$ are computed once in set-up runs and then stored on disk as random access data-sets to be used in the main D_2 program.

Rotation of the quantities A(k) and $B_\alpha(k)$ implies multiplication of vectors with (often sparse) rotation matrices. In cubic systems this means nothing but interchange of vector indices and sometimes sign changes of vector components. Only for the components of d-orbitals with e_g symmetry, the 2 x 2 rotation matrices are more complicated. Furthermore, in many systems the phase factors of Eq. (A47) turn out to be just powers of i. Odd powers of i imply - apart from sign changes - an interchange of real and imaginary part of vector components. Thus, in cubic systems we can avoid to a large extend the time-consuming multiplication operations. Instead we may use much faster schemes which only interchange components of vectors.

All this program optimization leads to considerable savings of computing time. For instance, in the A15 calculations about a factor 10 has been gained between the most optimized program version and the first, where only the A vectors are rotated (by rotation matrices) and the γ_α matrices are generated each time needed. Even with the optimized version, the D_2 run for a general q point in Nb_3Sn takes $t \approx 90$ min. on the IBM 370/3033 (10 bands around E_F with altogether 55 band-to-band transition were included and a mesh factor N = 10 was used). When carrying out the summation

$$\sum_{\substack{k \\ k' = k+q}}$$

in Eq. (2.14 or A26), we take advantage of any symmetry of the phonon wavevector q. This symmetry is described by G_q, the group of q. G_q consists of all those point group operations of the full point group G, which leave q unchanged. For instance, in a cubic system, there are 8 out of 48 operations which leave a vector $q = (\xi,0,0)$ along the line Δ unchanged. Thus, the dimension of G_Δ is m = 8. Apart from G_q we need the "complimentary" group G^q (of dimension n), defined that

$$G_q \times G^q = G \tag{A50}$$

This means that G^q is the group which multiplied on G_q generates the full point group (of dimension m.n).

Be

$$h_i \in G_q \quad , \quad i = 1 \dots m$$

and

$$h^j \in G^q \quad , \quad j = 1 \dots n \tag{A51}$$

Now we take

$$k = R(h_1.h^j)k^\star \quad , \quad j = 1 \dots n \tag{A52}$$

where R(h) is the rotation matrix associated with the operation

$$h = h_i.h^j$$

of the full group. $k^\star$ is a mesh point in the irreducible wedge of the BZ and should have general symmetry. Then we calculate $\chi(k, k+q)$ and $g_{k,k+q}$ (for j = 1 in Eq. (A52) we have $k = k^\star$, of course). Now we take the vectors

$$\bar{k} = R(h_i h^j)k^\star \quad , \quad i = 2 \dots m \tag{A53}$$

It is easy to see that

$$\chi(\bar{k}, \bar{k} + q) = \chi(k, k + q)$$

Further the vectors $g_{\bar{k},\bar{k}+q}$ are found by applying the appropriate space group operation on $g_{k,k+q}$. This treatment is very similar to the one used for generating the vectors $A(k)$ and $B_\alpha(k)$ from $A(k^\star)$ and $B_\alpha(k^\star)$.

In the summation over the star of k, $\chi(k, k + q)$ and $g_{k,k+q}$ need to be calculated only n instead of n.m times. Note that some care has to be exercised when $k^\star$ has more than general symmetry.

In the case of high symmetry phonons the exploitation of the symmetry means an enormous saving of time. For instance in the case of A15 calculations, we get $t \approx 10$ mins for the Γ and R point, and $t \approx 20$ min for a phonon along Δ, compared to $t = 90$ min for a general symmetry phonon.

The electron-phonon coupling matrices Γ (Eq. 2.18) are obtained in a very similar way as D_2. Γ is given by

$$\Gamma(\kappa\alpha,\kappa'\beta|q) = \sum_{\substack{k\mu\mu' \\ k'=k+q}} \Delta\sigma_{k\mu}\, \Delta\sigma_{k'\mu'}\, g^{\kappa\alpha}_{k\mu,k'\mu'}\, g^{\kappa'\beta}_{k'\mu',k\mu} \qquad \text{(A54)}$$

Here

$$\Delta\sigma_{k\mu} = \int_{\substack{FS \\ \text{micro-unit}}} \frac{dS_k}{|v_k|}$$

is the Fermi surface element of the μ-th band in the k-th micro-unit. There are of course much fewer contributions to Γ than to D_2, as Γ is a Fermi surface integral while D_2 is a volume integral. As a consequence, numerical stability of Γ is achieved only using denser grids with smaller micro-units than required for D_2.

The BZ summation part of the D_2 program including the various group-theoretical operation subroutines consists of $\approx$ 3000 FORTRAN statements, so that the total length of the D_2 code is about 6000 statements. We could not formulate the D_2 program in the same generality as the NTB energy code. It has become evident from the above discussion that any change in the crystal structure, of the sites or the sorts of atoms in the unit cell etc. means modifications of some parts of the D_2 program. We have succeeded in confining these modifications to a few subroutines.

A7.3. The Phonon Programs

Since in our theory, the short range forces ($D_0 + D_1$) are parametrized by near neighbor Born-and-von-Kármán force constants, we have also developed a code for setting up a dynamical matrix for a Born-and-von-Kármán model in an arbitrary crystal structure. We have included options
i) to add D_2 or not and ii) to use axially symmetric or general forces between given pair of atoms.

The code has many similarities to the NTB code, likewise there exist two versions, one for simply calculating ω_{qi} (as well as the phonon linewidth γ (Eq. 2.17)), and the other for determining the force constants by fits to experimental phonon frequencies.

The fitting procedure of the phonon program is the same as the one used in the NTB code. Altogether, the phonon programs have about 3000 FORTRAN statements. In the following we discuss some special features of these programs.

Employing axially symmetric (or central) forces between the atoms is analogous to using the two-center approximation in the tight-binding scheme. The force constant matrix ϕ between atoms $(\ell\kappa)$ and $(\ell'\kappa')$ then read as

$$\phi(\ell\kappa\alpha,\ell'\kappa'\beta) = -A(\ell\kappa,\ell'\kappa')\,\hat{r}_\alpha(\ell\kappa,\ell'\kappa')\,\hat{r}_\beta(\ell\kappa,\ell'\kappa') - B(\ell\kappa,\ell'\kappa')[\,\delta_{\alpha\beta} - \hat{r}_\alpha(\ell\kappa,\ell'\kappa')\,\hat{r}_\beta(\ell\kappa,\ell'\kappa')] \tag{A55}$$

Here $\hat{r}(\ell\kappa,\ell'\kappa')$ is the distance vector normalized to unity. A and B are the "longitudinal" and "transverse" force constants, respectively. If we assume that they are derived from two-body potentials $\Phi_{\kappa\kappa'}(r)$, then

$$A(\ell\kappa,\ell'\kappa') = (\partial^2/\partial r^2)\,\Phi_{\kappa\kappa'}(r)\Big|_{\vec{r}=\vec{R}(\ell\kappa,\ell'\kappa')} \tag{A56}$$

and

$$B(\ell\kappa,\ell'\kappa') = 1/r(\partial/\partial r)\,\Phi_{\kappa\kappa'}(r)\Big|_{\vec{r}=\vec{R}(\ell\kappa,\ell'\kappa')}$$

The quantities A and B are then assumed to be equal for all pairs (κ,κ') of the same distance $|R(\ell\kappa,\ell'\kappa')|$. As is evident from Eq. (A55), the force constant matrix ϕ is symmetric in the Cartesian indices α,β.

There is an almost complete analogy of central forces and the tight-binding case with p-orbitals. The quantities A_i and B_i correspond to the $(pp\sigma)_i$ and $(pp\pi)_i$ energy integrals, and thus the functional form of the dynamical matrix $D(\kappa\alpha,\kappa'\beta|q)$ is the same as that of the Hamiltonian matrix $H(\kappa\alpha,\kappa'\beta|k)$. The only difference is that the equivalent of the orbital energies $E_{\kappa m}$ does not exist, instead, the on-site force constant matrix is found from the requirement of infinitesimal translational invariance (see Maradudin et al., Eq. 2.1.13)

$$\phi(\ell\kappa\alpha,\ell\kappa\beta) = -\sum_{\substack{\ell' \\ \ell'\neq\ell}} \phi(\ell\kappa\alpha,\ell'\kappa'\beta)$$

In general, the matrix ϕ has more than two independent elements. Their maximum number is determined by the symmetry of the crystal. ϕ need not be symmetric in α and β, thus there may appear as many as nine independent elements. In tight-binding, the analogy to general forces is to incorporate the three center integrals. The maximum number of independent matrix elements in ϕ can be found by applying group-theoretical techniques (see Maradudin et al., p. 10ff).

To find the general form of the force constant matrices in a crystal, we have set up an auxiliary program. This constructs a dynamical matrix D(q) for a wave factor q of general symmetry

$$q = n_1 G_1 + n_2 G_2 + n_3 G_3$$

with

$$0.5 > n_1 > n_2 > n_3 > 0$$

using axially symmetric forces of sufficient range. This means each atom $(\ell\kappa)$ is coupled to all other atoms (ℓ,κ') and also to near neighbor atoms $(\ell'\kappa)$. Then, the eigenvalue equation (A5) is solved and the quantity

$$U(\ell\kappa\alpha,\ell'\kappa'\beta) = \sum_{\{q\}} e(\kappa\alpha|qj)\, e^{\star}(\kappa'\beta|qj) \times \exp[i\vec{q}\,\vec{R}(\ell\kappa,\ell'\kappa')] \tag{A57}$$

is calculated for one particular j. {q} means summing over the star of q. The quantity U (which is related to the equal time displacement correlation function Eq. 2.4.24 of Maradudin et al. has the same symmetry properties as ϕ. By analyzing U we are then able to determine the form of ϕ and the number of independent matrix elements in ϕ.

In more complicated crystal structures such as the A15 structure (see Fig. 35), there appear various kinds of generalizations of the central force model which are unexpected at first glance. For instance, the general forces for 1st nearest neighbors (A atom intrachain coupling) exhibit two different transverse force constants. These are interchanged when going in the opposite direction along the chain. Furthermore, there are two different kinds of 7th nearest neighbor pairs : One consists of A atoms situated at two adjacent, parallel chains, the other is made up by two A atoms of the same chain.

For cubic systems this method of determining the symmetry properties of ϕ seems to be fool-proof. However, in hexagonal (and probably also in trigonal) symmetry there is an additional problem. When we examine force constants matrices between atoms $(\ell\kappa)$ and $(\ell'\kappa')$ in the hexagonal plane, it appears as if matrices ϕ with $\hat{r}(\ell\kappa,\ell'\kappa') = (1,0,0)$ and $(1/2, \sqrt{3}/2, 0)$ do not have the same number of independent elements (the latter matrix has a non-zero xy component). However, the two matrices are transformed into each other by the appropriate symmetry operations. This means that, in hexagonal symmetry, special care has to be exercised to find out the smallest number of independent elements in ϕ.

A8. Change of the Energy Zero in NTB

For computational purposes it is advantageous to put the energy zero at E_F. Then for most cases we can neglect the term $\varepsilon\nabla S$ in Eq. (2.16 or A32). This leads to a considerable decrease of computing time. The most physical choice however is to put the vacuum energy

$$E_{vac} = E_F + \phi = 0$$

where ϕ is the work function. Then, NTB matrix elements and their gradients of different systems can be compared most simply.

The change of the energy zero implies not only a constant shift in the orbital energy $E_{\kappa m}$ of Eq. (A43) but also a change of the off-diagonal matrix elements $H(\ell\kappa m,\ell'\kappa'm')$. As a consequence, also the derivatives $\nabla_\alpha H$ are affected.

Be

$$HA = SAE \tag{A11}$$

the NTB eigenvalue equation. We now shift all energies by Δ to get new energies

$$E' = E + \Delta$$

E' is the eigenvalue matrix of a new NTB equation

$$H'A' = S\,A'\,E' = H'\,A' = S\,A'(E + \Delta)$$

Here S is the same as above, since the overlap does not depend on the zero of energy. Comparing

$$(H' - \Delta S)A' = S\,A'\,E \quad \text{with} \quad H\,A = S\,A\,E$$

We can equate

$$A' = A$$

and

$$H' = H + \Delta S \tag{A58}$$

As the matrices have to be equal element by element, this leads to

$$H'(\ell\kappa m,\ell'\kappa'm') = H(\ell\kappa m,\ell'\kappa m') + \Delta S(\ell\kappa m,\ell'\kappa'm')$$

or

$$(mm\mu)_i' = (mm\mu)_i + \Delta[mm\mu]_i \tag{A59}$$

We now define

$$\alpha = \frac{\partial}{\partial r}(mm\mu)_i/(mm\mu)_i$$

$$\beta = \frac{\partial}{\partial r}[mm\mu]_i/[mm\mu]_i \tag{A60}$$

$$\alpha' = \frac{\partial}{\partial r}(mm\mu)_i'/(mm\mu)_i'$$

Then

$$\alpha' = \alpha + (\alpha - \beta)\,.\,\frac{\Delta[mm\mu]_i}{(mm\mu)_i'} \tag{A61}$$

A9. Long range forces in D_0

In the case of "neutral objects", i.e., for elemental crystals, charge neutrality provides the cancellation of the long range part of the difference of the ion-ion and electron-electron Coulomb forces. This cancellation is strictly true only under the assumption that the electronic charge densities around the ion cores move rigidly with the displaced ions. There is however a displacement-induced charge density δn, whose leading term is

$$\delta n(q) \approx \sum_{\substack{k,\mu\mu' \\ k'=k+q}} \sum_m A^*(m|k\mu)\,A(m|k'\mu')\,|\phi_m(r - R_\ell)|^2 \times \frac{f_{k\mu}(1 - f_{k'\mu'})}{\varepsilon_k - \varepsilon_{k'\mu'}}\, g^\alpha_{k\mu,k'\mu'}\, u^\alpha_q \tag{A62}$$

The induced charge density $\delta n(q)$ thus consists of expressions very similar to those entering D_2. Therefore one may suspect, that interaction terms $\delta n(q)\ v_{e-e}\ \delta n(q)$ may contribute considerably to the long range forces in a metal, thereby causing something like a "screening resonance" (Hanke et al., 1976). However there is strong evidence that these induced electron-electron interactions are not very important for a phonon calculation. We first quote results of a study by Evans (1975) who investigated the various contributions to the pair potential in Na. He directly calculated the single particle or band-structure part corresponding to our $D_1 + D_2$, and the neutral object part corresponding to D_0 (see Fig. 50). In Na, long range forces are due to the Friedel oscillations, or in phonon language, due to Kohn anomalies. These oscillations are clearly visible at large distances, yet their dominant contributions (80 - 90%) arise from the band-structure part.

We have carried out a similar estimate for Nb by calculating $\delta n(q)$ in a way analogous to the D_2 calculations (see Sect. 3.1.2). Then $\delta n(q)\ v_{e-e}(q)\ \delta n(q)$ has been estimated by approximating v_{e-e}

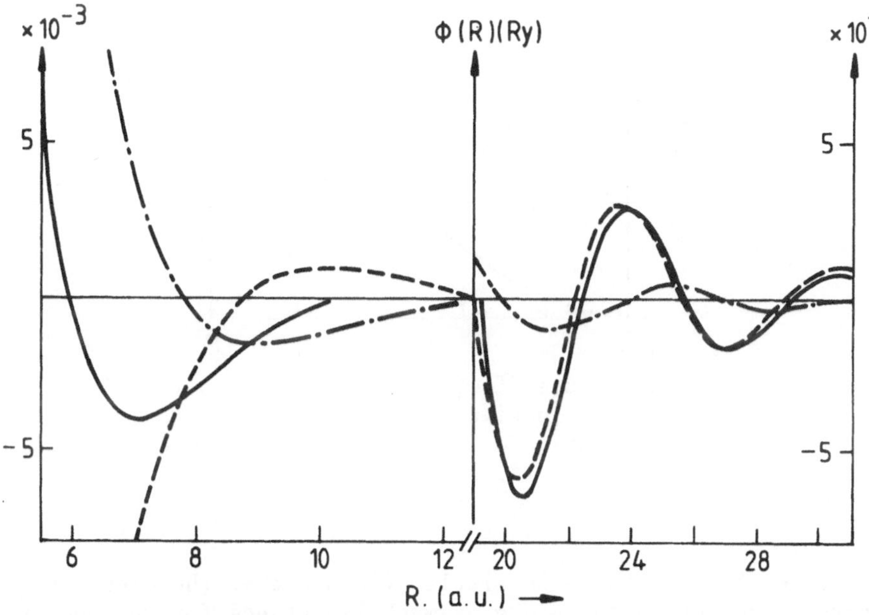

Fig. 50. Different contributions to the pair potential of Na. The full line is the total potential calculated using the Singwi-Vashishta exchange and correlation function. This potential is the sum of the band-structure contribution (dashed line) and the neutral object part (dashed-dotted) (Evans, 1975).

by the bare Coulomb interaction of point charges at the ion sites.

The resulting longitudinal component along the Δ line is shown in Fig. 51. The magnitude of $\delta n \, v_{e-e} \, \delta n$ (obtained from contributions of all bands), is less than $D_2^<$, the intraband contribution near E_F causing the phonon anomaly. In addition, the Coulomb term is rather smoothly varying and does not show much structure at the anomalous wavelengths where $D_2^<$ has a pronounced peak. The main reason why $\delta n \, v \, \delta n$ behaves much smoother is : In Eq. (A62) the form factors g^α (which may differ in sign) are first added, and then squared in $\delta n \, v \, \delta n$, while to compute D_2 (Eq. 2.14), the form factors g^α are squared before they are added up.

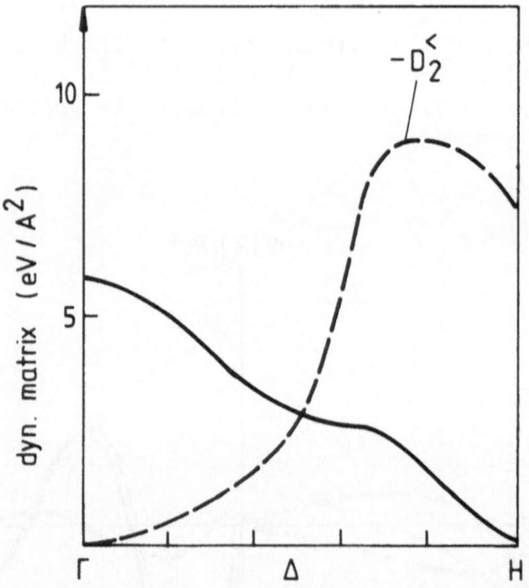

Fig. 51. Estimate along the $(\xi,0,0)$ line of the contribution $\delta n v_{e-e} \delta n$ to the longitudinal part of the dynamical matrix of Nb (full line). For comparison, the term $D_2^<$ is also shown.

In conclusion, there is little evidence that the electron-electron interaction terms are of much importance for the phonon anomalies in transition metals. This result is consistent with the observations of Cooke (1978, 1980), who also found that there is no screening resonance at phonon anomalies.

A10 Tables of NTB Matrix Elements and their Derivatives

Table A10.1 : NTB parameters for Nb energy bands. Energies are given in Rydberg. Model I (r.m.s error ≈ 15 mRy) has been used the calculations of Varma and Weber 1977, 1979, and Varma et al. 1978, 1979. Model II (Mattheiss and Weber 1979) has a much better r.m.s error of 3 mRy and has been employed in the control calculations of Richter and Weber (1983). Model I is obtained by assuming $E_F = -0.3$ Ry, model II by choosing $E_F = 0$. In model IIs the NTB energy parameter of model II are given by assuming $E_F = -0.3$ Ry (see Appendix A8). The rows D denote the derivatives of the NTB matrix elements as defined in Eq. (2.25).

(a) Orbital energies

	E_s	E_p	$E_d(t_{2g})$	$E_d(e_g)$
I	0.1302	0.4119	-0.2224	-0.2612
II	0.2227	0.6721	0.08556	0.0282
IIs	-0.0776	0.3718	-0.21474	-0.2721

(b) Crystal-field parameters

	$(dd)_1$
I	0.0145
II	0.0215

$E_d(t_{2g}) = E_d(e_g) + 8/3\ (dd)_1$

(c) Two-center energy parameters

		(ssσ)	(spσ)	(sdσ)	(ppσ)	(ppπ)	(pdσ)	(pdπ)	(ddσ)	(ddπ)	(ddδ)
1NN	I	-0.1077	0.1115	-0.1038	0.1153	-0.0216	-0.1074	0.0409	-0.1000	0.0776	-0.0287
	II	-0.1077	0.1004	-0.0743	0.0933	-0.0017	-0.0641	0.0465	-0.0723	0.0685	-0.0130
	IIs	0.1222	0.1200	-0.0984	0.1252	-0.0174	-0.0912	0.0637	-0.0972	0.0875	-0.0162
	D	-0.217	-	-0.434	-	-	-0.325	-0.373	-0.651	-0.747	-0.890
2NN	I	-0.0430	0.0744	-0.0557	0.1287	-0.0287	-0.0964	0.0260	-0.0722	0.0236	0.0
	II	-0.0224	-0.0026	-0.0524	0.0153	-0.0007	-0.0224	-0.0026	-0.0524	0.0153	-0.0007
	IIs	-0.0498	0.040	-0.0761	0.0700	-0.0120	-0.0652	0.0036	-0.0817	0.0272	-0.0011
	D	-0.217	-	-0.434	-	-	-0.325	-0.373	-0.651	-0.747	-0.890

(d) Overlap parameters

		[ssσ]	[spσ]	[sdσ]	[ppσ]	[ppπ]	[pdσ]	[pdπ]	[ddσ]	[ddπ]	[ddδ]
1NN	I	-0.001	-0.0089	0.0093	-0.0653	0.0122	0.0680	-0.0259	0.0708	-0.0549	0.0097
	II	0.0482	-0.0631	0.0805	-0.1063	0.0524	0.0904	-0.0574	0.0829	-0.0632	0.0106
	D	-0.193	-	-0.230	-	-	-0.128	-0.240	-0.256	-0.481	-0.701
2NN	I	0.0681	-0.0949	0.0590	-0.1323	0.0295	0.0822	-0.0222	0.0511	-0.0167	0.0
	II	0.0913	-0.1422	0.0790	-0.1831	0.0371	0.1427	-0.0208	0.0976	-0.0397	0.0012
	D	-0.193	-	-0.230	-	-	-0.128	-0.240	-0.256	-0.481	-0.701

Table A10.2 : NTB parameters for the energy bands of NbC, NbN and VN, fitted to the APW calculations of Schwarz (1977, 1971) and Neckel et al. (1976), respectively. Energies are given in Rydberg, $E_F \approx 0$ in all three cases. The r.m.s values are ≈ 10 mRy. The subscripts 1 and 2 denote NM and TM atoms, respectively.

(a) Orbital energies

	E_s	E_p	$E_d(t_{2g})$	$E_d(e_g)$
NbC	-0.7163	-0.1349	-0.0187	-0.0234
NbN	-1.0171	-0.2362	0.0103	-0.0057
VN	-1.1047	-0.2929	-0.0576	-0.0753

(b) Crystal field parameters

	$(dd)_1$
NbC	-0.0016
NbN	-0.0053
VN	-0.0059

$$E_d(e_g) = E_d(t_{2g}) + 3(dd)_1$$

(c) Two-center energy parameters

	$(ss\sigma)_{1-1}$	$(sp\sigma)_{1-1}$	$(sd\sigma)_{1-2}$	$(pp\sigma)_{1-1}$	$(pp\pi)_{1-1}$	$(pd\sigma)_{1-2}$	$(pd\pi)_{1-2}$	$(dd\sigma)_{2-2}$	$(dd\pi)_{2-2}$	$(dd\delta)_{2-2}$
NbC	-0.0189	0.0322	-0.1755	0.0549	-0.0156	-0.1640	0.0948	-0.0740	0.0259	0.0019
NbN	-0.0126	0.0235	-0.1920	0.0439	-0.0130	-0.1728	0.0915	-0.0764	0.0308	-0.0023
VN	-0.0149	0.0267	-0.1591	0.0481	-0.0082	-0.1501	0.0826	-0.0445	0.0188	-0.0016

(d) Overlap parameters

	$[ss\sigma]_{1-1}$	$[sp\sigma]_{1-1}$	$[sd\sigma]_{1-2}$	$[pp\sigma]_{1-1}$	$[pp\pi]_{1-1}$	$[pd\sigma]_{1-2}$	$[pd\pi]_{1-2}$	$[dd\sigma]_{2-2}$	$[dd\pi]_{2-2}$	$[dd\delta]_{2-2}$
NbC	0.0	-0.0007	0.0879	-0.0232	0.0066	0.1199	-0.0693	0.0222	-0.0078	0.0
NbN	0.0	0.0	0.0786	-0.0286	0.0085	0.1102	-0.0584	0.0176	-0.0071	0.0005
VN	0.0	0.0	0.0641	0.0	0.0	0.0909	-0.0500	0.0	0.0	0.0

Table A10.3 : Derivatives of NTB parameters for NbC, NbN, and VN, as defined by Eq. (2.25). The derivatives have been obtained, using atomic wavefunctions and potentials. For NbN the derivatives have also been found from fits to the APW energy band calculations of Schwarz (1971).

(a) Derivatives of two-center energy parameters

	$(ss\sigma)_{1-1}$	$(sp\sigma)_{1-1}$	$(sd\sigma)_{1-2}$	$(pp\sigma)_{1-1}$	$(pp\pi)_{1-1}$	$(pd\sigma)_{1-2}$	$(pd\pi)_{1-2}$	$(dd\sigma)_{2-2}$	$(dd\pi)_{2-2}$	$(dd\delta)_{2-2}$
NbC	- 1.25	- 1.16	- 0.90	- 0.84	- 1.16	- 0.75	- 1.10	- 0.60	- 0.75	- 0.87
NbN	- 1.25	- 1.16	- 0.90	- 0.84	- 1.16	- 0.75	- 1.10	- 0.60	- 0.75	- 0.81
VN	- 1.20	- 1.40	- 0.72	- 1.20	- 1.40	- 0.90	- 1.05	- 0.80	- 1.0	- 1.20

(b) Derivatives of overlap parameters

	$[ss\sigma]_{1-1}$	$[sp\sigma]_{1-1}$	$[sd\sigma]_{1-2}$	$[pp\sigma]_{1-1}$	$[pp\pi]_{1-1}$	$[pd\sigma]_{1-2}$	$[pd\pi]_{1-2}$	$[dd\sigma]_{2-2}$	$[dd\pi]_{2-2}$	$[dd\delta]_{2-2}$
NbC	- 0.82	- 0.63	- 0.34	- 0.42	- 0.69	- 0.31	- 0.49	- 0.20	- 0.40	- 0.63
NbN	- 0.82	- 0.63	- 0.34	- 0.42	- 0.69	- 0.31	- 0.49	- 0.20	- 0.40	- 0.63
VN	- 1.o	- 1.2	- 0.50	- 1.0	- 1.2	- 0.70	- 0.85	- 0.60	- 0.80	- 1.0

Table A10.4 : NTB parameters for V_3Si (from Mattheiss and Weber 1982)

(a) Orbital energies

Atom	E_s	E_p	E_d
Si	−0.253 092	0.178 333	
V	0.039 765	0.583 616	0.091 792

(b) Crystal-field parameters

Neighbor shell i	Orbitals at	Potential at	$(sd)_i$	$(pp)_i$	$(pd)_i$	$(dd)_i$
1	V	V (1NN)	0.016 789	−0.032 249		−0.007 949
2	V	Si (1NN)			−0.035 954	−0.009 763
3	V	V (2NN)				−0.016 574

(c) Two-center energy parameters

	$(ss\sigma)$	$(sp\sigma)$	$(sd\sigma)$	$(pp\sigma)$	$(pp\pi)$	$(pd\sigma)$	$(pd\pi)$	$(dd\sigma)$	$(dd\pi)$	$(dd\delta)$
Si-Si 1NN	−0.004 883	0.002 452		0.009 730	−0.001 112					
Si-Si 2NN	0.001 958	−0.004 166		−0.011 106	−0.000 878					
Si-V 1NN	−0.106 114	0.120 220	−0.080 812	0.129 453	0.000 701	−0.077 472	0.024 823			
V-V 1NN	−0.080 477	0.032 545	−0.050 754	−0.025 599	−0.000 940	0.003 568	0.044 083	−0.065 828	0.068 481	−0.018 092
V-V 2NN	−0.080 316	0.100 794	−0.064 728	0.165 591	0.005 426	−0.083 842	0.018 449	−0.051 206	0.031 298	−0.001 166

(d) Two-center overlap parameters

	$[ss\sigma]$	$[sp\sigma]$	$[sd\sigma]$	$[pp\sigma]$	$[pp\pi]$	$[pd\sigma]$	$[pd\pi]$	$[dd\sigma]$	$[dd\pi]$	$[dd\delta]$
Si-Si 1NN	0.016 818	−0.013 573		−0.034 458	0.001 473					
Si-Si 2NN	0.015 390	−0.023 651		0.028 521	−0.008 374					
Si-V 1NN	0.056 561	−0.087 683	0.007 892	−0.017 725	0.090 742	0.048 162	−0.076 874			
V-V 1NN	0.092 248	−0.160 134	0.121 320	−0.242 265	0.030 957	0.210 416	−0.028 848	0.102 535	−0.078 158	−0.007 642
V-V 2NN	0.045 886	−0.039 154	0.035 794	−0.073 341	0.048 427	0.063 376	−0.029 011	0.044 994	−0.029 882	0.009 525

Table A10.5 : NTB parameters for Nb_3Sn (from Mattheiss and Weber 1982)

(a) Orbital energies

Atom	E_s	E_p	E_d
Sn	−0.420 818	0.228 746	
Nb	−0.023 456	0.430 981	0.034 732

(b) Crystal-field parameters

Neighbor shell i	Orbitals at	Potential at	$(sd)_i$	$(pp)_i$	$(pd)_i$	$(dd)_i$
1	Nb	Nb (1NN)	0.024 369	0.032 368		0.029 123
2	Nb	Sn (1NN)			−0.003 692	0.008 592
3	Nb	Nb (2NN)				0.010 334

(c) Two-center energy parameters

	$(ss\sigma)$	$(sp\sigma)$	$(sd\sigma)$	$(pp\sigma)$	$(pp\pi)$	$(pd\sigma)$	$(pd\pi)$	$(dd\sigma)$	$(dd\pi)$	$(dd\delta)$
Sn-Sn 1NN	−0.001 375	0.003 471		0.006 523	−0.005 589					
Sn-Sn 2NN	0.000 155	0.003 791		0.015 307	0.004 772					
Sn-Nb 1NN	−0.105 491	0.109 671	−0.093 074	0.110 960	−0.046 378	−0.109 061	0.041 343			
Nb-Nb 1NN	−0.093 903	0.014 173	−0.073 108	−0.045 513	−0.000 594	−0.009 266	0.039 936	−0.084 767	0.087 438	−0.012 375
Nb-Nb 2NN	−0.086 096	0.083 639	−0.065 681	0.111 346	−0.028 797	−0.084 164	0.027 815	−0.066 939	0.035 609	−0.004 452
Nb-Nb 3NN								0.000 108	0.000 184	
Nb-Nb 4NN[a]						−0.004 075	−0.002 533	−0.003 344	−0.001 090	−0.000 098
Nb-Nb 4NN								0.002 670	−0.000 544	

(d) Two-center overlap parameters

	$[ss\sigma]$	$[sp\sigma]$	$[sd\sigma]$	$[pp\sigma]$	$[pp\pi]$	$[pd\sigma]$	$[pd\pi]$	$[dd\sigma]$	$[dd\pi]$	$[dd\delta]$
Sn-Sn 1NN	0.007 203	0.001 495		−0.003 328	−0.000 649					
Sn-Sn 2NN	0.006 935	−0.000 922		0.038 359	0.005 640					
Sn-Nb 1NN	0.080 305	−0.131 234	0.064 031	−0.122 238	0.014 811	0.071 135	−0.011 263			
Nb-Nb 1NN	0.122 506	−0.226 220	0.069 779	−0.292 044	0.037 271	0.182 336	−0.084 092	0.117 837	−0.120 906	0.034 874
Nb-Nb 2NN	0.074 542	−0.119 859	0.089 340	−0.163 097	0.013 950	0.102 806	−0.027 607	0.063 673	−0.041 072	−0.001 212
Nb-Nb 3NN								−0.001 824	0.002 443	
Nb-Nb 4NN[a]						−0.004 912	−0.003 072	−0.013 845	−0.002 694	0.001 487
Nb-Nb 4NN								0.011 179	−0.004 776	

[a]Denotes 4th neighbor Nb-Nb interactions along the Nb-atom chains.

Table A10.6 : Derivatives of the NTB parameters for the $D_2(q)$ calculation for Nb_3Sn and Nb_3Sb. Only the derivatives of the most important parameters (1NN and 2NN Nb-Nb two-center energy matrix elements) have been included in the calculation. Note the positive sign of the $(pd\sigma)_{1NN}$ derivative. This is in accordance to the fact that $(pd\sigma)_{1NN} \ll (pd\sigma)_{2NN}$ (see Tables A10.4 and A10.5).

	$(ss\sigma)$	$(sp\sigma)$	$(sd\sigma)$	$(pp\sigma)$	$(pp\pi)$	$(pd\sigma)$	$(pd\pi)$	$(dd\sigma)$	$(dd\pi)$	$(dd\delta)$
1NN	- 0.1	0.0	- 0.2	0.0	0.0	0.8	- 0.3	- 0.3	- 0.7	- 0.9
2NN	- 0.1	0.0	- 0.2	0.0	0.0	0.0	- 0.3	- 0.6	- 0.7	- 0.9

REFERENCES

Allen, P.B. (1980), in "Dynamical Properties of Solids", Vol.3 (Eds. Horton, G.K., and Maradudin, A.A.), North Holland, Amsterdam, p.95.
Andersen, O.K. (1984), this volume.
Anderson, J.R., Papaconstantopoulos, D.A., McCaffrey, J.W., and Schirber, J.E. (1973), Phys.Rev.B 7, 5115.
Appelbaum, J.A., and Hamann, D.R. (1978) in "Transition Metals 1977" (Eds. Lee, M.J.G., Perz, J.M. and Fawcett, E.), Institute of Physics, Bristol, p. 111.
Ashkenazi, J., Dacorogna, M., Peter, M., Talmor, Y., Walker, E. and Steinemann, S. (1978), Phys.Rev.B 18, 4120.
Ashkenazi, J. and Dacorogna, M. (1981), J. de Physique 42, C6-355 (Proc. Int. Conf. Phonon Physics, Bloomington, 1981).
Axe, J.D. and Shirane, G. (1978), Phys.Rev.B 18, 3742.
Baym, G. (1961), Ann. Phys. 14, 1.
Bhatt, R.N. and McMillan, W.L. (1976), Phys.Rev.B. 14, 1007.
Bhatt, R.N. (1977), Phys.Rev.B. 16, 1915.
Bilz, H. (1958, Z.Phys. 153, 338.
Bohnen, K.P., Weber, W. and Hamann, D.R. (1980), Prog.Rep. IAK I, KfK-Report 3051, Kernforschungszentrum Karlsruhe, p.40.
Brovman, E.G. and Kagan, Yu.M. (1974) in "Dynamical Properties of Solids", Vol. 1 (Eds. Horton, G.K. and Maradudin, A.A.), North Holland, Amsterdam, p. 191.
Born, M. and v. Kármán, Th. (1912), Phys.Zeitschrift 13, 297.
Born, M. and Huang, K. (1954), "Dynamical Theory of Crystal Lattices" Clarendon Press, Oxford.
Butler, W.H. (1977), Phys.Rev.B. 15, 5267.
Butler, W.H., Smith, H.G. and Wakabayashi, N. (1977), Phys.Rev.Lett. 39, 1004.
Butler, W.H., Pinski, F.J. and Allen, P.B. (1979), Phys.Rev.B. 19, 3708.
Butler, W.H. (1981), in "Treatise on Materials Science and Technology", Vol.21, (Ed. Fradin, F.Y.), Academic Press, New York, p. 165.
Christensen, A.N., Dietrich, O.W., Kress, W., Teuchert, W.D. and Currat, R. (1979), Solid State Commun. 31, 795.
Clogston, A.M. and Jaccarino, V. (1961), Phys.Rev. 121, 1357.
Cochran, W. (1959), Proc.Roy.Soc. Ser. A 253, 260.
Cochran, W. (1972), CRC, Critical Reviews in Solid State Sciences 2, 1.
Cooke, J.F. (1978), Solid State Div. Progr. Rep., Oak Ridge Nat. Lab. Rep. No. 5486, p. 13.
Cooke, J.F. (1980), Solid State Div. Progr. Rep., Oak Ridge Nat. Lab. Rep. No. 5640, p. 15.
Crabtree, G.W., Dye, D.H., Karim, D.P., Koelling, D.D. and Ketterson, J.B. (1979), Phys.Rev.Lett. 42, 390.

Diamond, J. (1971), in "Computational Methods in Band Theory" (Eds. Marcus, P.M., Janak, J.F. and Williams, A.R.), Plenum Press, New York, p. 347.

Dick, B.G. and Overhauser, A.W. (1958), Phys.Rev. 112, 90.

Eschrig, H. (1973), Phys.Stat. Solidi (b) 56, 197.

Evans, R. (1975), unpublished results.

Fischer, K., Bilz, H., Haberkorn, R. and Weber, W. (1972), Phys. Stat. Solidi (b) 54, 295.

Friedel, J. (1969), in "The Physics of Metals" (Ed. Ziman, J.M.) Cambridge University Press, England.

Fröhlich, H. (1966) in "Perspectives of Modern Physics", Marshak, R.E., ed.) Interscience, New York.

Fujii, Y., Hastings, J.B., Kaplan, M., Shirane, G., Inada, Y. and Kitamura, N. (1982), Phys.Rev.B. 25, 364.

Geerk, J., Gläser, W., Gompf, F., Reichardt, W. and Schneider E. (1975), in "Low Temperature Physics - LT14", Vol.2, (Eds. Krusius, M. and Vuorio, M.), North Holland, Amsterdam, p. 411.

Geerk, J. and Langguth, K.-G. (1977), Solid State Commun. 23, 83.

Geerk, J., Gurvitch, M., McWhan, D.B. and Rowell, J.M. (1982), Physica 109 , 110 B, 1775.

Gilat, G. and Raubenheimer, L.J. (1966), Phys.Rev. 144, 390.

Glötzel, D., Rainer, D. and Schober, H.R. (1979), Z.Phys. B. 35, 317.

Glötzel, D., Segall, B. and Andersen, O.K. (1980), Solid State Commun. 36, 403.

Gompf, F., Richter, W., Scheerer, B. and Weber, W. (1981), Physica 108 B, 1337.

Gompf, F. (1984), to be published.

Gor'kov, L.P. (1973), Pis'ma Zh. Eksp. Teor. Fiz. 17, 525 (JETP Lett. 17, 379 (1973)).

Guiliano, E.S., Gyorffy, B.L., Ruggeri, R. and Stocks, G.M. (1978) in "Transition Metals 1977" (Eds. Lee, M.J.G., Perz, J.M. and Fawcett, E.), Inst. of Physics, Bristol, p. 410.

Gupta, M. and Freeman, A.J. (1976a), in "Superconductivity in d- and f-band Metals", (Ed. Douglass, D.H.), Plenum Press, New York, p. 313.

Gupta, M. and Freeman, A.J. (1976b), Phys.Rev.Lett. 37, 364.

Gyorffy, B.L. (1978) private communication.

Hamann, D.R., Schlüter, M. and Chiang, C. (1979), Phys.Rev.Lett. 43, 194.

Hanak, J.J. and Berman, H.S. (1967), J.Phys.Chem.Solids 28, 249.

Hanke, W. (1973), Phys.Rev.B. 8, 4585.

Hanke, W., Hafner, J. and Bilz, H. (1976), Phys.Rev.Lett. 37, 1560.

Harmon, B.N. and Sinha, S.K. (1977), Phys.Rev.B. 16, 3919.

Harmon, B.N., Weber, W. and Hamann, D.R. (1981), J. de Physique 42, C6-628 (Proc. Int. Conf. Phonon Physics, Bloomington, 1981).

Harmon, B.N., Weber, W. and Hamann, D.R. (1982), Phys.Rev.B. 25, 1109.

Herman, F. (1959), J.Phys.Chem.Solids 8, 405.

Ho, K.M., Pickett, W.E. and Cohen, M.L. (1979), Phys.Rev.B. 19, 1751.

Ho, K.M., Fu, C.L., Harmon, B.N., Weber, W., Hamann, D.R. (1982), Phys.Rev.Lett. 49, 673.
Horton, G.K. and Maradudin, A.A. (1974), "Dynamical Properties of Solids", Vol. 1, North Holland, Amsterdam.
Horton, G.K. and Maradudin, A.A. (1980), "Dynamical Properties of Solids", Vol. 3, North Holland, Amsterdam.
Jarlborg, T. and Arbmann, G. (1977), J.Phys.F. 7, 1635.
Jepsen, O. and Andersen, O.K. (1971), Solid State Commun. 9, 1763.
Jørgensen, J.-E., Axe, J.D., Corliss, L.M. and Hastings, J.M. (1982), Phys.Rev.B. 25, 5856.
Klein, B.M., Papaconstantopoulos, D.A. and Boyer, L.L. (1976a), in "Superconductivity in d- and f-Band Metals", (Ed. Douglass, D.H.), Plenum Press, New York, p. 339.
Klein, B.M., Boyer, L.L. and Papaconstantopoulos, D.A. (1976b), Solid State Commun. 20, 937.
Klein, B.M., Boyer, L.L., Papaconstantopoulos, D.A. and Mattheiss, L.F. (1978), Phys.Rev.B. 18, 6411.
Klein, B.M. and Boyer, L.L. (1980), in "Superconductivity in d- and f-Band Metals" (Eds. Suhl, H. and Maple, M.B.), Academic Press New York, p. 455.
Kleppmann, W.G. and Weber, W. (1979), Phys.Rev.B. 20, 1669.
Kobbelt, M., Nücker, N., Reichardt, W. and Scheerer, B. (1982), in "Superconductivity in d- and f-Band Metals 1982" (Eds. Buckel, W. and Weber, W.), Kernforschungszentrum Karlsruhe, p. 119.
Kohn, W. (1959), Phys.Rev.Lett. 2, 393.
Kohn, W. and Sham, L.J. (1965), Phys.Rev. 140, A1133.
Kress, W., Rödhammer, P., Bilz, H., Teuchert, W.D. and Christensen, A.N. (1978), Phys.Rev.B. 17, 111.
Kunc, K. (1983), Helvetica Physica Acta, 56, 559.
Kwo, J. and Geballe, T.H. (1982), Physica 109 , 110 B, 1665.
Labbé, J. and Friedel, J. (1966), J.Phys.Radium 27, 153.
Lee, T.K., Birman, J.L. and Williamson, S.J. (1977), Phys.Rev.Lett. 39, 839.
Lehmann, G., Rennert, P., Taut, M. and Wonn, H. (1970), Phys.Status Solidi 37, K27.
Lehmann, G. and Taut, M. (1972), Phys.Status Solidi (b), 54, 469.
Maradudin, A.A., Montroll, E.W., Weiss, G.H. and Ipatov , I.P., (1971), in "Theory of Lattice Dynamics in the Harmonic Approximation", Academic Press, New York.
Martin, R.M. (1969), Phys.Rev. 186, 871.
Maschke, K. and Baldereschi, A. (1979) in "Physics of Semiconductors 1978" (Ed. Wilson, B.L.H.), Institute of Physics, Bristol, p. 673.
Mattheiss, L.F. (1965), Phys.Rev. 138, A 112.
Mattheiss, L.F. (1970), Phys.Rev.B. 1, 373.
Mattheiss, L.F. (1975), Phys.Rev.B. 12, 2161.
Mattheiss, L.F. and Weber, W. (1979), unpublished results
Mattheiss, L.F. and Weber, W. (1982), Phys.Rev.B. 25, 2248.
McMillan, W.L. (1968), Phys.Rev. 167, 331.
Miller, P.B. and Axe, J.D. (1967), Phys.Rev. 163, 924.

Moruzzi, V.L., Janak, J.F. and Williams, A.R. (1978), "Calculated Electronic Properties of Metals", Pergamon Press, New York.
Neckel, A., Rastl, P., Eibler, R., Weinberger, P. and Schwarz, K. (1976), J.Phys.C. 9, 579.
Overhauser, A.W. (1960), Phys.Rev.Lett. 4, 415.
Pandey, K.C. and Phillips, J.C. (1975), Phys.Rev.B. 13, 750.
Peter, M., Klose, W., Adam, G., Entel, P. and Kudla, E. (1974), Helv.Phys. Acta 47, 807.
Peter, M., Ashkenazi, J. and Dacorogna, M. (1977), Helv.Phys.Acta 50, 267.
Phillips, J.C. (1968a), Phys.Rev. 168, 905.
Phillips, J.C. (1968b), Phys.Rev. 166, 832.
Pick, R.M., Cohen, M.H. and Martin, R.M. (1970), Phys.Rev.B. 1, 910.
Pick, R.M. (1973), in "Elementary Excitations in Solids, Molecules and Atoms" (Eds. Devreese, J.T., Kunz, A.B. and Collins, T.C.), Plenum Press, New York, Part B, p.25.
Pickett, W.E. and Allen, P.B. (1974), Phys.Lett. A48, 91.
Pickett, W.E. and Gyorffy, B.L. (1976), in "Superconductivity in d- and f-Band Metals" (Ed. Douglass, D.H.), Plenum Press, New York p. 251.
Pickett, W.E., Ho, K.M. and Cohen, M.L. (1979), Phys.Rev.B. 19, 1734.
Pintschovius, L. (1979), unpublished data.
Pintschovius, L., Smith, H.G., Weber, W., Reichardt, W., Wakabayashi, W., Webb, G., Fisk, Z., Chang, Y.K., Aker, E. and Politis, C. (1982), in "Superconductivity in d- and f-Band Metals 1982" (Eds. Buckel, W. and Weber, W.), Kernforschungszentrum Karlsruhe, p.9.
Pintschovius, L., Smith, H.G., Wakabayashi, N., Reichardt, W., Weber, W., Webb, G.W. and Fisk, Z. (1983a), Phys.Rev.B., submitted.
Pintschovius, L., Reichardt, W., Smith, H.G., Aker, E. and Politis, C. (1983b), to be published.
Powell, B.M., Martell, P. and Woods, A.D.B. (1968), Phys.Rev. 171, 727.
Rath, J. and Freeman, A.J. (1975), Phys.Rev.B. 11, 2109.
Rehwald, W., Rayl, M., Cohen, R.W. and Cody, G.D. (1972), Phys. Rev.B. 6, 363.
Reichardt, W. and Scheerer, B. (1979), in Prog.Rep.Inst. f. ang. Kernphysik I, KfK Rep. 2881, p. 4, Kernforschungszentrum Karlsruhe.
Richter, W. and Weber W. (1983), to be published.
Ries, G. and Winter, E. (1980), J.Phys.F. 10, 1.
Rumyantsev, A.Yu., Zemlyanov, M.G., Parshin, P.P., Chernoplekov, N.A., Usov, O.A. and Marchenko, V.A. (1977), Fiz. Tverd. Tela (Leningrad) 19, 1715 (Sov.Phys. Solid State 19, 100 (1977)).
Schicktanz, S., Kaiser, R., Spengler, W. and Seeber, B. (1978), Solid State Commun. 28, 935.
Schicktanz, S., Kaiser, R., Schneider, E. and Gläser, W. (1980), Phys.Rev.B. 22, 2386.
Schneider, U. and Geerk, J. (1984), to be published.

Schopohl, N. (1982), private communication.
Schröder, U. (1966), Solid State Commun. 4, 347.
Schwarz, K. (1971), Monatsh. Chem. 102, 1400.
Schwarz, K. (1975), J.Phys.C. 8, 809.
Schwarz, K. (1977), J. Phys.C. 10, 195.
Schweiß, B.P., Renker, B., Schneider, E. and Reichardt, W. (1976), in "Superconductivity in d- and f-Band Metals" (Ed. Douglass, D.H.), Plenum Press, New York, p. 189.
Sham, L.J. (1969), Phys.Rev. 188, 1431.
Sham, L.J. (1971), Phys.Rev.Lett. 27, 1725.
Shirane, G., Axe, J.D. and Birgeneau, R.J. (1971), Solid State Commun. 9, 397.
Shirane, G. and Axe, J.D. (1973), Phys.Rev.B. 8, 1965.
Simons, A.L. and Varma, C.M. (1980), Solid State Commun. 35, 317.
Simons, A.L., Varma, C.M. and Weber, W. (1981), Phys.Rev.B. 23, 2431.
Sinha, S.K. (1969), Phys. Rev. 177, 1256.
Sinha, S.K., Gupta, R.P. and Price, D.L. (1971), Phys.Rev.Lett. 26, 1324.
Sinha, S.K. (1973), CRC, Critical Reviews in Solid State Sciences, 3, 273.
Sinha, S.K. and Harmon, B.N. (1975), Phys.Rev.Lett. 35, 1515.
Sinha, S.K. (1980), in "Dynamical Properties of Solids", vol. 3 (Eds. Horton, G.K. and Maradudin, A.A.), North Holland, Amsterdam, p.1.
Slater, J.C. and Koster, G.F. (1954), Phys.Rev. 94, 1498.
Smith, H.G. and Gläser, W. (1970), Phys.Rev.Lett. 25, 1611.
Smith, H.G. and Gläser, W. (1971), in "Phonons" (Ed. Nusimovici, M.A.), Proceedings of the Rennes Conference, Fammarion, p. 145.
Smith, H.G. (1972), Phys.Rev.Lett. 29, 353.
Smith, H.G., Pintschovius, L., Wakabayashi, N. and Chang, Y.K. (1983) to be published.
Splettstösser, B. (1977), Z.Phys.B. 26, 151.
Testardi, L.R., Bateman, T.B., Reed, W.A. and Chirba, V.G. (1965), Phys.Rev.Lett. 15, 250.
Testardi, L.R. (1973), in "Physical Acoustics" (Eds. Mason, W.P. and Thurston, R.N.), Academic Press, New York, Vol. X, p. 193.
Testardi, L.R. (1975), Rev.Mod.Phys. 47, 637.
Van Kessel, A.T., Myron, H.W. and Mueller, F.M. (1978), Phys.Rev. Lett. 41, 181.
Van Kessel, A.T., Myron, H.W., Mueller, F.M., Arko, A.J., Crabtree, G. and Fisk, Z. (1980), in "Superconductivity in d- and f-Band Metals" (Eds. Suhl, H. and Maple, M.B.), Academic Press, New York, p. 121.
Varma, C.M. and Dynes, R.C. (1976), in "Superconductivity in d- and f-Band Metals" (Ed. Douglass, D.H.), Plenum Press, New York, p. 507.
Varma, C.M. and Weber, W. (1977), Phys.Rev.Lett. 39, 1094.
Varma, C.M., Vashishta, P., Weber, W. and Blount, E.I. (1978), Solid State Commun. 27, 919.

Varma, C.M., Blount, E.I., Vashishta, P. and Weber, W. (1979), Phys.Rev.B. 19, 6130.
Varma, C.M. and Weber, W. (1979), Phys.Rev.B. 19, 6142.
Weber, W., Bilz, H. and Schröder, U. (1972), Phys.Rev.Lett. 29, 373.
Weber, W., (1973), Phys.Rev.B. 8, 5082.
Weber, W. (1974), Phys.Rev.Lett. 33, 371.
Weber, W. (1977), Phys.Rev.B. 15, 4789.
Weber, W., Rödhammer, P., Pintschovius, L., Reichardt, W., Gompf, F. and Christensen, A.N. (1979), Phys.Rev.Lett. 43, 868.
Weber, W. (1980a), in "Superconductivity of d- and f-Band Metals" (Eds. Suhl, H. and Maple, M.B.), Academic Press, New York, p. 131.
Weber, W. (1980b), in "Physics of Transition Metals 1980" (Ed. Rhodes, P.), Institute of Physics, Bristol, p. 455.
Weber, W. and Mattheiss, L.F. (1981), Physica 170B, 263.
Weber, W. and Mattheiss, L.F. (1982), Phys.Rev.B. 25, 2270.
Weber, W. (1982), in "Superconductivity in d- and f-Band Metals" (Eds. Buckel, W. and Weber, W.), Kernforschungszentrum Karlsruhe, p. 15.
Weber, W. (1983), to be published.
Weger, M. (1964), Rev.Mod.Phys. 36, 175.
Weiss, L. and Rumyantsev, A.Y. (1981), Phys.Stat.Sol. (b) 107, K75.
Wendel, H. and Martin, R.M. (1978), Phys.Rev.Lett. 40, 950.
Wendel, H. and Martin, R.M. (1979), Phys.Rev.B. 19, 5251.
Wimmer, E., Podloucky, R., Herzig, P., Neckel, A. and Schwarz, K. (1982), J.Phys.Chem.Sol. 43, 439.
Winter, H., Rietschel, H., Ries, G. and Reichardt, W. (1978), J. de Physique 39, C6-474.
Winter, H. (1981), J.Phys.F. 11, 2283.
Wipf, H., Klein, M.V., Chandrasekhar, B.S., Geballe, T.H. and Wernick, J.H. (1978), Phys.Rev.Lett. 41, 1752.
Yin, M.T. and Cohen, M.L. (1980), Phys.Rev.Lett. 45, 1004.
Zeyher, R. (1975), Phys.Rev.Lett. 35, 174.

A FIRST PRINCIPLES APPROACH TO THE BAND THEORY OF RANDOM METALLIC ALLOYS

G. Malcolm Stocks

SERC Daresbury Lab
Daresbury
Warrington WA4 4AD
England

Oak Ridge National Lab.
Oak Ridge
Tennessee 37830
U.S.A.

Herman Winter

Kernforschungszentrum Karlsruhe
D-7500 Karlsruhe
German Federal Republic

PROLOGUE

In 1976 and again 1978 Balazs Gyorffy and myself presented lectures at Advanced Study Institutes in Ghent on the developments towards obtaining a first principles theory of electronic states in random alloys. The 1976 lectures were concerned mainly with the formal theory whereas the 1978 lectures were concerned with the nature of the solution of the basic equations ; with calculations on real systems and with the comparison of these results with experiment. Since 1978 there have been many further advances in our understanding of random alloys both from the theoretical and from the experimental point of view. It is these developments with which we will be mainly concerned. However, in order to make these lecture notes self-contained I have not hesitated to repeat some of the formal aspects covered in the previous notes. Since the scope of these lecture notes is wider than the previous ones, here we include the discussion of a fully self-consistent scheme, the discussion of the formal aspects of the theory is necessarily abbreviated. With this in mind the reader is referred to the previous sets of notes by Gyorffy and myself in "Electrons in Finite and Infinite Structures" Eds. P. Phariseau and L. Scheire (Plenum 1977) and "Electrons in Disordered Metals and at Metallic Surfaces" Eds. P. Phariseau, B.L. Gyorffy and L. Scheire (Plenum 1979).

I. INTRODUCTION

From the Schrödinger Equation to the Rolling Mill

At the outset it is worthwhile to state what we have in mind as our ultimate objective in studying condensed metallic systems. Our objective may be far away and only vaguely defined, yet it influences the theoretical and experimental approaches which we bring to bear in studying these systems. Here our objective is lofty indeed. We have in mind the attempt to understand the phase stability and the physical properties of metallic alloy systems in terms of the basic building blocks of matter, nuclei and electrons. Thus we are interested in theories which are first principles, that is, we wish to be able to discuss the properties of these systems without recourse to adjustable parameters. Ideally the only inputs to our theories should be the atomic numbers of the constituent species. We then proceed, by controlled approximations on the Schrödinger equation for this many particle system to the calculation of physical observables. Armed with this detailed understanding we may then wish to construct simple models which contain the important physical mechanisms and which we can use to extrapolate to systems for which detailed calculations have not been performed.

For pure metals and ordered alloys the advances of recent years mean that the above scheme can be carried to a high degree of completion[1,2]. Indeed, most of the other lectures in this summer school are concerned with showing just how far one can go with first principle methods for ever increasingly complex systems. In all of this work the fundamental underpinning is provided by the density-functional-theory of Hohenberg and Kohn[3,4] described by Dr. Von Barth at this summer school. However, great as these advances have been, if we were to tell a metallurgist, as testiment to our theoretical ingenuity and computational prowess 'Yes, if at - 273°C you produce pure Cu in your rolling mill it will have the face-centred-cubic crystal structure and will have a lattice parameter of 6.76 a.u.' we ought not to be surprised if he were profoundly uninterested. (In fact the reliable prediction of the T = 0°K crystal structure is apparently still beyong first principles methods even for the case of the pure metals). The materials scientist is generally interested in complicated alloy systems the properties of which he can refine to meet specific ends. He is interested in the whole phase diagram of alloy systems. The pure metal end points and the ordered phases which may form give only a limited amount of information. In order to fill in the picture we need also to consider the possible solid solutions[5]. That is we need to extend the first principles methods to alloys in which translational invariance does not exist.

The fact that a system does not possess translational invariance does not affect (at least in principle) the applicability of the density functional methods, it does however mean that the great armoury of band theory methods developed to solve the Schrödinger equation for periodic systems are no longer useful. During the last few years, however, methods have been developed for calculating the electronic properties of random alloys which, though computationally difficult, allow us to treat them on roughly the same footing as pure metals and ordered alloys[6,10]. It is the central concern of these lectures to acquaint you with these methods.

The theory of random alloys is not a slightly reworked version of the band theory of pure metals it is profoundly different. The fact that translational invariance is lost means that we can no longer classify states according to their Bloch wave-vectors, energy bands no longer exist the concept of a Fermi surface becomes fuzzy.

The central problem in dealing with random alloys is that the effective one electron crystal potential function entering the Schrödinger equation is random. The study of the eigenvalue spectra of random Hamiltonians has a long history which has resulted in the development of approximations of known standing which can be used to treat specific problems. The major development of concern here was that of the so-called coherent potential approximation (CPA)[11,14]. The CPA allows the treatment of the effects of disorder at a level which makes it worthwhile to attempt to use it as the basis of a first principles theory of random alloys. As well as our previous lecture notes in this series, Gyorffy and Stocks (1976)[6], Gyorffy and Stocks (1979)[7] there are excellent review articles by Elliott et al (1974)[13] and Ehrenreich et al (1976)[14] which review the theory of disordered systems and the early development of the CPA, and by Faulkner[5] who brings the subject more or less up to date.

In these notes we will show how the CPA can be applied to a muffin-tin-potential description of random alloys, (See II) and how concepts such as energy bands and Fermi surfaces, familiar in the band theory of ordered systems, can be generalized, through the use of spectral functions, to random alloys (See III). We will show how the CPA can be embedded within local density functional schemes to provide an entirely parameter free self-consistent field theory of the electronic states of random alloys (See IV). At all times we will keep in mind the contact which can be made with experiment. In particular we will show how spectroscopic studies can be used to measure the band structure of random alloys, how positron annihilation experiments can be used to measure the

'Fermi surface' (Sec V). We will also show how the results of these first principles calculations can be used to understand the concentration variation of the electrical resistivity in Ag_cPd_{1-c} alloys. In Sec VI we offer some conclusions on what has been accomplished and some speculations on future directions.

The Density Functional Approach and First Principles Band Theory

Although in general we are interested in much more than the ground state properties of alloys and defected metals we take over as the fundamental underpinning of our first principles theories the density functional (DF) theory[3,4]. Since DF-theory has been described in great detail by Dr. Von Barth, here we need only summarize the final results of the so-called local density functional approach in order to delineate the steps involved in carrying out a first principles calculations.

The major result of DF-theory is that the total energy E of a system of electrons in an external potential V_{ext} (the potential due to the ions) is a functional of the electron density $\rho(r)$ and that the ground state charge density, which minimizes E, can be obtained by solving the set of one particle, Hartree like self consistency equations[3,4,15].

$$[-\nabla^2 + V_{eff}(\vec{r}) - \varepsilon_n]\varphi_n(\vec{r}) = 0 \qquad \text{I.1}$$

where the effective potential V_{eff} is related to the charge density through

$$V_{eff} = \int d^3r \, \frac{\rho(\vec{r})}{|\vec{r}-\vec{r}'|} + V_{ext} + V_{xc} \qquad \text{I.2}$$

and the charge density is obtained from the solutions of Eq. I.1 through

$$\rho(\vec{r}) = \sum_{n\ occ} |\varphi_n(\vec{r})|^2 \qquad \text{I.3}$$

where the sum runs over the lowest N occupied states. In Eq. I.2 V_{xc} is the exchange correlation potential and is obtained from the exchange correlation energy through the functional derivative

$$V_{xc}(\vec{r}) = \frac{\delta E_{xc}[\rho(\vec{r})]}{\delta\rho(\vec{r})} \qquad \text{I.4}$$

In the local approximation the unknown and complicated exchange correlation energy E_{xc} for the interacting inhomogeneous electron gas is approximated by

$$E_{xc} = \int d^3r \, \rho(\vec{r})\varepsilon_{xc}[\rho(\vec{r})] \qquad \text{I.5}$$

where ε_{xc} is the exchange correlation energy per particle for an interacting but homogeneous electron gas having the (local)density $\rho(\vec{r})$. Having made this approximation advantage can be taken of the countless person hours which have been expended in obtaining accurate values of the exchange correlation energy for the homogeneous electron gas. The form of the local-density-functional (LDF) theory exchange correlation potential μ_{xc}

$$\mu_{xc} = \frac{\delta\rho(\vec{r})\dot{\varepsilon}_{xc}[\rho(\vec{r})]}{\delta\rho(\vec{r})} \qquad \text{I.6}$$

which is most commonly used is that of Hedin and Lundqvist[16]. In terms of the parameter r_s where

$$r_s = \{\frac{3}{4\pi} \cdot \frac{1}{\rho(\vec{r})}\}^{1/3} \qquad \text{I.7}$$

Hedin and Lundqvist write $\mu_{xc}(r_s)$ in the form

$$\mu_{xc}(r_s) = \beta(r_s)\mu_x(r_s) \qquad \text{I.8}$$

where

$$\mu_x = -\frac{1.22}{r_s} \qquad \text{I.9}$$

is the usual ($\rho^{1/3}$) exchange potential and the correlation effects are included through the function $\beta(r_s)$ which is taken to be of the form

$$\beta(r_s) = 1 + A\, r_s\, \ell n(1 + \frac{B}{r_s}) \qquad \text{I.10}$$

The values of constants A and B which were obtained by Hedin and Lundqvist by fitting to the numerical values of E_{xc} obtained by Singwi et al[17] are 0.0368 and 21.0 respectively although other values have also been used in some calculations[18].

In the absence of externally applied magnetic and electric fields the only external potential to which the electrons have to adjust themselves is that due to the nuclei which comprise the system. If these nuclei have atomic numbers Z_i and are situated at the points $\vec{R}_i$ then the external potential at the point $\vec{r}$ is

$$V_{ext}(\vec{r}) = -\sum_i \frac{Z_i}{|\vec{r} - \vec{R}_i|} \qquad \text{I.11}$$

and the total effective potential in the LDF theory is given by

$$V_{eff}(\vec{r}) = -\sum_i \frac{Z_i}{|\vec{r} - \vec{R}_i|} + \int d^3r' \frac{\rho(r')}{|r - r'|} + \mu_{xc}[\rho(r)] \qquad \text{I.12}$$

Thus within this local density approach Eqs. (I.1) and (I.3) and (I.12) are a parameter free set of equations which have to be solved self-consistently for the ground state charge density corresponding to the particular ionic configuration which was specified in Eq. (I.12). Once these equations have been solved the total energy corresponding to the specified ionic arrangement is then given by

$$E\{Z_i,R_i\} = \sum_{n\ \text{occ.}} \varepsilon_n - \frac{1}{2}\int d^3r \int d^3r' \frac{\rho(\vec{r})\rho(\vec{r}')}{|\vec{r} - \vec{r}'|} +$$

$$\int d^3r\ \rho(\vec{r})\{\varepsilon_{xc}[\rho(\vec{r})] - \mu_{xc}[\rho(r)]\} \qquad \text{I.13}$$

In Eq. I.13 the first term gives the contribution to the total energy from the single particle eigenvalue sum, the second term corrects this for what in simple Hartree approximation would be called double counting (the fact that in the eigenvalue sum the electron Coulomb self interaction has been counted twice), the third term gives the local density approximation to the exchange correlation energy.

Thus within LDF theory a self-consistent first principles calculation of the electronic structure of a system in which nuclei of atomic number Z_i are situated at points in space $\vec{R}_i$ consists of the following steps.

(i) Construction of some initial input guess of the charge density $\rho_{in}(\vec{r})$. This initial guess is often obtained according to the so-called Mattheiss prescription[14] which involves the overlap of neutral atom charge densities.
(ii) Calculation of the effective crystal potential from Eq. (I.13) using the input charge density.
(iii) Solution of the effective single particle Schrödinger equation (I.1).
(iv) Calculation of the output charge density from Eq. (I.3), ρ_{out}.
(v) Return to step (ii) using an admixture of ρ_{in} and ρ_{out} as the new ρ_{in} if the output charge density is different from the input charge density by more than some small tolerance.

The steps (ii) → (v) then comprise a self-consistency loop for the charge density (and effective potential), accordingly electronic structure calculations which have been carried out to this degree of sophistication are generally referred to as being self-consistent. It will be taken as read that in such calculations we will evaluate the relevant equations in particular the effective Schrödinger equation with the minimum of numerical compromises. If this involves us in heavy numerical work then this is simply the price which must be paid for having a first principles theory. However, once the numerical difficulties are overcome the full power of such a theory comes into play in that when the results of the calculations are compared with experiment the focus of attention is then the effective potential function and therefore our understanding of exchange correlation effects. Effects which have only been treated in an approximate manner in LDF theory[20].

As we have already pointed out, for pure crystalline metals and for ordered compounds the above system of equations can be solved essentially exactly (save for the muffin-tin-approximation) and the total energy calculated[1,2]. The equilibrium lattice spacing, cohesive energy, bulk modules can then be found by repeating the above calculation for several lattice spacings and locating the lattice spacing at which the total energy is a minimum.

In these lecture notes I will explore the extent to which similar calculations can be carried out in systems where the translational invariance of V_{eff} is broken by the formation of a concentrated solid solution. Regardless of whether or not translational invariance exists at the heart of performing first principles calculations is the devising of efficient numerical algorithms for solving the Schrödinger equation (I.1). Once the Schrödinger equation has been solved it is generally a much less exacting task to calculate the charge density and a new guess of V_{eff}.

Because of the complexities involved in solving the Schrödinger equation for the alloy problem it is only recently that fully self-consistent calculations have been carried out[21,22] Indeed at the time of writing these notes there are still no reliable calculations of the total energies in such systems which would allow calculation of for example the equilibrium lattice spacing of random alloys and the formation energies for disordered alloys. Thus until recently it has been conventional to construct V_{eff} by some <u>ad-hoc</u> procedure, to solve the Schrödinger equation and from these solutions to go directly to a comparison with experiment, without iterating the potential to self-consistency. This has been particularly true of the development of the first principles coherent-potential-approximation[8,9] theory for random alloys where the theoretical and computational difficulties associated with solving the Schrödinger are worst. However, a decade's experience with similar non-self-consistent band structure calculations on ordered systems had given rise to <u>ad-hoc</u> prescriptions for construction of V_{eff} which gave good descriptions of the electronic structure. Both the Mattheiss prescription[14] and the renormalized atom[23] picture have been extended to random alloys[24]. In the extended Mattheiss prescription[7,25] on which some of the results shown in later sections are based, V_{eff} is constructed from the overlap of neutral atom potentials and charge densities and the exchange correlation potential is taken to be of the form suggested by Slater[26] and often called full Slater exchange namely

$$\mu_{xc}^{Slater} = \frac{3}{2} \mu_x \qquad \text{I.14}$$

Using this prescription apparently reliable potential functions appropriate to Cu_cNi_{1-c}[27], Ag_cPd_{1-c}[25], Nb_cMo_{1-c}[28], Cu_cZn_{1-c}[29], Ni_cPt_{1-c}[30], Au_cPt_{1-c}[31] alloys have been constructed and have been used as a basis for gaining experience with the theoretical and computational tools.

The Non Overlapping-Muffin-Tin Model of the Alloy Crystal Potential

Before going on to discuss in detail the models which we use to treat random alloys and the theoretical and computational tools which we bring to bear on their solution we should begin by stating precisely how we visualize a random alloy and what aspects of the problem we include.

In fig. I.1 we show our idealization of a random alloy for the case of a binary A_cB_{1-c} solid solution. It is assumed that the two atomic species occupy the sites $\vec{R}_i$ of an underlying regular lattice. It is further assumed that the site occupancies are uncorrelated, thus the probability that any site is occupied by the A-species is $c_A (\equiv c)$ the concentration of the A species and the probability that any site is occupied by the B-species is $c_B[\equiv(1-c)]$ the concentration of the B-species.

For such a system it is sufficiently general to write the effective one electron crystal potential $V(\vec{r})$ which enters Eq. I.1 as a sum of potential wells $v_i(\vec{r} - \vec{R}_j)$ centred on the lattice sites $\vec{R}_i$,

$$V(\vec{r}) = \sum_i v_i(\vec{r} - \vec{R}_i) \qquad \text{I.15}$$

In general, even for the case of a binary alloy, all the v_i are different since the surroundings of each site in a random alloy is different from all the others. If we perform the thought experiment of making up the random alloy by bringing together neutral atoms then roughly speaking the potential on any site will be made up of the atomic like potential of the species occupying the site plus the tails of atomic potentials from the neighbouring sites. Clearly it is only to the extent that sum of the tails of the atomic potentials corresponding to the different surrounding atomic arrangements are different that the potentials corresponding to different A(B) sites will be different from one another. In general these differences will be small on the average. Accordingly we will ignore them for the moment and assume that there are only two different site potentials $V_A(\vec{r} - \vec{R}_i)$ and $V_B(\vec{r} - \vec{R}_i)$ corresponding to the two different atomic species.

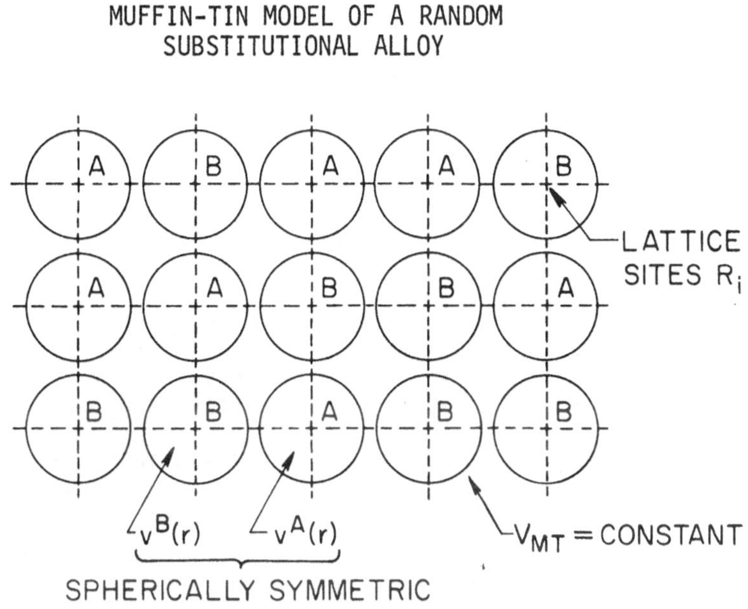

Fig. I.1. Schematic representation of the alloy crystal potential in the non-overlapping muffin-tin model.

On the basis of our thought experiment is should also be clear that for close packed systems that in regions close to the nuclei the potential will be roughly spherically symmetric and in interstitial regions where the potential is made up only from tails it will be approximately constant. Thus for computational purposes we will approximate the full site potentials by the so-called muffin-tin form where

$$v_{A(B)}(\vec{r}-\vec{R}_i) = \begin{cases} v_{A(B)}(|\vec{r}-\vec{R}_i|) & |\vec{r}-\vec{R}_i| < r_{MT} \\ v_{MT} = \text{const.} & |\vec{r}-\vec{R}_i| > r_{MT} \end{cases} \qquad \text{I.16}$$

the muffin-tin radius r_{MT} is taken to be half the nearest neighbour distance in order that the maximum volume possible is enclosed within the muffin-tin spheres. This generalized muffin-tin model is what is illustrated in fig. I.1. In summary the crystal potential inside the spheres is assumed spherically symmetric being either $V_A(r)$ or $V_B(r)$ according as the site is occupied by the A or B species, in the region between the spheres it is assumed to have a constant value V_{MT} which is usually set to be the zero of energy.

Thus for this muffin-tin model the total crystal potential can be written

$$V(\vec{r}) = \sum_i \{\xi_i v_A(\vec{r} - \vec{R}_i) + (1 - \xi_i) v_B(\vec{r} - \vec{R}_i)\} \qquad \text{I.17}$$

where ξ_i is a random occupation variable such that $\xi_i = 1$ if the site $\vec{R}_i$ is occupied by the A-species and $\xi_i = 0$ if $\vec{R}_i$ is occupied by the B species. Thus in a completely random binary A_cB_{1-c} alloy ξ_i takes the values 0 or 1 with the probabilities c_A and c_B.

Implicit in our idealization of a random alloy is that the atoms occupy a set of sites which form an ordered lattice. Since all the atoms in the alloy are surrounded by different configurations of other atoms presumably there are slight relaxations about the ideal-lattice positions $\vec{R}_i$ for all sites in the crystal. However, that sharp Bragg maxima occurs in diffraction patterns taken from random alloys is testiment to the fact that these local distortions are in general small and that the concept of an underlying regular lattice is a good one.

In the above idealization of a random alloy it is the approximation that the site occupancies are statistically uncorrelated which is the most problematical, nearly all alloys show some tendency either for like atoms to cluster together as in the Cu_cNi_{1-c} [32] system or to show signs of incipient superlattice formation as in the Cu_cPd_{1-c} [33] alloy system. Indeed such clustering and short range ordering effects can play an important role in determining the metallurgical properties of alloys. However, we shall take the view that we will begin by treating the idealized random system as completely as possible in order to use it as a basis for judging the importance of these effects and as a base onto which these effects can be built.

The Calculation of Observables

Within the model of the previous section a particular alloy configuration is specified by a set (ξ) of values $\xi_1, \xi_2, \ldots\ldots, \xi_N$ of the site occupation variable ξ_i. For such a configuration the corresponding Schrödinger equation takes the form

$$[-\nabla^2 + \sum_i \{\xi_i v_A(\vec{r} - \vec{R}_i) + (1 - \xi_i) v_B(\vec{r} - \vec{R}_i)\}]\varphi_n(\vec{r}, \xi_i, \ldots, \xi_N) =$$

$$\varepsilon_n(\xi_i, \ldots, \xi_N)\varphi_n(\vec{r}, \xi_i, \ldots, \xi_n) \qquad \text{I.18}$$

Clearly the solution of this equation for the configuration dependent eigenfunctions $\Phi_n(\vec{r},\xi_1,\ldots,\xi_N)$ and eigenvalues $\varepsilon_n(\xi_1,\xi_2,\ldots,\xi_N)$ is an impossible task for a system containing $\sim 10^{23}$ sites. Both from the point of view of constructing observables in general and in particular the charge density $\rho(r,\xi_1,\xi_2,\ldots,\xi_N)$, required in the solution of self-consistent LDF theory equations, it is also a pointless task. The reason for this is that we are not directly interested in the wavefunctions and eigenvalues corresponding to some particular configuration rather we are interested in calculating physical observables which have been averaged over all possible configurations of the alloy consistent with the concentrations of the A and B species being c_A and c_B[34]. It is these configurational averages which are actually measured in experiments on random alloys. Consider the matrix element $Q_{nn'}(\xi_1,\xi_2,\ldots,\xi_n) = \langle\varphi_n|Q|\varphi_{n'}\rangle$ of some operator Q, then the point is that the only information which is obtained about $Q_{nn'}$ from experiment is about its configurational averaged value $\langle Q_{nn'}\rangle$

$$\langle Q_{nn'}\rangle = \sum_{\{\xi\}} P(\xi)\ \langle\varphi_n|Q|\varphi_{n'}\rangle \qquad \text{I.19}$$

where $P(\xi)$ is the probability of the occurence of the configuration $\{\xi\}$ and the sum is taken over all configurations consistent with the concentration of the A and B species. Thus rather than calculating the individual φ_n's and ε_n's it is obviously more efficatious to attempt to calculate the configurationally averaged observables directly.

This task is greatly expedited if rather than considering the solution of the Schrödinger equation (I.18) we consider the solution of the inhomogeneous differential equation

$$[-\nabla^2 + V(\vec{r}) - \varepsilon]\ G(\vec{r},\vec{r}\,',\varepsilon) = -\delta(\vec{r}-\vec{r}\,') \qquad \text{I.20}$$

where the Green's function $G(\vec{r},\vec{r}\,',\varepsilon)$ describes the propogation of an electron of energy ε in the field defined by Eq. I.17. If all the eigenfunctions and eigenvalues of Eq. I.18 are known then $G(\vec{r},\vec{r}\,',\varepsilon)$ can be constructed directly from

$$G(\vec{r},\vec{r}\,',\varepsilon) = \lim_{\eta\downarrow} \sum_n \frac{\varphi_n(\vec{r})\varphi_n(\vec{r}\,')}{\varepsilon - \varepsilon_n + i\eta} \qquad \text{I.21}$$

The rotation $\eta\downarrow$ implies that the limit is to be taken as η approach zero from a positive value. An expression for the single particle density of states $n(\varepsilon)$ in terms of this Green's function is easily obtained by setting $\vec{r} = \vec{r}\,'$ in Eq. I.21 and integrating over all space to yield

$$n(\varepsilon) = -\frac{1}{\pi}\,\text{Im}\int d^3r\ G(\vec{r},\vec{r}\,',\varepsilon) = \sum_n \delta(\varepsilon - \varepsilon_n) \qquad \text{I.22}$$

The configurationally averaged density of states $<n(\varepsilon)>$ the quantity of interest is given in terms of the configurationally averaged Green's function by

$$<n(\varepsilon)> = -\frac{1}{\pi}\,\mathrm{Im}\int d^3r <G(\vec{r},\vec{r}',\varepsilon)> \qquad \text{I.23}$$

Thus it is the task of alloy theory to find approximations schemes in which $<G(\vec{r},\vec{r}',\varepsilon)>$ can be calculated directly. Once a reliable and accurate scheme has been obtained other quantities of interest can be obtained in a similar fashion. The average charge density $<\rho(\vec{r})>$ is given by

$$<\rho(\vec{r})> = -\frac{1}{\pi}\int_{-\infty}^{\varepsilon_F} d\varepsilon\, <G(\vec{r},\vec{r},\varepsilon)> \qquad \text{I.24}$$

More usefully the average charge density in the neighbourhood of some particular species, which as we shall see later is required for the self-consistent solution of the LDF-theory equations, is given by

$$<\rho(\vec{r})>_{A(B)} = -\frac{1}{\pi}\int_{-\infty}^{\varepsilon_F} d\varepsilon\, <G(\vec{r},\vec{r},\varepsilon)>_{A(B)} \qquad \text{I.25}$$

where the angle brackets $< >_{A(B)}$ imply that the averaging has been taken over all configurations of the alloy consistent with there being a A(B) site at the origin.

Obviously ensuring

$$<\rho(\vec{r})> = c_A<\rho(\vec{r})>_A + c_B<\rho(\vec{r})>_B \qquad \text{I.26}$$

is a non-trivial requirement on any approximate calculation of the exact averages. It is precisely for the calculation of such quantities that the coherent-potential-approximation (CPA) was devised.

<u>The Coherent-Potential-Approximation</u>

Obtaining methods by which reasonable solutions can be obtained for Hamiltonians having random potentials has a long and complex history[13,14]. However, in the present context, it was only with the introduction of the coherent-potential-approximation by Soven[11] for calculating electronic states in random alloys and by Taylor[12] in the analogous phonon problem that an approximation of sufficient flexibility and sophistication to be worthy of serious application to real alloy systems was achieved. Following the discovery of the CPA a deluge of papers appeared in which the CPA was rederived in a host of ways and its usefulness emphasized usually on the basis of calculations performed for Kronig-Penney and simple tight-binding models. Of particular importance was a paper by Velicky et al[35] which clarified the standing of the CPA as being the best possible approximation within a hierarchy of so-called single-site approximations. We will not go through these arguments in detail rather we will try to motivate the CPA physically, before deriving, in the next section, the relevant equations in the context of the muffin-tin Hamiltonian.

It is the task of alloy theory to find approximations for the configurationally averaged Green's function $\langle G(\vec{r},\vec{r}',\varepsilon)\rangle$. If we define the free particle Green's function $G_0(\vec{r},\vec{r}',\varepsilon)$ through the solution of

$$[-\nabla^2 - \varepsilon]\, G_0(\vec{r},\vec{r}',\varepsilon) = -\,\delta(\vec{r}-\vec{r}') \qquad \text{I.27}$$

then formally the Green's function for a particular arrangement of atomic sites is given from Eq. I.20 as

$$G(\vec{r},\vec{r}',\varepsilon) = G_0(\vec{r},\vec{r}',\varepsilon) + \int d^3r_1 G_0(\vec{r},\vec{r}_1,\varepsilon)V(\vec{r}_1)G(\vec{r}_1,\vec{r}',\varepsilon) \qquad \text{I.28}$$

and formally the averaged Green's function is given by

$$\langle G(r,r',\varepsilon)\rangle = G_0(\vec{r},\vec{r}',\varepsilon) + \langle \int d^3r_1 G_0(\vec{r},\vec{r}_1,\varepsilon)V(\vec{r}_1)G(\vec{r}_1,\vec{r}',\varepsilon)\rangle \qquad \text{I.29}$$

A re-occurring theme in the developments in alloy theory which ultimately led to the discovery of the CPA was to obtain approximations $\tilde{G}(\vec{r},\vec{r}',\varepsilon)$ to the averaged Green's function

$$\tilde{G}(\vec{r},\vec{r}',\varepsilon) \approx \langle G(\vec{r},\vec{r}',\varepsilon)\rangle \qquad \text{I.30}$$

which have the full periodicity of the lattice

$$\tilde{G}(\vec{r} + \vec{R}_n,\, \vec{r}' + \vec{R}_n,\varepsilon) = \tilde{G}(\vec{r},\vec{r}',\varepsilon) \qquad \text{I.31}$$

Thus it is implicitly assumed that there is some effective ordered system with a corresponding effective potential function which can be placed on every site in the alloy and used in Eqs. I.20 or I.28 for the purpose of calculating the approximation to the configurationally averaged Green's function. In the CPA some effective (unknown) potential $v_i(\vec{r},\vec{r}',\varepsilon)$ is placed on every site in the lattice, the "best" choice for v_i the coherent-potential is then obtained by satisfying some constraint.

To see how this goes let us first define the total t-matrix $T(\vec{r},\vec{r}',\varepsilon)$ for the system through

$$G(\vec{r},\vec{r}',\varepsilon) = G_0(\vec{r},\vec{r}',\varepsilon) + \int d^3r_1 \int d^3r_2 G_0(\vec{r},\vec{r}_1,\varepsilon)T(\vec{r}_1,\vec{r}_2,\varepsilon)G_0(\vec{r}_2,\vec{r}',\varepsilon) \qquad \text{I.32}$$

As we will see more clearly later $T(\vec{r},\vec{r}',\varepsilon)$ describes the scattering of an incident wave by the assembly of potential wells. Evidently the relationship between the t-matrix and the potential is

$$T(\vec{r},\vec{r}',\varepsilon) = V(\vec{r})\delta(\vec{r}-\vec{r}') + V(\vec{r})\int d^3r_1 G(\vec{r},\vec{r}_1,\varepsilon)T(\vec{r},\vec{r}',\varepsilon) \quad \text{I.33}$$

Clearly the configurationally averaged Green's function is given by

$$\langle G(r,r',\varepsilon)\rangle = G_0(\vec{r},\vec{r}',\varepsilon) + \int d^3r_1\int d^3r_2 G_0(\vec{r},\vec{r}_1,\varepsilon)\langle T(\vec{r}_1,\vec{r}_2,\varepsilon)\rangle G_0(r_2,r_1',\varepsilon) \quad \text{I.34}$$

Thus finding approximations for $\langle T(\vec{r}_1,\vec{r}_2,\varepsilon)\rangle$ is completely equivalent to finding them for $\langle G(\vec{r},\vec{r}',\varepsilon)\rangle$.

In terms of the t-matrix $T_v(\vec{r},\vec{r}',\varepsilon)$ which describes the scattering from the ordered array of coherent potentials $\sum_i v_i$ the CPA Green's function $G_v(r,r',\varepsilon)$ is given by

$$G_0(\vec{r},\vec{r}',\varepsilon) = G_0(\vec{r},\vec{r}',\varepsilon) + \int d^3r_1\int d^3r_2 G_0(\vec{r},\vec{r}_1,\varepsilon)T_v(\vec{r}_1,\vec{r}_2,\varepsilon)G_0(\vec{r}_2\vec{r}',\varepsilon) \quad \text{I.35}$$

The idea then is to choose v_i such that

$$G_v(\vec{r},\vec{r}',\varepsilon) \approx \langle G(\vec{r},\vec{r}',\varepsilon)\rangle \quad \text{I.36}$$

In the coherent-potential-approximation this is done by first constructing the Green's function for the system in which the coherent-potential is placed on every site except the central one on which the true A-potential v_A is placed. The Green's function $G_v^A(\vec{r},\vec{r}',\varepsilon)$ for this system is

$$G_v^A(\vec{r},\vec{r}',\varepsilon) = G_0(\vec{r},\vec{r}',\varepsilon) + \int d^3r_1\int d^3r_2 G_0(\vec{r},\vec{r}_1,\varepsilon)T_v^A(r_1,r_2,\varepsilon)G_0(\vec{r}_2,\vec{r}',\varepsilon) \quad \text{I.37}$$

There is a similar Green's function $G_v^B(\vec{r},\vec{r}',\varepsilon)$ and t-matrix $T_v^B(\vec{r},\vec{r}',\varepsilon)$ for a system having a B-site potential v_B at the origin. The 'best' coherent-potential is then chosen by requiring that replacement at a single site of the coherent-potential by either the true A- or B-site potential produces no further scattering on the average. In terms of the t-matrices defined above this requirement reads

$$c_A T_v^A(\vec{r},\vec{r}',\varepsilon) + c_B T_v^B(\vec{r},\vec{r}',\varepsilon) = T_v(\vec{r},\vec{r}',\varepsilon) \quad \text{I.38}$$

and is shown schematically in fig. I.2. Equation I.38 is an equation which determines v_i. Thus, in principle, the solution can then be substituted into Eq. I.35 and the CPA approximation to the full averaged Green's function can be calculated and from this the densities of states etc.

Prior to the development of the CPA many other approximations were developed two of these are worthy of mentioning if only for the sake of setting up 'straw men'. These are the 'virtual-crystal-approximation' (VCA) of Nordheim[36] and Muto[37] and the averaged-t-

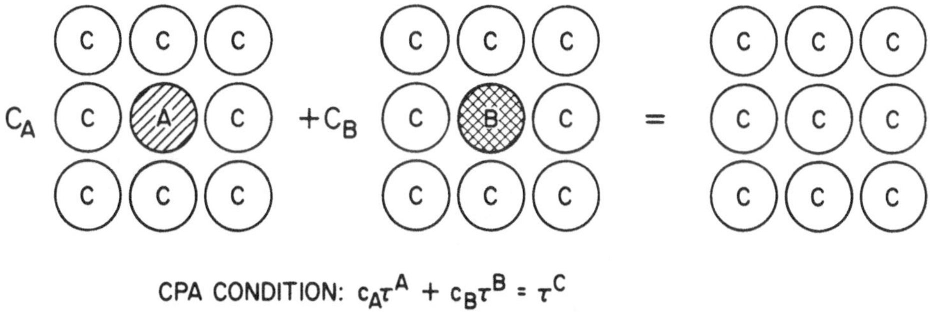

Fig. I.2. Schematic representation of the CPA condition.
The sites labelled 'c' are occupied by a coherent-potential.
The site labelled 'A' is occupied by the A potential v_A.
The site labelled 'B' is occupied by the B potential v_B.

matrix-approximation (ATA) of Korringa[38,39]. Along the lines of approximations involving replacing the disordered system by an effective ordered system the VCA is the simplest such approximation. On each site of the effective ordered system is placed the concentration weighted virtual crystal potential $v_{AV}(\vec{r}-\vec{R}_i)$

$$v_{AV}(\vec{r}-\vec{R}_i) = c_A v_A(\vec{r}-\vec{R}_i) + c_B v_B(\vec{r}-\vec{R}_i) \qquad \text{I.39}$$

Clearly the total crystal potential

$$V_{AV}(\vec{r}) = \sum_i v_{AV}(\vec{r}-\vec{R}_i) \qquad \text{I.40}$$

is real and periodic. Consequently the eigenvalues and eigenfunctions of Eq. I.1 and hence the densities of states etc., can be calculated by the normal methods of band theory. The disadvantage of this approximation is that it is correct only in the limit $v_A \to v_B$ and as a result it applies to no real alloy system of which I am aware.

The ATA is much more sophisticated. In the ATA it is the scattering properties of the individual v_A and v_B potentials which are averaged. These scattering properties are described by the t-matrix for an individual scatterer which is defined, in a way analogous to Eq. I.34 for the assembly of scatterers, as

$$t^{\alpha}(r,r',\varepsilon) = v_{\alpha}(\vec{r})\delta(\vec{r}-\vec{r}\,') + v_{\alpha}(\vec{r})\int d^3r G_0(\vec{r},\vec{r}_1)t^{\alpha}(r_1,r,\varepsilon) \qquad \text{I.41}$$

where α = A or B. The ATA consists of placing on every site of the effective ordered system a t-matrix t_{ATA} which is the concentration weighted average of the single scatter t-matrices t_A and t_B

$$t_{ATA}(r,r',\varepsilon) = c_A t^A(\vec{r},\vec{r}',\varepsilon) + c_B t^B(r,r',\varepsilon) \qquad \text{I.51}$$

The Green's function etc. are then to be calculated for this ordered array of t-matrices. Although the single-site potentials to which t_A and t_B correspond are real it turns out that the single-site potential to which t_{ATA} corresponds to is complex ; it is like an optical potential. Consequently Bloch states are no longer eigenstates of the system and any attempt to so describe them results in them having a complex energy[24] and therefore a finite lifetime. Thus the ATA has the effects of disorder are built in at a very basic level.

Unfortunately the ATA like the VCA has a number of problems and is inferior to the CPA. Rather than spend freams of paper trying to justify algebraically why the CPA is superior to both ATA and VCA I will take the view "The proof of the pudding is in the eating". In figs. I.3 and I.4 I show a comparison between the densities of states obtained on the basis of exact calculations with those obtained using variously VCA, ATA and CPA taken from the review article of Faulkner[10] for a simple two-level tight binding model of a random alloy. The Hamiltonian for this system takes the form

$$H = \sum_i \varepsilon_i |i\rangle\langle i| + \sum_{i \neq j} \omega_{ij} |i\rangle\langle j| \qquad \text{I.52}$$

the site energy ε_i takes the values $\varepsilon_{A(B)}$ depending on whether the site i is occupied by the A(B) species. The hopping integral ω_{ij} which allows electrons to hop from the site i to the site j is taken to be ω if i and j are nearest neighbours and zero otherwise. Figure I.3a shows the densities of states for four concentrations of an A_cB_{1-c} alloy having the simple cubic crystal structure for which the parameters of the model are taken to be $\varepsilon_A = -\varepsilon_B = 0.4$ and $\omega = 0.1667$. The CPA results are shown by the dotted curves. For this simple model it is possible to perform exact calculations and those of Alben et al[40] are shown as the solid line. The CPA gives a good account of the exact results. Clearly the CPA is a very different theory from the VCA. For this model the VCA simply corresponds to an ordered system in which the site energy at every site is $\varepsilon_{AV} = c_A\varepsilon_A + c_B\varepsilon_B$. In other words, the density of states curve for the VCA looks exactly like that for pure A (or B) except that the centre of gravity of the curve is at ε_{AV}. In Fig. I.3a the VCA is shown for $c_A = c_B = 0.5$ by the chain line. Obviously the VCA completely fails to reproduce the fact that for all of these alloys the total densities of states are split into a low lying 'host' sub-band and a high energy 'impurity' band. In Fig. I.3b the results[41] of calculations using the ATA are shown and again are compared with the exact results, clearly the ATA does much less well than the CPA.

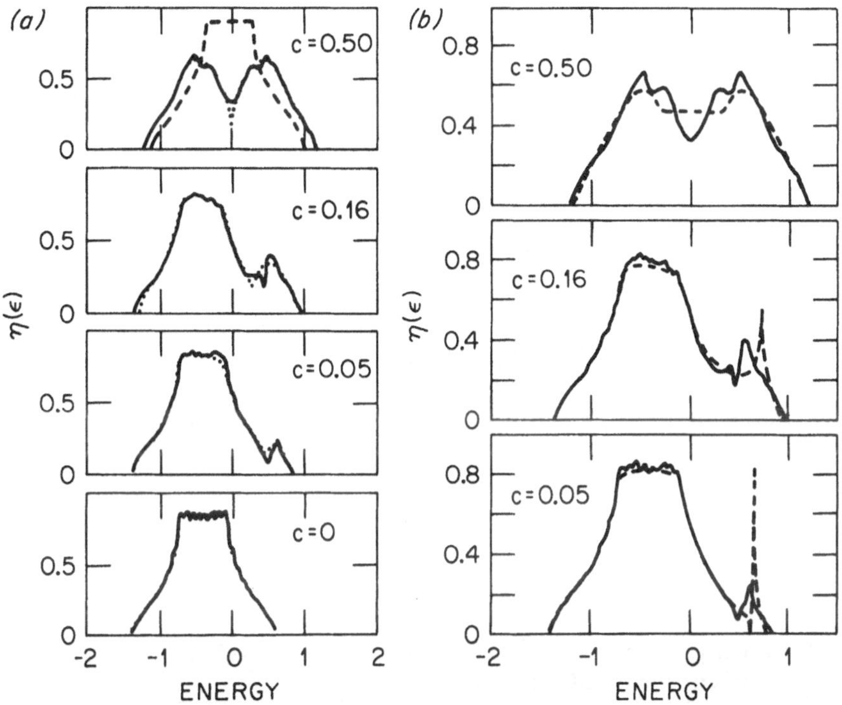

Fig. I.3. a) Comparison of the results of exact solutions (solid) of the densities of states for a random tight-binding Hamiltonian with those from the CPA (dots) and the virtual crystal approximation (dash). After ref. 5.

b) Comparison of exact calculation (solid), as in a) with those of the average t-matrix approximation (dash). After ref. 5.

In Fig. I.4 results are shown for an alloy for which the CPA does not do quite so well. For this alloy $\varepsilon_A = -\varepsilon_B = 0.75$. Whilst the CPA reproduces the fact that the impurity band is well split off from the host band it completely fails to reproduce the fine structure which is present in the exact calculation. The origin of this structure lies in the fact that even in a random alloy it is quite common to get small clusters of some specific configuration, say a single A atom completely surrounded by B atoms[42]. Since the CPA is a single-site theory all such 'molecular' effects

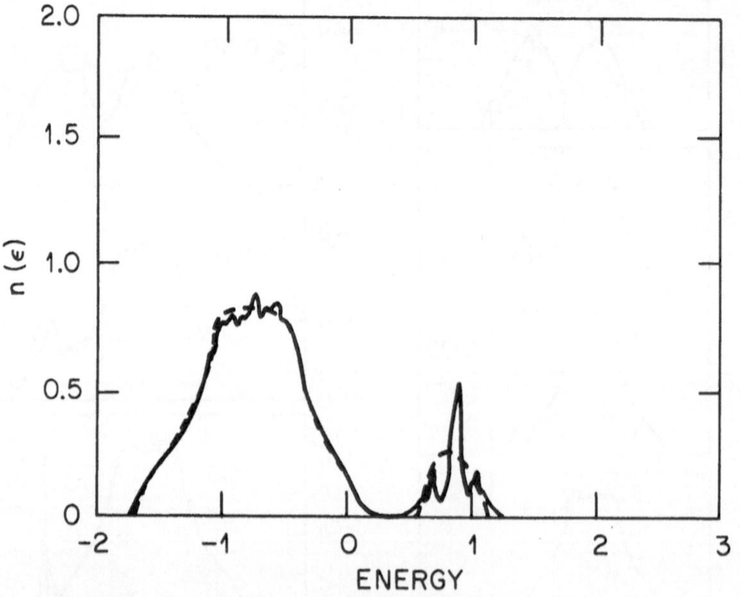

Fig. I.4. Comparison of exact calculations (solid), as in Fig. I.4a, with those of the CPA (dash).

are outside its scope. However, it is only in the extreme split band limit, i.e. when $\varepsilon_A - \varepsilon_B \gg$ band width of the pure constituents, as in Fig. I.4 but not in I.3a, that these effects become pronounced. Judging from the success of the full muffin-tin CPA which we will be describing it seems that such effects are difficult to observe experimentally even in systems where it might be expected they would occur.

II. MULTIPLE SCATTERING THEORY AND THE KORRINGA-KOHN-ROSTOKER COHERENT POTENTIAL APPROXIMATION

The central issue in obtaining a first principles theory of the electronic states in random alloys is to find approximation schemes for calculating the configurationally averaged Green's function. In order to proceed in this direction we must first be able to write down a general expression for the Green's function for some particular configuration. Having done this we will then be able to see simply how the coherent-potential-approximation fits in and how the equations of the first principles Korringa-Kohn-Rostoker-coherent-potential-approximation (KKR-CPA) are obtained. However before we can write down expressions for the Green's function for an arbitrary array of muffin-tin scatterers we must first include a brief review of the scattering properties of a single muffin-tin potential well. A more complete discussion is given in our previous lecture notes and in particular references 43 and 44.

The Scattering Properties of a Single Muffin-Tin Well

The Schrödinger equation for a single-muffin-tin well takes the form

$$[-\nabla^2 + v(\vec{r})]\Psi(\vec{r}) = \varepsilon\Psi(\vec{r}) \qquad \text{II.1}$$

where

$$v(\vec{r}) = \begin{cases} v(r) - V_{MT} & r \leqslant r_{MT} \\ 0 & r \geqslant r_{MT} \end{cases} \qquad \text{II.2}$$

In analogy with Eq. I.28 the Green's function $G'(r,r',\varepsilon)$ for a single muffin-tin potential well is given by

$$G'(\vec{r},\vec{r}',\varepsilon) = G_0(\vec{r},\vec{r}',\varepsilon) + \int d^3r_1 G_0(\vec{r},\vec{r}_1,\varepsilon)v(\vec{r}_1)G'(\vec{r}_1,\vec{r}',\varepsilon) \qquad \text{II.3}$$

The single scatterer t-matrix is defined through

$$G'(\vec{r},\vec{r}',\varepsilon) = G_0(\vec{r},\vec{r}',\varepsilon) + \int d^3r_1 \int d^3r_2 G_0(\vec{r},\vec{r}_1,\varepsilon). \\ \cdot (\vec{r}_1,\vec{r}_2,\varepsilon)G_0(\vec{r}_2,\vec{r}_1,\varepsilon) \qquad \text{II.4}$$

where G_0 is the solution of Eq. I.27, and

$$t(\vec{r},\vec{r}',\varepsilon) = \bar{v}(\vec{r})\delta(\vec{r} - \vec{r}') + \int d^3r_1 v(\vec{r})G_0(\vec{r},\vec{r}_1,\varepsilon)t(\vec{r}_1,\vec{r}',\varepsilon) \qquad \text{II.5}$$

In general the t-matrix is a complex quantity, it carries all the information about the scattering properties of the potential $v(r)$. However, for the purposes of the present discussion, where the potential function is of finite range, the only information that is required about the scattering event is how an incident wave of a given energy is transformed into a scattered wave of the same energy. These so-called 'on-the-energy-shell' matrix elements[45] of the t-matrix t_{LL} are defined as

$$t_{LL'}(\varepsilon) = \int d^3r_1 \int d^3r_2 j_\ell(\kappa r_1)Y_L(\hat{r})t_1(\vec{r}_1,\vec{r}_2,\varepsilon)Y_{L'}(\hat{r}_2)j_{\ell'}(\kappa r_2) \qquad \text{II.6}$$

Here $Y_L(\hat{r}) \equiv Y_{\ell,m}(\hat{r})$ is a (real) spherical harmonic, $j_\ell(\kappa r)$ is a spherical Bessel function and $\kappa = \sqrt{\varepsilon}$. For the spherically symmetric muffin-tin potential being considered here the on-energy-shell t-matrix, reduces to the diagonal form $t_L \delta_{LL'}$,[43] where t_L is related to the scattering phase shifts $\eta_\ell(\varepsilon)$ through

$$t_L(\varepsilon) = -\frac{1}{\kappa} \sin\eta_\ell(\varepsilon) e^{i\eta_\ell(\varepsilon)} \delta_{mm'} \qquad \text{II.7}$$

The phase shifts η_ℓ are obtained from the solutions

$$Z_L(\vec{r},\varepsilon) = z_\ell(r,\varepsilon) Y_L(\hat{r}) \qquad \text{II.8}$$

of Eq. II.1 which are regular at the origin ($z_\ell(r,\varepsilon) \sim r^\ell$ as $r \to 0$) by the requirement that this solution joins smoothly to

$$z_\ell(r,\varepsilon) = j_\ell(\kappa r)\eta_\ell(\varepsilon) - i\kappa h_\ell^+(\kappa r) \qquad \text{II.9}$$

at $r = r_{MT}$, where $h^+(\kappa r)$ is a spherical Hankel function [$\equiv j_\ell(\kappa r) \mp i\eta_\ell(\kappa r)$, where $\eta_\ell(\kappa r)$ is a spherical Neuman function] and

$$m_\ell = t_\ell^{-1} = -\kappa\cot\delta_\ell + i\kappa \qquad \text{II.10}$$

If $\gamma_\ell = \frac{1}{z_\ell}\frac{dz_\ell}{dr}\Big|_{r_{MT}}$ is the logarithmic derivative at $r = r_{MT}$ then clearly

$$\cot\eta_\ell = \frac{\kappa\eta_\ell'(\kappa r_{MT}) - \gamma_\ell(\varepsilon)\eta_\ell(\kappa r_{MT})}{\kappa j_\ell'(\kappa r_{MT}) - \gamma_\ell(\varepsilon) j_\ell(\kappa r_{MT})} \qquad \text{II.11}$$

The $\gamma_\ell(\varepsilon)$ are easily obtained by numerical integration of the radial Schrödinger equation

$$\frac{d^2}{dr}[rz_\ell(r)] - [v(r) - \varepsilon + \frac{\ell(\ell+1)}{r^2}]\, rz_\ell(r) = 0 \qquad \text{II.12}$$

Clearly the phase shifts are a very convenient way of describing the scattering proporties of a single muffin-tin. The wave function and Green's function outside the range of the potential are entirely specified in terms of them. As we shall see later they also are sufficient to specify the energy band structure of the ordered array of muffin-tin potentials corresponding to a pure metal or ordered alloy[44,45]. They do not however contain sufficient information to specify the Green's function hence the charge density etc. within the range of the potential.

It is possible to write down many different equations for the Green's function II.4 which are valid throughout all space. Of particular utility is the form used by Faulkner and Stocks namely

$$G'(r,r',\varepsilon) = \sum_{LL'} \{Z_L(\vec{r},\varepsilon)t_{LL'}(\varepsilon)Z_{L'}(\vec{r}',\varepsilon) - Z_L(\vec{r},\varepsilon)J_L(\vec{r}',\varepsilon)\delta_{LL'}\} \qquad \text{II.13}$$

The function $J_L(\vec{r}',\varepsilon) \equiv \tilde{j}(r,\varepsilon)Y_L(\hat{r})$ in this equation is the solution of Eq. II.1 which is irregular at the origin and joins smoothly to $j_\ell(\kappa r)Y_L(\hat{r})$ at $r = r_{MT}$. This particular representation of the Green's function has several convenient features, firstly it is valid for all values of $\vec{r}$ and $\vec{r}'$. Secondly, as will be shown later, this equation has a particularly nice generalization to an arbitrary array of muffin-tin scatterers i.e. to our random alloy. Finally it can be shown that wave functions $Z_L(\vec{r},\varepsilon)$ and $J_L(\vec{r},\varepsilon)$ are both real thus greatly facilitating the calculation of charge densities and densities of states. In particular the density of states inside the muffin-tin sphere is given by

$$n'_{MT}(\varepsilon) = -\frac{1}{\pi}\,\mathrm{Im}\int d^3r\, G'(r,r,\varepsilon) \qquad \text{II.14a}$$

$$= -\frac{1}{\pi}\sum_{LL'}\int_{|\vec{r}|\leq r_{MT}} d^3r\, Z_L(\vec{r},\varepsilon)Z_{L'}(\vec{r},\varepsilon)\,\mathrm{Im}\, t_{LL'}(\varepsilon) \qquad \text{II.14b}$$

$$= \frac{1}{\pi}\sum_{\ell}(2\ell+1)\sin^2\delta_\ell \int_0^{r_{MT}} dr\, z_\ell^2(r,\varepsilon) \qquad \text{II.14c}$$

Thus to obtain the density of states etc. we do not in fact require the irregular solution of the Schrödinger equation $J_L(r,\varepsilon)$.

Before closing it is worthwhile to investigate the physical content of Eq. II.14c for the density of states in the neighbourhood of a single scatterer since it already contains many of the features of the density of states for an assembly of such scatterers. In figure II.1 we show the phase shifts appropriate to Li and Mg muffin-tin wells in a $Li_{0.8}Mg_{0.2}$ alloy[47] (Fig. II.1a) and to Ag and Pd muffin-tin wells in a $Ag_{0.5}Pd_{0.5}$ alloy[25] (Fig. II.1b). A moment's thought about the electronic structure of the isolated atoms provides a simple interpretation of the corresponding phase shifts. As a general feature, the energy range where the phase shifts are large reflect the energies of the outermost occupied bound states of the isolated atom. The fact that the s-wave ($\ell = 0$) phase shift are large for the Mg muffin-tin well in the Li-Mg alloy is that the Mg atomic 3s level is no longer a bound state of the muffin-tin potential owing to the overlap of the potential with the tails of Li and Mg atomic potentials from neighbouring sites, however, it almost is and consequently the Mg muffin-tin potential is a strong scattering centre for energies just above the muffin-tin zero. Similar reasoning applies to the d-wave phase shift for the Ag and Pd muffin-tin-potentials except that the corresponding occupied atomic level which gives rise to these d-resonances is

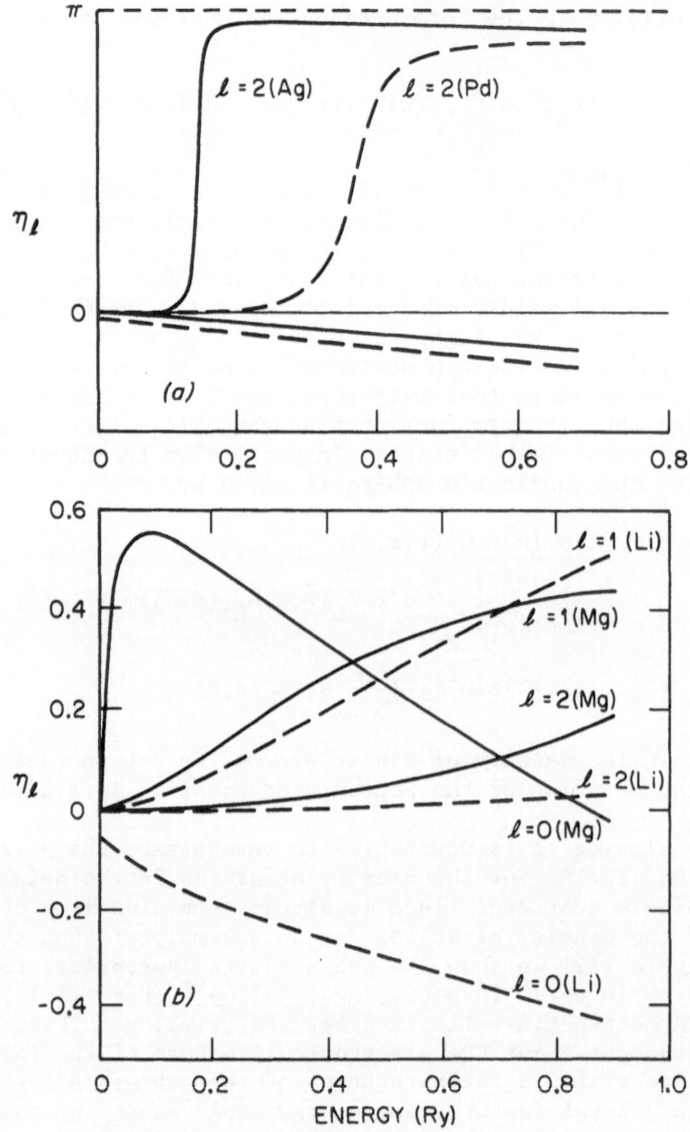

Fig. II.1. a) Upper. Phase shifts ($\ell = 0$ and $\ell = 2$) for single Ag and Pd muffin-tin wells. The potentials are appropriate to a $Ag_{0.2}Pd_{0.8}$ random alloy (fcc).
b) Lower. As above except for a $Li_{0.8}$ $Mg_{0.2}$ random alloy (bcc).

the 4d. A resonant phase shift is one which rises from 0 to π in some fairly narrow energy range. The resonance energy ε_r being defined by $\delta(\varepsilon_r) = \pi/2$. I will leave it as a exercise to explain the behaviour of the Mg $\ell = 1$ and Li $\ell = 0$ phase shifts. In going through the above analysis it should not have escaped your notice that the structure in $\eta'(\varepsilon)$ comes largely from the $\sin^2\delta_\ell$ term in Eq. II.14c. The reason for thus is that differential scattering amplitude for scattering of an electron of energy ε through an angle θ is given by

$$f(\varepsilon,\theta) = \sum_\ell (2\ell+1) f_\ell(\varepsilon) P_\ell(\cos\theta)$$

where the scattering amplitude $f_\ell(\varepsilon)$ is given by

$$f_\ell(\varepsilon) = -\sqrt{\varepsilon}\, t_\ell = \sin\eta_\ell\, e^{i\eta_\ell} = \frac{1}{2i}\left(e^{2i\eta_\ell} - 1\right) \qquad \text{II.15}$$

and the total scattering cross section is given by

$$\delta(\varepsilon) = \frac{4\pi}{\varepsilon} \sum_\ell (2\ell+1)\sin^2\eta_\ell \qquad \text{II.16}$$

Thus the structure in $\eta'(\varepsilon)$ is simply a reflection of the scattering cross section of the single muffin-tin well for electrons of energy ε. The quantity $\varepsilon\int_0^{r_{MT}} dr\, z_\ell^2(r,\varepsilon)$ then has little structure and simply serves as a weighting factor between different angular momentum channels.

As we will see later the energy region in which the phase shift is large and hence the single scatterer density of states is large corresponds to the energy regime in which the corresponding partial wave density of states in a metal or alloy is large. Of course the width of such regions will be radically different because of multiple scattering effects.

Multiple Scattering Theory

In this section we will write out expressions for the total t-matrix and Green's function for an arbitrary array of muffin-tin potential functions that will allow us to write out the CPA condition Eq. I.38 and expressions for the average densities of states and other quantities within the CPA. The development followed is that given by Faulkner and Stocks[46] to which the reader is referred for a more detailed discussion.

We can take the crystal potential function V(r) to be of the general form

$$V(\vec{r}) = \sum_{i=1}^{N} v_i(\vec{r}_i) \qquad \text{II.17}$$

where the vectors $\vec{r}_i$ are defined by $\vec{r}_i = \vec{r} - \vec{R}_i$ and is assumed to be zero outside the bounding sphere of radius r_{MT}.

The Green's function Eq. I.32 for such a system of scatterers can be written in operator notation as

$$G = G_0 + G_0TG_0 \qquad \text{II.18}$$

The real space representation is obtained in the obvious manner $G(\vec{r},\vec{r}',\varepsilon) = \langle\vec{r}|G|\vec{r}'\rangle$ etc. The t-matrix for the system can be decomposed according to

$$T = \sum_{i,j} \tau^{ij} \qquad \text{II.19}$$

The quantities τ^{ij} were first introduced by Gyorffy and Stott[48] who called them <u>scattering-path</u> operators. The scattering path operators are defined by

$$\tau^{ij} = t^i\delta_{ij} + \sum_{k\neq i} t^iG_0\tau^{kj} \qquad \text{II.20}$$

where

$$t^i = v_i(1 + G_0t^i) \qquad \text{II.21}$$

is the t-matrix for a single potential well of the previous section (Eq. II.5). If we recall that the role of the t-matrix t^i is to generate the scattered wave from the incident wave for a single potential well v_i and the role of the t-matrix T is to do the same for the assembly of potential wells $\sum_i v_i$ then the physical content of Eqs. II.20 and II.21 is clear. τ^{ij} generates the scattered wave at site i which results from the incident wave at site j. Evidently, $\sum_{ij} \tau^{ij}$ takes the incident waves arriving at all the sites in the lattice and adds up all the scattered waves. This is precisely what T does, hence the equality of Eq. II.19. Equation II.20 is then a self-consistency condition on the τ^{ij} which says that the scattered wave at site i which results from an incident were at site j is made up of two parts a direct part and an indirect part. The direct part simply consists of the conversion of the incident wave at site i into the scattered wave at site i. The indirect part consists of summing over all possible processes which involve a wave incident on site k being converted into scattered waves at sites k (k≠i) which are then propagated to site i according to the free electron propagator G_0 at which they are then the incident wave. This incident wave is then converted into the scattered wave at site i by the application of t-matrix t^i.

The importance of this formulation of the problem to the crystal potential which consists of non-overlapping muffin-tin wells is that for non-overlapping potentials the "on the energy shell" part of Eq. II.20 decouples from the rest and this gives rise to particularly powerful tools for calculating the energy bands and other physical properties of the system of scatterers.

By analogy with Eq. II.6 it is possible to define the 'on-energy-shell' matrix elements of the scattering path operator as

$$\tau^{ij}_{LL'}(\varepsilon) = \int d^3r_i \int d^3r'_j \; Y_L(\hat{r}_i) j_\ell(\kappa r) \tau^{ij}(\vec{r}_i,\vec{r}_j;\varepsilon) j_{\ell'}(\kappa r'_j) Y_{L'}(\hat{r}'_i) \qquad \text{II.22}$$

where L is the angular momentum about the site R_i and L' is that about the site $\vec{R}_{i'}$, $\vec{r}_i = \vec{r} - \vec{R}_i$ and $\vec{r}'_j = \vec{r}' - \vec{R}_j$. An expression for $\tau^{ij}_{LL'}(\varepsilon)$ in terms of the on-energy-shell matrix elements of t_L of the previous section can be found by putting Eq. II.20 'on-the-energy' shell. The result is

$$\tau^{ij}_{LL'}(\varepsilon) = t^i_L(\varepsilon)\delta_{ij}\delta_{LL'} + \sum_{k\neq i}\sum_{L''} t^i_L(\varepsilon) g^{ij}_{LL''}(\varepsilon)\tau^{kj}_{L''L'}(\varepsilon) \qquad \text{II.23}$$

where

$$g^{ij}_{LL''}(\varepsilon) = 4\pi\kappa i^{\ell-\ell'+1} \sum_{L''} C^{L''}_{LL'} i^{\ell''} h^+_{\ell''}(\kappa|\vec{R}_i - \vec{R}_j|) Y_{L''}(\widehat{R_i - R_j}) \qquad \text{II.24}$$

are the 'on-energy-shell' matrix elements of the free particle propagator $G_0(\vec{r},\vec{r}',\varepsilon)$. We will often refer to these latter quantities as real space structure constants; they depend only on the spacial arrangement of the scatterers. In II.24 the $C^{L''}_{LL'}$ are the so-called Gaunt numbers defined by

$$C^{L''}_{LL'} = \int d\Omega Y_L(\Omega) Y_{L''}(\Omega) Y_{L'}(\Omega)$$

Equation II.23 is the fundamental multiple scattering equation, it gives $\tau^{ij}_{LL'}$ in terms of $t^i_L(\varepsilon)$, i.e. the phase shifts, and the real space structure constants $g^{ij}_{LL'}$. This equation is valid for any arrangement of potentials, even if at each site we have a different scatterer. Thus, it is a good starting point to discuss pure metals and liquids as well as random alloys. The only limitation on its validity is the requirement that the potential wells at the different sites may not overlap.

Before going on, in the next subsection, to write down the CPA condition for the muffin-tin model of a random alloy we will obtain a general expression for the Green's function $G(\vec{r},\vec{r}',\varepsilon)$ for the assembly of scatterers which will be of use in later sections when we wish to obtain expressions for observables within the CPA. To this end it is helpful to consider two cases separately. Firstly the case when both $\vec{r}$ and $\vec{r}'$ are near some site n, secondly the case where $\vec{r}$ is near some site n and $\vec{r}'$ is near some other site m'.

For the case where both $\vec{r}$ and $\vec{r}'$ are near the site n Eq. II.18 can be rewritten as

$$G = G^n + G^n T_{nn} G^n \qquad \text{II.25}$$

where G^n is the single scatterer Green's function Eq. II.4 for the particular site n, and

$$T_{nn} = \sum_{i \neq n} \sum_{j \neq n} \tau^{ij} \qquad \text{II.26}$$

Using these expressions together with Eq. II.13 and the fact that $G^1(\vec{r},\vec{r}\,',\varepsilon)$ can also be written in the form

$$G'(\vec{r},\vec{r}\,',\varepsilon) = i\kappa \sum_{LL'} Z_L(\vec{r},\varepsilon) t_{LL'}(\varepsilon) h_{\ell'}(\kappa r') Y_{L'}(\hat{r}') \qquad \text{II.27}$$

where $r'_n > r_n$ and $r'_n > r_{MT'}$. Faulkner and Stocks[48] obtained

$$G(\vec{r},\vec{r}\,',\varepsilon) = \sum_{LL'} \{Z^n_L(\vec{r}_n,\varepsilon)\tau^{nn}_{LL'}(\varepsilon) Z^n_{L'}(\vec{r}\,'_n,\varepsilon) - Z^n_L(\vec{r}_n,\varepsilon) J^n_L(\vec{r}\,'_n,\varepsilon)\delta_{LL'}\} \qquad \text{II.28}$$

where τ^{nn} is given by the solution of Eq. II.23 for the case $i = j = n$. This expression is valid for any value of $\vec{r}$ and $\vec{r}\,'$ so long as do not lie in any of the bounding spheres excepting the nth.

To obtain an expression for the Green's function when $\vec{r}$ is near the site n and $\vec{r}\,'$ is near the site m it is best to decompose G in the form

$$G = G^{nn} + G^n T_{nm} G^m \qquad \text{II.29}$$

where

$$G^{nm} = (G_0 t^n + 1)(1 + t^m G_0) \qquad \text{II.30}$$

and

$$T_{nm} = \sum_{i \neq n} \sum_{j \neq m} \tau^{ij} \qquad \text{II.31}$$

from which it can be shown[48]

$$G(\vec{r},\vec{r}\,',\varepsilon) = \sum_{LL'} Z^n_L(\vec{r}_n,\varepsilon)\tau^{nm}_{LL'} Z^m_{L'}(\vec{r}_m,\varepsilon) \qquad \text{II.32}$$

which is valid so long as $\vec{r}$ is within no bounding sphere other than the nth and $\vec{r}\,'$ is within no bounding sphere other than the mth. Equations II.28 and II.31 can be combined into the general expression

$$G(\vec{r},\vec{r}',\varepsilon) = \sum_{LL'} \{Z_L^n(\vec{r}_n,\varepsilon)\tau_{LL'}^{nm}(\varepsilon)Z_{L'}^m(\vec{r}_m',\varepsilon) - Z_L^n(\vec{r}_n,\varepsilon)J_L^n(\vec{r}_n,\varepsilon)\delta_{LL'}\delta_{nm}\} \qquad \text{II.33}$$

which again is valid for all n and m including n = m with the same restrictions as Eq. II.32.

Evidently Eq. II.32 is the generalization to an assembly of scatterers of the single site formula Eq. II.13 like its single scatter counterpart this particular form of the Green's function has a number of nice features. Firstly it is valid for any array of non overlapping potentials and as such is an excellent starting point for the calculation of the average densities of states in a random alloy. Secondly the wavefunction dependent information contained in the Z^n is completely separated from the multiple scattering information contained in the τ^{nm}. As we shall see this also greatly facilitates the calculation of averaged values of observables. Again, as with the single scatter formula, because the functions Z_L^n and J_L^n are real it is a trivial matter to take this imaginary part. A final point worthy of note is that formulas for observables derived from this form of the Green's function usually turn out to be extremely easy to evaluate numerically, usually involving simple integrals over the unit cell of quantities which are simple functions of the Z^n.

The Korringa-Kohn-Rostoker-coherent-potential-approximation (KKR-CPA)

Having obtained the expressions for on-energy-shell-matrix elements of the t-matrix and the scattering path operator in the previous section it is now a simple matter, for the muffin-tin model, to convert the formal CPA self-consistency requirement of chapter I into a practical computational scheme. The equations of this section were first written down by Gyorffy[49]. Equations formally equivalent to these were previously written down by Soven[50] and by Shiba[51]. However, it is the form written down by Gyorffy which has provided the basis for practical calculations.

It should now be clear from the preceding section that if we implicitly assume that the coherent-potential which we use in order to describe the effective ordered system within the CPA is of the non-overlapping muffin-tin form then it is not really necessary to consider the actual 'coherent-potential' at all. It is sufficient to know the effective t-matrix t_C corresponding to an individual 'coherent-potential'-well, and to solve Eq. I.38 on-the-energy-shell. In general the coherent potential will be complex and energy dependent.

In terms of the on-energy-shell matrix elements of τ^{ij} the

CPA condition Eq. I.38 reads

$$c_A \underline{\tau}^{A,ij} + c_B \underline{\tau}^{B,ij} = \underline{\tau}^{c,ij} \qquad \text{II.34}$$

where the matrices τ^{ij} are to be regarded as matrices in the angular momentum index $\overline{L}$. The $\tau^{c,ij}$ then is to be calculated for an ordered array of $\underline{t}^{c}$'s situated at the lattice sites $\vec{R}_i$ and $\tau^{ij}_{A(B)}$ is to calculated for the same array except that at the site i there is an 'impurity' $t^{A(B)}$ the t-matrix corresponding to a single $v_{A(B)}$ muffin-tin well. In practice it is only necessary to satisfy the self-consistency condition for the special case j = i = 0 (say)

$$c_A \underline{\tau}^{A,00} + c_B \underline{\tau}^{B,00} = \underline{\tau}^{c,00} \qquad \text{II.35}$$

which is the CPA condition worked out by Gyorffy, since it can be shown that implicit in the solution of this case is the solution of Eq. II.34.

From Eq. II.23 we have for an ordered lattice of t_c's

$$\tau^{ij}_{LL'}(\varepsilon) = t^c_L(\varepsilon)\delta_{ij}\delta_{LL'} + \sum_{k\neq i}\sum_{L''} t^c_L(\varepsilon) g^{ik}_{LL''}(\varepsilon)\tau^{c,kj}_{L''L'} \qquad \text{II.36}$$

which can be solved by the method of lattice Fourier transforms. If we define $\underline{\tau}^c(\vec{k},\varepsilon)$ and $g(\vec{k},\varepsilon)$ by the relations

$$\underline{\tau}^c(\vec{k},\varepsilon) = \frac{1}{N}\sum_{ij} e^{i\vec{k}.(\vec{R}_i - \vec{R}_j)}\, \underline{\tau}^{c,ij}(\varepsilon) \qquad \text{II.37}$$

and

$$\underline{g}(\vec{k},\varepsilon) = \frac{1}{N}\sum_{ij} e^{i\vec{k}.(\vec{R}_i - \vec{R}_j)}\, \underline{g}^{ij} \qquad \text{II.38}$$

we find

$$\underline{\tau}^{c,00}(\varepsilon) = \frac{1}{\Omega_{BZ}} \int d^3k [\,\underline{m}^c(\varepsilon) - \underline{g}(\vec{k},\varepsilon)]^{-1} \qquad \text{II.39}$$

$$\equiv \frac{1}{\Omega_{BZ}} \int d^3k \; \underline{\tau}^c(\vec{k},\varepsilon) \qquad \text{II.40}$$

where $\underline{m}^c = (\underline{t}^c)^{-1}$ where $\underline{t}^c$ is the matrix whose elements are $t^c_{LL'}$. The matrix elements $g_{LL'}(\vec{k},\varepsilon)$ of the matrix $\underline{g}(\vec{k},\varepsilon)$ are the 'structure constants' familiar in the KKR-band theory method and can be obtained from efficient numerical algorithms[52,53]. In fact the KKR[44,45] method for finding the energy band structure of a pure metal involves no more than finding the energies $\varepsilon_{\vec{k}}$ for which the determinant of the KKR-matrix $[\underline{m} - \underline{g}(\vec{k},\varepsilon)]$ is zero, where the matrix $\underline{m}$ is obtained from the (real) muffin-tin potential corresponding to the pure metal. Evidently at the eigenvalues ε_k, $\underline{\tau}^{c,00}$ and thence T and G are singular which is the correct condition for having an eigen-solution. In order to obtain the quantities $\underline{\tau}^{A,00}$ and $\underline{\tau}^{B,00}$ we have to solve Eq. II.23 for the case that the site at the origin is occupied by the A(B) species whilst at all other sites there is a $\underline{t}_c$. It is a matter of simple matrix algebra to show that the solution to these impurity problems are

$$\underline{\tau}^{A,00} = \underline{D}_c^A \, \underline{\tau}^{c,00} \qquad \text{II.41}$$

where

$$D_c^A = [\underline{1} + \underline{\tau}^{c,00} (\underline{m}^A - \underline{m}^c)]^{-1} \qquad \text{II.42}$$

with similar expressions for $\underline{\tau}^{B,00}$. The matrix $\underline{m}^A$ is inverse of the single site t-matrix whose matrix elements are $t_{LL'}^A$.

Equations II.35, II.39, II.41 and II.42 comprise a set of equations which have to be solved self-consistently for the effective t-matrix $\underline{t}_c$. These are the so-called KKR-CPA equations. The CPA condition Eq. II.35 can be manipulated into a number of equivalent forms with the aid of Eqs. II.41 and II.42. Most useful of these are

$$c_A \underline{D}_c^A + c_B \underline{D}^B = \underline{1} \qquad \text{II.43}$$

and

$$\underline{m}^c = c_A \underline{m}^A + c_B \underline{m}^B + (\underline{m}^A - \underline{m}^c) \underline{\tau}^{c,00} (\underline{m}^B - \underline{m}^c) \qquad \text{II.44}$$

The latter form is that given in Gyorffy's original derivation.

Solution of the KKR-CPA Equations

The individual elements which comprise a KKR-CPA calculation are now clear. Given some input muffin-tin potentials $v^A(\vec{r}_i)$ and $v^B(\vec{r}_i)$ and hence phase shifts $\eta^A(\varepsilon)$ and $\eta^B(\varepsilon)$ and corresponding single site t-matrices $\underline{t}^A(\varepsilon)$ and $\underline{t}^B(\varepsilon)$. An initial guess of the matrix elements of the effective t-matrix $\underline{t}_c$ is made and the integral Eq. II.39, is evaluated. The resulting values of $\underline{\tau}^{c,00}$ together with the input $\underline{t}_c$ and the values of $\underline{t}_A$ and $\underline{t}_B$ are substituted into Eq. II.35 (say) to see if the input $\underline{t}_c$ is a solution. If it is not a new guess is made, say, by a generalized Newton-Ralphson method. The process is repeated until a solution is found. A sensible starting guess for the effective t-matrix is the average t-matrix approximation.

$$\underline{t}_c = c_A \underline{t}_A + c_B \underline{t}_B \qquad \text{II.45}$$

If we retain terms in Eq. II.35 having $\ell \leqslant 2$, all the matrices involved in the calculation are 9 × 9. However if we take full advantage of the crystal symmetry, for the fcc and bcc crystal structures the matrices $\underline{t}_c$ and $\underline{\tau}^c$ are diagonal and have only four independent elements one each corresponding to the four irreducible representations Γ_1 (s-like), Γ_{15} (p-like) $\Gamma_{25'}$ and Γ_{12} (d-like). Thus rather than being a matrix equation the CPA equation reduces to a set of four coupled non-linear equations, this greatly simplifies its solution.

From what has been said about the calculation of the phase shifts it should come as no surprize that in the above scheme the limiting step is the evaluation of the Brillouin zone integral in Eq. 35. The nature of the difficulty involved in the evaluation of this integral can be seen if we consider the form of the scattering amplitude f_L^c to which a complex 'coherent-potential' will give rise. In general we can write the scattering amplitude corresponding to a complex potential and hence phase shift $\eta_\ell = \eta_\ell^R + i\eta_\ell^I$ in the form, c.f. Eq. II.15, $f_\ell(\varepsilon) = 1/2i\ (\alpha_\ell \exp(2i\eta_\ell^R) - 1)$ where η_ℓ^R is a real phase shift and $\alpha_\ell = \exp(-2\eta_\ell^I)$ is a measure of the inelasticity in the 'coherent potential' ; $\alpha_\ell = 1\ (\eta_\ell^I = 0)$ corresponds to a real potential ; $\alpha < 1$ corresponds to a complex potential which is a sink. Decomposing the inverse of f_ℓ into real and imaginary parts as $\underline{f}^{-1} = (\underline{f}^{-1})_R - i(\underline{f}^{-1})_I$ and using the fact that we can write $G_{LL'}(k,\varepsilon) = B_{LL'}(\vec{k},\varepsilon) + i\kappa\delta_{LL'}$ Eq. (II.35) becomes

$$\tau^{c,00}(\varepsilon) = -\frac{1}{\Omega_B}\int_{BZ} d^3k\ \{\kappa(\underline{f}^{-1})_R + \underline{B}(\vec{k},\varepsilon) - i\kappa[(\underline{f}^{-1})_I - \underline{1}]\}^{-1}$$

Now if $\alpha_\ell = 1$ for all ℓ, then $(\underline{f}^{-1})_I$ is the unit matrix and the integrand reduces to the inverse of the KKR matrix which is singular for $\varepsilon = \varepsilon_{\vec{k}}$ the Bloch state energies, i.e., for energies such that Det $|-\kappa \cot\delta_\ell\, \delta_{LL'} - B_{LL'}(\vec{k},\varepsilon)| = 0$ which is the KKR condition[44,45]. If $\alpha_\ell < 1$, then $(f_L^{-1})_I \neq 1$ and the integral is intrinsically complex, the corresponding KKR determinant will have zeros only at the complex energies $\varepsilon = \varepsilon_{\vec{k}}^R + i\varepsilon_{\vec{k}}^I$. Clearly, if $\varepsilon_{\vec{k}}^I$ is small, the weight of the integrand in Eq. II.35 will be distributed on some, possibly complicated, constant energy surface in the Brillouin zone. Fortunately nature works somewhat in our favour. In energy regions where these constant energy surfaces have their most complicated behaviour, for example in the middle of the d-bands in a transition metal alloy, the disorder broadening is largest with the result that the weight of the integrand is distributed over a large part of the Brillouin zone.

The algorithms we have used to evaluate Eq. II.35 and other integrals of a similar type are based on what are called 'directional methods'[53]. In these methods a set of N-directions ($\hat{k}_i$, i = 1...N) emanating from the Brillouin zone centre is defined, the line integral $\underline{\tau}^c_{\hat{k}_i}(\varepsilon)$ of $\underline{\tau}^c(\vec{k},\varepsilon)$ along each direction

$$\underline{\tau}^c_{\hat{k}_i}(\varepsilon) = \int_0^{k_i^{BZ}} dk\ \omega(k)\underline{\tau}^c(\vec{k},\varepsilon) \qquad \text{II.46}$$

is formed, where k_i^{BZ} denotes the Brillouin zone boundary along the direction $\hat{k}_i$, and the required integral is then approximated by a sum of the line integrals

$$\underline{\tau}^{c,00}(\varepsilon) = \sum_{i=1}^{N} \omega_i\ \tau^c_{\hat{k}_i}(\varepsilon) \qquad \text{II.47}$$

where ω_i is again some appropriate weight function. In the 'prism method'[153] the set of directions are chosen to be the centres of mass lines of the set of prisms. The prisms are chosen so as to exactly divide the Brillouin zone, the weight function $\omega(k)$ used in the one dimensional integrations is taken to be the cross sectional area of the prism, consequently the weight function $\omega_i = 1$. For the fcc and bcc crystal structures we typically divide the irreducible 1/48 of the Brillouin zone into 36-prisms. Typically of the order of 100-200 k-points are used along each direction in the evaluation of the 1-dimensional integrals. For special purposes, the accurate determination of the Fermi energy and Fermi energy densities of states, we generally use 136-prisms and have on occasions used 518[29].

Another choice of directions are the 'special directions' of Bansil[54] and of Fehlner and Vosko[55]. Here $\omega(k) = k^2$ and ω_i are Gaussian weights, the particular set of directions is chosen so as to exactly reproduce the first n-moments of the integral when expanded in terms of the cubic harmonics. Though we have often used this method we find it to offer few advantages and indeed have often found the convergence in the number of directions used erratic and less reliable than the prism method.

In fig. II.2 we show plots of the self-consistent scattering amplitudes which are solutions of the CPA equations for three Ag_cPd_{1-c} alloys having concentrations C = 0.2, 0.5 and 0.8. The information is plotted in the form of an Argand diagram Ref_ℓ vs Im f_ℓ for the ℓ = 2 channel only. As we argued in the previous section, in transition metals it is the position and shape of the ℓ = 2 phase which determines the position and the width of the d-band complex. Similarly is the position and width of resonances in the scattering amplitude which determines the band structure of the alloys. A moments consideration of general form of the scattering amplitude discussed in the last section will convince you that for a real potential (α_ℓ = 1) having a resonant phase shift implies that the scattering amplitude for that ℓ will go once

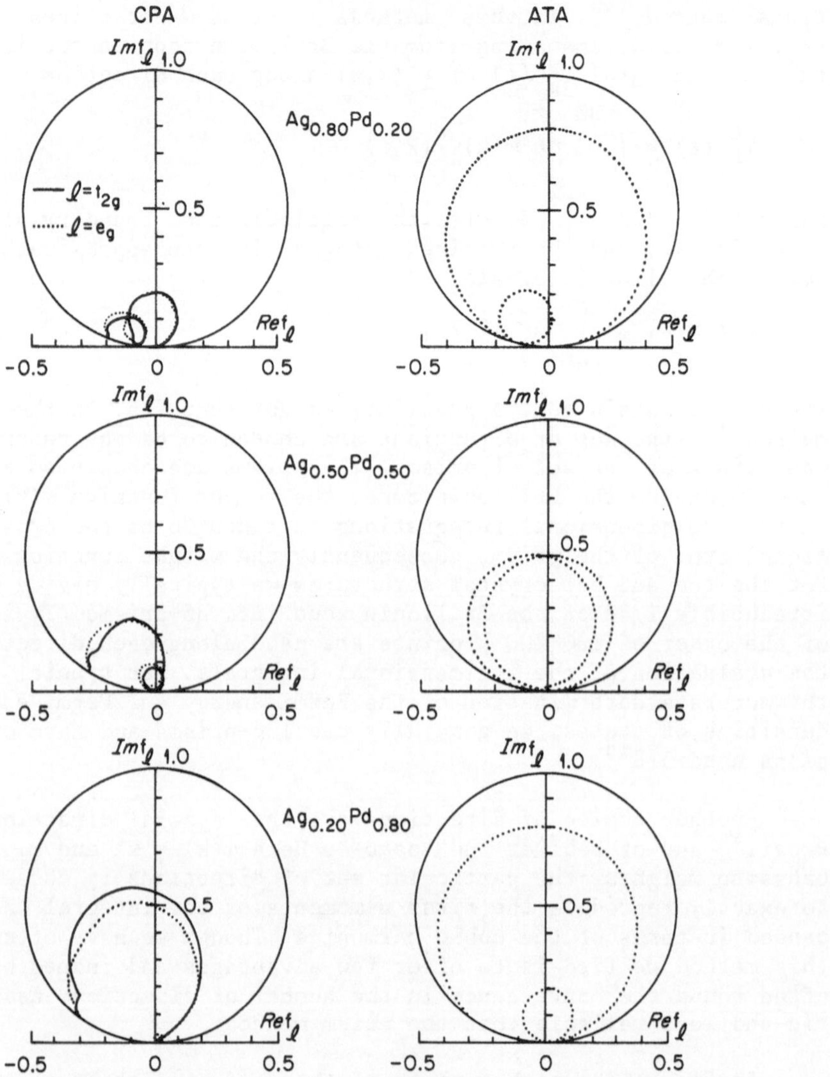

Fig. II.2. Argand diagrams corresponding to CPA (left) and ATA d-wave effective scattering amplitudes, $f_2(\varepsilon)$. The upper two curves are for $Ag_{0.8}Pd_{0.2}$, the middle pair for $Ag_{0.5}Pd_{0.5}$ and the low pair for $Ag_{0.2}Pd_{0.8}$ random alloys.

around the unitarity circle (the large circle centre (0,1/2) and radius 1/2 shown in fig. II.2). Inelastic scatterers which are sinks have scattering amplitudes which lie inside the unitarity circle, and resonances appear as loops within the unitarity circle. Evidently the scattering amplitudes of all three alloys correspond to inelastic scatterers which have two resonances. Also shown in fig. II.2 are the scattering amplitudes corresponding to the ATA. Though topologically similar in that they also have two loops, it is clear that the CPA scattering amplitudes are very different from those of the ATA. It should be noted that CPA d-scattering amplitude has two components this is because the coherent potential only has the point group symmetry of the lattice and therefore splits into a t_{2g} and an e_g component. The effective potential corresponding to the ATA on the other hand has spherical symmetry and therefore has only a single d-component. Given that each component of the CPA effective scattering amplitude has two resonances would seem to imply that the alloy will have two sets of (split-) d-bands. As we will see later this is in fact the case.

In terms of our parametrization of $f_\ell^c(\varepsilon)$ the degree of elasticity of the scattering amplitude is contained in the parameter α_ℓ. In fig. II.2 $\alpha_\ell/2$ is the distance of the scattering amplitude from the centre of the unitarity circle. The scattering cross-section of the effective scattering centre is contained in the real effective phase shift η_ℓ^R.

The effective phase shift η_ℓ^R and the absorption parameter α_ℓ are shown in fig. II.3 corresponding to the effective scattering amplitudes of fig. II.2 for the $Ag_{.5}Pd_{.5}$ alloy. For most energies the CPA effective phase shifts stay closer to 0 or π than the corresponding ATA. This implies that the CPA effective scatterer is less effective in deflecting electrons. In the ATA there are two well formed resonances. These differences in phase shifts will give rise to different band positions. On the other hand the CPA scatter absorbes over a wider energy range. From these pictures it is clear that in the ATA only bands near the two resonances energies will have short lifetimes while in the CPA the smearing of the bands will be more widely spread in energy.

III. KKR-CPA DENSITIES OF STATES AND BLOCH SPECTRAL DENSITIES

In this section we will show how the expressions for the Green's function for an arbitrary array of non-overlapping muffin-tin potentials can be used to obtain formulas for calculating total and component densities of states and Bloch spectral functions for random alloys with the KKR-CPA scheme. We will also show results for these quantities, for a number of systems, which illustrate the physical content of these formulas and of the KKR-CPA.

The derivations of the formulas for the densities of states from the Green's function of the previous section illustrate the logical steps required to formulate expressions for any observable with the KKR-CPA. These steps are further illustrated in Sec. VI

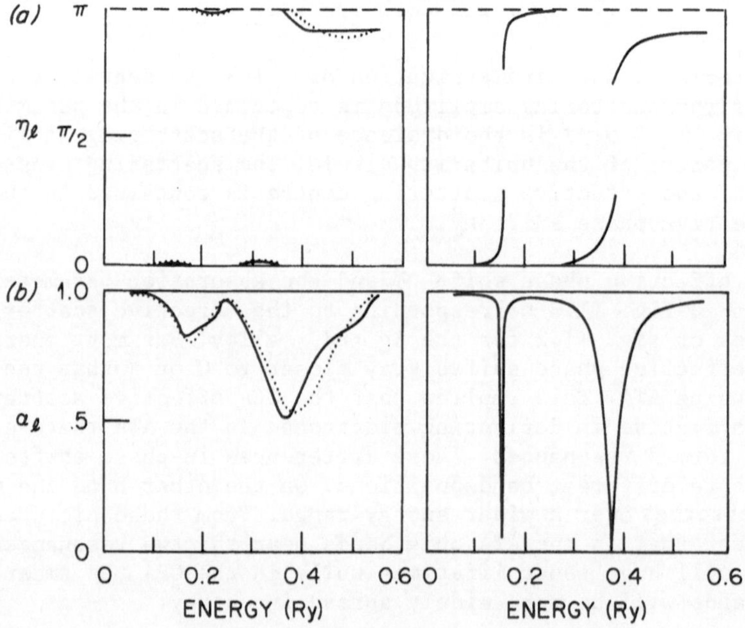

Fig. II.3. Upper. Effective d-wave phase shifts for a $Ag_{0.5}Pd_{0.5}$ alloy corresponding to CPA (left) and ATA (right) effective scatterers.

Lower. Absorption coefficient corresponding to the above effective scatterers.

where a formula for calculating the momentum density is derived. This quantity being relevant to the interpretation of positron annihilation experiments. In the lecture notes of Dr. Durham[56] similar methods will be used to derive expressions for the spectrum of emitted X-rays in soft X-ray experiments on random alloys[57,58] and for the intensity of photoemitted electrons in the angle resolved photoemission experiments[59,60,61]. The former experiment probes the local densities of states in the neighbourhood of the emitting atom, the latter experiment measures the remnants of the electronic energy band structure present in the alloy which has its realization in terms of the Bloch spectral function.

The Configurationally Averaged Densities of States

In order to calculate the average densities of states Eq. I.28 we must first obtain an expression for the average Green's function $\langle G(\vec{r},\vec{r}',\varepsilon)\rangle$ when $\vec{r} = \vec{r}'$. This is easily achieved with the help of the site diagonal form of the Green's function for an arbitrary array of scatterers Eq. II.28. The averaging process proceeds then in two stages. First the average is taken over all members of the ensemble of alloys in which the occupancy of some particular site $\vec{R}_i$ is held fixed. Next the average is taken over the possible occupancies [A(B) for the case of an A_cB_{1-c} random alloy] of this site. The result of this process is[46]

$$\begin{aligned}\langle G(r,r',\varepsilon) = \sum_{LL'} [\, & c_A Z_L^A(\vec{r}_n,\varepsilon)\langle\tau_{LL'}^{nn}\rangle_A\, Z_{L'}^A(\vec{r}_n,\varepsilon) + \\ & + c_B Z^B(\vec{r}_n,\varepsilon)\langle\tau_{LL'}^{nn}\rangle_B\, Z^B(\vec{r}_n,\varepsilon) - \{c_A Z_L^A(\vec{r}_n,\varepsilon) J_L^A(r_n,\varepsilon) + \\ & + c_B Z_L^B(\vec{r}_n,\varepsilon) J_L^B(r_n,\varepsilon)\}\delta_{LL'}] \end{aligned} \qquad \text{III.1}$$

The quantities $\langle\tau_{LL'}^{nn}\rangle_A$ and $\langle\tau_{LL'}^{nn}\rangle_B$ are the conditional averages of Eq. II.23 for the case that there is diffinitely an A or B on the site n but the occupancy of all other sites has be averaged over. Equation III.1 is an exact expression for the average site diagonal Green's function, since as defined above the $\langle\tau_{LL'}^{nn}\rangle_{A(B)}$ are the exact restricted averages.

In order to obtain a formula for the site diagonal Green's function with the CPA we must find an approximation to the exact restucted averages $\langle\tau_{LL'}^{nn}\rangle_{A(B)}$. Bearing in mind the single site nature of the CPA and the schematic representation embodied in fig. I.2, these exact averages are approximated by

$$\begin{aligned}\underline{\tau}^{A,nn} &\approx \langle\underline{\tau}^{nn}\rangle_A \\ \underline{\tau}^{B,nn} &\approx \langle\underline{\tau}^{nn}\rangle_B\end{aligned} \qquad \text{III.2}$$

A formal derivation of the correctness of these replacements within the CPA is given in the paper of Faulkner and Stocks[46]. Thus the CPA approximation $G_c(\vec{r},\vec{r}',\varepsilon)$ to the average Green's function $\langle G(\vec{r},\vec{r}',\varepsilon)\rangle$ is obtained by substituting the approximations Eq. III.2 into Eq. II.1. The CPA approximation $\bar{n}(\varepsilon)$ to the exact average densities of states is then given by

$$\bar{n}(\varepsilon) = -\frac{1}{\pi}\,\mathrm{Im\ tr}\ \underline{F}^c\underline{\tau}^{c,00} \qquad \text{III.3}$$

where

$$\underline{F}^c = \int_\Omega d^3r\, F^c(\vec{r},\vec{r}) = c_A\underline{F}^A\underline{D}_c^A + c_B\underline{F}^C\underline{D}_c^B \qquad \text{III.4}$$

where the matrix elements of $\underline{F}^A$ and $\underline{F}^B$ are given by

$$F^\alpha_{LL'}(\varepsilon) = \int_\Omega d^3r\, Z^\alpha_L(\vec{r},\varepsilon)Z^\alpha_{L'}(\vec{r},\varepsilon) \qquad \text{III.5}$$

where the integrals in Eqs. III.4 and III.5 are over a single unit cell. From these equations it is clear that we can resolve the total densities of states $\bar{n}(\varepsilon)$ into components $\bar{n}_B(\varepsilon)$ and $\bar{n}_B(\varepsilon)$, which can be thought of as the average densities of states on a A-type site and B-type site in the alloy, according to

$$\bar{n}(\varepsilon) = c_A\bar{n}_A(\varepsilon) + c_B\bar{n}_B(\varepsilon) \qquad \text{III.6}$$

where $\bar{n}_\alpha(\varepsilon)$, α = A,B are given by

$$\bar{n}_\alpha(\varepsilon) = -\frac{1}{\pi}\,\mathrm{Im\ tr}\ \underline{F}^\alpha\underline{D}^\alpha\underline{\tau}^{c,00} \qquad \text{III.7}$$

The above formulae for the total and component densities of states are those derived by Faulkner and Stocks[46]. These equations are simple to evaluate once the CPA equations have been solved. They involve the $\underline{t}_c$ and $\underline{\tau}^{c,00}$ which are obtained during a CPA calculation and the wavefunctions $Z^A_L(r,\varepsilon)$ and $Z^B_L(r,\varepsilon)$ and phase shifts for a single A and B type muffin-tin well.

In the right hand column of fig. III.1 we show the calculated densities of states for a series of Ag_cPd_{1-c} alloys. The potentials used in these calculations are based on the Mattheiss prescription. As expected from our earlier consideration of the CPA scattering amplitude the alloy d-bands are split between a low lying Ag sub-band and a high energy Pd sub-band. A particularly striking feature of each sub-band is that its position in energy relative to the Fermi energy is independent of energy. In the centre column are shown the XPS results of Hüfner et al[62]. If we interpret this experiment as roughly measuring the density of states then these experiments confirm the features just described. Both the UPS experiments and the KKR-CPA calculations are in sharp disagreement with the results obtained using the VCA which are shown in the left

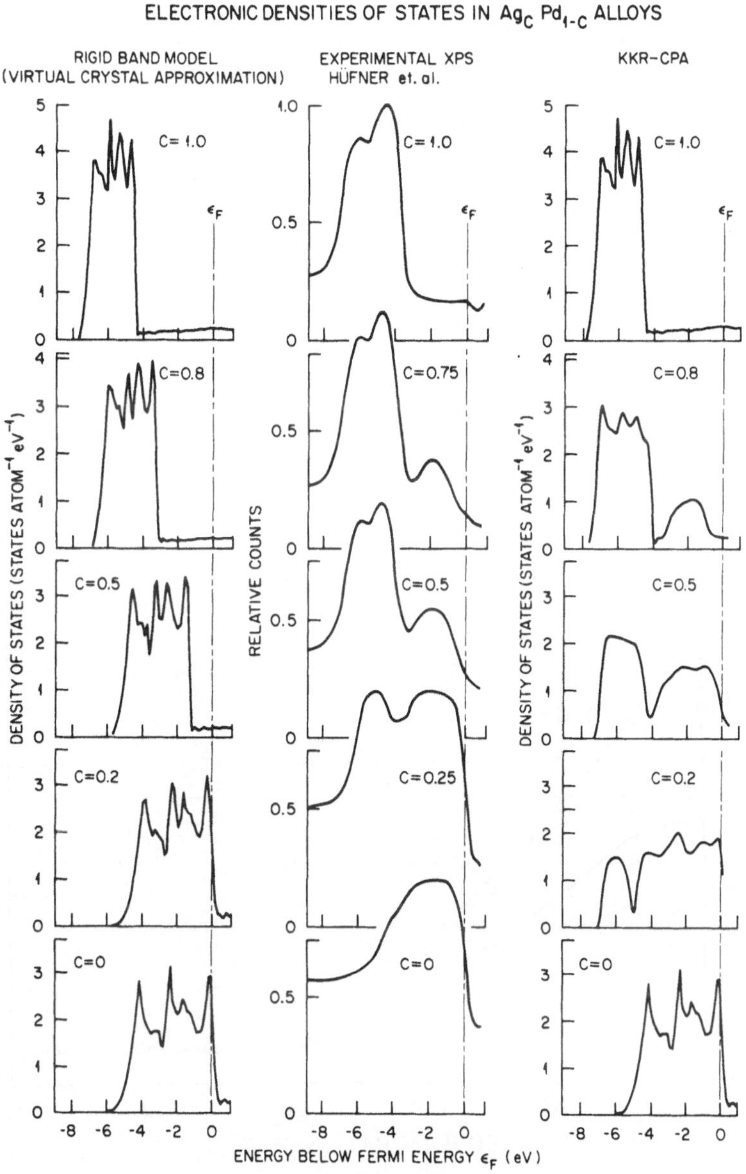

Fig. III.1. Densities of states for Ag_cPd_{1-c} alloys.
Left column : Densities of states calculated according to the virtual crystal approximation.
Centre column : Experimental XPS spectra {Hüfner et al}.
Right column : Densities of states calculated according to the KKR-CPA.

hand column of fig. III.1. The VCA gives only a single d-band, the position of which changes linearly as a function of concentration, in keeping with the results of the model calculations discussed earlier.

In fig. III.2 we show plots for the Ag_cPd_{1-c} alloys system of the calculated[63] and measured[64] low temperature specific heat coefficient. As is well known γ is related to the density of states at the Fermi energy $n(\varepsilon_F)$ through

$$\gamma = \gamma_0(1 + \lambda) \qquad \text{III.8}$$

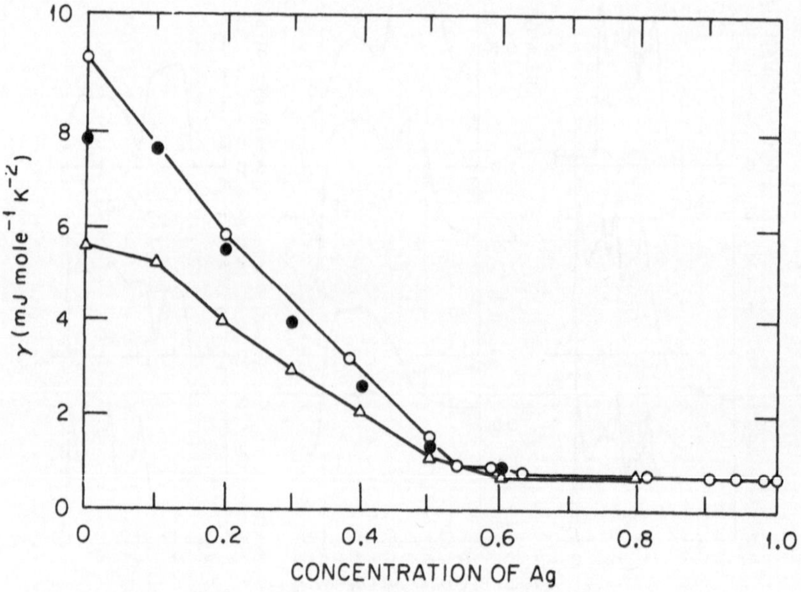

Fig. III.2. Calculated and measured low temperature {electronic} specific heat coefficient in Ag_cPd_{1-c} alloys.

Δ : Base electronic specific heat coefficient calculated using the KKR-CPA

○ : Electron-phonon enhanced specific heat coefficient obtained from KKR-CPA.

● : Experimentally measured specific heat coefficient according to Montgomery et al.

where

$$\gamma_0 = \frac{\pi^2}{3} k_B^2 \, n(\varepsilon_F) \qquad \text{III.9}$$

The quantity γ_0 in the unenhanced low temperature specific heat coefficient and the factor λ accounts for enhancements of the bare $n(\varepsilon_F)$ arising primarily from the electron phonon interaction. In fig. III.2 the triangles give the calculated values of γ_0. The solid circles give the calculated values of γ where an enhancement factor to account for the electron-phonon interaction has been included and which was also calculated using the results of the same KKR-CPA calculations[63]. The open circles give the experimental results of Montgomery et al[64]. Apart from very rich Pd-alloys where it is known that there is another contribution to λ from spin-fluctuations the KKR-CPA results give an excellent account of the experiment.

In fig. III.3 we show similar results for the Cu_cZn_{1-c} alloy system for $0 \leqslant c \leqslant 0.3$[29]. This alloy system is of particular interest since it is one of the so-called Hume-Rothery electron compounds[65,66]. For this class of alloys (e.g. Cu_cZn_{1-c}, Cu_cGa_{1-c}, Cu_cGe_{1-c}) the phase boundaries which separate equilibrium crystal structures occur at particular values of electron concentration,ξ, i.e. average number of conduction electrons per atom (ignoring the d-electrons). Furthermore many of the properties of the α-phase alloys ($e/a < 0.3$) also scale with ξ. The early interpretations of this behaviour were based on the rigid band model. This was obviously a sensible strategy since the only variable of the rigid band model is ξ. However, on the basis of modern measurements it has become clear that for the low temperature specific heat coefficient the results for the various systems neither scale with ξ nor agree with the predictions of the rigid band model. In fig. III.2a the experimental values of γ (open triangles and circles) are compared with the prediction of the rigid band model (solid line) and with the KKR-CPA results (solid circles). Clearly the KKR-CPA when properly applied gives a very good account of the experimental behaviour. As with the Ag_cPd_{1-c} system it was necessary to properly account for the concentration variation of λ (Fig. III.2c) enhancement of the bare specific heat (fig. III.2b).

In fig. III.4 we show the densities of states of a $Li_{0.2}Mg_{0.8}$ bcc random alloy calculated using the KKR-CPA. Such alloys are normally treated on the basis of pseudopotential theory where it is assumed that we are in the weak scattering limit. However, if we recall the phase shifts of fig. II.1b it is clear that for the Mg s-wave the scattering is far from weak. In fact there is an incipient s-wave resonance which makes the use of pseudopotential theories inappropriate. The effect of this Mg s-wave resonance is manifested in the peak at low energies in the Mg s-component of the densities of states. Clearly the Mg potential is almost suf-

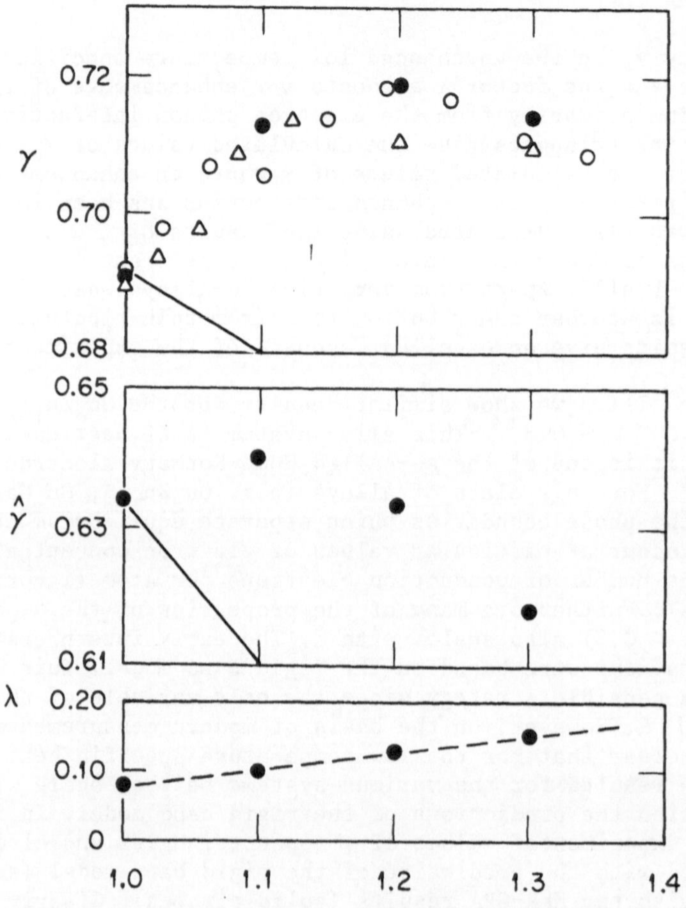

Fig. III.3. a) Upper : Low temperature specific heat coefficient for α-phase Cu_cZn_{1-c} random alloys. • calculated on the basis of KKR-CPA ; ○ and Δ experiment. The solid line gives the prediction of the rigid band model.

b) Middle : • - Bare electronic specific heat calculated using KKR-CPA. The solid line gives the same for the rigid band model.

c) Lower : Electron-phonon mass enhancement factor extracted from resistivity measurements.

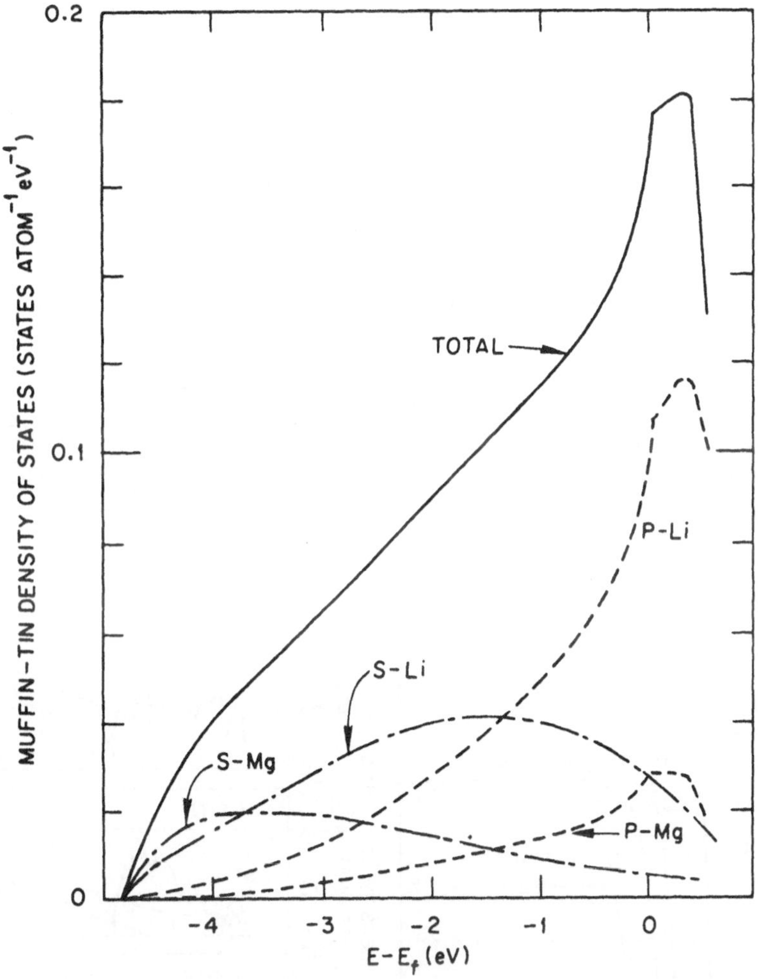

Fig. III.4. Total and component densities of states for $Li_{0.8}Mg_{0.2}$ according to the KKR-CPA.

ficiently attractive to form a separate Mg s-derived band corresponding to the Mg atomic 3s-level when embedded in the Li-matrix. This structure in the Mg s-band can be seen quite clearly in soft X-ray emission experiments[47,69,70]. Since these experiments are discussed by Dr. Durham in his lectures I will pursue the issue no further.

For alloy systems in which one or more of the constituent metals have atomic numbers > 50 relativistic corrections become of a magnitude that can no longer be ignored. A fully relativistic extension of the KKR-CPA described here has been developed by Weinberger and Staunton[30,71]. As well as deriving and programming the extension to the KKR-CPA condition itself they have also developed formulas for calculating the densities of states and Bloch spectral functions which are the relativistic generalizations of these given in this section. Inclusion of relativistic effects greatly increases the computational effort required since the size of all the matrices involved is doubled. However calculations on two alloy systems Ni_cPt_{1-c} and Au_cPt_{1-c} have been made. In fig. III.5 we show results for the Au_cPt_{1-c} system taken from the thesis of Staunton[71]. In fig. III.5a are plotted the d-wave phase shifts appropriate to the Au and Pt muffin-tin potentials, the spin orbit splitting between the d-scattering resonances associated with the $d^{3/2}$ and $d^{5/2}$ components is of the same order as that between the elements. In fig. III.4b,c and d are shown the total density of states and the d-densities of states at Pt and Au sites respectively for a $Au_{0.5}Pd_{0.5}$ alloy. The d-densities of states are resolved into $d^{3/2}$ and $d^{5/2}$ contributions. Unlike Cu_cNi_{1-c} and Ag_cPd_{1-c} alloy systems which are iso-electronic with the Au_cPt_{1-c} the

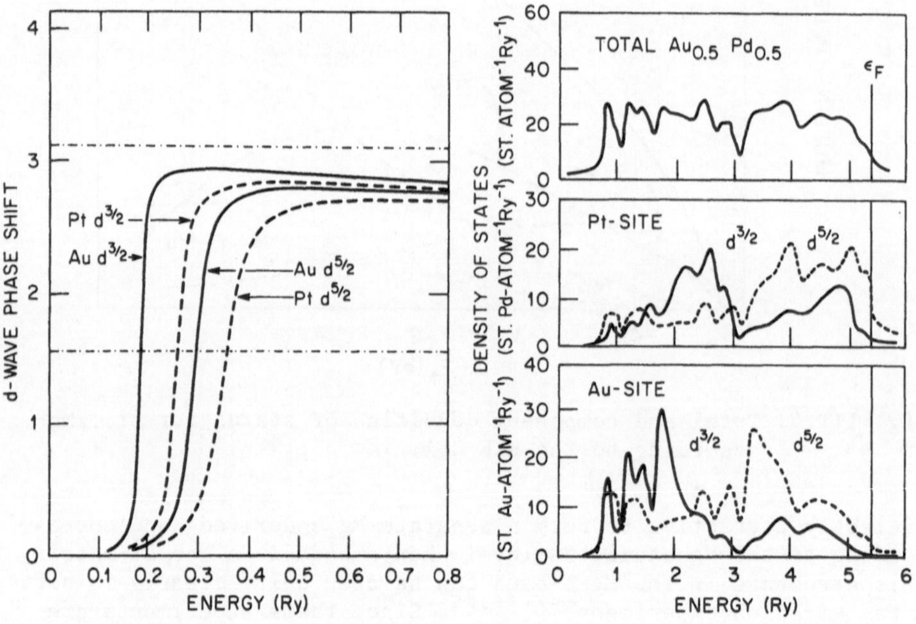

Fig. III.5. a) Left : Calculated relativistic $d^{3/2}$ and $d^{5/2}$ phase shifts for single Au and Pt muffin-tin potentials.

b) Right : Calculated total and component describes of states of a $Au_{0.5}Pd_{0.5}$ alloy calculated according to the relativistic KKR-CPA.

d-bands arising from the Au and Pd-sites completely overlap one another giving rise to much more of a common band picture for this alloy as opposed to the split band behaviour of the Cu_cN_{1-c} and Ag_cPd_{1-c} alloy systems.

The KKR-CPA method has also been extended to systems having more than one atom per unit cell (complex lattices). Pindor and Temmerman[72,73] have performed calculations for the $Ag_cPd_{1-c}H_x$ system. The underlying lattice for this system is CsCl with the Ag and Pd atoms occupying one sub-lattice with hydrogen and vacancies occupying the other. Once again this extension doubles the size of all the matrices involved in the calculation. However, at least at the level of non-change self-consistent calculations the calculation is still tractable. In Fig. III.6 are shown the densities of

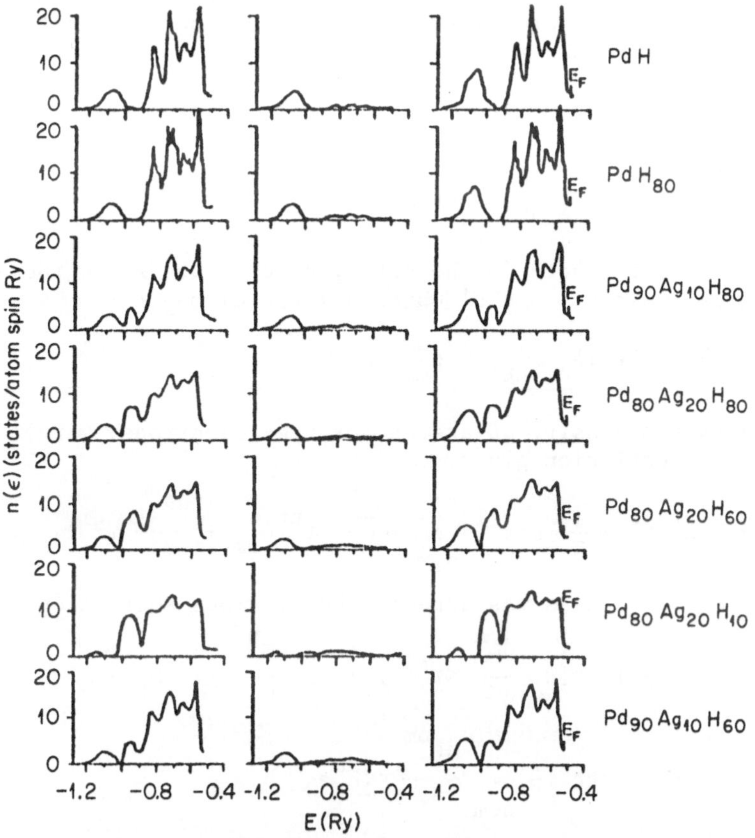

Fig. III.6. Total and component densities of states for Ag_cPd_{1-c} hydrides.
Left column : Contribution from metal sites (Ag and Pd)
Centre column : Contribution from non-metal sites (H and vacancies)
Right column : Total density of states.

states for a number of AgPd-hydrides. For all of the hydrides the hydrogen (premarily) s-band can be seen below the structure due to the metal d-bands. One particularly interesting feature of these results is that the presence of the hydrogen induces a substantial number of states (primarily d) associated with the metal atoms in the energy range of the hydrogen s-band.

A formula for the Integrated Density of States and a Cautionary Tale

An apparent alternative to calculating the density of states from the Green's function as in the previous section is to begin with the formula for the density of states of an arbitrary array of nonoverlapping muffin-tin potentials due to Lloyd[74]. This formula appears as the natural generalization to an array of scatterers of Friedel's formula[75] for the density of states in the neighbourhood of a single scatterer embedded in a free electron gas. It has been shown by Faulkner[76] that all such formulas can be derived from a fundamental theorem due to Krein[77].

According to Lloyd's formula, the density of states for an arbitrary array of muffin-tin scatterers is given by

$$n(\varepsilon) = n^0(\varepsilon) - \frac{1}{\pi N}\frac{d}{d\varepsilon}\,\mathrm{Im}\,\ell n\,\mathrm{Det}\,\underline{M} \qquad \text{III.10}$$

where $\underline{M}$ is a matrix both in the site indices n and m and the angular momentum indices L and L' which has the elements

$$M^{nm}_{LL'}(\varepsilon) = m^n_L(\varepsilon)\delta_{nm}\delta_{LL'} - g^{nm}_{LL'}(\varepsilon) \qquad \text{III.11}$$

and $n^0(\varepsilon)$ is the density of states for free electrons. Carrying out the differentiation gives

$$n(\varepsilon) = n^0(\varepsilon) - \frac{1}{\pi}\,\mathrm{Imtr}\,\frac{1}{N}\sum_{n,m}\left\{\frac{d\underline{m}^n}{d\varepsilon}\,\underline{\tau}^{nn}\delta_{nm} + \frac{d\underline{g}^{nm}}{d\varepsilon}\,\tau^{nm}\right\} \qquad \text{III.12}$$

Averaging Eq. III.12 as was done in the previous subsection gives

$$n(\varepsilon) = n^0(\varepsilon) - \frac{1}{\pi}\,\mathrm{Imtr}\left\{c_A\frac{d\underline{m}^A}{d\varepsilon}\langle\underline{\tau}^{00}\rangle_A + c_B\frac{d\underline{m}^B}{d\varepsilon}\langle\underline{\tau}^{00}\rangle_B\right\}$$

$$+\frac{1}{\pi}\,\mathrm{Imtr}\left\{\frac{1}{N}\sum_{n,m}\frac{d\underline{g}^{nm}}{d\varepsilon}\langle\underline{\tau}^{nm}\rangle\right. \qquad \text{III.13}$$

If we once again make the identifications embodied in Eq. III.2 together with

$$\underline{\tau}^{c,nm} \approx \langle\underline{\tau}^{nm}\rangle \qquad \text{III.14}$$

it can be shown that the average density of states in the CPA is given by

$$\bar{n}(\varepsilon) = n^0(\varepsilon) - \frac{1}{\pi} \mathrm{Imtr}\{d^3k[c_A \frac{d\underline{m}^A}{d\varepsilon} D_c^A + c_B \frac{d\underline{m}^B}{d\varepsilon} D_c^B - \frac{d\underline{g}}{d\varepsilon} (\vec{k},\varepsilon)]\underline{\tau}^c(\vec{k},\varepsilon) \qquad \text{III.15}$$

where $\underline{\tau}_c(\vec{k},\varepsilon)$ is defined in Eq. II.39. This equation can be manipulated into a number of equivalent forms with the help of the CPA condition, Eq. II.34. In particular it can be manipulated into the form of a complete derivative

$$\bar{n}(\varepsilon) = \frac{d}{d\varepsilon} \bar{N}(\varepsilon) = \frac{d}{d\varepsilon}\{N^0(\varepsilon) + \frac{1}{\pi} \mathrm{Im}[\frac{1}{\Omega_{BZ}} \int_{BZ} d^3k \,\ln \mathrm{Det}\, \underline{M}^c(\vec{k},\varepsilon) \qquad \text{III.16}$$

$$- c_B \ln \mathrm{Det} \left(\frac{\underline{m}^A - \langle\underline{m}\rangle}{\underline{m}^A - \underline{m}^c}\right) - c_A \ln \mathrm{Det} \left(\frac{\underline{m}^B - \langle\underline{m}\rangle}{\underline{m}^B - \underline{m}^c}\right)]\}$$

where $\underline{M}^c = [\underline{m}^c - g(\vec{k},\varepsilon)]$, $\langle\underline{m}\rangle = c_A\underline{m}^A + c_B\underline{m}^B$. $\bar{N}(\varepsilon)$ is the integrated density of states and $N^0(\varepsilon)$ is the integrated density of states for free electrons. Equation III.14 is the form first given by Gyorffy and Stocks[78].

In terms of calculating the densities of states Eq. III.14 is less convenient then Eq. III.3. The reason for this has to do with the fact that the structure constants $g_{LL'}(k,\varepsilon)$ are singular at the free-electron energies which causes problems when performing the necessary Brillouin zone integration. This singularity simply gives a contribution to the density of states which exactly cancels against the free-electron term $n(\varepsilon)$. Furthermore if turns out to be numerically more stable to calculate $n(\varepsilon)$ directly from Eq. III.3 than to first calculate the integrated density of states $N(\varepsilon)$ from Eq. III.14 and then obtain $n(\varepsilon)$ by numerical differentiation.

The expresiion for the integrated density of states Eq. III.14, however, is extremely useful. The point is simply that it allows the accurate determination of the Fermi energy, ε_F, without evaluation of the density of states throughout the band. This is possible because ε_F is determined by the requirement $N(\varepsilon_F) - N(\varepsilon_B) = c_A z_A + c_B z_B$ where z_A and z_B are the number of conduction electrons associated with the A and B alloying species and $N(\varepsilon_F)$ and $N(\varepsilon_B)$ are the integrated density of states at ε_F and at ε_B, the bottom of the band. In determining ε_F we are forced to evaluate $N(\varepsilon_B)$ because, in Eq. III.14 $N(\varepsilon)$ is obtained from the phase of a complex number which is arbitrary to multiples of 2π.

From the way I have introduced the calculation of the averaged densities of states both from the averaged Green's function and from the Lloyd's formula you might be forgiven for thinking that it must somehow be obvious that they are equivalent ways of calculating n(ε). This in fact is not the case. That use of the Green's function method is the correct way to do this is not in doubt. Thus what remains is to prove that Eq. III.13 is equivalent to Eq. III.3. This has been done by Faulkner and Stocks[46]. However, the proof required the multiple use of the CPA condition Eq. II.34. Thus it would appear that Eq. III.15 and III.16 are correct if and only if t_c satisfies the CPA condition. Furthermore it appears that Eq. III.15 cannot be broken up even for the CPA into contributions from the different species as was done by Schwartz and Bansil[79]. The consequences doing this are shown in fig. III.7 for the case of a 1-dimensional model[81]. The upper curves show total and component densities of states calculated on the basis of the Green's function formulas of this section while the lower curve were calculated using the formula for component densities of states proposed by Schwartz and Bansil. Not only are Schwartz and Bansil component densities of states different from those obtained from the Green's function formulas they are also pathological being negative in certain energy regions.

Schwartz and Bansil and later Bansil suggested the formulas for calculating densities of states on the basis of the ATA could be obtained beginning from the Lloyd formula. Correct formulas for calculating n(ε) within the ATA can be obtained trivially by replacing $\underline{t}_c$ by the ATA quantity $\underline{t}_{ATA} = c_A \underline{t}_A + c_B \underline{t}_B$ in the Green's function formulas of this section. In fig. III.8 we show, again for a 1-dimensional model ; a comparison of densities of states calculated using the CPA and ATA on the basis of the Green's function formulas with the Lloyd formula based expressions for the ATA densities of states calculated according to the formulas of Schwartz and Bansil[79] (ATA1) and Bansil[80] (ATA2). Though different from the CPA the Green's function based ATA results are sensible. Those labelled ATA1 are clearly pathological, those labelled ATA2 while not pathological for the total density of states turn out to be so for the component densities of states. Clearly the moral of this story is that the Lloyd formula is a dangerous place to start a calculation of the average densities of states in disordered systems.

The Bloch Spectral Function

The Bloch spectral density $A^B(k,\varepsilon)$ was first introduced by Soven[82] in connection with the one-level tight-binding model of an alloy. The Bloch spectral-density function has the property that for a perfect periodic solid it can be expressed as a sum of δ-functions

$$A^B(\vec{k},\varepsilon) = \sum_{\nu} \delta(\varepsilon - \varepsilon_{\vec{k},\nu}) \qquad \text{III.17}$$

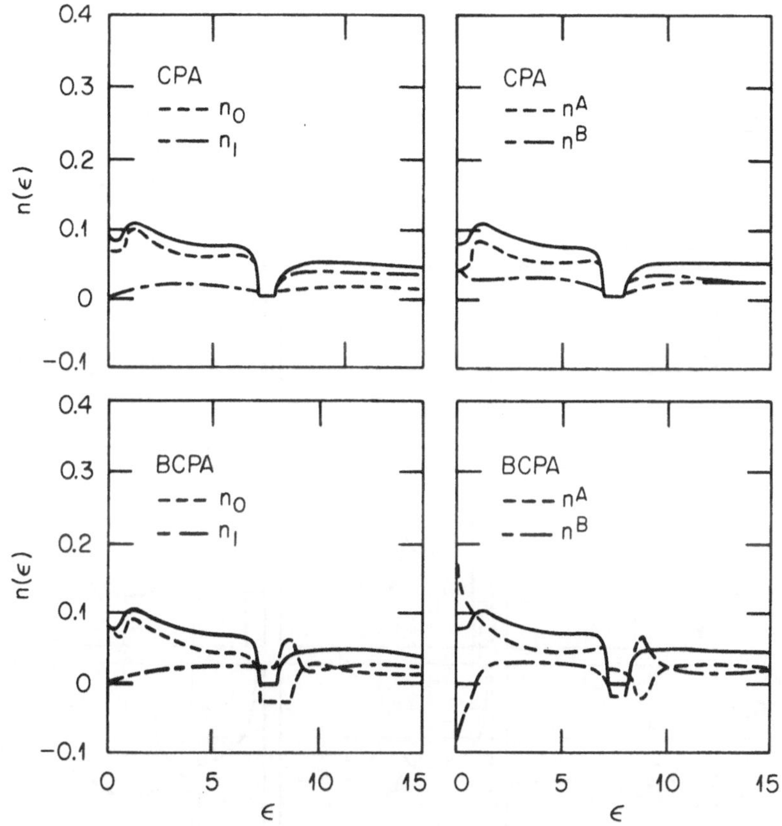

Fig. III.7. Total and component CPA densities of states for a 1-dimensional muffin-tin model alloy calculated according to various formulas.

Upper : Angular momentum decomposed (left) and site decomposed (right) densities of states calculated according to the Green's function formulas of Faulkner and Stocks.

Lower : Same as upper except calculated according to the formulas of Schwartz and Bansil[77].

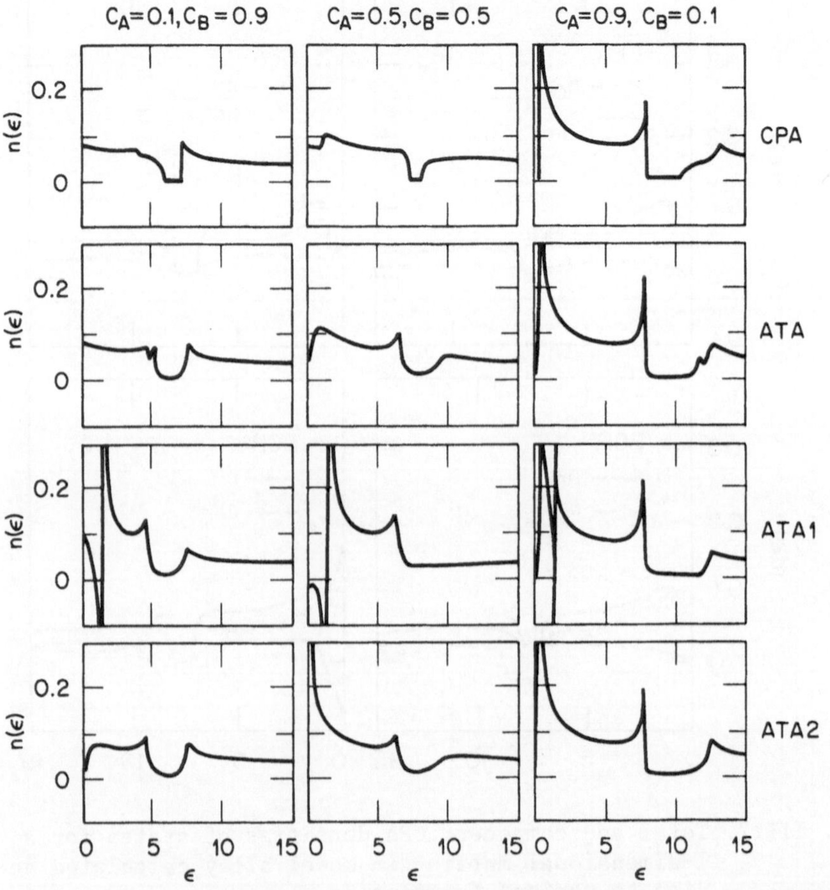

Fig. III.8. Total densities of states for the same model as Fig. III.7 for three different concentrations using the CPA Green's function formulas (upper) the ATA Green's function formulas (upper middle) the Lloyd formula based ATA expressions of Schwartz and Bansil (lower middle) and of Bansil (lower).

where the $\varepsilon_{\vec{k},\nu}$ are the energy eigenvalues obtained from a band-theory calculation. For a given $\vec{k}$, $A^B(\vec{k},\varepsilon)$ is a function of ε that is zero except for the energies $\varepsilon_{\vec{k},\nu}$ where it can be looked upon as having an infinitely sharp peak.

From both intuition and model calculations it is clear that when the system is disordered these peaks will broaden ; roughly speaking, the greater the disorder the greater the broadening. This broadening can be related to the lifetime of an electron in a Bloch state and can be measured in experiments such as the Haas-van Alphen effect where it shows up as a Dingle temperature. It can also be measured in more concentrated alloys using angular-resolved photoemission techniques, it also gives rise to the residual resistivity. It is the Bloch spectral function which for a random alloy carries the information which is contained in plots of the ε vs $\vec{k}$ relation for a pure metal. The above general features of the Bloch spectral function for a pure transition metal and a dilute random alloy are illustrated in fig. III.9. The central frame shows a band structure typical of a fcc-transition metal along the (001) direction in the Brillouin zone. The frames to the left and right show the Bloch spectral density for the pure metal as horizontal bars and the alloy as the dashed line for the zone centre Γ-point and the zone boundary X-point. For a dilute alloy most of the structure in the alloy spectral function will be related to the positions of eigenvalues in the pure metal, however, it is likely that there will also be structure related to the impurity at energies associated with the band structure of the impurity element.

The CPA Green's function $G_c(\vec{r},\vec{r}',\varepsilon)$ can be put in momentum representation by the transformation

$$G_c(\vec{r},\vec{r}',\varepsilon) = \frac{1}{N\Omega} \int d^3r \int d^3r' \; e^{i(\vec{p}.\vec{r} - \vec{p}'\vec{r}')} G_c(\vec{r},\vec{r}',\varepsilon) \qquad \text{III.18}$$

The fact that $G_c(\vec{r},\vec{r}',\varepsilon)$ has the full periodicity of the underlying lattice means that, if we write $\vec{p} = \vec{k} + \vec{K}_n$ and $\vec{p}' = \vec{k} + \vec{K}_n$ where $\vec{k}$ is a vector in the first Brillouin zone and $\vec{K}_n$ is a vector of the reciprocal lattice, we can define a function

$$\tilde{G}_c(\vec{k},\vec{k}',\varepsilon) = \sum_n G_c(\vec{k} + \vec{K}_n, \vec{k}' + \vec{K}_n,\varepsilon) \qquad \text{III.19}$$

which has all the information contained in $G_c(p,p',\varepsilon)$ folded back into the first Brillouin zone. It can be shown that $G_c(\vec{k},\vec{k}',\varepsilon)$ is given by

$$\tilde{G}_c(\vec{k},\vec{k}',\varepsilon) = \delta(\vec{k} - \vec{k}') \sum e^{i\vec{k}.\vec{R}_n} \int_{\Omega_0} d^3r \; G_c(\vec{r},\vec{r} + \vec{R}_n,\varepsilon) \qquad \text{III.20}$$

where the integral is over the central unit cell only.

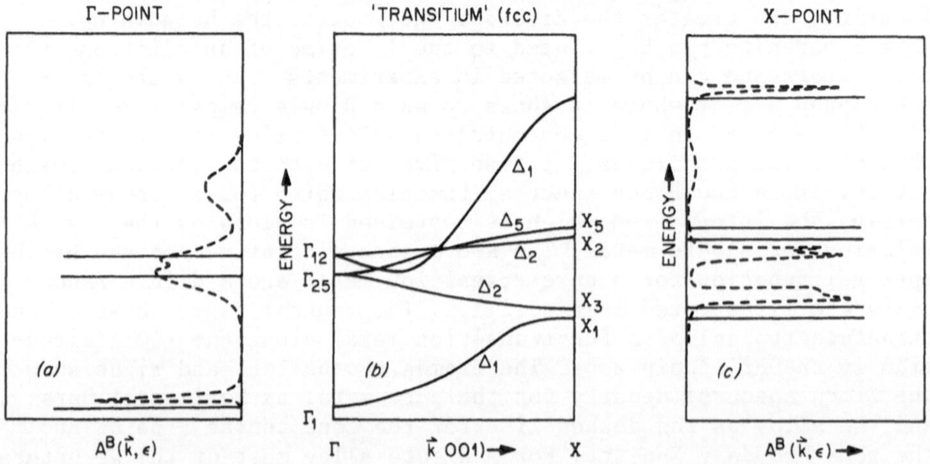

Fig. III.9. Schematic representation of the Bloch spectral function for a pure (canonical) transition metal (solid) and a disordered alloy (dash) at the Γ-point (left) and the X-point (right). The band structure, along Γ-X, of the hypothetical transition metal is shown in the center panel.

We now assert that the Bloch spectral-density function is given by

$$A^B(\vec{k},\varepsilon) = -\frac{1}{\pi}\,\mathrm{Im}\,\tilde{G}_c(\vec{k},\vec{k},\varepsilon) \qquad \text{III.21}$$

For an ordered crystal it is a trivial matter to show that

$$A^B_{ord}(\vec{k},\varepsilon) = \sum_\nu \delta(\varepsilon - \varepsilon_{\vec{k},\nu}) \qquad \text{III.22}$$

where the $\varepsilon_{k,\nu}$ are the band energies, ν being the band index.

In order to obtain an expression for the Bloch spectral function in the CPA it is clear from Eq. III.20 that we require an expression for the CPA approximation to the average Green's function $\langle G(r,r',\varepsilon)\rangle$ for the case where $\vec{r}$ is in the neighbourhood of the origin and $\vec{r}'$ is in the neighbourhood of the site at $\vec{R}_n$. That is we need to consider averages of the non-site-diagonal Green's function.

The ensemble average of the Green's function for the non-site diagonal (NSD) case can be obtained starting from Eq. II.31. First the average is taken over the subset of the ensemble that leaves the potential in cells n and m fixed :

$$\langle G(\vec{r},\vec{r}',\varepsilon)\rangle_{nm} = \mathrm{tr}\, \underline{F}^{nm}(\vec{r}_n,\vec{r}'_m)\ \langle \tau^{nm}\rangle_{nm} \qquad \text{III.23}$$

Here, the matrix $\underline{F}^{nm}(\vec{r}_n,\vec{r}'_n)$ has elements

$$F^{nm}_{LL'}(\vec{r}_n,\vec{r}'_n) = Z^n_L(\vec{r}_n,\varepsilon)Z^m_{L'}(\vec{r}'_m,\varepsilon) \qquad \text{III.24}$$

The matrix $\langle \underline{\tau}^{nm}\rangle_{nm}$ is the restricted average in which the potential on site n is known to be $v_n(\vec{r}_n)$ and the one on m is $v_m(\vec{r}_m)$. The final averaging over the possible occupants of sites n and m leads to

$$\langle G(\vec{r},\vec{r}',\varepsilon) = \mathrm{tr}[c_A^2\underline{F}^{AA}(\vec{r}'_n,\vec{r}'_m)\langle\underline{\tau}^{nm}\rangle_{AA} + c_Ac_B\underline{F}^{AB}(\vec{r}_n,\vec{r}'_m)\langle\underline{\tau}^{nm}\rangle_{AB}$$

$$+ c_Bc_A\underline{F}^{BA}(\vec{r}'_n,\vec{r}'_m)\langle\underline{\tau}^{nm}\rangle_{BA} + c_B^2\underline{F}^{BB}(\vec{r}_n,\vec{r}'_m)\langle\underline{\tau}^{nm}\rangle_{BB}] \qquad \text{III.25}$$

The various quantities $\langle\tau^{nm}\rangle_{\alpha\beta}$ are the restricted averages of $\underline{\tau}^{nm}$ such that the dite n is definitely occupied by the α species whilst the site m is definitely occupied by the β species.

Thus in order to find a CPA approximation to the non-site diagonal average Green's function we must obtain expressions for the two site restricted averages $\langle\underline{\tau}^{nm}\rangle_{\alpha\beta}$. Within the CPA these averages are much more problematical than the ones considered in the calculation of $n(\varepsilon)$. The CPA medium was obtained by satisfying a self-consistency condition involving only a single site condition, the fact that we are now asking for quantities which involve averages where two sites are fixed would appear to go beyond the CPA. In one sense this is true, however it is possible to obtain CPA expressions for these two site averages the derivation of which requires us only to make approximations of the type which were made in the original CPA[46,83]. The CPA approximation for the two site averages is

$$\underline{D}_c^{\alpha}\, \underline{\tau}_c^{nm}\, \underline{\tilde{D}}_c^{\beta} \approx \langle \underline{\tau}^{nm} \rangle_{\alpha\beta} \qquad \text{III.26}$$

where $\underline{\tilde{D}}_c^{\beta}$ is the transpose of $\underline{D}_c^{\beta}$. Thus the CPA approximation to the ensemble averaged Green's function for the non-site diagonal case can be written

$$G_c(\vec{r},\vec{r}\,',\varepsilon) = \mathrm{tr}\, \underline{F}^{cc}(\vec{r}_n,\vec{r}\,'_m)\underline{\tau}_c^{nm} \qquad \text{III.27}$$

where

$$\underline{F}^{cc}(\vec{r}_n,\vec{r}\,'_m) = c_A^2 \underline{\tilde{D}}^A \underline{F}^{AA}(\vec{r}_n,\vec{r}\,'_m)\underline{D}^A + c_A c_B \underline{\tilde{D}}^B \underline{F}^{AB}(\vec{r}_n,\vec{r}\,'_m)\underline{D}^A$$

$$+ c_A c_B \underline{\tilde{D}}^A \underline{F}^{BA}(\vec{r}_n,\vec{r}\,'_m)\underline{D}^B + c_B^2 \underline{\tilde{D}}^B \underline{F}^{BB}(\vec{r}_n,\vec{r}\,'_m)\underline{D}^B \qquad \text{III.28}$$

An expression for the Bloch spectral function can now be obtained. Inserting the above expression for the non-site-diagonal Green's function and the corresponding expression for the site diagonal Green's function implicit in Eq. III.1 and III.2 into Eq. III.20 yields

$$\tilde{G}_c(\vec{k},\vec{k},\varepsilon) = \mathrm{tr}\, \underline{F}^{cc} \sum_{n=0} e^{i\vec{k}.\vec{R}_n}\, \underline{\tau}^{c,0n} + \mathrm{tr}\, \underline{F}^c \underline{\tau}^{c,00} - Q(\varepsilon) \qquad \text{III.29}$$

In this expression $Q(\varepsilon)$ is an integral over the unit cell of the second term in Eq. III.1 and is real, and $\underline{F}^{cc}$ is defined in analogy with Eq. III.4 as

$$\underline{F}^{cc}(\varepsilon) = \int_\Omega d^3r\, F^{cc}(\vec{r},\vec{r},\varepsilon) \qquad \text{III.30}$$

Using Eq. III.29 with III.30 and II.37 it can be shown that the Bloch spectral function of a disordered alloy is given by

$$A^B(\vec{k},\varepsilon) = -\frac{1}{\pi}\, \mathrm{Imtr}\, \underline{F}^{cc}\underline{\tau}^c(\vec{k},\varepsilon) + \Delta(\varepsilon) \qquad \text{III.31}$$

where

$$\Delta(\varepsilon) = -\frac{1}{\pi}\, \mathrm{Imtr}\, (\underline{F}^c - \underline{F}^{cc})\underline{\tau}^{c,00} \qquad \text{III.32}$$

Equation II.39 can be used to prove

$$\frac{1}{\Omega} \int d^3k\, A^B(\vec{k},\varepsilon) = \bar{n}(\varepsilon) \qquad \text{III.33}$$

as it should be.

In fig. III.10 we show in the lower frame the Bloch spectral density for a $Ag_{.2}Pd_{.8}$ random alloy plotted as a function of energy for six $\vec{k}$-points along the (001)-direction. In the upper frame we

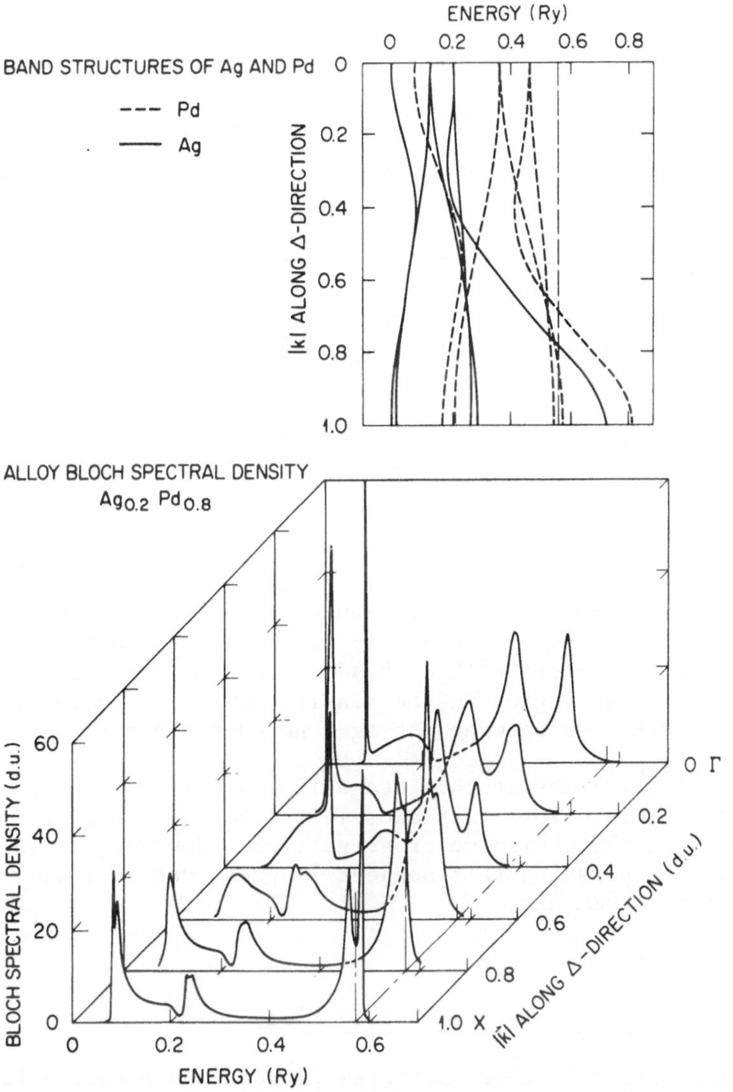

Fig. III.10. Upper : The band structures of pure Ag and Pd along Γ-X. The potentials used in these calculations are those corresponding to Ag and Pd sites in a random $Ag_{0.2}Pd_{0.8}$ alloy.

Low : Bloch spectral function plotted as a function of energy at various $\vec{k}$-points along the Γ-X direction.

show the energy band structures for a pure Ag and pure Pd having the lattice spacing of the alloy. Evidently the Bloch spectral function for a real alloy can be a very complicated quantity. Some states are very smeared out by disorder while some remain sharp. However, comparison of the spectral functions with the band structures above reveals order in the chaos. For example, at the Γ-point the low energy δ-function spike in $A^B(\vec{k},\varepsilon)$ can be identified with the states which form the bottom of the sp-band in the pure metals. This state is virtual crystal like in that its position is intermediate between the position of the state in pure Ag and the position of the state in pure Pd. The two high energy large peaks are clearly identified with the two d-zone-centre states in pure Pd. The small peak just above the low lying δ-function peak is the start of the formation of the two Ag related zone-centre d-states; recall that Ag is the minority component. Thus these states behave in a split band way, they appear once for each constituent.

At the zone boundary the low energy two peaked structure delineates the bottom of the Ag d-band whilst the similar structure at high energy delineates the top of the Pd sub-band, both of these sets of states have clear origins in the pure metal band structures. It is perhaps surprizing that even though the Ag constitutes only 20% of the alloy that the Ag derived states at the bottom of the band are so well formed. The structure at intermediate energy apparently arises from the zone boundary states at the bottom of the Pd band. For all $\vec{k}$-vectors $A^B(\vec{k},\varepsilon)$ has a trough at ~ 0.2 ryd this corresponds to the region in the density of states curve of fig. III.1 where $n(\varepsilon) \sim 0$ between the Ag- and Pd-sub-bands.

While $A^B(\vec{k},\varepsilon)$ may at this stage look a fairly esoteric object structure in this quantity is closely related to structure in angle-resolved photoemission cross-sections. However, since Dr. Durham will be pursuing this subject in great detail I will again forego further comments.

IV. SELF-CONSISTENT KKR-CPA

The Self-Consistent Single Site Potential

At this stage in these lectures it would be possible to use the equations of the KKR-CPA and the numerical methods which have been developed to solve them and to calculate a host of physical observables and to make contact with experiment. Indeed historically this is how the KKR-CPA developed. KKR-CPA calculations based on ad-hoc potential functions constructed according to the Mattheiss prescription were performed for a number alloy systems by the Oak Ridge/Bristol/Messina/Daresbury/Vienna axis, similar calculations based on potential functions constructed according to the renormalized atom method were performed by the North Eastern/ Brandeiss axis[84,86]. However, because of the lack of charge self-consistency, these calculations were confined to systems consisting of similar constituents e.g. Ag_cPd_{1-c}, Cu_cNi_{1-c}, Nb_cMo_{1-c}, Ni_cPt_{1-c} etc. That for these systems reliable potential should be constructed is bourne out by using the results obtained, densities of states, local densities of states, Bloch spectral functions etc. as a basis of the interpretation of experimental measurements of, for example, low temperature specific heat coefficient, electron-phonon-coupling parameters, Fermi surface dimensions obtained from position annihilation experiments, soft x-ray emission experiments, angle integrated UPS and XPS data, angle resolved photoemission cross sections etc.

However, successful as these calculations have been it is still the case that self-consistency in the alloy potential in the sense of density functional theory even for systems like those cited above is not only desirable in order to improve the results for these quantities, but also in order to be able to calculate quantities which are likely to be very sensitive to the finer details of the electronic structure, e.g. the cohesive energy, short range order effects etc. Only if charge self-consistency has been achieved can we establish conclusively if the KKR-CPA-method itself is good enough to deal with these more subtle problems. Of paramount importance, however, is self-consistency for alloys consisting of atoms dissimilar in their atomic numbers and (or) their numbers of valence-electrons. Without self-consistency it is difficult to make a theory of their physical properties. Thus, in this section we show how the KKR-CPA method can be embedded in the local density functional LDF scheme laid out in the introduction, and how numerical algorithms of sufficient sophistication, speed and accuracy can be developed to allow for the implementation of the resulting self-consistent-field (SCF)-KKR-CPA.

In order to establish a formalism which is suitable to obtain self-consistency with respect to the alloy potentials we have to give a prescription for the construction of the potentials out of configurationally averaged single-site charge densities obtained in the CPA. In contrast to the ordered case this is not straightforward, because again some kind of configurational averaging is involved in this step. Obviously the configurational averaging involved in the potential construction step should be on the same level as that applied to the scattering equation. For this purpose it is useful to devise a hierarchy of approximations to the alloy problem within which the CPA may be considered as the lowest-order one.

This is done in the following way. The alloy is subdivided into two regions, fig. IV.1. Region I contains the sites whose properties are going to be calculated, region II consists of more distant sites the properties of which are going to be averaged over in the calculation of the properties of the site in the central

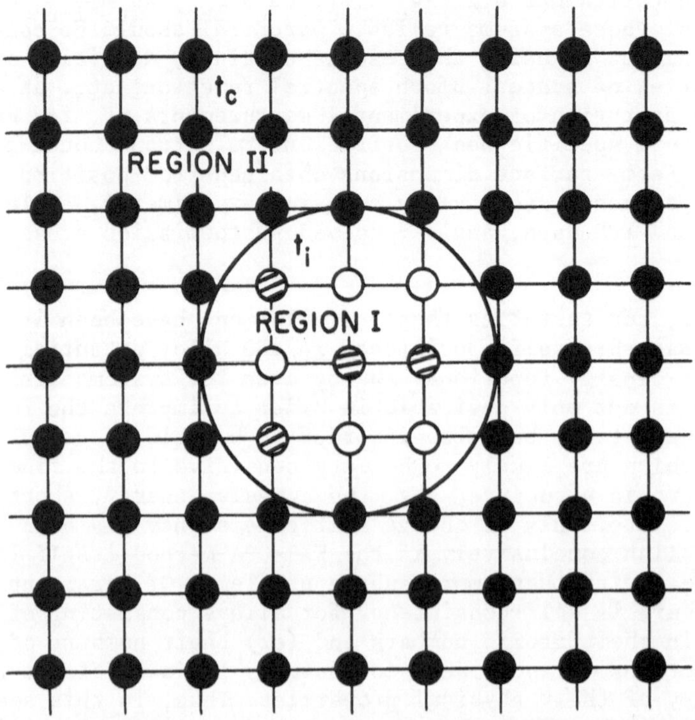

Fig. IV.1. Schematic of selfconsistency procedure in a random alloy. In region I the sites are occupied by the real A and B scatterers. In region II the sites are occupied by an "average" atom. In the single site CPA the average atom is determined by the CPA.

region I. Whereas in region I a particular configuration of the atoms is considered, the atoms of region II are repaced by the scatterers of some effective medium. The CPA consists of restricting region I to just the central site and is the lowest order in a hierarchy of schemes which consist in using different sizes for region I.

The prescription for the construction of the potential fitting into this scheme, now comes in a natural way. It is based on the following charge densities. Within site n of region I the true charge density, $\rho_I^n(r)$, is used. Of course for the general case this would have to be calculated within some extention of the CPA method which would allow us to calculate the properties, charge density etc, of a small cluster of scatterers rather than just the single scatterer averages obtained in the CPA. In region II we ascribe the concentration weighed average $\overline{\rho}(\vec{r})$, of the charge densities $\rho^A(\vec{r})$ and $\rho^B(\vec{r})$ at the average A- and B-atoms where this average is taken over all sites including those contained in the cluster.

$$\overline{\rho}(\vec{r}) = c_A\rho^A(\vec{r}) + c_B\rho^B(\vec{r}) \tag{IV.1}$$

In terms of $\rho^n(r)$ and $\overline{\rho}(r)$ the potential V^n at the site n in region I corresponding to a given configuration is from Eq. I.12.

$$\begin{aligned} V^n(\vec{r}) = {} & \sum_{m\subset I} \left\{\frac{\int d^3r'\rho^n(\vec{r}')}{|\vec{r} + \vec{R}_n - \vec{r}' - \vec{R}_m|} - \frac{Z^m}{|\vec{r} + \vec{R}_n - \vec{R}_m|}\right\} \\ & + \sum_{m\subset II} \left\{\frac{\int d^3r'\overline{\rho}(\vec{r}')}{|\vec{r} + \vec{R}_n - \vec{r}' - \vec{R}_m|} - \frac{\overline{Z}}{|\vec{r} + \vec{R}_n - \vec{R}_m|}\right\} \end{aligned} \tag{IV.2}$$

Here Z^m is the nuclear charge at site m and $\overline{Z}$ is the average nuclear charge

$$\overline{Z} = c_A Z^A + c_B Z^B \tag{IV.3}$$

Since we require the site potential functions in the muffin-tin form we approximate $\rho^n(r)$ and $\overline{\rho}(r)$ in the following manner.

$$\rho^n(\vec{r}) = \begin{cases} \rho^n_s(r) : \text{the spherically averaged charge density} & \text{for } \vec{r} \text{ within the muffin tin sphere of site n} \\ \langle\rho^n\rangle : \text{the spatially averaged interstitial charge density within the Wigner-Seitz-cell of site n} & \text{for } \vec{r} \text{ outside the muffin-tin sphere but within the Wigner-Seitz-cell n} \end{cases} \tag{IV.4}$$

$$\overline{\rho}(\vec{r}) = \begin{cases} \overline{\rho}_s = c_A\overline{\rho}^A_s(r) + c_B\overline{\rho}^B_s(r) \text{ with } \overline{\rho}^A_s \text{ and } \overline{\rho}^B_s \text{ the spherically averaged charge densities } \overline{\rho}^A \text{ and } \overline{\rho}^B \text{ respectively.} & \text{for } \vec{r} \text{ within the muffin-tin sphere of some site in region II} \\ \langle\overline{\rho}\rangle = c_A\langle\overline{\rho}^A\rangle + c_B\langle\overline{\rho}^B\rangle \text{ with } \langle\overline{\rho}^A\rangle \text{ and } \langle\overline{\rho}^B\rangle \text{ the spatial averages over } \overline{\rho}^A \text{ and } \overline{\rho}^B \text{ respectively in the interstitial regions} & \text{for } \vec{r} \text{ in the interstitial parts or region II} \end{cases} \tag{IV.5}$$

Having specified how the site potentials v^n are to be reconstructed from charge densities which are output from the alloy theory the formalism is now complete for the fully self-consistent treatment of random alloys. Of course we begin by restricting region I to a single site and use the CPA to obtain the average A and B site charge densities required in the reconstruction of the site potentials. In the alloys we have studied so far Ag_cPd_{1-c} and Cu_cPd_{1-c} this approximation seems sufficient for studying most of the properties of the systems. If it were not, or one is especially interested in short range-order-effects, the size of region I would have to be increased.

For the case where region I is restricted to a single site Eq. II.2 can be evaluated fairly straight forwardly for the two possible occupancies of the central site yielding

$$v^n(r) = \begin{cases} -\frac{2Z^n}{r} + 8\pi \int_0^r \left(\frac{r_1^2}{r} - r_1\right) \rho_s^n(r_1)dr_1 \\ \quad + 8\pi \int_0^{r^n_{MT}} r_1\rho_s^n(r_1)dr_1 + V_{xc}[\rho_s^n(r)] & r < r_{MT} \\ \quad - V_{xc}[<\rho>] + \frac{e}{a} <\bar{\rho}> (\bar{\Omega}_{ws} - \bar{\Omega}_{MT}) \\ 0 & r > r_{MT} \end{cases} \qquad \text{IV.6}$$

Here n is either an A- or a B-atom. In Eq. IV.6, $\bar{\Omega}_{ws}$ is the volume of the average Wigner-Seitz-cell in the alloy, a is the lattice constant and c, the Madelung constant, is given by a number, which depends on the crystal structure only. In the case of the fcc-structure, its value is 4.5848756[15]. An average site-independent constant interstitial potential is used in eq. IV.6.

We now require, for the KKR-CPA an expression for the configurationally averaged A- and B-site charge densities $\rho^A(r)$ and $\rho^B(r)$. As in the self-consistent band theory of ordered systems it is convenient to break the charge density up into a contribution $\rho^\alpha_{core}(r)$ from the low lying core electron plus a contribution ρ^α_{cond} from the electrons in the conduction band

$$\rho^\alpha(\vec{r}) = \rho^\alpha_{core}(\vec{r}) + \rho^\alpha_{cond}(\vec{r}) \qquad \text{IV.7}$$

Using the CPA expression for $G_c(\vec{r},\vec{r},\varepsilon)$ of the preceding section and the definition embodied in Eq. I.25 the contribution to the charge densities resulting from the conduction electrons is given by

$$\rho^{\alpha}_{cond}(\vec{r}) = -\frac{1}{\pi}\sum_{LL'}\int_{\varepsilon_B}^{\varepsilon_F} d\varepsilon Z^{\alpha}_{L}(\vec{r},\varepsilon)Z^{\alpha}_{L'}(\vec{r},\varepsilon)\mathrm{Im}\tau^{\alpha}_{LL'}(\varepsilon) \qquad \text{IV.8}$$

for α = A or B. The energy ε_B denotes the bottom of the conduction band. The core contribution to the charge density is obtained as in SCF-band theory either by taking it to be equal to the core charge density calculated for an isolated atom, the so-called frozen core approximation, or by finding the core eigenvalues and eigenfunctions for the isolated v^A and v^B muffin-tin potential and thereby reconstructing the core charge density at each iteration. Calculations performed in this latter mode are generally referred to as being relaxed core.

Fast Cluster Method for Solving the KKR-CPA Equations

Much more numerical labour is involved in obtaining the self-consistent electronic structure of a substitutionally disordered alloy than for an ordered system. This is because the self-consistency requirements, which have to be met simultaneously for both the single-site-scattering matrix, t_c, of the effective medium and the alloy potentials require us to run through the two interrelated self-consistency loops. The inner loop involves solving the KKR-CPA equations for given input potentials and constructing the charge densities of Eq. IV.7, we will refer to this as the CPA loop. The outer loop involves the iteration of these charge densities until the input and output charge densities for a particular CPA loop are the same i.e. we have solved the LDF-theory equations of section 1, we refer to the outer loop as the charge self-consistency loop. In practical calculations it is the CPA loop which is the limiting step and in particular the calculation of $\underline{\tau}^{c,00}$ for a particular input guess of $\underline{t}_c$ by evaluation of the integral in Eq. II.38. Therefore, it would be desirable to perform at least part of the steps involved by using a method considerably faster than the 'full' KKR-CPA. Henceforth we will refer to KKR-CPA calculations where $\underline{\tau}^{c,00}$ is calculated from Eq. II.38 without further approximation as full-KKR-CPA calculations.

An apparent alternative to solving Eq. II.35 by the lattice Fourier transform method and the evaluation of $\underline{\tau}^{c,00}$ from Eq. II.38 is to regard $\underline{\tau}^{c}$ as a matrix in both site and angular momentum indices and to solve Eq. II.35 directly by matrix inversion.

$$\tau^{c,00}_{LL'} = [\, m^c_{i,L}\delta_{ij}\delta_{LL'} - g^{ij}_{LL'}]^{-1}_{LL'} \quad i=j=0 \qquad \text{IV.9}$$

where on the r.h.s. of Eq. IV.9 we have made the dependence of the matrix elements of $\underline{m}$ and $\underline{g}$ on site and angular momentum indices explicit. Of course the difficulty with Eq. IV.9 is that we have to invert a matrix whose dimension is infinite since it involves the site index. Equation III.9 does however offer the possibility that, for the purposes of calculating the matrix elements of the scattering path operator for the central site, it may be sufficient to consider approximating the infinite array of t_c's by a small cluster of t_c's surrounding the central site as in fig. IV.2. Such a finite cluster approach was first treated out by Guiliano et al. In order to obtain any meaningful results for the transition-metal-alloy-systems we are aiming at, the cluster must not be too small. Dealing with just one shell of atoms, surrounding the central site is not good enough. It is also inappropriate to neglect s- and p-wave-scattering even in systems where d-wave-scattering is highly dominant as was done in the work of Guiliano et al.

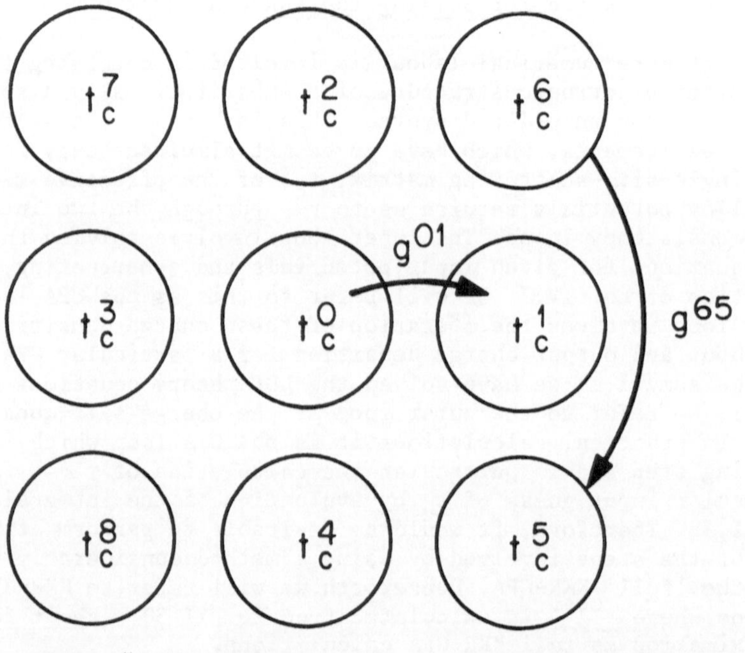

Fig. IV.2. Schematic of cluster method for solving the CPA equations. On each site in the finite cluster there is an effective CPA t-matrix, t_c^1. An electron propagates between sides according to the free particle Green function g^{ij}.

If we truncate the L index at $\ell = \ell_{max} = 2$ and we include N-sites in the cluster the size of the matrix we are required to invert is $N(\ell_{max} + 1)^2$ by $N(\ell_{max} + 1)^2$ i.e. 9N by 9N. Attractive as the cluster approach seems at first site, it soon becomes clear that if applied directly it offers no numerical saving over the full KKR-CPA method for the cluster sizes (> 40 sites) necessary to treat fcc and bcc systems. However, by transcribing the fast symmetrized cluster approach developed by Winter and Reiss[88] to the alloy problem it becomes possible to deal with 100 and more sites. This makes the cluster method quite attractive for the type of problem we are concerned with here. The superiority of that approach results from the fact that the scattering equation for the effective medium has the full point symmetry of the crystal lattice. Using this symmetrized cluster approach for solving the KKR-CPA equations and for performing the CPA loop it is possible, by including > 40 sites, to obtain single-site charge densities which are essentially identical to those obtained from the full KKR-CPA at a fraction of the computational cost (~ $1/50^{th}$).

Because the ability to handle clusters consisting of about 100 sites is the key to obtaining self-consistency, it is worthwhile to show in detail how the speeding up of the computations is achieved. Most important in this connection is the choice of the most appropriate basis set of single-particle-wavefunctions necessary to cast the scattering equation, Eq. II.20, into matrix form. If, instead of the free-particle angular-momentum eigenfunctions centered around the different lattice sites $j_\ell(\kappa r_n)Y_L(r_n)$ used in Eqs. II.21 and II.23, we use wavefunctions adapted to the point group symmetry of the system then the matrix elements of $\underline{\tau}^c$ between states belonging to different representations of the point group, as well as between different basis sates of the same representation, vanish. Using these basis states, the matrix :

$$[\underline{\tau}^c]^{-1} = \underline{t}_c^{-1} - \underline{g} \qquad \text{IV.10}$$

is therefore block-diagonal and $\underline{\tau}^c$ can be constructed by inverting the different symmetry blocks separately. The effect of this change of basis can be seen by taking a 79-site cluster employed to represent the fcc structure.The effective medium has O_h-point-symmetry, possessing 10 irreducible representations, M. Because the matrix elements of τ^c are only needed at the central site and it is sufficient to take $\ell_{max} = 2$ we have to include the blocks of 4 representations only. Their symbols and the dimensions of the corresponding blocks are displayed in table IV.1. In the non-symmetrized treatment of the problem on the other hand the dimension of the matrix to be inverted would have been 711.

The symmetry-adapted states can be constructed in the following way. The atoms are grouped into shells surrounding the point-symmetry-centre of the system. The positions of the atoms in a given shell are related to each other by symmetry operations. Each symmetrized state consists of a linear combination of single-site states, involving the sites of a particular shell, s, only and may be written as:

Table IV.1. The dimensions of the blocks in the symmetrized representation of scattering-path-operator

Symbols of the representations			Dimension of the corresponding block in the T-matrix
1	Ag	Γ_1	24
2	T_{1u}	Γ_{15}	52
3	E_g	Γ_{12}	38
4	T_{2g}	Γ_{25}	45

$$|M\nu\kappa\rangle = \sum_{i \subset s_\mu} C^{M\nu\kappa}_{\mu,i} |\lambda\mu i\rangle \qquad \text{IV.11}$$

Here $|M\nu\kappa\rangle$ is a basis state, ν, of the irreducible representation, M. The index κ is a further label necessary to account for the multiplicity of the basis states. $|\lambda\mu i\rangle$ is an unsymmetrized state centered at site i and characterized by the two quantum numbers λ and μ. The $|\lambda\mu i\rangle$'s contained in a particular symmetrized state all carry the same quantum number λ. The states $|\lambda\mu i\rangle$, employed in the cases of O_h-symmetry, are the cubic harmonics. By applying a Schmidt-orthogonalization-procedure to the $|M\nu\kappa\rangle$, the expansion coefficients, C, are constructed in such a way as to satisfy orthogonality relations.

Use of the basis sates IV.11 not only means that the equations determining $\underline{\tau}^c$ decouple in the way detailed above but also that the amount of time required to set up the matrices to be inverted can be greatly reduced. The point is simply that the matrix elements of the free particle Green's function can be written in the form

$$\langle M\nu\kappa|g|M\nu\kappa\rangle = \sum_{\ell, i' \subset s'} B_{\kappa\kappa'}(\ell, i)\sqrt{\varepsilon}\, h^+_\ell(\sqrt{\varepsilon}\, R_{ii'}) \qquad \text{IV.12}$$

and the energy- and lattice-constant-independent (structural) quantities $B_{\kappa\kappa'}(\ell, i')$ can be calculated once for each crystal structure and cluster size and stored permanently. This greatly speeds up the setting up of the matrix elements of the free particle Green's function which is the most time consuming step in setting up the matrices to be inverted.

Despite our advocacy of using this cluster method of treating the CPA loop of our SCF-method we should not lose site of the fact that it is an approximate way of solving the CPA equations. Its value rests on its ability to duplicate the charge density that would have been obtained if we had in fact used the full KKR-CPA.

In fig. IV.3 we show the single site charge densities $\rho^{Ag}(r)$ and $\rho^{Pd}(r)$ obtained for a $Ag_{.2}Pd_{.8}$ random alloy. The input potentials to these calculations were based on the Mattheiss prescription. We show the charge density obtained on the basis of the cluster method outlined above for clusters having 43 sites (4-shells of neighbours) and 79 sites (6-shell of neighbours) as well as those obtained using the full KKR-CPA. Clearly the 4-shell, 6-shell and full-KKR-CPA charge densities are identical to one another. Thus it appears that, provided we keep the size of the cluster reasonably large, use of the cluster method in the CPA loop does not affect the final converged self-consistent field potentials since the only quantities from the use cluster method in the CPA loop which are used in charge self-consistency loop are the single-site charge densities.

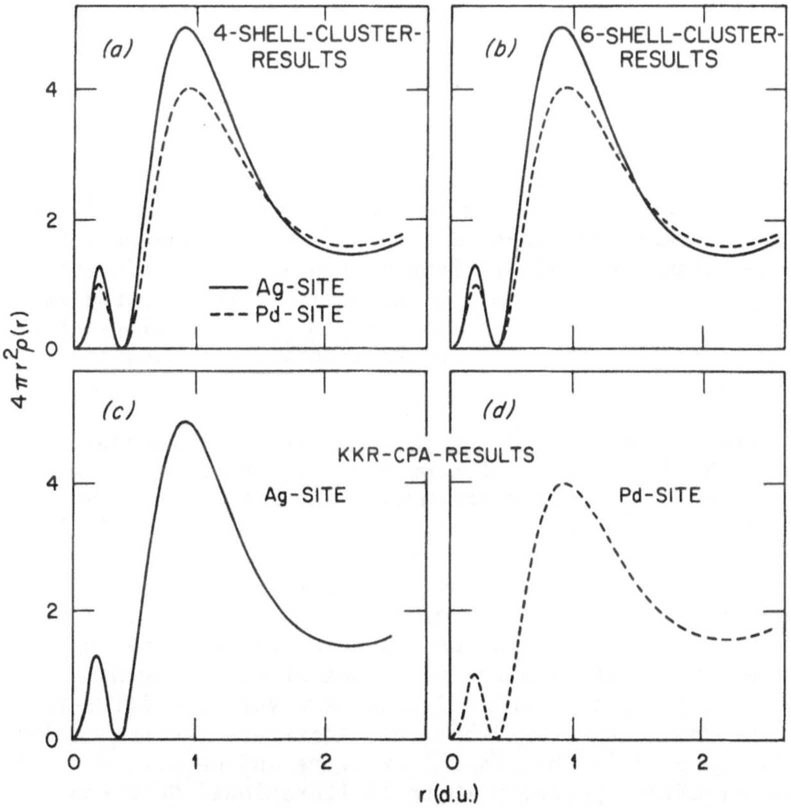

Fig. IV.3. Average Ag and Pd site charge densities calculated using the KKR-CPA. Frames a) and b) give the charge densities calculated using the cluster method with 4-shells and 6-shells of neighbours respectively. Frames c) and d) give the results obtained using the full KKR-CPA.

The densities of states corresponding to the charge densities shown in fig. IV.3 are shown in fig. IV.4. These densities of states were calculated from Eq. III.3 with $\underline{\tau}^{c,00}$ being obtained from the 4-shell, 6-shell cluster and the full KKR-CPA. Unlike the charge density, the densities of states are quite dependent on the cluster size the larger the cluster the better the approximation to the full KKR-CPA results. The fact that the charge density is quite insensitive to these differences in the densities of states is simply because it involves an integration over energy. However, the differences seen between the cluster densities of states and the full KKR-CPA densities of states do mean that we cannot rely on the fast cluster method alone for solution of the KKR-CPA equations. Having obtained the self-consistent alloy potentials it is necessary to perform one last CPA loop using the full-KKR-CPA for the purposes of calculating densities of states, Bloch spectral functions and non-energy integrated observables. If we in addition calculate the single site charge densities it also provides one final check that nothing went wrong during the SCF calculation.

SCF-KKR-CPA Calculations

So far as we are aware the only SCF-KKR-CPA calculations which have been performed as random alloys are our own. So far we have performed calculations in two alloy systems Ag_cPd_{1-c} and Cu_cPd_{1-c}. The calculations on the former wer performed largely to check out our methods since we already had good ad-hoc potentials for this system. The calculations on the latter were performed because potentials constructed on the basis of the generalized Mattheiss prescription do not allow even the beginnings of a description of the experimental information which is available on that system.

In general we applied the self-consistency procedure to clusters of both 43 (4 shells of atoms) and 79 (6 shells of atoms) sites. The maximum angular momentum taken into account was ℓ_{max} = 2 for each site ; f-scattering could easily be included if necessary. As starting potentials we used potentials obtained from the Mattheiss-prescription. The input potentials for the ith iteration-loop are obtained by admixing the output-charge-densities of the $i\text{-}1^{th}$ iteration with the input-charge-densities of the $i\text{-}1^{th}$ iteration. Although the positions of extremely narrow d-resonances present in these systems are very sensitive to the details of the charge distributions, admixture factors as large as 0.2 could be used in the final iterations and we attained self-consistency after typically 10 to 15 iterations. Self-consistency has been considered achieved as soon as the r.m.s. difference between the output-potentials of the two subsequent iterations fell below 10^{-5} ryd. This amounts to a shift in the d-resonances, which are the most sensitive quantities, of less than 0.001 ryd. We treated the core charge density using the frozen core approximation

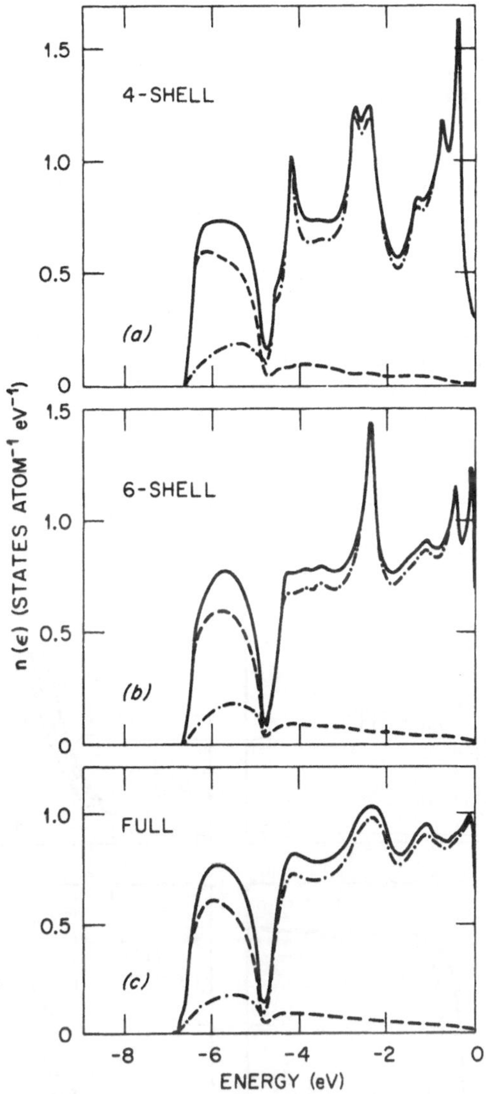

Fig. IV.4. Total and component densities of states for a $Ag_{0.2}Pd_{0.8}$ alloy calculated using the cluster CPA with 4 (upper) and 6 (middle) shells of neighbours and using the full KKR-CPA (lower). Total n(ε)-solid lines ; Ag-site component-dash line ; Pd-site component-dash/dot line.

although this could be relaxed at the expense of little additional effort. In order to stabilize the self-consistency procedure the scattering-path-operator for about 180 energy points throughout the occupied conduction band had to be determined.

As expected for the case of the Ag_cPd_{1-c} alloy system the differences between the self-consistent results and the results based on the Mattheiss prescription potentials are small, but they are not zero. In fig. IV.5 we show the d-wave phase shifts corresponding to the Ag and Pd muffin-tin potentials both for the SCF potentials and the Mattheiss prescription potentials for the three alloys having compositions $Ag_{.2}Pd_{.8}$, $Ag_{.5}Pd_{.5}$ and $Ag_{.8}Pd_{.2}$. At each concentration the major effect of self-consistency is to move the Ag-site d-scattering resonance to higher energy by approximately 0.05 ryd while the position of the Pd-site d-resonance is little affected. A moments thought should convince you that this should result in a decrease in the Ag-Pd d-band splitting from that seen in fig. II.1. That this the case can be seen clearly in fig. IV.6 where for the $Ag_{.2}Pd_{.8}$ alloy we plot the full-KKR-CPA densities of states corresponding to both the self-consistent and the non-self-consistent potentials.

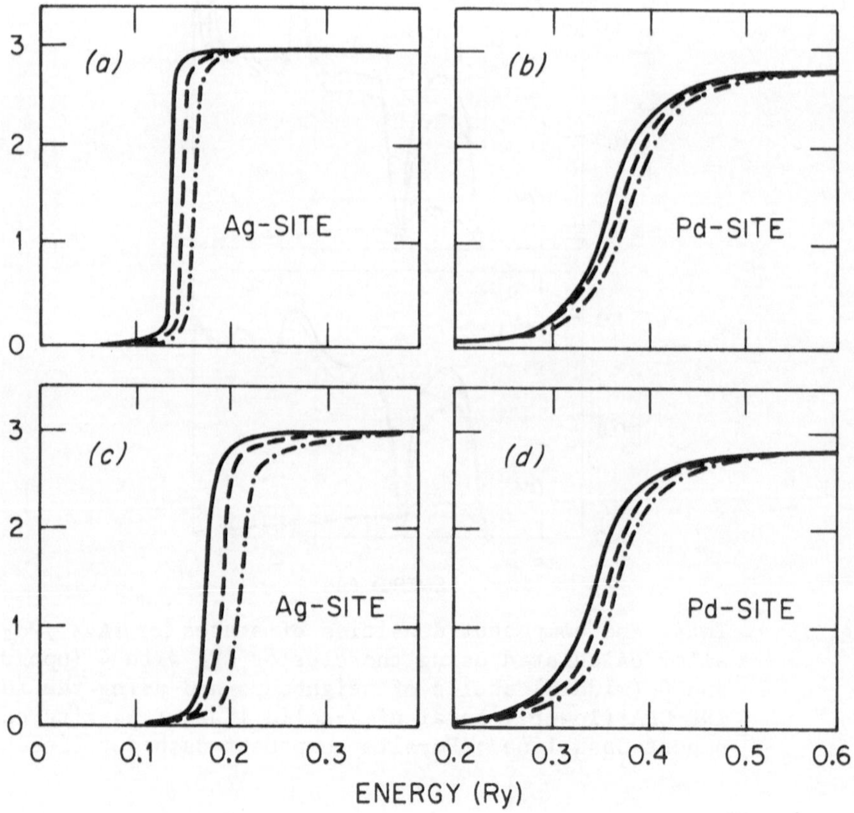

Fig. IV.5. Non-self consistent (upper two frames) and self-consistent d-wave phase shifts for three Ag_cPd_{1-c} alloys. $Ag_{0.2}Pd_{0.8}$ (solid) ; $Ag_{0.5}Pd_{0.5}$(dash) ; $Ag_{0.8}Pd_{0.2}$ (dash/dot).

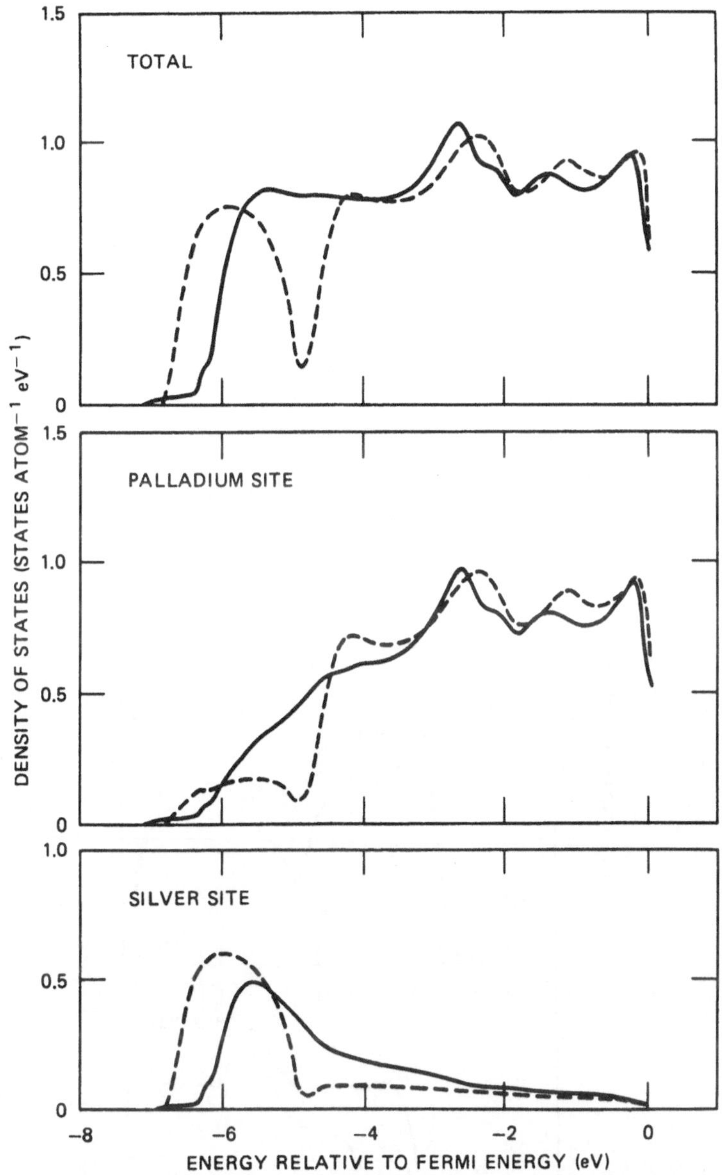

Fig. IV.6. Self-consistent (solid) and non-self-consistent total and component densities of states for $Ag_{0.2}Pd_{0.8}$.

The increased tailing of the Ag-site component densities of states on the high energy side of the Ag sub-band and the corresponding increased tailing of the Pd-site component densities of states on the low energy-side in the self-consistent calculations reflects the increased Ag-Pd hybridization which results from the decrease in the separation between the Ag and Pd d-scattering resonances. In fig. IV.7 we collect together the self-consistent total and component densities of states for the three alloys considered, the corresponding non-self-consistent results are shown in fig. II.1. Insofar as the positions of the Pd and Ag sub-bands are concerned it appears that the self-consistent results are in slightly better agreement with the UPS results displayed in fig. II.1.

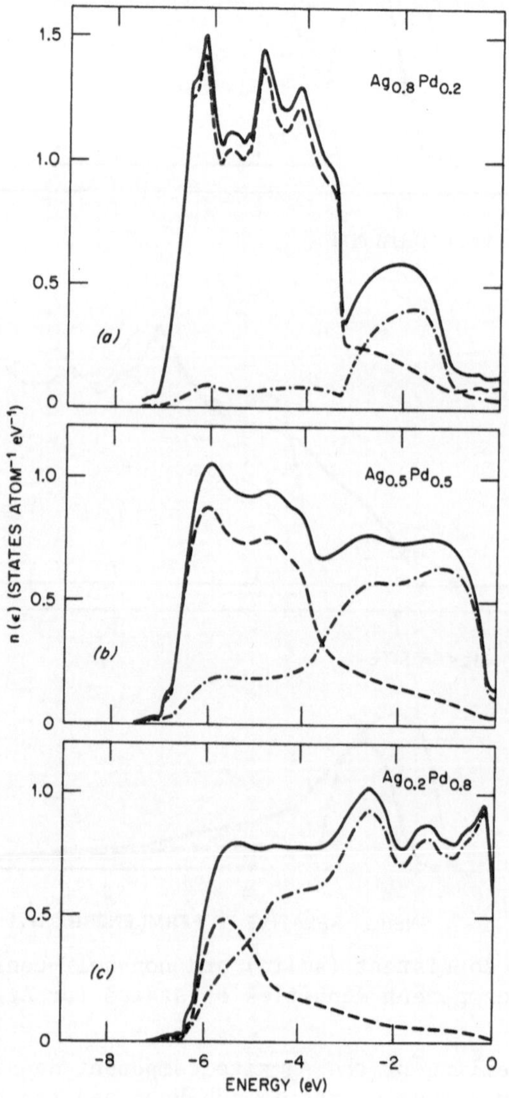

Fig. IV.7. Total and component densities of states for three Ag_c Pd_{1-c} alloys calculated using the self-consistent KKR-CPA. The key is as in Fig. IV.4.

However, the dip seen in the UPS spectrum between the Ag and Pd sub-bands for Pd-rich alloys is no longer present in the total densities of states. Thus in this respect the self-consistent densities of states are in worse agreement with experiment that the non self-consistent results where the dip is clearly seen, if somewhat exaggerated. We believe that the reason for the discrepancy largely has to do with matrix elements effects, which are ignored when we directly compare UPS spectrum with total densities of states curves. Results of preliminary calculations[89] seem to substantiate this view.

Before leaving consideration of the densities of states it is worthwhile to draw attention to just how good and how bad the cluster method can be for calculating the densities of states. In fig. IV.8 and fig. IV.9 we show the self-consistent total and component densities of states for the $Ag_{.5}Pd_{.5}$ and the $Ag_{.2}Pd_{.8}$ alloys calculated using the 4- and 6-shell clusters as well as the full KKR-CPA. The reason that both the 4- and 6-shell clusters give good densities of states for the $Ag_{.5}Pd_{.5}$ alloy is that the d-wave CPA effective scattering amplitude $f_{c,2}$ for this alloys is relatively inelastic $\alpha_2 \ll 1$ which means in turn that the sharp structure normally seen in cluster calculations in the d-band complex is largely washed out by the disorder. On the contrary, particularly in the energy range of the Ag sub-band, for the $Ag_{.8}Pd_{.2}$ the d-wave effective scattering amplitude is quite elastic $\alpha_2 \approx 1$. The consequence of this is that sharp structure typical of the cluster calculations persists even in the alloy.

In fig. IV.10 we show the SCF conduction band charge densities as out-putted from a full-KKR-CPA calculation using the SCF-potentials as input for all three alloys considered. We have not shown the corresponding cluster charge densities since they are identical to the full-KKR-CPA charge densities. The charge-densities both at the silver- and the palladium-sites vary smoothly throughout the concentration-range. With increasing Ag-concentration and increasing lattice spacing the main-peaks of the valence charge densities at the silver- and at the palladium-sides are slightly shifted towards the ion-cores, whereby their peak-heights show a small increase. This effect accompanies the shift of the d-resonances to lower energies as one approaches the Ag-rich-side of the concentration-range as shown in fig. IV.5.

In table IV.2 we summarize the charges within the muffin-tin-spheres and the Wigner-Seitz-cells surrounding Ag and Pd-sites both for the self-consistent-potentials and the Mattheiss prescription potentials. Of particular note is that, both for the self-consistent and non-self-consistent potentials, the individual Ag and Pd cells are essentially charge neutral. For the self-consistent potentials the average Pd site is slightly negatively charged while a Ag site is correspondingly net positively charged. The opposite is

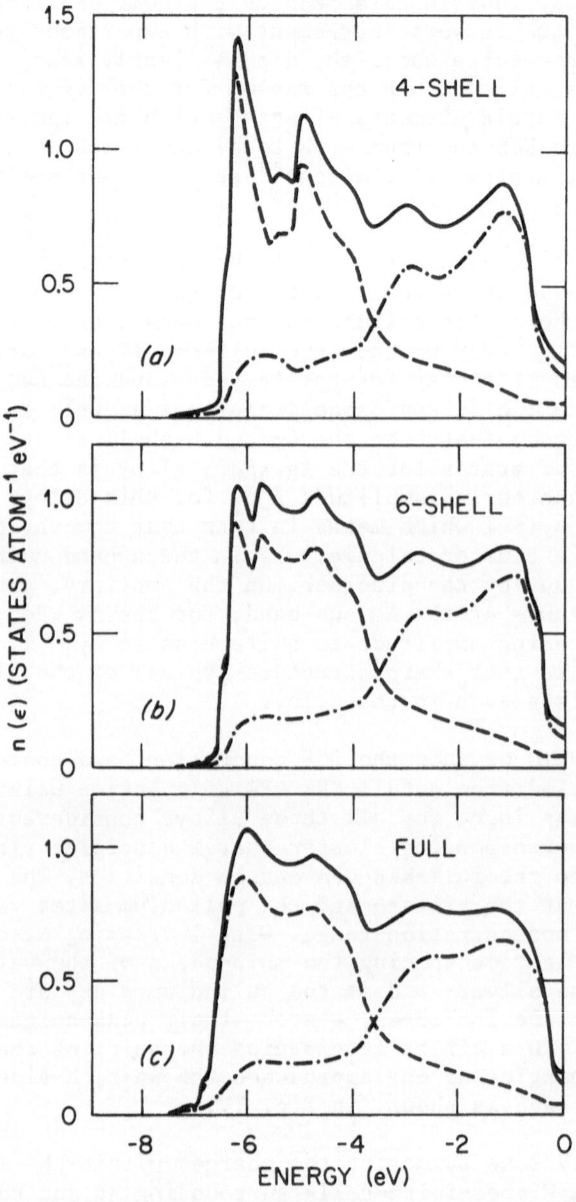

Fig. IV.8. Total and component densities of states for $Ag_{0.5}Pd_{0.5}$ calculated using the self-consistent potentials and the 4-shell, 6-shell and full KKR-CPA. The key is as in Fig. IV.4.

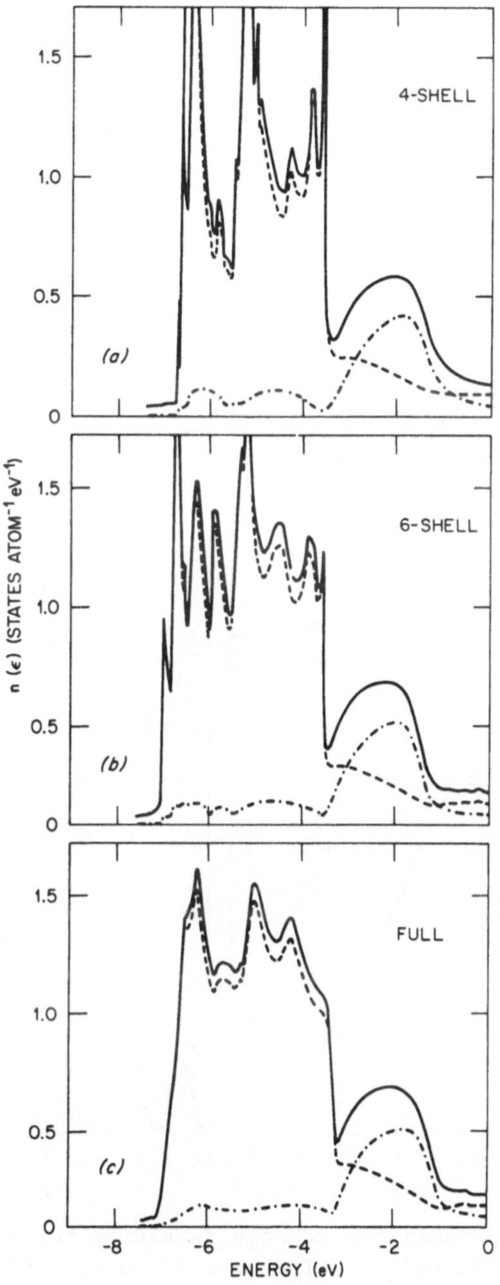

Fig. IV.9. As Fig. IV.8 except for $Ag_{0.8}Pd_{0.2}$.

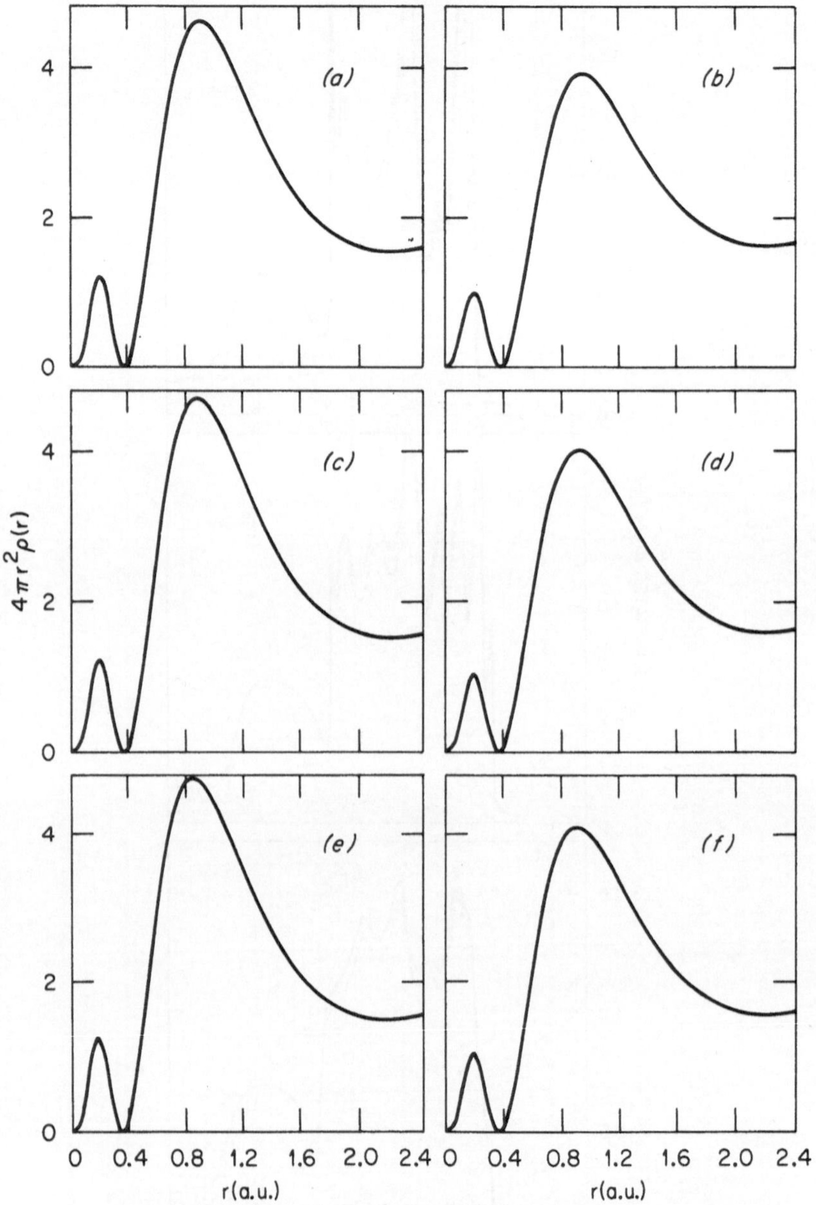

Fig. IV.10. Self-consistent charge densities for Ag (left column) and Pd (right column)-sites in $Ag_{0.2}Pd_{0.8}$ (upper) $Ag_{0.5}Pd_{0.5}$ (middle) and $Ag_{0.8}Pd_{0.2}$ (lower) random alloys.

Table IV.2. Some results of the selfconsistent calculations versus those based on the non-selfconsistent potentials. A detailed comparison to the experiments of Hüfner et al. requires the inclusion of matrix-element-effects

composition	lattice constant (a.u.)	Fermi-energy (Ry)		Positions of the d-bands (eV)					
		nonselfconsistent	selfconsistent	nonselfconsistent		selfconsistent		experiment	
				Ag-site	Pd-site	Ag-site	Pd-site	Ag-site	Pd-site
$Ag_{.2}Pd_{.8}$	7.415	.577	.571	5.9	2.4	5.5	2.2	5.4	2.2
$Ag_{.5}Pd_{.5}$	7.518	.544	.552	5.5	1.9	5.4	1.9	5.4	2.0
$Ag_{.8}Pd_{.2}$	7.630	.542	.554	5.4	1.7	5.4	1.9	5.5	1.9

composition	valence-charge in the muffin-tin sphere				valence-charge in the Wigner-Seitz-cell			
	nonselfconsistent		selfconsistent		nonselfconsistent		selfconsistent	
	Ag-site	Pd-site	Ag-site	Pd-site	Ag-site	Pd-site	Ag-site	Pd-site
$Ag_{.2}Pd_{.8}$	10.394	9.206	10.222	9.230	11.097	9.983	10.926	10.017
$Ag_{.5}Pd_{.5}$	10.397	9.284	10.224	9.308	11.015	9.985	10.943	10.057
$Ag_{.8}Pd_{.2}$	10.49	9.393	10.278	9.416	10.998	10.01	10.962	10.151

true of the non-self-consistent potentials for the $Ag_{.2}Pd_{.8}$ and $Ag_{.5}Pd_{.5}$. For the $Ag_{.8}Pd_{.2}$ alloy the charge transfer is in the same direction as the self-consistent-results. For the self-consistent potentials the Pd-site becomes increasingly negatively charged with increasing Ag content. The opposite is true of the non-self-consistent potentials. From this it should be clear that if one is interested in details of charge transfer in random alloys, as in the case in many semi-empirical theories of phase stability from Hume-Rothery's to Miedema's, this self-consistency is of paramount importance.

In figs. IV.11, IV.12 and IV.13 we displace the Bloch spectral density functions based on the self-consistent-potentials for five $\vec{k}$-points in the (100)-direction. For the $Ag_{.2}Pd_{.8}$ these can be compared with the non-self-consistent Bloch spectral functions shown in fig. II.10 and in the work of Pindor et al[25]. An overall comparison of the self-consistent and non-self-consistent Bloch spectral functions reveals a number of marked differences.

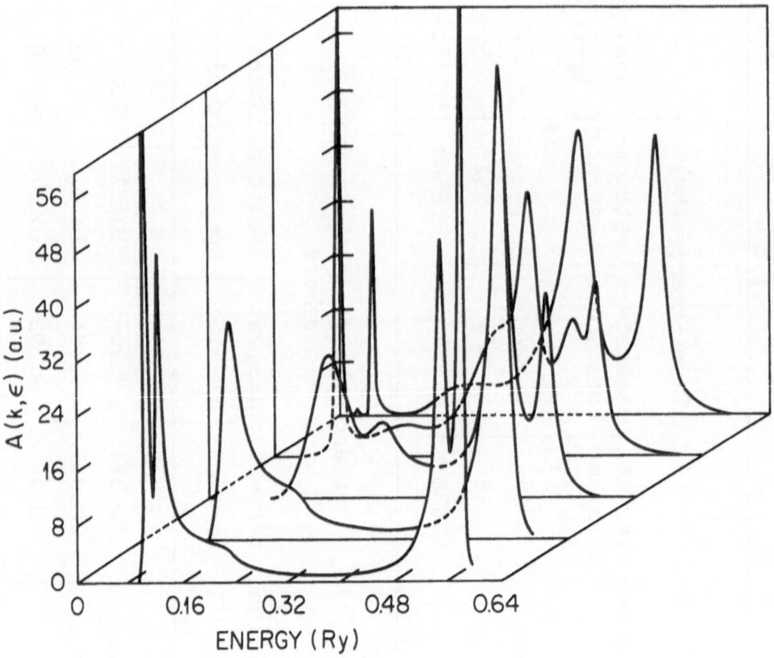

Fig. IV 11. Bloch spectral density as a function of energy at five equally spaced $\vec{k}$-points along Γ-X for $Ag_{0.2}Pd_{0.8}$ calculated on the basis of self-consistent potentials.

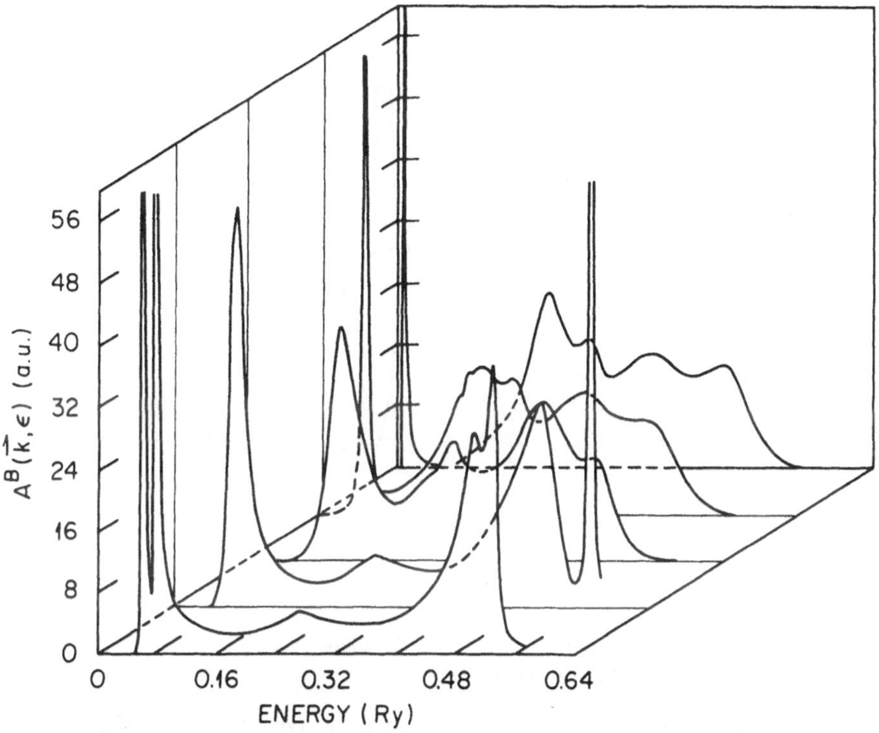

Fig. IV.12. Same as Fig. IV.11 except for $Ag_{0.5}Pd_{0.5}$.

Whereas the non-self-consistent results exhibit gaps between the low-energy-structures caused by the silver-bands and the higher-energy-structures due to the palladium-bands, the self-consistent results do not. Furthermore in the silver-rich alloys the unstructured peak indicating the Pd-impurity-band is narrower and more pronounced in the self-consistent case. The self-consistent results show more structures for the $Ag_{.5}Pd_{.5}$ alloy than the non-self-consistent ones. Furthermore, self-consistency almost obliterates the features found near the gaps of the Mattheiss-prescription results. In addition we observe slight shifts in the positions of the peaks. The examples cited above clearly show that self-consistency causes changes in the broadened bands, which could be probed by performing angle-resolved photoemission-experiments of the type described by Dr. Durham in this summer school.

Let us now turn our attention to an alloy system, Cu_cPd_{1-c}, where the effects of self-consistency are much more pronounced. Our earlier attempts to understand the available experimental

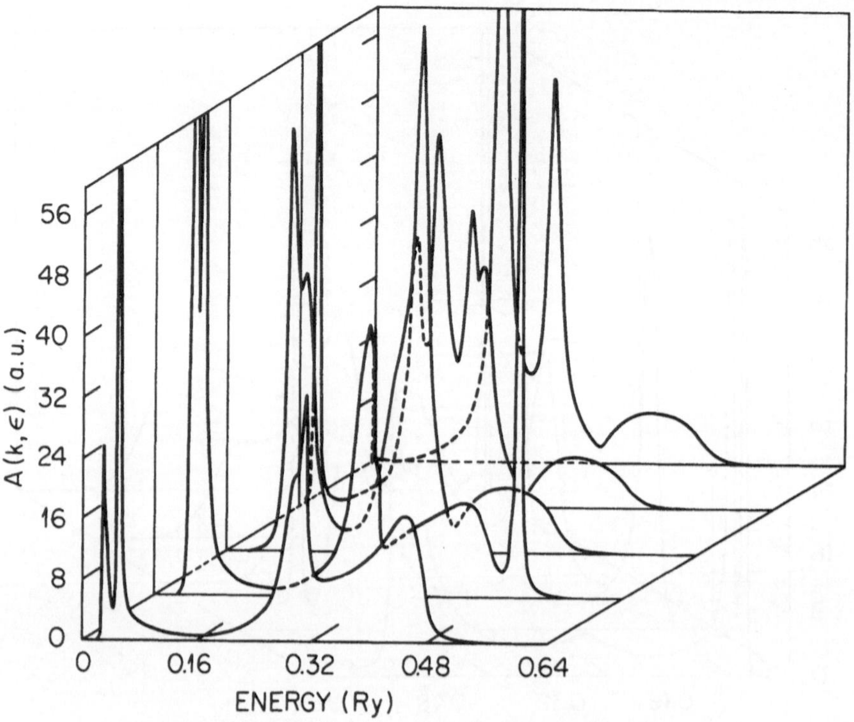

Fig. IV.13. Same as Fig. IV.11 except for $Ag_{0.8}Pd_{0.2}$.

information on this very interesting alloy system on the basis of KKR-CPA calculations always came to nought because of our inability to construct reliable potential functions using the Mattheiss prescription. Whilst one could take the view that the potential function could be adjusted to obtain agreement with some experiment and then regard the potential as having some semi-empirical foundation this is clearly an unsatisfactory situation. It may be that one could be adjusting the potentials to fix up what could be a fundamental breakdown of the CPA itself.

In fig. IV.14 we show the total and component densities of states based on non-selfconsistent Mattheiss prescription potentials and on self-consistent potentials. The densities of states were calculated using the 6-shell cluster $\underline{\tau}^{c,00}$'s and are, as a consequence, unduly spikey. Let us consider first the non-self-consistent densities of states. These densities of states are in some respects similar to those of Cu_cNi_{1-c} and Ag_cPd_{1-c} alloys having similar noble metal content. The structure in the densities

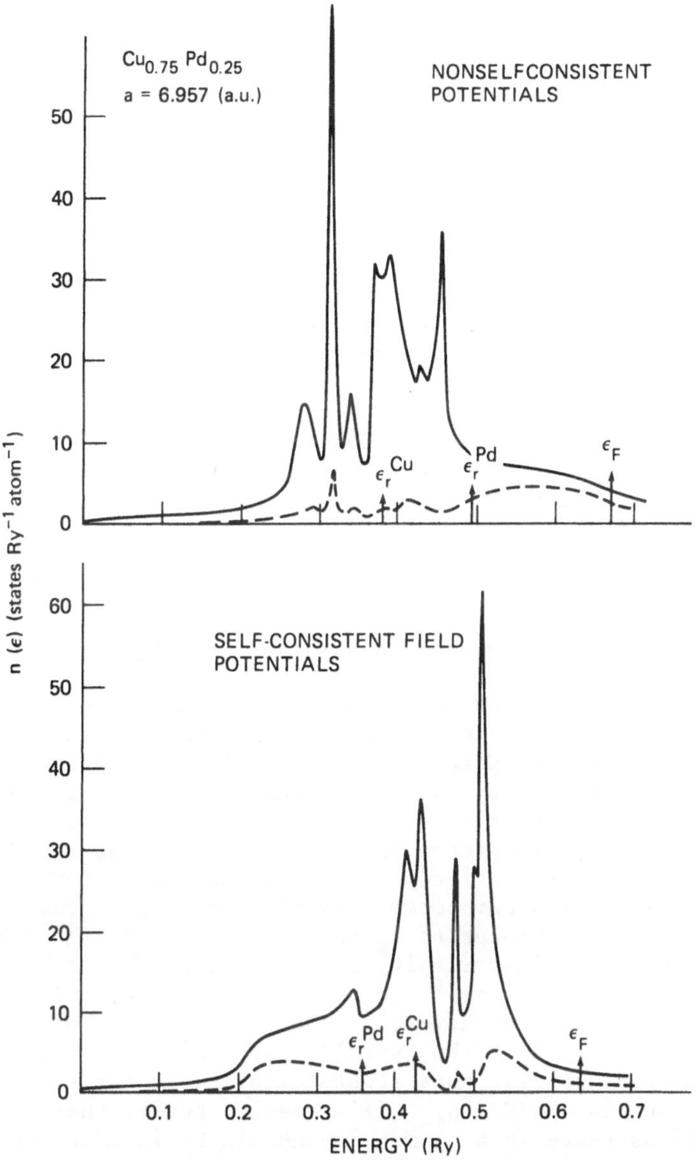

Fig. IV.14. Upper : Total (solid) and palladium component (dash) densities of states for $Cu_{0.75}Pd_{0.5}$ for non-selfconsistent potentials. The densities of states were calculated using the 6-shell cluster method.

Lower : Same as upper except for the self-consistent potentials.

of states from the minority late transition metal is in the form of a 'virtual bound state' like structure which lies above the d-band structure of the majority noble metal. The difference between $Cu_{.75}Pd_{.25}$ alloy on the one hand and the Cu_cNi_{1-c} and Ag_cPd_{1-c} alloys on the other is that the Pd-virtual bound state in the CuPd alloy is extremely wide. The consequence of this is that, in order to accommodate the relevant number of electrons below the Fermi energy ε_F has to rize substantially with respect to the top of the Cu d-band complex compared with its position in pure Cu. This does not occur for Cu_cNi_{1-c} alloys, neither does it occur for Ag_cPd_{1-c} discussed previously as can be seen from figs. III.1 and IV.7. Despite this rise in ε_F, it is still the case that the Pd-virtual bound state is only partially occupied. In table IV.3 we display the calculated changes on the Cu and Pd sites. For the non SCF potentials the Pd site only has ~ 9 electrons/atom. Clearly for a metallic system such charge transfer is nonsense. Use of such output charge densities to construct a new input potential would result in the Pd-site being positively charged, thus making the net potential very attractive to electrons and the d-scattering resonance being pulled down in energy towards the core level. The reverse would be true of the Cu-site. This effect is the cause of the reversal of the ordering of the Cu and Pd d-scattering resonances, which are marked in Fig. IV.14, between the selfconsistent and non-selfconsistent results.

In Fig. IV.15 we show the densities of states for $Cu_{0.95}Pd_{0.05}$ and $Cu_{0.75}Pd_{0.25}$ calculated using the full KKR-CPA. This figure is taken from the work of Weightman et al[90]. The densities of states show typical 'virtual crystal' type behaviour. This is apparently because the Pd d-scattering resonance lies well inside the Cu d-band complex and vice-versa (see fig. IV.14). Thus the Pd d-densities of states lies entirely inside the Cu d-band (or vice versa). For these self-consistent potentials the top of the common d-band complex is slightly closer to ε_F than is the top of the d-band complex in pure Cu. This finding seems to be supported by the XPS results of Weightman et al[90] and by preliminary angle-resolved photoemission results of Meyers[91]. A further feature of the densities of states is that the overall d-band width is greater than in pure Cu. Further self-consistent calculations for several other concentrations in the Cu_cPd_{1-c} alloy series reveal that for all concentrations there is a common d-band. As Pd is added to Cu the top of the d-band complex gradually approaches the Fermi energy finally intersecting with it at mid-concentration range. At the same time the d-band width increases from its value in pure Cu to its value in pure Pd.

At this stage it would be nice to make contact with the 'rolling mill' to the extent of having calculated the total energy and hence the equilibrium lattice spacing, heats of solution etc. and their concentration variation, for the random alloys. In fact

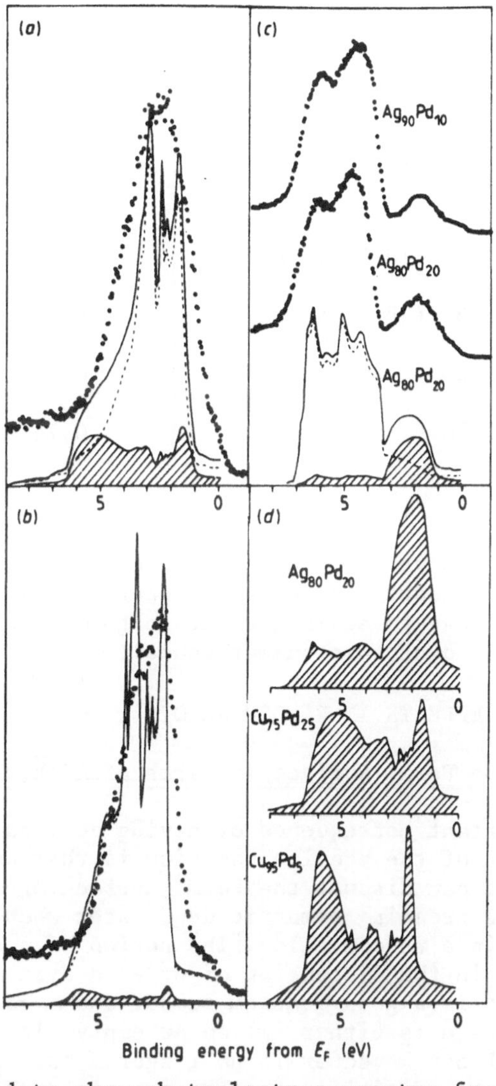

Fig. IV.15. The dots show photoelectron spectra from the conduction bands of $Cu_{82}Pd_{18}$ (a), $Cu_{95}Pd_5$(b), $Ag_{90}Pd_{10}$ and $Ag_{80}Pd_{20}$(c). The full curves show the total density of states given by SCF-KKR-CPA calculations for $Cu_{75}Pd_{25}$ (a), $Cu_{95}Pd_5$(b) and $Ag_{80}Pd_{20}$(c). The contributions to the total densities of states made by Pd are shown shaded and the contributions from Cu and Ag are shown by the broken curves. Figure V.2(d) shows the calculated Pd densities of states per Pd site.

Table IV.3. Charges on Cu and Pd sites in a $Cu_{0.75}Pd_{0.25}$ alloy for both self-consistent and non-self-consistent potentials calculated with the KKR-CPA.

Total charges : $(Cu_{.75}Pd_{.25})$

	non-self-consistent	self-consistent
Q^{Cu}	11.36 (10.71 in MT)	11.05 (10.40 in MT)
Q^{Pd}	8.93 (8.17 in MT)	9.84 (9.10 in MT)
$Q^{Cu}_{interst.}$	0.65	0.65
$Q^{Pd}_{interst.}$	0.76	0.74

the ansatz that we have used in the self-consistency procedure implies a similar ansatz for the total energy. However, such calculation will have to await our return to the CRAY-1S at Daresbury at the end of this summer school.

V. FERMI SURFACE EFFECTS IN RANDOM ALLOYS

Description of the Fermi Surface in Terms of $A^B(\vec{k},\varepsilon_f)$

A most important consequence of having defined the Bloch spectral function of the previous section is that we now have a means by which we can discuss the Fermi surface of a random alloy. Obviously, we are treading semantic deep water when we talk about a Fermi surface in a random alloy. The notion of a Fermi surface arises naturally in the discussion of ordered systems where the Bloch wave vector is a good quantum number and at zero temperature a particular $\vec{k}$-state is either filled or empty. In a disordered alloy where the Bloch $\vec{k}$-vector is no longer a good quantum number there is no strict definition of a Fermi surface. However, this does not prevent us from taking a sensible view of the situation and continuing to refer to the Fermi surface in an alloy if the smearing of the Fermi surface which results from the disorder is small compared with the dimensions of the Fermi surface.

The extent to which the Fermi surface in an alloy exists can be displayed nicely by plotting the Bloch spectral function at the Fermi energy $A^B(\vec{k},\varepsilon_F)$, along a series of $\vec{k}$-directions emanating from the Brillouin zone centre and terminating at the zone boundary.

This is done in fig. V.1 for the three Ag_cPd_{1-c} alloys considered previously. The potentials for which these Fermi surface are plotted are the non-self consistent Mattheiss prescription potentials of Pindor et al. The Fermi surfaces corresponding to the self-consistent potentials, as might be expected from the discussions of the previous section, are little different from these.

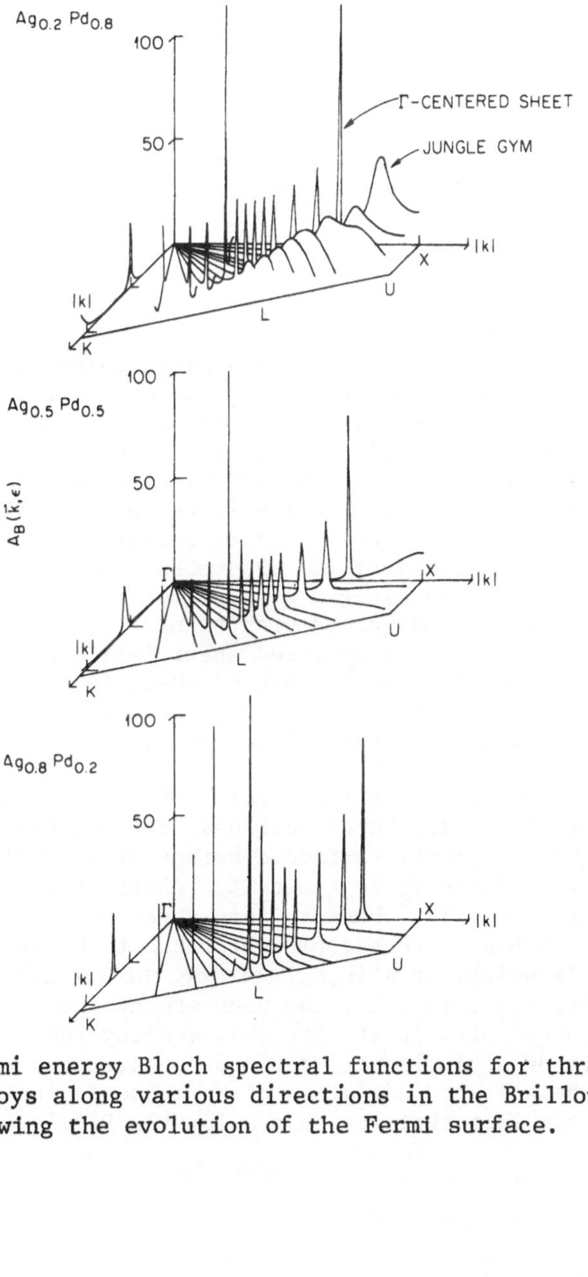

Fig. V.1. Fermi energy Bloch spectral functions for three Ag_cPd_{1-c} alloys along various directions in the Brillouin zone showing the evolution of the Fermi surface.

The set of directions are in the ΓLK-plane and the ΓLX-plane of the fcc structure which define two of the bounding planes of the irreducible 1/48th of the Brillouin zone. For all the alloys $A^B(\vec{k},\varepsilon_F)$ is sharply peaked around certain $\vec{k}$-vectors. If we recall that for a pure metal $A^B(\vec{k},\varepsilon_F) = \sum_\nu \delta(\varepsilon - \varepsilon_{\vec{k},\nu})$ a plot similar to those in fig. V.1 for a pure metal would consist of a series of δ-function spikes the locus of which would define the Fermi surface. Clearly in all of these alloys the δ-function spikes have been broadened but for the most part the broadening is still small on a scale set by the Brillouin zone dimensions. Thus it make sense to define an alloy Fermi surface by the locus of the peak positions in $A^B(\vec{k},\varepsilon_F)$. The width $\gamma_{\vec{k}}$ of the peaks in $\vec{k}$ defines the coherence length $\ell_{\vec{k}} \sim h/\gamma_{\vec{k}}$ of electrons on the Fermi surface.

Before trying to understand the topology and the smearing of the alloy Fermi surface it is useful to recall the Fermi surfaces corresponding to the pure metal and points. The Ag Fermi surface consists of a single multiply connected electron surface centered on the Γ-point and touching the Brillouin zone boundary near to the L-point ; it is topologically equivalent to the well known Fermi surface of Cu. If we refer back to the band structure plots corresponding to Ag and Pd in the upper frame of Fig. III.9, along the ΓX-direction the Ag Fermi surface arises from the intersection of the uppermost band with the Fermi energy, which is shown as a horizontal dashed line . The Fermi surface of Pd which is shown in Fig. V.2 is more complicated, it consists of several sheets.Chief of these are a Γ-centered closed electron sheet and a multiply connected hole-jungle-gym,the arms and pipes of which intersect at the X-point and run along the (001)-directions. As with Ag, along the ΓX-direction the Pd Γ-centred sheet results from the intersection with the Fermi energy of the uppermost band which is plotted in fig. III.9. The Pd jungle-gym arises from the intersection of the uppermost flat d-bands of Pd with ε_F near to the X-point. In Ag the volume beneath the Γ-centered sheet is 1.0 electron/atom. In Pd the volume beneath the Γ-centered sheet is about 0.4 electron/atom ; the volume of the hole surface is then approximately 0.4 holes/atom since the total volume has to be zero.

Armed with a picture of the Fermi surface of Ag and Pd let us now consider the alloy Fermi surfaces. For the $Ag_{0.8}Pd_{0.2}$ alloy it is clear that the Fermi surface consists of a single well defined sheet, the smearing $\gamma_{\vec{k}}$ is typically 1/30th of the Brillouin zone dimension. The volume beneath this sheet of Fermi surface is almost exactly 0.8 electrons/atoms which has to be the case if there is little weight in $A^B(\vec{k},\varepsilon_F)$ outside the smeared δ-function peaks. Clearly, for this alloy the neck around the L-point has just about pinched off. In the $Ag_{0.5}Pd_{0.5}$ alloy the Fermi surface again consists of a single well defined sheet, although near to the X-point there is an indication of an increase in $A^B(\vec{k},\varepsilon_F)$ which results from the fact that for this alloy the Fermi energy is in

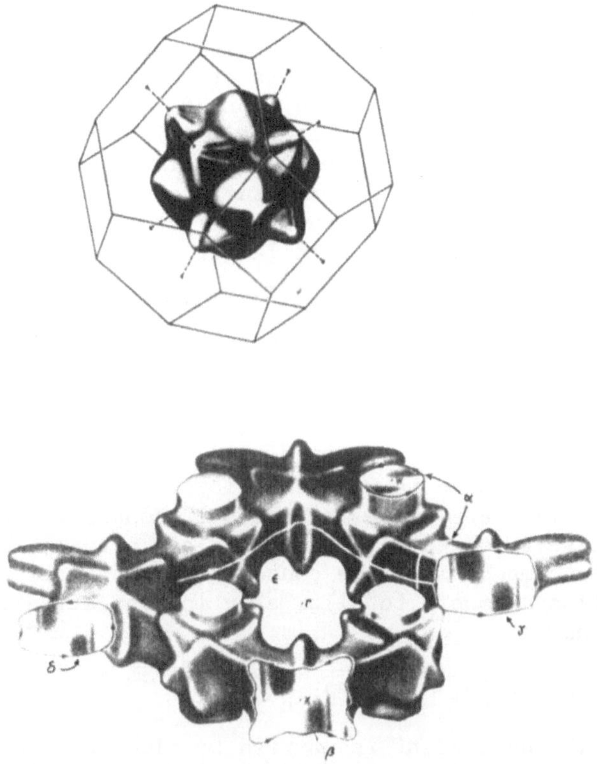

Fig. V.2. Reconstruction of the major sheets of Fermi surface in Pd.
Upper : Γ-centred sheet. The electrons are primarily d-electrons that have high velocities.
Lower : Jungle gym. The electrons are primarily d-electrons that have low velocities.

very close proximity to the top of the Pd-sub-band (recall fig. IV . 7). For this alloy there is no contact of the Γ-centered sheet of Fermi surface with the Brillouin zone boundary. The volume beneath this sheet is almost exactly 0.5 electrons/atoms as might be expected. For the $Ag_{0.2}Pd_{0.8}$ alloy there are evidently two sheets of Fermi surface a well defined closed Γ-centered sheet and a very smeared out sheet which is actually forming along the XW-directions. Obviously this latter sheet is the precurser to the jungle-gym of pure Pd. The reason that this sheet of Fermi surface is so smeared out is because it arises from the intersection of the flat d-bands at the top of the Pd-sub-band with ε_F. Since the angle of intersection is so acute even a small smearing of the bands in energy will give rise to a large smearing in $\vec{k}$. The concentration evolution of the various sheets of Fermi surface is now clear. If we begin with pure Ag, the addition of Pd results in the gradual shrinkage of the Γ-centered sheet to its size in pure Pd, however, at all concentrations it remains well defined. At a concentration of about 50% Pd the sheet of Fermi surface which will eventually expand into the jungle-gym of pure Pd first appears. At all times $\gamma_{\vec{k}}$ on the jungle-gym is very much larger than for the Γ-centered sheet.

In fig. V.3 we show the concentration variation in the ΓXKXΓ-plane of the Ag_cPd_{1-c} Γ-centered sheet of Fermi surface across the complete alloy diagram. This clearly illustrates the fact that this sheet shrinks on the addition of Pd, furthermore, this sheet goes from being convex in the Γ K-direction in Ag-rich alloys to being concave in Pd-rich alloys. The Fermi surfaces shown in fig. IV.3 are those for the self-consistent potentials. Although the self-consistent potentials are actually calculated at only three compositions the concentration variation is sufficiently small and smooth to allow easy interpolation to other concentrations.

Residual Resistivities of Ag_cPd_{1-c}

In metals the current is carried by electrons on the Fermi surface. Thus the residual resisitivity ρ_R, the resistivity of a random alloy at T = 0°K gives a direct measure of the degree to which the Fermi surface is smeared out by the disorder in the system. Since at each concentration it is a single number the information about the topology and lifetimes of the electrons on the highly convoluted. However, as we shall see for the Ag_cPd_{1-c} alloy the residual resistivity measurements confirm in great detail our picture of the behaviour of the electrons at the Fermi surface.

The study of the residual electrical resistivity random alloys has a long history. In a series of papers published in the mid 30's Mott applied the concepts of the then new quantum mechanics to obtaining an understanding of the electronic states and transport properties of transition metals and alloys. In these papers, he proposed both the rigid band model (RBM) (a close relative of the VCA) for discussing the concentration dependence of the electronic densities of states functions in random alloys and the

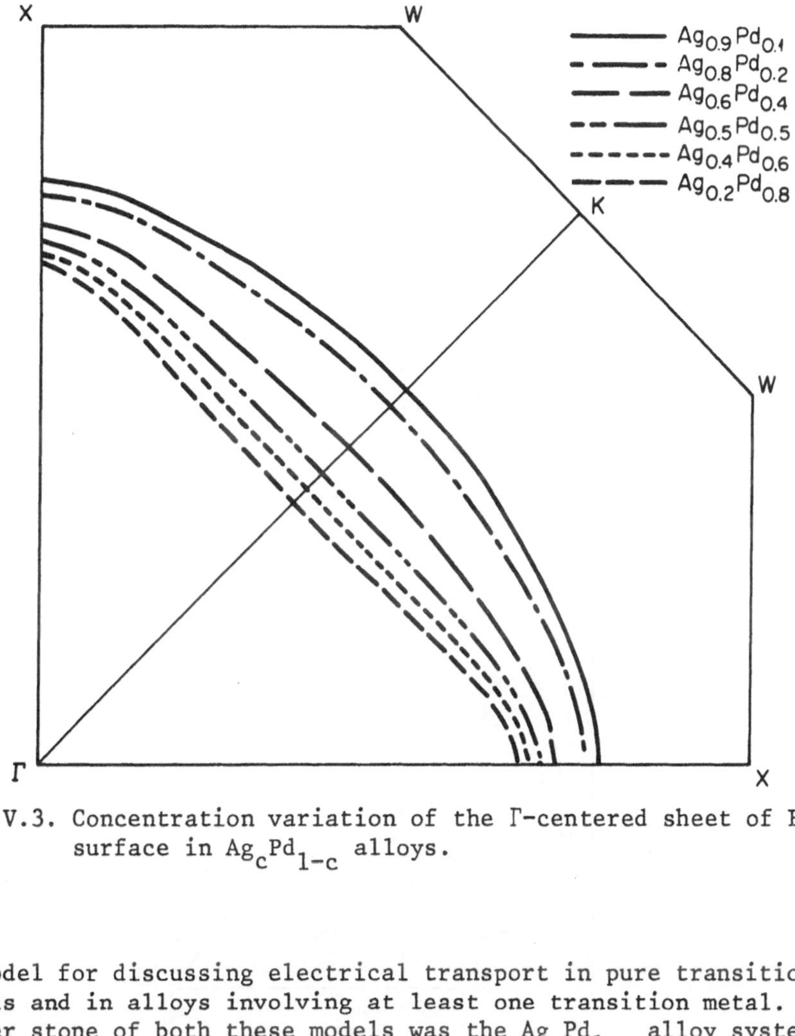

Fig. V.3. Concentration variation of the Γ-centered sheet of Fermi surface in Ag_cPd_{1-c} alloys.

sd-model for discussing electrical transport in pure transition metals and in alloys involving at least one transition metal. A corner stone of both these models was the Ag_cPd_{1-c} alloy system. Various experimental results on this system, in addition to the residual resistivity were correlated using these models ; magnetic susceptibility, electronic specific heat coefficient, thermo-electric powere etc. The rigid band model as conceived by Mott for the Ag_cPd_{1-c} alloy system is illustrated in fig. V.4a. Mott pictured transition metals as having two kinds of electrons s-electrons which form a broad low density of states conduction band and d-electrons which form a narrow high density of states band which may overlap the s-band. Mott pictured, correctly, Pd to have an incompletely filled, high density of states d-band complex superimposed on a low density of states sp-band, the Fermi energy ${}_F$ lying at the upper edge of the d-band complex leaving ~ 0.5 d-holes per Pd site. He pictured Ag to have a low lying, filled, d-band and a sp-band which contained one electron. On adding Ag to Pd he

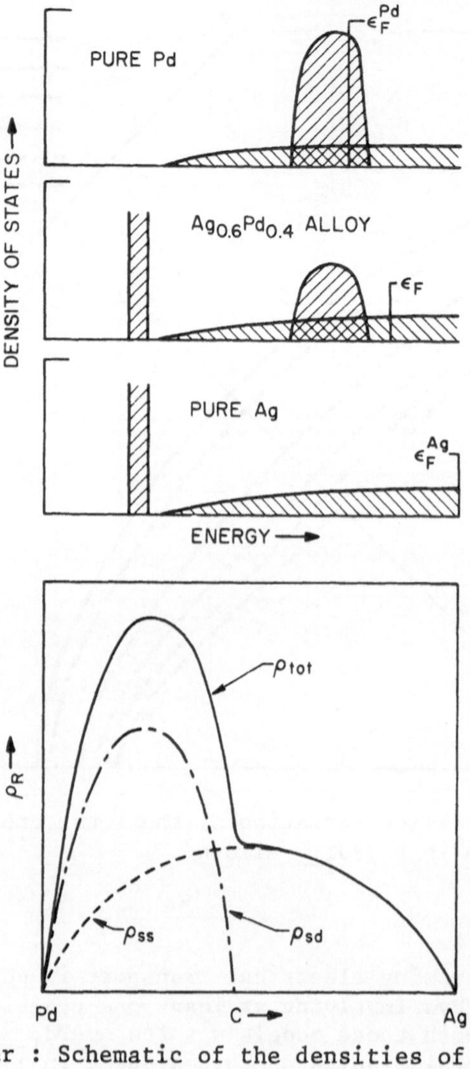

Fig. V.4. a) Upper : Schematic of the densities of states in Pd, Ag and alloys of them according to Mott's rigid band model.
b) Lower : The various contribution to the residual resistivity of Ag_cPd_{1-c} alloys according to the sd-model.

pictured, incorrectly, that the extra sp-electron per Ag site fills the Pd d-holes resulting in the experimentally observed rapid decrease in the density of states at the Fermi energy $n(\varepsilon_F)$, which continues until all the d-holes are filled at which time ε_F is in the low density of states sp-band and $n(\varepsilon_F)$ is typical of a simple metal. Beyond this concentration, $n(\varepsilon_F)$ remains roughly constant and ε_F rises rapidly away from the d-band complex. It is worth stressing that this model of Ag_cPd_{1-c} is not the common band model often attributed to Mott. Even in 1935 Mott, unlike many people who followed him, was sufficiently clever to realise that the tightly bound d-electrons of the different alloy constituents would not form a common virtual crystal like band. He only assumed that the s-electron would form a common band. Whilst having some of the basic elements of the KKR-CPA calculations the RBM is incorrect in several important respects , in particular it predicts that once the Pd-band is filled a further increase in Ag content will result in the centres of gravities of the Ag and Pd d-bands rapidly receeding from ε_F. This is clearly inconsistant with both UPS experiments and the KKR-CPA. A further problem with the model is that it totally ignores the effects of sd-hybridization. In the sd-model used to describe the transport properties, Mott postulated that the current is carried entirely by the s-electrons. The residual resistivity then arises from the scattering of these electrons by potential fluctuations either into other mobile s-states or when available into the d-holes at Pd-sites which are assumed to act as traps. If it is assumed that the s-electrons propogated in the average virtual crystal potential then the scattering potentials seen at A and B sites are :

$$v_{A(B)} - v_{Av} = \begin{cases} (1-c)(v_A - v_B) & \text{A-site} \\ -c(v_A - v_B) & \text{B-site} \end{cases} \qquad \text{V.1}$$

The scattering cross-section at the A and B sites are then given by $(1-c)^2M^2$ and c^2M^2 respectively where M is the matrix element $\langle\psi_f|v_A - v_B|\psi_i\rangle$ for scattering from the initial to the final state. Thus the contribution ρ_{ss}, to the resistivity which results from scattering from a s-state into another s-state is given by

$$\rho_{ss} \propto c(1-c)^2\, M^2\, n_s^A(\varepsilon_F) + (1-c)c^2\, M^2\, n_s^B(\varepsilon_F) \qquad \text{V.2}$$

where n_s^A and n_s^B are the densities of s-states at the Fermi energy at the A and B sites respectively. Since the s-band is assumed common $n_s^A = n_s^B = n_s$ the total s-density of states at ε_F and

$$\rho_{ss} = A\, c(1-c)\, M^2\, n_s(\varepsilon_F) \qquad \text{V.3}$$

where A is a constant of proportionality. The contribution ρ_{sd} to the resistivity which results form the scattering of a s-electron into a d-hole is given by an equation exactly analogous to Eq. V.2 except n_d^A and n_d^B the densities of states of the d-holes at the A and B sites respectively replace n_s^A and n_s^B. However, in Mott's rigid band model for Ag_cPd_{1-c} alloys, where it is assumed that there are never any d-holes at Ag sites ρ_{sd} is given by

$$\rho_{sd} = B\,c^2(1-c)M^2\,n_d(\varepsilon_F) \qquad \text{V.4}$$

where $n_d(\varepsilon_F)$ is the density of d-states at ε_F and B is another constant of proportionality. The total resistivity ρ_{tot} from both mechanisms is then given by

$$\rho_{tot} = \rho_{ss} + \rho_{sd} = Ac(1-c)M^2 n_s(\varepsilon_F) + Bc^2(1-c)M^2 n_d(\varepsilon_F) \qquad \text{V.5}$$

The contribution ρ_{ss} is simply Nordheims rule. If it is assumed that $n_s(\varepsilon_F)$ is fairly independent of concentration then ρ_{ss} gives a contribution proportional to $c(1-c)$. This contribution is shown as the dotted curve of fig. V.4b. If it is assumed that the $n_d(\varepsilon_F)$ falls from a lager value in Pd to zero at $c \sim 0.5$, as in fact can be inferred from the low temperature specific measurements of fig. III.2, then ρ_{sd} gives a contribution to the resistivity which has the form shown by the dashed curve in fig. V.4b. The total resistivity then has the form of the solid line, which as we shall see later is the correct shape. In fact using this model Dugdale and Guenault obtained an excellent fit to experimental data by choosing suitable values of the constant A and B and by inferring the concentration variation of $n_s(\varepsilon_F)$ and $n_d(\varepsilon_F)$ from the low temperature specific heat data (unfortunately not corrected for the electron phonon-enhancement which as we have already seen is substantial). Using the same model they were also able to understand the room temperature coefficient of resistance, which becomes negative in mid-concentration range, as well as the early (though certainly not the most recent and complete) measurements of the thermoelectric power. Thus we are in the interesting position of having a semi-empirical theory which works very well but which has little or no theoretical underpinning. Thus, given the very detailed description of the Fermi surfaces provided through $A^B(\vec{k},\varepsilon)$ it is interesting to attempt to reinterpret the experimental results in terms of these.

In the presence of an applied electric field $\vec{E}$ the Boltzmann equation describing the electron distribution function take the form[92]

$$-\frac{\partial f^0_{\vec{k}}}{\partial \varepsilon}\,\vec{v}_{\vec{k}}.e\vec{E} = \sum_{k'} \{g_{\vec{k}}\,\Gamma_{\vec{k}\vec{k}'} + g_{\vec{k}'}\,\Gamma_{\vec{k}'\vec{k}}\} \qquad \text{V.6}$$

where $f^0_{\vec{k}}$ is the equilibrium distribution function and $g^0_{\vec{k}}$ describes the deviation from equilibrium and $\vec{v}_{\vec{k}}$ is the electron velocity. The first term on the right of Eq. V.6 is called the 'scattering out' term and describes the scattering from a state $\vec{k}$ to a state $\vec{k}'$ which, in the present problem, results from the disorder. The second term describes the scattering back from the state $\vec{k}'$ to the state $\vec{k}$ and is often called the 'scattering in' term. In calculations of the temperature variation of the electrical resistivity of pure Pd, where the scattering term on the r.h.s. of Eq. V.6 results from the electron-phonon interaction Pinski, Butler and Allen reached two conclusions which are of great importance for the present study. These were : a) most of the current (> 80%) is carried by electrons on the Γ-centered sheet, b) the 'scattering in' term gives only a small contribution to the resistivity. If we now assume that these two conclusions also apply when the scattering results from disorder and neglect the scattering in term and, when performing Fermi surface integrations, we perform them only over the Γ-centered sheet of Fermi surface, it is then a simple matter to solve the Boltzmann equation and to calculate the resistivity. Since the jungle-sym sheet of Fermi surface is rapidly broadened by disorder and in addition shrinks away on the addition of Ag, a) is probably more true in the alloys than in pure Pd. However, in Ag-rich alloys where there is only a single sheet of Fermi surface and scattering can only arise between different points on the same sheet of Fermi surface, b) is certainly less the case in the alloys than in pure Pd ; we will return to this point later. Neglecting the scattering in term and writing $\Gamma_{\vec{k}\vec{k}'} = 1/\tau_{\vec{k}}\,\delta_{\vec{k}\vec{k}'}$, where $\tau_{\vec{k}}$, is the electron lifetime the residual resistivity is given by[92]

$$1/\rho_R = \delta_R = \frac{e^2}{12\pi^3}\int \tau_{\vec{k}} v_{\vec{k}} dS \qquad \text{V.7}$$

where the integral is over the Fermi surface.

By obtaining constant energy surfaces in the form of a dense mesh of points in k-space for $\varepsilon = \varepsilon_F$ and $\varepsilon = \varepsilon_F \pm \delta\varepsilon$ it is possible to calculate $v_{\vec{k}}$. The electron life-time can be calculated by setting $\vec{k}$ to a point on the Fermi surface and then calculating $A^B(k,\varepsilon)$ as a function of ε. The width in energy $\gamma^{\varepsilon}_{\vec{k}}$ of the peak in is then related to the electron life-time $\tau_{\vec{k}}$ by $\gamma^{\varepsilon}_{\vec{k}} = X/\tau_{\vec{k}}$. If the peaks in $A^B(k,\varepsilon)$ are Lorentzian, which is the case for the Γ-centred sheet in Ag_cPd_{1-c} alloys, the width of the peak in $\vec{k}$, $\gamma^{k}_{\vec{k}}$ is related to width of the peak in ε, $\gamma^{\varepsilon}_{\vec{k}}$ through $\gamma^{\varepsilon}_{\vec{k}} = \gamma^{k}_{\vec{k}}\, d\varepsilon/dk$, thus obviating the need to evaluate $A^B(\vec{k},\varepsilon)$ as a function of ε for all points on the Fermi surface.

The values of ρ_R calculated using Eq. V.7 are shown[93] in fig. V.5. The input potentials to these calculations were the self-consistent-KKR-CPA potentials of the previous section. Thus the remarkable agreement between theory and experiment has to be viewed in the light of the fact that the only inputs to the calculation were the two atomic numbers, the measured lattice spacing and crystal structure. The characteristic asymmetry, which so intrigued Mott is properly reproduced and for Pd rich alloys the calculated values of ρ_R are within a few percent of experiment. For Ag-rich alloys the agreement is worse (~ 30%). This is largely due to the neglect of the scattering in term. The effect of this

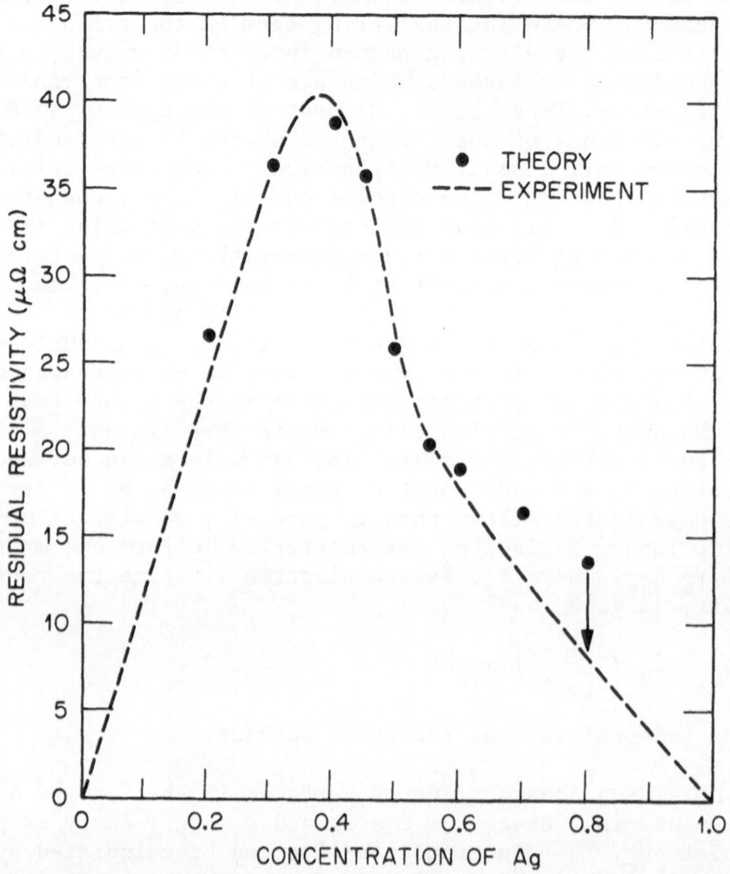

Fig. V.5. Experimental and theoretical residual resistivities of Ag_cPd_{1-c} alloys.

omission can be estimated on the basis of a formula appropriate to a low concentration of scatterers in a simple or noble metal where the ratio of transport life-time τ^t to the quasi-particle life-time τ^{qp} is given by

$$\frac{\tau^{qp}}{\tau^t} = \sum_\ell \frac{(\ell+1)\sin^2(\Delta\eta_{\ell+1} - \Delta\eta_\ell)}{\sum_\ell (2\ell+1)\sin^2\Delta\eta_\ell} \qquad \text{V.8}$$

where $\Delta\eta_\ell = \eta_\ell^{Ag} - \eta_\ell^{Pd}$. Using phase shifts appropriate to $Ag_{0.2}Pd_{0.8}$

$\tau^{qp}/\tau^{t} = 0.63$. When this correction factor is applied to the calculated ρ_R for Ag-rich alloys the results shown by 'bars' in fig. V.5 are obtained.

Having seen that the experimental results can be so nicely reproduced on the basis of the Fermi surfaces obtained from KKR-CPA calculations it is interesting to look at the details of the calculations in order to see what if anything survives of the sd-model beyond the obvious connection that the s-electrons of Mott are electrons on the Γ-centred sheet (which actually have a majority d-component ; 10%-s ; 10%-p ; 80%-d but nonetheless have high velocities) and that the d-electrons of Mott are electrons on the 'jungle-gym'.

In the upper two frames of fig. V.6 we have plotted (solid lines) as functions of concentration the Fermi surface averages γ^{ε} and $\langle v^2\rangle^{1/2}$ of the scattering rate $\gamma^{\varepsilon}_{\vec{k}}$ and velocity $v_{\vec{k}}$. For comparison we have also plotted (dashed curves) the form of γ^{ε} implied by the sd-model and obtained on the basis of the RBM. Clearly the values obtained in the KKR-CPA bear little resemblance to the sd-model predictions. The KKR-CPA calculations show that for low concentrations ($c \leqslant 0.2$ and $c \geqslant 0.8$) the effects on γ^{ε} of adding Pd impurities to Ag or Ag impurities to Pd are similar i.e., $|(d\gamma^{\varepsilon}/dc)_{c=0}| \simeq |(d\gamma^{\varepsilon}/dc)_{c=1}|$. If it is to agree with experiment, the Mott model requires $|(d\gamma^{\varepsilon}/dc)_{c=0}| \approx 2|(d\gamma^{\varepsilon}/dc)_{c=1}|$ since $\langle v^2\rangle^{1/2}$ is assumed constant and $|(\partial\rho_R/\partial c)_{c=0}| \simeq 2|(\partial\rho_R/\partial c)_{c=1}|$. In our model, it is the fact that $(\langle v^2\rangle^{1/2})_{c=0} \approx 2(\langle v^2\rangle^{1/2})_{c=1}$ which accounts for the differences in ρ_R at the two ends of the concentration range. This in turn is related to the change in character, from <u>d</u> to <u>sp</u>, of the electrons on the Γ-centered sheet as Ag is added to Pd.

A further striking feature of our calculated γ^{ε} and $\langle v^2\rangle^{1/2}$ curves is the sharp dip near $c = 0.5$ (the density of states on the Γ-centered sheet has a corresponding hump). These features do not appear, however, in the ratio $\gamma^{\varepsilon}_{\vec{k}}/v_{\vec{k}} = \gamma^{k}_{\vec{k}}$ which is physically the inverse of the mean-free path and which is shown in the lower frame of fig. V.6. Features similar to those in the concentration variation of γ^{ε} and $\langle v^2\rangle^{1/2}$ also occur in the energy variation of $\gamma^{\varepsilon}_{\vec{k}}$ and $v_{\vec{k}}$ for energies near the top of the Pd d-band complex. An example of this behaviour is shown in fig. V.7 where we show in the upper frame the energy variation of $(d\varepsilon/dk)$ in the neighbourhood of ε_F for the $Ag_{0.4}Pd_{0.6}$ alloy, an alloy just to the left of the structure in fig. V.7. For this alloy ε_F lies just below the structure in $(\partial\varepsilon/\partial k)$. If we imagine the effect of adding more Ag to the alloy as simply increasing ε_F then it is clear that the structure in $\langle v^2\rangle^{1/2}$ is related to this structure. Also shown in the upper frame of fig. V.7 is the magnitude of the smearing $\gamma^{k}_{\vec{k}}$ in $\vec{k}$ of $A^B(\vec{k},\varepsilon_F)$. In fig. V.7 all the quantities are evaluated

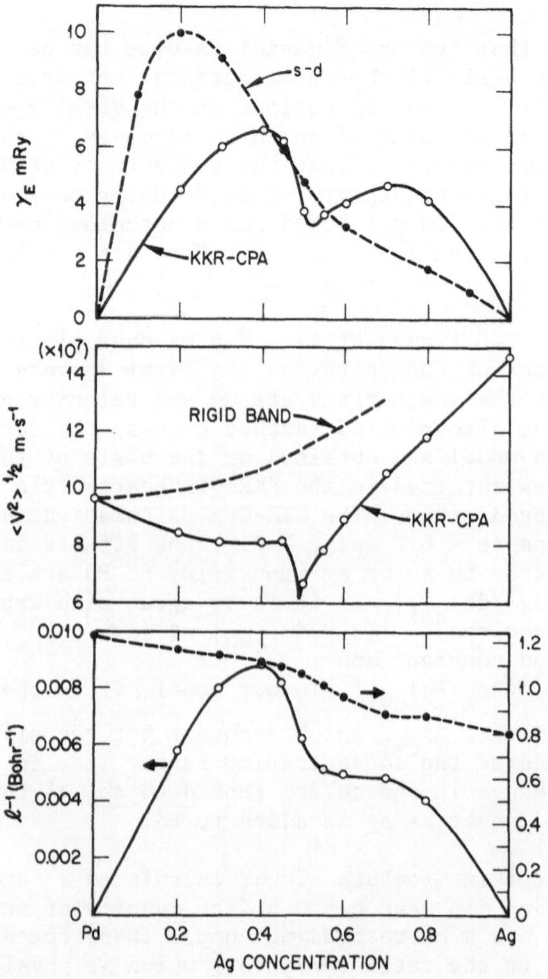

Fig. V.6. Fermi-surface averages of various quantities involved in calculating the resistivity of Ag_cPd_{1-c} alloys. Upper frame : $\langle\gamma^{\varepsilon}\rangle$ obtained in our KKR-CPA calculation ; also shown is the form of γ^{ε} obtained for the Mott model. Middle frame : rms Fermi velocity obtained from our KKR-CPA calculation (solid line), and from the rigid-band model (dashed line). Lower frame : inverse of mean free path (solid line) and surface area of Γ-centered sheet (dashed line).

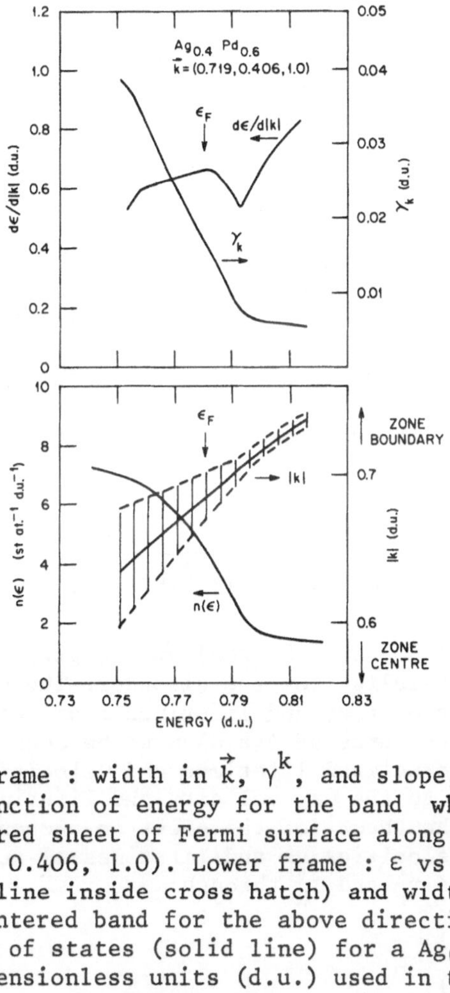

Fig. V.7. Upper frame : width in $\vec{k}$, γ^k, and slope, $v_{\vec{k}} = |d\varepsilon/dk|$, as a function of energy for the band which forms Γ-centered sheet of Fermi surface along the direction (0.719, 0.406, 1.0). Lower frame : ε vs $\vec{k}$ relation (solid line inside cross hatch) and width (cross hatch) of Γ-centered band for the above direction and total density of states (solid line) for a $Ag_{0.45}Pd_{0.55}$ alloy. The dimensionless units (d.u.) used in this figure may be converted to atomic units by multiplying by $(2\pi/a)^p$ where a is the lattice constant (7.494 a.u.) and p = 1 for k, γ^k, and dε/dk;p = -2 for n(ε).

for $\vec{k}$-vectors along a direction joining the Brillouin zone centre to the point (0.719, 0.406, 1.0). Like the mean free path plotted in the lower frame of fig. V.6 $\gamma_{\vec{k}}^{k}$ does not have the structure seen in $(d\varepsilon/dk)$ etc. If we recall that $\gamma^{\varepsilon}_{\vec{k}} = \gamma^{k}_{\vec{k}}(\partial\varepsilon/dk)$ then it is also clear where the structure in γ^{ε} in fig. V.6 comes from. In the lower frame of fig. V.7 are plotted the densities of states and, by the line in the centre of the cross hatched area, a section of the 'energy band' which gives rise to the Γ-centered sheet of Fermi surface ; the cross hatch gives $\gamma^{k}_{\vec{k}}$. Clearly the structure in $(\partial\varepsilon/dk)$ etc. occurs at energies where the densities of states is falling rapidly i.e. where the jungle-gym structure is just about to disappear. Thus the effects seen in $(d\varepsilon/dk)$ can be viewed as hybridization effects having to do with a slight flattening of the free-electron derived sp-band as it begins to interact with the d-band complex in a random alloy. The large values of $\gamma^{k}_{\vec{k}}$ at energies where $n(\varepsilon)$ is large clearly shows that the short mean-free path of electrons on the Γ-centered sheet in Pd rich alloy is related to the availability of states on the 'jungle-gym' into which it is possible to scatter. From what has been said it should be clear that these effects are peculiar to random alloys and that they are quite general features which will manifest themselves whenever ε_F lies in the vicinity of the top of a d-band complex. Another way to see this is to consider a disordered system Green's function of the form

$$G(\vec{k},\varepsilon) = \frac{1}{\varepsilon - \varepsilon_k - [\Delta(\vec{k},\varepsilon) + i\gamma(\vec{k},\varepsilon)]} \qquad \text{V.9}$$

where $\Delta(\vec{k},\varepsilon)$ and $\gamma(\vec{k},\varepsilon)$ are the real and imaginary parts of the self-energy which results from the disorder. The point is that when calculating velocities and lifetimes from such a Green's function the energy dependence of $\Delta(\vec{k},\varepsilon)$ must be accounted for, thus giving rise to renormalised lifetimes and velocities in a manner entirely analogous to the way velocity and lifetime renormalisation effects occur in the theory of the electron-phonon interaction. The renormalised velocities $v_{\vec{k}}$ and lifetimes $\tau_{\vec{k}}$ are related to unrenormalised ones $v^{0}_{\vec{k}}$ and $\tau^{0}_{\vec{k}}$ through

$$\begin{aligned} v_{\vec{k}} &= v^{0}_{\vec{k}}/(1+\lambda) \\ \tau_{\vec{k}} &= \tau^{0}_{\vec{k}}(1+\lambda) \end{aligned} \qquad \text{V.10}$$

where the enhancement parameter $\lambda = -\,\partial\Delta(\vec{k},\varepsilon)/\partial\varepsilon$. It is then the rapid variation in $\Delta(\vec{k},\varepsilon)$ as a function of energy which in turn arises from the rapid fall off of the density of states at the top of the d-band complex which gives rise to the structure in $v_{\vec{k}}$ and $\tau_{\vec{k}}$ calculated in the KKR-CPA. It is the renormalised velocities and lifetimes which are generated automatically by the KKR-CPA. The reason that these renormalisation effects do not show up in the resistivity (and mean-free-path) is now clear, $\rho_R \sim v_{\vec{k}}\tau_{\vec{k}}$ and the normalisation effects cancel in the product and so not affect the residual resistivity.

Alloy Fermiology by Positron Annihilation

Having gained considerable confidence in the existence of Fermi surfaces in random alloys, one can now reasonably search for methods by which the topology and sharpness of the Fermi surface can be measured more directly. Such Fermi surface studies played a very important role in the development of one electron theories of pure metals. The fact that Fermi surface topologies and dimensions could be measured and that these compared favourably with Fermi surfaces computed on the basis of one electron band theory, gave band theory practitioners confidence that, despite being constantly harangued by many body theorists, they were not completely out of court. Unfortunately, from the point of view of confirming our predictions about alloy Fermi surfaces most of the methods used in Fermiology experiments on pure metals, de Haas-van Alphen effect, magneto-acoustic effect , de Haas-Shubnikov effect etc. require that the electron-mean-free path be extremely long which is not the case for random alloys, except in the extreme dilute limit. In the de Haas-van Alphen experiment for reasonably sized magnetic fields resistivity ratios, $\rho_{273}/\rho_{4.2}$, the ratio of the resistivity at 273°K to the resistivity at 42°K, of ~ 30 are required. For most concentrated alloys $\rho_{273}/\rho_{4.2} \sim 1$. Thus, if we are to study the Fermi surface of random alloys we must turn to experimental techniques which do not require a long mean-free-path. Out of a range of possible experimental techniques which include angle-resolved photoemission, Compton profile and positron annihilation, it is the positron annihilation experiment which offers the best possibility at the present time. The use of positron annihilation to measure Fermi surfaces has a long history. However, when the major focus of interest was in pure metals, where good single crystals with high resistivity ratios could be prepared easily, positron annihilation could not complete with the more conventional methods because of its lack of resolution.

When a fast positron is injected into a sample it is rapidly thermalized. The thermalized positron then annihilates with an electron yielding two 2γ-rays as the most probable process in a bulk solid free from defects. In this 2γ-annihilation the two photons are not quite colinear, the deviation from colinearity resulting from the fact that the momentum of the annihilating electron is in general finite. Thus, measurement of the departure from colinearity can be used to infer the electron momentum. This is what is done in 2γ-ACAR (angular correlation of annihilation radiation)-experiments. These experiments and the basic theory have been fully covered in a number of excellent review articles[94,95,96].

The basic geometry of what is called a 2-dimensional ACAR experiment is shown in fig. V.8, which is taken from the review article by Berko[95]. Coincidences are counted in the planes c_1 and c_2 and the two components (p_y, p_z) of the momentum are measured,

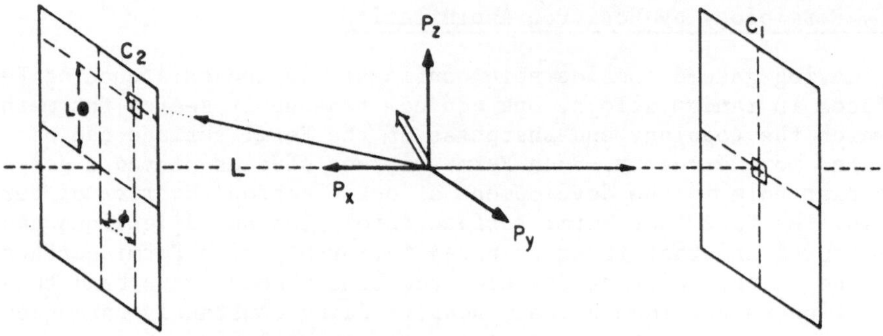

Fig. V.8. Geometry of 2γ-ACAR experiments. Taken from Berko[95].

the third component of momentum appears as a Doppler shift and is in general not measured. Thus the experiment measures a coincidence rate which is proportional to the 2-D electron-positron momentum density $N^{2\gamma}(p_y,p_z)$ which is in turn related to the full 3-D electron-positron momentum density by

$$N^{2\gamma}(p_y,p_z) = \int_{-\infty}^{\infty} N^{2\gamma}(\vec{p})dp_x \qquad V.11$$

Currently there are several 2-D ACAR machines in operation and these have been used to measure $N^{2\gamma}(p_y,p_z)$ in a number of systems including some alloy systems. Some of this mountinous information has been back and convoluted[97] to give actual sectors through the full 3-D $N^{2\gamma}(\vec{p})$. It is this quantity which contains the clearest information about the Fermi surface. In principle it is the one positron-many electron many body system for which the momentum density is measured. In an independent particle model this simply reduces to[94,95]

$$N^{2\gamma}(\vec{p}) = \frac{1}{N}\sum_{n\ occ} |\int d^3r\ e^{i\vec{p}.\vec{r}}\ \psi_n^e(\vec{r})\psi^p(\vec{r})|^2 \qquad V.12$$

where ψ^e and ψ^p are the wavefunctions corresponding to the electrons and the positron respectively. In Eq. V.12 the sum extends over all the occupied electron states. If we now postulate that the position is equally likely to be anywhere in the crystal, $\psi^p(\vec{r})$ = constant, then Eq. V.12 is simply related to the electron momentum density.

$$n(\vec{p}) = \frac{1}{N}\sum_{n\ occ} |\int d^3r\ e^{-i\vec{p}.\vec{r}}\ \psi_n^e|^2 \qquad V.13$$

Thus to a first approximation 2-D ACAR measures the electron momentum distribution. In fig. V.9a we show $n(\vec{p})$ corresponding to metallic Pd for $\vec{p}$ along the 001-direction calculated by the KKR method[98]. Apart from the general decrease in $n(\vec{p})$ with increasing $\vec{p}$ the most striking features are the breaks which occur whenever $\vec{p} = \vec{k}_F + \vec{K}_n$ where $\vec{k}_F$ is a wave vector corresponding to a point on the Fermi surface of Pd and $\vec{K}_n$ is a reciprocal lattice vector. These Fermi surfaces break in $n(\vec{p})$ and can be understood simply in terms of the band structure of Pd (see inset). Each band below ε_F gives a contribution to $n(\vec{p})$ thus as a particular band crosses ε_F we loose its contribution resulting in a discontinuous drop in $n(\vec{p})$. In fact for $\vec{p}$ along the 001-direction only the bands having Δ_1 - symmetry (solid) can contribute to $n(\vec{p})$ on symmetry grounds. Thus along this direction we only see breaks in $n(\vec{p})$ corresponding to the Γ-centered sheet of Fermi surface and we do not see any breaks at all at values of $\vec{p}$ corresponding to the k-vector where the bands which give rise to the jungle-gym cross the Fermi energy.

From what has already been said about the Fermi surface in random alloys it should be clear that, for the parts of the Fermi surface which remain sharp eg. the Γ-centered sheet in Ag_cPd_{1-c}, $N_{1-c}Cu_{1-c}$ etc. the form of the momentum density should be quite similar to that of the pure metal. This is indicated schematically in fig. V.9b where the shading on the inset band structure indicates the smearing of the bands resulting from disorder. Obviously the breaks at the values of $\vec{p}$ corresponding to Fermi wave vectors will no longer be sharp but will have a width related to $\gamma_{\vec{k}}^{k}$ the smearing of the Fermi surface in $\vec{k}$.

In terms of the Greens functions discussed previously the electron momentum density is given by

$$n(\vec{p}) = -\frac{1}{N\pi}\int_0^{\varepsilon_F} d\varepsilon \int d^3r \int d^3r' \; \mathrm{Im}\; e^{i\vec{p}\cdot(\vec{r}-\vec{r}')}\; G(\vec{r},\vec{r}',\varepsilon) \qquad \text{V.14}$$

Thus in a random alloy the configurationally averaged momentum density is given by

$$\overline{n}(\vec{p}) = -\frac{1}{\pi N}\,\mathrm{Im}\int_{-\infty}^{\varepsilon_F} d\varepsilon < \int d^3r \int d^3r' \; e^{i\vec{p}\cdot(\vec{r}-\vec{r}')} G(\vec{r},\vec{r}',\varepsilon)>\} \qquad \text{V.15}$$

$$= \int_{-\infty}^{\varepsilon_F} d\varepsilon\; \overline{A}(\vec{p},\varepsilon) \qquad \text{V.16}$$

where

$$\overline{A}(\vec{p},\varepsilon) = -\frac{1}{\pi N}\,\mathrm{Im}<G(\vec{p},\vec{p}',\varepsilon)> \qquad \text{V.17}$$

is the momentum spectral function. Evidently the Bloch spectral function of Sec. III is related to the momentum spectral function through

$$A^B(\vec{k},\varepsilon) = \sum_{K_n} \overline{A}(\vec{k} + \vec{K}_n,\varepsilon) \qquad \text{V.18}$$

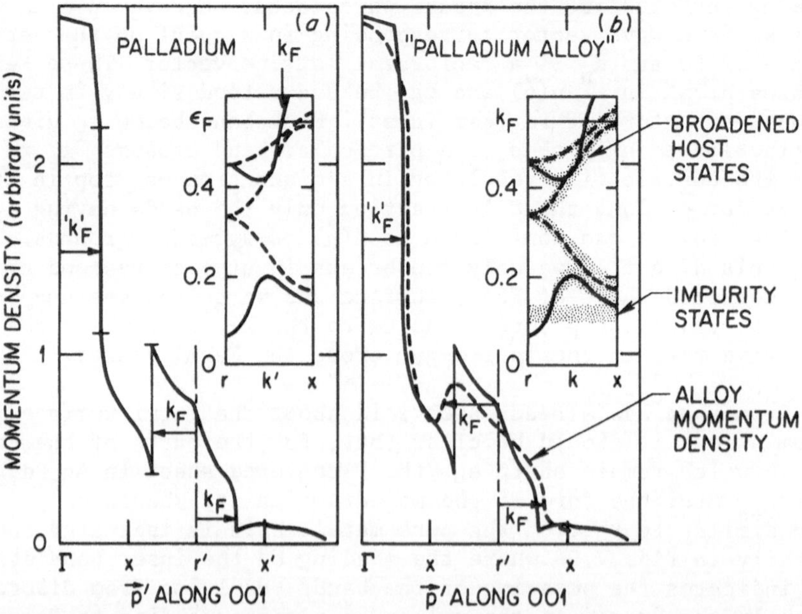

Fig. V.9. a) Electron momentum density of Pd along the 001 direction and the band structure of pure Pd along Γ-X showing the bands (solid) which contribute to the momentum density.
b) The same as a) except for a hypothetical random alloy.

Following the method outlined in the derivation of the Bloch spectral function it is possible to derive an expression for the momentum spectral function within the KKR-CPA the result is[99]

$$\overline{A}(\vec{p},\varepsilon) = -\frac{1}{\pi}\,\mathrm{Im}\sum_{LL'}\{F^{cc}_{LL'}(\vec{p},\varepsilon)\tau^{c}_{LL'}(\vec{p},\varepsilon) + [F^{c}_{LL}(\vec{p},\varepsilon) - F^{cc}_{LL}(\vec{p},\varepsilon)] \cdot \tau^{c,00}_{LL}(\varepsilon)\delta_{LL'}\} \qquad \text{V.19}$$

where $\tau^{c}_{LL'}(p,\varepsilon)$ and $\tau^{c,00}_{LL}$ have their usual definitions and can be obtained from the results of a KKR-CPA calculation. In analogy with Eqs. III.3 and III.17 the matrix elements $F^{c}_{LL'}(\vec{p},\varepsilon)$ and $F^{cc}_{LL'}(\vec{p},\varepsilon)$ have the form

$$F^{c}_{LL'}(\vec{p},\varepsilon) = \sum_{\alpha,L_1} c_\alpha Z_L(\vec{p},\varepsilon)Z_{L_1}(-\vec{p},\varepsilon)D^{\alpha}_{L_1,L'}(\varepsilon) \qquad \text{V.20}$$

and

$$F^{cc}_{LL'}(\vec{p},\varepsilon) = \sum_{\alpha,\beta}\sum_{L_1,L_2} c_\alpha c_\beta D^\alpha_{LL_1}(\varepsilon)Z^\alpha_{L_1}(\vec{p},\varepsilon)Z^\beta_{L_2}(-\vec{p},\varepsilon)D^\beta_{L_2,L'}(\varepsilon) \qquad \text{V.21}$$

where, as previously, α = A or B and β = A or B, c_A and c_B are the concentrations of the A and B species and the matrix elements $D^A_{LL'}(\varepsilon)$ and $D^B_{LL'}(\varepsilon)$ are given by Eq. II.41. The matrix elements $Z^\alpha_L(\vec{p},\varepsilon)$ are the Fourier transforms of the $Z^\alpha_L(\vec{r},\varepsilon)$ defined previously Eq. II.8 and are given by

$$Z^\alpha_L(\vec{p},\varepsilon) = 4\pi \sum_{L'} i^{\ell'} Y_{L'}(\hat{p})\int d^3r Z^\alpha_L(\vec{r},\varepsilon) j_{\ell'}(pr)Y_L(\hat{r})Y_{L'}(\hat{r}) \qquad \text{V.22}$$

Evidently all the information which is required to calculate $\overline{A}(\vec{p},\varepsilon)$ is readily available at the end of a KKR-CPA calculation. The configurationally arranged momentum density within the CPA $\overline{n}(\vec{p})$ can then be obtained from $\overline{A}(\vec{p},\varepsilon)$ by Eq. V.16.

A comparison between the formulas for the momentum spectral function and Bloch spectral function reveals that $\overline{A}(\vec{p},\varepsilon)$ merely takes the peaks of $A_B(\vec{k},\varepsilon)$ apart and scatters the bits about in momentum space (with weights that add up to unity). The expressions for $A_B(\vec{k},\varepsilon)$ and $\overline{A}(\vec{p},\varepsilon)$ have the same form but different matrix elements, the peaks in both quantities come from corresponding peaks in $\underline{\tau}^c(\vec{k},\varepsilon)$ which is periodic in $\vec{k}$. However, in $A^B(\vec{k},\varepsilon)$ the structure in $\underline{\tau}^c(\vec{k},\varepsilon)$ is modulated by matrix elements $\underline{F}^{cc}(\vec{p},\varepsilon)$ which are not periodic. As expected, these matrix elements decrease when $|\vec{p}|$ becomes large. Thus, the relative heights of corresponding peaks of $\overline{A}(\vec{p},\varepsilon)$ in different Brillouin zones contain interesting wave-function information. Moreover, as for ordered systems these matrix elements impose powerful selection rules on the peaks in $\overline{A}(\vec{p},\varepsilon)$ when $\vec{p}$ is along a high symmetry direction.

That $\overline{A}(\vec{p},\varepsilon)$ decomposes $A^B(\vec{k},\varepsilon)$ throughout momentum space is illustrated in fig. V.10 where we have plotted in the uppermost frame the Bloch spectral function at the X-point for an $Ag_{0.2}Pd_{0.8}$ random alloy and in the frames beneath the momentum spectral function for points in momentum space which fold back to the X-point i.e. values of $\vec{p}$ such that $\vec{p}$ = (0,0,1) + $\vec{K}_n$. The actual $\vec{p}$-points shown are those near to the origin, $\vec{p}$ = (0,0,1), (1,1,0), (0,2,1). In $A^B(\vec{k},\varepsilon)$ the two upper peaks correspond to the broadened X_2 and X_5 states of pure Pd which delineate the top of the Pd d-band complex. The two lowest peaks are derived from the X_1 and X_3 states of Ag, states which form the bottom of Ag d-band complex. The peak at $\varepsilon \sim 0.25$ Ry is the remnant of the bottom of the Pd d-band complex (X_1 & X_3 -states). The contributions to $A^B(\vec{k},\varepsilon)$ from $A(\vec{p},\varepsilon)$ for each of the $\vec{p}$-points shown are marked by bars for the lowest X_1-derived peak and two uppermost X_2- and X_5-derived peaks. The

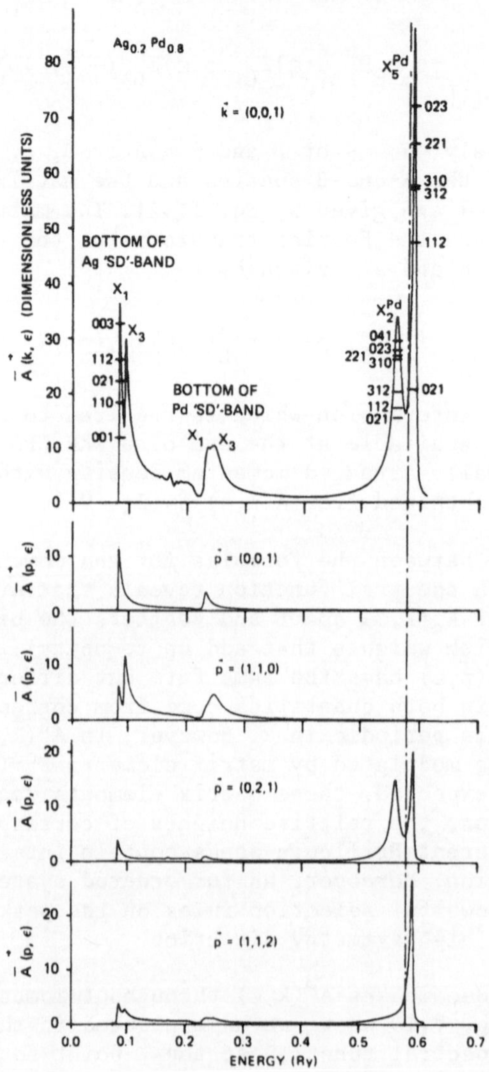

Fig. V.10. Bloch spectral densities $A(\vec{k},\varepsilon)$ and momentum spectra $A(\vec{p},\varepsilon)$ as a function of energy in a $Ag_{0.2}Pd_{0.8}$ alloy. (a) $A(\vec{k},\varepsilon)$ for $\vec{k}$ = (0,0,1) ; (b)-(e), $A(\vec{p},\varepsilon)$ for $\vec{p}$ = (0,0,1), (1,1,0), (0,2,1) and (1,1,2) respectively. The Fermi energy is marked by the chain line. For the lowest and the two uppermost peaks in (a) we have indicated by the hatch marks the contribution to those peaks from $A(\vec{p},\varepsilon)$ at the values of p indicated (e.g. 021 implies p = (0,2,1). The values of $A(\vec{p},\varepsilon)$ have been multiplied by a factor n to account for the number of equivalent contributions to $\overline{A}(\vec{k},\varepsilon)$. For the four values of $\vec{p}$ above n=2,8,8,8.

first major contribution to the lowest X_1-peak comes from the first Brillouin zone p = (0,0,1), however, there are substantial contributions form the X-points of more distant Brillouin zones. As might be expected, from the fact that the matrix elements decrease with increasing $\vec{p}$, the majority of the weight in $A^B(\vec{k},\varepsilon)$ comes from parts of momentum space near the origin. However, it is clear that this convergence is in general slow. The one major exception to this is $\vec{p}$ = (0,0,0) where about 75% of $A^B(\vec{k},\varepsilon)$ results from $A(\vec{p},\varepsilon)$ for p = (0,0,0). The first significant contributions to the d-like high energy X_2- and X_5-derived peaks of $A^B(\vec{k},\varepsilon)$ arise at $\vec{p}$ = (0,2,1) and (1,1,2).

In fig. V.11a we show the momentum density along (0,0,1) for an $Ag_{.2}Pd_{.8}$ alloy[99], as expected it is very similar to the curve for pure Pd. The major difference being a shift in the position of the Fermi surface break in $n(\vec{p})$ corresponding to an expansion of about 10% in Γ-centered sheet coupled with a general rounding off of the structure. The rounding off of the Fermi surface break is related to the degree of smearing of the Γ-centered sheet of Fermi surface. In fact if we assume Lorentzian smearing as in the previous subsection it can be shown that $(dn(p)/dp)p_F = \ell p_F$ the mean-free-path. Fig. V.11b shows curves similar to those of fig. V.11a except for a $Cu_{0.6}Ni_{0.4}$[100] which further demonstrate the similarities between the Ag_cPd_{1-c} and Cu_cNi_{1-c} alloy systems. In fig. V.12 we plot the momentum spectral functions along the (001)-direction for several p-points (not evenly spaced) between $\vec{p}$ = (0,0,0) and p = (0,0,1) for the $Ag_{0.2}Pd_{0.8}$ alloy. A comparison between this figure and the earlier plots of the Bloch spectral function illustrates the selection rule which prevents any state which does not have Δ_1 type symmetry being seen along the (001)-direction. For p near the Brillouin zone centre $\bar{A}(p,\varepsilon)$ has only a single spike corresponding to the low lying Δ_1-band. For larger values of $\vec{p}$ structure from the remnants of higher energy Δ_1-band arises. For p = 0.66 it is the upper Δ_1-band which dominates and it is the crossing of ε_F by this band which results in the rapid decrease in $\bar{n}(\vec{p})$ seen in fig. V.11a. Having located structures in $\bar{n}(\vec{p})$ associated with the Γ-centered sheet of Fermi surface it is interesting to look in the case of the $Ag_{.2}Pd_{.8}$ for structure associated with the high density of states precursors of the 'jungle-gym' seen in fig. V.1.

Unfortunately, $\bar{n}(\vec{p})$ does not reflect this structure in the first Brillouin zone because of selection rules and matrix element effects. The 'jungle-gym' near the X-point results from the intersection with ε_F of the bands which form the Δ_5 band along 001 and which result in the uppermost peak in $A^B(\vec{k},\varepsilon)$ seen in fig. III.10. This structure is not seen in $\bar{A}(\vec{p},\varepsilon)$ at p = (0,0,1). Clearly from the plots of $\bar{A}(\vec{p},\varepsilon)$ in fig. V.10 the value of $\vec{p}$ where this structure is most easily seen is (1,1,2), therefore, the most likely region of $\vec{p}$ space to see the 'jungle-gym' is in the neigh-

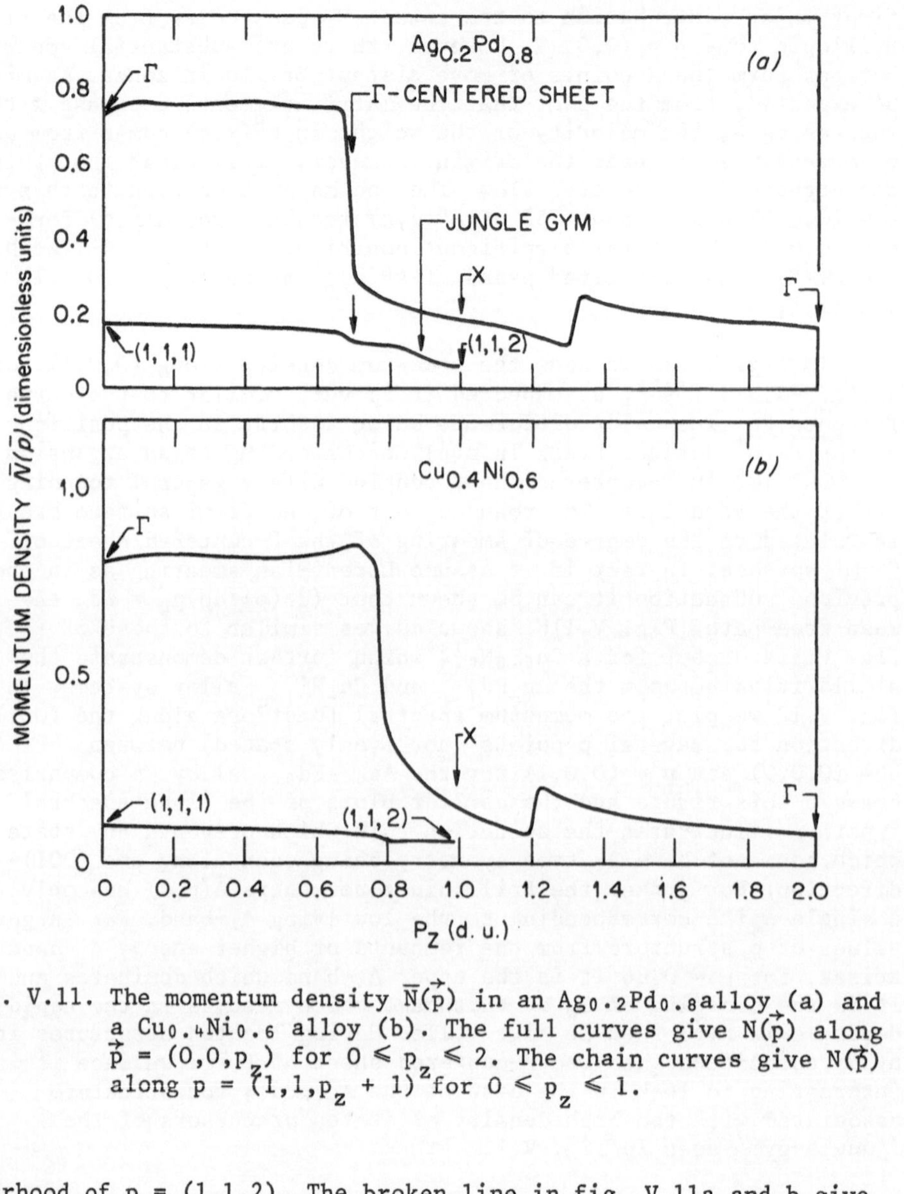

Fig. V.11. The momentum density $\overline{N}(\vec{p})$ in an $Ag_{0.2}Pd_{0.8}$alloy (a) and a $Cu_{0.4}Ni_{0.6}$ alloy (b). The full curves give $N(\vec{p})$ along $\vec{p} = (0,0,p_z)$ for $0 \leqslant p_z \leqslant 2$. The chain curves give $N(\vec{p})$ along $p = (1,1,p_z + 1)$ for $0 \leqslant p_z \leqslant 1$.

bourhood of p = (1,1,2). The broken line in fig. V.11a and b give $\underset{\sim}{N}(p)$ along the (001) direction in the Brillouin zone centered on $\vec{K}_n = (1,1,1)$. The gentle break in $\overline{n}(1,1,p + 1)$ is all that is seen of the 'jungle-gym'. This break in $n(\vec{p})$, which begins at $\vec{p}_z = 0.80$ (dimensionless units, du) and ends at $p_z = 0.955$ du, results from the passage of the Δ_5 'band' through the Fermi energy as $\vec{p}$ is increased. The gentleness of the break results from the fact that the 'band' in the neighbourhood of the X-point, is very flat i.e. ℓp_F is very short.

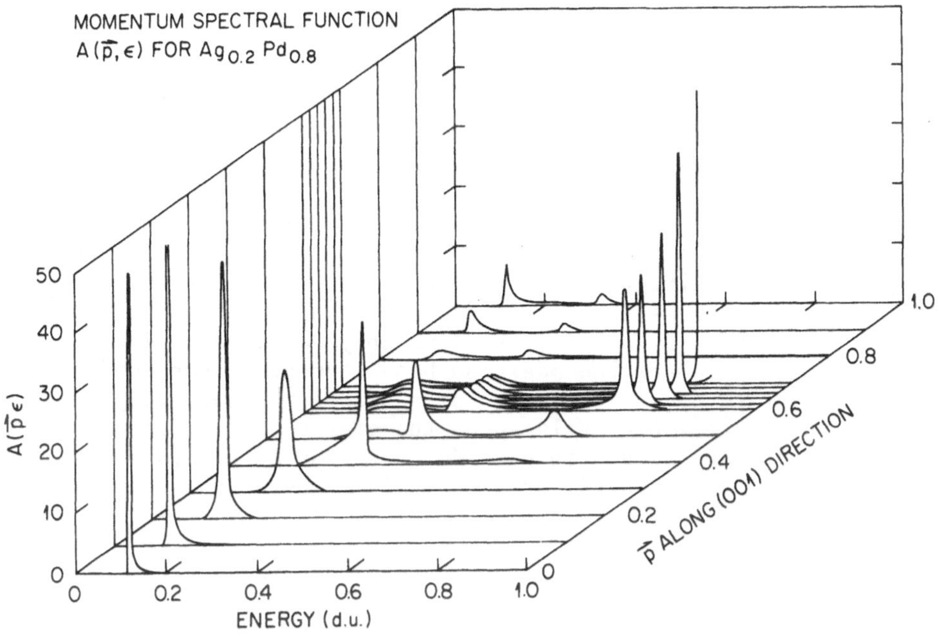

Fig. V.12. The momentum spectral function plotted as a function of energy at several $\vec{p}$-points along the (001)-direction. Near where the Δ_1-band crosses the Fermi energy the spectral functions are plotted at closely space intervals in $\vec{p}$.

So far we have totally ignored the role of the positron and, rather than calculate the full configuration averaged electron-positron momentum density $N^2\gamma(\vec{p})$ we have calculated just the electron-momentum density $\overline{n}(\vec{p})$. The approximation that the positron samples all regions of the crystal equally is clearly not valid, the positron will avoid the regions of high electron density around the ion cores. This will then result in a modulation of the momentum matrix elements and of the momentum density. In a random alloy the positron density will preferentially accumulate on one species or the other. Both of these effects can be studied quite straight forwardly by considering the problem of a positron in a random alloy. If we neglect the many body interactions between the single positron and the many electrons, under the assumption that the effects of these interactions can be accounted for by enhancing $\overline{N^2\gamma(\vec{p})}$ in the neighbourhood $\vec{p}_f$ by a known function as is done in calculations on pure metals[101,102], then the potential seen by a positron in a random alloy will just be the negative of the potential seen by the electron excepting that there will be no exchange-correlation contribution. Given these potentials we are then in a position to study the charge distribution, spectral function etc. of the positron by the now familiar methods of the KKR-CPA.

In fig. V.13 are plotted the phase shifts corresponding to positron potentials corresponding to $Ag_{.2}Pd_{.8}$. Phase shifts for some other system are shown in Reg. 100. For all the potentials the dominant phase shift is the s-wave. Furthermore, for alloy systems consisting of atoms adjacent in the periodic table there is little difference between the s-wave phase shifts. This means that so far as the positrons are concerned alloys like Cu_cNi_{1-c}, Ag_cPd_{1-c}, Nb_cMo_{1-c} etc. are only weakly disordered. In fact, if we calculate the positron band structures corresponding to the separate constituents in these alloys they are very similar to one another[103]. Fig. V.14 shows the energy band structure for positrons in pure Ag and in pure Pd using the Ag and Pd positron potentials appropriate to the alloys. From these phase shifts and band structures it should not be difficult to guess what the band structure of the 'positron'-alloy looks like. It will be virtual crystal like and the corresponding Bloch spectral function for

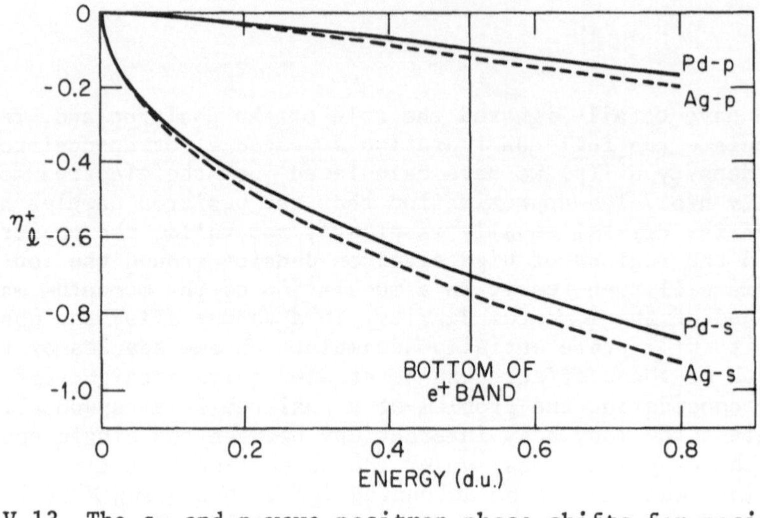

Fig. V.13. The s- and p-wave positron phase shifts for positron potentials corresponding to Ag and Pd sites in an $Ag_{0.2}Pd_{0.8}$ alloy.

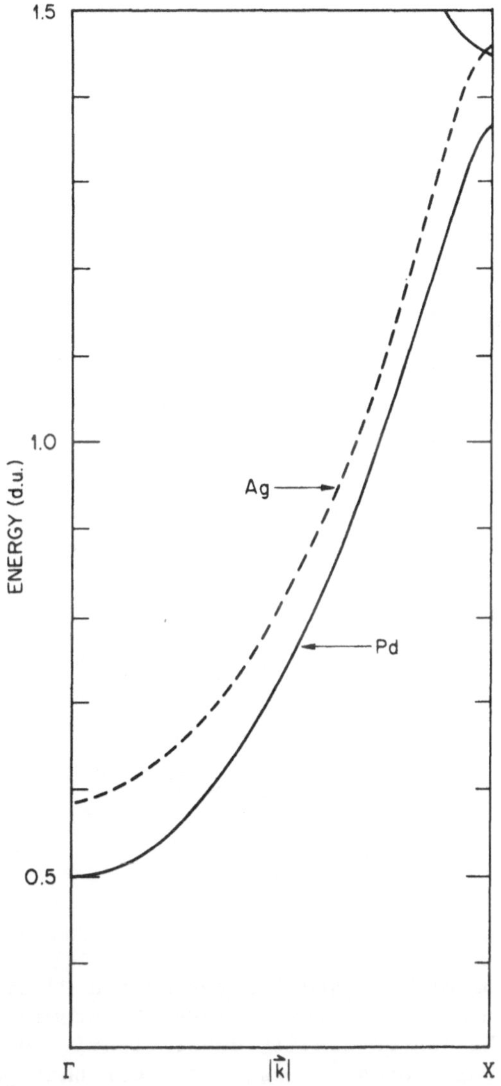

Fig. V.14. Positron band structure for pure Ag and Pure Pd. The positron potentials used were there appropriate to Ag and Pd sites in a $Ag_{0.2}Pd_{0.8}$ alloy.

the positron will be δ-function like. This is shown in fig. V.15 where we show both the electron and positron Bloch spectral functions at the Γ-point for a $Ag_{.2}Pd_{.8}$ alloy, a similar plot for $Cu_{.6}Ni_{.4}$ is shown in ref.100. The positron Bloch spectral function is a single δ-function. This is not to say that the effects of disorder are entirely zero in fig. V.16, we show the charge densi-

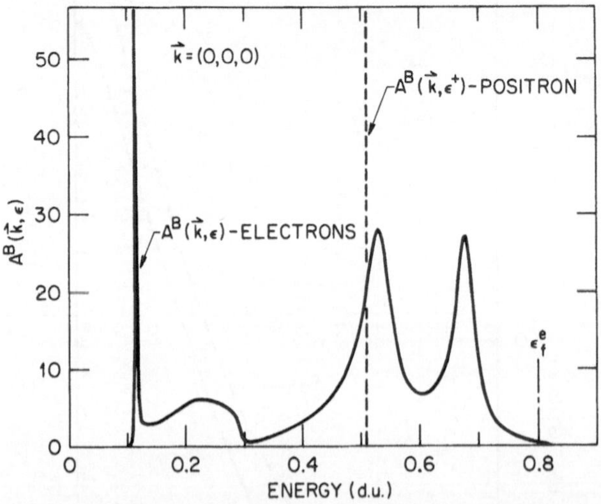

Fig. V.15. Electron (solid) and positron (dash) Bloch spectral functions at the Γ-point for a $Ag_{0.2}Pd_{0.8}$ alloy.

ties at the bottom of the band for the Ag and Pd sites in the $Ag_{0.2}Pd_{0.8}$ random alloy. In these plots the atomic sites are at the centres of the charge density cups. Clearly the positrons stay away from the atomic cores, it is also clear that positrons have a slight preference for the Pd-sites.

In order to see how these effects show up in the electron-positron momentum density let us consider the direct calculation of $N^2\Upsilon(\vec{p})$ within the Green's function fomulism. $N^2\Upsilon(\vec{p})$ can be written in the form

$$N^2\Upsilon(\vec{p}) = \frac{1}{N}\int_{-\infty}^{\varepsilon_F^p}\frac{d\varepsilon^+}{\pi}\int_{-\infty}^{\varepsilon_F^e}\frac{d\varepsilon}{\pi}\int d^3r\int d^3r'\; e^{-i\vec{p}(\vec{r}-\vec{r}\,')}\; \mathrm{Im}G^p(\vec{r},\vec{r}\,',\varepsilon)\; \mathrm{Im}G^e(\vec{r},\vec{r}\,',\varepsilon') \qquad V.23$$

Where all quantities having a 'p' on them refer to the positron

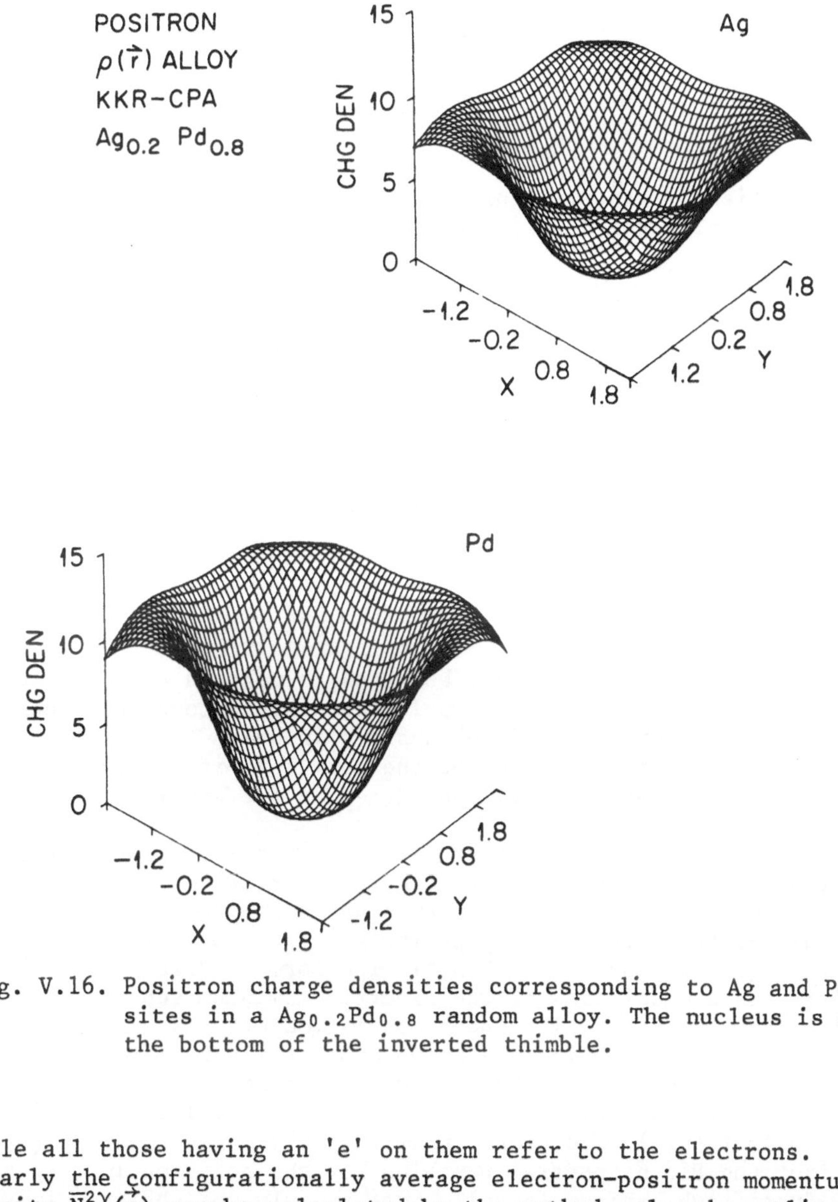

Fig. V.16. Positron charge densities corresponding to Ag and Pd sites in a $Ag_{0.2}Pd_{0.8}$ random alloy. The nucleus is at the bottom of the inverted thimble.

while all those having an 'e' on them refer to the electrons. Clearly the configurationally average electron-positron momentum density $\overline{N}^{2\gamma}(\vec{p})$ can be calculated by the methods already outlined.

The result of averaging is[100]

$$\overline{N}^{2\gamma}(\vec{p}) = \sum_{L_1L_1'L_2L_2'} \int_{-\infty}^{\varepsilon_F^e} \frac{d\varepsilon'}{\pi} \int_{-\infty}^{\varepsilon_F^p} \frac{d\varepsilon}{\pi} \{\sum_{\alpha} c_\alpha Z^{\alpha}_{L_1L_2}(\vec{p},\varepsilon,\varepsilon') Z^{\alpha}_{L_1'L_2'}(-\vec{p},\varepsilon,\varepsilon') .$$

$$. \langle \mathrm{Im}\tau^{e,nn}_{L_1L_1'}(\varepsilon)\, \mathrm{Im}\tau^{p,nn}_{L_2L_2'}(\varepsilon') \rangle_{n=\alpha} \quad +$$

$$\sum_{m\neq n'}\sum_{\alpha,\beta} c_\alpha c_\beta Z^{\alpha}_{L_1L_2}(\vec{p},\varepsilon,\varepsilon')Z^{\beta}_{L_1'L_2'}(-\vec{p},\varepsilon,\varepsilon')\,.$$

$$\cdot\ \langle \mathrm{Im}\tau^{e,nm}_{L_1L_1'}(\varepsilon)\,\mathrm{Im}\tau^{p,nm}_{L_2L_2'}(\varepsilon')\rangle_{\substack{n=\alpha\\ m=\beta}} \qquad \text{V.24}$$

The quantities $Z^{\alpha}_{LL'}(\vec{p},\varepsilon,\varepsilon')$ are given by an expression like Eq. V.22 for the electron case except that it now includes the positron wave function

$$Z^{\alpha}_{LL'}(\vec{p},\varepsilon,\varepsilon') = 4\pi\sum_{L_1} i^{\ell_1}Y_{L'}(\hat{p})\int d^3r Z^{e,\alpha}_{L}(\vec{r},\varepsilon)Z^{p,\alpha}_{L'}(\vec{r},\varepsilon)j_{\ell_1}(pr)Y_{L_1}(\hat{r}) \qquad \text{V.25}$$

Clearly in order to obtain a formula for $\bar{N}^{2\gamma}(\vec{p})$ within the CPA we have to be able to calculate the restricted averages $\langle \mathrm{Im}\underline{\tau}^{e,nm}\,\mathrm{Im}\underline{\tau}^{p,nm}\rangle_{n=\alpha,n=\beta}$ etc. Within the KKR-CPA this can in fact be done. However, as a first approximation we can consider the decoupling scheme

$$\langle\underline{\tau}^{e,nm}\underline{\tau}^{p,nm}\rangle_{\substack{n=\alpha\\ m=\beta}} \Rightarrow \langle\underline{\tau}^{e,nm}\rangle_{\substack{n=\alpha\\ m=\beta}}\ \langle\underline{\tau}^{p,nm}\rangle_{\substack{n=\alpha\\ m=\beta}} \qquad \text{V.26}$$

since within the CPA we already know how to calculate restricted averages of the type $\langle\underline{\tau}^{nm}\rangle_{n=\alpha,m=\beta}$. In fact this decoupling scheme will probably be quite good for systems in which the disorder is not too great, as in the case for the positron, if not the electron system, in the alloys we have considered in the section.

We will not go into further detail on the inclusion of the positron except to make a few remarks concerning the likely outcome of calculations in which the positron has been properly included. In an alloy, the effect of the positron is seen both in the matrix elements Eq. V.25 and in Im $\underline{\tau}^{p}(\vec{p},\varepsilon)$. The former effects will only give rise to a modulation of the electron momentum density which we have shown in fig. V.11, these modulation effects will be little different from those which are known about for pure metals[96]. These effects introduce no extra broadening of the Fermi-break. Such broadening effects could only arise from the convolution of the peaks in Im $\underline{\tau}^{p}(\vec{p},\varepsilon)$ with those in Im $\underline{\tau}^{e}(\vec{p},\varepsilon)$. However, as we have seen from the Bloch spectral function for the positron Im $\underline{\tau}^{p}(\vec{p},\varepsilon)$ has δ-function behaviour (recall there is only one positron in the system with $\vec{p}$ = (0,0,0)) thus the introduction of the positron will, for Cu_cNi_{1-c}, Ag_cPd_{1-c} etc, introduce no further broadening in the Fermi break.

In summary the situation with respect to being able to make direct contact between theoretic alloys computed $\bar{N}^{2\gamma}(\vec{p})$ and those deconvoluted from 2-D ACAR data means that, in the not too distant future, we should be able to obtain direct experimental verification (or otherwise) of the Fermi surfaces obtained on the basis of KKR-CPA calculations.

VI. CONCLUSIONS AND NEW FRONTIERS

Although the development of the first principles theory of the electronic properties of random alloys has been much slower than that for pure metals and ordered alloys, the methods described in these lecture notes to a large extent bring it up to the same level. Of course the solution of the Schrödinger equation for a pure metal or alloy provided by band theory is exact whilst at best the CPA is an approximation. However, it now appears, that when the CPA is used within the context of a first principles theory it is capable of giving a description of the electrons that is of a very high order indeed. The degree of agreement between the results of KKR-CPA calculations and experimental measurements on alloys is certainly as good as that obtained for pure metals and ordered alloys. Indeed as the experimental probes of the electronic structure of alloys such as angle resolved photoemission become more and more refined it will be interesting to see at what level the CPA fails and specifically multisite effects have to be included within the theory.

Far as the developments of the last decade have taken us towards an understanding of the properties of random alloys it is important to remember that important as it is to understand the intrinsic properties of the conduction electrons in an alloy and to offer detailed explanations of the results of photoemission, low-temperature specific-heat, soft x-ray, positron annihilation experiments etc., this is not the major goal of theoretical studies of alloys. Having gained confidence in the correctness of our theoretical methods, we must then use this knowledge to understand how the electron glue that holds on alloys atomic building blocks together determines the structural arrangement of the atoms in an alloy and, hence, its metallurgical properties.

Even in an alloy which is nominally random there is usually some residual order in the way the different species arrange themselves with respect to one another. In some alloys, for example Cu_cNi_{1-c}, like-atoms tend to cluster together, while in others, like Cu_cPd_{1-c}, atoms of one species prefer to have near-neighbors of the opposite species. Alloys of the former type are said to show clustering, while the latter are said to exhibit short-range order (SRO). At the lowest temperatures, alloys which cluster tend to separate into regions of pure A and pure B - they are immiscible, while alloys which show SRO tend to form ordered compounds.

Recently, using the KKR-CPA methodology as a basis, we have begun to study the forces driving SRO and clustering in alloys. To this end we are developing two rather different though in many respects complementary methods.

In the first approach, in collaboration with Dr. Gyorffy we have developed a new first principles electronic model for the forces driving clustering and SRO[104]. This theory which is based on a classical analogue of density functional theory, has recently been used to explain the existance of concentration waves in $Cu_c Pd_{1-c}$ alloys. A concentration wave may be thought of as a long-lived periodic variation in the site concentration about the average homogeneous value. For a completely random, $A_c B_{1-c}$, binary alloy the probability of occupancy of each site by the A species is c. In a system with a concentration wave along some direction, the average concentration of the A species at sites along that direction is different from c and varies in a periodic manner. The period of this variation need not be related to the periodicity of the lattice ; such a concentration wave is said to be incommensurate with the lattice.

Experimentally, concentration waves show up in conventional x-ray, electron, and neutron diffraction experiments. For an alloy in which there is a concentration wave diffraction patterns not only show the conventional Bragg maxima, from which the crystal structure and lattice spacing can be determined, but also extra diffuse scattering peaks. The position of these diffuse scattering spots is trivially related to the wave length of the concentration wave.

Electron diffraction pictures from $Cu_c Pd_{1-c}$ clearly reveal the existence of concentration waves, Fig. VI.1. Furthermore, it is clear from the variation with concentration of the separation between the diffuse scatttering spots that the wavelength of the concentration waves varies substantially with alloy concentration. The conventional argument for describing this concentration variation is based on the Krivoglaz-Clapp-Moss[105,106] theory for the SRO parameter $\alpha(\vec{k})$, which is related to the lattice Fourier transform of a pairwise interchange potential. Our new theory enables us to obtain an exact expression for $\alpha(\vec{k})$. Using this expression we are able to obtain a mean field approximation for $\alpha(\vec{k})$ based on the use of the CPA. We find that concentration-dependent diffuse scattering peaks can arise at values of $\vec{k}_0$ that connect well-defined parallel flat pieces of Fermi surface. In Fig. VI.2 we show the calculated Fermi surfaces for three $Cu_c Pd_{1-c}$ in the ΓXWK plane of the Brillouin zone. Clearly as the concentration of Pd is increased in these alloys the Fermi surface becomes increasingly flat normal to 110-directions. It is these pieces of flat Fermi surface that give rise to the concentration waves in $Cu_c Pd_{1-c}$ alloys. Also shown in Fig. VI.2 for a number of $Cu_c Pd_{1-c}$ alloys are calculated and measured[106] values for the separation between the diffuse scattering spots. Clearly the Fermi surface mechanisms for the driving force behind these concentration wave is capable of providing a detailed explanation of this concentration dependence.

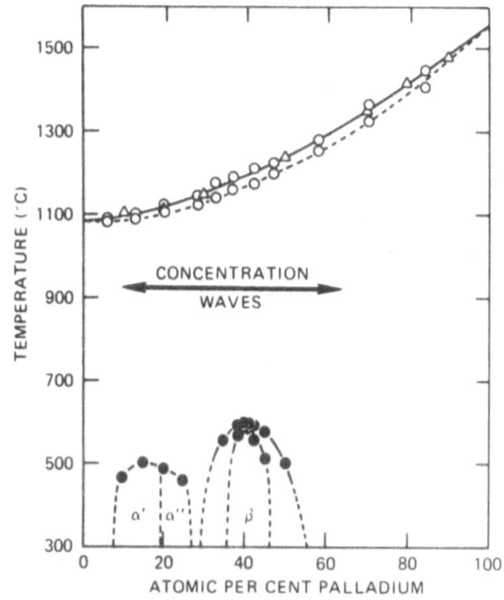

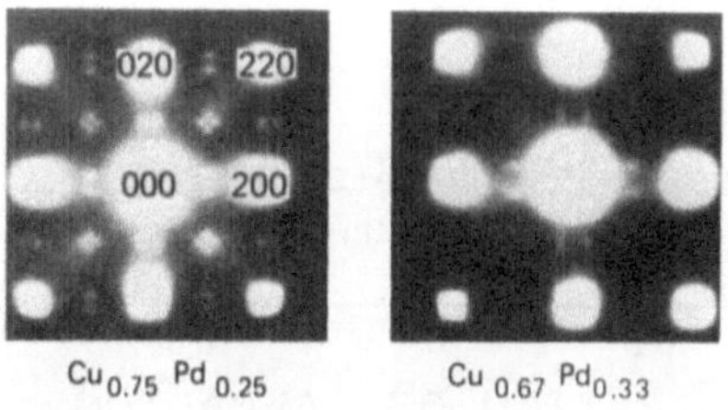

Fig. VI.1. Upper : Phase diagram of Cu_cPd_{1-c} alloys. The concentration range over which concentration waves are seen is indicated by the arrows.

Lower : Electron diffraction patterns for two Cu_cPd_{1-c} alloys taken from the work of Oshima and Watanabe. The incident beam was parallel to [001]. The alloys were quenched from 1000°C.

In the second approach we have taken to the question of ordering in alloys, which we call a cluster method, we attempt to calculate directly the energy associated with particular local arrangements of atoms in the otherwise random alloy. These arrangements are chosen to represent various situations ranging from clustering to SRO. The arrangement which the actual system will prefer to adopt is the one of lowest energy. Using this method developed in collaboration with Tony Gonis of Northwestern University and Bill Butler of ORNL we were able to calculate the electronic

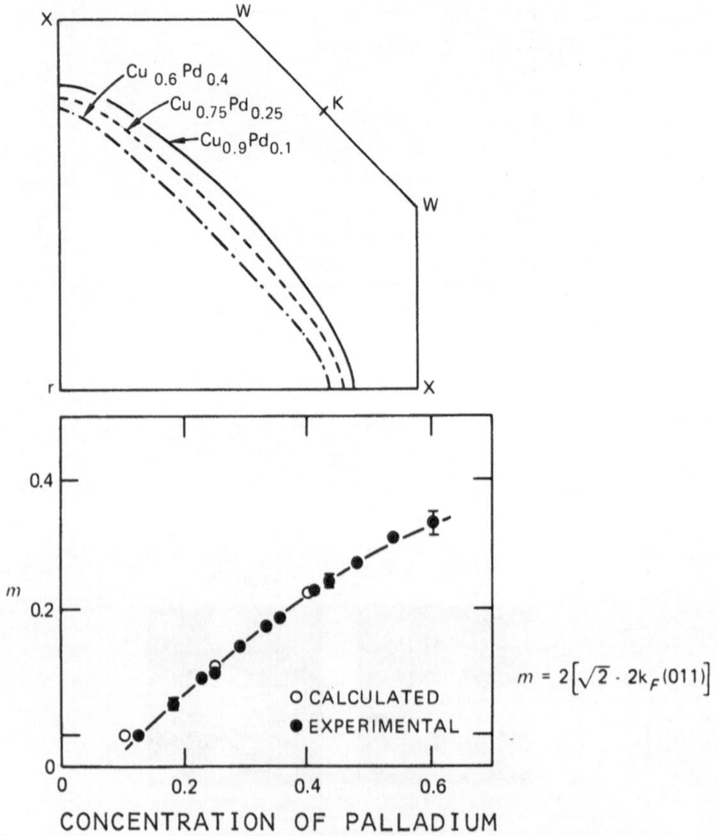

Fig. VI.2. Upper : Concentration variation of the Fermi surfaces of Cu_cPd_{1-c} alloys.
Lower : Comparison of experimental and theoretical values of the diffuse scattering spot separation m, for Cu_cPd_{1-c} alloys.

structure of specific near-neighbor clusters of atoms in an $Ag_{0.5}Pd_{0.5}$ alloy[107]. For the first time within a parameter free theory these calculations go beyond the single site CPA. In Fig. VI.3 we show calculated densities of states for palladium and silver sites at the center of particular 13-atom clusters in an $Ag_{0.5}Pd_{0.5}$ random alloy compared with single-site CPA calculations. Clearly, the cluster density of states can deviate substantially from those obtained in the single-site CPA. In particular, the density of states $n(\varepsilon)$ for the central site of an all-palladium cluster (solid line) resembles that of pure palladium, but $n(\varepsilon)$ for a palladium atom surrounded by 12 silver atoms displays a single peak (dotted line) characteristic of an impurity. The densities of states corresponding to a silver atom on the central site show similar behaviour. On the basis of a calculation of the contribution

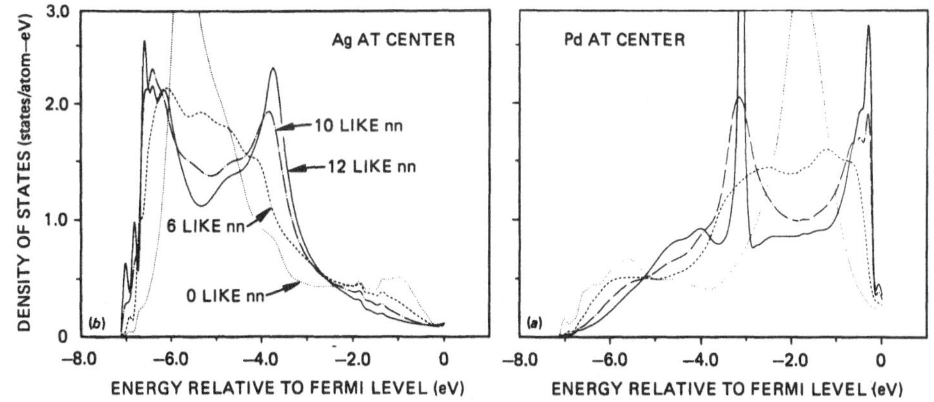

Fig. VI.3. Central site densities of states in $Ag_{0.5}Pd_{0.5}$ for a cluster consisting of a central site plus 12-near-neighbours. Left : A silver site is at the centre of the cluster. Right : A palladium site is at the centre of the cluster.

to the total energy which comes from the sum of the energy of single particle states it proved possible to predict that the $Ag_{0.5}Pd_{0.5}$ alloy system will exhibit SRO. The results of measurements of various thermodynamic properties[108] taken together with careful measurements of the variation of electrical resistivity with heat treatment[109] made by Bob Williams at ORNL are very suggestive that this is the case.

Far as the developments of the last decade have taken us towards a detailed theoretical understanding of the properties of metallic alloys, it is clear that we have as yet hardly scratched the surface. Just to be able to predict as a function of temperature the stable equilibrium phases still requires many further theoretical developments. Not only do we need to be able to calculate the energies associated with disordered solid solutions and the various ordered compounds, we need to be able to do this with sufficient precision to be able to distinguish between the various possible crystal structures. Even for pure metals this is still beyond the reach of first principles calculations since these energy differences are, in general, extremely small. Clearly alloy theory still has a long way to go.

ACKNOWLEDGEMENTS

This work was supported in part by the Division of Materials Sciences, US Department of Energy, under contract number

W-7405-erg-26. One of us would like to thank the UK-SERC and the Daresbury Laboratory for this support and hospitality during a one year stay, under the auspecies of CCP-9 in which period these notes were assembled, and for the pleasures which occured from using their CRAY-1S computer.

REFERENCES

1. V.L. Moruzzi, J.F. Janak and A.R. Williams, 'Calculated Electronic Properties of Metals', Pergamon, New York (1978).
2. A.R. Williams, J. Kübler and C.D. Gelatt, Jr., Phys.Rev. B19, 6094 (1979).
3. P. Hohenberg and W. Kohn, Phys.Rev. 136, B864 (1964).
4. W. Kohn and L.J. Sham, Phys.Rev. 140, A1133 (1965).
5. L. Kaufman and H. Bernstein, "Computer Calculation of Phase Diagrams" Academic Press, New York (1970).
6. B.L. Gyorffy and G.M. Stocks, "Electrons in Finite and Infinite Structures" Ed. P. Phariseau and L. Scheire, Plenum Press NATO ASI Series Physics B24 (1977).
7. B.L. Gyorffy and G.M. Stocks, "Electrons in Disordered Metals and at Metallic Surfaces" Ed. P. Phariseau, B.L. Gyorffy and L. Scheire, Plenum Press NATO ASI Series Physics B42 (1979).
8. G.M. Stocks, W.M. Temmerman and B.L. Gyorffy, Phys.Rev.Lett. 41, 339 (1978).
9. A. Bansil, Phys.Rev.Lett. 41, 1670 (1978).
10. J.S. Faulkner, Progress in Materials Science 27, 1 (1982).
11. P. Soven, Phys.Rev. 156, 809 (1967).
12. D.W. Taylor, Phys.Rev. 156, 1017 (1967).
13. R.J. Elliott, J.A. Krumhansl and P.L. Leath, Rev.Mod.Phys. 46, 465 (1974).
14. H. Ehrenreich and L. Schwartz, "Solid State Physics" Ed. H. Ehrenreich, F. Seitz and D. Turnbull, Academic Press (1976).
15. J.F. Janak, Phys.Rev. B9, 2985 (1974).
16. L. Hedin and B.I. Lundqvist, J.Phys. C4, 2064 (1971).
17. K.S. Singwi, A. Sjölander, M.P. Tosi and R.H. Land, Phys.Rev. B1, 1044 (1970).
18. U. von Barth and L. Hedin, J.Phys. C5, 1629.
19. L.F. Mattheiss, Phys.Rev. B1, 373 (1970).
20. U. von Barth, this volume.
21. G.M. Stocks and H. Winter, Z.Phys. B46, 95 (1982).
22. H. Winter and G.M. Stocks, Phys.Rev. B27, 882 (1983).
23. R.E. Watson, H. Ehrenreich and L. Hodges, Phys.Rev.Lett. 24, 829 (1970).
24. A. Bansil, H. Ehrenreich, L. Schwartz and R.E. Watson, Phys.Rev. B9, 445 (1974).
25. A.J. Pindor, W.M. Temmerman, B.L. Gyorffy and G.M. Stocks, J.Phys.F. 10, 2617 (1980).
26. J.C. Slater, Phys.Rev. 81, 385 (1951).

27. W.M. Temmerman, B.L. Gyorffy and G.M. Stocks, J.Phys.F. 8, 2461 (1978).
28. E.S. Guiliano, R. Ruggeri, B.L. Gyorffy and G.M. Stocks, "Transition Metals 1977", Ed. M.J.G. Lee, J.M. Perz and E. Fawcett, The Institute of Physics, Bristol 410 (1978).
29. J.S. Faulkner and G.M. Stocks, Phys.Rev. B23, 5628 (1981).
30. J. Staunton, P. Weinberger and B.L. Gyorffy, J.Phys.F. 13, 779 (1983).
31. P. Weinberger, J. Staunton and B.L. Gyorffy, J.Phys.F. 12, 2229 (1982).
32. W. Wagner, R. Poerschke and H. Wollenberger, J.Phys.F. 12, 405 (1982).
33. K. Oshima and D. Watanabe, Acta Crystallogr. Sec. A 29, 520 (1973).
34. B.L. Gyorffy, "Fundaments Di Fisica Dello Stati Solido", Ed. F. Fumi, University of Genova (1970).
35. B. Velicky, S. Kirkpatrick and H. Ehrenreich, Phys.Rev. 175, 747 (1968).
36. L. Nordheim, Ann.Physik 9, 607 (1931), 9, 641 (1931).
37. T. Muto, Sci. Papers Inst. Phys.Chem.Res. (Toyko) 34, 377 (1938).
38. J. Korringa, J.Phys.Chem.Solids 7, 252 (1958).
39. J.L. Beeby, Phys.Rev. 135, A130 (1964).
40. R. Alben, M. Blume and M. McKeown, Phys.Rev.B. 16, 3829 (1977).
41. When using the ATA there is some liberty in chosing the reforme medium. If the VCA is taken as the reference the results are generally superior to those obtained when the reference is taken to be the free particle propagator.
42. R. Mills and P. Ratanavararaksu, Phys.Rev.B. 18, 5291 (1978).
43. P. Lloyd and P.V. Smith, Adv.Phys. 21, 69 (1972).
44. J. Korringa, Physica 13, 392 (1947).
45. W. Kohn and N. Rostoker, Phys.Rev. 94, 1111 (1954).
46. J.S. Faulkner and G.M. Stocks, Phys.Rev.B. 21, 3222 (1980).
47. G.M. Stocks, J.A. Tagle, T.A. Callcott and E.T. Arakawa "Inner-Shell and x-ray Physics of Atoms and Solids, Ed. D.J. Fabian, H. Kleinpoppen and L.M. Watson, Plenum 619 (1980).
48. B.L. Gyorffy and M.J. Stott, "Band Structure Spectroscopy of Metals and Alloys", Ed. D.J. Fabian and L.M. Watson, Academic Press (1972).
49. B.L. Gyorffy, Phys.Rev. B 5, 2382 (1972).
50. P. Soven, Phys.Rev.B 2, 4715 (1970).
51. H. Shiba, Prog.Theoret.Phys. (Kyoto) 16, 77 (1971).
52. H.L. Davis, "Computational Methods in Band Theory", Eds. Marcus, Janak and Williams, Plenum Press 183 (1971).
53. G.M. Stocks, W.M. Temmerman and B.L. Gyorffy, "Electrons in Disordered Metals and at Metallic Surfaces", Ed. P. Phariseau, B.L. Gyorffy and L. Scheire, Plenum Press NATO ASI Series Physics B42 (1979).
54. A. Bansil, Solid State Comm. 16, 885 (1975).
55. W.H. Fehlner and S.H. Vosko, Can.J.Phys. 54, 2159 (1976).

56. P.J. Durham, this volume.
57. P.J. Durham, B.L. Gyorffy, G.M. Stocks and W.M. Temmerman, "Transition Metals 77", Inst.Phys.Conf.Series 39, Inst. Physics London (1978).
58. P.J. Durham, D. Ghaleb, B.L. Gyorffy, C.F. Hague, J.M. Mariot, G.M. Stocks and W.M. Temmerman, J.Phys.F. 9, 1719 (1979).
59. B.L. Gyorffy, R. Jordan, D.R. Lloyd, C.M. Quinn, N.V. Richardson, G.M. Stocks and W.M. Temmerman, "Transition Metals 77" Inst. Phys.Conf.Series 39, Inst.Phys. London (1978).
60. P.J. Durham, J.Phys.F. 11, 2475 (1981).
61. N.K. Allen, P.J. Durham, B.L. Gyorffy and R.G. Jordan, J.Phys. F. 13, 223 (1983).
62. S. Hüfner, G.K. Wertheim and J.H. Wernick, Phys.Rev.B. 8, 4511, (1973).
63. B.L. Gyorffy, A.J. Pindor and W.M. Temmerman, Phys.Rev.Lett. 43, 1343 (1979).
64. H. Montgomery, G.P. Pells and E.M. Wray, Proc.R.Soc. A301, 261, (1967).
65. W. Hume-Rothery, G.W. Mabbott and K.M. Chanel Evans, Philos. Trans. R. Soc. (London) Sec. A233, 1 (1934).
66. T.B. Massalski and U. Mizutani, Prog.Mater.Sci. 22, 151 (1978).
67. B.W. Veal and J.A. Rayne, Phys.Rev. 130, 2156 (1963).
68. U. Mizutani, S. Noguchi and T.B. Massalski, Phys.Rev.B. 5, 2057 (1972).
69. T.A. Callcatt, J.A. Tagle, E.T. Arakawa and G.M. Stocks, App. Optics 19, 4035 (1980).
70. R.S. Crisp, "Inner-Shell and x-ray Physics of Atomis and Solids" Eds. D.J. Fabien, H. Kleinpoppen and L.M. Lewis, Plenum Press 625 (1980).
71. J.B. Staunton, Ph.D. Thesis, Univ. Bristol (1982).
72. W.M. Temmerman and A.J. Pindor, J.Phys.F. (in press).
73. W.M. Temmerman and A.J. Pindor, J.Phys.F. (in press).
74. P. Lloyd, Proc.Phys.Soc. (London) 90, 207 (1967).
75. J. Friedel, Nuovo Cimento Suppl. 7, 287 (1958).
76. J.S. Faulkner, J.Phys.C. 10, 4661 (1977).
77. M.G. Krein, Mat.Sb. 33, 597 (1953).
78. B.L. Gyorffy and G.M. Stocks, J.Phys. (Paris) 35, C4-75 (1974).
79. L. Schwartz and A. Bansil, Phys.Rev.B. 10, 3261 (1974).
80. A. Bansil, Phys.Rev.B. 20, 4025 (1979), 20, 4035 (1979).
81. A. Gonis and G.M. Stocks, Phys.Rev.B. 25, 659 (1982).
82. P. Soven, Phys.Rev. 151, 539 (1966).
83. P.J. Durham, B.L. Gyorffy and A.J. Pindor, J.Phys.F. 10, 661 (1980).
84. R. Prasad, S.C. Papadopoulos and A. Bansil, Phys.Rev.B. 23, 2607 (1981).
85. R. Prasad and A. Bansil, Phys.Rev.Lett. 48, 113 (1982).
86. A. Bansil, "Positron Annihilation", Eds. P.G. Coleman, S.C. Sharna and L. Diana, North-Holland (1983).
87. G.M. Stocks, B.L. Gyorffy, E.S. Guiliano and R. Ruggeri, J.Phys.F. 7, 1859 (1977).

88. G. Ries and H. Winter, J.Phys.F. 9, 1589 (1979).
89. P.J. Durham, G.M. Stocks and H. Winter, J.Phys.F. (in press).
90. P. Weightman, P.T. Andrews, G.M. Stocks and H. Winter, J.Phys. F. 16, 181 (1983).
91. Private Communication.
92. J.M. Ziman, "Electrons and Phonons", Oxford (1960).
93. G.M. Stocks and W.H. Butler, Phys.Rev.Lett. 48, 55 (1982).
94. R. West, Advances in Physics 22, 263 (1973).
95. S. Berko, "Compton Scattering" Ed. B.W. Williams, McGraw-Hill London (1977).
96. P.E. Mijnarends, "Positrons in Solids", Ed. P. Hautojarvi, Springer Verlag (1978).
97. A.A. Manuel, Phys.Rev.Lett. 49, 1525 (1982).
98. R. Podlouky, P. Lasser, E. Wimmer and P. Weinberger, Phys.Rev. 19, 4999 (1979).
99. B.L. Gyorffy and G.M. Stocks, J.Phys.F. 10, 1321 (1980).
100. Z. Szotek, B.L. Gyorffy, G.M. Stocks and W.M. Temmerman, "Positron Annihilation", Eds. P.G. Coleman, S.C. Sharma and L.M. Diana, North-Holland (1982).
101. B. Chakravarty, Phys.Rev.B. 24, 7423 (1983).
102. S. Kahana, Phys.Rev. 129, 1622 (1963).
103. Z. Szotek, private communication.
104. B.L. Gyorffy and G.M. Stocks, Phys.Rev.Lett. 50, 374 (1983).
105. M.A. Krivoglaz, "Theory of X-Ray and Thermal Neutron Scattering by Real Crystals", Plenum (1969).
106. S.C. Moss, Phys.Rev.Lett. 22, 1108 (1969).
107. A. Gonis, W.H. Butler and G.M. Stocks, Phys.Rev.Lett. 50, 1482 (1983).
108. O. Kubaschewski and J.A. Catterall, "Thermochemical Data of Alloys", Pergamon Science Series vol.3 (1956).
109. R.K. Williams, Scripta Metallurgica 16, 683 (1982).

ALLOY PHASE DIAGRAMS FROM FIRST PRINCIPLES

J.W.D. Connolly* and A.R. Williams†

IBM T.J. Watson Research Center
Yorktown Heights
New York 10598

The first applications of density-functional theory to disordered systems are described. The usual difficulty of solving the single-particle Schrödinger equation for systems lacking long-range order is circumvented by using the cluster-expansion method of Sanchez and deFontaine to describe the alloy, and deducing the coefficients appearing in this expansion from energy-band calculations for several ordered compounds. The cluster expansion obtained in this way describes the alloy as a continuous function of stoichiometry, short-range order and volume. We have used the technique, in combination with an approximate description of the entropy of mixing, to calculate the dominant features of the phase diagrams of 28 transition-metal alloy systems. Agreement with measured phase diagrams is generally very good. Not tested in these first applications of the technique is its ability to describe properties of disordered systems other than the total energy, properties such as the state density. Also not yet tested is the ability of the technique to directly predict the degree of short-range order present in the system, by minimization of the free energy.

---

* Permanent address : National Science Foundation, Washington, DC 20550.

† This paper describes one of five lectures presented by A.R. Williams at the NATO Advanced Study Institute "The Electronic Structure of Complex Systems. References to the material presented in the other four lectures can be found in Ref. 24.

It is by now well-accepted that the density functional theory[1,3] provides an accurate description of the electronic structure of small systems, such as simple molecules[4,5] and systems, such as solids[6,7] that are effectively small due to their symmetry. Particularly relevant to the present discussion is the fact that applications of this theory to ordered intermetallic compounds[8] have shown that it is capable of predicting their formation energies with errors on the order of only 0.01 eV/atom. Despite the development of methods based on the coherent-potential approximation[9] the application of density-functional methods to disordered materials has proceeded less rapidly, because of the great difficulty posed by both the solution of the single-particle Schrödinger equation (an essential ingredient of density-functional theory) and the accurate treatment of the electrostatics (Poisson's equation) for disordered systems[10]. The present development is based on the observation that the source of difficulty in the usual approach to disordered systems (the absence of long-range order) is relatively unimportant to many properties of such systems, and on the complementary observation that the source of simplicity in the study of compounds (the presence of long-range order) is relatively unimportant to many of the calculated results. The important specification of an energy-band calculation is, in this view, the local arrangement of atoms ; the periodic replication of this arrangement is better regarded as a mathematical convenience than as an essential property of the physics system. Similarly, the important difference between ordered and disordered alloys, in this view, is not the presence or absence of long-range order, but rather the presence of a single versus the presence of several local atomic arrangements. To the extent that this view is correct, we might hope that an appropriate synthesis of the results of several energy-band calculations will provide an accurate description of a disordered alloy. The cluster-expansion formalism of Sanchez and deFontaine[11] constitutes a concise mathematical formulation of this point of view. In this formalism, properties of an alloy, such as its total energy, are assumed to depend only on the probability P_n with which various local arrangements of atoms (denumerated by n) appear in the alloy. The total energy, considered as a function of the volume Ω , is written as

$$E(\Omega) = \sum_n P_n \varepsilon_n(\Omega) \tag{1}$$

where again the index n denumerates the various local arrangements (clusters), such as pairs, triangles etc., and the energy coefficients $\{\varepsilon_n(\Omega)\}$ are assumed to be transferable from one system to another.

The practical utility of the cluster expansion rests on rapid convergence of the expansion with increasing cluster complexity. Measured alloy properties provide a useful indication here. The fact that, for the hypothetical case of complete disorder, the cluster probability P_n for an n-atom cluster is an n-th order polynomial in the concentration suggests a general correlation between cluster complexity and the rapidity with which the quantities of interest vary with concentration. The rather slow variation of most measured concentration dependences therefore constitutes evidence that the expansion (Eq. 1) should converge rapidly.

Even with rapid convergence, however, there remain a large number of cluster energy coefficients to be determined, particularly if the atomic positions are entirely free to vary. This is why the cluster-expansion idea is particularly well suited to the description of "chemical" disorder, the assumption that the atoms all reside on a single crystal lattice. This assumption changes the set of possible clusters from continuous to discrete, greatly reducing the number of clusters that must be considered. In the applications described below, we assume the disorder to be exclusively chemical. An important point to emerge from the applications described below is that the freedom to vary the volume (change the lattice constant) can alter qualitative inferences drawn from measured phase diagrams (see below).

In addition to rapid convergence of the expansion and the restriction to chemical disorder, a third important factor in reducing the quantity of information required to implement the cluster expansion is the fact that the cluster probabilities $P_n(\Omega)$ appearing in Eq. 1 are not independent. The simplest example of this dependence is the fact that, in an A-B alloy, P_A and P_B are not independent because $P_A + P_B = 1$. Similarly, pair probabilities are linked, as for example by the relation $P_{AB} + P_{AA} = P_A$. Sanchez and deFontaine[11] have developed an elegant formalism for identifying and manipulating these relations. When these dependencies are properly accounted for, the basic cluster expansion, Eq. 1, is written most compactly as follows

$$E(\Omega) = \sum_n \xi_n \, v_n(\Omega) \qquad (2)$$

While Eq. 2 has the same form as the original cluster expansion (Eq. 1), the quantities appearing in Eq. 2 are defined somewhat differently and, more importantly, they are far less numerous. The multisite correlation functions ξ_n are defined as follows

$$\xi_n = \frac{1}{N_n} \sum_{\{p_i\}} \sigma_{p_1} \sigma_{p_2} \cdots \sigma_{p_n} \qquad (3)$$

where σ_p is a spin-like variable that takes the values +1 or -1, depending on whether the lattice point p is occupied by an A or B atom. The sum extends over all the n-th order clusters of a given type. The coefficients of the correlation functions, the $\{v_n(\Omega)\}$, are differences of the cluster energies $\varepsilon_n(\Omega)$, appearing in the original cluster expansion (Eq. 1). For example

$$v_4(\Omega) \propto \varepsilon_{AAAA} - 4\varepsilon_{AAAB} + 6\varepsilon_{AABB} - 4\varepsilon_{ABBB} + \varepsilon_{BBBB} \qquad (4)$$

(We have suppressed the Ω dependence of each term in Eq. 4 to simplify the notation). The virtue of the transformation of the cluster expansion from Eq. 1 to Eq. 2 is that, by accounting for the interdependencies among the P_n's, a very detailed statistical description of the local chemical environments occurring in an alloy is provided by a cluster expansion containing only a few terms. For example, the applications described below are based on a five-term cluster expansion which accounts for all the points, pairs, triangles and tetrahedra involving exclusively nearest neighbors on an fcc lattice.

The fact that such a detailed description of the local atomic environment is provided by a five-term cluster expansion is the basis of the present study. We write the five-term cluster expansion for five different systems composed of the same constituents :

$$E^i(\Omega) = \sum_{n=1}^{5} \xi_n^i \, v_n(\Omega) \; ; \; i = 1...5 \qquad (5)$$

where $E^i(\Omega)$ is the total energy of the i-th system, and ξ_n^i is the (known) probability of finding the n-th cluster in the i-th system. In the present study, we have taken these five systems (the i's in Eq. 5) to be the two pure elements and the three compounds A_3B, AB and AB_3. Using the fcc structure for the elements, the CuAu structure for the AB compound and the Cu_3Au structure for the A_3B and AB_3 compounds means that these five systems represent different "chemical" arrangements on an fcc lattice. Because, for these ordered structures, the probability functions ξ_n^i are known (see Table I), Eq. 5 can be inverted to obtain the interaction potentials $v_j(\Omega)$.

Consider now the application of the formalism described above to real materials. We feel that a sensible first test of the theory is to see if it accurately describes broad chemical trends. We therefore have considered the dominant features of the phase diagrams of a large number of intermetallic compounds, all 28 binary compounds that can be formed from the 4d transition metals Y through Pd. The measured phase diagrams for these systems exhibit considerable variety. We set as our objective a sensible categorization of these phase diagrams into characteristic types and the comparison of these types with the predictions of the parameter-free theory described above. The 4d transition series was chosen over the 3d or 5d series because the absence of both magnetic and relativistic complications makes the comparison with experiment least ambiguous.

Table I. Cluster correlation functions for five structures on an f.c.c. lattice. ξ_0 represents the structure-independent term, $\xi_1 = x_A - x_B$ is the "point" correlation function, and ξ_2,ξ_3 and ξ_4 correspond to n. n. pairs, n. n. triangles and n. n. tetrahedra respectively.

Formula	Structure	Correlation functions				
		ξ_0	ξ_1	ξ_2	ξ_3	ξ_4
A	fcc	1	1	1	1	1
A_3B	$L1_2$	1	1/2	0	- 1/2	- 1
AB	$L1_0$	1	0	- 1/3	0	1
AB_3	$L1_2$	1	- 1/2	0	1/2	- 1
B	fcc	1	- 1	1	- 1	1

Our calculated results fall into two fundamental categories. First, the cluster energy parameters (the $v_n(\Omega)$ in Eqs. 2 and 5) make a great deal of intuitive sense and provide a valuable perspective from which to view earlier work. Second, by combining the cluster expansion of the total energy with an approximate expression for the entropy, we are able to estimate the quantities that distinguish the various phase-diagram types. We show that the phase-diagram types given in this way describe very well the chemical trends in the phase diagrams of these systems. By constructing and inverting the cluster expansion of the total energy for the two pure constituents and the three compounds for each of the 28 4d/4d systems at several volumes near the theoretical equilibrium, we have constructed the cluster energy parameters and their volume dependence for all 28 systems. These interaction potentials are entirely consistent with chemical intuition. (We note that it would not have been possible to perform the enormous number of self-consistent energy-band calculations required for such a chemical trend study without a technique possessing the efficiency and reliability of the Augmented-Spherical-Wave method[12]).

The most important aspect of the interaction potentials we have obtained is that they decrease rapidly with cluster complexity ($|v_2| < |v_3| < |v_4|$), indicating a convergence of the cluster expansion that is consistent with our expectations. Another indication that the inversion of the cluster expansion yields reasonable interaction potentials is the fact that the triangle and tetrahedron potentials v_3 and v_4 are very small for systems such as PdAg and CuAg, in which directional bonding is relatively unimportant and are larger in systems such as YPd for which the average number of d electrons indicates an approximately half-filled d shell and directional bonding.

As mentioned above, the fact that our procedure allows the cluster interaction potentials to be volume dependent has interesting implications. The potentials $v_n(\Omega)$ are frequently taken to be independent of volume and concentration. In this approximation, there is a simple relation between the interaction potentials and the compound heats of formation. For example, this quantity for the $L1_0$ structure would be just $\Delta H(L1_0) = -4v_2/3$, as can be seen from equation (1) and Table I, whereas there is a more complicated relationship involving all of the v_n when the potentials are Ω-dependent. This implies that v_2 is always positive when an ordered compound forms, i.e., $\Delta H < 0$, whereas there is no such connection in the general case. Also, the significance of volume independent v_n is that they describe an n-th order dependence of E on concentration, i.e., v_2 alone would give a parabolic concentration dependence, v_3 introduces a cubic asymmetry, and v_4 is a fourth-order term. This concentration dependence remains in the present formulation, with additional terms dependent on derivatives of v_n.

We turn now to the calculation of phase diagrams. The first step in our efforts to interpret the chemical trends in alloy phase diagrams is to define what we mean by chemical trends in this context. We find that the most important qualitative features of the phase diagrams for these systems are determined by the relative magnitudes of three total energies, the concentration-weighted average of the constituent total energies, the total energy of the disordered alloy, and that of the ordered compound. We denote the difference between the first two by ΔH_D, the heat of formation of the disordered alloy, and the difference between the first and third by ΔH_O, the heat of formation of the ordered compound. For example, when ΔH_O is more negative than ΔH_D, ordered compounds will form at low temperatures. As the temperature rises, entropy causes a transition to the disordered state at a temperature given roughly by

$$T_D \sim (\Delta H_D - \Delta H_O)/S_D \tag{6}$$

where S_D is the entropy of mixing. The crystal melts at the temperature T_M. If $T_M < T_D$, then the disordering transition is not observed. A different type of phase diagram occurs when both ΔH_D and ΔH_O are positive. In this case, we get phase separation (clustering) at low temperatures, but entropy again drives a transition to a disordered-alloy state at a temperature T_D given approximately by

$$T_D \sim \Delta H_D/S_D \tag{7}$$

The different qualitative phase-diagram types that can arise depending on the relative magnitudes of the quantities ΔH_O, ΔH_D, T_D and T_M are shown in Fig. 1.

The energy-band calculations give us values of ΔH_O for the specific concentrations 25%, 50% and 75%. We use the cluster expansion Eq. 2 to calculate the total energy of the disordered alloy, required for ΔH_D. The cluster expansion requires the cluster interaction potentials, $\{v_n(\Omega)\}$, whose evaluation we described above, and the multisite correlation functions $\{\xi_n(\Omega)\}$ which are known for the case of complete randomness to be simply $\xi_{n,D} = (x_A - x_B)^n$.

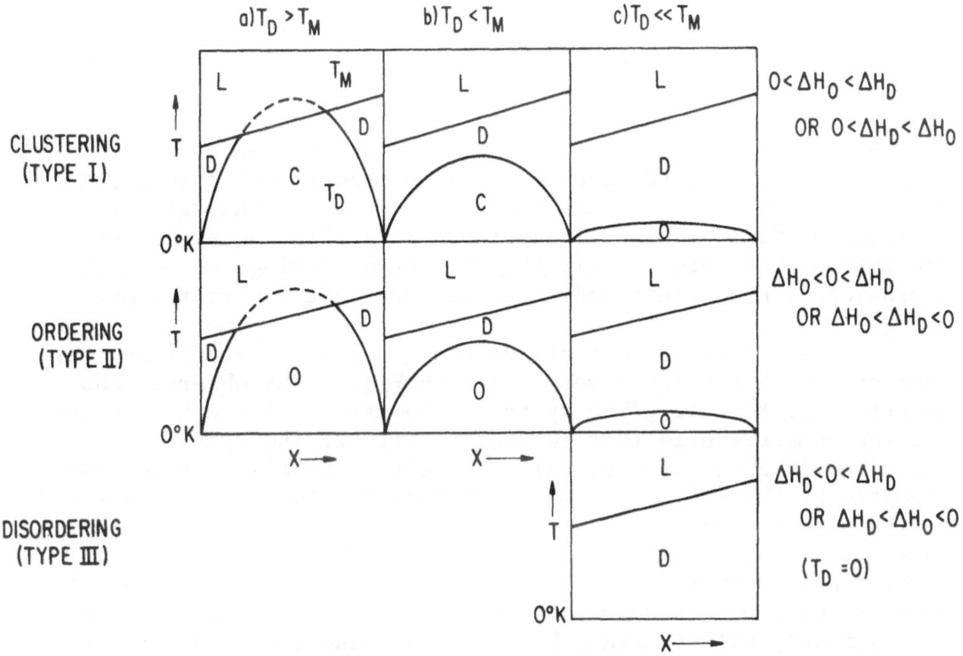

Fig. 1. Schematic phase diagrams for the three basic types. x is the concentration (= x_B), T is the absolute temperature, T_D is the disordering temperature, and T_M is the experimental melting point. L denotes the liquid phase, C means clustering (phase segregation), D is for disordering (solid solutions) and O is a region where ordered compounds occur. In order to compare with the experimental phase diagrams, we distinguish among three sub-types ; (a) where $T_D > T_M$ for most concentrations ; (b) where $T_D \lesssim T_M$ and (c) where $T_D \ll T_M$. We note that types I(c), II(c) and III will in most cases be experimentally indistinguishable.

(Note that the inversion of the multisite correlation functions appearing in Eq. 5 can be viewed as yielding a set of interpolation coefficients with which the results of calculations for the ordered alloys should be combined at each concentration and degree of short-range order. These interpolation coefficients can be applied to properties other than the total energy, the state density for example[13]. All that remains, therefore, is to estimate the entropy. For this purpose, we use the following simple meanfield expression[14]

$$S_D \sim -k(X_A \log X_A + X_B \log X_B) \quad (8)$$

where k is the Boltzmann constant and X_A and X_B are the concentrations of A and B atoms. An indication of the accuracy of this expression for the entropy is provided by Table II where we compare the disordering temperatures for a particular expression for the total energy that result from increasingly sophisticated treatments of the entropy[15,17]. Table II indicates that the disordering temperature T_D we obtain by using Eq. 8 will be too low when $\Delta H_D < \Delta H_O$ and too high when $\Delta H_O < \Delta H_D$, but we think Eq. 8 is quite adequate for our present purposes. Table II also suggests that the simple linear relations we use (Eqs. 6 and 7) to deduce the transition temperatures from the heats of formation and the entropy are slightly more accurate than mean-field theory. For example, Eq. 7 gives approximately the same dependence of T_D on the ratio v_4/v_2 as found by Monte Carlo calculations of Ref. 19. Note that Eq. 8 provides as estimate of only the entropy of mixing; we neglect vibrational, electronic and magnetic contributions to the entropy.

Figure 2 compares our calculated results to experiment. For each of the 28 compounds we compare in Fig. 2 the observed phase-diagram type to that given by the calculations. The notation used for the comparison is that defined in Fig. 1. The comparison is organized in a two-dimensional array with each alloy system designated by the average atomic number of the two constituents and by the atomic-number difference. This organization is useful because it reveals the tendency expected on the basis of recent theoretical analysis of the heat of formation of these alloys[8,22]. That is, we find a tendency toward ordering for systems with an approximately half-filled d band ($\bar{z} \sim 42$) and a tendency toward clustering for systems with a nearly full or nearly empty d band. Disordering occurs in the transition region. In comparing our results to experiment, it should be borne in mind that we have constrained the atoms to lie on an fcc lattice which is inappropriate for systems (elements or compounds) with an approximately half filled d shell. Another aspect of the same restriction to fcc is that vibrational entropy, which we have neglected, is larger in the bcc crystal structure. None the less, with a few exceptions, some of which call the experimental data into question, the theory presented here reproduces these trends quite well.

Table II. Comparison of disordering temperatures[a] for the nearest neighbor fcc Ising model in three dimensions.

	c = 1/2	c = 1/2	c = 1/4
	$v_2 < 0$	$v_2 > 0$	$v_2 > 0$
Bragg-Williams[b]	1.000	0.333	0.273
CVM[c]	0.835	0.158	0.160
Exact	0.816[d]	e	e
Monte Carlo	0.813[f]	0.145[g]	0.155[g]
Linear approx.[h]	0.722	0.240	0.222

[a] expressed in units of $2|v_2|/k$. Except where noted, these numbers are taken from Ref. 14.

[b] Mean field theory.

[c] Cluster Variation Method with clusters up to, and including nearest neighbor tetrahedra.

[d] Ref. 18.

[e] Not available.

[f] Ref. 19.

[g] Ref. 20.

[h] Using Eqs. 6 and 7.

To summarize, the cluster expansion provides a means by which the local-density approximation to exchange and correlation effects and modern energy-band methods, such as the Augmented-Spherical-Wave method, can be used to study disordered systems. The first applications of this idea described above are encouraging, but mucht more experience with the technique is required to establish its full strength and its limitations. For example, it seems clear that the technique is best suited to properties that are intrinsically local in character. Now, there is a general correlation between locality of a property and the fraction of the system involved in its definition. For example, the total energy involves all the occupied states of a system and is therefore expected to reflect primarily local properties of the system. The total electron density is another example of a very local quantity, as can be seen vividly in the range of the electron-density disturbance associated with a vacancy in a semiconductor[23]. Note that the semiconductor-vacancy exhibits this locality principle, despite the

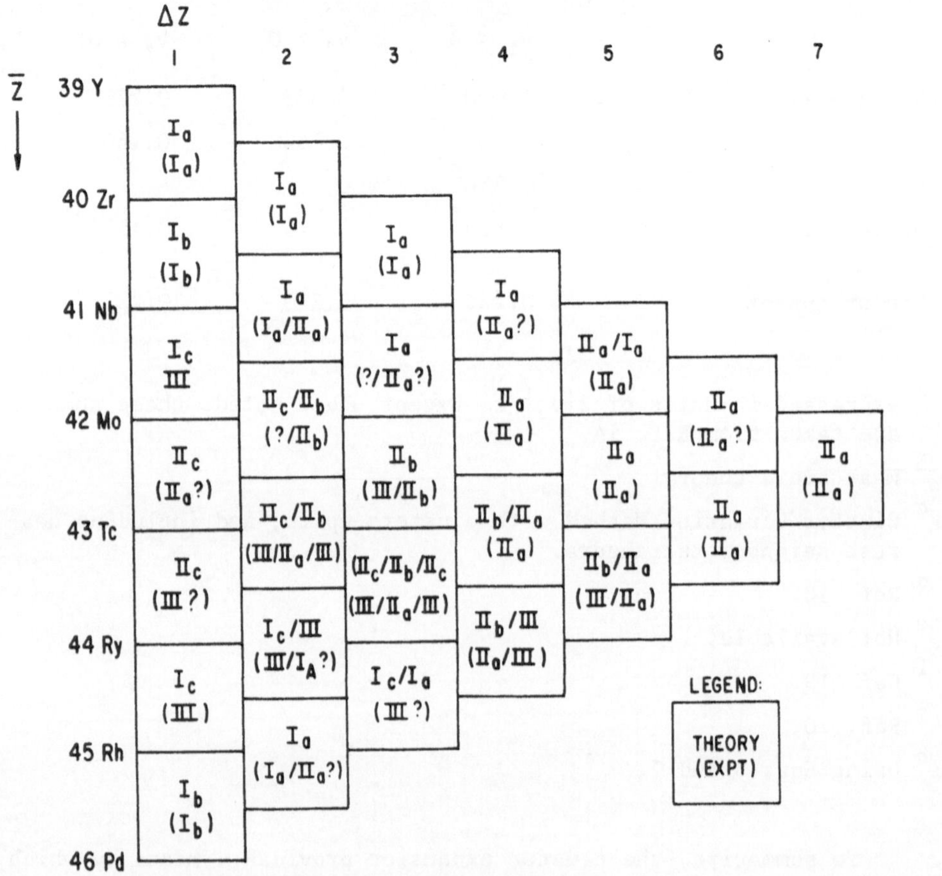

Fig. 2. Comparison of theory and experiment. A plot of phase diagram types for the 4d transition series, according to the classification of Figure 1. The horizontal coordinate is Δz, the difference between the atomic numbers of the constituents, and the vertical coordinate $\bar{z}$ is their average. The upper type is that corresponding to the present calculations. The lower (in parentheses) is the experimental type as published in the References 21. The notation I/II means that the alloy is type I at the left of the phase diagram (smaller Z) and type II to the right, etc. (?) indicates incomplete (or questionable) experimental data. Note that types I_c, II_c and III are experimentally indistinguishable.

absence of metallic screening. Less integrated properties, such as the electron density associated with particular electronic states, are generally less local. The electron density associated with states induced by semiconductor vacancies[23] serves as an illustrative example.

ACKNOWLEDGMENTS

We wish to thank V.L. Moruzzi for performing the computer calculations used to derive the total energies of the ordered compounds of this study. We are also grateful to J.M. Sanchez for helpful discussions.

REFERENCES

1. W. Kohn and L.J. Sham, Phys.Rev. 140, A1133 (1965).
2. A.R. Williams and U. von Barth, "Applications of Density-Functional Theory to Atoms, Molecules and Solids" in "The Theory of the Inhomogeneous Electron Gas", N.H. March and S. Lundqvist Eds. Plenum, New York (1982).
3. M. Schlüter and L.J. Sham, Physics Today 35, 36 (1982).
4. O. Gunnarsson, J. Harris and R.O. Jones, J.Chem.Phys. 67, 3970 (1977).
5. B.I. Dunlap, J.W.D. Connolly and J.R. Sabin, J.Chem.Phys. 71, 4993 (1979).
6. V.L. Moruzzi, J.F. Janak and A.R. Williams, "Calculated Electronic Properties of Metals", Pergamon Press, New York (1978).
7. M.T. Yin and M.L. Cohen, Phys.Rev.Lett. 45, 1004 (1980).
8. A.R. Williams, C.D. Gelatt and V.L. Moruzzi, Phys.Rev.Lett. 44, 429 (1980).
9. H. Ehrenreich and L. Schwartz, Solid State Physics 31, 149 (1976).
10. F. Gautier, F. Ducastelle and J. Giner, Phil.Mag. 31, 1373 (1975).
11. J.M. Sanchez and D. deFontaine, "Theoretical Prediction of Ordered Superstructures in Metallic Alloys", in Structure and Bonding in Crystals, vol. II, M. O'Keeffe and A. Navrotsky Eds. (Academic Press 1981), p. 117.
12. A.R. Williams, J. Kübler and C.D. Gelatt,Jr., Phys.Rev.B 19, 6094 (1974). The aspect of reliability that is particularly important for the study of chemical trends is the ability of the computer programs that implement the Augmented-Spherical-Wave method to run unattended during the night. This means that both the self-consistent-field iteration and the iteration to find the crystal volume that minimizes the calculated total energy have been successfully automated (by V.L. Moruzzi).
13. If one substitutes the values of v_n of (3) into the expression for E_D one finds that the coefficients of E_M (m = the structure index for the five structures of Table I) are the same as in the density of states expression used by C.B. Sommers

et al., Solid State Comm. 37, 761 (1981).
14. D. deFontaine, Solid State Physics 34, 73 (1979).
15. R. Kikuchi, Phys.Rev. 81, 988 (1951).
16. R. Kikuchi and D. deFontaine, NBS Publ. SP-496, 967 (1978).
17. J.M. Sanchez and D. deFontaine, Phys.Rev. B21, 216 (1980).
18. M.E. Fisher and R.J. Burford, Phys.Rev. 156, 583 (1967).
19. O.G. Mouritsen, S.J. Knak Jensen and B. Frank, Phys.Rev. B24, 347 (1981).
20. K. Binder, J.L. Lebowitz, M.K. Phani and M.H. Kalos, Acta.Met. 29, 1655 (1981).
21. M. Hansen, Constitution of Binary Alloys (McGraw-Hill 1958) ; R.P. Elliott, Constitution of Binary Alloys, First Supplement (McGraw-Hill 1965) ; F.A. Shunk, Constitution of Binary Alloys, Second Supplement (McGraw-Hill 1969) ; W.G. Moffatt, The Handbook of Binary Phase Diagrams (General Electric Co. 1977, latest update Nov. 1981).
22. D.G. Pettifor, Phys.Rev.Lett. 49, 846 (1979).
23. J. Bernholc, N.O. Lipari and S.T. Pantelides, Phys.Rev.Lett. 41, 895 (1978) ; Phys.Rev. B21, 1545 (1980).
24. The subjects of the other four lectures were : Metallic Cohesion, Transition-Metal Compound Formation, The Bonding of Transition Metals to Non-Transition Metals, and Magnetism in Transition Metals and Their Compounds. Some information on all of these subjects can be found in the book chapter cited as Ref. 2. The Augmented-Spherical-Wave energy-band method, which is the fundamental tool used to obtain much of the information discussed in the lectures is described in detail in Ref.12. The qualitative aspects of pure-metal cohesion are discussed in A.R. Williams, C.D. Gelatt Jr. and J.F. Janak, in Theory of Alloy Phase Formation L.H. Bennett editor (The Metallurgical Society of AIME, New York, 1980). Transition-metal compound formation is discussed in Ref. 8. The bonding of transition metals to non-transition metals is discussed in C.D. Gelatt Jr., A.R. Williams and V.L. Moruzzi, Phys.Rev. B 1982, (in press). Magnetism in transition-metal compounds is discussed in A.R. Williams, R. Zeller, V. Moruzzi, C.D. Gelatt Jr. and J. Kübler, J.Appl.Phys. 52, 2067 (1981) ; A.R. Williams, V.L. Moruzzi, C.D. Gelatt Jr., J. Kübler and K. Schwarz, J.Appl.Phys. 53, 2019 (1982) ; and A.R. Williams, V.L. Moruzzi, C.D. Gelatt Jr. and J. Kübler, J.Mag. and Mag. Mater. (proceedings of the ICM 82 Conf., Kyoto 1982).

ON THE THEORY OF FERRO-MAGNETISM OF TRANSITION METALS AT FINITE TEMPERATURES

B.L. Gyorffy[1], J. Kollar[2], A.J. Pindor[3], G.M. Stocks[4], J. Staunton[1] and H. Winter[5]

[1] H.H. Wills Physics Laboratory, University of Bristol U.K.
[2] Max-Planck Institute fur Festkorperforschung Stuttgart, F.R.G.
[3] Cavendish Laboratory, University of Cambridge, U.K.
[4] S.E.R.C., Daresbury Laboratory, Warrington, U.K., and Oak Ridge National Laboratory, Oak Ridge, Tenn., U.S.A.
[5] Kernforschungszentrum, Karlsruhe F.R.G.

1. ON THE NATURE OF THE PROBLEM

1) Introduction

The nature and origin of magnetic order in metals is one of the most ellusive problems in solid state physics. Fifty years after the pioneering work of Bloch (1929), Mott (1935), Stoner (1936) and Slater (1936) it is still not as well understood as the apparently more esoteric phenomenon of superconductivity (Parks 1969). While this subject has never lacked devotees at the moment we are witnessing a particularly active period due to the fact that recent progress seems to have brought the possibility of a fairly complete microscopic quantitative theory within sight. The purpose of these lectures is to sketch the outlines of the emerging picture and the prospects for doing realistic first principles calculation for metallic magnets at finite temperatures.

The spin-polarized band-theory described elsewhere in this volume is a theory of the magnetic ground state. As such it is very successful : used with care it can predict that Fe, Co, Ni are ferromagnetic but V and Pd are not (Gunnarsson 1976), the exchange splittings of the spin ↑ and ↓ bands, Δ, are roughly right (Δ_{Ni} = .65 eV, Δ_{Fe} = 1.5 eV, Δ_{Co} = 1.3 eV : Eastman et al. 1979) and it can even handle the complex spin-density-wave ground state of Cr (Rath and Callaway 1973). However, a straightforward generalization of this theory to finite temperatures results in a first principles version of the Stoner-Wohlfarth model and fails miserably : since the Curie temperature T_c works out to be $\sim 1/k_B\ \Delta$ the predicted $T_c \sim 10000$(for $\Delta \sim 1$ eV) is much too high ($T_c^{Fe} = 1043$ K, $T_c^{Co} = 1388$ K, $T_c^{Ni} = 637$ K^0). There is no Curie-Weiss law, $\chi_0^c(T) = C/T-T_c$ where χ_0 is the zero field susceptibility, and no moment even on the Fe atoms for $T > T_c$ in contradiction with experiments. Although the theoretical framework can be stretched to include spin-waves (Herring and Kittel 1951), indeed spin diffusion constants have been calculated with some success (Wang and Callaway 1976, Cook 1980), there is no way for reproducing the experimentally observed spin-waves above T_c (Mook and Nicklow 1973, Lynn, J.W. 1975).

The reasons for these failures of the theory have to do with the subject of much debate from the 1930's onwards under the heading of 'bonds versus bands'. A penetrating account of these early arguments are to be found in the monograph of Herring (1966) or in the recent book by Mattis (1982).

In short, from the point of view of spin-polarized band theory the missing ingredient is an account of 'spin fluctuations". Namely, since in such calculations all unit cells must be considered to be the same it is difficult to incorporate a proper statistical mechanics of spin configurations into the theory. Roughly speaking the magnetization goes to zero as $T \to T_c$ because the exchange splitting Δ, and therefore the magnetic moment on each site, goes to zero and not because the direction of these moments becomes randomized.

In the light of the above remarks the band picture appears quite untenable and one is tempted to attempt a description based on the Heisenberg model which is a good description for insulating magnets. The attractive feature of this approach is that one dispenses with the orbital motion of the electrons at once and one starts with a spin Hamiltonian : $H = \sum_{i,j} J_{ij}\vec{s}_i.\vec{s}_j$ where $\vec{s}_i$ is the spin (operator) at the i-th lattice point and J_{ij} is an exchange interaction integral (Heitler London Bond). In this model one gets a Curie-Weiss law without any trouble and in the paramagnetic state there is a temperature independent moment, μ, on each site albeit they point in random directions. However, following this line of thought creates more problems than it solves. For instance, in an external magnetic field, H, this model predicts that the magnetization saturates once all the spins are lined up. Experimentally, the magnetization, M, does not saturate so easily. This is in accord with the band model for which M can increase with H even at T = 0. (Callaway and Chatterjee 1978). Moreover, according to the Heisenberg model the magnetic moment per site, μ, is an integral multiple of the Bohr magneton μ_B while experimentally practically any value can be found between 0 and 10 μ_B

($\mu^{Fe} = 2.2\ \mu_B$, $\mu^{Ni} = .6\mu$, $\mu^{Gd} = 7.1\ \mu_B$). Even the central virtue of this picture : the Curie-Weiss law, is suspect. In the mean field approximation the Curie constant $C = N4\mu_B^2 q(q+1)/3k_B$ where k_B is the Boltzman constant, N is the number of sites and q is the number of magnetic carriers (Bohr magneton per site - see P. Rhodes and Wohfarth 1963). Evidently q also should be determined by the zero temperature magnetization μ per site since $q_s = \mu/\mu_B$. However, as was stressed by Rhodes and Wohlfarth (1963), q from the measured values of C, is usually different from q_s. This illustrated by their plot of q/q_s vs. T_c shown in Fig. 1.

In summary the Heisenberg Hamiltonian approach gives a good description of the thermal fluctuations in the relative orientation of the individual atomic "spins" but fails to recognise that the moment formation is a collective phenomena involving itinerant electrons and the pair-wise exchange bond, J_{ij}, with its implicit assumption of localized electrons, is also an inadequate description of the exchange forces which line up the moments once they have been formed. Evidently a complete theory would incorporate the correct features of both pictures.

Over the past 10 years or so there has been much progress towards such a synthesis. From where I stand the most visible efforts are Edwards (1970, 1980, 1982), Korenmann, Murray and Prange (1977), Korenman and Prange (1979), Moriya and Takahashi (1978) and Moriya 1979, Hubbard (1979a and b), Hasegawa (1979, 1980). A common feature of these is that they incorporate into the band theory picture the formation of localized moments whose orientation and magnitude are allowed to fluctuate. The rest of this lecture will be devoted to a simple derivation of their central results.

2) The Stoner Wohlfarth Theory at finite temperature

It is economical to base the argument on the Hubbard Hamiltonian

$$H = \sum_{i,j,\sigma} (\varepsilon_0\delta_{ij} + t_{i,j})a^+_{i\sigma}a_{i\sigma} + \frac{1}{2}I\sum_{i,\sigma} a^+_{i\sigma}a_{i,\sigma}a^+_{i,-\sigma}a_{i,-\sigma} \qquad \text{I-1}$$

where $a^+_{i,\sigma}, a_{i,\sigma}$ create and annihilate electrons with spin eigenvalue σ in the orbital state $\phi_0(\vec{r} - \vec{R}_i)$ centred on the lattice point located at R_i, ε_0 and t_{ij} are the site energy and the overlap integral of the underlying one electron tight-binding Hilbert space, and I is an effective intra-atomic electron-electron interaction parameter.

It is not difficult to include more than one band and Hund's rule exchange interaction in this model. Without this it would be a non-starter for Fe and Co where there are more than one hole per site on the average. However, for the purposes of illustration

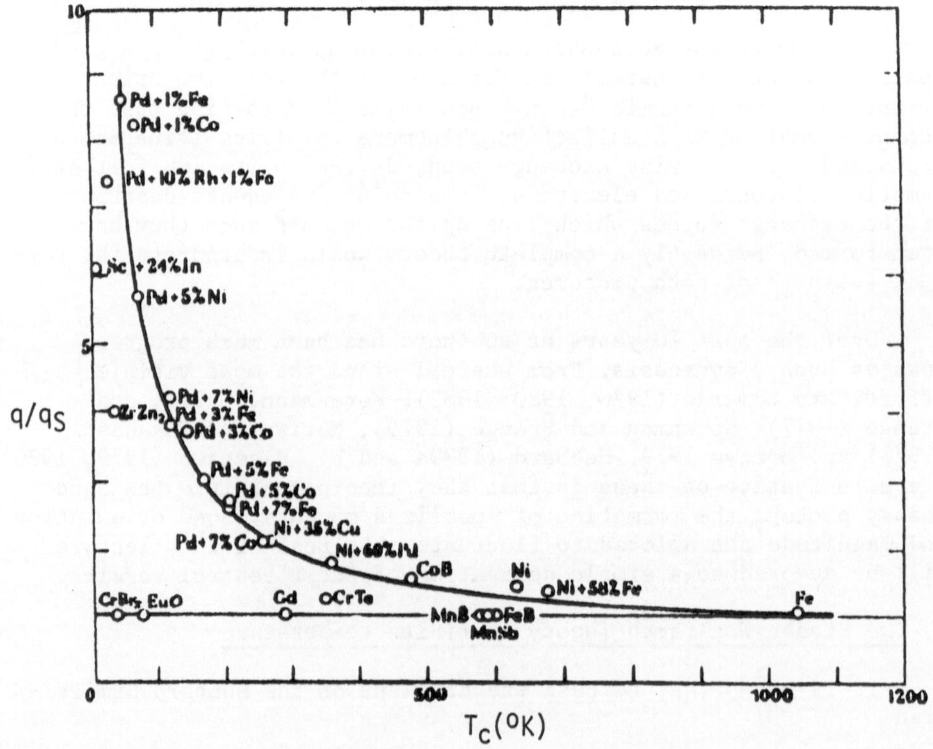

Fig. 1. Ratio q/q_S as function of Curie temperature. q, number of carriers deduced from Curie-Weiss constant C ; q_S, number of carriers deduced from saturation magnetization ; T_C , Curie temperature.

the simple form in Eq. 1 will suffice. Moreover, we will treat H in the Hartree-Fock approximation and assume that an important part of the correlations, which keep holes with parallel spins apart, have been included in I (Kanamori 1963, Hubbard 1963). Then the model is directly relevant to Ni with only .2 holes per site. In any case as a semi-phenomenological framework the Hamiltonian in Eq. 1, within the Hartree-Fock approximation, is a basis for much useful discussion (Edwards, 1979, Shimizu 1981).

One of the virutes of the Hubbard Hamiltonian is that it identifies very clearly the basic reason for the tendency of the electronic spins to line up. Parallel spins do not feel the Coulomb repulsion and hence the spins will want to line up as best they can. Locally, this leads to moment formation : if there is an up spin at a site the down spin electrons will avoid that site in order to avoid paying the energy penalty I. Since the electrons are mobile, they hop from site to site with the probability amplitude t_{ij}, the size and direction of the moments will fluctuate. To describe the statistical mechanics of these fluctuations and their correlations from site to site in a fully quantum mechanical way is the aim of the solution to be presented here and in the next section.

We seek the Hartree Fock solution by linearizing the Hamiltonian in Eq. 1

$$H_{HF} = \sum_{i,j,\sigma} [(\varepsilon_0 + I\bar{n}_{i,-\sigma})\delta_{i,j} + t_{i,j}]\, a^+_{i\sigma} a_{j\sigma} \qquad \text{I-2}$$

where

$$\bar{n}_{i,\sigma} = \frac{1}{Z_H} \operatorname{tr}\{e^{-\beta H_{HF}}\, a^+_{i,\sigma} a_{i,\sigma}\} = \langle a^+_{i,\sigma} a_{i,\sigma}\rangle$$

Then, the Fourier transform of the time dependent one particle Green's function defined as $G_{\sigma\sigma}(i,j;t) = -\,i\langle T\{a_{i,\sigma}(t)a^+_{j,\sigma}(0)\}\rangle$, where T is the Dyson time ordering operator, satisfies the following one electron-like equation

$$\sum_k [(\varepsilon^+ - \varepsilon_0 - I\bar{n}_{i,-\sigma})\delta_{i,k} - t_{i,k}]\, G_{\sigma\sigma}(k,j;\varepsilon) = \delta_{i,j} \qquad \text{I-3}$$

with the self-consistency condition

$$\bar{n}_{i,\sigma} = -\frac{1}{\pi} \operatorname{Im} \int_{-\infty}^{\infty} d\varepsilon f(\varepsilon) G_{\sigma\sigma}(i,i;\varepsilon) \qquad \text{I-4}$$

Here ε is the energy variable conjugate to t and $\varepsilon^+ = \varepsilon + i\eta$ where η is the usual positive infinitesimal quantity. Eq. I-3 is derived in Appendix 1.

The thermodynamics of the system is described by the grand potential $\Omega(T,V,\mu_e) = -\,1/\beta \,\ell n\, Z$ where $Z = \operatorname{tr}\{\exp(-\beta(H-\mu_e N))\}$.

Namely, from Ω we can calculate the pressure $P = -\frac{1}{V}\Omega(T,V,\mu_e)$, the entropy $S = -(\partial\Omega/\partial T)_{\mu_e,V}$ and the total number of particles $\overline{N}(\mu_e) = -(\partial\Omega/\partial\mu_e)_{T,V}$.

To find Ω_{HF} we integrate

$$\overline{N}(\mu_e) = \sum_{i,\sigma} \overline{n}_{i,\sigma} = -\frac{1}{\pi}\,\mathrm{Im}\sum_{i,\sigma}\int d\varepsilon f(\varepsilon) G_{\sigma,\sigma}(i,i;\varepsilon) \qquad \text{I-5}$$

By direct differentiation of

$$\Omega_{HF}(T,V,\mu_e) = \frac{1}{\pi}\,\mathrm{Im}\sum_{\sigma}\int d\varepsilon f(\varepsilon)\ln\left\|(\varepsilon^+ - \varepsilon_0 - I\overline{n}_{i,-\sigma})\delta_{ij} - t_{ij}\right\|$$

$$- \frac{1}{2} I \sum_{i,\sigma} \overline{n}_{i,\sigma}\overline{n}_{i,-\sigma} \qquad \text{I-6}$$

with respect to μ_e we get Eq. 5 and hence this expression is Hartree-Fock approximation to Ω_{HF}. Evidently, the second term is there to cancel out the contribution which arises when we take the derivative of the 'potential' $I\overline{n}_{i,-\sigma}$ in the first term. It is the usual correction for double counting.

Let us now change variables from $\overline{n}_{i\uparrow},\overline{n}_{i\downarrow}$ to the charge density $\overline{n}_i = \overline{n}_{i\uparrow} + \overline{n}_{i\downarrow}$ and the magnetic moment $\overline{\mu}_i = (\overline{n}_{i\uparrow} - \overline{n}_{i\downarrow})$ per site in units of μ_B. Then, factoring out the matrix $[G_0^{-1}]_{i\sigma,j\sigma} = (\varepsilon^+ - \varepsilon_0)\delta_{ij} - t_{ij}$ in the argument of the $\ln$ in Eq. 6 we obtain

$$\Omega_{HF} = \Omega_0 + \frac{1}{\pi}\,\mathrm{Im}\sum_{\sigma}\int d\varepsilon f(\varepsilon)\ln\left\|\delta_{ij} + \frac{1}{2} I\mu_i\sigma G_{\sigma\sigma}(i,j;\varepsilon)\right\|$$

$$- \frac{1}{4} I \sum_i (\overline{n}_i^2 - \overline{\mu}_i^2) \qquad \text{I-7}$$

where, the unperturbed grand potential

$$\Omega_0 = \frac{2}{\pi}\,\mathrm{Im}\int d\varepsilon f(\varepsilon)\ln\left\|(\varepsilon^+ - \varepsilon_0)\delta_{ij} - t_{ij}\right\| \qquad \text{I-8}$$

In the magnetic state $\overline{\mu}_i = \overline{\mu}$ for all i and near a magnetic instability $\overline{\mu}$ is small. Expanding Eq. 7 in powers of $\overline{\mu}$ gives

$$\Omega_{HF} = \Omega_0 - \frac{1}{4} I \sum_i \overline{n}_i^2 + \frac{N}{4} I\overline{\mu}^2(1 - I\chi_0) \qquad \text{I-9a}$$

where the susceptibility

$$\chi_0 = \frac{1}{\pi N}\,\mathrm{Im}\sum_{i,j}\int d\varepsilon f(\varepsilon) G^0(i,j;\varepsilon) G^0(j,i;\varepsilon) \qquad \text{I-9b}$$

Clearly the magnetic contribution to Ω_{HF} is negative if $(1 - I\chi_0) < 0$. Thus the paramagnetic state becomes unstable when

$$1 - I\,\chi_0(T_c) = 0 \qquad \text{I-10}$$

This is the finite temperature Stoner condition. At T = 0, $\Omega = E_0 - \mu_e\overline{N}$ where E_0 is the ground state energy and

$$\chi_0 = \frac{1}{\pi N}\,\text{Im}\sum_{i,j}\int d\varepsilon f(\varepsilon)G^0(i,j;\varepsilon)G^0(j,i;\varepsilon)$$

$$\cong -\frac{1}{\pi N}\,\text{Im}\sum_i\int_0^{\varepsilon_F} d\varepsilon\,\frac{\partial}{\partial\varepsilon}G^0(i,i;\varepsilon) = n(\varepsilon_F) \qquad \text{I-11}$$

the density of states at the Fermi energy ε_F. Thus we obtain the usual Stoner condition $1 - In(\varepsilon_F) < 0$ for the occurrence of the magnetic ground state.

Consider a system which favours a magnetic ground state e.g. $1 - In(\varepsilon_F) < 0$. As T increases χ_0 decreases and at T_c, defined by Eq. I-10, the paramagnetic state becomes the equilibrium state of the system : it will cost energy to produce a moment $\overline{\mu}$. As mentioned earlier, this T_c will be too high ($\Delta \sim I\mu \sim k_BT_c$). The technical reason for this is that the Fermi factor in Eq. 9 will not change sufficiently rapidly since the magnetic energy, $I\overline{\mu}$ is small on the scale of electronic energies $\sim \varepsilon_0$. A more physical way of putting this is to note that in this model the entropy is due only to excitations of the electron hole pairs described by χ_0. This under-estimates the entropy of the paramagnetic state and allows it to have higher free energy, $\Omega = U - TS + \mu_e\overline{N}$, than the magnetic state to higher temperatures. What is missing is the entropy associated with randomness in the local moments μ_i. Evidently, the disorder may be due to μ_i taking values $\pm\,\mu$ in a random fashion, transverse (Ising) spin fluctuations, or to a distribution in the magnitude of μ_i, longitudinal spin fluctuations. In general both of these effects will operate.

3) A simple theory of spin fluctuations

It is useful to regard Ω_{HF} in Eq. 6 as a function of the variables n_i,μ_i

$$\tilde{\Omega}_{HF}(\{n_i,\mu_i\}) = \frac{1}{\pi}\,\text{Im}\sum_\sigma\int d\varepsilon f(\varepsilon)\ell n\left|\left|(\varepsilon^+ - \varepsilon_0 - \tfrac{1}{2}In_i + \tfrac{1}{2}I\mu_i\sigma)\delta_{ij} - t_{ij}\right|\right|$$

$$- \frac{1}{4}\,I\sum_i(n_i^2 - \mu_i^2) \qquad \text{I-12}$$

Then, one may notice that minimizing $\Omega_{HF}(\{n_i,\mu_i\})$ with respect to independent variations in n_i and μ_i leads to Eq. 3 and 4. Namely the equilibrium values $\overline{n}_i,\overline{\mu}_i$ defined as the solution of the equations

$$\frac{\partial}{\partial n_i}\tilde{\Omega}_{HF}\bigg|_{\{\overline{n}_i,\overline{\mu}_i\}} = 0 \quad \frac{\partial}{\partial \mu_i}\tilde{\Omega}_{HF}\bigg|_{\{\overline{n}_i,\overline{\mu}_i\}} = 0 \qquad \text{I-13}$$

satisfy Eq. 3 and 4. This is the well-known stationary property of the Hartree-Fock approximation expressed in somewhat unusual terms (Anderson 1975).

The above remark suggests that $\tilde{\Omega}_{HF}(\{n_i,\mu_i\})$ may be used not only to calculate the grand potential $\Omega_{HF}(T,V,\mu_e)$ in thermal equilibrium by evaluating $\tilde{\Omega}_{HF}(\{\bar{n}_i,\bar{\mu}_i\})$ but also for associating a grand potential with any arbitrary configurations $\{n_i,\mu_i\}$. $\tilde{\Omega}_{HF}(\{n_i,\mu_i\})$ may be regarded as the work done on the system to bring about a specified charge $\{n_i\}$ and magnetic configuration $\{\mu_i\}$. This allows us to discuss the fluctuations about $\bar{\mu}_i,\bar{n}_i$.

Following the general arguments of Landau and Lifshitz (1975) we write the probability that a certain configuration $\{n_i,\mu_i\}$ occurs as

$$P(\{n_i,\mu_i\}) = \frac{1}{Z}\exp(-\beta\tilde{\Omega}_{HF}\{n_i,\mu_i\}) \qquad \text{I-14}$$

where Z is the normalization constant. Clearly, $\tilde{\Omega}_{HF}$ plays the role of an effective Hamiltonian and the thermodynamic value of the grand potential is $\Omega(T,V,\mu_e) = -\frac{1}{\beta}\ell n Z$.

If we evaluate the integrals in

$$Z = \prod_i \int dn_i \int d\mu_i \exp(-\beta\tilde{\Omega}_{HF}(\{n_i,\mu_i\})) \qquad \text{I-15}$$

in the saddle point approximation, e.g. we approximate it by the value of the integrand at the stationary point $\{\bar{n}_i,\bar{\mu}_i\}$ defined in Eq. 13, then

$$Z = \exp(-\beta\tilde{\Omega}_{HF}(\{\bar{n}_i,\bar{\mu}_i\})) \quad , \quad \Omega = \tilde{\Omega}(\{\bar{n}_i,\bar{\mu}_i\}) = \Omega_{HF} \qquad \text{I-16}$$

That is to say we have recovered the Hartree-Fock approximation discussed previously. Doing the integral better will describe the fluctuations in n_i and μ_i as we have aimed to do. Clearly, in this generalized theory the equilibrium values of n_i and μ_i are their averages with respect to the probability distribution given in Eq. 14. For example,

$$\bar{\mu}_i = \prod_j \int dn_j \int d\mu_j \; \mu_i \; P(\{n_\ell,\mu_\ell\}) \qquad \text{I-17}$$

Before discussing ways of evaluating Ω, $\bar{\mu}_i$ and $\bar{n}_i$ in this new theory we wish to make a few changes of variables to aid physical intuition. Note that the argument of the ℓn in Eq. 13 is the determinant of the inverse of the one particle Green's function $G_{\sigma\sigma}(i,j,\varepsilon)$ in Eq. 3. Evidently, 1/2 I $\bar{n}_i$ is a Coulomb field in which the electron described by $G_{\sigma\sigma}(i,j,\varepsilon)$ moves. Therefore we define a 'potential' variable $i\phi_i = \frac{1}{2} In_i$ where the $i = \sqrt{-1}$ was introduced for computational convenience to become evident later. Similarly, it is convenient to introduce the exchange field $\nu_i = \frac{1}{2} I\mu_i$. Then, using an obvious notation

$$(\varepsilon \underset{\approx}{1} - \underset{\approx}{H}^{eff}_{\sigma,\sigma}(\{\phi_i,\nu_i\})\underset{\approx}{G}_{\sigma\sigma}(\varepsilon) = \underset{\approx}{1},\ H^{eff}_{i\sigma,j\sigma} = (\varepsilon_0 + i\varphi_i - \nu_i\sigma)\delta_{ij} + t_{ij} \qquad \text{I-18a}$$

and

$$Z = \prod_i \int d\phi_i \int d\nu_i \exp(-\frac{\beta}{I}\sum_i(\varphi_i^2+\nu_i^2) - \beta\Omega^{eff}(\{\phi_i,\nu_i\})) \qquad \text{I-18b}$$

where Ω^{eff} is the grand potential for a set of non interacting electrons in an external field as described by Eq. I-18a. Namely

$$\Omega^{eff} = -\frac{1}{\beta}\ell n Z_{eff} = \frac{1}{\pi} \mathrm{Im} \sum_\sigma \int d\varepsilon f(\varepsilon) \ell n ||\varepsilon \underset{\approx}{1} - H^{eff}_{\sigma\sigma}|| \qquad \text{I-19}$$

and

$$Z = \langle Z_{eff} \rangle$$

where < > means averaging over the Gaussian distribution $\exp(\frac{-\beta}{I}\sum(\phi_i^2 + \nu_i^2))$. It is now clear that we have used a complex potential $i\phi_i$ in order that this Gaussian measure would be well behaved. Doubts about this procedure will be allayed in the next section.

The theory now has a very appealing physical meaning : an electron is moving in an external Coulomb and exchange fields which are due to other electrons. To get the form of Eq. 18 and 19 from a first principles theory we must assume that these fields ϕ_i, ν_i fluctuate slowly compared with the hopping time and therefore the adiabatic theorem can be used. The energy stored in these fields is proportional to $\sum_i(\phi_i^2 + \nu_i^2)$ and hence the averaging process. Our procedure has managed to average out the fast electronic motions but unlike the previous Hartree-Fock theory retained slow fluctuations on the time scale $\hbar\beta$. It is hoped that these will describe the spin fluctuations which were missing in the Hartree-Fock theory. This is precisely the theory derived by Hubbard (1976, 1979) and Hasegawa (1979, 1980) in what they called the static approximation.

II. AN OUTLINE OF THE MODERN SYNTHESIS

1) A general theory of spin-fluctuations

The theory described in the previous lecture might be thought to be too intuitive, lacking a conceptual framework within which its range of validity may be investigated. In this section we will derive it in a more systematic fashion.

The authors whose works are most central to the current discussions all base their argument on a functional integral representation of the partition function Z. This is achieved by the formal device often referred to as the Hubbard-Stratonovitch transformation (Hubbard 1959, Stratonovitch 1957). Like many other famous mathematical tricks, such as Feynman diagrams, the power of this approach lies in the way it breaks down a complicated problem into simpler components. If these correspond sufficiently closely to actual stages of the physical process described then our intuition about the latter may help us to make headway with the complicated mathematics.

In the interest of simplicity we shall continue to work with the single band Hubbard Hamiltonian given in Eq. I-1. In the interaction representation

$$Z = \langle T\{\exp(-\int_0^\beta d\tau\, H_{int}(\tau))\}\rangle_0 \qquad \text{II-1}$$

where $H_{int} = \frac{I}{2}\sum_{i\sigma} a^+_{i\sigma}a_{i\sigma}a^+_{i-\sigma}a_{i-\sigma}$ the evolution in complex time τ is described by $H_{int}(\tau) = \exp(\tau H_0)H_{int}\exp(-\tau H_0)$ and the average $\langle\ \rangle$ is with respect to the density matrix $(\exp(-\beta H_0))/\mathrm{Tr}\exp(-\beta H_0)$. As before we introduce the new (operator) variables $\hat{n}_i$ and $\hat{\mu}_i$ by the relations

$$n_i = \sum_\sigma a^+_{i\sigma}a_{i\sigma}\ ,\ \hat{\mu}_i = \sum_\sigma \sigma\, a^+_{i\sigma}a_{i\sigma}$$

Following the usual practice in path-integral formulations of problems (Feynman and Hibbs 1961) we discretize τ on the interval $0 \to \beta$ and replace the integral $\int_0^\beta d\tau H_{int}(\tau)$ by $\sum_n^N \frac{\beta}{N} H_{int}(\tau_n)$ where $\tau_n = \frac{\beta}{N} n$. Then

$$T\{\exp(-\int_0^\beta d\tau H_{int}(\tau))\} \cong T\{\exp(-\frac{1}{4} I \sum_{i,n}^{N} \frac{\beta}{N}(\hat{n}_i^2(\tau_n) - \hat{\mu}_i^2(\tau_n))\} \qquad \text{II-2}$$

For large enough N the exponential of the sum can be converted into products of exponentials. Each of these may be written as an integral using the identities

$$e^{\hat{\theta}^2} = \int_{-\infty}^{\infty} dx\, e^{-\pi x^2 + 2\sqrt{\pi}\hat{\theta}x}\ ,\ e^{-\hat{\theta}^2} = \int dy\, e^{-\pi y^2 + i2\sqrt{\pi}\hat{\theta}y} \qquad \text{II-3}$$

The integrand of the resulting multiple integral is a product of exponentials which can be converted back into an exponential of a sum. The result is that the right hand side of E. II-2 is equal to

$$\prod_{i,n} \int dx_{in}\int dy_{in} \exp\{-\pi \sum_{in}^{N} (x_{in}^2 + y_{in}^2) + \sqrt{\frac{I\beta\pi}{N}}(iy_{in}\hat{n}_i(\tau_n) + x_{in}\hat{\mu}_i(\tau_n)\} \qquad \text{II-4}$$

We now return to the continuum picture and replace the discrete variables x_{in}, y_{in} by the functions $x_i(\tau)$ and $y_i(\tau)$. The final result is

$$Z = Z_0 \langle T\{\exp(- \int_0^\beta d\tau H_{int}(\tau))\}\rangle_0 = \prod_i \int Dx_i(\tau) \int Dy_i(\tau)$$

$$\exp(- \int_0^\beta d\tau[\, \pi \sum_i (x_i^2(\tau) + y_i^2(\tau)) + \Omega^{eff}(\{x_i(\tau), y_i(\tau)\})] \qquad \text{II-5}$$

where $\int Dx_i(\tau)$ and $\int Dy_i(\tau)$ are functional integrals over all functions $x_i(\tau)$, $y_i(\tau)$ defined on the interval $0 \leqslant \tau \leqslant \beta$ and Ω^{eff} is the effective grand potential for independent particles moving in the combined fields of H_0 and $\{x_i(\tau), y_i(\tau)\}$. This is to say

$$\exp(-\beta\Omega^{eff}(\{x_i(\tau), y_i(\tau)\}))$$

$$= tr\{\exp(-\int_0^\beta d\tau\, [H_0 + \sqrt{I\pi} \sum_i (iy_i(\tau)\hat{n}_i(\tau) + x_i(\tau)\hat{\mu}_i(\tau))]\} \qquad \text{II-6}$$

Eq. II-5 and II-6 are the functional integral representation we sought. As it stands it is an exact rewrite of the conventional many-body theory based on the Hubbard Hamiltonian. It is easy to show that in perturbation theory it leads to the same diagrammatical expansion. If the functional integrals are evaluated by the saddle-point method we get HF. If Ω^{eff} is expanded to second order in the deviations of $x_i(\tau)$ and $y_i(\tau)$ from their Hartree-Fock values and the resulting Gaussian integrals are fully evaluated the result is the usual RPA. If the higher order terms are retained in a "Hartree" approximation we have the mode-mode coupling theory of Hertz and Klenin (1974). Evidently, the formulation of the problem is very close to the result of the more intuitive approach summarized in I-18 and I-19. Simple scaling factors apart iy_i may be identified with the Coulomb field ϕ_i which is coupled to the charge density n_i and x_i plays the role of the exchange field. The principle difference is that x_i and y_i now depend on the imaginary time τ. These 'dynamical' fluctuations have been left out of our simpler theory. Thus, it is seen as a static approximation within the more general framework described above. Clearly, it is a useful simplification if the most relevant field configurations change slowly and on the time scale $\hbar\beta \sim 1/\omega_{sw}$ where ω_{sw} a typical spin-wave frequency.

It might be useful to recall, at this point, that what we mean by a local moment and alignment of such moments is very much a matter of time scales. On time scales less than the nopping time to the nearest neighbour, $\hbar/t_{ij}$, there is usually a moment even in Na. This is the moment due to the spin of the resident unpaired electron. However, in Na, as this electron leaves and another with different spin polarization arrives, the moment averages out. If the spin polarization is maintained for a time, longer than $\hbar/t_{ij}$, but less than $\hbar\beta \sim 1/\omega_{sw}$, by a succession of electrons visiting the site, then we can say that we have a local moment which can flip on the $\hbar\beta$ time scale. Nevertheless, on thermodynamic time scales even these moments will average out in the paramagnetic state.

The extent to which local moments form varies from one magnetic material to another. There are good, large μ, moments in Fe where features of the Heisenberg model can be very relevant (Edwards 1970) but moment is small, or if it exists at all it is not associated with a site but larger regions, in materials like Ni or $ZrZn_2$. It is common to refer to the former as strong ferromagnets while the latter as weak. (There is an older parlance according to which Ni is a strong ferromagnet. We are not using the word 'strong' in that sense here). Given this situation a general theory should not be an expansion in the size of the moments. On the other hand taking into account only those degrees of freedom correctly which vary slowly on the time scale $\hbar\beta$ may not be a bad approximation. An example where the dynamics we are leaving out in going from Eq. II-5 to Eq. I-16 is of central importance is the Kondo effect (Anderson and Yuval 1969).

The aim of this short, formal discussion was to show how our picture of independent electrons moving in a fluctuating Coulomb, ϕ_i, and exchange, ν_i fields can be arrived at on a rigorous basis and thereby identify its link with the conventional many-body theory based on interaction Hamiltonians quartic in creation and annihilation operators. We now continue with the static approximation derived in Lecture I.

2) The CPA and the static approximation

In this section we will discuss an approximation scheme for carrying out the functional integral in Eq. I-16. This is a difficult problem in practice because : a) we have to solve the effective one electron problem defined by H^{eff} for an arbitrary potential field $i\phi_i - \sigma\nu_i$ at each site and b) we have to integrate over all such fields. The usual tight binding Hamiltonians have the full translational symmetry of the lattice and therefore are easily diagonalized by taking their lattice Fourier transform, (Bloch theorem). The presence of the random fields introduce all the difficulties encountered in the theory of electronic structure of disordered systems like random alloys (Faulkner 1982). In connection with b) the trouble is that only Gaussian integrals are easy to do and H^{eff} is not even a polynomial. Thus, even though we neglected the dynamics of spin fluctuations the problem is still rather intractible. Nevertheless, the new shape of the theory will be seen to be of great help because of our physical understanding of the problem can be used to render both a) and b) manageable.

Recall that evaluating the integrals in Eq. I-18b by the saddle point method leads to the Hartree-Fock approximation, e.g. the usual Stoner-Wohlfarth spin-polarized band theory. While this is an inadequate treatment of the spin orientation for T = 0 there is no reason to think that it is grossly wrong for the charge fluctuations (see, however, Moriya 1979). Therefore we shall evaluate the integrals over ϕ_i by evaluating the integrand at the stationary point ϕ_i of the exponent. To be specific

$$\varphi_i^0(\{\nu_i\}) = -\frac{1}{2} I \left(\frac{\partial\Omega^{eff}}{\partial\varphi_i}\right)\Bigg|_{\{\varphi_i^0\}} \qquad \text{II-7}$$

A second simplification is suggested by a consideration of the physical nature of the exchange field. Clearly ν_i is a scalar because, for simplicity, we are dealing with up and down (Ising) spins only. For $\nu_i > 0$ an electron scattering at the site i will see a more attractive potential if its spin is ↑ than if it is down ↓. For $\nu_i < 0$ it is the other way around. Thus the role of the exchange field is to line up the incoming electrons with itself and from the spin statistics point of view its most important characteristic is its sign. This corresponds to the fact that in the Hartree-Fock theory ν_i is determined by the moment of an electron already at the site i,$\nu_i \sim I\mu_i$. This also suggests that the most relevant values of $|\nu_i|$. If the low-energy states of the system, (these are the ones which dominate Z), are such that the sizes of the moments are roughly the same and only their orientation varies then it is reasonable to assume that $|\nu_i|$ will not fluctuate much about some fixed value ν_i^0. Hence we shall treat the variable as a classical, Ising, spin variable $\nu_i = \nu_i^0 s_i^z$ where s_i^z takes the values $\pm$ 1 and therefore determines the sign of ν_i. For the magnitude ν_i we shall take its Hartree-Fock value at a given $\{s_i^z\}$ configuration. Namely, ν_i^0 minimizes the exponent in Eq. I-16 and is given by

$$\nu_i^0 = -\frac{1}{2} I \left(\frac{\partial\Omega^{eff}}{\partial|\nu_i|}\right)\Bigg|_{\{\nu_i^0\}} \qquad \text{II-8}$$

Consequently, we replace the integral $\pi_i\int_{-\infty}^{\infty} d\nu_i$ by a sum over all spin configurations and an evaluation of the integrand at $\nu_i = \nu_i^0 s_i^z$. That is to say we evaluate the integrals over ν_i using the saddle point approximation with respect to the magnitude of ν_i but not its sign.

With these approximations Eq. I-18 becomes

$$Z = \mathrm{tr}\{\exp(-\beta H_{spin}(\{s_i^z\}))\} \qquad \text{II-9}$$

where the spin-only Hamiltonian H_{spin} is

$$H_{spin} = \frac{1}{I}\sum_i[\varphi_i^0(\{s_i^z\}))^2 + (\nu_i^0(\{s_i^z\}))^2]$$

$$+ \frac{1}{\pi}\, \mathrm{Im}\sum_\sigma\int d\varepsilon f(\varepsilon)\ell n||(\varepsilon^+ - \varepsilon_0 - i\varphi_i^0 + \nu_i^0 s_i^z\sigma)\delta_{ij} - t_{ij}|| \qquad \text{II-10}$$

where $\sigma(=\pm 1)$ is the eigenvalue of spin operator $\hat{\sigma}_z$ for the electrons.

To complete the picture we recall that ϕ_i^0 and ν_i^0 are determined by Eqs. II-7 and II-8. Working out the right hand side of each we obtain

$$i\varphi_i^0 = -\frac{1}{2} I \frac{1}{\pi} \mathrm{Im}\sum_\sigma \int d\varepsilon f(\varepsilon) G_{\sigma\sigma}(i,i;\varepsilon) = \frac{1}{2} I n_i^{eff}$$

II-11

$$\nu_i^0 = -\frac{1}{2} I \frac{1}{\pi} \mathrm{Im}\sum_\sigma \sigma s_i^z \int d\varepsilon f(\varepsilon) G_{\sigma\sigma}(i,i;\varepsilon) = \frac{1}{2} I |\mu_i^{eff}|$$

where $G_{\sigma\sigma}^{eff}(i,j;\varepsilon)$ is the Green's function of an electron in the field $V_{i\sigma}^{eff} = + i\phi_i^0 - \nu_i^0 s_i^z \sigma$, n^{eff}, μ_i^{eff} are the corresponding charge density and local moment at i. Note that Eq. II-11 is a pair of self-consistent equations for ϕ_i^0, ν_i^0. In principle, they may be solved by iteration for each spin configuration $\{s_i^z\}$.

We have now reduced the problem to that of statistical mechanics based on a spin Hamiltonian : Eq. II-9 albeit for a very complicated, non pairwise, spin-spin interaction defined by Eqs. II-10, II-11. As the next step we shall solve this in the mean field approximation. This is certainly consistent with the spirit of treating σ_i^0 and ν_i^0 in the Hartree-Fock, saddle-point, approximation.

An efficient way of doing a mean field theory for an arbitrary spin Hamiltonian $H(\{s_i^z\})$ is to use the Feynman inequality : (Feynman 1972, Hubbard 1979b, Takahashi 1981)

$$F \leqslant F_0' + \langle H - H_0 \rangle_0 \qquad \text{II-12}$$

where $F_0 = -\frac{1}{\beta} \ln \mathrm{tr}\{e^{-\beta H_0}\}$, the free energy corresponding to an arbitrary trial Hamiltonian and $\langle\ \rangle_0$ is an average with respect to the density matrix $\rho_0 = 1/Z_0\ e^{-\beta H_0}$, as a basis for a variational calculation. In this approach one guesses a simple form for H_0 with some variational parameters and determines these by minimizing the right hand side of Eq. II-12.

As an illustration of this method let us consider the Ising Hamiltonian $H = -\sum_{ij} J_{ij} s_i^z s_j^z$ and take $H_0 = -\sum_i h_i s_i^z$ where h_i is a fictitious local magnetic field to be determined variationally. Denoting the right hand side of Eq. II-12 by $\hat{F}$ we find

$$\hat{F} = -\frac{1}{\beta}\sum_i \ln(2\cosh\beta h_i) + \sum_i h_i \langle s_i^z \rangle - \sum_{i,j} J_{i,j} \langle s_i^z \rangle \langle s_i^z \rangle \qquad \text{II-13}$$

and

$$\langle s_i^z \rangle = \tanh(\beta h_i) = m_i$$

It is convenient to use m_i, the magnetization at the site i, as the variational parameter. To change variables we note that

$$h_i = \frac{1}{2\beta} \ln\left(\frac{1 + m_i}{1 - m_i}\right) \qquad \text{II-14}$$

substituting this result into Eq. II-13 leads to

$$\hat{F} = \frac{1}{\beta} \sum_i \frac{1}{2}(1 + m_i)\ln \frac{1}{2}(1 + m_i) + \frac{1}{2}(1 - m_i)\ln \frac{1}{2}(1 - m_i) - \sum_{i,j} J_{i,j}\, m_i m_j \qquad \text{II-15}$$

The first term is, evidently, - T times the entropy due to the spin disorder.

$\hat{F}$ is minimized by the configuration $\{m_i\}$ which satisfy the Euler-Lagrange equation

$$\left.\frac{\partial \hat{F}}{\partial m_i}\right|_{\{\bar{m}_i\}} = \frac{1}{2\beta} \ln\left(\frac{1 + m_i}{1 - m_i}\right) - \sum_j J_{ij} m_j = 0 \qquad \text{II-16}$$

This may be recast as

$$\bar{m}_i = \tanh \sum_j J_{ij} m_j \qquad \text{II-17}$$

which is the usual mean field theory result for the Ising model.

Let us now return to our problems as summarized in Eqs. II-9-10-11. In this case the Feynman inequality reads

$$\Omega(T,V,\mu_e) \geqslant \Omega_0 + \langle H_{spin} - H_0 \rangle \equiv \hat{\Omega} \qquad \text{II-18}$$

Again we take $H_0 = -\sum_i h_i s_i^z$ and hence

$$\hat{\Omega} = \frac{1}{\beta} \sum_i \left[\frac{1}{2}(1 + m_i)\ln \frac{1}{2}(1 + m_i) + \frac{1}{2}(1 - m_i)\ln \frac{1}{2}(1 - m_i)\right] + \langle H_{spin} \rangle \qquad \text{II-19}$$

The Euler-Lagrange equation which determines the thermal equilibrium average $\bar{m}_i = \langle s_i^z \rangle_0$ is

$$\left.\frac{\partial \hat{\Omega}}{\partial m_i}\right|_{\{\bar{m}_i\}} = \frac{1}{2\beta} \ln\left(\frac{1 + \bar{m}_i}{1 - \bar{m}_i}\right) + \left.\frac{\partial}{\partial m_i} \langle H_{spin} \rangle\right|_{\{\bar{m}_i\}} = 0 \qquad \text{II-20}$$

This leaves us with the problem of calculating $\frac{\partial}{\partial m_i} \langle H_{spin} \rangle_0$.

The average $\langle \ \rangle_0$ is with respect to the density matrix $\rho_0 = 1/Z \exp(+\beta\Sigma h_i s_i^Z)$ which may be written as a product of probabilities $\prod_i^N P_i(s_i^Z)$ where

$$P_i(s_i^z) = \begin{cases} P_i(\uparrow) = \dfrac{e^{\beta h_i}}{e^{\beta h_i} + e^{-\beta h_i}} = \dfrac{1}{2}(1 + m_i) \\[2ex] P_i(\downarrow) = \dfrac{e^{-\beta h_i}}{e^{\beta h_i} + e^{-\beta h_i}} = \dfrac{1}{2}(1 - m_i) \end{cases} \qquad \text{II-21}$$

Thus we must average over an inhomogeneous distribution of independent spins : at each site the spin is ↑ with probability $1/2(1 + m_i)$ and ↓ with probability $1/2(1 - m_i)$ and m_i, in principle, varies from site to site.

If we regard an ↑ site as an A atom and a ↓ site spins then the problem of $\langle H_{spin} \rangle$ is formally the same as encountered in studying the electronic structure of random binary alloys. This suggests that we evaluate $\langle H_{spin} \rangle$ using the well tried method of the Coherent Potential Approximation (CPA) which has been described in the lectures of Dr. Stocks. As was shown by Schwartz and Sigga (1972) the CPA may be regarded as a mean-field theory of disorder. In view of the fact that we have made several mean-field like approximations for ϕ_i^0, ν_i^0 and s_i^z it is singularly appropriate that electronic motion in the random exchange field is also treated in the same spirit.

It follows from Eq. II-10 that to find $\langle H_{spin} \rangle$ we must calculate $\langle (\phi_i^0)^2 \rangle_0, \langle (\nu_i^0)^2 \rangle_0$ and $\langle \Sigma \ell n || [(\varepsilon^+ - \varepsilon_0)\delta_{ij} - v_{i,\sigma}^{eff}]\delta_{ij} - t_{ij} || \rangle$. As the theory stands ϕ_i^0 and ν_i^0 not only depend on s_i^z the spin orientation at i, but also on s_j^z for $j \neq i$, i.e. the environment. In order to proceed we make the further assumption that there are only two kinds of potentials $V_{i\sigma} = i\,\phi_i^0(\{s_i^Z\}) + \nu_i^0(\{s_i^Z\}) s_i^Z \sigma$ e.g. $V_{i\uparrow,\sigma}$ on an ↑ site and $V_{i\downarrow,\sigma}$ on a ↓ site. Of course, these two kinds of potentials are distributed in a random fashion. Clearly, the above statement is equivalent to the assumption that $\phi_i^0(\{s_i\}) = \overline{\phi}_i^0(s_i^z)$ and $\nu_i^0(\{s_i^Z\}) = \overline{\nu}_i^0(s_i^z)$ only depend on s_i^Z. This leads to a self-consistent theory if we take

$$i\varphi_{i\uparrow}^0 = -\frac{I}{2\pi} \operatorname{Im} \sum_\sigma \int d\varepsilon f(\varepsilon) \langle G_{\sigma\sigma}^{eff}(i,i;\varepsilon) \rangle_{i\uparrow} = \frac{I}{2} \overline{n}_{i\uparrow}$$

$$\nu_{i\uparrow}^0 = -\frac{I}{2\pi} \operatorname{Im} \sum_\sigma \sigma \int d\varepsilon f(\varepsilon) \langle G_{\sigma\sigma}^{eff}(i,i;\varepsilon) \rangle_{i\uparrow} = \frac{I}{2} \overline{\mu}_{i\uparrow} \qquad \text{II-22}$$

with similar expressions for $i\phi_{i\downarrow}^0$ and $\nu_{i\downarrow}^0$. The notation $\langle \ \rangle_{i\uparrow \text{or} \downarrow}$

means an average over all configurations such that the spin at i is ↑ or ↓ respectively. For clarity we stress that in $n^{eff}_{i\uparrow}$ following site index i means the orientation of the moment at the site and σ denotes the spin of the electron which is scattering at that site.

Eq. II-23 becomes a self-consistency condition to be solved for and $\phi^0_i(s^Z_i), \overline{\nu}^0_i(s^Z_i)$ if we can calculate $\langle G^{eff}_{\sigma\sigma}(r,r;\varepsilon)\rangle$ in terms of $\overline{\phi}^0_i(s^Z_i)$, $\nu^0_i(s^Z_i)$. This is what the CPA is good for.

Briefly, the CPA strategy is as follows : one assumes that the fully averaged Green's function

$$\langle G^{eff}_{\sigma\sigma}(i,j;\varepsilon)\rangle = c_i \langle G^{eff}_{\sigma\sigma}(i,j;\varepsilon)\rangle + (1 - c_i)\langle G^{eff}_{\sigma\sigma}(i,j;\varepsilon)\rangle_{i\downarrow} \qquad \text{II-23}$$

where $c_i = 1/2(1 + m_i)$ the concentration of ↑ spin sites, is given by the Green's function $G^c_{\sigma\sigma}(i,j,\varepsilon)$ for an electron moving in a medium specified by the self energy $\Sigma^i_{\sigma\sigma}\delta_{ij}$. If at the site i, $\Sigma^i_{\sigma\sigma}(\varepsilon)$ is removed and replaced by $V_{i\uparrow,\sigma}$ the scattering of an electron with spin σ from the difference $V_{i\uparrow,\sigma} - \Sigma^i_{\sigma\sigma}(\varepsilon)$ will be described by the t-matrix

$$t_{i\uparrow,\sigma} = (1 - (V_{i\uparrow,\sigma} - \Sigma^i_{\sigma\sigma})G^c_{\sigma\sigma}(i,i;\varepsilon))^{-1}(V_{i\uparrow,\sigma} - \Sigma^i_{\sigma\sigma}) \qquad \text{II-24}$$

The medium is determined by the condition that on the average such impurities do not disturb the electrons moving in that medium. Namely $\Sigma^i_{\sigma\sigma}(\sigma)$ must satisfy the condition

$$c_i\, t_{i\uparrow,\sigma} + (1 - c_i)t_{i\downarrow,\sigma} = 0 \qquad \text{II-25}$$

The solution of Eq. II-24 and 25 determines not only $\Sigma^i_{\sigma\sigma}(\varepsilon)$ and therefore $G^c_{\sigma\sigma}(i,i;\varepsilon)$ but also gives

$$\langle G^{eff}_{\sigma\sigma}(i,i;\varepsilon)\rangle_{i\uparrow} = (1 - (V_{i\uparrow,\sigma} - \Sigma^i_{\sigma\sigma})G^c_{\sigma\sigma}(i,i;\varepsilon))^{-1}G^c_{\sigma\sigma}(i,i;\varepsilon) \qquad \text{II-26}$$

and a similar expression for $\langle G^{eff}_{\sigma\sigma}(i,i,\varepsilon)\rangle_{i\downarrow}$. Thus Eq. II-24-25 and 26 together with Eq. II-22 constitutes a theory for $\overline{\phi}^0_i(s^Z_i)$ and $\overline{\nu}^0_i(s^Z_i)$. Furthermore, the required averages, are given by

$$\langle(\nu^0_i)^2\rangle = c_i(\overline{\varphi}^0_{i\uparrow})^2 + (1 - c_i)(\overline{\phi}^0_{i\downarrow})^2$$

$$\langle(\nu^0_i)^2\rangle = c_i(\overline{\nu}^0_{i\uparrow})^2 + (1 - c_i)(\overline{\nu}^0_{i\downarrow})^2 \qquad \text{II-27}$$

To find the average of the last term in Eq. II-10, the all important single particle contribution, we rewrite the $\langle \ell n|| \ldots ||\rangle$ factor in the integrand to read

$$\ell n||G^{c,-1}_{\sigma,\sigma}(i,j;\varepsilon)|| + \langle \ell n||\delta_{ij} + (\overline{v}_{i\sigma}(s^z_i) - \Sigma^i_{\sigma\sigma})G^c_{\sigma,\sigma}(i,j;\varepsilon)||\rangle \qquad \text{II-28}$$

where all symbols stand for the appropriate site space matrices. Within the CPA one can neglect the contribution coming from the site off diagonal parts of $G^c(i,j,\varepsilon)$ in the second term. This allows us to write it as

$$\langle \ell n||\delta_{ij} + (\overline{v}_{i,\sigma}(s^z_i) - \Sigma^i_{\sigma\sigma})G^c(i,j;\varepsilon)||\rangle \cong \sum_i$$

$$c_i \ell n(1 + (v_{i\uparrow,\sigma} - \Sigma^i_{\sigma,\sigma}(i,i;\varepsilon)G^c(i,i;\varepsilon) +$$

$$+ (1 - c_i)\ell n(1 + (v_{i\downarrow\sigma} - \Sigma^i_{\sigma\sigma})G^c(i,i;\varepsilon)) \qquad \text{II-29}$$

Collecting the above results leads to

$$\langle H_{spin}\rangle = + \overline{I} \sum_i [\, c_i(\overline{\varphi}^{02}_{i\uparrow} + \overline{\nu}^{02}_{i\uparrow}) + (1 - c_i)(\overline{\varphi}^{02}_i + \overline{\nu}^{02}_i)$$

$$+ \frac{1}{\pi} \operatorname{Im} \sum_\sigma \int d\varepsilon f(\varepsilon)|\ell n||G^{c,-1}(i,j;\varepsilon)|| + \sum_i$$

$$+ c_i \ell n(1 + (v_{i\uparrow,\sigma} - \Sigma^i_{\sigma\sigma})G^c(i,i;\varepsilon))$$

$$+ (1 - c_i)\ell n(1 + (v_{i\downarrow,\sigma} - \Sigma^i_{\sigma\sigma})G^c(i,i;\varepsilon))] \qquad \text{II-30}$$

It should be stressed that the CPA theory described above is very different from the usual calculations (Faulkner 1982) on account of the fact the 'concentration' $c_i = 1/2(1 + m_i)$ varies from site to site. Namely Eq. II-26 has to be satisfied at each site separately. Consequently, the self-energy, $\Sigma^i_{\sigma\sigma}(\varepsilon)$ will be site dependent. Such an inhomogeneous CPA has been made use of by Ducastelle and Gautier (1976) and Gyorffy and Stocks (1982) in the context of random alloys. Of course, it is a formal theory only since it can not be solved for an arbitrary 'concentration' configuration $\{c_i\}$. However, we need it only for the formal purpose of taking the derivative $\partial/\partial m_i \langle H_{spin}\rangle$. Since we are interested in the homogeneously magnetized state, we wish to solve the Euler-Lagrange equation, Eq. II-20, for $m_i = \overline{m}$ for all i. Thus having derived an analytic expression for $\partial/\partial m_i \langle H_{spin}\rangle_0$ we can evaluate it for a homogeneous CPA.

If $\overline{\phi}^0_i$ and $\overline{\nu}^0_i$ i.e. the charge at a site and the local moment, do not depend much on the magnetic structure of the equilibrium state as it is described by the probability c_i that the spin is ↑ at $\overline{R}_i$, then $\partial/\partial m_i \langle H_{spin}\rangle$ takes a particularly simple form. Under this circumstances only derivatives with respect to c_i's which appear explicitly in Eq. II-30 need to be taken. This is so because derivatives such as $\partial\Sigma^i_{\sigma\sigma}/\partial m_i\ \partial/\partial\Sigma^i_{\sigma\sigma}$ do not contribute since in the CPA the last term in H_{spin}, the grand potential for the electrons, is stationary with respect to variations of the self energies. (Ducastel,). Consequently, we find

$$\frac{\partial}{\partial m_i}\langle H_{spin}\rangle\Big|_{m_i=\hat{m}} = \frac{1}{2}\frac{\partial}{\partial c_i}\langle H_{spin}\rangle_{c_i=c} \quad \text{for all}$$

$$= +\frac{1}{2I}\left[(\bar{\varphi}^0_{i\uparrow})^2 - (\bar{\varphi}^0_{i\downarrow})^2 + (\bar{\nu}^0_{i\uparrow})^2 - (\bar{\nu}^0_{i\downarrow})^2\right]$$

$$+ \frac{1}{2\pi}\,\mathrm{Im}\sum_\sigma \int d\varepsilon f(\varepsilon)\left[\ell n(1 + (\bar{v}_{\uparrow\sigma} - \Sigma_{\sigma\sigma})G^c_{\sigma\sigma}(i,i;\varepsilon))\right.$$

$$\left. - \ell n(1 + (\bar{v}_{\downarrow\sigma} - \Sigma_{\sigma\sigma})G^c(i,i;\varepsilon))\right] \qquad \text{II-31}$$

where now $\Sigma^i_{\sigma\sigma}$ is the same on all sites and $G^c_{\sigma\sigma}(i,i,\varepsilon)$ the solution of the ordinary, homogeneous CPA Green's function.

In a good moment system the majority moment $\bar{\mu}_\uparrow$ should be the same as the minority moment $\bar{\mu}_\downarrow$. This implies by the virtue of Eq. II-11, that $\bar{\nu}^0_{\uparrow\sigma} = \bar{\nu}^0_{\downarrow\sigma}$. If the size of the moment does not depend on which way it points then it is likely that the charge density and therefore the Coulomb potential is also the same on ↑ sites and ↓ sites. For such systems the first in Eq. II-31 does not contribute. Fe may be an example.

Noting that formally

$$\frac{1}{\pi}\,\mathrm{Im}\ell n(1 + (V_{\uparrow\sigma} - \Sigma_{\sigma\sigma})G^c_{\sigma\sigma}(i,i;\varepsilon)) = N^\uparrow_\sigma(\varepsilon)$$

is the Friedel sum excess integrated density of states for electrons with spin eigenvalue σ due to a spin up impurity in the $\Sigma_{\sigma\sigma}$ medium, we may finally write

$$\frac{1}{2\beta}\,\ell n\,\frac{1+\bar{m}}{1-\bar{m}} + \frac{1}{2}\sum_\sigma \int d\varepsilon f(\varepsilon)(N^\uparrow_\sigma(\varepsilon) - N^\downarrow_\sigma(\varepsilon)) = 0 \qquad \text{II-32}$$

This is then our final answer : the theory which includes some of the consequences of spin-fluctuations and is to replace the Hartree-Fock theory discussed in lecture I.

As is clear from its derivation the first term in Eq. II-31 will be discussed again in the next lecture.

In summary one must solve the basic CPA equation, II-25, $c_i = c = 1/2(1+\bar{m})$ for all i, to determine Σ and G^c, then one must calculate $\bar{\phi}^0_i(s^z_i)$ and $\bar{\nu}^0_i(s^z_i)$ using Eq. II-22 and Eq. 26 ; if these quantities are the same as their starting values then the calculation is self-consistent ; if not the procedure must be repeated with a linear combination of the old and new values of $\overline{\phi}_i^0(s_i^Z)$ and $\overline{\nu}_i^0(s_i^Z)$ until convergence ; finally, one must repeat the whole calculation for several values of $\overline{m}$ at a given temperature until $\overline{m}$ is such that Eq. II-33 is satisfied.

The magnetization per atom at this temperature is $M(T) = \mu_B\overline{\mu}^0\overline{m}$. The input parameters are the site energy ε_0, the overlap integral t_{ij} (these two represent the description of the non-magnetic band structure), and the effective exchange integral I. The magnetic properties predicted by this model are strongly influenced by all three.

Although the derivation is somewhat different, the theory presented in this lecture is the same as that of Hasegawa (1979). In Fig. 2 we show some of his results for model parameters appropriate to Fe. It is satisfying to note that the magnetization M(T) ($\equiv m_0$) goes to zero very much like a Brillouin function for s = 1/2, as one would expect it to do for a Ising model, while the local moment $\overline{\mu}^0(\equiv m_i)$ remains non zero for $T > T_c$ as observed. The temperature scale is in units of half a bandwidth which is taken to be 5 Ry. In ordinary units T_c = 800 K° which is to be compared with the experimental value T_c = 1043 K. In light of the fact that for the same parameters the Hartree-Fock theory would give 5280 K° the prediction that T_c = 800°K is highly encouraging. Prompted by these and similar results by Hubbard (1979a,b) Holden et al. (1982) and You and Heine (1982), we have begun to study the possibility of formulating the above theory in the context of a more realistic and more first-principles description of metallic electrons. The headway we have made will be reported in the next two lectures.

III. A SPIN-DENSITY FUNCTIONAL THEORY WITH UNALIGNED SPINS AT T > 0

1) The Spin Hamiltonian : By now it should be clear that to understand magnetic order in metals at finite temperatures we must treat both the spin and the orbital degrees of freedom adequately. The Stoner-Wohlfarth theory mistreats the former by assuming that all the moments point in the same direction and the Heisenberg model fails to take proper account of the itinerant nature of the electrons. The simplifying feature which offers the possibility of a synthesis has to do with the fact that these two aspects of the problem manifest themselves on different time scales. The formation of the local moment takes place on a time scale long compared with the hopping time ($\hbar\, t_{ij}^{-1} \sim \hbar x$ (the inverse band width) $\hbar$/5eV i 10^{-15} sec.) but short compared with the periods of spin waves $\hbar$/.1eV 5.10^{-13} sec. or $\sim \hbar\beta_c$. The moment may then flip on this longer scale. Thus, a theory which is to give a full account of the phenomena must be suitably taylored in each regime. The way to do this has been suggested in the previous lecture. Here we wish to generalize this approach incorporating a fuller description of the electronic structure.

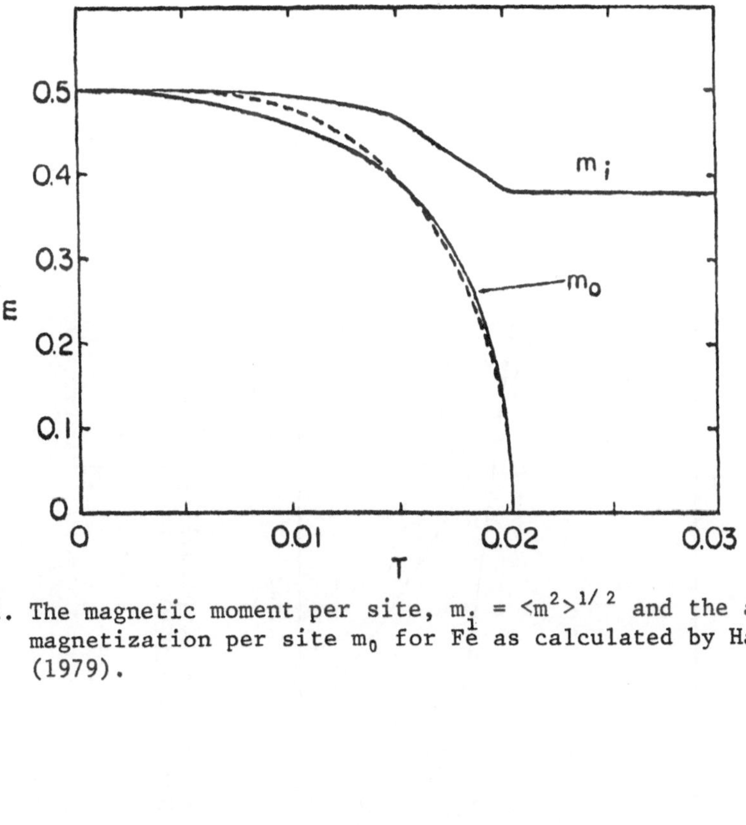

Fig. 2. The magnetic moment per site, $m_i = \langle m^2 \rangle^{1/2}$ and the average magnetization per site m_0 for Fe as calculated by Hasegawa (1979).

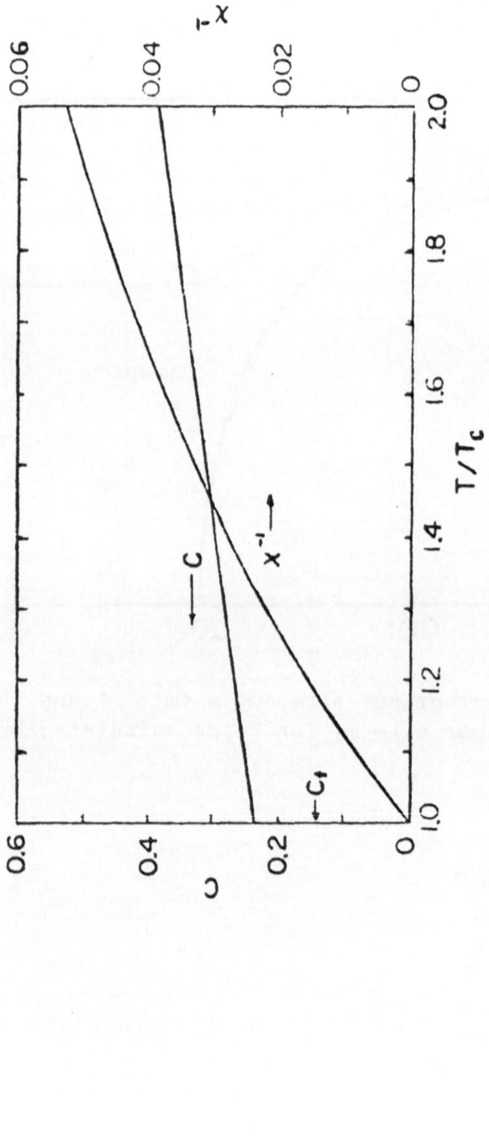

Fig. 3. The inverse susceptibility χ^{-1} and the Curie constant C for Fe as calculated by Hasegawa (1979).

Let us begin by assuming that on the long time scale $\hbar\beta$ moments have been formed and the state of the system is described by their orientation. For simplicity we shall consider only up (↑) and down (↓) (Ising) specification only, although the vector and quantum mechanical character of the moments may be important. Including these complicates the discussion but does not seem to present an unsurpassible barrier to the calculations we shall propose. The grand potential for the system may then be written as

$$\Omega(T,V,\mu_e) = -\frac{1}{\beta}\ln Z \ , \ Z = \sum_{\{s_i^z\}} e^{-\beta\Omega_e(\{s_i^z\})} \qquad \text{III-1}$$

where the coarse-grained (averaged over time scale $\hbar\, t_{ij}^{-1}$) grand potential Ω_e is that of the electron system with its moments oriented according to the configuration $\{s_i^z\}$. Thus Ω_e acts as an effective spin Hamiltonian. Clearly, our first problem is how to find a useful prescription for calculating Ω_e.

An attractive possibility is to use the finite temperature Spin-Density functional theory in the local Spin Density (LSD) approximation (Rajagopal, 1980). In this theory one constructs a functional of the charge, $n(\vec{r})$, and spin, $M(\vec{r})$, densities $\Omega[n(r)\ M(r)]$ whose minima with respect to arbitrary variations in $M(\vec{r})$ and $n(\vec{r})$ is the thermodynamic grand potential of the electrons in the external field of the atomic nuclei. Furthermore, the values of $n(\bar{r})$ and $M(\vec{r})$ at these minima, $n_0(\vec{r})$ and $M_0(\bar{r})$, are the equilibrium charge and spin densities. The suggestion here is to use the functional $\Omega[n(\bar{r}),M(\bar{r})]$ to associate a grand potential with the non-equilibrium state of the system specified by a moment orientation configuration $\{s_i^z\}$.

What makes this a viable proposition is the fact that there is an $\Omega_{e,\{s_i^z\}}[n(\bar{r}),M(\bar{r})]$ with an absolute minimum for all specified symmetries (Williams and von Barth, 1982) and these minima may be interpreted as the grand potentials of the system with those symmetries. The specification of a configuration $\{s_i^z\}$ imposes a certain symmetry on the system. Thus we may define $\Omega_e(\{s_i^z\})$ as the minimum of $\Omega_{c,\{s_i^z\}}[n(r),M(r)]$ with that symmetry. This allows us to find $\Omega_{e,\{s_i^z\}}$ by the usual procedure which yields a set of effective one-electron like Euler-Lagrange equations. In fact we find that $\Omega_{e,\{s_i^z\}}[n(\vec{r}),M(\vec{r})]$ is minimized for a given configuration $\{s_i^z\}$ by $n(\vec{r})$ and $M(\bar{r})$ defined by the Schrödinger equation

$$\left(-\frac{\hbar^2}{2m}\nabla^2 + V_\sigma^{eff}(\vec{r};[n,M,\{s_i^z\}])\right)\varphi_{\nu,\sigma}(\vec{r}) = \varepsilon_{\nu,\sigma}\varphi_{\nu,\sigma}(\vec{r}) \qquad \text{III-2}$$

for the spin dependent potential V_σ^{eff} and wave-function $\varphi_\sigma(r)$. As in other uses of the SD functional theory

$$n(\vec{r}) = \sum_{\nu,\sigma} |\varphi_{\nu,\sigma}(\vec{r})|^2 f(\varepsilon_{\nu,\sigma})$$
$$M(\vec{r}) = \sum_{\nu,\sigma} \sigma|\varphi_{\nu,\sigma}(\vec{r})|^2 f(\varepsilon_{\nu,\sigma}) \qquad \text{III-3}$$

At this stage it is natural to opt for the LD approximation and a muffin-tin form the effective potential

$$V_\sigma^{eff}(\vec{r}) = \sum_i (v_i^0(\vec{r} - \vec{R}_i) + v^z(\vec{r} - \vec{R}_i)\sigma s_i^z) \qquad \text{III-4}$$

where

$$v_i^0(\vec{r} - \vec{R}_i) = V^{ext}(r) + e^2\int dr^3 \frac{n(r')}{|\vec{r} - \vec{r}'|\pi} + \frac{\delta}{\delta n(r)} \Omega^{xc}_{e,\{s_i^z\}} \qquad \text{III-5}$$

$$v_i^z(\vec{r} - \vec{R}_i) = \frac{\delta}{\delta\mu_i(r)} \Omega^{xc}_{e,\{s_i^z\}} \qquad \text{III-6}$$

and Ω^{XC} is the usual LSD functional of n(r) and M(r) and $\mu_i(r)$ is the magnetization in the i-th unit cell stripped of its sign i.e. the local moment distribution. Of course we can contemplate only a 'gedanken' solution to Eq. III-2,3,4,5 and 6. However, it is sufficient for the present purpose.

The grand potential $\Omega_{e\{s_i^z\}}$ corresponding to the solution to the above equations is given by the usual expression (Rajagopal 1980)

$$\Omega_{e,\{s_i^z\}} = \int dr^3 V^{ext}(\vec{r}) + \frac{1}{2} e^2 \int dr^3 \int dr'^3 \frac{n(\vec{r})n(\vec{r}')}{|\vec{r} - \vec{r}'|} - \mu_e \int dr n(\vec{r})$$
$$+ G_s[n,M] + \Omega_e^{xc}[n,M]$$

where $G_s[n,M]$ is given in terms of the kinetic energy $T_s[n,M]$ and entropy $S_s[n,M]$ functional for non-interacting electrons as

$$G_s[n,M] = T_s[n,M] - T\,S_s[n,M]$$

and

$$\Omega_e^{xc}[n,M]$$

is the exchange and correlation contribution to the Gibbs free-energy and n(r) and $M(r) = \mu_B \sum \mu_i(\overline{r}) s_i^Z$ are the self-consistent solutions of the minimization procedure mentioned above.

Clearly $n(\bar{r})$ and $\mu_i(\bar{r})$ depend on configuration $\{s_i^Z\}$ which specifies the symmetry.

This is then our first principles prescription for the spin Hamiltonian. It supposed to describe the formation of moments. Note that it does not commit us to local moments. In general there will be large moments on some sites small moments on others and in fact, there will be some sites with no moment at all. If the distribution of moments for a large number of low $\Omega_{e\{s_i^z\}}$ configurations are such that they have a single peak centered on $\mu_i(r) = 0$ that is to say the no moment situation at a site is the most likely, then we may interpret this as a case where, on the time scale $\hbar\beta$, there are no local moments, only some magnetic fluctuations. On the other hand strong ferromagnetism would correspond to a peak at some finite $\bar{\mu}(r)$. It may also happen that some or all the moments turn out to be negative. This must be interpreted as implying that we did not find a locally stable minimum of the grand potential functional with the prescribed symmetry : $\Omega_{\{s_i^z\}}[n(r),M(r)]$.

2) The statistical mechanics of spin-configurations

a) The Curie Temperature T_c

We now return to the evaluation of $\Omega(T,V,\mu_c)$ in Eq. III-1. As in Lecture II we use the mean field theory based on the Feynman inequality. Following the same arguments we find

$$\tilde{\Omega} = -\sum_i m_i h_i + \frac{1}{\beta}\sum_i \left[\frac{1}{2}(1 + m_i)\ell n\,\frac{1}{2}\,(1 + m_i)\right]$$

$$+ \frac{1}{2}(1 - m_i)\ell n\,\frac{1}{2}\,(1 - m_i) + \langle\Omega_e(\{s_i^z\})\rangle \qquad \text{III-7}$$

where the average $\langle\ \rangle_0$ is to be taken with respect to the inhomogeneous product distribution function given in Eq. II-22 and we have added to a term of the form $-\Sigma h_i s_i^z$ which represents the coupling of the "spins", s_i^Z, to a fictitious dimensionless 'magnetic' field, h_i. This latter quantity have been introduced for strictly formal reasons and will be set esual to zero at the end of the calculation.

As before, the thermal equilibrium values m_i and Ω are obtained by minimizing $\hat{\Omega}$ in Eq. III-7. Taking the derivative of $\hat{\Omega}$ with respect to m_i leads to the equation of state.

$$\frac{i}{2\beta}\sum_i \left(\frac{1 + m_i}{1 - m_i}\right) + \frac{\partial}{\partial m_i}\langle\tilde{\Omega}_e(\{s_i^z\})\rangle - h_i = 0 \qquad \text{III-8}$$

To see how the theory works, we set $h_i = 0$ and solve Eq. III-8 for m_i :

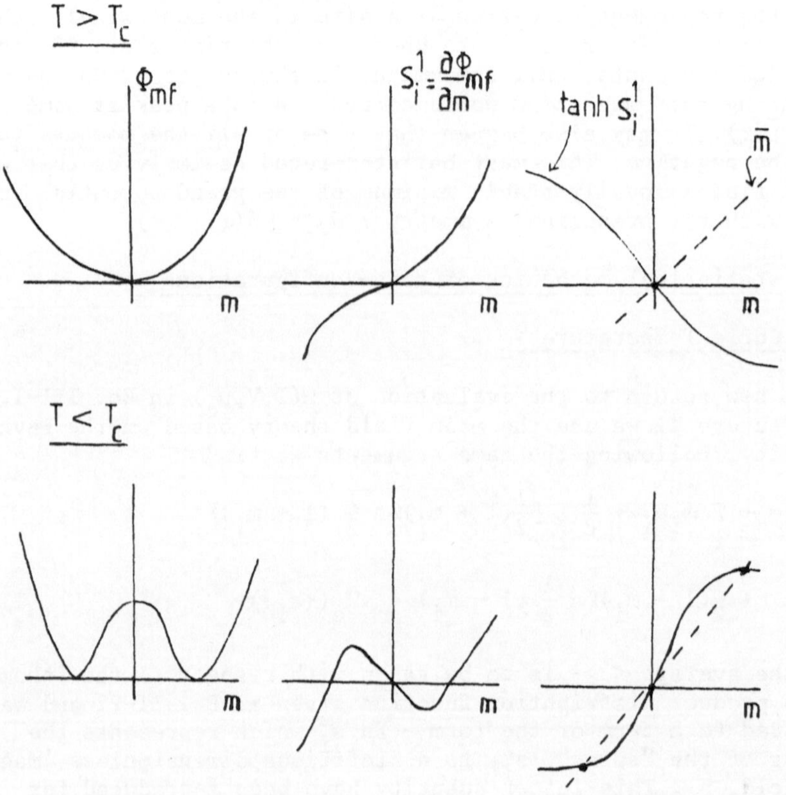

Fig. 4. A graphical scenario of how a solution to Eq. III-9 may arise.

$$\bar{m}_i = -\tanh \beta \left(\frac{\partial}{\partial m_i} \langle \Omega_{\varepsilon 0} \rangle \Big|_{\{\bar{m}_i\}}\right) \qquad \text{III-9}$$

A very satisfactory aspect of this central result is its similarly to the mean field approximation for the Ising model in Eq. II-18. Evidently $s_i^{(1)} = \partial/\partial m_i \langle \Omega_e \rangle_0 \Big|_{\{\bar{m}_i\}}$ is the mean exchange field at i analogous to $\sum_j J_{ij} \bar{s}_j^Z$. Note that this similarity is a reflection of the fact that we have incorporated into the theory the entropy associated with disorder in the spin orientations. At the same time we have maintained a full description of the itinerant nature of the electrons. This part of the problem will be handled in evaluating, $s_i^{(1)}$. Here we merely stress that the local exchange field, $s_i^{(1)}$, is a complicated function of the full configuration $\{\bar{m}_i\}$.

Now the magnetic state arises in the same way as for spin Hamiltonians. For a homogeneous state one looks for a solution to Eq. III-9 in the form $\bar{m}_i = \bar{m}$ for all i. Then whether $\bar{m} = 0$ or not depends on the functional form of $s^{(1)}(\bar{m}) = \partial/\partial m_i \langle \Omega_e \rangle |-$ as depicted in Fig. 4. To find the Curie temperature, T_c, we take the derivative of Eq. III-9 with respect to $\bar{m}$, using $\partial/\partial\bar{m} = \sum_j \partial/\partial m_j$. The result is

$$1 = -\frac{1}{\beta}(1 - \bar{m}^2) \sum_j s_{ij}^{(2)} \qquad \text{III-10}$$

where

$$s_{ij}^{(2)} = \left(\frac{\partial s_i^{(1)}}{\partial m_j}\right)\Big|_{\{\bar{m}_i\}} = \frac{\partial^2}{\partial m_i \partial m_j} \langle \Omega_e \rangle_0 \Big|_{\{\bar{m}_i\}}$$

As $T \to T_c$, $\bar{m} \to 0$ hence

$$k_B T_c = -\sum_j s_{ij}^{(2)} \qquad \text{III-11}$$

b) <u>The Curie-Weiss low</u>

As we shall show presently, the above theory is also suitable for a discussion of the static susceptibility. Let us begin with a consideration of the 'spin-spin' correlation function.

Using $\Omega_{e(\{s_i^s\})} - \sum_i h_i s_i^z$ as spin Hamiltonian

$$(G = -\frac{1}{\beta} \ln Z,\ Z = \mathrm{tr}\{\exp(-\beta(\Omega_e - \sum_i h_i s_i^z))\})$$

it is easy to show that $\tilde{\Omega}$ in Eq. III-7, with respect to h_i, is a generator $G \cong \tilde{\Omega}$ of the 'spin-spin' correlation functions :

$$<s_i^z> \equiv \bar{m}_i = -\left(\frac{\partial G}{\partial h_i}\right)_T \cong -\left(\frac{\partial \tilde{\Omega}}{\partial h_i}\right)\Bigg|_{\{\bar{m}_i\}}$$

$$<s_i^z s_j^z> - <s_i^z><s_j^z> \equiv q_{i,j} = -\frac{1}{\beta}\left(\frac{\partial^2 G}{\partial h_i \partial h_j}\right) = \frac{1}{\beta}\frac{\partial \bar{m}_i}{\partial h_i} \qquad \text{III-12}$$

$$\cong -\frac{1}{\beta}\left(\frac{\partial^2 \tilde{\Omega}}{\partial h_i \partial h_i}\right)\Bigg|_{\{\bar{m}_i\}}$$

etc.. where $\cong$ means the mean-field approximation defined by $\tilde{\Omega}$ in Eq. III-7, and $-1/\beta \, \ell n Z$ has been referred to G to recall the fact that this quantity is the Gibbs free energy of the spin system in an external magnetic field h_i.

It turns out that the derivatives of $<\Omega_{e,(\{s_i^z\})}>$ with respect to m_i also form a useful set of correlation functions

$$s_i^{(1)} = \left(\frac{\partial <\Omega_{e(\{s_i^z\})}>}{\partial m_i}\right)\Bigg|_{\{\bar{m}_i\}}$$

$$s_{ij}^{(2)} = \left(\frac{\partial^2 <\Omega_{e(\{s_i^z\})}>}{\partial m_i \, \partial m_i}\right)\Bigg|_{\{\bar{m}_i\}} \qquad \text{III-13}$$

etc... Evidently, $s_i^{(1)}$ is the exchange field discussed in connection with Eq. III-9 and $s_{ij}^{(2)}$ plays the role of the exchange integral (J_{ij}) in Eq. III-11. Moreover, $s_{ij}^{(2)}$ is related to the 'spin-spin' correlation function in a very interesting way.

To find the explicit form of this relation take the derivative of Eq. III-8 with respect to m_j. The result is

$$\frac{1}{2\beta}\frac{1}{(1-\bar{m}_i^2)}\delta_{ij} + s_{ij}^{(2)} - \frac{\partial h_i}{\partial m_j} = 0 \qquad \text{III-14}$$

where $\{h_i\}$ is that field which gives rise to $\{\bar{m}_i\}$ and hence h_i is being considered as a function of $\{\bar{m}_i\}$. From Eq. III-12 it follows that $\partial h_i/\partial m_i = 1/\beta (q^{-1})_{ij}$. Substituting this result into Eq. III-14 and inverting the resulting equation we find

$$q_{ij} = (1-\bar{m}_i^2)\,\delta_{ij} - \beta(1-\bar{m}_i^2)\sum_{\ell} s_{i\ell}^{(2)} q_{\ell j} \qquad \text{III-15}$$

Thus $s_{ij}^{(2)}$ is seen to be a kind of self-energy for the correlation function q_{ij}. In a homogeneous state where $\overline{m}_i = \overline{m}$ for all i and q_{ij} and $s^{(2)}_{ij}$ depend only on the spatial separation of the sites i and j, namely $\vec{R}_i - \vec{R}_j$. Eq. III-15 may be solved by taking its lattice Fourier Transform and we obtain

$$q(\vec{k}) = \frac{1 - \overline{m}^2}{1 + \beta(1 - \overline{m}^2)s^{(2)}(\vec{k})} \qquad \text{III-16}$$

In the hope that it might clarify matters for the reader we note that if $q(R_{ij})$ is regarded as a radial distribution function of the lattice gas model represented by our spin-spin system then $s^{(2)}(R_{ij})$ is very closely related to the corresponding Orstein-Zernicke direct correlation function (Fisher 1964). For an Ising model with pairwise exchange forces :

$$H_{Ising} = - \sum_{i,j} J_{ij} s^z_i s^z_j$$

in the Random Phase Approximation (RPA), $s^{(2)}(\vec{k}) = 2\, J(\vec{k})$. This is an other way of seeing that $s^{(2)}(\vec{k})$ plays the role of exchange interaction. However, it should be stressed that in the present theory $s^{(2)}_{ij}$ is defined by Eq. III-13 which makes no reference to pairwise spin-spin interactions. Indeed, it contains a fully itinerant description of the electrons and, moreover, it depends on the magnetic state of the system. Hence, in the language of spin Hamiltonians the theory contains many-spin interactions to all orders.

If the magnetic moment associated with each spin s^z_i were μ_B then the susceptibility χ_{ij} would be $\mu_B^2 q_{ij}$. However, the moments in our theory are the μ_i's and they are dependent on the magnetic state of the system e.g. $\{\overline{m}_i\}$. Consequently, they depend on the external magnetic field. Under these circumstances we must consider the interaction of the electrons with an external magnetic $\vec{H}(\vec{r})$ more carefully.

To begin with recall that in the spin-density functional theory $\Omega[n(\vec{r}),M(\vec{r})] = -\int dr^3 H(\vec{r})M(\vec{r}) + \Omega_e[n(\vec{r}),M(\vec{r})]$ where Ω_e is, for a given electrostatic potential and a prescribed symmetry, a unique functional of the charge density $n(\vec{r})$ and magnetization density $M(\vec{r})$. In fact Ω_e is given by Eq. III-4,5,6 with

$$V_\sigma(\vec{r}) = \sum_i v^0_i(r - \vec{R}_i) + v^z_i(\vec{r} - R_i)\sigma s^z_i$$

Thus, in the presence of a magnetic field $\tilde{\Omega}$ is of the form of Eq. III-7 h_i is replaced by $h_i = \mu_B \int \Omega_i dr \mu_i(r) H(\vec{r})$. If H varies from unit cell to unit cell but in each it is more or less constant, H_i, we may write $\mu_i H_i$. Then on minimizing $\tilde{\Omega}$ with respect to $\{m_i\}$ gives

$$\frac{1}{2\beta}\frac{1}{(1-\overline{m}_i^2)}\delta_{ij} + s_{ij}^{(2)} - \mu_B\mu_i\frac{\partial H_i}{\partial \overline{m}_j}$$

$$-\mu_B\sum_{\ell}\left(\frac{\partial \mu_\ell}{\partial \overline{m}_j}H_\ell + \mu_\ell\frac{\partial H_\ell}{\partial \overline{m}_j}\right)\overline{m}_\ell = 0 \qquad \text{III-17}$$

Moreover, the susceptibility may be defined as

$$\chi_{ij} = \frac{\partial m_i}{\partial H_j} = \frac{\partial}{\partial H_j}\left[(1+\overline{m}_i)\mu_{i\uparrow} + (1-\overline{m}_i)\mu_{i\downarrow}\right] \qquad \text{III-18}$$

where m_i, $\mu_{i\uparrow}$ and $\mu_{i\downarrow}$ all depend on H and $\mu_{i\uparrow} \neq \mu_{i\downarrow}$. Thus, for a general H and at a general temperature T,χ_{ij} is to be found from Eq. III-17 and Eq. III-18. Although this is a fairly straightforward matter it is lengthy and will not be gone into here. For our purposes it will be sufficient to consider only the zero field ($H_i = 0$) paramagnetic ($T > T_c$ and $\overline{m}_i = 0$ for all i) susceptibility.

Setting $H_i = 0$ and $\overline{m}_i = 0$ in Eqs. III-17 and III-18 we find

$$\chi_{ij} = \mu_B\mu_i\frac{\partial \overline{m}_i}{\partial H_i} + \frac{\partial}{\partial H_i}(\mu_{i\uparrow} - \mu_{i\downarrow})\Big|_{\overline{m}_i=0}$$

$$\frac{\partial H_i}{\partial \overline{m}_i} = \mu_B\mu_i\,\delta_{ij} - \beta\sum_\ell s_{i\ell}^{(2)}\frac{\partial H_\ell}{\partial \overline{m}_j} \qquad \text{III-19}$$

The term involving the field dependence of $\mu_{i\uparrow} - \mu_{i\downarrow}$ is a non spin-flip itinerant contribution (Edwards, 1982) and will be neglected here. Thus, in the homogeneous case $\mu_i = \overline{\mu}$ for all i, we have that

$$\chi(k,T) = \mu_B^2\overline{\mu}^2(T)\,\frac{1}{1+\beta s^{(2)}(\vec{k},T)} \qquad \text{III-20}$$

where we have emphasized the various sources of temperature dependences.

Note now that

$$s^{(2)}(\vec{k}=0,T) = \sum_j \frac{\partial^2\langle\Omega_{e,\{s_i^z\}}\rangle}{\partial m_i\,\partial m_i}$$

and hence, from Eq. III-11, $s^{(2)}(0,T_c) = -k_BT_c$. This is a very satisfactory result since it implies that

$$1 + \frac{1}{k_BT_C}s^{(2)}(0,T_c) = 0 \qquad \text{III-21}$$

and $\chi(k = 0,T_c) \cong \infty$. Thus Eq. III-21 is our finite temperature Stoner condition.

Assuming that $\overline{\mu}(T)$ varies more slowly with T then the denominator in Eq. III-20 we expand the latter about T_c

$$\chi(\vec{k} = 0,T) = \mu_B^2\overline{\mu}(T) \frac{1}{1 + \frac{1}{k_B}(\frac{\partial s^{(2)}}{\partial T})_{T_c}} , \frac{T_c}{T - T_c} \text{ for } T > T_c \qquad \text{III-22}$$

Clearly, if the moment $\overline{\mu}(T)$ varies slowly with temperature, then for some range of temperatures, determined by the validity of the above expansion, we have a Curie-Weiss Law inspite of the fact that our discussion is based entirely on a pure itinerant model. Of course, in general Eq. III-20 will not predict a straight line $\chi^{-1}(\vec{k},T)$. Fortunately, some deviation is in accord with the experimental facts. An other attractive feature of the above result is that any Curie constant extracted from the linear dependence of $\chi^{-1}(\vec{k},T)$ need not have anything to do with the saturation magnetization at T = 0 in contradiction to the Heisenberg and Ising models.

Having obtained a theory which, at least formally, has that right features now we shall turn to the problem of studying its quantitative consequences : namely that of calculating the direct correlation functions $s_i^{(1)}$ and $s_{ij}^{(2)}$.

IV. EVALUATION OF THE DIRECT CORRELATION FUNCTIONS $s_i^{(1)}$ AND $s_{ij}^{(2)}$ USING THE KKR-CPA

1) Restatement of the problem

To make the formal structure described in the previous lecture into a useful first principles theory we must now devise a tractible scheme for evaluating $<\Omega_{e,\{s_i^z\}}>$ and hence the correlation functions $s_i^{(1)}$ and $s_{ij}^{(2)}$. Since our model Hamiltonian is of the same general form as that used by Stocks and Winter in their lecture for developing a self-consistent band-theory for random binary alloys within the CPA it is natural to adopt their KKR-CPA and generalize it to suit the problem at hand. Evidently, the notion of A and B sites will be replaced by that of spin ↑ and spin ↓ sites. The only major change in the formal structure will come from incorporating the inhomogeneity of the distribution of ↑ and ↓ sites. However, as was the case in connection with the Hubbard model dealt with in Lec. II, this will not introduce conceptual difficulties. The theory emerging from such an approach will be entirely analogous to that of Hasegawa described earlier, in lec. II. The hoped for improvement over his and similar model calculations (Hubbard 1979a, 1979b, Holden and You 1982, You and Heine 1982, Wang et al. 1981) will be two fold : Firstly, the exchange and correlation will be described by the local spin-density functional approximation, eliminating the need for the adjustible parameter I. Secondly, the vastly more general description of the one electron like band structure efforded by the KKR aspect of the KKR-CPA will provide a, much needed, more realistic picture of the magnetic fluctuations.

At this stage it is advisable to restate the problem briefly. In principle we need to

a) Solve the self-consistent Schrödingers equations given by Eq. III-2,3,4 and 5 for a specific 'spin' configuration $\{s_i^z\}$.

b) Find the corresponding grand potential by evaluating

$$\Omega_{e,\{s_i^z\}} = -\frac{1}{\beta}\sum_n \ell n(1 + e^{-\beta(\varepsilon_n-\mu)}) - \frac{e^2}{2}\int dr^3 \int dr'^3 \frac{n(\vec{r})n(\vec{r})}{|r - r'|} + \int dr^3 n(\vec{r})(\varepsilon^{xc}(n(r) - V^0_{xc}(n(\vec{r})) + \sum_i \int drv_i^z(\vec{r} - \vec{R}_i)s_i^z M(\vec{r}) \qquad \text{IV-1}$$

for $n(\vec{r})$ and $M(\vec{r})$ obtained in a).

c) Find the average of $\Omega_{e,\{s_i^z\}}$ over the ensemble of 'spin' configurations described by the distribution function $p(\{s_i^z\}) = \prod_i p_i(s_i^z)$ with $p_i(\uparrow) = 1/2(1 + m_i)$ and $p_i(\downarrow) = 1/2(1 - m_i)$.

d) Take the derivatives $\partial/\partial m_i \langle\Omega_{e,\{s_i^z\}}\rangle$ and $\partial^2/\partial m_i \partial m_i \langle\Omega_{e,\{s_i^z\}}\rangle$ and, if we wish to study the homogeneous equilibrium state, evaluate these for $\bar{m}_i = \bar{m}$ for all i.

e) Solve the equation of state

$$\frac{1}{2\beta}\ell n\left(\frac{1+\bar{m}}{1+\bar{m}}\right) + \left(\frac{\partial}{\partial m_i}\langle\Omega_{e,\{s_i^z\}}\rangle\right)\Big|_{\{\bar{m}\}} = 0 \qquad \text{IV-2}$$

for $\bar{m}$ using the result of d).

f) Calculate $\chi(\vec{q})$ defined by Eq. III-20 for $T > T_c$ or, using Eq. III-17 and III-18, for $T < T_c$.

We shall now show how these formidable tasks can be rendered tractible by the formalism of the inhomogeneous KKR-CPA theory.

2) The Charge and Spin Selfconsistent Inhomogeneous KKR-CPA

Evidently, the selfconsistent equations given in Eq. 2,3,4,5 can not be solved exactly for an arbitrary configuration $\{s_i^z\}$ even numerically. The technical difficulty is that all the unit cells are different and therefore the Bloch's theorem, which is the bedrock of all band theory calculations, is not applicable.

Fortunately, we need only quantities like $<\Omega_{e,\{s_i^z\}}>$, $s_i^{(1)}$ and $s_{ij}^{(2)}$ which are averages over all configurations. Under this circumstances we can use the methods developed for the very similar problem of random binary alloys. As we have noted already we shall be adopting the selfconsistent KKR-CPA method described by Stocks and Winter elsewhere in this summer school. This alloy analogy' is well known in the literature of the Hubbard model (Rado and Suhl 1968) and indeed we have exploited it briefly in Lec. II. However, its use in the context of the density functional theory is novel and, while it undoubtedly opens up new visitas in spin polarized band theory, its foundations are no more then the intuitive arguments presented in Lec. III ; and its consequences are, as yet, largely unexplored. Therefore, in what follows we shall be more explicit about the details of this theory then we were in connection with Hasegawa's calculation based on the Hubbard model in Lec. II.

Firstly, let us consider the potential function in the Schrödingers equation given by Eq. III-4. By construction it is of the form of non overlapping, spherically symmetry potentials wells with a constant, V_{MTZ}, interstitial potential. V_{MTZ} is obtained by averaging the potential function over all the interstitial region. It should also be clear that a spin ↑ electron, $\sigma^z = 1$, sees two kinds of wells at the site i : $v_{i\uparrow\uparrow}(\vec{r} - \vec{R}_i)$ with probability $1/2(1 + m_i)$ and $v_{i\downarrow\uparrow}(\vec{r} - \vec{R}_i)$ with probability $1/2(1 - m_i)$. These two potential functions vary slightly from site to site because of their dependence on the environment. They are analogous to the two different kinds of scattering centers seen by an electron in a random binary alloy. This is the meaning of the 'alloy analogy'. However, here there is also an other Schrödingers equation for a ↓ spin electron. This sees also two kinds of wells $v_{i\uparrow\downarrow}(\vec{r} - \vec{R})$ with probability $1/2(1 + m_i)$ and $v_{i\downarrow\downarrow}(\vec{r} - \vec{R}_i)$ with probability $1/2(1 - m_i)$. Only in the homogeneous paramagnetic state where the concentration of the ↑ and ↓ sites are the same : $1/2(1 + m_i) = 1/2(1 - m_i) = 1/2$, e.g. $m_i = 0$ for all i, are $v_{i\uparrow\uparrow}(\vec{r} - \vec{R}_i) = v_{i\downarrow\downarrow}(\vec{r} - \vec{R}_i)$ and $v_{i\uparrow\downarrow}(\vec{r} - \vec{R}_i) = v_{i\downarrow\uparrow}(\vec{r} - \vec{R}_i)$ by symmetry. In a general homogeneous state $m \neq 0$ there is a prefered direction and the ↑ moment may be different from the ↓ moment and hence $v_{i\uparrow\uparrow} \neq v_{i\downarrow\downarrow}$ and $v_{i\uparrow\downarrow} \neq v_{i\downarrow\uparrow}$. Of course, apart from falling into one of these four qualitatively different categories in an inhomogeneous state all the potential wells will be different on account of the fact that in general each site will have a different environment.

For a given set of potential functions the two 'alloy problems' can be solved separately because there is no spin-flip term, containing σ^x or σ^y, in the Hamiltonian. However, the they are coupled if the calculation is carried on to selfconsistency because the results of both are needed to calculate the charge densities and the magnetization densities from which the new potentials are constructed. This will be explained more clearly later on when we have specified the sense in which our calculation will be selfconsistent.

Although our problem is different from that of Stocks and Winter in that the probability of a site being occupied by a given species varies from site to site we may proceed along the same line to our next stage of simplification. This is to treat the environment of each site in a mean field approximation, as is appropriate to the CPA idea which we shall use to carry out the averaging. Namely, we take, in place of V_σ^{eff} in Eq. III-2 the partially averaged crystal potential

$$\overline{V}_\sigma^{eff}(\vec{r}) = \sum_i [\overline{v}_{i\uparrow,\sigma}(\vec{r} - \vec{R}_i;\{m_i\})\tfrac{1}{2}(1 + s_i^z) + \overline{v}_{i\downarrow,\sigma}(\vec{r} - \vec{R}_i;\{m_i\})\tfrac{1}{2}(1 - s_i^z)] \qquad \text{IV-3}$$

where we have indicated that $v_{i\uparrow,\sigma}$ and $v_{i\downarrow,\sigma}$ depend on the averaged values $<s_i^z>$ at $j \neq i$ and not on the s_j^z's themselves which specify a given configurations. Note that this approximation is the same as the one made in connection with model calculation discussed in Lec. II where it II-23. To be explicit

$$\overline{v}_{i\uparrow,\sigma}(\vec{r}_i) = v_i^{ext}(\vec{r}_i) + e^2\int_{v_i} dr'^3 \frac{1}{|\vec{r}_i - \vec{r}_i'|} \overline{n}_{i\uparrow}(\vec{r}_2') \qquad \text{IV-4}$$

$$+ e^2\sum_j \int_{v_j} dr_j \frac{1}{|\vec{r}_i - \vec{r}_j|} \overline{n}_i(\vec{r}_i) + V_{i\uparrow,\sigma}^{xc}(\overline{n}_{i\uparrow}(r_i),\overline{\mu}_{i\uparrow}(r_i))$$

where $\vec{r}_i = \vec{r} - \vec{R}_i$, v_i is the volume of the unit cell centered on $\vec{R}_i$, $\overline{n}_{i\uparrow}$ is the partially averaged charge density, which is the sum of its two component $\overline{n}_{i\uparrow\uparrow}$ and $\overline{n}_{i\uparrow\downarrow}$, given that the moment at the site is pointing up, $\overline{n}_i$ is the fully averaged charge density : $1/2(1 + m_i)\overline{n}_{i\uparrow} + 1/2(1 + \overline{m}_i)\overline{n}_{i\downarrow}$, $\overline{\mu}_{i\uparrow}(r_i)$, $\overline{\mu}_{i\downarrow}(\vec{r}_i)$ and $\overline{\mu}_i(\vec{r})$ are the corresponding partially and fully averaged magnetization densities and $v^{ext}(\vec{r}_i)$ is the electrostatic potential due to the nucleus and the 'frozen' core electrons at $\vec{R}_i$. Clearly $\overline{n}_{i\uparrow}$, $\overline{n}_{i\downarrow}$, $\overline{n}_i$ and $\overline{\mu}_{i\uparrow}$, $\overline{\mu}_{i\downarrow}$, $\overline{\mu}_i$ all depend on the full specification of the ensemble $\{m_i\}$.

Then the problem is defined by the Schrödinger's

$$(\varepsilon - \frac{\hbar^2}{2m}\nabla^2 - \sum_i \overline{v}_{i,\sigma}^{eff}(r_i;s_i^z)G_{\sigma\sigma}(\vec{r},\vec{r};\varepsilon) = \delta(\vec{r} - \vec{r}') \qquad \text{IV-5}$$

the probability distribution $P(\{s_i^z\}) = \prod_i P_i(s_i^z)$ and the selfconsistency prescription

$$\begin{aligned} \overline{n}_{i\uparrow}(\vec{r}_i) &= -\frac{1}{\pi}\sum_\sigma \int d\varepsilon f(\varepsilon)\,\mathrm{Im}<G_{\sigma,\sigma}(\vec{r}_i,\vec{r}_i;\varepsilon)>_{i\uparrow} \\ \overline{\mu}_{i\uparrow}(r_i) &= -\frac{1}{\pi}\sum_\sigma \sigma\int d\varepsilon f(\varepsilon)\,\mathrm{Im}<G_{\sigma,\sigma}(\vec{r}_i,\vec{r}_i;\varepsilon)>_{i\uparrow} \end{aligned} \qquad \text{IV-6}$$

where $<\ >_{i\uparrow}$ denotes the average over $P(\{s_i^z\})$ with the restriction that the moment at $\vec{R}_i$ is ↑. The corresponding expressions for $\bar{n}_{i\downarrow}$ and $\bar{\mu}_{i\downarrow}$ are given by the obvious modifications of Eq. IV-6. Within the same approximation

$$
\begin{aligned}
<\Omega_{e,\{s_i^z\}}> = &-\frac{1}{\beta}\int d\varepsilon\bar{n}(\varepsilon)\ell n(1 + e^{-\beta(\varepsilon - \mu_e)}) \\
&- e^2 \sum_{i\neq j} \int_{V_i} dr_i^3 \int_{V_j} dr_j^3 \frac{\bar{n}_i(\vec{r}_i)\bar{n}_j(\vec{r}_j)}{|\vec{r}_i - \vec{r}_j|} \\
&- e^2 \sum_{i\alpha} c_{i,\alpha}\int_{V_i} dr_i^3 \int_{V_i} dr_i'^3 \frac{\bar{n}_{i\alpha}(r_i)\bar{n}_{i\alpha}(r_i')}{|r_i - r_i'|} \\
&+ \sum_{i\alpha} c_{i\alpha} \int_{V_{ii}} dr_i^3 [\varepsilon_{xc}(\bar{n}_{i\alpha},\bar{\mu}_{i\alpha}) - V^{xc}(\bar{n}_{i\alpha},\bar{\mu}_{i\alpha})]\bar{n}_{i\alpha}
\end{aligned}
\qquad \text{IV-7}
$$

where α = ↑ and ↓, $c_{i\uparrow} = 1/2(1 + m_i)$, $c_{i\downarrow} = 1/2(1 - m_i)$, ε_{xc} and V^{xc} are the local-density-approximations to the exchange and correlation energy density and potential respectively. For easy reference later we shall denote the first term, the so called particle contribution, by $<\Omega_{e,sp}>$ and the rest, the correction for double counting by $<\delta\Omega_e>$ e.g.

$$<\Omega_{e,\{s_i^z\}}> = <\Omega_{e,sp}> + <\delta\Omega_e> \qquad \text{IV-8}$$

Note that $<\Omega_{e(\{s_i^z\})}>$, or any part of it, depend on all the local 'concentrations' of up, $c_{i\uparrow}$, spin sites and therefore, at this stage, the derivatives $s_i^{(1)i\uparrow}$ and $s_{ij}^{(2)}$ see Eq. III.13, are well defined. Finally, we note that the partial average of the potential function at each site i makes our theory, in the parlance of the field which deals with Schrödinger like equations with random potential (Elliott et al. 1974, Ehrenreich and Schwartz 1976), a single site theory. Since the CPA is known to be the best possible single site theory, the selfconsistent equations, Eq. IV-3,4,5,6 only make sense if the partial averages like $<\ >_{i\uparrow}$ are calculated in the CPA. We now proceed to the CPA scheme appropriate to our inhomogeneous ensemble of configurations and the nonoverlapping muffin-tin form of our crystal potential.

The multiple scattering theory basis of the KKR-CPA approach requires that we begin by describing the scattering of an electron from a single potential well. In this respect a new feature of the present theory is that the potential functions $v_{i\uparrow,\sigma}(\vec{r} - \vec{R}_i)$ etc.. dependent on the spin orientation, σ, of the incident electron. Due to the fact that, for simplicity, we have chosen to work with Ising like ↑ and ↓ specification of the coupling between the electrons and the scattering centers (no spin flips) only, no complicated spin algebra will enter our discussion. Nevertheless, for the sake of completeness, we describe briefly the relevant parts of scattering theory for spin 1/2 particles with spin dependent targets in Appendix 2. Here, we merely quote the result that for potential functions which depend only on the z-component of the vector spin operator $\vec{\sigma}$, this is the case for $v_{i\uparrow,\sigma}(\vec{r} - \vec{R}_i)$ and $v_{i\downarrow,\sigma}(\vec{r} - \vec{R}_i)$ in our theory, the partial wave scattering amplitudes $f_{i\uparrow;L\sigma,L\sigma'}$, (L stands for both the polar, l, and azimuthal m, quantum numbers) are diagonal in spin space. For clarity we note that σ is the eigenvalue of the σ^z operator which refers to a common axis of spin quantization at each site. Of course $f_{i\uparrow;L\sigma,L\sigma'}$ is diagonal in angular-momentum space because the spatial part of the potential functions is taken to be spherically symmetric. Hence, a scattering event is completely described by the diagonal components

$$f_{i\uparrow,L\sigma}(e) = \frac{1}{2i}\left(e^{i2\delta^{i\uparrow}_{\ell,\sigma}(\varepsilon)} - 1\right) \text{ etc.} \qquad \text{IV-9}$$

where $\delta^{i\uparrow}_{\ell,\sigma}(\varepsilon)$ is the partial wave scattering phase-shift to be calculated from the potential function $v_{i\uparrow,\sigma}(\vec{r})$ as in spin dependent scattering problems. Evidently,

$$v_{i\uparrow\uparrow}(r) = v^0_i(\vec{r}) + v^z_i(r) \quad \text{and} \quad v_{i\downarrow\downarrow}(r) = v^0_i(\vec{r}) - v^z_i(\vec{r})$$

As mentioned previously, the inhomogeneous of the ensemble causes no conceptual difficulties in implementing the CPA idea (Goutier and Ducastelle, 1976, Gyorffy and Stocks, 1982). As in the case of the homogeneous KKR-CPA we seek a lattice of effective scattering centers amongst which the motion of an electron resembles the configurationally averaged motion of an electron on the random lattice of interest. The new feature is that each scattering center is different. This is due to the fact that the probabilities, $1/2(1 + m_i)$ and $1/2(1 - m_i)$, of a given site, i, being occupied by one (↑) or an (↓) of the scattering species of the original random lattice, vary from site to site. Thus, the effective lattice, as seen by an electron with spin-projection σ along the z-axis common to all the sites, is described by a set of effective scattering amplitudes $f_{ic;L\sigma,L'\sigma}(\varepsilon)$. Because the sites do not possess point group symmetry, the $f_{ic;L\sigma,L'\sigma'}$'s are not diagonal in the angular momentum index L(≡ ℓ,m). However, as $f_{i\uparrow;L\sigma}$ and $f_{i\downarrow,L\sigma'}$, they are diagonal in σ since the averaging does not introduce any spin-flip scattering.

In fact, it is more convenient to work the angular momentum components of the 'on the energy shell' t-matrices :

$$t_{i\uparrow;L\sigma} = -\frac{1}{\sqrt{\varepsilon}} f_{i\uparrow;L\sigma}, \quad t_{i\downarrow;L\sigma} = -\frac{1}{\sqrt{\varepsilon}} f_{i\downarrow;L\sigma}$$

and

$$t_{ic;L\sigma,L'\sigma'} = -\frac{1}{\sqrt{\varepsilon}} f_{ic;L\sigma L'\sigma}$$

Consider now the 'impurity' problem which arises when at the site i the effective scatterer, corresponding to $t_{ic;L\sigma,L'\sigma'}$ is replaced by an ↑ 'impurity' described by $t_{i\uparrow;L\sigma}$. The total scattering at i is described by $\tau^{\uparrow,ii}_{L\sigma,L'\sigma}$ of multiple scattering equations in Appendix 2 for the 'impurity problem'. Solving these is a simple matter of matrix algebra (Durham et al. 1981) and we find

$$\tau^{\uparrow,ii} = \frac{1}{1 + (t^{-1}_{i\uparrow} - t^{-1}_{i,c})\tau^{c,ii}} \tau^{c,ii} \qquad \text{IV-10}$$

The 'scattering path' matrix $\tau^{c,ij}$, whose site-diagonal components appear in the above relation, is the solution of the 'on the energy shell' multiple scattering equations for the effective lattice :

$$\sum_{\ell} [t^{-1}_{i,c}(\varepsilon)\delta_{i\ell} - G(\vec{R}_i - \vec{R}_\ell ;\varepsilon)] \tau^{c,\ell j}(\varepsilon) = \delta_{ij} \qquad \text{IV-11}$$

where $G(R_i - R_\ell ;\varepsilon)$ is the usual, real-space, KKR structure constant matrix in angular momentum space and, to be easy on the eye, we have suppressed the angular momentum (L) and spin (σ) indecies. Namely, the symbols 1, t^{-1}_{ic}, $t^{-1}_{i\uparrow}$, $\tau^{c,ij}$ and $G(R_i - R_j;\varepsilon)$ in the above formula are all matrecies in the angular momentum space spanned by L and carry an appropriate spin label. The implied divisions and multiplications are also those of matrix algebra. In what follows we shall often make use of this convention leaving it to the context to imply the full meaning of the formal expression.

Then the obvious generalization of the usual CPA condition is to require that

$$\frac{1}{2}(1 + m_i)\tau^{\uparrow,ii} + \frac{1}{2}(1 - m_i)\tau^{\downarrow,ii} = \tau^{c,ii} \qquad \text{IV-12}$$

Multiplying both sides of this equation from the left by $(1 + (t^{-1}_{\uparrow} - t^{-1}_{c})\tau^{c,ii}$ and from the right by $(1 + (t^{-1}_{\downarrow} - t^{-1}_{c})\tau^{c,ii}$ and, then, rearranging the terms leads to the fundamental equation of the inhomogeneous KKR-CPA for the present problem

$$t^{-1}_{i,c,\sigma} = \frac{1}{2}(1 + m_i)t^{-1}_{\uparrow\sigma} + \frac{1}{2}(1 - m_i)t^{-1}_{\downarrow\sigma} + (t^{-1}_{i,c,\sigma} - t^{-1}_{\uparrow\sigma})\tau^{c,ii}_{\sigma\sigma} (t^{-1}_{i,c,\sigma} - t^{-1}_{\downarrow\sigma}) \qquad \text{IV-13}$$

Evidently, we have a separate equation for each site and they are coupled through the $\tau^{c,ii}$ which depend on all the effective scattering amplitudes according to Eq. IV-10. This set of coupled equations determine all the $t_{i,c,\sigma;LL'}$'s.

As the next step we must now develop this theory further to facilitate the calculation of the partially averaged Greens functions which can be used in Eq. IV-6 for evaluating the partially averaged charge and magnetization densities : $\overline{n}_{i\uparrow}$ etc... and $\overline{\mu}_{i\uparrow}$ etc... respectively. Following Faulkner and Stocks (1981) and Stocks and Winter (1982) we find

$$\mathrm{Im}\langle G_{\sigma\sigma}(\vec{r}_i,\vec{r}_i;\varepsilon)\rangle_{i\uparrow} = \sum_{L,L'} Z_{i\uparrow,L\sigma}(\vec{r}_i;\varepsilon)Z_{i\uparrow,L'\sigma}(\vec{r}_i;\varepsilon)\mathrm{Im}\tau^{\uparrow,ii}_{L\sigma,L'\sigma} \qquad \text{IV-14}$$

where $Z_{i\uparrow,L\sigma}(\vec{r}_i;\varepsilon)$ is the solution of the Schrödingers equation for the potential well at $\vec{R}_i$ with the interstitial constant potential (V_{MT}) extended to infinity in every direction. It is regular at the origin and matches smoothly to an out going wave solution as this is defined in Appendix 2.

Equations IV-3,4,5,9,13,14 are now form a closed, selfconsistent set. In principle it can be solved by iteration and it yields the local charge and magnetization densities $\overline{n}_{i\uparrow}(\vec{r}_i)$, $\overline{n}_{i\downarrow}(\vec{r}_i)$, $\mu_{i\uparrow}(\vec{r}_i)$, $\mu_{i\downarrow}(\vec{r}_i)$ respectively. Of course they all depend on $\{m_i\}$. This is our selfconsistent inhomogeneous KKR-CPA theory. For clarity we stress that selconsistency here means with respect to charge and magnetization as well as disorder. This is the part of the problem which was solved by the saddle point approximation in the case of the Hubbard model discussed in Sec. II.

Let us now return to our primary aim of evaluating $\langle\Omega_e(\{s_i^z\})\rangle$, given in Eq. IV-7. Consider first the single particle contribution $\langle\Omega_{sp}\rangle$ defined in Eq. IV-8. Integrating by parts leads to

$$\langle\Omega_{sp}\rangle = \mu_e Z^V - \int d\varepsilon\, \overline{N}(\varepsilon)f(\varepsilon) \qquad \text{IV-15}$$

where Z^V is the number of valence electrons per atom and $\overline{N}(\varepsilon)$ is the configurationally averaged integrated density of states. The generalization of the usual KKR-CPA formula for $\overline{N}(\varepsilon)$ (Gyorffy and Stocks 1979, see also Eq. II-74 of Stocks and Winter in this volume) reads as follows

$$\overline{N}(\varepsilon) = N_0(\varepsilon) - \frac{1}{\pi n}\,\mathrm{Im}\sum_\sigma \ell n\,||t^{-1}_{c,i,\sigma}\,\delta_{ij} - G(\vec{R}_i,\vec{R}_j;\varepsilon)||$$

$$- \frac{1}{\pi n}\,\mathrm{Im}\sum_{i\sigma}\,[\tfrac{1}{2}(1 + m_i)\ell n||D^{i\uparrow}_\sigma|| + \tfrac{1}{2}(1 - m_i)\, n||D^{i\downarrow}_\sigma||] \qquad \text{IV-16}$$

where $D^{i\uparrow}_{LL'}(\varepsilon)$ is defined as

$$D^{i\uparrow}_{LL'}(\varepsilon) = [(1 + (t^{-1}_{\uparrow\sigma} - t^{-1}_{ic,\sigma}))^{-1}]_{LL'} \qquad \text{IV-17}$$

As we have mentioned before the above inhomogeneous theory can not be implemented computationally. Its purpose is purely formal. It allows us to take the derivatives of $<\Omega_e(\{s^z_i\})>$ and $<\Omega_{sp}>$ which define the correlation functions $s^{(1)}_i$ and $s^{(2)}_{ij}$. Once we have a formal expression for $s^{(1)}_i$ and $s^{(2)}_{ij}$ we shall evaluate it in the limit where $m_i = \overline{m}$ for all i. In a way we are studying inhomogeneous fluctuations about the homogeneous KKR-CPA state and the correlation functions have to do with its linear response. As we shall show presently these response functions are accessible by the computational means developed in the KKR-CPA studies of random alloys.

3) The mean exchange field $s^{(1)}_i$

We are now in the position to derive a formula for $s^{(1)}_i = \partial/\partial m_i <\Omega_e>$. Eq. IV-16 and 17 form a particularly convenient starting point for treating the single particle contribution. The reason for this is the remarkable property of $\overline{N}(\varepsilon)$ that it is stationary with respect to variations in $t_{ic,L,L',\sigma}(\varepsilon)$. For the homogeneous KKR-CPA this, most useful fact was first noted by Lloyd (1974) and it is easy to show that it is also true in the present inhomogeneous theory. One merely has to take the derivative of Eq. IV-16 and after some straight-forward matrix algebra note that the condition $(\partial/t_{ic,LL',\sigma})\overline{N}(\varepsilon) = 0$ implies Eq. IV-13. Thus

$$\frac{\partial}{\partial m_i} <\Omega_{e,sp}>_0 = \frac{1}{2}\frac{\partial}{\partial c_i} <\Omega_{e,sp}> = \frac{1}{2}\frac{\partial}{\partial c_i} <\Omega_{e,sp}>\Big|_{t_{ic,LL',\sigma}\ \text{fixed}} \qquad \text{IV-18}$$

and hence

$$S^{(1)}_{i,sp} = -\sum_\sigma \int d\varepsilon f(\varepsilon)\,(N_{i\uparrow,\sigma} - N_{i\downarrow,\sigma}) \qquad \text{IV-19}$$

where

$$N_{i\uparrow,\sigma}(\varepsilon) = -\frac{1}{\pi}\,\text{Im}\,||\underset{\sim}{1} + (t^{-1}_{i\uparrow,\sigma} - t^{-1}_{ic,\sigma})\tau^{c,ii}_{\sigma\sigma}|| \qquad \text{IV-20}$$

and we have again used our convention that the symbols within the determinant sign $||\ \ ||$ stand for angular momentum matrices.

It is reassuring to note that Eq.19 is of the same for as our earlier result in Eq. II.33 for the Hubbard model. We would like to stress that although we named the right hand side of Eq. IV-19 $N_{i\uparrow,\sigma}$ suggesting that it is the excess integrated density of states due to an ↑ spin in our effective medium it is not in fact that. Because the effective scattering amplitudes $t^{-1}_{ic,\sigma}$ correspond to energy dependent potentials there is a correction to the simple

Friedel sum on the right hand side of Eq. IV-19 for the true impurity integrated density of states. Nevertheless, we used this notation because we do not expect $N^{true}_{i\uparrow,\sigma}$ to differ much from $N_{i\uparrow,\sigma}$ and for qualitative discussion we may disregard this difference. Thus, the forgoing reservations notwithstanding we interpret Eq. IV-19 to mean that the local exchange field $s^{(1)}_{i,sp}$ is the free energy (grand potential) difference between and ↑ and ↓ impurity at $\vec{R}_i$ in our inhomogeneous medium. Evidently this is a pleasingly physical result.

Calculating the contributions to $s^{(1)}_i$ due to the double counting correction in Eq. IV-8 from Eq. IV-7 we find

$$s^{(1)}_{i,\alpha} \equiv \frac{\partial}{\partial m_i} \langle\delta\Omega\rangle = \int dr_i^3 [\varepsilon_{xc}(\bar{n}_i,\bar{\mu}_{i\uparrow}) - \varepsilon_{xc}(\bar{n}_i,\bar{\mu}_{i\downarrow})]\bar{n}_i(\vec{r}_i)$$

$$- (v^{xc}(\bar{n}_i,\bar{\mu}_{i\uparrow}) - V^{xc}(\bar{n}_i,\bar{\mu}_{i\downarrow})\bar{n}_i(\bar{r}_i) + \text{terms involving}$$

$$\frac{\partial n_i}{\partial c_j}\ \frac{\partial \bar{\mu}_{i\uparrow}}{\partial c_j},\ \frac{\partial \bar{\mu}_{i\downarrow}}{\partial c_j}\ \text{etc.} \qquad \text{IV-21}$$

Let us now return to our equation of state

$$\frac{1}{2\beta}\ \ell n(\frac{1+\bar{m}_i}{1-\bar{m}_i}) + s^{(1)}_i = 0 \qquad \text{IV-22}$$

where $s^{(1)}_i$ is to be determined by the selfconsistent inhomogeneous KKR-CPA as discussed above. Since $s^{(1)}_i$ depends on the full equilibrium orientation configuration, e.g. $\{\bar{m}_i\}$. In general Eq. IV-21 is a very complicated equation. However, with only minor loss in generality we may seek the equilibrium state in a particular form: ferromagnetic, antiferromagnetic or spin density wave state to name but a few. Namely we may specify a certain relation between the m_i's leaving only a limited number of parameters to be determined by Eq. 21. For instance if we expect a ferromagnetic state we may take as a trial solution $\bar{m}_i = \bar{m}$ for all i. Then the $N_{i\uparrow,\sigma}$ in Eq. IV-19 will be the same for all i. That is to say

$$\frac{1}{2\beta}\ \ell n\ (\frac{1+\bar{m}}{1-\bar{m}}) - \sum_\sigma \int d\varepsilon f(\varepsilon)(N_{\uparrow,\sigma}(\varepsilon) - N_{\downarrow\sigma}(\varepsilon)) = 0 \qquad \text{IV-23}$$

where, for simplicity we have discarded the contribution from the double counting terms. Evidently, $N_{\uparrow,\sigma}$ and $N_{\downarrow\sigma}$ in this relation is to be calculated by solving a homogeneous, albeit selfconsistent and spin-polarized, KKR-CPA problem. Fortunately, such calculations can be, actually, performed by the methods described in the lectures of Drs. Stocks and Winter. Thus, as we shall show presently, the quantitative consequence of the above theory can be explored without further essential simplifications.

We conclude this discussion by noting that for $\overline{m} = 0$ $N_{\uparrow\sigma} = N_{\downarrow -\sigma}$. This is a consequence of the fact that the CPA preserves the symmetry of the 'alloy problem' under the interchange of species and the corresponding concentrations ($v_\uparrow \to v_\downarrow$, $c = 1/2(1 + \overline{m}) \to 1 - c = 1/2(1 - \overline{m})$ and $1 - c = 1/2(1 - m) \to c = 1/2(1 + m)$. Consequently, $s_i^{(1)}$ as a function of $\overline{m}$ must pass through 0 at $\overline{m} = 0$ ($\overline{m} = 0$ is always a solution to Eq. 21 or 22). If $s_i^{(1)}$ rises from negative values to positive values as $\overline{m}$ goes from -1 to +1 then at som low enough temperature there will always be an other solution corresponding to a more stable ferromagnetic $\overline{m} \neq 0$ equilibrium state. This situation is sketched in Fig. 5. Alternatively, $s_i^{(1)}$ may fall from some positive values to some negative values for increasing $\overline{m}$. Then, as is clear from the topology of Fig. 5, there will always be only one paramagnetic $\overline{m} = 0$, solution. In that case we must conclude that there is no transition to a ferromagnetic state. However, there might be transitions to phases with more complicated magnetic structure. In order to investigate such possibilities we would have to study $s_i^{(1)}$ for the appropriate symmetry. As an example consider the possibility of an antiferromagnetic phase. Then, we would have to take $\overline{m}_i$ to be $\overline{m}_1$ on one sublattice and $\overline{m}_2$ on an other sublattice. Then $s_i^{(1)}$ would take on two different values $s_1^{(1)}$ and $s_2^{(1)}$ on the two different sublattices. These could be found by solving the two sublattice versions of Eq. IV-13. The methods for doing this have been developed recently by Pindor et al. (1982) in the context of two sublattice random alloys. Once $s_1^{(1)}$ and $s_2^{(1)}$ have been calculated as a function of the staggered magnetization variables $\overline{m}_1, \overline{m}_2$ the appropriate version of Eq. 21 needs to be checked for solutions.

4) The Direct Spin-Spin Correlation Function : $s_{i,j}^{(2)}$

To keep the discussion centered on the most interesting features of the problem we shall deal with the single particle contributions only. It follows from Eq. III-13 and Eq. IV-19 that it is given by

$$S_{ij}^{(2)} = -\frac{1}{2}\sum_{\sigma}\int d\varepsilon f(\varepsilon)\left(\frac{\partial}{\partial m_j} N_{i\uparrow,\sigma}(\varepsilon) - \frac{\partial}{\partial m_j} N_{i\downarrow,\sigma}(\varepsilon)\right) \qquad \text{IV-24}$$

With the help of the definitions given in Eq. IV-20 we can carry out the differentiation with respect to m_j. The result is

$$\frac{\partial}{\partial m_j}(N_{i\uparrow,\sigma} - N_{i\downarrow,\sigma}) = -\frac{1}{\pi}\,\mathrm{Im}\sum_{\ell\neq i} \mathrm{tr}\{(D_{i\uparrow} - D_{i\downarrow})\,(\tau^{c,ii})^{-1}$$

$$(\tau^{c,i\ell}\,\frac{\partial t^{-1}_{\ell,c}}{\partial m_j}\,\tau^{c,\ell i})\} \qquad \text{IV-25}$$

where tr is over the angular-momentum matrecies and we had made use of the relations

$$D_{i\uparrow}(t_{i\uparrow}^{-1} - t_{ic}^{-1})\,\tau^{c,ii} = 1 - D_{i\uparrow}$$

$$\frac{\partial}{\partial m_j}\tau^{c,ii} = -\sum_{\ell}\tau^{c,i\ell}\,\frac{\partial t_{\ell,c}^{-1}}{\partial m_j}\,\tau^{c,\ell i} \qquad \text{IV-26}$$

The next step is to derive an equation for the response function

$$\Lambda_{L,L'}^{i,j}(\varepsilon) = \frac{\partial}{\partial m_j}\,t_{i,c;LL'}^{-1}(\varepsilon) \qquad \text{IV-27}$$

To do this we take the derivative of the fundamental inhomogeneous KKR-CPA equations given in Eq. IV-13 with respect to m_j. After some rearranging of the terms this leads to

$$(1 + (t_{i\uparrow}^{-1} - t_{ic}^{-1})\tau^{c,ii})\Lambda^{i,j}(\tau^{c,ii}(t_{i\downarrow}^{-1} - t_{ic}^{-1}) + 1) = \frac{1}{2}\,(t_{i\uparrow}^{-1} - t_{i\downarrow}^{-1})\delta_{ij}$$

$$-\sum_{\ell\neq i}(t_{i\uparrow}^{-1} - t_{ic}^{-1})\tau^{c,i\ell}\Lambda^{\ell,j}\tau^{c,\ell i}(t_{i\downarrow}^{-1} - t_{ic}^{-1}) \qquad \text{IV-28}$$

This expression may be rewritten as

$$\Lambda_{L_1L_2}^{i,j}(\varepsilon) = \Lambda_{L_1L_2}^{0;i,j}(\varepsilon) - \sum_{\ell\neq i}\sum_{L_3L_4} I_{L_1L_2;L_3L_4}^{i,\ell}(\varepsilon)\Lambda_{L_3L_4}^{\ell,j}(\varepsilon) \qquad \text{IV-29}$$

where

$$\Lambda_{L_1L_2}^{0;i,j}(\varepsilon) = \frac{1}{2}\,(D_{i\uparrow}(t_{i\uparrow}^{-1} - t_{i\downarrow}^{-1})D_{i\downarrow})_{L_1L_2}\,\delta_{ij}$$

$$I_{L_1L_2;L_3L_4}^{i,\ell}(\varepsilon) = \sum_{L'L''}(D_{i\uparrow}(t_i^{-1} - t_{ic}^{-1})_{L_1L'}((t_{i\downarrow}^{-1} - t_{ic}^{-1})D_{i\downarrow})_{L''L_2}$$

$$\cdot\; X_{L'L'';L_3L_4}^{i\ell}(\varepsilon)$$

$$X_{L'L'';L_3L_4}^{i\ell}(\varepsilon) = \tau_{L'L_3}^{c,i\ell}(\varepsilon)\;\tau_{L_4L''}^{c,\ell i} \qquad \text{IV-30}$$

As we have emphasized repeatedly the inhomogeneous CPA was introduced to facilitate the formal evaluation of the derivatives $s_i^{(1)}$, $s_{i,j}^{(2)}$. In order to study the properties of the ferromagnetic state we can evaluate these within the homogeneous KKR-CPA, e.g. for $m_i = \overline{m}$ for all i. Then t_{ic}^{-1}, $D_{i\uparrow}$ and $D_{i\downarrow}$ are independent of i and, for cubic systems when the highest l retained in the angular momentum expansion is $\ell_{max} = 2$, are diagonal in L. Moreover, $X^{i\ell}$ and therefore $I^{i,\ell}$ depend on the difference coordinate $\vec{R}_i - \vec{R}_\ell$. Consequently, Eq. IV-30 can be simplified by taking its lattice Fourier Transform. This yields

$$\Lambda_{L_1L_2}(\vec{k};\varepsilon) = \Lambda^0_{L_1}(\varepsilon)\delta_{L_1L_2} - \sum_{L_3L_4} I_{L_1L_2;L_3L_4}(\vec{k};\varepsilon)\Lambda_{L_3L_4}(\vec{k};\varepsilon) \qquad \text{IV-31}$$

where

$$\Lambda^0_L(\varepsilon) = \frac{1}{2} D_{\uparrow L,L}(\varepsilon)\ (t^{-1}_{\uparrow,L} - t^{-1}_{\downarrow,L})\ D_{\downarrow,LL}(\varepsilon)$$

$$I_{L_1,L_2;L_3L_4}(\vec{k};\varepsilon) = D_{\uparrow L_1,L_1}\ (t^{-1}_{\uparrow L_1} - t^{-1}_{c_1L_1})(t^{-1}_{\downarrow,L_2} - t^{-1}_{c,L_2})D_{\downarrow L_2L_2} \cdot X_{L_1L_2;L_3L_4}(\vec{k};\varepsilon)$$

$$X_{L_1L_2;L_3L_4}(\vec{k};\varepsilon) = \frac{1}{\Omega_{BZ}}\int d^3k'\tau_{L_1L_3}(\vec{k}' + \vec{k};\varepsilon)\tau_{L_4L_2}(\vec{k}';\varepsilon) - \tau^{c,ii}_{L,L_3}(\varepsilon)\tau^{c,ii}_{L_4L_1}(\varepsilon) \qquad \text{IV-32}$$

To solve Eq. IV-31 is now only a matter of matrix algebra at each value of energy ε and wave vector k.

Returning to $s^{(2)}_{i,j}$ in Eq. IV-24 we take its lattice Fourier Transform and our final result is

$$S^{(2)}(\vec{k}) = \frac{1}{\pi}\ \mathrm{Im}\ \sum_{\sigma}\int d\varepsilon f(\varepsilon) \sum_{LL'L''} D_{\uparrow,LL}(t^{-1}_{\uparrow,L} - t^{-1}_{\downarrow L})D_{\downarrow,LL}\Lambda_{L',L''}(\vec{k}\varepsilon)\ X_{LL'L''}(\vec{k};\varepsilon) \qquad \text{IV-33}$$

Like KKR band theory the KKR-CPA converges rapidly in L (Faulkner Davis and Joy 1967, Gyorffy and Stocks 1979). Thus, it is likely that in evaluating the above formulae for Fe,Co and Ni there will be no need to go beyong taking the maximum azimuthal quantum number $\ell_{max} = 2$. Then Eq. IV-31 involves 9 X 9 metrices and hence its solution requires only a modest computational effort. However, the evaluation of the Brillouin zone integral in Eq. IV-32 is a major undertaking comparable to that solving the KKR-CPA equations selfconsistently.

Fortunately, the last of these remarks do not apply at $\vec{k} = 0$. $X_{L_1L_2;L_3L_4}(k = 0;\varepsilon)$ can be readily evaluated using routines which are part and parcel of a KKR-CPA calculation. This means that first principles calculations of $c^{(2)}$ (k = 0,T) and therefore of T_c from Eq. III-21 should be available soon.

In this connection we hasten to add that the above contribution to $S^{(2)}(k)$ is due to only those spin fluctuations in which the moments turn over rigidly. However, $\mu_\uparrow \neq \mu_\downarrow$ in general and we expect that the $\partial/\partial H(\mu_\uparrow - \mu_\downarrow)$ term in Eq. III-19 will play a role even in Fe. Fortunately, the theory of this contribution leads to an equation very similar to Eq. IV-31 and can be handled accordingly.

5) Exchange Splitting and Local Moments at Finite Temperatures

As yet, the consequences of the foregoing first-principles theory remain largely unexplored. Nevertheless, in this last section we are able to report on our first calculations concerning the formation of local moments in Iron above the Curie temperature T_c.

Using the computational techniques described by Dr. Stocks we have solved the homogeneous KKR-CPA equations, Eq. 13 with $m_i = 0$, for Fe at various temperatures. That is to say the concentration of ↑ sites, $1/2(1 + m_i) = 1/2$, was taken to be the same as the concentration of ↓ sites, and hence the system described was in the paramagnetic state. The temperature entered the calculation only through the Fermi factor in Eq. IV-6 which came into play when the charge densities and the magnetic moment densities were calculated at the end of each self-consistency cycle. The spin dependent potential functions were calculated using the local-spin-density approximation as in the book of Maruzzi, Janak, and Williams (1977) in each iteration.

The phase-shifts for a spin ↑ electron corresponding to the spin ↑ and the spin ↓ selfconsistent potentials in the T = 2000 K calculation are shown in Fig. 5. Because of symmetry the phase shifts for a spin ↓ electron are given by the same curves but $\delta_{\ell\uparrow}(\varepsilon)$ is renamed as $\delta_{\ell\downarrow}(\varepsilon)$ and $\varepsilon_{\ell\uparrow}(\varepsilon)$ as $\delta_{\ell\downarrow}(\varepsilon)$. The splitting between the two $\ell = 2$ resonances is about 1 eV which is comparable to the average exchange-splitting of bands in the T = 0 spin-polarized band theory calculations of Maruzzi et al. (1977). This is a good sign since it indicates that exchange splitting has taken place in spite of the fact that the moments point up and down in a random fashion with no net magnetization. The densities of states for a ↑ electron, $n_{\uparrow\uparrow}(\varepsilon)$ and $n_{\uparrow\downarrow}(\varepsilon)$ are shown in Fig. 6a. For clarity we also show the densities of states for a ↓ electron in Fig. 6b where the curves $n_{\downarrow\downarrow}(\varepsilon)$ and $n_{\uparrow\downarrow}(\varepsilon)$ are the same as $n_{\uparrow\uparrow}(\varepsilon)$ and $n_{\downarrow\uparrow}(\varepsilon)$ respectively in Fig. 6a on account of symmetry. The physical content of these results can now be read off with ease.

Consider the states at the energy ε_1 picked for the purposes

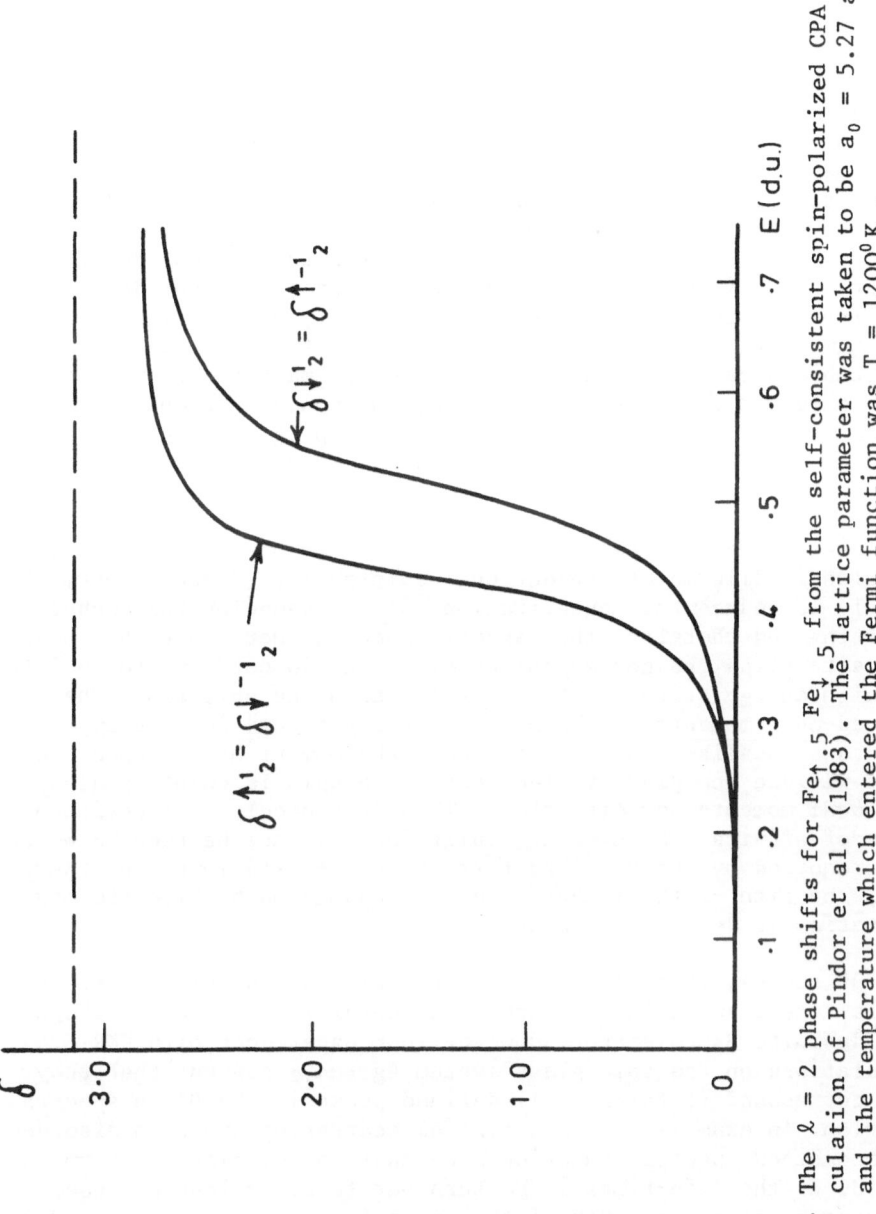

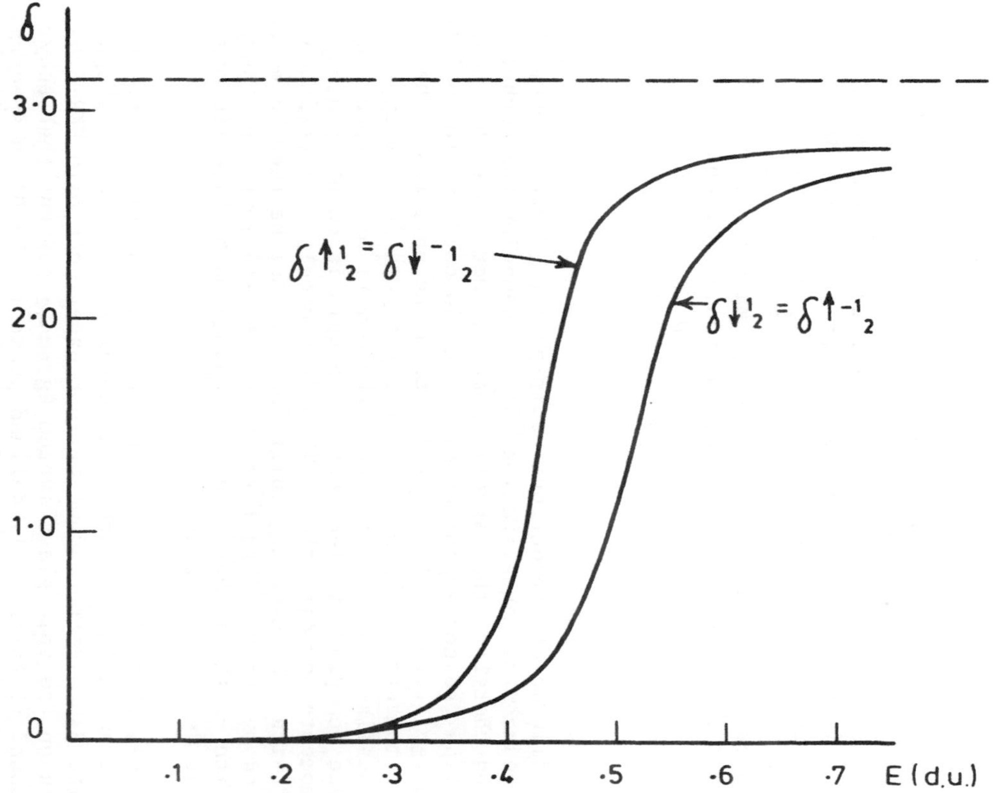

Fig. 5. The $\ell = 2$ phase shifts for $Fe_{\uparrow .5}\,Fe_{\downarrow .5}$ from the self-consistent spin-polarized CPA calculation of Pindor et al. (1983). The lattice parameter was taken to be $a_0 = 5.27$ au and the temperature which entered the Fermi function was $T = 1200^0$K .

of illustration only. From Fig. 6a we can see that the states for spin ↑ electrons have much larger amplitudes on the ↑ sites than on the ↓ sites ; on the other hand it follows from Fig. 6b that spin ↓ states have larger amplitudes on the ↓ sites than on the ↑ sites since $n_{\downarrow\downarrow}(\varepsilon_1) < n_{\uparrow\downarrow}(\varepsilon_1)$. Thus the CPA produced two sets of states at a given energy : those for spin up electrons which are localized on the ↑ sites and those for spin ↓ electrons which have substantial amplitudes only on the ↓ sites. Similar analysis applies at each energy although the degree of localization varies. This selfconsistently achieved spatial separation of the spin ↑ and spin ↓ down electrons is the mechanism of the moment formation. Evidently, the system has lowered its free energy, at this temperature, by this arrangement. As T increases the entropy due to the Stoner excitations increases and the size of the moment decreases. This is shown in Fig. 7. These are the first, ever, calculations of the magnetic moments in metals at finite temperatures without adjustable parameters. In order to find the magnetization we would have to repeat the above calculations for concentrations of upsites greater than 1/2. Such calculations are in progress and will be reported elsewhere. (Pindor et al. 1983). However, before bringing these notes to a close we wish to make two final remarks.

Firstly we note the very interesting way our theory resolved the classic dilemma of conventional spin-polarized band-theory which is able to reduce the magnetization only by reducing the exchange splitting and therefore the magnetic moments. Here when the moment on a site flips the corresponding electrons do not have to go into the high energy states dictated by the exchange splitting. The flipped-over moments of the present theory create low energy states which are localized on the site where the moment has flipped and these will be occupied by electrons whose spin is pointing along the local moments new directions. This is a nicely selfconsistent state of affairs. The exchange splitting need not be reduced beyond that required by the Fermi factors, Stoner excitations, and therefore, the size of the moments need not change much while the magnetization is definitely being reduced.

Of course, these localized states will not be eigenstates of the lattice translation operator and therefore will not correspond to a definite Bloch vector. However, our experience with KKR-CPA calculations on the spin-glass system $Ag_{.8}Mn_{.2}$ suggest that they will correspond to fairly well defined peaks in the Bloch spectral function. An example of the potential scattering and spin-disordered broadened spectral function from that calculation is shown in Fig. 8. If the life-times in Fe turn out to be as long as those corresponding to the width of the peaks in Fig. 8 then we could expect, even above T_c, a Fermi surface which show substantial features corresponding to exchange splitting. That this is roughly the case is shown in Fig. 9. A direct observation by positron annihilation measurements (Berko 1979), this electronic cause of the

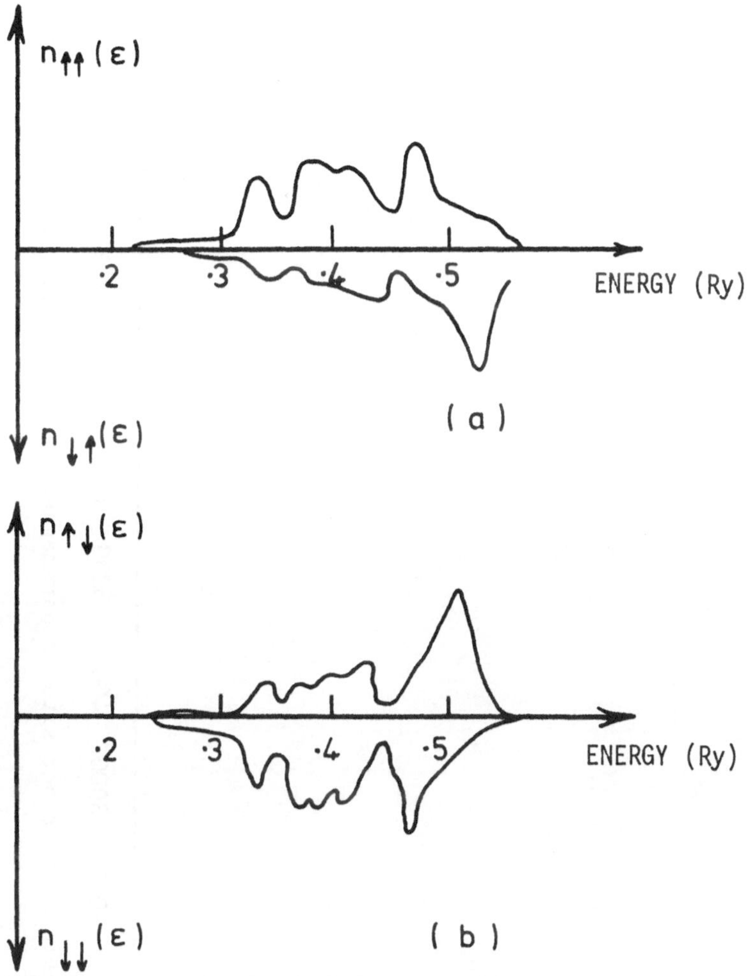

Fig. 6. The partially averaged density of state for spin-↑ (a) and spin-↓ (b) electrons on an ↑ and ↓ site as obtained in the self-consistent and spin polarized KKR-CPA calculation of Pindor et al. (1983).

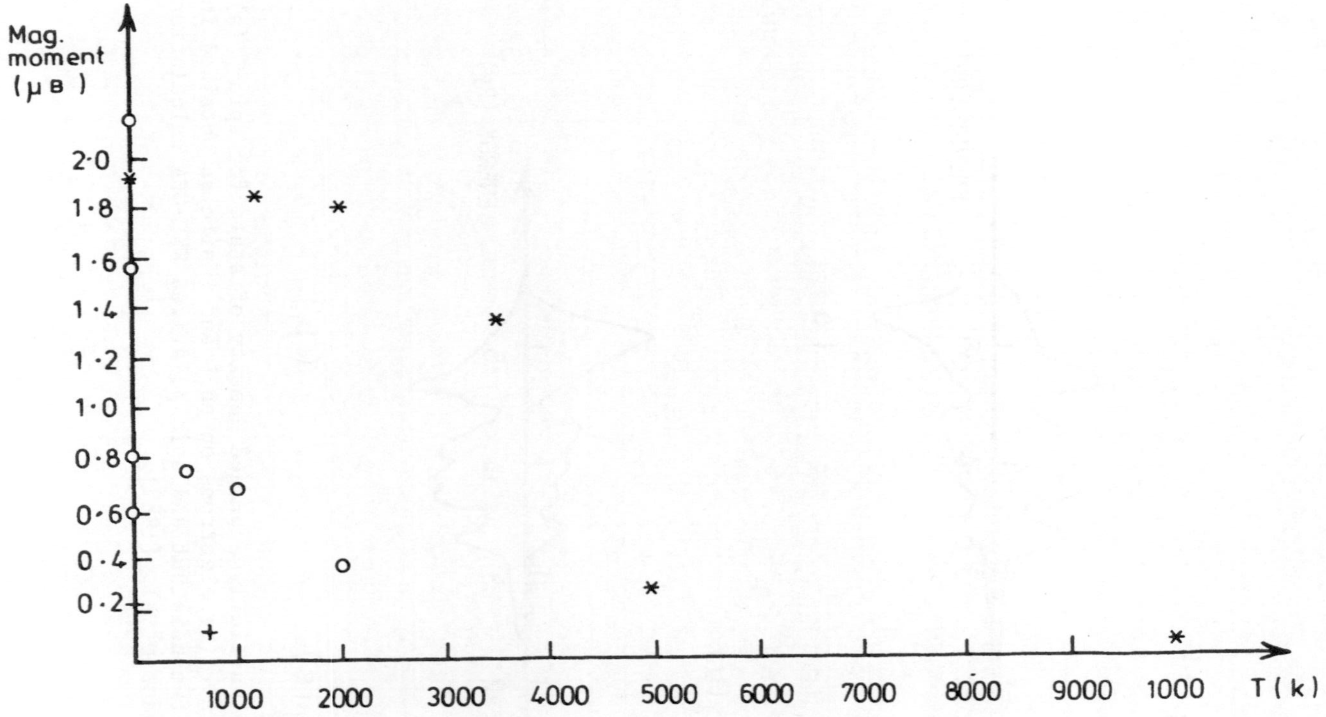

Fig. 7. The magnetic moments per site in paramagnetic Fe(*) Co (o) and Ni (+) as calculated by Pindor et al.

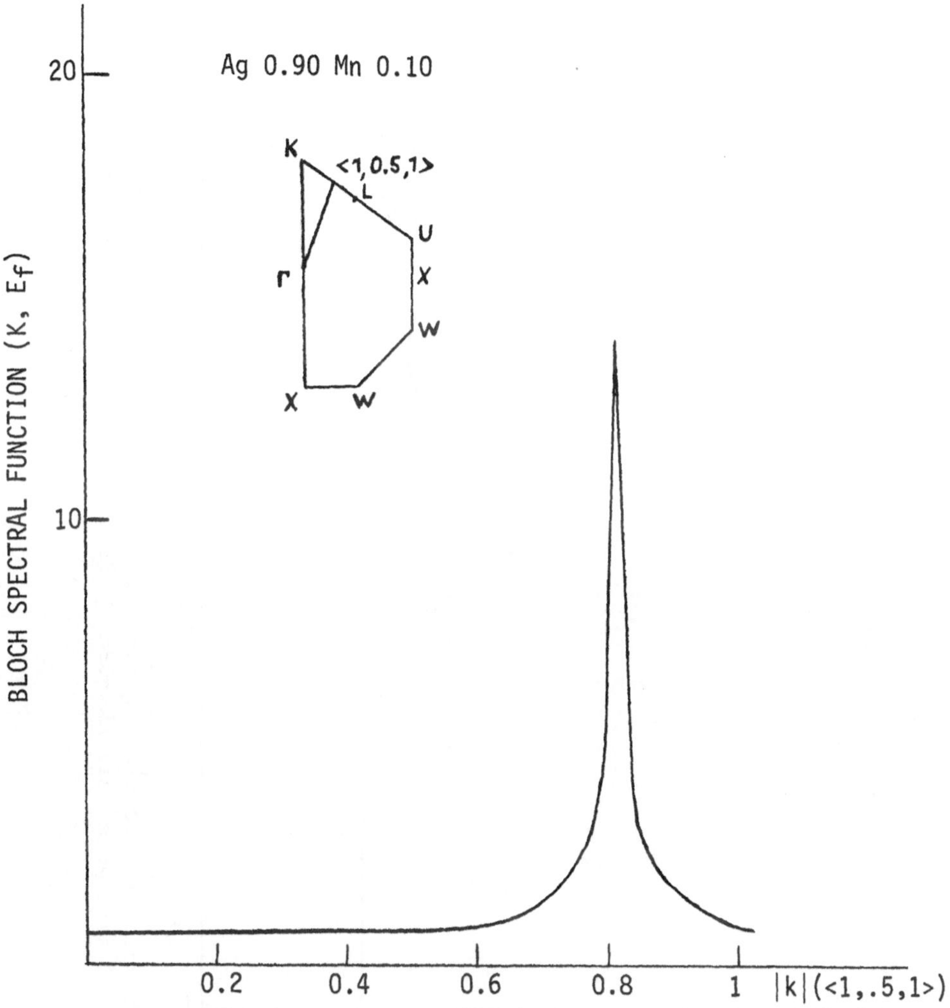

Fig. 8. The Bloch spectral function $\bar{A}_B(\vec{k},\varepsilon_F)$ as calculated by Munoz et al. (1983) for $Ag_{.90}Mn_{.10}$. The direction of $\vec{k}$ is indicated in the inset.

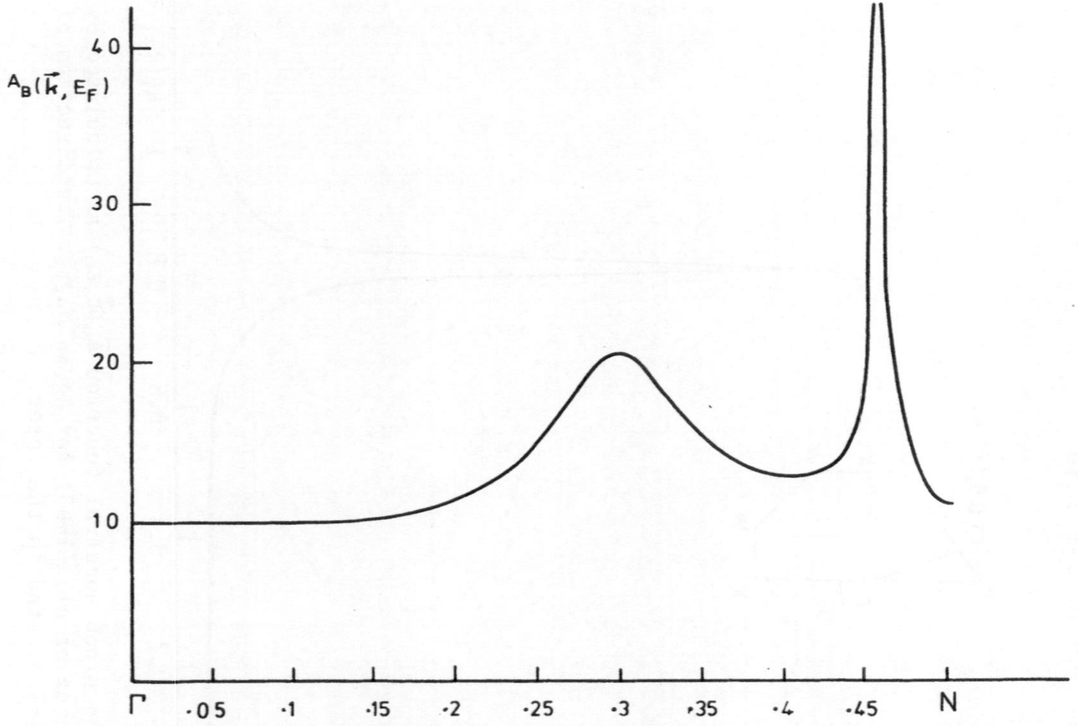

Fig. 9. The Bloch spectral function $A_B(\vec{k},\epsilon_F)$ for $Fe_{\uparrow .5}\ Fe_{\downarrow .5}$ as calculated in the self-consistent and spin-polarized KKR-CPA by Pindor et al.[5] (1983) at T = 1200K.

moment above T_c would enhance the credibility of our story so far.

Finally, it is worth noting that the CPA method of a averaging the electronic structure over all spin configurations has played a very important role in producing the results discussed above. Splitting bands in energy and localizing states on a subset of sites are the things that the CPA does much better than other simpler approximations designed to solve the same problem. For instance the Virtual Crystal Approximation would not have split the bands at all while the Average T-matrix Approximation would have done so rather unreliably (Ehrenreich and Schwartz 1976). We believe that to be consistent with doing a mean-field approximation for the spin system one must deal with the effect of random spin configurations on the electronic structure within the CPA.

Appendix I

In order to help the reader unacustomed to dealing with many body Hamiltonians, in this appendix, we derive the equation of motion for the one particle Greens function of the Hubbard Hamiltonian in the Hartree-Fock approximation. For further details and or more sophisticated treatment of this model the reader is refered to Doniach and Sondheimer (1974).

The Hamiltonian in I-1 defines a non-trivial many-body problem because the interaction term is quadratic in the number operator $\hat{n}_{i\sigma} = a^+_{i\sigma}a_{i\sigma}$. To linearize it, and thereby render the problem tractible by one-body means, we expand $\hat{n}_{i\sigma}$ in powers of its fluctuations about its mean : $\hat{n}_{i\sigma} - \bar{n}_{i\sigma}$. To linear order

$$H_{int} = \frac{1}{2}\sum_{i,\sigma} \hat{n}_{i\sigma}\hat{n}_{i-\sigma} = \frac{1}{2} I \sum_{i,\sigma} [\bar{n}_{i,\sigma} + (\hat{n}_{i,\sigma} - \bar{n}_{i,\sigma})]$$

$$[\bar{n}_{i-\sigma} + (\hat{n}_{i-\sigma} - \bar{n}_{i,-\sigma})] \underset{HF}{\cong} \frac{1}{2} I \sum_{i,\sigma} (\bar{n}_{i,\sigma}\bar{n}_{i,-\sigma} + \bar{n}_{i\sigma}$$

$$(\hat{n}_{i-\sigma} - \bar{n}_{i,-\sigma}) + (\hat{n}_{i,\sigma} - \bar{n}_{i,\sigma})\bar{n}_{i,-\sigma}) = -\frac{1}{2} I \sum_{i,\sigma}$$

$$\bar{n}_{i\sigma}\bar{n}_{i-\sigma} + I \sum_{i,\sigma} \hat{n}_{i,\sigma}\bar{n}_{i,-\sigma} \qquad \text{A-1}$$

where we also used the trivial identity

$$\sum_{i,\sigma} \bar{n}_{i\sigma}\hat{n}_{i-\sigma} = \sum_{i\sigma} \bar{n}_{i,-\sigma}\hat{n}_{i,\sigma}$$

Adding this result to the first term in Eq. I-1 leads to

$$H \cong \sum_{i,j,\sigma} [(\varepsilon_0 + I\bar{n}_{i,-\sigma})\delta_{ij} + t_{ij}] a^+_{i,\sigma}a_{j,\sigma} - \frac{1}{2} I \sum_{i,\sigma} \bar{n}_{i,\sigma}\bar{n}_{i,-\sigma} \qquad \text{A-2}$$

and dropping the operator independent last term in Eq. A-2 yields H_{HF} in the text, Eq. I-2.

To find the equation of motion for $G_{\sigma,\sigma}(i,j;t)$ we note that

$$G_{\sigma,\sigma}(i,j;t) = - i<T\{a_{i,\sigma}(t)a^+_{j,\sigma}(0)\}> = - i\theta(t)<a_{i,\sigma}(t)a^+_{j,\sigma}(0)>$$

$$+ i\theta(-t)<a^+_{j,\sigma}(0)a_{i,\sigma}(t)> \qquad \text{A-3}$$

where $\theta(t)$ is the Heavyside step function. Hence,

$$i\hbar \frac{\partial}{\partial t} G_{\sigma,\sigma}(i,j;t) = \ddot{\hbar} \delta(t)(<a_{i,\sigma}(0)a^+_{j,\sigma}(0)> + <a^+_{j,\sigma}(0)a_{i,\sigma}(0)>)$$

$$- i<T\{i\hbar a_{i,\sigma}(t)a^+_{j,\sigma}(0)\}> \qquad \text{A-4}$$

The coefficient of $\delta(t)$ in the above expression is 1 because of the commutation relations :

$$[a^+_{i,\sigma},a_{j,\sigma'}]_+ = a^+_{i,\sigma}a_{j,\sigma'} + a_{j,\sigma'}a^+_{i,\sigma} = \delta_{ij}\delta_{\sigma\sigma'} \qquad \text{A-5}$$

In the Hartree-Fock approximation

$$i\hbar \, a_{i,\sigma} = [a_i,H_{HF}] \qquad \text{A-6}$$

$$= \sum_{\ell,k,\sigma'} [(\varepsilon_0 + I n_{\ell,-\sigma'})\delta_{\ell,k} + t_{\ell,k}][a_{i,\sigma'}a^+_{\ell,\sigma'}a_{k,\sigma'}]_+$$

Moreover, using Eq. A-5 it is easy to show that

$$[a_{i,\sigma'}a^+_{\ell,\sigma'}a_{k,\sigma'}] = \delta_{i,\ell}\delta_{\sigma,\sigma'}a_{k,\sigma} \qquad \text{A-7}$$

Therefore

$$i \hbar \, a_{i,\sigma}(t) = \sum_k [\varepsilon_0 + I\bar{n}_{i,-\sigma})\delta_{ik} + t_{i,k}]a_{k,\sigma} \qquad \text{A-8}$$

Now, substituting Eq. A-5 and A-8 into Eq. A-4 leads to

$$i \hbar \frac{\partial}{\partial t} G_{\sigma,\sigma}(i,j;t) = \hbar \delta_{i,j} + \sum_k [(\varepsilon_0 + I\bar{n}_{i,-\sigma})\delta_{i,k} + t_{i,k}]G_{\sigma\sigma}(k,j;t)$$
$$\text{A-9}$$

Finally, we define the Fourier transform $G_{\sigma,\sigma}(i,j;\varepsilon)$ of $G_{\sigma\sigma}(i,j;t)$ by

$$G_{\sigma,\sigma}(i,j;t) = \frac{1}{\hbar} \int_{-\infty}^{\infty} \frac{d\varepsilon}{2\pi} e^{-i\frac{1}{\hbar}\varepsilon t} G_{\sigma,\sigma}(i,j;\varepsilon) \qquad \text{A-10}$$

and take the Fourier transform of Eq. A-9. The result is

$$\sum_k [(\varepsilon - \varepsilon_0 - I\bar{n}_{i,-\sigma})\delta_{ik} - t_{i,k}]\, G_{\sigma,\sigma}(k,j;\varepsilon) = \delta_{ij} \qquad \text{A-11}$$

as in the text : Eq. I-3.

Appendix II

Scattering of electrons from spin polarized targets

1) Scattering from a single potential well

As is customary in scattering theory (Newton, '66) instead of the Schrödingers equation we begin with the corresponding integral equation, the Lippman-Schwinger equation, which already incorporates the outgoing wave boundary condition

$$\psi_{\vec{k},\nu}(\vec{r};\sigma) = \varphi_{\vec{k},\nu}(r;\sigma) + \sum_{\sigma'} \int d'^3 G^0(r,\sigma;r'\sigma;k)V(r';\sigma,\sigma)\psi_{\vec{k},\nu}(\vec{r}';\sigma') \qquad \text{A2-1}$$

where the incident wave $\varphi_{\vec{k},\nu}$, specified by the wave vector $\vec{k}$ and the spin index ν, is the solution of the free particle Schrödingers equation, $G^0(r,\sigma;r'\sigma';k)$ is the free particle Greens function at the energy k^2 diagonal in the spin variables σ,σ', and $V(r;\sigma,\sigma')$ is the scattering potential which is an arbitrary function of σ and σ' but depends only on the magnitude of the position vector $\vec{r}$. Like the muffin-tin potential of band theory it is of finitie range, $v = 0$ for $r > r_{MT}$, the muffin-tin radius, and, as indicated by the fact that it depends only on the magnitude of $\vec{r}$, spherically in real space. However, in spin space it corresponds to a tensor force.

In order to investigate the behavior of $\psi_{\vec{k},\nu}(\vec{r},\sigma)$ far from the scattering potential well centered on the origin, $\vec{r} = 0$ we note that

$$G^0(r,\sigma;r'\sigma';k) = \sum_{L,\nu} g_\ell(r,r';k)Y_L(\hat{r})Y_L\chi_\nu(\sigma)\chi_\nu(\sigma') \qquad \text{A2-2}$$

where the $Y_L(\hat{r})$'s are the real spherical harmonics, eigen-functions of L^2 and L_z corresponding to the eigenvalues $\ell(\ell+1)$ and m respectively, the $\chi_\nu(\sigma)$'s are a complete set of spin functions, and the radial Greens function $g_\ell(r,r';k)$ satisfies the free particle radial equation

$$\left(k^2 + \frac{1}{r^2}\frac{d}{dr}\left(r^2\frac{d}{dr}\right) - \frac{\ell(\ell+1)}{r^2}\right) g_\ell(r,r';k) = \frac{1}{r^2}\,\delta(r - r') \qquad \text{A2-3}$$

Furthermore, we expand the incident plane wave in angular momentum partial waves as follows

$$\psi_{\vec{k},\nu}(\vec{r};\sigma) = e^{i\vec{k}.\vec{r}}\chi_\nu(\sigma) = 4\pi \sum_L i^\ell j_\ell(kr)Y_L(\hat{r})Y_L(\hat{k})\chi_\nu(\sigma) \qquad \text{A2-4}$$

where $j_\ell(kr)$ is a spherical Bessel function.

Evidently, $\psi_{\vec{k},\nu}(\vec{r};\sigma)$ is a function of $\vec{k},\vec{r}$ and σ. With respect to the direction of $\hat{k}$, $\hat{k}$ it has the expansion $\psi_{\vec{k},\nu}(\vec{r},\sigma) = \sum_L \psi_{L,\nu}(\vec{r};\sigma)Y_L(\hat{k})$. Because L^2 and L^z commutes with $V\psi_{L,\nu}(\vec{r};\sigma)$ must correspond to the same angular momentum eigen-function $Y_L(\hat{r})$ as the incident wave. Namely $\psi_{\vec{k},\nu}(\vec{r};\sigma)$ should depend only on the relative direction of $\vec{k}$ and $\vec{r}$. However, $\chi_\nu(\sigma)$ is not, in general, an eigenfunction of V and hence $\psi_{L,\nu}$ will contain not only $\chi_\nu(\sigma)$ of the incident wave but also all the other spin functions. Thus we may write

$$\psi_{\vec{k},\nu}(\vec{r};\sigma) = 4\pi \sum_{L,\nu'} i'^{\ell} R_{L,\nu',\nu}(r;k)Y_L(\hat{r})Y_L(\hat{k})\chi_{\nu'}(\sigma) \qquad \text{A2-5}$$

where $R_{L;\nu',\nu}(r;k)$ is a as yet undetermined, radial function. It satisfies the radial equations

$$R_{L;\nu',\nu}(r;k) = j_\ell(kr)\delta_{\nu'\nu} + \sum_{\nu''}\int^\infty dr'r'^2 g_\ell(r,r';k)V_{\nu'\nu''}(r')R_{L;\nu''\nu}(r';k) \qquad \text{A2-6}$$

where $V_{\nu'\nu''}(r')$ is the matrix element $\langle\chi_{\nu'}|V|\chi_{\nu''}\rangle$, as can be readily shown substituting Eq. A2-2, A2-4 and A2-5 into Eq. A2-1.

Now, we recall that for $r > r'$

$$g_\ell(r,r';k) = -\,ik\, h_\ell^+(kr)j_\ell'(kr) \qquad \text{A2-7}$$

where $h_\ell^+(kr)$ is a spherical Hankel whose asymptotic form is $-\,i(-i)^\ell\, 1/r\, e^{ikr}$. Then, from A2-6 we can deduce that

$$\lim_{r\to\infty} R_{L;\nu',\nu}(r;k) = \lim_{r\to\infty} (j_\ell(kr)\delta_{\nu\nu'} + ikf_{\ell,\nu'\nu}(k)h_\ell^+(kr) \qquad \text{A2-8}$$

where

$$f_{\ell;\nu'\nu}(r) = -\sum_{\nu''}\int_0^\infty dr'r'^2 j_\ell(kr')V_{\nu'\nu''}(r')R_{L;\nu'',\nu}(r';k) \qquad \text{A2-9}$$

Substituting Eq. A2-8 into Eq. A2-5 we find

$$\lim_{r\to\infty} \psi_{\vec{k},\nu}(r;\sigma) = 4\pi \sum_{L\nu'} [\, i^\ell j_\ell(kr)\delta_{\nu',\nu} + f_{\ell;\nu',\nu}(k)\frac{e^{ikr}}{r}\,]\; Y_L(\hat{r})Y_L(\hat{k})\chi_{\nu'}(\sigma) \qquad \text{A2-10}$$

This identifies the coefficients $f_{\ell;\nu\nu'}(k)$ as the partial wave scattering amplitudes. Due to the fact that the potential is off-diagonal in the spin index ν the scattering amplitude $f_{\ell;\nu',\nu}$ is also off-diagonal, as one might have expected.

To facilitate the calculation of $f_{\ell;\nu',\nu}$ we convert Eq. A2-6 into a radial differential equation by operating on both sides from the left with

$$\left(k^2 + \frac{1}{r^2}\frac{d}{dr}\left(r^2 \frac{d}{dr}\right) - \frac{\ell(\ell+1)}{r^2}\right)$$

and using Eq. A2-3. This leads to

$$\left(k^2 + \frac{1}{r^2}\frac{d}{dr}\left(r^2 \frac{d}{dr}\right) - \frac{\ell(\ell+1)}{r^2}\right)R_{\ell;\nu',\nu}(r;k) + \sum_{\nu''} V_{\nu'\nu''}(r)R_{\ell;\nu''\nu}(r;k)=0 \qquad \text{A2-11}$$

Note that the second spin index on $R_{\ell;\nu'\nu}$ referes to the spin state of the incident wave, e.g. the boundary condition. That is to say if the $\chi_\nu(\sigma)$'s are eigenfunctions of $\hat{\sigma}^z$, with respect to some arbitrary axis of quantization, then the functions $R_{\ell;\uparrow\nu}(r;k)$ and $R_{\ell;\downarrow\nu}(r;k)$ satisfy the same set of coupled differential equations for all ν, but for each incident wave specification, ν, they will satisfy different boundary conditions

$$\begin{pmatrix} R_{\ell;\uparrow\uparrow}(r;k) \\ R_{\ell;\downarrow\uparrow}(r;k) \end{pmatrix} \Rightarrow \begin{pmatrix} j_\ell(kr) + if_{\ell;\uparrow\uparrow}h^+_\ell(kr) \\ if_{\ell;\downarrow\uparrow}h^+_\ell(kr) \end{pmatrix}$$

$$\begin{pmatrix} R_{\ell;\uparrow\downarrow}(r;k) \\ R_{\ell;\downarrow\downarrow}(k;k) \end{pmatrix} \Rightarrow \begin{pmatrix} f_{\uparrow\downarrow}h^+_\ell(kr) \\ j_\ell(kr) + if_{\ell;\downarrow\downarrow}h^+_\ell(kr) \end{pmatrix} \qquad \text{A2-12}$$

Clearly, Eq. A2-11 determines the scattering amplitudes $f_{\ell;\nu'\nu}(k)$ by matching smoothly its solutions to the form given in Eq. A2-12 at the muffin tin radius r_{MT}.

In the spinor representation, where $\chi_\uparrow(\sigma) = \begin{pmatrix}1\\0\end{pmatrix}$ and $\chi_\downarrow(\sigma) = \begin{pmatrix}0\\1\end{pmatrix}$ and ↑ and ↓ are with respect to the z-axis of what we shall refer to as the lab frame, the most general potential we may be interested in can be written as

$$\underset{\approx}{V}(r) = V^0(r)\,\underset{\approx}{1} + \vec{V}(\vec{r}).\vec{\underset{\approx}{\sigma}} \qquad \text{A2-13}$$

where the unit matrix $\underset{\approx}{1}$ and the Pauli spin-matrices σ^x, σ^y, σ^z are defined in the usual way :

$$\underset{\approx}{1} = \begin{pmatrix}1 & 0\\0 & 1\end{pmatrix},\ \underset{\approx}{\sigma}^x = \begin{pmatrix}0 & 1\\1 & 0\end{pmatrix},\ \underset{\approx}{\sigma}^y = \begin{pmatrix}0 & i\\-i & 0\end{pmatrix},\ \underset{\approx}{\sigma}^z = \begin{pmatrix}1 & 0\\0 & -1\end{pmatrix} \qquad \text{A2-14}$$

Thus the full potential $\underset{\approx}{v}(r)$ is defined by four spherically symmetric potentials $V^0(r)$, $V^x(r)$, $V^y(r)$ and $V^z(r)$. In general, these will give rise to four scattering amplitudes $f_{\ell;\uparrow\uparrow}$, $f_{\ell,\uparrow\downarrow}$, $f_{\ell,\downarrow\uparrow}$ and $f_{\ell,\downarrow\downarrow}$ for each azimuthal quantum number ℓ.

To do full justice to the quantum nature of the fluctuating moments in the problem of itinerant magnetism, along the line of attack we have discussed in these lectures, we would have to use the full machinery of the above scattering theory to describe scattering events at each site. However, since we have considered classical moments only the calculation of the scattering amplitudes may be simplified considerably. In fact we may write $\underset{\approx}{V}$ in the form

$$\underset{\approx}{V}(r) = V^0(r)\,\underset{\approx}{1} + V^{\hat{e}}(r)\hat{e}.\vec{\underset{\approx}{\sigma}} \qquad \text{A2-15}$$

where the unit vector $\hat{e}$ denotes the orientation of the target. In the applications we have in mind $\hat{e}$ would specify the orientation of the local magnetic moment.

Note now that $\underset{\approx}{V}(r)$ as given in Eq. A2-15 is diagonal in the representation where the axis of quantization is along $\hat{e}$. Namely, if the eigen functions χ_+, χ_- which span the spin space, are such that $\vec{\sigma}.\hat{e}\chi_\pm = \pm\chi_\pm$ then

$$\underset{\approx}{V}(\vec{r}) = \begin{pmatrix} V_{++} & 0 \\ 0 & V_{--} \end{pmatrix} = \begin{pmatrix} V_0+V^{\hat{e}} & 0 \\ 0 & V_0-V^{\hat{e}} \end{pmatrix} \equiv \begin{pmatrix} V_{\hat{e},\uparrow} & 0 \\ 0 & V_{\hat{e},\downarrow} \end{pmatrix}$$

where $V_{\hat{e},\uparrow}(r)$ is the potential seen by an electron whose spin is parallel to $\hat{e}$ and $V_{\hat{e},\downarrow}(r)$ is that seen by an electron whose spin is antiparallel to $\hat{e}$. In this representation the radial equations given in Eq. A2-11 decouple and the scattering amplitudes may be taken as diagonal

$$\underset{\approx}{f}_\ell = \begin{pmatrix} f_{\hat{e};\ell\uparrow} & 0 \\ 0 & f_{\hat{e};\ell\downarrow} \end{pmatrix} \qquad \text{A2-17}$$

Evidently, we now have two separate scattering problem. Matching the logarithmic derivatives of $j_\ell(kr) + if_{\hat{e};\ell\uparrow}(k)h_\ell^+(kr)$ and the regular solution $R_{\hat{e};\ell\uparrow}(r;k)$ at r_{MT} we find

$$\cot\delta^{\hat{e}}_{\ell,\uparrow}(k) = \frac{kn_\ell'(kr_{MT}) - \gamma^{\hat{e}}_{\ell,\uparrow}(k)n_\ell(kr_{MT})}{\sqrt{\varepsilon}j_\ell'(kr_{MT}) - \gamma^{\hat{e}}_{\ell,\uparrow}(k)j_\ell(kr_{MT})} \qquad \text{A2-18}$$

where n_ℓ is a spherical. Neumann function, $\gamma^{\hat{e}}_{\ell,\uparrow}(k)$ is the logarithmic derivative corresponding to $R_{\hat{e},\ell\uparrow}(r;k')$ and the phase shifts $\delta^{\hat{e}}_{\ell,\uparrow}(\varepsilon)$ determine the scattering amplitudes by the usual relation

$$f_{\hat{e};\ell\uparrow}(k) = \frac{1}{2i}\left(e^{i2\delta^{\hat{e}}_{\ell,\uparrow}(k)} - 1\right) \qquad \text{A2-19}$$

with similar expressions for $f_{\hat{e};\ell\downarrow}(k)$.

Let us now rotate the spin frame of reference from the z-axis beeing along $\hat{e}$ to coinciding with that of the lab frame. The result most easily seen if we rewrite Eq. A2-17 as

$$\underset{\approx}{f}_\ell = \frac{1}{2}(f_{\hat{e};\ell\uparrow} + f_{\hat{e};\ell\downarrow})\, \underset{\approx}{1} + \frac{1}{2}(f_{\hat{e};\ell\uparrow} - f_{\hat{e};\ell\downarrow})\underset{\approx}{\sigma}^z \qquad \text{A2-20}$$

where $\sigma^z = \begin{pmatrix} 1 & 0 \\ 0 & -1 \end{pmatrix}$ refers to the spin projection along $\hat{e}$. On rotating from the local to the lab frame $\underset{\approx}{\sigma}^z$ becomes $\vec{\underset{\approx}{\sigma}}.\hat{e}$ and hence

$$\underset{\approx}{f}_\ell(\hat{e};k) = \frac{1}{2}(f_{\hat{e};\ell\uparrow} + f_{\hat{e};\ell\downarrow})\underset{\approx}{1} + \frac{1}{2}(f_{\hat{e};\ell\uparrow} - f_{\hat{e};\ell\downarrow})\vec{\underset{\approx}{\sigma}}.\hat{e} \qquad \text{A2-21}$$

Thus in the lab frame, for an arbitrary orientation of $\hat{e}$, the scattering amplitude $\underset{\approx}{f}_\ell(\hat{e};k)$ is determined by $f_{\hat{e};\ell\uparrow}(k)$, $f_{\hat{e};\ell\downarrow}$ and the direction cosines of $\hat{e}$.

2) Multiple Scattering

Consider now a lattice of scattering centers, of the kind described above. The lattice sites are located at the positions $\vec{R}_i$ and the orientation of the scatterer at $\vec{R}_i$ is denoted by $\hat{e}_i$. It is straightforward to generalize the spin independent multiple scattering theory described in the lectures of Dr.Stocks to treat the problem of the propagation of an electron on such lattice. Since the potentials are, again, non-overlapping it is sufficient to consider multiple scattering on the energy shell only. Then the problem reduces to that of finding the scattering path matrix $\underset{\approx}{\tau}^{i,j}_{L,L'}(\varepsilon)$ which is now a matrix in spin-space as well as in site-space and angular momentum-space. The real space KKR equations for $\underset{\approx}{\tau}^{i,j}_{L,L'}(\varepsilon)$ may be written down by inspection

$$\sum_{k,L''} (\underset{\approx}{t}^{-1}_{i;L}(\hat{e}_i;\varepsilon)\delta_{ik}\delta_{L,L''} - G_{L,L''}(\vec{R}_i - \vec{R}_k;\varepsilon)\underset{\approx}{1})\underset{\approx}{\tau}^{k,j}_{L'',L'}(\varepsilon) = \underset{\approx}{1}\delta_{ij}\delta_{L,L'} \qquad \text{A2-22}$$

where $t^{-1}_{i;L}(\hat{e}_i;\varepsilon)$ is the inverse of the spin-space "on the energy shell" t-matrix defined by $\underset{\approx}{t}_{i;L}(\hat{e}_i;\varepsilon) = -1/\sqrt{\varepsilon}\, \underset{\approx}{f}_{i;L}(\hat{e};\varepsilon)$ and the other symbol's carry the same meaning as before.
The suffix i on $\underset{\approx}{t}_{i;L}$ means that the potential function is allowed to vary from site to site. For an arbitrary configuration of potential functions, of the form described in A2-17, the integrated density of states are given by the obvious generalization of the Lloyd formula (Faulkner 1977) :

$$N(\varepsilon) = 2N^0(\varepsilon) - \frac{1}{\pi}\, \mathrm{Im\,tr}\ell n \| \underset{\approx}{t}^{-1}_{i;L}(\hat{e}_i;\varepsilon)\delta_{i,j}\delta_{L,L'} - G_{L,L'}(\vec{R}_i - \vec{R}_j;\varepsilon)\underset{\approx}{1} \| \qquad \text{A2-23}$$

where $G_{LL'}(R_i - R_j;\varepsilon)$ is a real space KKR structure constant, the determinant is to be taken with respect to the site, i, and angular momentum, L, indices and the trace operation is to be carried out in spin-space.

Furthermore, for $\vec{r}_i = \vec{r} - \vec{R}_i$ and $r_i' = \vec{r}' - \vec{R}_i$ within the muffin-tin sphere surrounding the site i the spin matrix Greens function is given by

$$\underset{\approx}{G}(\vec{r}_i,\vec{r}_j;\varepsilon) = \sum_{L,L'} \underset{\approx}{Z}_{i;L}(\vec{r}_i,\varepsilon)\underset{\approx}{\tau}^{ii}_{L,L'}(\varepsilon)\underset{\approx}{Z}_{i,L}(\vec{r}_i';\varepsilon)$$

$$+ \sum_{LL'} \underset{\approx}{Z}_{i,L}(\vec{r}_i;\varepsilon)\underset{\approx}{J}_{i,L'}(\vec{r}_i';\varepsilon) \qquad \text{A2-24}$$

where $\underset{\approx}{Z}_{i,L}(\vec{r}_i;\varepsilon) = \underset{\approx}{R}_{i,L}(r_i;\varepsilon)Y_L(\hat{r}_i;\varepsilon)$ and $\underset{\approx}{J}_{i,L}(\vec{r}_i;\varepsilon)$ is the irregular solution of the single potential well radial Schrödingers equation of which $\underset{\approx}{R}_{i;L}(r_i;\varepsilon)$ is the regular solution. The site off-diagonal Greens function, for which $r_i = \vec{r} - \vec{R}_i$ is within the i-th muffin-tin sphere and $\vec{r}_j = \vec{r}' - \vec{R}_j$ is the j-th muffin-tin sphere, is given by

$$\underset{\approx}{G}(\vec{r}_i;\vec{r}_j;\varepsilon) = \sum_{L,L'} \underset{\approx}{Z}_{i;L}(\vec{r}_i;\varepsilon)\underset{\approx}{\tau}^{ij}_{L,L'}(\varepsilon)\underset{\approx}{Z}_{j;L'}(\vec{r}_j';\varepsilon) \qquad \text{A2-25}$$

As is shown in the lectures of Dr. Stocks, in the spin independent cases, this quantity is needed to calculate the spectral function $\overline{A}_B(\vec{r};\varepsilon)$. In the theory of the main text the spectral function is a spin matrix defined by

$$\underset{\approx}{\overline{A}}_B(k\varepsilon) = \sum_j e^{i\vec{k}.(R_j-R_i)} \int_{V_i} dr_i^3 \langle \mathrm{Im}\, \underset{\approx}{G}(r_i + R_j;r_i;\varepsilon)\rangle \qquad \text{A2-26}$$

where the average < > is with respect the ensemble of crystal potentials. Using Eq. A2-25 and the KKR-CPA described in the text an expression analogous to that used by Dr. Stocks can be derived for $\overline{A}_B(\vec{r};\varepsilon)$.

In the text we have made used of the special case of the above theory where the orientation vectors were allowed to be only up, ↑ or down ↓, with respect to what we called here the lab frame. That is to say the Ising "spin" system considered corresponds to taking the axis of quatization : to be parallel on all sites. Then we may use the simplified notation

$$\underset{\approx}{f}_{i,L}(\hat{e}_i = \uparrow;\varepsilon) = \underset{\approx}{f}_{i\uparrow,L}(\varepsilon) = \begin{pmatrix} f_{i\uparrow,L\uparrow}(\varepsilon) & 0 \\ 0 & f_{i\uparrow,L\downarrow}(\varepsilon) \end{pmatrix}$$

$$\underset{\approx}{f}_{i,L}(\hat{e}_i = \downarrow, \varepsilon) = \underset{\approx}{f}_{i\downarrow,L}(\varepsilon) = \begin{pmatrix} f_{i\downarrow,L\uparrow} & 0 \\ 0 & f_{i\downarrow,L\downarrow}(\varepsilon) \end{pmatrix} \qquad \text{A2-27}$$

We can now dispense with the spin matrices and use the symbols $f_{i\uparrow,L\sigma}$ and $f_{i\downarrow,L\sigma}$ as in the text. In this limit the spectral function $\underset{\approx}{\bar{A}}_B$ is also diagonal and $\bar{A}_{B,\sigma}(\vec{k};\varepsilon)$ is the same as that for binary alloys in the lectures of Dr. Stocks with the transcription that $A \rightarrow \uparrow$ and $B \rightarrow \downarrow$.

In conclusion we stress that the arrows, ↑ and ↓, following the site index i of $f_{i\uparrow,L\uparrow}$ and $f_{i\uparrow,L\downarrow}$ referes to the orientation of the local moments while to the right of the angular-momentum symbol L to the spin of the scattering electron.

3) The RKKY-Interaction

In order to illustrate the physical content of these formulae we shall now work out the interaction energy of two scattering centers immersed in an otherwise non-interacting, free electron system. (Rudderman and Kittel 1956, Koshuya 1957, Yoshida 1958). The energy difference between the presence and absence of two scattering centers at $\vec{R}_1$ and $\vec{R}_2$ with magnetic moments in the $\hat{e}_1$ and $\hat{e}_2$ directions, E_{12}, is given by

$$E_{1,2} = -\sum_{\sigma} \int_0^{\varepsilon_F} d\varepsilon N_\sigma^{1,2}(\varepsilon) \qquad \text{A2-28}$$

where $N_\sigma^{1,2}(\varepsilon)$ is the excess integrated density of states due to the presence of the two scattering centers for an electron with a spin projection along a common, lab frame, axis of quantization. It follows from Eq. A2-22 that

$$E_{1,2} = +\frac{1}{\pi} \int_0^{\varepsilon_F} d\varepsilon \ \mathrm{tr}\ \mathrm{Im}\ \ell n \left\| \begin{matrix} \underset{\approx}{t}_{1,L}^{-1}(\hat{e}_i;\varepsilon)\delta_{LL'} & G_{L,L'}(\vec{R}_{12};\varepsilon)\underset{\approx}{1} \\ G_{LL'}(-\vec{R}_{12};\varepsilon)\underset{\approx}{1} & \underset{\approx}{t}_{2,L}^{-1}(\hat{e}_2;\varepsilon)\delta_{L,L'} \end{matrix} \right\| \qquad \text{A2-29}$$

where we have made use of the fact that the structure constants are defined to be 0 for i = j. Factoring out the matrix

$$\begin{pmatrix} \underset{\approx}{t}_{1,L}^{-1}(\hat{e}_1;\varepsilon)\delta_{L,L'} & 0 \\ 0 & \underset{\approx}{t}_{1,L}^{-1}(\hat{e}_2;\varepsilon) \end{pmatrix}$$

from the argument of the determinant in Eq. A2-24 we find

$$E_{1,2} = -\sum_{L,i=1,2} \operatorname{Im} \int^{\varepsilon_F} d\varepsilon \; \operatorname{tr} \{\ell n t^{-1}_{i,L}(\hat{e}_i;\varepsilon)\}$$

$$- \operatorname{Im} \int^{\varepsilon_F} d\varepsilon \; \operatorname{tr} \begin{Vmatrix} \underset{\approx}{1} & \underset{\approx}{t}_{1,L}(\hat{e}_1;\varepsilon) G_{L,L'}(\vec{R}_{12};\varepsilon) \\ t_{2,L}(\hat{e}_2;\varepsilon) G_{L,L'}(-R_{12};\varepsilon) & \underset{\approx}{1} \end{Vmatrix} \qquad \text{A2-30}$$

In the first term we carry out the 'tr' operation in the local frame. Recalling that

$$t^{-1}_{i,L} = -\sqrt{\varepsilon} f^{-1}_{i,L} = \frac{\sqrt{\varepsilon}}{\sin\delta_{i,L}} e^{-i\delta_{i,L}}$$

we find the contribution

$$E^0_{1,2} = -\frac{1}{\pi} \sum_{\substack{i=1,2 \\ \ell}} (2\ell + 1) \int_0^{\varepsilon_F} d\varepsilon (\delta_\ell^{\hat{e}_i,\uparrow}(\varepsilon) + \delta_\ell^{\hat{e}_i,\downarrow}(\varepsilon)) \qquad \text{A2-31}$$

Now, we note that the excess integrated density of states due to one impurity is given by the Friedel sum :

$$\delta N(\varepsilon) = \frac{1}{\pi} \sum_{\ell,\sigma} (2\ell + 1) \delta^\sigma_\ell(\varepsilon)$$

and hence recognize that $E^0_{1,2}$ is the sum of the self-energies of two independent impurities. Thus, we may identify the second term in Eq. A2-25 as the interaction energy. Note that it depends on the relative distance $\vec{R}_{12} = \vec{R}_1 - \vec{R}_2$ as it should. This is the full, including near field correction, RKKY interaction.

As $R_{1,2} \to \infty$, in the asymptotic regime, we may expand in powers of $G_{L,L'}(\vec{R}_{1,2};\varepsilon)$. The first term in such an expansion is given by

$$E^{RKKY}_{1,2} = -\sum_{L,L'\sigma} \int_0^{\varepsilon_F} d\varepsilon \; \operatorname{tr}\{\underset{\approx}{t}_{1,L'} G_{LL'}(\vec{R}_{12};\varepsilon) \underset{\approx}{t}_{2,L'} G_{L',L}(-\vec{R}_{12};\varepsilon)\} \qquad \text{A2-32}$$

where, asymptotically,

$$G_{L,L'}(R_{1,2};\varepsilon) = 4\pi \sqrt{\varepsilon} \sum_{L''} C^{L''}_{L,L'} \, i^{\ell''} h^+_{\ell''}(\sqrt{\varepsilon} R_{12}) Y_{L''}(\hat{R}_{12})$$

$$- i \sqrt{4\pi} \frac{e^{i\sqrt{\varepsilon} R_{12}}}{R_{12}} \delta_{LL'} \qquad \text{A2-33}$$

with $C^{L''}_{L,L'}$ standing for the Gaunt numbers : $\int d\Omega Y_L(\Omega) Y_{L''}(\Omega) Y_{L'}(\Omega)$. Using the spin-matrix identities

$$\operatorname{tr}\{\sigma.\hat{e}_1 \sigma.\hat{e}_2\} = \hat{e}_1.\hat{e}_2 \quad ; \quad \operatorname{tr}\{\sigma.\hat{e}_1\} = 0 \qquad \text{A2-34}$$

we arrive at that final result that

$$E^{RKKY}_{1,2} = - \sum_{\ell}(2\ell + 1)\,\mathrm{Im} \int_0^{\varepsilon_F} \frac{d\varepsilon}{\varepsilon} (f_{1,e_1;\ell\uparrow} - f_{1,e_1;\ell\downarrow})$$

$$(f_{2,\hat{e}_2;\ell\uparrow} - f_{2,\hat{e}_2;\ell\downarrow}) \frac{e^{i2\sqrt{\varepsilon}R_{1,2}}}{R^2_{12}} \hat{e}_1,\hat{e}_2 \qquad \text{A2-35}$$

As expected the interaction energy is isotropic in spin-space, e.g. it only depend on the rotationally invariant scalar $\hat{e}_1.\hat{e}_2$. In the simplest discussions of this interaction the scattering amplitudes are treated only in the Born approximation :

$$f_{\ell,\sigma}(\varepsilon) = - \sqrt{\varepsilon} \int_0^{\infty} dr\; r^2 j_{\ell}'^{\,2}(\sqrt{\varepsilon}r) v_{\sigma}(r)$$

for a δ-function $v^0_{\sigma} \frac{1}{r^2} \delta(r)$ and hence $f_0 = - v^0_{\sigma} \sqrt{\varepsilon}$. In this approximation the energy integral in Eq. A2-29 can be performed analytically. The results for $R_{12} \to \infty$ is

$$E^{RKKY}_{1,2} \cong - 2(V^0_{\uparrow} - V^0_{\downarrow})^2\, k_F \frac{\cos 2k_F R_{12}}{R^3_{12}} \hat{e}_1.\hat{e}_2 \qquad \text{A2-36}$$

where, to conform with the usual notation we have denoted $\sqrt{\varepsilon_F}$ by k_F. For further discussion of this interaction and its physical significance the reader is referred to Kasuya (1966). Evidently, the interaction energies in the inhomogeneous KKR-CPA considered in the text are generalizations of the above pair-wise interactions to the case of arbitrarily many site interactions.

ACKNOWLEDGEMENT

One of us, B.L.G., would like to thank the Instituto de Fisica "Gleb Wataghin", Universidade Estadual de Campinas for the gracious hospitality during which these notes have taken their present form, the CNP9 and the Royal Society for making the visit to Campinas possible and Mrs. Maria Inês Costa Soares for the efficient typing of the manuscript.

REFERENCES

Anderson, P.W., 1964 "Concepts in Solids" (W.A. Benjamin Inc.).

Anderson, P.W. and Yuval, G., 1969 Phys.Rev.Lett. 89, 723.

Berko, S., 1979, "Electrons in Disordered Metals and at Metallic Surfaces" Eds. P. Phariseau, B.L. Gyorffy, L. Scheire (Plenum Press 1979).

Bloch, F., 1929, Z.Phys. 57, 545, 1930, Z.Phys. 61, 206.

Callaway, J., Laurent, O., Wang, C.S., 1977, "Transition Metals 1977", Eds. Lee, M.J.G., Perz, J.M. and Fawcett, E., IP Conference Series 39, p.41.

Callaway, J. and Chatterjee, A.K., 1978, J.Phys.F : Metal Physics 8, 2569.

Capelman, H., 1974, J.Phys.F : Metal Physics, 1466.

Capelman, H., 1979, Z.Phys. B34, 29.

Cook, J.F., 1979, J.Appl.Phys. 50, (11).

Doniach, S. and Sondheimer, E.H., 1974 "Green's Functions for Solid State Physicists" (W.A. Benjamin, Inc. 1974).

Ducastelle, F. and Gautier, F., 1976, J.Phys.F : Metal Physics 6, 2039.

Eastman, D.E., Janck, J.F. and Williams, A.R., 1979, J.App.Phys. 50, 7423.

Edwards, D.M., 1970, Phys.Lett. A33, 183 and 1973 Lecture Notes at the Mont Tremblant International Summer School.

Edwards, D.M., 1978, Proc.Int.Conf. on Transition Metals, Toronto 1977, IP Conf. Series No. 39.

Edwards, D.M., 1979, "Electrons in Disordered Metals and Metallic Surfaces", Ed. P. Phariseau, B.L. Gyorffy and L. Scheire, ASI Series B42, p. 355, (Plenum Press 1979).

Edwards, D.M., 1980, J.Magn.Magn.Mat. 15-18, 262.

Edwards, D.M., 1982, J.Phys.F : Metal Physics 12, 1789.

Ehrenreich, H. and Schwartz, L., 1976, "Solid State Physics" Ed. H. Ehrenreich, F. Seitz and D. Turnbull (Academic Press).

Elliott, R.J., Krumhansl, J.A. and Leath, P.L., 1974, Rev.Mod.Phys. 46, 465.

Faulkner, J.S., 1977, J.Phys.C : Solid State Physics, 10, 4661.

Faulkner, J.S., Davis, H.L., Joy, W.H., 1967, Phys.Rev. 161, 656.

Faulkner, J.S. and Stocks, G.M., 1980, Phys.Rev. B21, 3222.

Faulkner, J.S., 1982 "Progress in Materials Science" 27 ed. T. Massalski (Oxford : Pergamon).

Feynman, R.P., 1972, "Statistical Mechanics" (Benjamin), Peierls proved this theorem in Phys.Rev. 54, 918, 1938.

Feynman, R.P. and Hibbs, A.R., 1965, "Quantum Mechanics and Path Integrals" (McGraw-Hill Book Company), Wohlfarth, E.P., 1980, "Physics of Transition".

Fisher, M.E., 1964, J.Math.Phys. 5, 944.

Gordon, B.E.A., Temmerman, W.M., Gyorffy, B.L., 1981, J.Phys.F : Metal Physics 11, 821.

Gunnarsson, O., 1976, J.Phys.F : Metal Physics 6, 587-605.

Gyorffy, B.L., and Stocks, G.M., 1979 "Electrons in Disordered and at Metallic Surfaces", Eds. P. Phariseau, B.L. Gyorffy, L. Scheire, ASI Series B42 (Plenum Press).
Gyorffy, B.L. and Stocks, G.M., 1982, Phys.Rev.Lett.
Herring, C. and Kittel, C., 1951, Phys.Rev. 81, 869.
Herring, C., 1966, "Magnetism" ed. G. Rado, H. Suhl (Academic Press, New York 1966).
Hertz, J.A. and Kleuin, H.A., 1974, Phys.Rev. B10, 1084.
Holden, A.J. and You, M.V., 1982, J.Phys.F : Metal Physics 12, 195.
Hohenberg, P. and Kohn, W., 1964, Phys.Rev. 136B, 864.
Hubbard, J., 1959, Phys.Rev.Lett. 3, 77.
Hubbard, J., 1963, Proc.Roy.Soc. A276, 238, 1979.
Hubbard, J., 1979a, Phys.Rev. B19, 2626, 1979b Phys.Rev. B20, 4584.
Kanamori, J., 1963, Prog.Theor.Phys. 30, 275.
Kasaya, T., 1956, Progr.Theoret.Phys. (Kyoto) 16, 45 and 58.
Kasuya, T., 1966, "Magnetism" Ed. G.T. Rado and H. Suhl (Academic Press 1966) p.215.
Korenmann, V., Murry, J.L. and Prange, R.E., 1977, Phys.Rev. B16, 4032, B16, 4058.
Korenmann, V. and Prange, R.E., 1979, Phys.Rev. B19, 4698.
Landau, L.D. and Lifshitz, 1980, "Statistical Physics" 3rd Ed.Part 1 (Pergamon Press).
Lloyd, P. and Smith, P.V., 1972, Adv.Phys. 21, 69.
Lloyd, P. and Best, P.R., 1975, J.Phys.C : Solid State Physics 8, 3752-66.
Lynn, J.W., 1975, Phys.Rev. B11, 2624.
Mattis, D.C., 1981, "The Theory of Magnetism" I, (Springer-Verlag 1981).
Mook, H.A. and Nicklow, R.M., 1973 Phys.Rev. 336.
Moriya, T. and Takashi, 1978, J.Phys.Soc., Japan 45.
Moriya, T., 1979, J.Magn.Mat. 14, 1.
Newton, R.G., 1966, "Scattering Theory of Waves and Particles" (McGraw-Hill Book Co. 1966).
Parks, R.D. (Editor), 1969 "Superconductivity" (Marcel Dekker, Inc; New York).
Pindor, A.J., Temmerman, W.M. and Gyorffy, B.L., 1983, submitted to J.Phys.F. : Metal Physics.
Pindor, A.J., Staunton, J., Stocks, G.M. and Winter, H., J.Phys.F. Metal Physics 1983 (to be published).
Rajagopal, A.K., 1980, Adv.Chem.Phys. 41, Eds. I. Prigogine and S.A. Rice (Wiley).
Rhodes, P. and Wohlfarth, E.P., 1963, Proc.Roy.Soc. A273, 247.
Ruderman, M.A. and Kittel, C., 1954, Phys.Rev. 96, 99.
Schwartz, L. and Siggia, E., 1972, Phys.Rev. B5, 383.
Shimizu, M., 1981, Rep.Prog.Phys. 44, 329.
Slater, J.C., 1936, Phys.Rev. 49, 537, 931.
Stocks, G.M. and Winter, H., 1982, Z.Phys. B46, 95.
Stocks, G.M. and Winter, H., 1983, in this volume.
Stoner, E.C., 1936, Proc.Roy.Soc. A154, 656, 1938, Proc.Roy.Soc. A165, 372.

Takahashi, M., 1981, J. of Phys.Soc.Japan 50, 1854.
Von Barth, U. and Williams, A.R., 1982 "Theory of the Inhomogeneous Electron Gas", Eds. S. Lundqvist, N.H. March, (Plenum Press, New York, 1982).
Wang, C.S. and Callaway, J., 1976, Solid St. Commun. 20, 255.
Wang, C.S., Prange, R.E. and Korenman, V., 1982, Phys.Rev. B25, 5766.
Wohlfarth, E.P., 1980, "Physics of Transition Metals 1980", Ed. P. Rhodes, IP Conference Series 55 (1981).
Yosida, K., 1957, Phys.Rev. 106, 893.
You, M.V. and Heine, V., 1982, J.Phys.F : Metal Physics 12, 177.

RELATING ELECTRON CONFIGURATION, CRYSTAL STRUCTURE AND BAND STRUCTURE

John A. Wilson

H.H. Wills Physics Laboratory
University of Bristol
Bristol BS8 1TL, England

The following article is not designed as a research article ; nor indeed as a 'true' review article. Rather, as befits a Summer School, it is aimed at teaching ; at passing on the gist of a large amount of material yet to find its way into any single text-book. A reader new to the field probably will find the material very condensed, but he might at least be able to read all the words at one sitting. Hopefully besides appearing short enough, it is also interesting enough to read more than once.

Basically the article is about materials and as such is constructed around a representative progression of materials. To be of some longterm value to the reader a materials-cum-reference index is provided to facilitate retrieval of information. The references in the main, are to band structure related work, or where this is lacking, to structural work. Just sufficient references are given to allow access into each topic in question. Accordingly the majority of references are post-1975. It is because of the wealth of data appearing from an ever-expanding array of systems that a broader look into some of the work of the past decade is not amiss. The article, materials-wise, is deliberately broad-based and as a necessity for those who have not yet found it worth memorizing I supply a Periodic Table.

I am conscious of the fact that, because of the way in which the paper was written, the sub-headings do not entirely embrace what is contained below them. However they should provide some guide and a breathing point or two.

An informed reader will soon pick up that my early perspectives were set by the work of Slater, Goodenough, Jorgensen, Hulliger

Periodic Table

IA	IIA	IIIA	IVA	VA	VIA	VIIA	VIIIᵃA	VIIIᵇA	VIIIᶜA	IB	IIB	IIIB	IVB	VB	VIB	VIIB	O
Li^{3} $2s$	Be^{4} $2s^{2}$											B^{5} $2s^{2}2p$	C^{6} $2s^{2}2p^{2}$	N^{7} $2s^{2}2p^{3}$	O^{8} $2s^{2}2p^{4}$	F^{9} $2s^{2}2p^{5}$	Ne^{10} $2s^{2}2p^{6}$
Na^{11} $3s$	Mg^{12} $3s^{2}$											Al^{13} $3s^{2}3p$	Si^{14} $3s^{2}3p^{2}$	P^{15} $3s^{2}3p^{3}$	S^{16} $3s^{2}3p^{4}$	Cl^{17} $3s^{2}3p^{5}$	Ar^{18} $3s^{2}3p^{6}$
K^{19} $4s$	Ca^{20} $4s^{2}$	Sc^{21} $3d$ $4s^{2}$	Ti^{22} $3d^{2}$ $4s^{2}$	V^{23} $3d^{3}$ $4s^{2}$	Cr^{24} $3d^{5}$ $4s$	Mn^{25} $3d^{5}$ $4s^{2}$	Fe^{26} $3d^{6}$ $4s^{2}$	Co^{27} $3d^{7}$ $4s^{2}$	Ni^{28} $3d^{8}$ $4s^{2}$	Cu^{29} $3d^{10}$ $4s$	Zn^{30} $3d^{10}$ $4s^{2}$	Ga^{31} $4s^{2}4p$	Ge^{32} $4s^{2}4p^{2}$	As^{33} $4s^{2}4p^{3}$	Se^{34} $4s^{2}4p^{4}$	Br^{35} $4s^{2}4p^{5}$	Kr^{36} $4s^{2}4p^{6}$
Rb^{37} $5s$	Sr^{38} $5s^{2}$	Y^{39} $4d$ $5s^{2}$	Zr^{40} $4d^{2}$ $5s^{2}$	Nb^{41} $4d^{4}$ $5s$	Mo^{42} $4d^{5}$ $5s$	Tc^{43} $4d^{6}$ $5s$	Ru^{44} $4d^{7}$ $5s$	Rh^{45} $4d^{8}$ $5s$	Pd^{46} $4d^{10}$ –	Ag^{47} $4d^{10}$ $5s$	Cd^{48} $4d^{10}$ $5s^{2}$	In^{49} $5s^{2}5p$	Sn^{50} $5s^{2}5p^{2}$	Sb^{51} $5s^{2}5p^{3}$	Te^{52} $5s^{2}5p^{4}$	I^{53} $5s^{2}5p^{5}$	Xe^{54} $5s^{2}5p^{6}$
Cs^{55} $6s$	Ba^{56} $6s^{2}$	La^{57} $5d$ $6s^{2}$	Hf^{72} $4f^{14}$ $5d^{2}$ $6s^{2}$	Ta^{73} $5d^{3}$ $6s^{2}$	W^{74} $5d^{4}$ $6s^{2}$	Re^{75} $5d^{5}$ $6s^{2}$	Os^{76} $5d^{6}$ $6s^{2}$	Ir^{77} $5d^{9}$ –	Pt^{78} $5d^{9}$ $6s$	Au^{79} $5d^{10}$ $6s$	Hg^{80} $5d^{10}$ $6s^{2}$	Tl^{81} $6s^{2}6p$	Pb^{82} $6s^{2}6p^{2}$	Bi^{83} $6s^{2}6p^{3}$	Po^{84} $6s^{2}6p^{4}$	At^{85} $6s^{2}6p^{5}$	Rn^{86} $6s^{2}6p^{6}$
Fr^{87} $7s$	Ra^{88} $7s^{2}$	Ac^{89} $6d$ $7s^{2}$															

Ce^{58} $4f^{2}$ $6s^{2}$	Pr^{59} $4f^{3}$ $6s^{2}$	Nd^{60} $4f^{4}$ $6s^{2}$	Pm^{61} $4f^{5}$ $6s^{2}$	Sm^{62} $4f^{6}$ $6s^{2}$	Eu^{63} $4f^{7}$ $6s^{2}$	Gd^{64} $4f^{7}$ $5d$ $6s^{2}$	Tb^{65} $4f^{8}$ $5d$ $6s^{2}$	Dy^{66} $4f^{10}$ $6s^{2}$	Ho^{67} $4f^{11}$ $6s^{2}$	Er^{68} $4f^{12}$ $6s^{2}$	Tm^{69} $4f^{13}$ $6s^{2}$	Yb^{70} $4f^{14}$ $6s^{2}$	Lu^{71} $4f^{14}$ $5d$ $6s^{2}$
Th^{90} – $6d^{2}$ $7s^{2}$	Pa^{91} $5f^{2}$ $6d$ $7s^{2}$	U^{92} $5f^{3}$ $6d$ $7s^{2}$	Np^{93} $5f^{4}$ $7s^{2}$	Pu^{94} $5f^{6}$ $7s^{2}$	Am^{95} $5f^{7}$ $7s^{2}$	Cm^{96} $5f^{7}$ $6d$ $7s^{2}$	Bk^{97}	Cf^{98}	Es^{99}	Fm^{100}	Md^{101}		Lw^{103}

Mooser, Mott, Phillips and Cohen. As for band structure work itself (where I am, of course, not a direct practioner) my research has brought me most closely into contact with the Cambridge group, with Mattheiss, Schlüter and Zunger, and with Bullett, Gyorffy and Temmerman. That doubtless shows in the references too.

It has been my intention to present something of wider interest than just our present research work on T.M. chalcogenides and CDWs. I have not wished to duplicate in print what may be found in my earlier, more broadly based work on transition metal compounds and the Mott transition which may be found in refs. 47 and 36b ; on rare earth compounds and interconfiguration fluctuation behaviour in refs. 156a and 156b ; and on layer compounds and charge density waves in refs. 10b, 114, 34, 100, 106, 102 etc., etc.

Those who know my work will be surprised to find a paper with no new figure. I hope the words are sufficiently graphic in the first instance, but do recommend that reference be made to Hulliger's excellent structural figures in refs. 10a and 68. The not widely known book by Clark (ref. 246) is also very useful for the present class of compound. In this line references 7 and 10e need little introduction.

§ 1. INTRODUCTORY DISCUSSION OF BAND STRUCTURE REFINEMENT AND STRUCTURE PARAMETRIZATION

The continuous development of band structure techniques over the past 12 years in improving the quality and speed of the calculations has been most impressive[1]. These improvements satisfy two instincts. First, one may for a basic material like NaCℓ[2], or Si[3] or Ni[4] aim at ever upgrading the form of the starting potential, especially in regard to its exchange and correlation aspects, or in its non-locality. Self-consistency procedures may also be upgraded. Further to this, one can probe the extent to which the results of a single calculation may ever be able closely to cover both ground state and excited state properties[5]. In such studies it is now becoming possible at the high level of precision necessary (0.05 eV) to perform total energy calculations that permit (from a presented choice of simple structure alternatives) identification of the normal ground state structure, and of transformations from that phase under pressure[6].

Structure choice has for many years been monitored more or less closely by radius ratio rules, by classical electronegativity plots (like Mooser-Pearson plots[7]) or by extraction of average band structure characteristics for each particular solid (as after Phillips and van Vechten[8]). The structure-transferable, hard-core-pseudopotential, radial parametrization of structure choice recently made by Zunger, Cohen and coworkers[9] has provided a major breakthrough here. It deals as readily with alloys as with semiconducting compounds ; with T.M. as with non-T.M. materials. The smooth atomic-like characteristics of the Periodic Table, long used intuitively in solid state work, are brought into a quantitative basis. The reason for the similarity of Be with Aℓ, B with Si, As with Te, S with I in solid state behaviour is made explicit. From a table of characteristic, ℓ-dependent, atomic, hardcore radii (r) it has proved possible to form a meaningful 2D-parametrization of the vast majority of known AB materials into isostructural domains. To the extent that different domain boundaries are needed for super-octet materials (e.g. GaS, PbS) as compared with sub-octet (e.g. LiO) and normal octet materials (e.g. ZnS) means that a more awkward 3D-parametrization is really necessary. Projection to the 2D plot leads to quite unrelated families like CrB (B33) and TℓSe (B27) occupying the same region. It remains nice, however, to see a clean separation being made of, for example, the NaCℓ-(B1) and NiAs-($B8_1$) structured materials. Although the presently selected form of parametrization for these plots works so well and is so simple ; namely (R_σ,R_π) where

$$R_\sigma^{AB} = |(r_p^A + r_s^A) - (r_p^B + r_s^B)| \quad \text{'size mismatch'}$$
$$R_\pi^{AB} = |r_p^A - r_s^A| + |r_p^B - r_s^B| \quad \text{'orbital non-locality index'}$$

it is not clear that it is the 'best' parametrization, and in line with this a clear casual route to structure selection remains lacking. Why should R_σ = 0.765 a.u., R_π = 0.335 a.u. mean a NiAs- and not a NaCℓ-structured material ?

Structural choice particularly among transition metal compounds clearly has a myriad facets. What is the role of the d-electrons in making the NiAs structure unknown outside T.M. compounds ? The deviation of structure from NiAs in NiS (millerite)[10] is very specific, as is the false divalency of CuS ($\equiv Cu_2^{1+}S_2Cu^{2+}S$)[10]. The square-planar coordination and semiconductivity of PdS and PtS[10] is a common ligand-field imposed feature of d^8. It recurs in semiconducting $Ni^{2+}P_2$[11](and semimetallic $Au^{3+}GeAs$)[12], but is lost with PtP_2[13] which achieves its semiconductivity through octahedral coordination and a 'quadrivalent' t_{2g}^6 platinum (pyrite structure - see later). The above approach will in all likelihood yield a less satisfactory division for structures of AX_2 stoichiometry than it provided for AX. However in playing down the real space aspects of electronegativity, i.e. of 'soft' radii and charge transfer, I am sure the new approach is providing a valuable service. Attempting to put those notions on a quantitative basis is not a 'proper' pursuit, as noted long ago by Slater[14]. This is reflected in Mattheiss' successful 7/6 prescription for APW calculations[15].

Trustworthy charge distribution maps naturally prove invaluable for perceiving structural character. Eigen-function information is rapidly improving, though always lagging on that for eigen-values. Besides total valence band plots, the charge distribution can be plotted band-by-band, or indeed k-point by k-point to help isolate significant structural information. It remains so to be clarified for d^2MoS_2 as to whether its trigonal prismatic coordination is stabilized over the octahedral coordination of d^0ZrS_2 by what happens within the 'd'-band or in the 'p'-band. We use labels of convenience here, these two bands being in reality heavily hybridized[16]. Just as for $\rho(r)$, so it is possible also to decompose the density of states band-by-band or state-by-state, according here to effective quantum number ℓ[17]. The fact that $\ell = 3$ may show up well away from the rare-earths or actinides reminds one to be careful not to assign too close a parentage to these decompositions, in a simple tight-binding, charge transfer sense.

Although neither ZrS_2 nor SnS_2 are quadrivalent in an ionic sense, it is, nevertheless, appropriate to label them as quadrivalent in regard to electron counting (or equivalently to Bloch state counting). Four remains, whether dealing with 'ionic' HfO_2 or semimetallic $TiSe_2$[17a] the number of electrons needed to complete the V.B.-forming double octet of bonding state electrons. Only for strongly semimetallic $TiTe_2$ is the 'rule of eight' compromised, although the AX_2 stoichiometry is not relinquished. Charge plots give a good feel for what the 'level of ionicity' is actually like in a compound (see ref. 18 for SnS_2), after we have learnt what to expect through a series like (CuBr), ZnSe, GaAs, Ge[19] as minimum and average bonding-antibonding bandgaps progressively close up (3.0/8.0, 2.3/7.0, 1.4/5.2, 0.7/4.3 eV)[20]. For each formula unit per unit cell one has, of course, 4 's-p' valence bands in a simple AX compound like ZnSe, CaO[21] or CsCℓ[22], independent of coordination of A by X, or vice versa. For AX_2 (e.g. ZrS_2)[16] the number becomes 8 ; for AX_3 (e.g. ReO_3)[23] 12 and for AX_4 16.

§ 2. THE OBTAINING OF BAND STRUCTURES FOR COMPLEX MATERIALS

We are at this point brought to the second route open to current band structure work. Namely to press on beyond the 'simple' materials to produce band structures for more complex materials, known already to present great potential for interesting physical properties. A glance at the formulae of the following semiconducting materials immediately reveals they cannot be simple in the above sense

NaS, ZnS_2, SiP_2, SiP, ZnSb, ZrP_2, ZrI_2, $ZrSe_3$, $ZrTe_5$, NbS_3, $CoAs_2$, $CoAs_3$.

Many materials here contain an odd number of electrons per formula unit ; some, like SiP_2 or $ZrSe_3$ appear 'sub-octet' ; others, like SiP or ZrI_2 are 'super octet'; The development of tight-binding methods by Bullett[1.d)] and of the ASA method by O.K. Andersen[1.f)]

have been particularly useful in attacking materials of some complexity, often containing 20 or even more atoms per unit cell (e.g. $NbSe_3$(24)[24], β-rhombohedral boron (105)[25], $Tl_2Mo_6Se_8$(16)[26]). Bullett[27] has made a check of his method against the thorough APW results of Mattheiss for the d^4 rutile metal RuO_2[15.a)] with a very satisfactory outcome. His results for the structurally complex and diverse T.M. trichalcogenide family (e.g. $ZrSe_3$, NbS_3, $NbSe_3$ and $TaSe_3$) similarly compare well with experiment[24,28].

The band structure of $NbSe_3$, as for many of those compounds in the list earlier, contains many bands ; in $NbSe_3$ there are 3 inequivalent Nb and 9 inequivalent Se sites per unit cell. It is most desirable in such circumstances to have an understanding of how to 'read' them : indeed how in the absence of a band calculation to produce a skeleton band structure starting from the crystal structure and electron configuration. For the rest of this lecture we shall review by liberal example the many facets dictating the form of band structures for materials beyond the 'simple'.

Many alloys have complex structures and stoichiometries as a glance at Pearson's book[7] will reveal. Personally I have no experience nor great insight into when a phase diagram will be simple (like Fe/Ni) or when littered with stoichiometric phases (like Y_2Ni_{17}, YNi_5, Y_2Ni_7, YNi_3, YNi_2, Y_4Ni_3 in the magnetically interesting Y/Ni system). The same goes for metal-rich metalloid phases, like Pd_8P, Pd_6P, Pd_5P, Pd_3P, Pd_5P_2, Pd_7P_3. Such complexity starts of course in the elements themselves, with In, Mn, U and especially boron. These problems I leave for others to discuss[29]. The vast majority of heavily metal-rich materials are naturally metallic, and this goes as much for sulphides (e.g. $Nb_{21}S_8$, Nb_2S) as for the phosphides above[10a, 242de]. It is even true for the structurally very simple layer halides Ag_2F[30], ScCl and ZrCl[31].

§ 3. CONTRASTING THE STRUCTURAL SITUATION OF OPEN s/p BANDS WITH OPEN d-BANDS

Compounds like the latter highlight the fact that by choosing the growth route with care a very large selection of unlikely looking materials may be prepared. Materials like these and those listed earlier look 'unlikely' because we have learnt not to expect systems which produce binary compounds with integrally open p- and s-bands (nor indeed a d-band quite as in ZrCl). Sometimes semimetallic overlap may fractionally open bands as in h.p. SiP_2[32] or $TiSe_2$[33], but this is normally of little consequence (though it is the origin of the low temperature distortion in $TiSe_2$)[34]. Simple compound materials having integrally open p- or s-bands and attendant metallic character are indeed few. SnAs[35] is one such metal. We shall discuss shortly the ways in which materials like SnP, GeP, Sn_2S_3, etc., avoid this state.

By contrast there do occur very many standardly structured metals which present an open d-band e.g. RuO_2 (d^4, rutile)[15.a], CoS_2 (d^7, pyrite)[36], VS (d^3,NiAs)[37]. Nonetheless in structurally favourable circumstances we know there is a fair drive to drop off filled d-sub-bands. This is able to occur in the d^1 compounds by forming discrete cation pairs, as in Ti_2O_3 [38](corundum structure, face-sharing octahedra) and in VO_2 [39](rutile structure ; edge-sharing octahedra), though not in ReO_3 [23](where the octahedra share corners and the structure is cubic). With two electrons or two holes per cation, sub-band dropping is often achieved by the formation of cation chains ; zig-zag in WTe_2[10.b](d^2, distorted CdI_2), linear in $FeAs_2$[40](t_{2g}^4, distorted marcasite). Unlike these two d^2MoO_2[41] continues to pair and is left a metal (distorted rutile). Interestingly d^2TiS being NiAs-structured cannot give c-axis chain formation, and shows indeed a very large c/a ratio[42].

A surfeit of two d electrons in layered $LiVO_2$[43]leads to triangular clusters, while in d^3ReS_2[10.b] ($E_g \geqslant 1$ eV) linked clusters of 4 Re atoms occur. These clusters are diamagnetic with saturated pair-bonding. However, the triangular clusters formed in Nb_3I_8[31] are not saturated (7e) and metallic conductivity is presumably only avoided because of the poor cluster overlap. Octahedral metal atom clusters in NbO[44] (doubly defect NaCℓ) do leave metallic conduction, as also they do in the superconducting Chevrel phases based on Mo_6Se_8[45]. Again the 19e cluster halide Nb_6I_{11} ($\equiv Nb_6I_8I_{6/2}$) is a Mott insulator having interesting magnetic properties[46].

In all cases of distortion under M-M d-bonding it must be remembered that significant changes occur at the same time in the main 's-p' valence bands, these being directly affected in their energies by the M-X bond lengths. We have already made this point for d^2MoS_2 and d^8PtS when discussing the semiconductivity arising there from the adoption of trigonal prismatic and square planar coordination respectively.

For some highly symmetric structures semiconductive gapping throughout the zone may not be possible, as the bands in question can become tied together at some high symmetry point. In particular for octahedral coordination, where low-spin t_{2g}^6 semiconductivity is such a common feature[47], e.g. $PtS_2(CdI_2)$, $PtO_2(CaC\ell_2)$, PtP_2 and OsS_2 (pyrite), it cannot occur for the rock-salt or NiAs structures, thereby precluding the existence of semiconducting RuO or RuS.

When gaps occur as in MoS_2 through coordination, or in VO_2 through distortion, these gaps often look very large in comparison with the various sub-band widths. For example the minimum gap in MoS_2 is $1\frac{1}{3}$ eV, which is virtually the same as the filled sub-band's width. This is the band often loosely labelled 'd_{z^2}'. However, as shown by Mattheiss[15.a] the bare d_{z^2} band would be considerably wider than this, and the narrowness of the filled sub-band finally dropped off from the main d-band is actually a consequence of heavy p/d hybridization.

§ 4. CLOSED 'LONE-PAIR' BANDS

We are brought at this point to consider what are known as 'lone-pair s^2' bands. In SnAs[35] with the extra electron beyond InAs (and a change in structure to rocksalt, which renders InAs itself semi-metallic at high pressures, and probably at ambient, given the situation in Cd_3As_2[48] and Melé's calculation for metallic rocksalt InSb[49]) we expect the Fermi level to lie in a quite broad s-like conduction band. With one more electron for rocksalt SnTe we suddenly however achieve semiconductivity[50], as for all other divalent IVB compounds from SiS through SnSe[51] to PbTe[52]. The narrowness of this closed band is again doubtless the product of hybridization. It certainly is heavily mixed with $\ell = 1$ states. For PbTe the weight of the $\ell = 0$ contribution is in the sub-band found just below the main p-valence band, but by PbI_2[53] and PbO[54] the latter p-band has been so stabilized by the principal binding that the 'lone-pair' weight has transferred to the top-most isolated V.B. Such a semiconductor is somewhat unusual for non-T.M. compounds in then having an intra-cation and strongly excitonic band gap, rather than customary 'charge transfer' band gap. [N.B. Because of high symmetry, there arises in $CsSnBr_3$[55](perovskite), a symmetry-induced zero-band gap situation at the R point].

The inversion of role above between bands of similar symmetry is reminiscent of what happens between copper[56] and silver-monohalides[57] with respect to the p-band and just closed d-band. For anybody wishing to compare the virtues and failings of various band structure techniques at a common point in time, they would do well to see the dozen or so calculations made for CuCℓ[58] at the time of the furore concerning high temperature, transient, Meissner-like effects in 'pressure-quenched' material[59]. Similar effects for CdS[60] did not produce such a spate[61]. The remarkable effects of course have nothing to do with CuCℓ or CdS as such, but with the nature of their decomposition product[62].

Lone-pair 's^2' compounds are encountered for III^{1+} and V^{3+} in addition to IV^{2+}, e.g. TℓBr, BiI_3. The wide principal gap in BiI_3[63] secures a free high-lying s-band, but in Tℓ monohalides it remains mixed in with the p-bands much as for PbS[52].

These lone-pair states provide unusually valenced compounds with one route to avoid an otherwise metallic product ; namely valence disproportionation to incorporate such a configuration. For example in Sn_2S_3[65], or in the TℓSe[66], both s^2 and s^0 cation sites can be identified in the complex structure through their characteristic sizes. BiTe[68] is likewise not what it might appear, being a layered mixture of Te_2 layers and the standard 's^2' layer compound sandwich Bi_2Te_3. Bismuth in Bi_2Te_3 takes regular coordination, as with other s^2 materials such as BiI_3[63], PbTe[52], PbI_2[53], PbF_2, TℓCℓ[64] for which the s^2 band is fairly clear of the p-band. In general, however, the stereochemistry of lone-pair materials is notoriously complex, as the following examples will illustrate, InCℓ[75], TℓF[68], $Tℓ_2S$[68], SnI_2[69], PbO[54], As_2Se_3[70]. In many cases the molecular unit becomes bent as a lone-pair orbital seems to develop. The tetragonal layer structure of red PbO typifies this : the Pb atoms are one the outside of the sandwich in contrast with the situation for otherwise isostructural AgI. In view however of the high coordination and structural complexities encountered in the halides of the alkaline earths (e.g. BaI_2, $SrBr_2$, etc.)[71] it is dangerous to apportion blame for these structures without a really wide ranging investigation.

§ 5. STRUCTURAL COMPLEXITIES ACCOMPANYING THE CLOSURE OF OPEN s/p BANDS

(a) Cationic superoctets

SnAs[35], we have noted in beginning sections 3/4, is most unusual in retaining its regular structure (NaCℓ) in an s^1 situation. For SnP and the high-pressure forms of GeP and GeAs, the M and X sublattices become displaced towards each other along one fourfold axis, though metallic or at least semi-metallic behaviour persists[72]. However, in the low pressure forms of GeP, SiP, GaTe, GaSe, InS, etc., completely non-metallic character is secured through the formation of cation-cation pairing (generally within layer structures[68], except for InS[10c]). Written $(Ga_2)S_2$, gallium sulphide becomes the direct structural analogue of MoS_2. The band-gap is again intra-cationic, but here is from the Ga-Ga bonding band to its corresponding anti-bonding band[73]. Charge maps are available to show the changes occurring between GaS and InSe[74] Mercurous materials like Hg_2S[10.c], $Hg_2Cℓ_2$[76] also contain an M-M pair.

(b) Anionic sub-octets

By contrast with the above 'super-octet' materials, 'sub-octet' materials to avoid an open p-band must show anion-anion bonding in some form or other. For some this may entail X-X pairs, (e.g. NaS, ZnS_2, SrP) ; for some X-X chains and rings, (e.g. NaP, NiP_2, $CoAs_3$) ; for others entire planes (e.g. $CaSi_2$). The variety is enormous and the potential for novel materials abounds. For example the calcium may be extracted from $CaSi_2$ to produce a layered silicon. Much of the crystal chemistry has been well reviewed by Hulliger[77, 10.a], especially as it pertains to layer compounds[68]. Some T.M. materials like NbS_3 and VS_4, show pairings on both sublattices. To penetrate this complexity we must look back at the band structures of the elements in groups V, VI and VII.

The usual semiconductive structures of elements such as As, Se and Br are of course expressive of the 'rule of eight' - just as is that of Ge. However, we must try to divorce the former electron-counting routine from the notion of directed-bonding often associated with sp^3 hybridization. Actually I believe that the latter notion is much overplayed. Except where symmetry demands it at the X point in the diamond structure, the 's' and 'p' valence bands are well separated. Even the X-point degeneracy is gone in the zincblende structure of GaAs or SiC[14]. The drive to four-fold coordination in such materials is principally geometric, as Zunger and Cohen's results confirm[1.c)]. No great benefit accrues from establishing four-fold coordination with very little difference in total energy between, say, rocksalt and zincblende CdS[61.b)]. We accordingly encounter sixfold coordination of Si and Ge in SiP_2[32] and (rutile) GeO_2[78] duly cut back to four in $SiAs_2$ and GeS_2[68], but up at six again in certain silicides and germanides (e.g. TiSi). Even carbon nearly presents us with graphite as its ground state[79]. It would be nice to see an analysis of charge density plots for say rocksalt and zincblende-structured AℓAs or GaAs, in order to assess just what action the presence of 'bond charges'[8,19] on the inter-atom cols has on structure selection. This should be done in both structures, firstly at the two 'appropriate' bond lengths, and secondly at common bond lengths and then volumes. Hopefully this would highlight what influences bond-charge formation, band-gap narrowing (minimum and average), and indeed volume collapse between the two structures.

Structures with coordinations of 4 or less (including the puckered layer structure of arsenic, the helical chains or rings of selenium, and the isolated pairs of bromine) so often are discussed in valence bond terms, but throughout it seems more credible, safer, to stick to a M.O. approach and simply identify bonding, non-bonding and antibonding bands. The number of occupied bands is determined solely by the atomic configurations and the atomic content of the unit cell. Clearly there has to be little sense in introducing d-orbitals when discussing an octahedrally coordinated SiP_2[32] (or indeed CdI_2-structured SnS_2[18]).Six-fold coordination, say in rocksalt CdO or CdS[61], is of course established with eight electrons and four orbitals, just as much as was four-fold coordination in ZnS[19] (or two-fold coordination is in α-HgS[80]). Electron counting must, we see, be positively hampered by any ball and stick model of the local geometry. Instead we should look only for the identification of the unit repeat in the structure, together with an appreciation from interatomic distances of the interactions current in that unit. It is in the latter aspect that SnS_2 is 'simple', and SiP_2 is not.

The only safe way to represent electron counting is in a

pseudo-ionic fashion without recourse to a diagram ; thus Cd(II)O, In(III)$_2$S$_3$, Sn(IV)S$_2$, Mo(IV)S$_2$ [d^2], Sc(III)P [d^0], Gd(II)S [f^7d^1], Pb(II)S ['s^2'], Sn(II)Sn(IV)S$_3$ ['s^2',s^0]. The square brackets give the number of electrons residual to the main bonding bands, plus a loose identification of character. When M-M bonding is present, as in V(IV)O$_2$[d^1] and Ti(III)$_2$O$_3$[d^1], these simple representations do not register the pairing. If we felt this necessary, one could write {V$_2$(VIII)}O$_4$ [d^2] and {Ti$_2$(VI)}O$_3$ [d^2]. Similarly for GaS we need to replace Ga(II)S ['s^1'] by {Ga$_2$(IV)}S$_2$ ['s^2'], to register the Ga-Ga pairing. We can now move forward along these lines to look at the band structures of X-X bonded elements like phosphorus and compounds like SiP_2 or $Zn_2(P_2S_6)$.

The case of tellurium has been studied in some detail, with band-by-band charge density maps[81]. The structure is a helical chain with 3 atoms per unit cell, the 120^0 pitch leading to D_3^4/D_3^6 symmetry. There are accordingly three s-bands per unit cell, slightly differentiated by bonding, non-bonding and antibonding character but all occupied. There follow nine p-bands, likewise differentiated though much more strongly so. The structure is formed with a complement of 4 p-electrons available per Te to place the Fermi level in the sizeable gap between non-bonding and antibonding bands. Integrated $\psi^2(\underline{r})$ plots for the three batches of dominently p-bands show in ascending order nodal, lone-pair and anti-nodal characteristics relative to the Te-Te axis. The bonding band as usual for short bonds in low coordination shows a charge hill on the nodal col. Under simulated pressure this so-called 'bond-charge' is found to evaporate as the helix is forced back to comply with a cube body-diagonal and a regular 6-fold coordination[81.a]. In this process the semiconductive gap is known experimentally to be lost for both Se and Te. The real process may however be considerably more complicated than was modelled in ref. 81a[82].

Solid iodine is formed from molecular pairs based on the single p-band hole state. Again the pairing can be progressively decoupled under pressure, to yield en route overlap of the nonbonding and anti-bonding p-bands and metallic behaviour[83].

Going the other way from group VI to group V, we find the low gap, semiconductive, layer structures of black phosphorus, arsenic, antimony and bismuth, plus the higher gap, molecular cluster states[68]. All express the bonding/anti-bonding activity now of all three p-electrons present per atom. The three-fold coordinate layer structures can again be related back to a simple cube, and high pressures again force them towards this and metallization. The recently published band structure work for the case of black phosphorus is to be found in refs. 84.

The minimum gaps in bromine and iodine are approximately 2.1 and 1.3 ev ; in selenium and tellurium about 1.6 and 0.33 eV ; in phosphorus and arsenic about 0.3 and 0.1 eV. The average non-bonding/anti-bonding and bonding/anti-bonding gaps are considerably greater than this, in tellurium being ∿ 3 eV and 5 eV respectively. For black phosphorus the corresponding values are around 5 and 7 eV, though the band roles are less clear out[84]. In due course we shall relate these numbers to the band structures of $MgTe_2$, SiP_2[32], FeS_2[26], etc.

The pyrite structure of the latter compounds is equivalent to that of rocksalt but with a simple anion replaced by the diatomic anion $(Te_2)^{2-}$ or $(P_2)^{4-}$ (giving their pseudo-ionic valences). In the cubic pyrite structure (Pa3, Z=4) these units are aligned equally along the four body diagonals (ref. 10b, Fig. 7). In the pyrites $MnTe_2$ and PtP_2, for which structural refinement has been made, the X-X pair length is 2.75 and 2.17 Å respectively[86]. These lengths when compared with closest approaches in the elements of 2.86 and 2.22 Å indicate the bonding to be very comparable in the two situations. The comparison is of course not 'direct' since the X-X length in question is now in the pair geometry of a pseudo-halogen. The following table clarifies such inter-relationships for less familiar cases, aided by actual examples to illustrate the stoichiometry appropriate to the discrete pair and to the infinitely extended chain and layer anions of this kind.

From the table one may readily see how 3d NiP_2 with its infinite phosphorus chains can display divalent Ni(II) in the square planar coordination appropriate to d^8, whilst for pyrite 5d PtP_2 the semiconducting t_{2g}^6 configuration of Pt(IV) obtains.

Although the relatively simple scheme of Table 1 embraces a large number of compounds, there are many more which are more complex and lie outside it - though often including some of its features. For example the linear chain compound $ZrSe_3$[68,24] mixes Se^{2-} and $(Se_2)^{2-}$ units to achieve standard Zr(IV)d^0 semiconductors tivity as $ZrSe(Se_2)$. Likewise to represent the situation in LaS_2[87] one can write $La_2S_2(S_2)$. The low pressure form of SiP_2 and $GeAs_2$ mixes single As atoms and continuous As chains to show quadrivalent semiconductivity, $GeAs^{3-}[As]^{1-}$.

In some cases the chains or layers in the structure are not infinite, which leads to further stoichiometries, e.g. SrS_3, TlP_5. Sometimes such fragments are of two types as in CeP_2, or interlinked as in LiGe, which makes the stoichiometry still less transparent[77]. The example of $Ca_{11}Sb_{10}$ quoted by Hulliger[77] gives an idea of the complexity possible. The latter material includes a square loop which is present also in the simpler skutterudite family of t_{2g}^6 semiconductors, e.g. $CoAs_3$[10a]. The full cell in $CoAs_3$ contains eight molecules, and contact may be made with the

Table 1. Pair, chain and layer anions based on the structures of the elements of groups V to VII. The valence of the unit is given, and illustrated underneath by a variety of insulating (or semi-metallic) examples of appropriate stoichiometries.

Sheets As-As	Chains -Se-Se-	Pairs (Br-Br)
[>Ge-As<] $^{1-}$	[-As-Se-] $^{1-}$	(Se-Br) $^{1-}$
[>Ge-Ge<] $^{2-}$	[-As-As-] $^{2-}$[-As-] $^{1-}$	(Se-Se) $^{2-}$
	[-Ge-As-] $^{3-}$	(As-Se) $^{3-}$
	[-Ge-Ge-] ^{4} [-Ge-] $^{2-}$	(As-As) $^{4-}$
?	?	?
$EuGe_2$, $CaSi_2$	red ZnP_2, NaSb	$V(S_2)_2$, ZnS_2, Na_2S_2
	AuGeAs	Co(AsSe)
	CaSi	SiP_2, $Ca_2(As_2)$

table by writing the contents as $[Co_4^{3+}(As_4)_3^{4-}]_2$.

Some clusters are polyhedral such as the P_7^{3-} unit of $Ba_3(P_7)_2$, or Si_4^{4-} in $BaSi_2$[77]. Many borides are like this, semi-conducting CaB_6[88] and metallic d^1 LaB_6[89] having the octahedral B_6^{2-} anion. TiB_2 has the AlB_2 structure with dense B-B bonded hexagonal sheets, and the four electrons per B_2 repeat which are available from the Ti atom all but yield semiconductivity, i.e. $Ti(IV)B_2$ with a 10e V.B.[90]. This is a particularly low V.B. content for an AX_2 compound, as was 18e per formula unit very low in $CoAs_3$. Large numbers of anti-bonding bands are being ejected above the Fermi level in these materials.

We have already found in the elements that the anti-bonding states are ejected several eV on average above the top of the V.B. We see similar action to this in a compound like ZnS_2 (pyrite) (or $MgTe_2$)[91], where the pale yellow colour of the compound indicates a band gap in excess of $2\frac{1}{2}$ eV. (Such a value is comparable with the main p → s gap in tetragonally coordinate ZnSe). In ZnS_2 the S - S bond is probably somewhat longer (i.e. weaker) than in particularly stable and dense t_{2g}^6 FeS_2 for which a (X - X) ejection of $4\frac{3}{4}$ eV had been estimated[36b], in good agreement with recent calculation[36a]. The S - S pairing through FeS_2, CoS_2 and NiS_2 actually shows remarkably little change with d-electron content despite these being semiconductor, ferromagnetic metal, and Mott insulator, at d^6, d^7 and d^8 respectively. For this sequence (completed by CuS$_2$ at d^9 and ZnS_2 at d^{10}) the e_g band is being filled. This is anti-bonding w.r.t. p-d M-X bonding, and through the sequence the lattice parameter itself increases strongly. The Te-Te pairs of $MgTe_2$ and high-spin $MnTe_2$ are likewise very similar.[91]

The absence of cation-cation pairing at e_g^1 (d^7) above, whether for the pyrite or marcasite forms of CoS_2 and $CoSe_2$, should be contrasted with the pairing which arises for the favorable t_{2g}-based orbital geometry at low-spin d^5 in $CoAs_2$, etc. (arsenopyrite, FeAsS)[92]. The semiconductivity of that family is in this way the product of pairing on both sub-lattices. This also is true of d^1 NbS_3[28] (≡ $NbS(S_2)$ and directly derived from ZrS_3).

Returning to d^7, we meet with a very interesting effect through the 3d, 4d, 5d sequence $CoSe_2$, $RhSe_2$, $IrSe_2$. Although 2 is a most acceptable valence for Co, 3 is more favoured for Rh and Ir, and so in α-$RhSe_2$ and $IrSe_2$ the marcasite structure becomes modified to achieve this[10b]. The Se-Se pairing already weakening slightly in β-$RhSe_2$ is for the new structure completely lost to half the Se content. The resulting structure is complex (Z=8), but can be written

$Ir_8^{3+}(Se^{2-})_8(Se_2)_4^{2-}$ to reveal its t_{2g}^6 semiconductivity.

The drive to achieve this semiconducting configuration is such that the defective pyrite $IrSe_3 \equiv Ir^{3+}_{2/3}Se_2$ occurs also[10a].

In several instances M-M or X-X bonding clearly is present in a structure but not of a sufficient strength or with the right geometry to see filled-band semiconductivity actually gained. The various strongly metallic distortions of the NiAs structure illustrate the case under M-M bonding (e.g. MnP, FeS)[10a], as likewise do the d^2 and d^3 rutiles MoO_2[41] and ReO_2. Strong X-X bonding we have seen was quite compatible with CoS_2 and CuS_2 being metallic, and the same is true too of $ZrTe_5$ ($\equiv ZrTe_3 : Te_2$)[93] and $NbTe_4 \equiv Nb(Te_2)_2$[10a]. NiP[10a] remains metallic despite pairing on both sublattices, as does $TiAs_2$[10a]. For NiP the P-P length of its pairs is relaxed to 2.43 Å compared[11] with 2.22 Å in the chains of NiP_2. Comparison of NiP and $TiAs_2$ respectively with NiSe and $TiTe_2$, suggests that semiconductivity was denied to these two compounds by the Fermi level being low enough to intersect the main p-based valence band.

§ 6. THE DETERMINING OF p-d OVERLAP

We can now move to a problem that has come much to the fore recently in band structure work on transition metal compounds, especially in connection with TiS_2/Se_2[33], and to a lesser extent the group III rock-salt pnictides like ScP[94,44]. It is the question of accurate calculation of the separation of the p and d bands. Unfortunately the poor stoichiometry of the above families does not help when trying to identify the point of p/d overlap unequivocally. Although particularly apparent at d^0 because of the passage from semiconductor to semimetal, the problem is known to be present in calculations on d^n materials. For example Mattheiss' APW calculations[15,16,21] on a wide range of compounds, despite giving a good description of the internal structure of the d-bands (e.g. Fermi surface detail, monitorable by de Haas-van Alphen measurement)[95] invariably place the d-band complex too high by 1 or more eV relative to the p-band (as assessed by optical and photoemission data)[47]. Though the majority of earlier calculations, like those of Mattheiss have tended to over-estimate the p-d separation (c.f. early calculations on TiS_2), the more recent 'self-consistent' calculations have tended to swing the other way. The gap now certainly appears under-estimated for TiS_2 in refs 33, and for ScP, etc. in ref. 94. In view of the large p-d gaps in TiO_2 and Sc_2O_3, it is most unlikely that they are shut right down in TiS_2 and ScP. It looks experimentally that this only becomes the case with $TiSe_2$ and with ScAs or YSb. Among other things it is clear that for the calculations to be discriminatory on this point they must be fully relativistic. Indeed it is well known that the semiconductivity of PbTe[14,52] or CsAu[96], or Yb[97] under pressure, only comes as a consequence of the reduced symmetry of the double-groups forcing gapping.

Relativistic calculations for the C.D.W.-induced superlattice condition in (d^0) $TiSe_2$[98] and (d^1) 2H-$TaSe_2$[99] are clearly going to be of much interest to the next round of band structure improvements. A point of particular importance emerging for these layer compound CDW's, as for $NbSe_3$, is the phasing with which the periodic structural distortion is placed on the underlying lattice[100]. Though the results of such a deterministic band calculation[101] for 2H-$TaSe_2$ (3 x 3) looked promising when made, they are now in question following new structural modelling to accommodate the recently revealed space group of Cmcm[102,103]. The new work suggests that the common origin of the triple-axis CDW should be shifted from a Ta atom to a Ta vacancy axis. We shall return to discuss CDW problems later. Relatively simple dHvA data does exist for 2H-$TaSe_2$ which should provide a powerful constraint on the acceptibility of future calculations[104]. For $TiSe_2$ there is almost complete loss of its semi-metallic free carrier content in the low temperature superlattice condition.

We now proceed to look at one of two more complicated d^0 materials which offer additional points of band structural interest. SiP_2 we have repeatedly mentioned already. The high pressure pyrite form is semi-metallic with $\sim$ 0.01 carriers/molecule and these appear to exhibit an excellent example of Baber e-h scattering to above room temperature ($\theta_D \sim 600^0$K)[105]. The new band structure calculation[32b] has not yet been analyzed to reveal whether the C.B. electrons are s-like, p-like or d-like. dHvA data exists to check the calculation.

Quadrivalent ThS_2[10a] from the other half of group IV, naturally has a band gap p-d that is very much greater. It does not take the CdI_2 structure shown by ZrS_2 or HfS_2, but the higher coordinated $PbCl_2$ structure, a representation of which appears in figure 1. The primary coordination is trigonal prismatic with three more S atoms at only marginally greater distance capping the vertical faces of the prisms. The prisms themselves form continuous chains. High coordination is often found for this side of group IV, e.g. in the baddeleyite form of ZrO_2. ZrP_2 and $ZrAs_2$[10a] (unlike $TiAs_2$, p17) continue the trend with a slight distortion of the $PbCl_2$ structure which comes close to yielding d^0 semiconductivity. The structure is sheared in such a way that half the anions move to form As chains. With Z=4 we may then write the cell contents $4[Zr^{4+}As^{3-}(-As-)^{1-}]$. The As-As bond length appears not quite short enough to avoid a semimetallic end product. All the Zr atoms in $ZrAs_2$ are equivalent.

A rather similar circumstance of much interest recently is that presented by $TaSe_3$[106]. The trigonal prismatic chains in this material are capped only on two prismatic walls, as in d^0ZrSe_3, for (as with $ZrSe_3$) the 'back' wall of the prism is X-X paired and

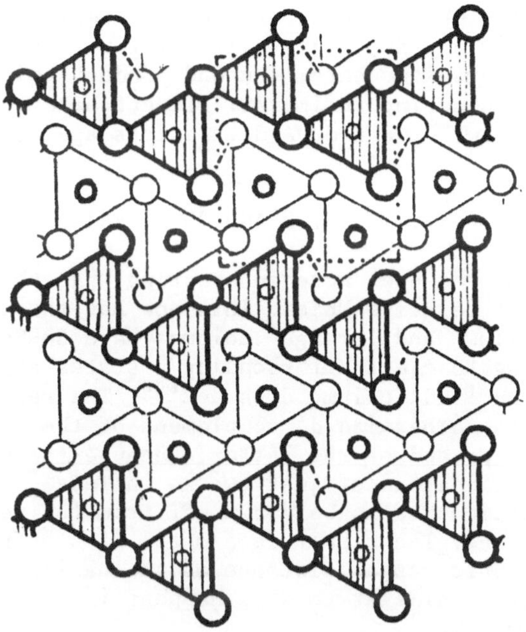

Fig. 1 : Structure of $PbCl_2$, and indication of As chain formation in $ZrAs_2$.[68]

hence much smaller. In $ZrSe_3$ (and in directly derived d^1, M-M paired , NbS_3) all the metal atoms were equivalent. However now in $TaSe_3$ we meet the interesting new circumstance that this is no longer so. Half of the prismatic columns (termed 'orange' in ref. 106) have moderately strong Se-Se bonding, but the other half (termed 'red') show very weak pairing that is insufficient to eject their anti-bonding p-states above the Fermi level. As a consequence the three red column -Se anions need 6 electrons to complete the p-band. With only 5 forthcoming from the red cations, the electron which is surplus to the $Ta(Se_2)Se$ situation of the orange column is drawn across. The result is that both Ta atom types end up with the $d^0Ta(V)$ configuration, though at somewhat different energies because of their different environments. The red column state is of course the less tightly bound. The cell contents are then $2[Ta^{(O)}(Se_2)Se + Ta^{(R)}Se_3]$. Actually, as for $ZrAs_2$, semiconductivity is not quite attained here, the end product being a 5×10^{20} cm^3 carrier, superconducting semimetal. Present band structure work supports this view[24a], and dHvA data exists against which to make improvement[107].

§ 7. PAIRING VS. CDWS. 1D VS. 2 AND 3D MATERIALS

What has proved even more interesting than the last material is its immediate relative $NbSe_3$. This now has three types of column. Both 'yellow' and 'orange' columns have Se-Se pairings sufficiently strong to eject anti-bonding p-states above the Fermi level, while the 'red' column does not[106]. The result is charge transfer from the otherwise d^1 complement of the former columns to complete the p-band of the latter. Because there are now three chains involved, instead of just two as in $TaSe_3$, only half an electron per cation has to be drawn off from the yellow and orange columns by the red. The result is a metal with a fractional electron count per unit repeat distance of the basic structure along the chain. That is similar to what happens in a 'charge transfer salt' like TTF-TCNQ.

Transfer from the yellow and orange cations in $NbSe_3$ is of course not quite identical. As I have discussed elsewhere at length the 'd_{z^2}' band corresponding to the orange chain becomes slightly more than half-full, the yellow slightly less[108,106]. The q-vectors of the two periodic structural distortions that develop in $NbSe_3$ at low temperatures reflect this. The preciseness and temperature independence of those wave-vectors suggests a pair of soliton lattices in otherwise alternatively paired pair-bond chains. The soliton periodicity marks the deviation from exactly one half for the electron counts per atom in the orange and yellow chains. A recent band structure analysis lends considerable support to this picture[24b]. What makes the two "CDWs" in $NbSe_3$ of particular interest is that they can, much more readily than others, be induced to slide and so carry current (e.g. by 5mV/cm d.c. field).

Moreover when sliding the CDWs generate sharp MHz frequency electrical noise. The CDW/PSDs are rapidly suppressed by pressure, then the chains being not so independent either electrically or elastically[109]. Elastically the properties are set by all the electrons of the valence bands ; while electronically the electrons of interest are ultimately those left in the "d_{z^2}" bands - 6 bands and 2 electrons per cell.

Two somewhat less differentiated chain types occur in the Nb_2Se_3 and Mo_2S_3 structure[110], and it is not surprising that two transitions have been reported in Mo_2S_3 (182 and 145K)[111]. A furhter weaker transition at 80K may represent coupling of these two to each other or to a lattice periodicity.

4Hb-$TaSe_2$ has two more strongly differentiated structural entities which experience an appreciable charge transfer ; namely complete MX_2 sandwiches. These alternately adopt octahedral and trigonal prismatic coordination. Both types of sandwich continue to show CDWs, 1T- or 2H-like, corresponding to their type of coordination[112]; strong in the octahedral layers and locking in at $\sqrt{13}a_0$; weak in the trigonal prismatic layers and related to $3a_0$. Charge transfer occurs from octahedral to trig. prism layers†. An attempt has been made to assess the transfer theoretically and to produce an appropriate band structure[113]. A transfer of 0.1 electrons per Ta has been deduced. This shift matches the drop in observed incommensurate wave-vector to below $1/\sqrt{13}$ for the octahedral CDW, and the passage to above 1/3 for the trigonal prismatic[112 114].

The above incommensurate CDW wavevectors can be nicely related to the Fermi-surface geometry, as was clearly shown in work on the substituational alloys 1T-$(Ta/Ti)S_2$[114]. Nesting of Fermi surface walls leads both to peaks in $\chi^0(q)$, the bare electronic susceptibility[115], and to strong reinforcement of $\chi(q)$ over the same range of wavevectors by the electron-phonon coupling matrix-elements[116,117]. Indeed, as suggested from the observed simple scaling of the transition metal dichalcogenide CDW onset temperatures with molecular weight[114], it is possible that the regular high temperature phases are only ever stabilized through anharmonic lattice fluctuations[117,118]. For CDW's unlike SDW's the lattice response and electronic system are quite inseparable[119]. In particular the effect of a discrete lattice with its set electron-count can be that the local environment and near-neighbour interactions begin to dictate the geometry, the form and the phasing, of the low temperature commensurate states[100,117] - and their carry through into higher temperature discommensurateness[102].

In a one-dimensional material it is quite difficult to make distinction between distortion wave-vectors impressed on a system

† appropriate to the ground state nature of the latter.

that signal, on the one hand, commensurateness from Fermi surface geometry in a 'nesting' condition, or on the other from short-range M-M bonding forces deriving directly from the 'surplus' electron count. Frequently in a strictly 1D situation the two 'superlattice' periodicities are the same, and then one has to turn to the magnitude and particularly the phasing of the PLD to provide clues about the seat of the distortion. Earlier we have attributed diamagnetic semiconductivity in '1D' d^1 compounds such as NbS_3, Nb_2Se_4, NbI_4, $NbC\ell_4$, $NbS_2C\ell_2$, $NbOC\ell_2$, VS_4 directly to M-M pair-bonding. For these the paired and ideal undistorted M-M lengths are

	NbS_3	NbI_4	$NbC\ell_4$	$NbS_2C\ell_2$	$NbOC\ell_2$	VS_4
pair	3.04	3.31	3.06	2.90	3.14	2.83 Å
ideal	3.36	3.83	3.41	3.51	3.93	3.03 Å
shift	0.16	0.26	0.17	0.30	0.40	0.10 Å
ref.	120	68	68	68	68	121

The third number here is the atomic shift into the pair configuration. The pairing displacement is seen to be rather variable, but possibly diminishing as the ionicity falls. For $NbSe_3$, the unusual properties of which were earlier presented in terms of alternate-pair-bonding[108], the shifts seem from the satellite intensities to be ~ 0.1 Å. This value, while comparable to the shift in VS_4, is appropriately smaller than that in fully-paired NbS_3. It would of course be particularly nice to know what the phasing of the PSD in $NbSe_3$ is. Present indications[108] are that it is indeed as sketched above. As long as it is possible that the distortion in $NbSe_3$ (and even NbS_3) might be taken as being Fermi surface determined, one should bear in mind that linear-chain Mott-insulators, such as $RuBr_3(d^5)$[122], $MoBr_3(d^3)$[123] and $NbOC\ell_2(d^1)$[68] also often show pairing. The values for $RuBr_3$ corresponding to those above are 2.73, 2.93 and 0.10 Å.

When one moves from 1D to 2D or 3D materials the periodicities from nesting and from pair-bonding no longer are so tied. We have already seen that the PSDs in the transition metal dichalcogenides, whether 1T or 2H, can definitely be accounted for in terms of the Fermi surface geometry[114] (via doping and tracking of the incommensurate q-vectors). Remember that an abnormal F.S. geometry is able to impress itself both through $\chi^0(q)$ and $\chi(q)$[116]. Where commensurate distortions are induced they are now often complex (e.g. the 4 x 4 cell of d^1VSe_2), and cannot be interpreted in terms of conventional pair-bonding (e.g. the two states reported for $NbTe_2$)[100]. Complementary to this, as noted at length above, M-M bonding still is prevalent in those 2D and 3D materials to present an appropriate structure and electron configuration. Corresponding to the above 1D pair data we can thus add the following data for the canted c-axis pairing (across shared octahedron edges) in the d^1 rutiles NbO_2 and VO_2 - 2.80, 3.00, 0.10 and 2.65, 2.88, 0.12 Å, respectively. Actually a short-lived attempt tried to relate these two distortions to Fermi surface effects and in particular $\chi^0(q)$, but it was not a feasible course to sustain[124,125]. Certainly strong nesting may dictate events in a 3D material, as the perovskite system Na_xWO_3 reveals[126], but neither NbO_2 nor VO_2 is such a case.

§ 8. PAIRBONDING, THE MOTT TRANSITION, AND DISORDER

NbO_2[127] and VO_2[128] are nonetheless very interesting materials. In particular they are representative of compounds where the M-M bonding can be destroyed thermally. In VO_2 this occurs in a strongly first order transition at only 340 K, but in NbO_2 the event is deferred to much higher temperatures and then is weaker. The d^1 corundum Ti_2O_3 similarly loses its pairing around 450 K, though this time in progressive fashion, there being no symmetry change[129]. In all three cases depairing is associated with an insulator-to-metal conversion and the materials have been extensively discussed in connection with the Mott transition (see ref. 47). As we shall see in a minute, there indeed is some interplay of phenomena, but it has to be pointed out straight away that many examples of temperature-dependent pair-bonding exist which have nothing to do either with CDWs or with the Mott transition, e.g. the h.t. loss of pairing in d^5 arsenopyrite $CoAs_2$[130] - or that in d^6 and d^3 MnP-structured CoAs, CrAs[131] and VS[132] where both conditions are metallic. By contrast with the latter materials, the prior mentioned oxides naturally are much closer to Mott, correlation-driven, localization of carriers (see Fig. 1 and Fig. 11 of ref. 47). Indeed for VO_2 there exists an alternative form having the diaspore structure where the compound is insulating (E_a = 0.12 eV) and strongly paramagnetic[133]. For this structure the rutile-like columns occur cross-linked in edge-sharing duos. This in fact causes the total volume to decrease (33.5 Å 3/molecule vs 36.9 Å ^{3}). One finds appropiately though that the crucial V-V distance within a column has risen to 2.92 Å from (an extrapolated) 2.85 Å in metallic rutile-structured $VO_2(R)$[134] (or an average of 2.87 Å in the monoclinic paired phase, (M_1)). It similarly is most revealing that under slight uniaxial pressure, over a narrow temperature range below T_0, half the chains in rutile VO_2 can be prevented from pairing[135]. Normally the pairs are canted, but now (phase M_2) those of the paired half show strong contraction with no cant (V-V = 2.54 Å), while the other half is not paired but strongly canted[135,128]. In the latter half the V-V distance is 2.93 Å, and as for the diaspore variety that leaves the material strongly paramagnetic and Mott insulating. Between this M_2 phase and the M_1 fully-paired l.t. phase a further transitional phase (T) exists ; it too naturally is insulating[128].

The above and related work on VO_2 serves to show several points. It reveals how close unpaired metallic VO_2 is to the Mott transition. It emphasizes that to study that 'delicate' transition a material must be chosen for which the opportunity for pair-bonding and cation displacement is absent. By these criteria the cubic NiS_2/Se_2 pyrite system is one of the very few to provide straightforward access[47,36.b)]. Finally it makes one aware, through the band structure calculations done on VO_2[39] (and now NiS_2[36.a]), of the one-electron band-width Δ, for a d-band at the point of delocalization. There the average Hubbard U intersite correlation energy becomes $\approx \Delta$, and calculation presents that value as $\sim 2\frac{1}{2}$ eV.

Many efforts have been made to unravel all the aspects of the electronic and structural transitions in VO_2, two of the more recent being refs. 136 and 137. The situation is possibly a little simpler in the much studied d^1 and d^2 corundums[138] Ti_2O_3[139] and V_2O_3(Cr)[140]. Magnetic ordering appears though as a complicating factor in the latter system. What all three systems uniformly bring to one's attention is that the exchange and correlation-driven enhancement of mass and the reduction of intersite charge fluctuation are so advanced as for doping of the insulating phases not to yield one-electron-like results. The electronic and structural disorder leads to carrier trapping and often to valence modification of the dopant. Thus metallic Sc-doping of paired Ti_2O_3 is not obtained[141]. By contrast V-doping of Ti_2O_3 does proceed fairly satisfactorily to give metallic products (they show spin-glass magnetic behaviour at low temperature[142]). However in the corresponding system $(V/Ti)O_2$ below T_d metallic conductivity is not gained even for very heavy doping[143]. The semi-insulating character there must result from a highly local mix of intact V-V pairs and isolated $Ti^{4+}(d^0)$ and $V^{4+}(d^1)$ ions, the latter being localized and strongly magnetic. Such disorder-induced localization is in evidence even for pure VO_2, where the photoconductivity gap of the M_1 phase is appreciably greater than the optical gap, implying the existence of strong mobility edge behaviour[144]. This quasi-amorphous view is supported by the log of the diminishing electrical conductivity below T_0 plotting as $T^{-1/4}$.

It is rather surprising that the VO_2:Ti[143] system has not been investigated as much as have other dopants since with Ti doping the valence of the dopant is unambiguously set at 4. Nb-doping by contrast[145] leads to the situation where the dopant sites more and more take up a valence of 5+, compensated by V sites moving to 3+. This charge transfer is registered by the development of a d^2 V-based magnetism along with degradation of the metallic conductivity. Conversely when doping with Cr[146], which firmly adopts a valence of 3+, it is the vanadium that is forced into the compensating 5+ condition. A variety of semi-ordered l.t. insulating phases appears. In $(V/Nb)O_2$ the strong linear depression of T_d at low x clearly shows that reducing the possibilities for pair formation can withhold the insulating phase despite the growing electronic and structural disorders. No one, I believe, has tried to form $Ti_{1/2}Cr_{1/2}O_2$ to see if a pseudo-VO_2 can be formed. d^2CrO_2[147] itself is still metallic, but now strongly magnetic (indeed ferromagnetic with T_c = 394°K). The system $(Cr/Mn)O_2$[148] though without the pairing problem again shows differentiation of site, Cr^{3+}/Mn^{5+}, and as expected a rapid loss of conductivity. Pure MnO_2 is a Mott insulator, as in $d^3Cr_2O_3$. Cr-doping indeed rapidly renders V_2O_3 insulating (1% at 300°K)[140]. However from what we have seen this system cannot (even at this low doping level) provide the ideal test system to monitor the Mott transition, especially since in the corundum structure the V atoms at the transition take advantage of the fact that their c-axis position is not a special point. Moreover much the same results are obtainable in the volume contracted system $(V/A\ell)_2O_3$[149] in which presumably the d-electrons avoid the Aℓ sites ; disorder as well as correlation clearly plays a significant role. Nonstoichiometry as V_2O_{3+x} gives a similar result[150]. Actually it would be nice to probe for narrow d-band metals cases where disorder provides the dominant factor : how significant then is the shift to localization ? It is not easy to identify two suitable isostructural partners which are of - and will remain of - like valence. We might suggest $d^1(Y/Lu)S$, $d^2(Mo/W)O_2$ or $d^4(Ru/Os)O_2$. CrO_2 and MoO_2 from 3d and 4d naturally mix 3+/5+[151].

§ 9. FURTHER ASPECTS OF DISORDER

It was in an attempt to accommodate both cation size and a matching of Fermi levels that the system $(Ta/Ti)S_2$ was selected (over say $(Ta/Zr)S_2$) to track the CDW wavevector through the $d^1 \rightarrow d^0$ occupancy range[114]. Even so it is amazing that a rigid-band approach worked so well. One has only to look at stages in the system $(Ti/V)Se_2$[152], where d^2 local moments appear, or in $(Ti/V)S_2$[153], where the resistivity rises with doping, to realize that the edge of the metallic range is being approached in these CDW-perturbed chalcogenides (see below).

Weaker localization is found in 'TiO' and 'VO'. These metallic oxides though CDW-free are closer to the Mott transition. Their negative α, disorder effects in ρ arise from a very high (∿ 15 %) spontaneous vacancy content on both sublattices. Surprisingly as these are partially squeezed out under pressure, the lattice parameter increases[155]. Recently an evaluation of the Mooij weak-localization criteria has been presented by McLachlan[154]. $\ell \simeq \lambda_F \simeq a_0$ where localization sets in.

Another batch of materials that may be regarded as 'self-disordering' are the rare-earth interconfiguration fluctuation compounds[156], TmSe, SmS (6 < P < 40 kB), SmB_6, etc. The lattice in these materials is attempting at the single site level to accommodate to the fluctuations between the two electronic configurations involved (e.g. for Sm(II),f^5d^1 and f^6). The volumes of these states in normal non-fluctuating conditions are very disparate. In I.C.F. conditions the fluctuation rate being comparable with phonon times ($\sim 10^{-13}$sec) permits only a partial relaxation[157], but this is quite sufficient to perturb the lattice and carriers continually, and so cause the resistivity once again to rise on cooling. It is possible to obtain a fair $T^{-1/4}$ plot[156a)], although at the lowest temperature ρ seems to fall away from this, possibly due to 'shorting' by a non-fluctuating semiconductive f^6d^0 surface layer! Sm metal itself shows a 'divalent' surface condition.

Perturbation of a lattice and its electron system by a static CDW similarly carries one towards localization problems. A rising resistivity results in $1T\text{-}TaS_2$ at low temperatures (though not in the selenide)[158]. This situation is made appreciably more marked by doping, even though that doping tends to reduce the amplitude and coherence of the CDW (e.g. l.t. lock-in)[159]. A particularly marked example is with the substitution of iron into $1T\text{-}TaS_2$[160]. This is not a magnetic effect, since here the iron is low-spin at low temperatures[161].

Proceeding from such randomly 'substituted' materials to fully ordered ternary materials, many of the above questions relating to band structural characteristics persist. Thus while there is not much of a problem set by a material like $LiTiSe_2$ (fully intercalated CdI_2)[33b)] or $LaVO_3$ (perovskite)[162], where there is more or less complete electron donation from the Li or La atoms to the T.M. atom d-band, (leaving the ionized Li s-band or La s/p bands well above E_F), the situation is automatically not so simple, for say $Fe_{1/3}TaS_2$[163] or $NiVO_3$[164](ilmenite). Magnetic properties then provide the clearest information about correlated site-electron contents, letting one, for example, know in the above compound that the iron is clearly FeII(d^6) and not FeIII(d^5).

The valence situation is less clear cut in $CuFeS_2$ (chalcopyrite)[165] or $CuCr_2S_4$ (spinel)[166]. For this pair there has been considerable discussion as to whether they basically are Cu(II) or Cu(I) compounds. The superconductivity of CuV_2S_4 and $CuRh_2S_4$ indicates the latter, as does photoemission[167]. The d-band in copper compounds, being upon the point of completion, is always low-lying and ready to accept electrons. This induces in the countercation a split-valence situation, e.g. formally $V^{3+}(d^2)/V^{4+}(d^1)$. The three low temperature transitions recently reported[168] for CuV_2S_4 in CDW terms are thus more likely to be due to valence ordering, as occurs in Fe_3O_4[169] and possibly also LiV_2O_4[170]. In binary CuS (covellite)[10a),171] electrons are actually 'abstracted' from the p-band to fill the d-band for 2/3 of the Cu sites ; structural examination reveals the situation as being $Cu^{I}_2(S_2)Cu^{II}S$. I have earlier given a general presentation of the relative positioning of bands for transition-metal compounds in reference 47 supported by reference 36b), and a similar presentation for rare-earth compounds in reference 156a) supported by reference 156b). I do not wish to repeat that here, beyond showing Figure 2.

In §4 we saw that some non-transition metal compounds will valence segregate, as in InTe - or $InTlSe_2$ (TlSe structure)[66], or again through the very interesting perovskite-based system $Ba^{2+}(Bi/Pb)^{4+}O_3$ (which at high Pb content can be made as a high T_c ceramic superconductor)[172]. Correspondingly now among transition metal compounds we can cite Co_3S_4, a direct spinel[10a)]. Here valence differentiation 2+/3+ occurs between spinel A and B sites ; the B site is octahedral, the A site tetrahedral. However with defect-NiAs materials each type of metal site is octahedral, but still site differention (as revealed by magnetic behaviour) remains pronounced, e.g. in $Cr_3S_4 \equiv Cr_{1/2}CrS_2$[173] and $V_5S_8 \equiv V_{1/4}VS_2$[174]. Both these materials are of considerable value toward understanding the stages of development in metallic magnetism. It is somewhat surprising that only the Japanese schools have seriously followed up what such materials have to offer in this direction.

Where, in a ternary, variable valence accompanies counter-ion disorder, as in $(La^{3+}/Sr^{2+})V^{3+/4+}O_3$[175], that disorder can manage to force Mott-Anderson localization despite in this case both end materials d^2LaVO_3 and d^1SrVO_3 being metals. $V^{3+/4+}$ mixtures acquire progressive structural order in the rutile-based Magneli phases V_nO_{2n-1}, e.g. V_4O_7[176]. The site ordering is more complicated than might have been expected. Furthermore, at least as gauged by careful X-ray analysis of site coordination dimensions, the charge differentiation is not integral. Admittedly in a multiply connected network it can be difficult to establish clear differentiation. However the 3+/4+ balance in the sites is clearly adjusted at the M/I transition when pairing occurs.

§ 10. LOCAL MOMENTS, MAGNETISM AND SPIN-POLARIZED BANDS

When we compare the optical spectrum of d^1 metallic 2H-NbS_2 with that of virtually isostructural d^2 semiconducting 2H-MoS_2, we see a strong increase in the level of detailed structure[10.b)]. The excitonic enhancement of critical point detail in MoS_2 is largely screened out by the free carriers in NbS_2. In 2H-MoS_2 there is sufficient detail present for the low energy part of the spectrum to be satisfactorily related to a joint optical density of states, coming from a one-electron band structure calculation[16], and covering both d → d (1.3 eV edge) and p → d transitions[177]. The optical spectra of 3d semiconducting $FeS_2(t_{2g}^6)$ looks as if it too can satisfactorily be related back[36.b)] to a band structure[36.a)] with a 0.9 eV $t_{2g} \rightarrow e_g$ edge. Likewise the optical spectra of d^0TiO_2[178] or TiS_2[179], for which the initial part of the spectrum is of the p → d charge transfer type, appears satisfactorily encompas-

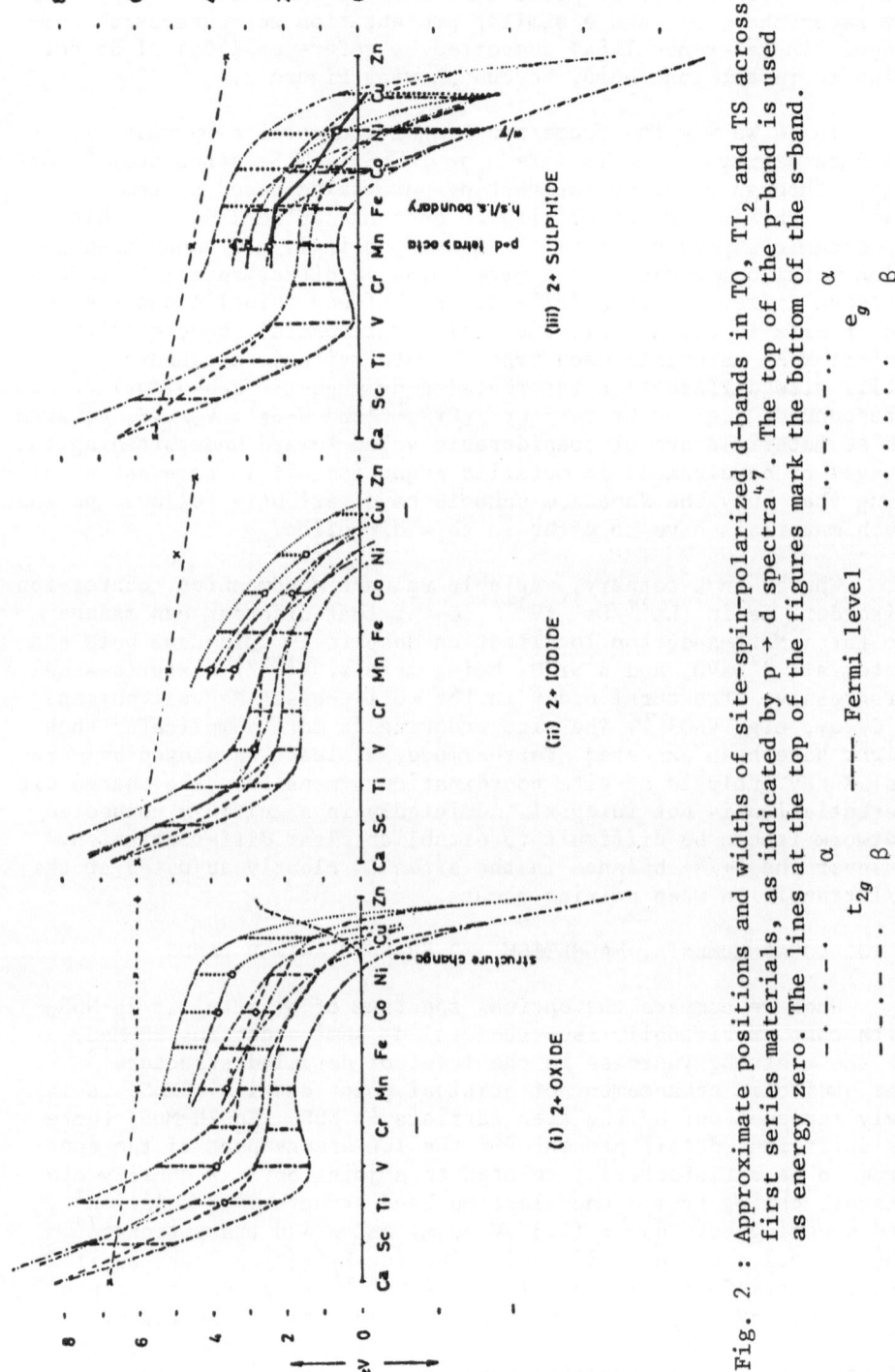

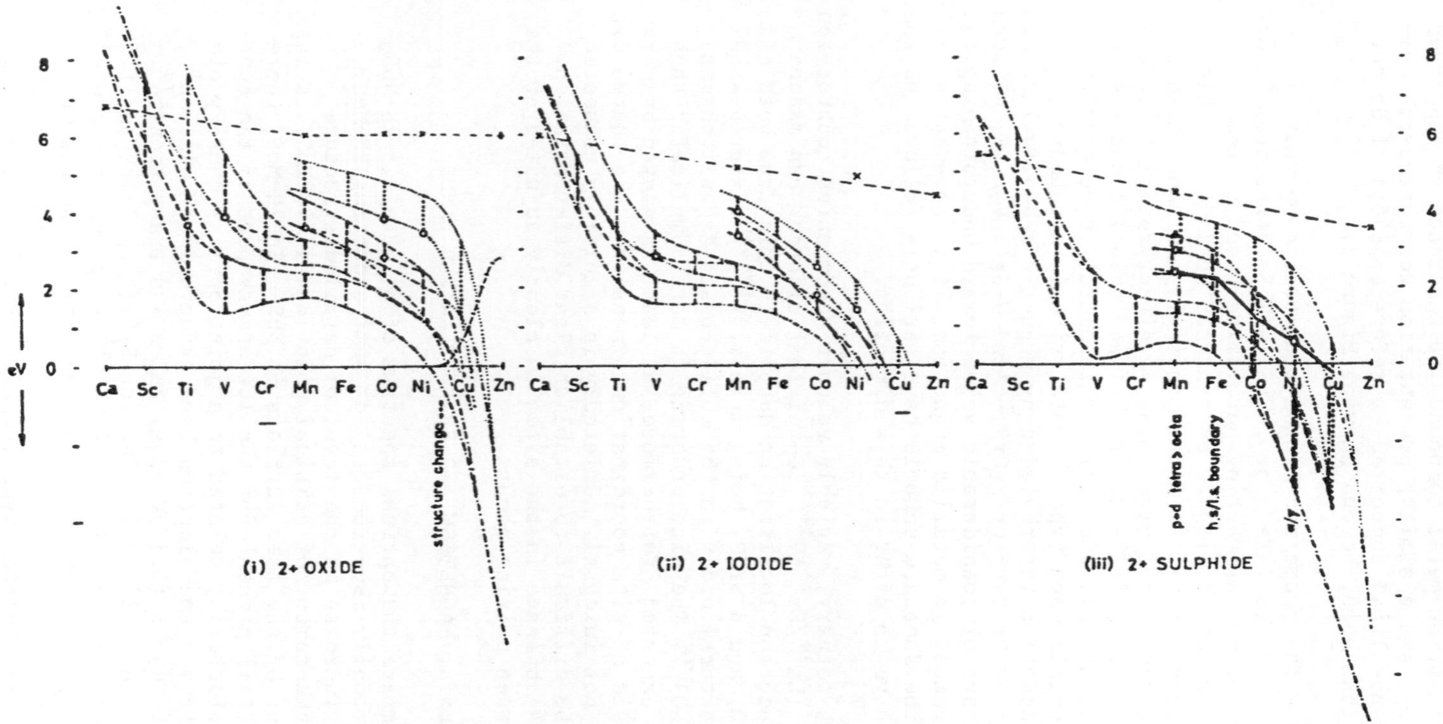

Fig. 2 : Approximate positions and widths of site-spin-polarized d-bands in TO, TI_2 and TS across first series materials, as indicated by p → d spectra[47]. The top of the p-band is used as energy zero. The lines at the top of the figures mark the bottom of the s-band.

-·-·-·- α / -·-·-·- β : t_{2g} ---- Fermi level -··-··-·· α / · · · · · · β : e_g

sed by the existing band structures[91,33]. When we remember that neighbouring CrO_2 or CoS_2 are metals which show Curie-Weiss moments (and are ferromagnetic at lower temperatures), and following these that MnO_2 and NiS_2 are Mott insulators, we see that p → d and even d → d spectra can in large measure be described in single particle terms right up to the Mott transition[47,36.b)]. At this point, as we have seen, the d-band is ∿ 2½ eV wide.

Considerably beyond the Mott transition, the p → d spectrum of say d^8NiBr_2[190] or d^5MnS[181] or d^3CrCl_3[182] continues to have the basic appearance of a one-electron spectrum. Such spectra have normally been discussed by chemists in molecular orbital terms[183]. Somewhere in this range there must be a complete break-down of k-selection rule behaviour, though as yet there has been no close examination of this matter. In VCl_4 clearly that point has passed for one finds a p → d spectrum little different between the solid and liquid phases[184] !

p → d spectra rarely show great detail of the sort apparent on p → f spectra. In actinide trihalides one sees for example[185] evidence of excited J states of the residual coupled configurations f^{n+1}. Likewise photoemission from SmS and EuS leaves the systems in the excited states of the f^5 and f^6 configurations respectively[186]. Because photoemission actually removes an electron from the system, and hence drastically changes the local screening, significant relaxation effects begin to be encountered in the photoemission process for d electrons in addition to f, and there is some evidence for excited V.B., on-site, final-state, configurations besides core-state multiplet production[187].

In rare earth f → d spectra (e.g. EuS) the detail which appears is exclusively associated with the f-electron set[188]. As we know from the metallic properties of d^1 LaS, the 5d-electron is well-delocalized (ω_p large)[190]. Correspondingly in f^2USe_2, p → d spectra are band-like whilst f → d spectra are still multipleted[191].

Of much weaker intensity than 'charge transfer' p → d or cationic f → d spectra are the on-site multiplet spectra internal to the coupled d^n or f^n sets. These are normally termed 'ligand field spectra'[183]. They are evident only where not overlayed by the charge transfer spectra, and might well persist unseen into the metallic regime. They appear still to be present in Mott insulating $ScVO_3(d^2)$[192] (-bixbyite not corundum structure). In materials like $CrCl_3$[182], MnO[193] and $NiCl_2$[194], detailed analysis of their ligand field spectra permits some assessment of the level of p - d co-valency, - hybridization which as we have seen so strongly builds up the 'd-band' width. The changes in ligand field spectra are most readily monitored through the Tanabe-Sugano diagrams[183] (at each d^n, for each coordination type). The diagrams plot E/B against

10Dq/B, where 10Dq is the crystal-field splitting of the p/d non-bonding and anti-bonding orientated d-wavefunctions (t_{2g} and e_g for octahedral coordination) and $B = F_2 - 5F_4$ is a composition of Slater integrals assessing d/d interaction.

Comparisons of the ℓ.f. spectra of 3d and the corresponding 4d material reveal considerable decrease in B for the latter, with increase in 10Dq. This shifts one to the right on the Tanabe-Sugano diagram while often increasing the excitation energies. The relative electron cloud expansion (nephelauxetic effect) for 4d over 3d is we know so highly developed that the stereochemistry of $3d^n$ and $4d^n/5d^n$ series compounds rarely are alike. For example at d^2 we find CdI_2-structured $TiC\ell_2/Br_2/I_2$[195], but MoS_2-like semiconductivity (trig.prism coordn.), Mo_6-like clusters, and WTe_2-like M-M chains hold in the 4d counterparts $ZrC\ell_2/Br_2/I_2$[196]. Similar though less marked wave-function expansions occur for a 3d compound under pressure, or when following a series of increasing covalency, e.g. $NiF_2/C\ell_2/Br_2/I_2$[47]. The appearance of the NiAs structure for MnTe, following the rocksalt and zincblende structures of MnO/S/Se is a manifestation of such d-function expansion.

d^5MnTe, and pyrites $MnTe_2$[197], remain in the high-spin ground-state $6A_1$ based on $t_{2g}^3 e_g^2$. The most spectacular effect associated with advancing covalency, and one of great importance with regard to magnetism, is the inducing of a high-spin/low-spin cross-over[198]. Between d^5FeCl_3 and $RuCl_3$[199] this occurs on the insulating side of the Mott transition and carries one from the $^6A_1$ h.s. ground-state based on $t_{2g}^3 e_g^2$ to t_{2g}^5 based on $^2T_2$. By the time that a metallic condition is gained in t_{2g}^5 IrO_2 local moment paramagnetism is gone[200]. However a moment still is presented by the more 'dilute' metal t_{2g}^5 $LaRuO_3$[201].

For d^6 the crossover is from $t_{2g}^4 e_g^2$-based $^5T_2$ to t_{2g}^6-based $^1A_1$. 3d FeS in the former h.s. state continues both metallic and magnetic[202]. The Neel point is 600 K with a spin-flip transition at 445 K, and with troilite triangular clustering of the NiAs basis occurring at 420 K. Already we have encountered low-spin and rather standardly semiconducting t_{2g}^6 FeS_2 ; this may be doped both n- and p-type, though it does possess a sizeable van Vleck paramagnetism[203]. Pressurizing the above monosulphide FeS cannot lead to a comparable semi-conducting t_{2g}^6 phase within the NiAs structure because of the symmetry reasons noted earlier (see ref. 204 for NiS). The actually observed non-magnetic form of FeS, mackinawite[205] adopts the tetragonal AgI/anti-PbO structure and has the larger volume (34.1 vs. 30.5 A^3 per molecule) ; Mackinawite has tetrahedrally coordinated Fe^{2+} (the tetrahedra share edges in sandwiches). Its diamagnetic semiconductivity apparently results from there being two molecules per unit cell, M-M interactions completely dropping off a filled t-type sub-band $(e^4.t_+^{\uparrow\downarrow})^2$. Band structure work is underway to test this view of mackinawite suggested by Goodenough[206].

Not apparent above for FeS because of the coordination change is the very large drop in volume normally associated with a h.s. → ℓ.s. transition, since anti-bonding e_g states are being depopulated for non-bonding t_{2g} ones. Fig. 10 of ref. 47 shows the 10 % reduction in Fe-S bond length present between h.s. FeS and ℓ.s.FeS_2. A reduction of this sort actually makes the volume of ℓ.s.$4d^5RuCl_3$ smaller than that of h.s.$3d^5FeCl_3$. The volume change is so considerable (as it was between f^6d^0 and f^5d^1 SmS)[156] that such transitions induced under pressure are often first-order, although in ternary systems where the lattice is braced a second-order change may arise (e.g. $(Mn/Fe)S_2$[207] and 1T-$(Ta/Fe)S_2$[208]).

Occasionally $Fe^{2+}d^6$ crossover proceeds via the intermediary configuration $t_{2g}^5 e_g^1$. If 'Mott-insulating' $CuFeS_2$[209] (tetrahedrally coordinated chalcopyrite) indeed contains Cu^{1+} and hence d^5Fe^{3+}, one might correspondingly understand the rather small Fe-S bond length there and the reduced moment (μ = 3.85 μ_B) in terms of the s = 3/2 intermediate configuration e^3t^2 (see Table 2). [T_N ~ 800°K].

A ℓ.s./h.s. change marks d^6 $LaCoO_3$[210] as an especially interesting perovskite, moving from semiconducting to metallic behaviour. A ℓ.s. → h.s. transition similarly marks metallic magnetic d^4MnAs : P[211]. At d^7 low spin (i.e. $t_{2g}^6 . e_g^1$) metallic magnetic behaviour is shown by CoS_2[212], though isostructural isoelectronic NiPS[247] has become momentless, and $LaNiO_3$[213] is very unusual.

When only one sub-band is open, as in $Co(S/Se)_2$ or in many d^1, d^2 and d^3 cases such as metallic ferromagnetic $t_{2g}^2 CrO_2$[147] or antiferromagnetic t_2^3 CrSb[214], it seems a natural extension of normal spin-polarized band structure calculations (e.g. ref. 215 for ferromagnetic metallic $CuCr_2Se_4$ spinel) to adhere to the same calculation above T_C or T_N. In a magnetic Mott insulator like d^3 $NaCrS_2$[216] it is known that there is only a small modification of the p → d spectra across the ordering temperature. The edge shifts blue by only 0.02 eV as the sample is cooled through T_N. This is because in very highly correlated materials the Curie or Neel point sees little change in magnitude of the on-site moment, but is simply the point of cooperative spatial ordering. In $e_g^1 CoS_2$ the ferromagnetically ordered moment is actually slightly less than the Curie-Weiss paramagnetic moment[212] ; a probable reflection of band narrowing in the disordered situation. When there is more than one magnetic electron, as in CrSb[214] or Cr_2Te_3[217], the magnetic moment appears cooperatively preserved to considerably greater degrees of p-d covalent hybridization than at d^1. At d^1 even TiI_3[195] and TiN/P[10.a] were momentless. A similar lack of moment occurs for one anti-bonding e_g hole in CuS_2, though one t_{2g} hole in ℓ.s. d^5FeP/As/Sb does hold a reduced moment[218]. $t_{2g}^5 Fe_2Te_3$ shows no moment by contrast with t_2^3 Cr_2Te_3[217].

Table 2. Spin states of di- and tri-valent iron.

octa. coord.	h.s.	int.	ℓ.s.
d^6Fe^{2+}	$t_{2g}^4 e_g^2$	$t_{2g}^5 e_g^1$	t_{2g}^6
s=	4/2	2/2	0
d^5Fe^{3+}	$t_{2g}^3 e_g^2$	$t_{2g}^4 e_g^1$	t_{2g}^5
s=	5/2	3/2	1/2

tetra. coord.	h.s.	int.	ℓ.s.
d^6Fe^{2+}	e^3t^3	–	e^4t^4
s=	4/2		2/2
d^5Fe^{3+}	e^2t^3	e^3t^2	e^4t^1
s=	5/2	3/2	1/2

4d and 5d series metals, as we have seen, normally do not achieve a moment, e.g. t_{2g}^4 RuO_2[200], t_{2g}^5 IrO_2[200], t_{2g}^2 ThI_2[219], though there is the occasional case like $LaRuO_3$[201]. Rh-doping of CoS_2 quite quickly suppresses the magnetism[220], as does Mo-doping of CrO_2[151].

The absence of a moment in CuS_2 (and also the Jahn-Teller effect) is fostered by the exceptionally heavy p-d hybridization as the Fermi level encounters the top of the p-band[36.b)]. Back up at d^8, NiS_2 has a reasonably standard Curie-Weiss moment, though the ordered moment that is displayed is become abnormally low (1.17 μ_B)[221]. Cu-substitution[222] or Se-substitution[223], both of which induce the insulator-metal Mott transition, further promote loss of the Ni moment. Fig. 3 drawn from ref. 47 indicates the locus of loss of C-W moment across the 3d compound array.

For NiAs-structured α-NiS[224] the close c-axis stacking of the Ni atoms (2.675 Å) puts it on the metallic side of the Mott transition. No trace of local moment is displayed for T > 300 K, it like $NiSe_2$[225] simply showing enhanced Pauli paramagnetism. However at 265 K α-NiS suddenly becomes antiferromagnetic and almost semiconducting. The moment that appears is almost in full alignment/of full magnitude for the two-electron e_g^2 situation. The entropy change at this transition is only about one quarter that of $R.\ln(2S + 1)^2$, the value for complete spatial disordering of a salt-like d^2 antiferromagnet that maintains a moment of constant magnitude. The transition is associated with a 0.03 Å c-axis increase in Ni-Ni spacing on cooling. Moment formation and ordering in α-NiS would seem to be precipitated by the possibility of opening up a semiconductive gap at d^8 under the lower symmetry of the antiferromagnetic condition (viz. D_{3d}^3), a process which is not permitted in the higher symmetry of the normal state (D_{6h}^4)[36.b),204]. It takes ∿ 20 kB at helium temperatures to suppress this 'AFI' state.

t_{2g}^2 V_2O_3 by contrast with e_g^2NiS shows a normal Curie-Weiss moment in its h.t. metallic phase. However, as for Mott insulating NiS_2, this diminishes to 1.2 μ_B when ordered antiferromagnetically[226]. It should be noted that in V_2O_3 the two electrons are not structurally equivalent in M.O. terms. The entropy change involved at 'T_N' though twice as great as for α-NiS is still only half the 'simple' value. The appearance of this magnetic order carries V_2O_3 into the Mott-insulating regime[140,150]. At helium temperatures 20 kBars of pressure again are required to recover metallic conductivity. No magnetic order then obtains[227], in contrast with the AFM phase holding in the system $Ni(S/Se)_2$ just beyond delocalization[228].

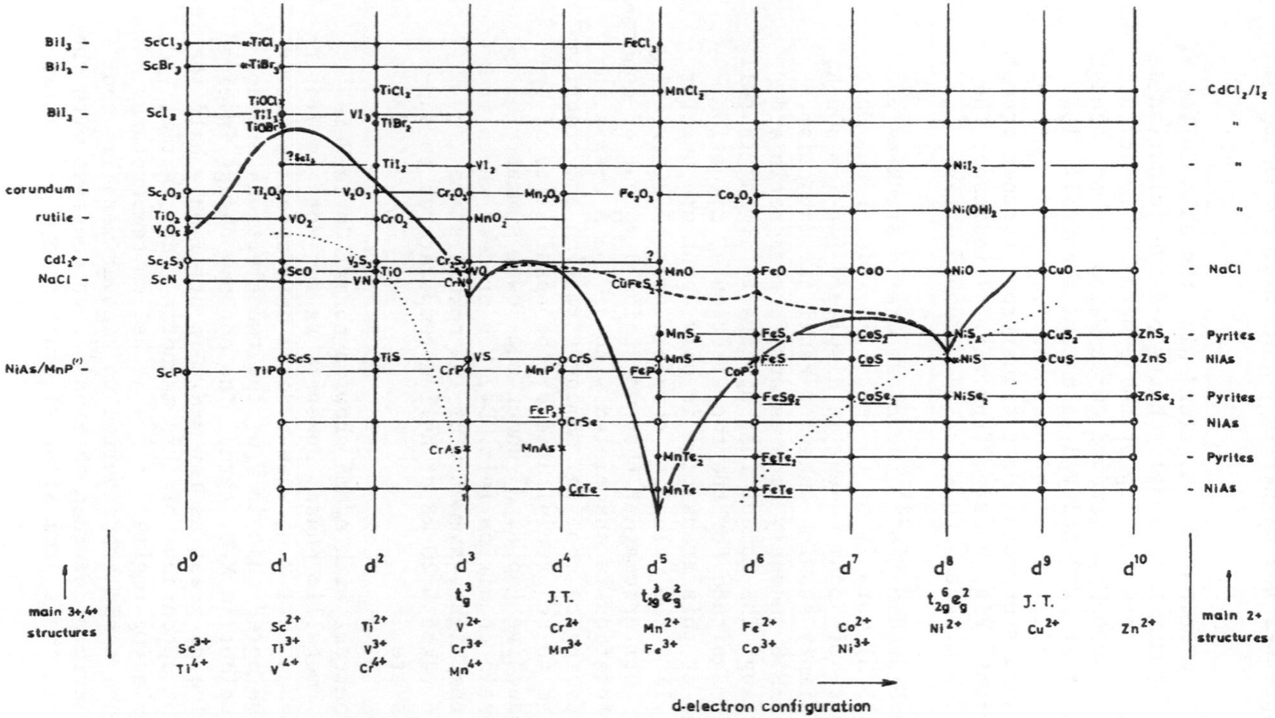

Fig. 3 : First series T.M. compounds of octahedral coordination arrayed by d^n number, and by anion electronegativity with deference to cation valence, and to structure type and mode of coordination unit linkage[47]. The ranking of compounds at each given n is separated into metals and Mott insulators by the heavy line, while within the metallic regime the dotted lines indicate the point of loss of magnetic moment. Note not all compounds are labelled. Open circles indicate a change of structure from the main run. The movements of the separator line indicate the effects of d-band filling, of crystal field effects and of p-d hybridization.

A progressive degradation of the Curie-Weiss moment occurs towards $NiSe_2$ within the metallic part of the e_g^2 system $Ni(S/Se)_2$, which is not encountered for the corresponding e_g^1 system $Co(S/Se)_2$[212]. The difference may be that in an e_g^1 system (unlike the e_g^2) electrons are not required to reverse spin during delocalized site-occupancy fluctuations. What in fact is found through the $Co(S/Se)_2$ is a change in Curie-Weiss θ_p from positive to negative, although a LRO AFM state is not ultimately attained in $CoSe_2$[229] NMR experiments indicate that the l.t. peak observed in χ for $CoSe_2$ may be due to weak CDW behaviour. This it seems arises too in CuS_2 and $CuSe_2$[230].

An antiferromagnetic SDW probably is responsible for the transitions observed in layered t_{2g}^2 $CrSe_2$[231] following up on the ferromagnetism CrO_2. The moment is lost in wider band divalent t_{2g}^2 TiS and also VS, where CDW activity was earlier suggested[132,37].

As has been said, the majority of these magnetic materials may be described in their overall behaviour above as below T_N or T_c by site-spin polarized band schemes of the type presented in Figs. 1 and 2 of ref. 36.b) for the pyrite sulphides. Spin becomes the dominant aspect in the correlated behaviour close to the Mott transition. However whether detailed band calculation following such a procedure can properly yield the Fermi Surface in say $CoSe_2$ or $CrSe_2$ remains open to question. Still more problematic is the type of site-spin polarized representation given in Fig. 16 of ref. 47 [reproduced here as Fig. 4] for a high-spin metal like FeS or MnAs, and especially as attempting to cover high-spin/low-spin transitions in single particle band-structural form.

Such diagrams were of course prompted by the fact that very similar diagrams are of relevence in discussing charge transfer $p \rightarrow d$ spectra, whether of Mott-insulating or of delocalized materials (see Fig. 2 ; also ref. 47). In particular one should appreciate the large shift to higher energies at d^5 compared with d^4 for the $p \rightarrow d$ edge, since now the transferring electron has to pass up into a β-spin orbital. Thus the edge in d^5 MnI_2 falls beyond 3 eV (and with very weak spin-forbidden l.f. peaks the material is left transparent). Just as significant are the shifts which come for octahedral coordination at d^3 and d^8. There the α/β spin separation requires that the transferring electron must be excited up into the corresponding e_g sub-band. For example the edge in d^2VCl_3 is lower than in d^3CrCl_3. Similarly it is lower in d^7CoCl_2 than in d^8NiCl_2, though greater for the anion $-CoCl_4^{2-}$ than $-NiCl_4^{2-}$ where the cation site geometry has become tetrahedral with e/t splitting.

As stated above these schemes reduce to k-independent molecular orbital modelling somewhere along the line, and probably well before the above chlorides are reached. The optical probe is a

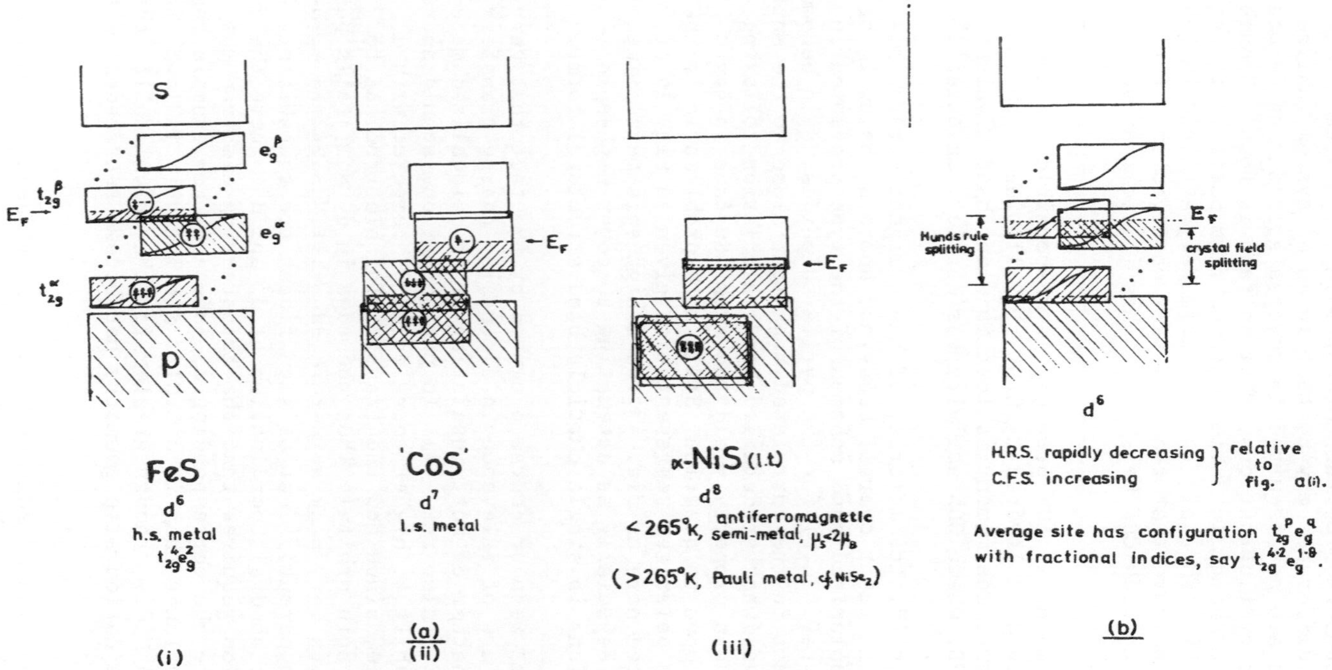

Fig. 4 : (a) Site spin polarized band scheme for h.s. FeS, hypothetical ℓ.s. CoS, and antiferromagnetic α-NiS, from ref. 47. Since ref. 47 dots have been added linking t_{2g} and e_g components to indicate that for a simple NiAs situation these sub-bands are not fully separable.

(b) The beginning of the h.s. → ℓ.s. transition. Compared to (a)i) the e_g/t_{2g} crystal-field splitting is augmented, the α/β spin splitting diminished.

rather severe one by which to investigate the existence and nature of band-like behaviour in highly correlated materials. It is not yet clear at what stage k-independent crystalline selection rules become irrelevant. It is notoriously difficult even to be sure of peak identifications in wide band semiconductors like ZnSe[232] or TiO_2[178]. More polarized, electro-reflectance studies are needed. On the metallic side of the Mott transition, a very sensitive probe of the quality/relevance of a band structure calculation is the de Haas-van Alphen type measurement of Fermi surface geometry. In this area the results on ReO_3[233] of RuO_2[15.a)] continue to match up but what the situation will be in more highly correlated $NiSe_2$, or $CoSe_2$, or V_2O_3 under pressure remains to be discovered. The ultimate in such work will be to uncover the situation in a high-spin metal like MnAs just prior to crossover. Presently crossover has only been dealt with in atomic terms, even with a metal like 1T-(Ta/Fe)S_2[208].

Of course all theoretical work in this very difficult area has need to be self-consistent at various levels, as both the magnitude and direction of the site local moments fluctuate due both to thermal effects and to the site occupancy fluctuations of the metallic state. Just as the local moments are maintained in magnitude and direction by their neighbourhood, under some level of short range order and coupling, so the fluctuations also will be cooperatively interactive. Metallic materials have been cited above which span the full range of moment stability from a highly established moment in $CuCr_2S_4$ or CrO_2 to a moment that only appears in conjunction with LRO as in α-NiS (or NiAsS, V_3S_4). Intermediate cases naturally exist, like V_2O_3 or $LaNiO_3$. It will be of interest to follow the behaviour of some of these under pressure, e.g. the C-W behaviour of $CoSe_2$. Co(S/Se_2) is a system known to show a marked metamagnetic response to an applied field. Unfortunately substitutional doping, as we saw for the Mott transition, does not produce the average environment looked for, but one strongly reflecting the specific local coordination, as the systems Co(S/Se)$_2$[212] and (Co/Rh)S_2[220] have nicely shown.

There is clearly much interesting experimental work to be done in this area. Theory is making great strides here too. In the past five years the work of Capellmann[234] and of Hubbard[235], of Korenmann and Prange[236], of Moriya and Hasegawa[237], of Heine and co-workers[238] has picked up on the earlier work by Goodenough[239] Evenson, Schrieffer and Wang[240], Cyrot[241], Julien, Gutzwiller, Sokoloff, Liu, Inagaki and many others, pulling together work on the Hubbard model, on spin fluctuations, on the Stoner model and spin waves, and on spin glasses and alloys. These however are matters of high theory and I believe Dr. Gyorffy will be addressing himself in this school as at home to these matters. Perhaps before long in such hands we shall be seeing band calculations which go well beyond simple spin-polarized calculations of the type introduced by Slater and his co-workers. It is as well to remember though that by being as close to real materials as to theory Slater had envisaged the sort of situation expressed in refs. 36.b) and 47, which only now is being so powerfully filled out.

§ 11. CONCLUSION

It is clear that things are now moving forward at a great pace. New materials are emerging under new preparative techniques. Some, like $NbSe_3$ and $Ni_2P_2S_6$, are relatively easy to prepare once the conditions are known. Others like CrO_2, NiP_2 or SiP_2 are somewhat more specialized. The low valence halides[242.a)] are one such growth area e.g. ScCl, Sc_7Cl_{12}, $CsScCl_3$. The ZrCl, YCl family offers possible extension to UCl and even BaCl. Many unusual R.E. compounds like $LaTe_2$ and $NdTe_3$ remain virtually unexplored[68]. Another big materials growth area is in polypnictides such as SiP_2, NiP_2, VP_2, the $CaAs_3$ family[243], and the very large divalent FeP_4 family[242.b)].

It would be nice to see one or two more band structures for mixed anion compounds like BiSeBr[244], or TiOCl, or $Ni_2(P_2S_6)$.

Especially relevant to say the Mott transition in MnO_2 or the very unusual magnetic properties of CeSb[242.c)] will be theoretically improved determination of p/d band separations.

Experimentally as theoretically a problem in this area remains the degree to which band calculations can describe both ground state and optical excitations. At what point too do k-selection rules become irrelevant to p → d spectra ? In the narrow band situation disorder plays a dominant role in magnetism (e.g. $(Co/Rh)S_2$), in the Mott transition (e.g. $(V/Al)O_3$) and in CDWs (e.g. 1T-$(Ta/V)S_2$). It will be of great interest to find just how well the Fermi surface really survives in the system 1T-$(Ta/Ti)S_2$, or are local bonding forces preeminent in determining events[245]. Determination of the absolute phasing of CDWs would reveal a lot.

Finally, with regard to magnetism we see that there exists a fine complementary array of compounds upon which to test ideas that presently the elements Fe, Co and Ni are forming the main forcus of attention for.

MATERIALS INDEX

NbS_2	681	10b,16	ReO_2	671	-
NbS_3	661,662,	24,28,	ReO_3	661,663,691	23,233
	665,670	106,120,	ReS_2	663	10b
Nb_2Se_3	675	110	$RhSe_2$	670	10b
$NbSe_3$	662,672,	24,28,	$RuBr_3$	676	122
	674,676	100,	$RuCl_3$	684,685	199
		106,	RuO/S	663	-
		108,	RuO_2	662,663,679,	15a,200
		109		687,691	
$NbTe_2$	676	10b,	Sb	667	68
		100	Sc halides	662,692	31,242a
$NbTe_4$	671	10a	ScAs	671	94
$NdTe_3$	691	68	Sc_2O_3	671	47
NiAs	660,671	1c,10a	ScP	667,671	44,94
NiAsS	691		$ScVO_3$	683	192
$NiBr_2$	683,684	47,180	Se	666,667,668	81,82
$NiCl_2$	683,684,	47,194	Si	659,666	3
	689		$SiAs_2$	666	68
NiP	671	11	SiC	666	14
NiP_2	660,665,	11	SiP	661,665	68
	668,671		SiP_2	661,666,667,	32,68
NiPS	685	247		668,669,672	105
NiS	660,684,	7,10,	SiS	664	10c
	687,690,	36b,	SmB_6	679	156
	691	204	SmS	679,683,685	156,186
NiS_2	670,678,	36,47,	SnAs	662,664,665	35
	683,687	221	SnI_2	665	69
$(Ni/Cu)S_2$	687	222	SnP	662,665	72
$Ni(S/Se)_2$	678,687,	223,	SnS_2	661,666,667	10b,18
	689,	228	Sn_2S_3	662,664,667	65
NiSe	671	10a	SnSe	664	51
$NiSe_2$	678,691	36b,	SnTe	664	52
		47,225	$SrBr_2$	665	71
$NiVO_3$	678	164	SrP	665	77
OsS_2	663,679	10	SrS_3	668	77
P(black)	667,	68,84	$SrVO_3$	681	175
$PbCl_2$	672,673	10c	IT-TaS_2	680	10b,100,
PbF_2	665	-			114,115,
PbI_2	664,665	53			117,158
PbO	664,665	54	IT-$(Ta/Fe)S_2$		
PbS	664,667	52		680,685,691	160,161,
PbTe	664,665,	14,52			163
	671		IT-$(Ta/Ti)S_2$		
Pd_xP	662	10a,		675,679,692	114,159
		242d	IT-$TaSe_2$		114,158
PdS	660	10	2H-$TaSe_2$	672	99,100,
PtO_2	663	47			101,102,
PtP_2	660,663,668	13,86			103,104,
PtS	660,663	10			114,116,
PtS_2	663	10			117,118

REFERENCES

1.a) A full bibliography to about 1970 can be found in - J.C. Slater, 'The Quantum Theory of Molecules and Solids, Vol.4' (New York : McGraw-Hill, 1974).
1.b) A recent review of SCF band calculations is provided by - D.D. Koelling, Reports on Progr. in Phys. 44, 139-212 (1981).
1.c) Pseudopotential work may be traced through the articles by Cohen and Zunger in - 'Structure and Bonding in crystals', Vol.1, Ed. M.O'Keefe and A. Navrotsky, (Academic 1981). M.L. Cohen, p.25-48 : A. Zunger, p. 73-136 respectively.
1.d) Recent work of the Cambridge school is summarized in - Solid state physics, vol.35, (Academic, 1980). For our present purposes note especially the article by D.W. Bullett p. 129-214 on improved T.B. methods.
1.e) The discrete variational method is discussed in - A. Zunger and A.J. Freeman, Phys.Rev. B15, 4716, 5049 (1977) B16, 906, 2901 (1977) ; B17, 1839 (1978).
1.f) The LMTO-ASA approach is introduced in - O.K. Andersen, Phys.Rev. B12, 3060 (1975), see J. Pure and Appl.Chem. 52, 93 (1979) for example of T.M. monoxides.
2. A.R. Williams, J. Kübler and C.D. Gelatt, Jr., Phys.Rev. B19, 6094 (1970).
3. D. Glötzel, B. Segall and O.K. Andersen, Sol.St.Comm. 36, 403 (1980). A. Zunger, Phys.Rev. B21, 4785 (1980).
4. J.R. Anderson, D.A. Papaconstantopoulos, L.L. Boyer, J.E. Schirber, Phys.Rev. B20, 3172 (1979).
5.a) W. Hanke, Adv. in Phys. 27, 287 (1978).
5.b) G. Strinati, H.J. Mattausch, W. Hanke, Phys.Rev. 25, 2867, (1982).
5.c) J.P. Perdew and A. Zunger, Phys.Rev. B23, 5048 (1981) - also ref. 61b.
5.d) A.B. Kunz, J.Phys.C.Sol.St. 14, L455 (1981) - case of NiO.
6. M.T. Yin and M.L. Cohen, Phys.Rev.Lett. 45, 1004 (1980), Sol.St.Comm. 38, 625 (1981).
7. W.B. Pearson, 'The Crystal Chemistry and Physics of Metals and Alloys' (Wiley, 1972).
8. J.C. Phillips, 'Bonds and Bands in Semiconductors' (Academic 1973).
9. Ref. 1c, and A. Zunger, Phys.Rev.Lett. 47, 1086 (1981).
10.a) A clear and comprehensive work on the structures of chalcogenides and pnictides of T.M. elements is - F. Hulliger, Structure and Bonding, 4, 83-229 (1968).
10.b) Additional discussion of some of these can be found in - J.A. Wilson and A.D. Yoffe, Adv. in Phys. 18, 193 (1969).
10.c) Further information on sulphide structures in general can be obtained from - F. Jellinek, Ch.19 (p.670-747), 'Inorganic Sulphur Chemistry' Ed. G. Nickless (Elsevier, 1968).
10.d) and - C.N.R. Rao and K.P.R. Pisharody, Prog.Sol.St.Chem. 10, 207 (1976).

10.e) Also of much value is ref.7 and - A.F. Wells 'Structural Inorganic Chemistry' 4th edition (Oxford University Press, 1975).
For NiS, CuS and PtS see (a) Figs. 49,50,51 and (b) p.675.
11. E. Larson, Arkiv. Kemi 23, 335 (1964) and ref. 10a, Fig.15. J.P. Odile, S. Soled, C.A. Castro, A. Wold., Inorg.Chem. 17, 283 (1978).
12. R.G. Vincent and J.W. Steeds, unpublished.
13. E. Dahl, Acta Chem. Scand. 23, 2677 (1969). D.H. Damon, R.C. Miller and A. Sagar, Phys.Rev. 138A, 636 (1965).
14. J.C. Slater 'Quantum Theory of Molecules and Solids, vol.2' (McGraw-Hill, 1965).
15. L.F. Mattheiss, (a) Phys.Rev. B13, 2433, (1976) (rutiles) (b) Phys.Rev. B12, 2162 (1975) (A15's) and refs. 16, 21, 23b.
16. L.F. Mattheiss, Phys.Rev. B8, 3719 (1973).
17.a) W. Temmerman, to be published (case of TiS_2 and $TiSe_2$).
17.b) A. Neckel, P. Rastl, R. Eibler, P. Weinberger and K. Schwarz, J.Phys.C.Sol.St. 9, 579 (1976) (metallic 3d rocksalts).
18. M. Schlüter and M.L. Cohen, Phys.Rev. B14, 424 (1976).
19. J.R. Chelikowsky and M.L. Cohen, Phys.Rev. B14, 556 (1980).
20. Ref.8, p.12, p.42.
21. L.F. Mattheiss, Phys.Rev., B5, 290, 306 (1972).
22. Y. Onodera, J.Phys.Soc.Jap., 25, 469 (1968).
23.a) H.W. Myron, R.P. Gupta, S.H. Liu, Phys.Rev. B8, 1292 (1973).
23.b) L.F. Mattheiss, Phys.Rev. 181, 987 (1969) ; see also B24, 692 (1981).
24. D.W. Bullett, J.Phys.C.Sol.St., 12, 277 (1979), and 15, 3069 (1982).
25. D.W. Bullett, J.Phys.C.Sol.St., to be published (1982).
26. O.K. Anderson, H. Nohl, P. Kelly, Europhysics Conf.Abs. 6A, 218 (1982) (Manchester EPS 11).
27. D.W. Bullett, unpublished (1980).
28. D.W. Bullett, J.Sol.St.Chem. 33, 13 (1980) ; also see J.Phys.C.Sol.St. 13, 1267 (1980).
29. To couple into the mass of literature concerning Miedema's parametrization of this topic see - A.R. Williams, G.D. Gelatt and A. Morizzi, Phys.Rev.Lett., 44, 429 (1980), and Phys.Rev. B25 (1982), May 15 issue.
30. H. Terauchi, N. Sakamoto, I. Shirotani, J.Phys.Soc.Jap. 38, 595 (1975).
31. J.F. Marchiando, B.N. Harmon, S.H. Liu, Physica 99B, 259 (1980). D.W. Bullett, Inorg.Chem. 19, 1780 (1980).
32. O.V. Farberovich and E.P. Domashevskaya, Phys.Stat.Sol.(b) 72, 661 (1975).
33. A great many band structures now exist for $TiSe_2$, the most recent of which are - G.A. Benesh and A.M. Woolley, Phys. Rev.Lett., to appear ; W. Temmerman, to appear ; H. Isomaki and J.V. Boehm, J.Phys.C.Sol.St. 14, L75 (1981), and to appear. The various methods have great difficulty

in reproducing the experimentally determined p-d overlap of about 0.1 eV.
34. J.A. Wilson, Phys.Stat.Sol., (b) 86, 11 (1978). K. Motizuki, Y. Yoshida and Y. Takaoka, Physica 105B, 357 (1981). R.H. Friend, D. Jerome and A.D. Yoffe, J.Phys.St. 15, 2183 (1982), and to appear.
35. V.G. Losev, S.S. Kabalkina, L.F. Vereshchagin, Sov.Phys.Sol. St. 16, 965 (1974).
36. a) W. Temmerman, unpublished. b) J.A. Wilson, 'Phase Transitions, 1973', Ed.: E. Cross (Pergamon 1973) p. 101-115.
37. W.B. England, S.H. Liu and H.W. Myron, J.Chem.Phys. 60, 3760 (1974).
38. J. Ashkenazi and T. Chuchem, Phil.Mag. 32, 763 (1975).
39. E. Caruthers, L. Kleinman and H.I. Zhang, Phys.Rev. B7, 3753, 3760 (1973). M. Gupta, A.J. Freeman, D.E. Ellis, Phys.Rev. B16, 3338 (1977).
40. A. Baghdadi and A. Wold, J.Phys.Chem.Sol. 35, 811 (1974), and ref. 13.
41. M.A.K.L. Dissanayake and L.L. Chase, Phys.Rev. B18, 6872 (1978).
42. D.K. Misemer and J. Nakahara, H. Franzen and D.K. Misemes (Ames Lab.), Amer.Chem.Soc., Las Vegas Meeting 1981, (Abs. 78 and 228).
43. K. Kobayashi, K. Kosuge and S. Kachi, Mat.Res.Bull. 4, 95 (1969). Comp. trigonal prismatic coord^{n}. of $NaNbO_2$, G. Meyer and R. Hoppe, Z.Anorg.Allg.Chem. 424, 128 (1976).
44. E. Wimmer, K. Schwartz, R. Podloucky, P. Herzig, A. Neckel, J.Phys.Chem.Sol. 43, 439 (1982).
45. D.W. Bullett, Phys.Rev.Lett. 39, 664 (1977). T. Jarlborg and A.J. Freeman, Phys.Rev.Lett. 44, 178 (1980).
46. J.J. Finley, R.E. Camley, E.E. Vogel, V. Zavin, E. Gmelin, Phys.Rev. B24, 1323 (1981).
47. see J.A. Wilson, Adv.in Phys. 21, 143 (1972).
48. P. Plenkiewicz and D. Dowgiallo-Plenkiewicz, Phys.Stat.Sol. (b) 95, 29 (1979) and refs. therein.
49. E.J. Mele and J.D. Joannopoulas, Phys.Rev. B24, 3145 (1981).
50. P.C. Kemeny and M. Cardona, J.Phys.C.Sol.St. 9, 1361 (1976).
51. G. Ciucci, A. Guarnieri, G.L. Masserimi, L. Quartapelle, Sol.St.Comm. 29, (1979).
52. G. Martinez, M. Schluter, M.L. Cohen, Phys.Rev. B11, 651, 3308 (1975).
53.a) I.Ch.Schluter and M. Schluter, Phys.Rev. B9, 1652 (1974).
b) M. Schluter and M.L. Cohen, Phys.Rev. B14, 424, (1976) ρ-plots.
54. J. Robertson, Phil.Mag. B43, 497 (1981).
55. D.E. Parry, M.J. Tricker, J.D. Donaldson, J.Sol.St.Chem. 28, 401 (1979).
56. E. Calabrese and W.B. Fowler, Phys.Stat.Sol.(b) 57, 135 1973).
57. H. Overhof, J.Phys.Chem.Sol. 38, 1214 (1977). J.Shy-Yih Wang, M. Schlüter, M.L. Cohen, Phys.Stat.Sol.(b) 77, 295 (1976).

58. e.g. A. Zunger and M.L. Cohen, Phys.Rev. B20, 1189 (1979).
H. Overhof, Phys.Stat.Sol. (b) 97, 267 (1980).
L. Kleinman and K. Mednick, Phys.Rev. B20, 2487 (1979).
N.J. Doran and A.M. Woolley, J.Phys.C.Sol.St. 12, L321 (1979).
A.B. Kunz and R.S. Weidman, J.Phys.C.Sol.St. 12, L371 (1970).
O.V. Farberovich, R.G. Akopdzhanov, S.I. Kurganskii, E.P. Domashevskaya, Sov.Phys.Sol.St. 21, 1691 (1980).
N.I. Kulikov and E.S. Alekseev, JETP Lett. 30, 665 (1979).
V.N. Borzunov, A.E. Kadyshevich, A.P. Rusakov, Phys.Stat. Sol.(b) 104, K113 (1981).
S.Ves, D. Glötzel, M. Cardona and H. Overhof, Phys.Rev. B24, 3078 (1981).

59. N.B. Brandt, S.V. Kuvshinnikov, A.P. Rusakov, V.M. Semenov, JETP Letters 27, 33 (1978).

60.a) E. Brown, C.G. Homan, R.K. MacCrone, Phys.Rev.Lett. 45, 478 (1980).

60.b) G.C. Vezzoli, M. Ottoni, P. Houser, L.W. Doremus and S. Krasner, Mat.Res.Bull. 17, 485 (1982). N.B. The suggested Cd-producing dissociation should read $2Cd^{II}S \rightarrow Cd^0 + Cd^{II}(S_2)$ pyrites (see ref. 10b).

61.a) A.B. Kunz, R.S. Weidman, T.C. Collins, J.Phys.C.Sol.St. 14, L581 (1981).

61.b) A. Zunger and A.J. Freeman, Phys.Rev. B17, 4850 (1978). (zincblende vs rocksalt).

62.a) J.A. Wilson, Phil.Mag. B38, 427 (1978).

62.b) G.C. Vezzoli and J. Bera, Phys.Rev. B23, 3022 (1981).

63. M. Schlüter, M.L. Cohen, S.E. Kohn and C.Y. Fong, Phys.Stat. Sol.(b) 78, 737 (1976).

64.a) J. Treusch, Phys.Rev.Lett. 34, 1343 (1975) (NaCl vs. CsCl strs.).

64.b) M. Schreiber and W. Schäfer, Phys.Rev. B21, 3571 (1980).

65. H.R. Chandrasekhar and D.G. Mead, Phys.Rev. B19, 932 (1979).

66. F.M. Gashimzade and G.S. Orudzhev, Phys.Stat.Sol.(b) 106, K67 (1981).

67. F. Borghese and E. Donato, Nuovo Cim. 53B, 283 (1968).

68. F. Hulliger, 'Physics and Chemistry of Materials with Layered Structures'. Vol.5 'Structural Chemistry of Layer Type Phases'. Ed. F. Levy (D. Reidel Pub.Co., Dordrecht, Holland, 1976).

69. E. Doni, G. Grosso, I. Ladiana, Physica 99B, 281 (1980).

70. D.W. Bullett, Phys.Rev. B14, 1693 (1976).

71. J.G. Smeggil and H.A. Eick, Inorg.Chem. 10, 1458 (1971).

72. P.C. Donohue, Inorg.Chem. 9, 335 (1970).
P.C. Donohue and H.S. Young, J.Sol.St.Chem. 1, 143 (1970).

73. M. Schlüter, J. Camassel, S. Kohn, J.P. Voitchovsky, Y.R. Shen, M.L. Cohen, Phys.Rev. B13, 3534 (1976).
J. Robertson, J.Phys.C.Sol.St. 12, 4777 (1979).

74.a) Y. Depeursinge, Nuovo Cim. 64B, 109 (1981).

74.b) also experimental X-ray maps just published. A. Kuhn, A. Bourdon, J. Rigoult, A. Rimsky, Phys.Rev. B24, 4081 (1981).
75. C.P.J.M. van der Vorst and W.J.A. Maaskant, J.Sol.St.Chem. 34, 301 (1980).
76. W. Dultz and E. Rehaber, J.Phys.C.Sol.St. 12, L137 (1979).
77. F. Hulliger, Ch. 26 in Vol.2 of 'Structure and Bonding in Crystals' Ed. M.O'Keefe and A. Navrotsky (Academic, 1981).
78.a) F.J. Arlinghaus and W.A. Albers, Jr., J.Phys.Chem.Sol. 32, 1455 (1971).
78.b) J.L. Jacquemin and G. Bordure, J.Phys.Chem.Sol. 36, 1081 (1975).
79.a) R.C. Tatar and S. Rabii, Phys.Rev. B25, 4126 (1982).
79.b) D.W. Bullett, J.Phys.C.Sol.St. 8, 2695, 2707 (1975).
80. The situation is somewhat complicated in cinnabar by there being 3 molecules per unit cell, see - E. Doni, L. Resca, S. Rodriguez and W.M. Becker, Phys.Rev. B20, 1663 (1979).
81.a) E. Mooser, I.Ch. Schlüter and M. Schlüter, J.Phys.Chem.Sol. 35, 1269 (1974).
81.b) H. Isomaki and J. von Boehm, J.Phys.C., Sol.St. 13, L485 (1980). The second paper reminds one to beware of ρ-plots deriving from pseudo-potential calculations.
82. G. Doerre and J.D. Joannopoulos, Phys.Rev.Lett. 43, 1040 (1979).
83. K. Takemura and S. Minomura, p. 131-135 in 'Physics of Solids under High Pressures' ed. J.S. Schilling and R.N. Shelton (N. Holland, 1981).
84.a) Y. Takao, H. Asahina and A. Morita, J.Phys.Soc.Jap. 50, 3362 (1981).
84.b) D. Schiferl, Phys.Rev. B19, 806 (1979).
85. A comprehensive catalogue of band gaps is available in W.H. Strehlow and E.L. Cook, J.Phys.Chem.Ref.Data 2, 163-199 (1973) - but not much discrimination was made on the quality of the data.
86. G. Brostigen and A. Kjekshus, Acta Chem.Scand. 24, 2993 (1970). full compilation of pyrite and marcasite lattice parameters.
87. J. Dugué, D. Carré and M. Guittard, Acta Cryst. B34, 403 (1978). By contrast YbS_2 is all paired with Yb(II), C.L. Teske, Zeit.Naturfors. 29B, 16 (1974).
88. A. Hasegawa and A. Yanase, J.Phys.C.Sol.St. 12, 5431 (1979). The magnitude of E_g is uncertain experimentally.
89. A.J. Arko, G. Crabtree, J.B. Ketterson, F.M. Mueller, P.F. Walch, L.R. Windmiller, Z. Fisk, R.F. Hoyt, A.C. Mota, R. Viswanathan, D.E. Ellis, A.J. Freeman and J. Rath, Int.J.Quant.Chem.Symp. 9, 569 (1975).
90. T. Tanaka and Y. Ishizawa, J.Phys.C.Sol.St. 13, 6671 (1980).
91. S. Yanagisawa, M. Tashiro and S. Anzai, J.Inorg.Nucl.Chem. 31, 943 (1969).
92. A. Kjekshus and T. Rakke, Acta Chem. Scand. A31, 517 (1977).
93. D.W. Bullett, to be published, see also GM 9,10,11,12 in Bull.Amer.Phys.Soc., 27, 283/4 (1982).

94.a) E. Wimmer, A. Neckel, K. Schwarz, R. Eibler, J.Phys.C.Sol.St. 12, 5441, 5453 (1979).
94.b) A. Hasegawa, J.Phys.C.Sol.St. 13, 6147 (1980).
94.c) A. Gusatinskii, G.I.Al'perovich, I.I.Geguzin and L.M. Monastyrskii, Sov.Phys.Sol.St. 22, 1805 (1980).
95.a) J.E. Graebner, E.S. Greiner and W.D. Ryden, Phys.Rev. B13, 2426 (1976).
95.b) G.K. Wertheim and H.J. Guggenheim, Phys.Rev. B22, 4680 (1980).
96. A. Hasegawa and M. Watabe, J.Phys.F.Metals 7, 75 (1977).
97.a) D.B.McWhan, T.M. Rice and P.H. Schmidt, Phys.Rev. 177, 1063 (1969).
97.b) G. Johansen and A.R. Mackintosh, Sol.St.Comm. 8, 121 (1970).
98. J.A. Wilson, A.S. Barker, F.J. DiSalvo, J.A. Ditzenberger, Phys.Rev. B18, 2866 (1978).
99. J.A. Wilson, Phys.Rev. B15, 5748 (1977).
100. J.A. Wilson, Phys.Rev. B17, 3880 (1978).
101. N.J. Doran and A.M. Woolley, J.Phys.C.Sol.St. 14, 4257 (1981).
102. K.K. Fung, S.McKernan, J.W. Steeds and J.A. Wilson, J.Phys. C.Sol.St. 14, 5417 (1981) and Physica scripta TI,74 (1982).
103. M.B. Walker and A.E. Jacobs, Phys.Rev. B24, 6770 (1981). W.L. McMillan, Phys.Rev., to appear.
104. J.E. Graebner, Sol.Stat.Comm. 21, 353 (1977), interpreted through band folding of early band structures in ref.99.
105. J.A. Wilson, unpublished.
106. J.A. Wilson, Phys.Rev. B19, 6456 (1979).
107. R.M. Fleming, R.A. Polo,Jr. and R.V. Coleman, Phys.Rev. B17, 1634 (1978).
108. J.A. Wilson, J.Phys.F. Metals, 12, 2469 (1982).
109. A Briggs, P. Monceau, M. Nunez-Regueiro, J. Peyard, M. Ribault and J. Richard, J.Phys.C.Sol.St. 13, 2117 (1980).
110. F. Kadijk, R. Huisman and F. Jellinek, Acta Cryst. B24, 1102 (1978). F. Kadijk, Ph.D. thesis, Groningen Univ. Netherlands (1969).
111. H. Rashid, V. Katkanant, D.J. Sellmyer and R.D. Kirby, Bull.Amer.Phys.Soc. 27, 283 (GM 8) (1982).
112. F.J. DiSalvo, D.E. Moncton, J.A. Wilson and S. Mahajan, Phys.Rev. B14, 1543 (1976).
113. N.J. Doran, G. Wexler and A.M. Woolley, J.Phys.C.Sol.St. 11, 2967 (1978).
114. J.A. Wilson, F.J. DiSalvo and S. Mahajan, Phys.Rev.Lett. 32, 882 (1974), Adv. in Phys. 24, 117 (1975).
115.a) H.W. Myron, J. Rath and A.J. Freeman, Phys.Rev. B15, 885 (1977).
115.b) N.J. Doran, B. Ricco, D.J. Titterington and G. Wexler, J.Phys.C.Sol.St. 11, 699 (1978).
115.c) A. Zunger and A.J. Freeman, Phys.Rev. B17, 1839 (1978).
116.a) W. Weber, Inst.Phys.Conf.Ser. 55, 495-504 (1980) (Physics of T.M. Metals, Leeds, UK).
116.b) N.J. Doran, J.Phys.C.Sol.St. 11, 1959 (1978).
117. J.E. Inglesfield, J.Phys.C.Sol.St. 13, 17 (1980).

118. W.L.McMillan, Phys.Rev. B16, 643 (1977).
119. S.K. Chan and V. Heine, J.Phys.F.Met. 3, 795 (1973).
120. J. Rijndsdorp and F. Jellinek, J.Sol.St.Chem. 25, 325 (1978).
121. R. Allmann, I. Bauman, A. Kutoglu, H. Rösch, E. Hellner, Naturwiss. 51, 263 (1964).
122. K. Brodersen, H.K. Breitbach and G. Thiele, Z.Anorg.Allg. Chem. 357, 162 (1968).
123. D. Babel, J.Sol.St.Chem. 4, 410 (1972).
124. M. Gupta, A.J. Freeman and D.E. Ellis, Phys.Rev. B16, 3338 (1977).
125. M. Posternak, A.J. Freeman and D.E. Ellis, Phys.Rev. B19, 6555 (1979).
126. W.A. Kamitakahara, B.N. Harmon, J.G. Taylor, L. Kopp, H.R. Shanks and J. Rath, Phys.Rev.Lett. 36, 1393 (1976). L.Kopp, B.N. Harmon and S.H. Liu, Sol.St.Comm. 22, 677 (1977).
127. R. Pynn, J.D. Axe and R. Thomas, Phys.Rev. B13, 2965 (1976). R. Pynn, J.D. Axe and P.M. Raccah, Phys.Rev. B13, 2196 (1978).
128. J.P. Pouget and H. Launois, J. de Physique, Coll. C4, 37, C4-49 (1976).
129. J.J. Capponi, M. Marezio, J. Dumas, C. Schlenker, Sol.St. Comm. 20, 293 (1976). (Note the pairing in Ti_2O_3 amounts to a reduction in the repulsive displacement from the centres of the c-axis pairs of octahedra that share a common face).
130. A. Kjekshus and T. Rakke, Acta Chem.Scand. A31, 517 (1977).
131. K. Selte and A. Kjekshus, Acta Chem.Scand. 27, 3195 (1973).
132. H.F. Franzen and G.A. Wiegers, J.Sol.St.Chem. 13, 114 (1975). S.H. Liu, W.B. England and H.W. Myron, Sol.Stat.Chem. 14, 1003 (1974).
133. J. Muller and J.C. Joubert, J.Sol.St.Chem. 11, 79 (1974). Structure figures : Amer.Min. 38, 1242 (1953) ; 40, 861 (1955).
134. D.B. McWhan, M. Marezio, J.P. Remeika and P.D. Dernier, Phys.Rev. B10, 490 (1974).
135. J.P. Pouget, H. Launois, D'Haenens, P. Merenda and T.M. Rice, Phys.Rev.Lett. 35, 873 (1975).
136. D. Paquet and P. Leroux-Hugon, Phys.Rev. B22, 5284 (1980).
137. F. Pintchovski, W.S. Glaunsinger and A. Navrotsky, J.Phys. Chem.Sol. 39, 941 (1978).
138. J. Ashkenazi, M.G. Vincent, K. Yvon and J.M. Honig, J.Phys. C.Sol.St. 14, 353 (1981).
139. C. Castellani, D. Feinberg and J. Ranniger, J.Phys.C.Sol.St. 12, 1541 (1979).
140. H. Kuwamoto, J.M. Honig and J. Appel, Phys.Rev. B22, 2626 (1980). C. Castellani, D. Feinberg and J. Ranniger, Phys.Rev. B18, 4945, 4967, 5001 (1978).
141. G.V. Chandrashekhar, L.L. van Zandt, J.M. Honig and A. Jayaraman, Phys.Rev. B10, 5063 (1974).

C.E. Rice and W.R. Robinson, Mat.Res.Bull. 12, 421 (1977).
142. H. Kuwamoto, H.V. Keer, J.E. Keem, S.A. Shivashanker, L.L. Van Zandt and J.M. Honig, J. de Phys. 37, (Colloq. 4), C4-35 (1976). J. Dumas, Phys.Rev. B22, 5085 (1980).
143. T. Horlin, T. Niklewski and M. Nygren, Acta Chem.Scand. A30, 619 (1976) J. de Phys. 37 (Colloq 4), C4-69 (1976). For $(Nb/Ti)O_2$ see T. Sakata and E. Fromm, Z. Anorg.Allg. Chem. 398, 129 (1973).
144. G. von Schulthess and P. Wachter, Sol.St.Comm. 15, 1645 (1974).
145. Ref. 128 and res. therein.
146. J.P. Pouget, H. Launois, T.M. Rice, P. Dernier, A. Gossard, G. Villeneuve and P. Hagemuller, Phys.Rev. B10, 1801 (1974). G. Villeneuve, M. Drillon, J.C. Launay, E. Marquestant and P. Hagenmuller, Sol.St.Comm. 17, 657 (1975). M. Marezio, D.B. McWhan, J.P. Remeika and P.D. Dernier, Phys.Rev. B5, 2541 (1972).
147. B.L. Chamberland, Crit.Rev.Sol.St. 7, 2 (1977).
148. B.L. Chamberland, W.H. Cloud, C.G. Frederick, J. Sol.St.Chem. 8, 238 (1973).
149. A.C. Gossard, A. Menth, W.W. Warren $ J.P. Remeika, Phys. Rev. B3, 3993 (1971).
150. Y. Ueda, K. Kosuge and S. Kachi, J.Sol.St.Chem. 31, 171 (1980).
151. Y. Shimony and L.Ben-dor, Mat.Res.Bull. 15, 227 (1980).
152. F.J. DiSalvo and J.V. Waszczak, Phys.Rev. B17, 380 (1978). C. Schlenker, C. Lander, R. Buder, F. Levy, J.Mag.Mag. Mat. 15-18, 91 (1980).
153. A.T. Chang, P. Molinie and M.J. Sienko, (LT 15) J.de Phys., Coll. 6, 39, C§-1070 (1978). L.E. Conroy and K.R. Pisharody, NBS.Sp.Publn. 364, 663 (1972), (Proc.5th Symp.Mat.Res./Sol.St.Chem.).
154. D.S. McLachlan, Phys.Rev. B25, 2285 (1981).
155. M.D. Banus, T.B. Reed and A.J. Strauss, Phys.Rev. B5, 2775 (1972).
156. J.A. Wilson, Structure and Bonding, 32, 57-91 (1977), and p. 427 in Conf. on 'Valence Instabilities and Related Narrow Band Phenomena', Ed. R.D. Parks (Plenum, 1977).
157. W. Kohn, T.K. Lee and Y.R. Lin-Liu, Phys.Rev. B25, 3557 (1982). Note also effect of B' to M phase transition in (Sm/Y)S, discussed on page 76 of ref. 156.
158. P.D. Hambourger and F.J. DiSalvo, Physica 99B, 173 (1980), and three following papers. H. Fukuyama and K. Yoshida, Physica 105B, 132 (1981) and preceding paper.
159. F.J. DiSalvo, J.A. Wilson, B.G. Bagley and J.V. Waszczak, Phys.Rev. B12, 2220 (1975).
160. F.J. Di Salvo, J.A. Wilson and J.V. Waszczak, Phys.Rev.Lett. 36, 885 (1976).

161. M. Eibschütz, M.E. Lines and F.J. DiSalvo, Phys.Rev. B15, 103 (1977), Phys.Rev.Lett. 36, 104 (1976).
162. L.F. Mattheiss, Phys.Rev. B6, 4718 (1973).
163. D.A. Whitney, R.M. Fleming and R.V. Coleman, Phys.Rev. B15, 3405 (1977). S.S.P. Parkin and R.H. Friend, Phil.Mag. B41, 65, 95 (1980).
164. B.L. Chamberland, J.Sol.St.Chem. 2, 521 (1970).
165. T. Teranishi and K. Sato, J.de Phys., Coll. 3, 36, C3-149 (1975).
166. V.V. Val'kov and S.G. Ovchinnikov, Sov.Phys.Sol.St. 22, 2000 (1980).
167. J.C.Th. Hollander, G. Sawatzky and C. Haas, Sol.Stat.Comm. 15, 747 (1974).
168. R.M. Fleming, F.J. DiSalvo, R.J. Cava and J.V. Waszczak, Phys.Rev. B24, 2850 (1981).
169. M. Iizumi and G. Shirane, Sol.St.Comm. 17, 433 (1975), this is a good deal more complicated than originally presumed.
170. H. Kessler and M.J. Sienko, J.Chem.Phys. 55, 5414 (1971).
171. C. Suguira, Y. Goshi and I. Suzuki, Phys.Rev. B10, 338 (1974).
172. D.E. Cox and A.W. Sleight, Sol.Stat.Comm. 19, 969 (1976).
T.M. Rice and L. Sneddon, Phys.Rev.Lett. 47, 689 (1981).
L.R. Gilbert, R. Messier and R. Roy, Mat.Res.Bull. 17, 467 (1982).
173. Y. Tazuke, J.Phys.Soc.Jap. 50, 413 (1980).
174. Y. Kitaoka and H. Yasuoka, J.Phys.Soc.Jap. 48, 1460 (1980).
175. P. Dougier and P. Hagenmuller, J.Sol.St.Chem. 15, 158 (1976).
176. J.L. Hodeau and M. Marezio, J.Sol.St.Chem. 29, 47 (1979), and refs. therein.
177. A.R. Beal and H.P. Hughes, J.Phys.C.Sol.St. 12, 881 (1979).
M. Tanaka, H. Fukutami, G. Kuwabara, J.Phys.Soc.Jap. 45, 1899 (1978). S. Saiki, M. Yoshimi, S. Tanaka, Phys.Stat. Sol. (b) 88, 607 (1978).
178. K. Vos, J.Phys.C.Sol.St. 10, 3893, 3917 (1977).
179. S.C. Bayliss and W.Y. Liang, J.Phys.C.Sol.St. 15, 1283 (1982).
H.M. Isomaki and J. Von Boehm, J.Phys.C.Sol.St. 14, L1043 (1981).
180. I. Pollini and G. Spinolo, J.Phys.C.Sol.St. 7, 2391 (1974).
181. H.H. Chou and H.Y. Fan, Phys.Rev. B10, 901 (1974).
H. Terasawa, T. Kambara, K.I. Kondaira, T. Teranishi, K. Sato, J.Phys.C.Sol.St. 13, 5615 (1980).
182. L. Nosenzo, E. Reguzzoni, G. Samoggia, G. Guizetti and I. Pollini, Sol.Stat.Comm. 29, 793 (1979). Also see Phys.Rev. B14, 4622 (1976).
183. A.B.P. Lever 'Inorganic Electronic Spectroscopy' (Elsevier, 1968).
184. R.J.H. Clark, J.Chem.Phys. 36, 633 (1962).
T. Parameswaran and D.E. Ellis, J.Chem.Phys. 58, 2088 (1973). Also : E.G.M. Tornqvist and W.F. Libby, Inorg. Chem. 18, 1792 (1979) for TiI_4.
185. L.J. Nugent and K.L. Van der Sluis, J.Chem.Phys. 59, 3440 (1973).

186. D.E. Eastman, F. Holtzberg, J. Freeouf, M. Erbudak, AIP Conf. Proc. 18, 1030 (1974).
M. Campagna, E. Bucher, G.K. Wertheim and L.D. Longinotti, Phys.Rev.Lett. 33, 165 (1974).
187. T. Yamaguchi, S. Shibuya and S. Sugano, J.Phys.C.Sol.St. 15, 2652, 2640 (1982).
188. G. Guntherodt, Phys.Kond.Mat. 18, 37 (1974).
190. V.P. Zhuse, M.G. Karin, D.P. Lukirskii, V.M. Sergeeva and A.I. Shelykh, Sov.Phys.Sol.St. 22, 1558 (1980).
191. J.A. Wilson, unpublished.
192. A.F. Reid and M.J. Sienko, Inorg. Chem. 6, 521 (1976).
193. M. Yokogawa, K. Taniguchi and C. Hamaguchi, J.Phys.Soc.Jap. 42, 591 (1976).
194. I. Pollini, G. Spinolo and G. Benedek, Phys.Rev. B22, 6369 (1980).
195. R. Colton and J.H. Canterford 'Halides of First Row Transition Metals' (Wiley-Interscience, 1969).
196. A. Cisar, J.D. Corbett and R.L. Daake, Inorg.Chem. 18, 836 (1979). H. Imoto, J.D. Corbett and A. Cisar, Inorg.Chem. 20, 145 (1981). J.D. Corbett and D.H. Guthrie, J.Sol.St. Chem. 37, 256 (1981).
197. O. Okada and T. Miyadai, J.Phys.Soc.Jap. 43, 343 (1977).
198. E. Konig and G. Ritter, Sol.St.Comm. 18, 279 (1976).
R. Zimmermann and E. Konig, J.Phys.Chem.Sol.38, 779 (1977).
S. Ohnishi and S. Sugano, J.Phys.C.Sol.St. 14, 39 (1981).
199. G. Guizzetti, E. Reguzzoni and I. Pollini, Phys.Lett. 70A, 34 (1979).
200. W.D. Ryden and A.W. Lawson, J.Chem.Phys. 52, 6058 (1970).
201. R.J. Bouchard, J.F. Weiher and J.L. Gillson, J.Sol.St.Chem. 6, 519 (1973).
202. J.L. Horwood, M.G. Townsend and A.H. Webster, J.Sol.St.Chem. 17, 35 (1976).
J.M.D. Coey and H.Roux-Buisson, Mat.Res.Bull. 14, 711 (1979).
203. P. Burgardt and M.S. Seehra, Sol.Stat.Comm. 22, 153 (1977).
204. L.F. Mattheiss, Phys.Rev. B10, 995 (1974).
205. E.F. Bertaut, P. Burlet and J. Chappert, Sol.St.Comm. 3, 335 (1965).
206. J.B. Goodenough, Annales de Chimie, to be published. (Proc.C.I.C.S.E.T. VIIe, Grenoble 1982).
207. C.B. Bargeron, M. Avinor and H.G. Drickamer, Inorg.Chem. 10, 1338 (1971).
208. M. Eibschütz and M.E. Lines, Phys.Rev.Lett. 39, 726 (1977). Also refs. 160/1.
209. T. Teranishi and K. Sato, J. dePhys. (C3) 36, C3-149 (1975). T. Oguchi, K. Sato and T. Teranishi, J.Phys.Soc.Jap. 48, 123 (1980). T. Kambara, K. Suzuki and K.I. Gondaira, J.Phys.Soc.Jap. 39, 764 (1975).

210. V.G. Bhide, D.S. Rajoria, G. Rama Rao and C.N.R. Rao, Phys. Rev. B6, 1021 (1972).
R. Marx and H. Happ, Phys.Stat.Sol. (b) 67, 181 (1975).

211. S. Haneda, Y. Yamaguchi, H. Watanabe and N. Kazama, J.Phys. Soc.Jap. 46, 802 (1979).

212. K. Adachi, M. Matsui and M. Kawai, J.Phys.Soc.Jap. 46, 1474 (1979). N. Inoue, H. Yasuoka, M. Matsui and K. Adachi, J.Phys.Soc.Jap. 50, 1180 (1981). G. Krill, P. Panissod, M. Lahrichi, M.F. Lapierre-Ravet, J.Phys.C.Sol.St. 12, 4269, 4281 (1979).

213. N.F. Mott 'Metal Insulator transitions' (Taylor-Francis, 1974). G. Thornton, A.F. Orchard and C.N.R. Rao, J.Phys. C.Sol.St. 9, 1991 (1976).

214. A. Kallel, H. Boller and E.F. Bertaut, J.Phys.Chem.Sol. 35, 1139 (1974). J.W. Allen and J.C. Mikkelson, Phys.Rev. B15, 2952 (1977).

215. F. Ogata, T. Hamajima, T. Kambara and K.I. Gondaira, J.Phys. C.Sol.St. 15, 3483 (1982).

216. K.W. Blazey and H. Rohrer, Phys.Rev. 185, 712 (1969).

217. T. Hamasaki, T. Hashimoto, Y. Yamaguchi and H. Watanabe, Sol.St.Comm. 16, 895 (1975).
P. Terzieff, Proc.C.I.C.S.E.T.VIIe (Grenoble, 1982).

218. G.P. Flecher, F.A. Smith, D. Bellavance and A. Wold, Phys. Rev. B3, 3046 (1971).
K. Selte, A. Kjekshus and A.F. Andresen, Acta Chem.Scad. 26, 3101 (1972).

219. J.R. Peterson, 10th R.E.Res.Conf. (vol.1), 4 (1973).

220. J. Covino, K. Dwight, A. Wold, R. Chiannelli, J. Passaretti, Inorg.Chem. 21, 1744 (1982).

221. K. Kikuchi, T. Miyadai, T. Fukui, H. Ito, K. Takizawa, J.Phys.Soc.Jap. 44, 410 ; 45, 444 (1978).

222. G. Krill, M.F. Lapierre, F. Gautier, C. Robert, G. Czjek, J. Fink, H. Schmidt, J.Phys.C.Sol.St. 9, 761 (1976).
P. Kwizera, A.K. Mabatah, M.S. Dresselhaus and D. Adler, Phys.Rev. B24, 2972 (1981).

223. J.A. Wilson and G.D. Pitt, Phil.Mag. 25, 625 (1972).
G. Czjezek, J. Fink, H. Schmidt, G. Krill, M.F. Lapierre, P. Panissod, F. Gautier and C. Robert, J.Mag and Mag.Mat. 3, 58 (1976) and refs. therein.
H. Takano and A. Okiji, J.Phys.Soc.Jap. 50, 3835 (1981).

224. R. Brusetti, J.M.D. Coey, G. Czjzek, J. Fink, F. Gompf and H. Schmidt, J.Phys.F.Met. 10, 33 (1980).
M.T. Hutchings, M.G. Townsend, A.H. Webster, Sol.Stat. Comm. 22, 123 (1977).

225. N. Inoue, H. Yasuoka and S. Ogawa, J.Phys.Soc.Jap. 48, 850 (1980).

226. Y.B. Yelon, S.A. Werner, S. Shivashankar and J.M. Honig, Phys.Rev. B24, 1818 (1981).

227. D.M.McWhan, J.P. Remeika, S.D. Bader, B.B. Triplet and N.E. Phillips, Phys.Rev. B7, 3079 (1973).

228. G. Krill, P. Panissod, M.F. Lapierre, F. Gautier, C. Robert, G. Czjzek, J. Fink, H. Schmidt, R. Kuentzler, J. de Phys. (Colloq. 4) 37, C4-23 (1976).
229. P. Panissod, M. Lahrichi, M.F. Lapierre-Ravet, Sol.St.Comm. 31, 273 (1979).
S. Waki, N. Kasai and S. Ogawa, Sol.St.Comm. 41, 835 (1982).
230. G. Krill, P. Panissod, M.F. Lapierre, F. Gautier, C. Robert and M. Nasr Eddine, J.Phys.C.Sol.St. 9, 1521 (1976).
G. Vanderschaeve and B. Escaig, Mat.Res.Bull. 11, 483 (1976).
231. C.F. van Bruggen, R.J. Haange, G.A. Wiegers and D.K.G. de Boer, Physics 99B, 166 (1980).
H.W. Myron, Physica 105B, 120 (1981).
232. C.S. Wang and B.M. Klein, Phys.Rev. B24, 3417, 3393 (1981).
233. J.E. Schirber and L.F. Mattheiss, Phys.Rev. B24, 692 (1981).
234. H. Capellmann, Z.Phys. B35, 269 (1979).
235. J. Hubbard, Phys.Rev. B19, 2626 ; B20, 4584 (1979).
236. V. Korenman, J.L. Murray, R.E. Prange, Phys.Rev. B16, 4032, 4048, 4058 (1977).
237. T. Moriya and H. Hasegawa, J.Phys.Soc.Jap. 48, 1490 (1980).
H.Hasegawa, J.Phys.Soc.Jap. 49, 178, 963 (1980).
238. V. Heine, J.H. Samson and C.M.M. Nex, J.Phys.F.Met. 11, 2645 (1981). M.V. You and V. Heine, J.Phys.F.Met. 12, 177 (1982).
A.J. Holden and M.V. You, J.Phys.F.Met. 12, 195 (1982).
239. J.B. Goodenough, 'Magnetism and the Chemical Bond' (Wiley-Interscience, 1963) and Prog.Sol.Stat.Chem. 5 (1971), chapter 4.
240. W.E. Evenson, J.R. Schrieffer and S.Q. Wang, J.Appl.Phys. 41, 1199 (1970).
241. M. Cyrot, Phil.Mag. 25, 1031 (1972).
242. Papers given at CICSET VIIe Grenoble (1982), to be published in Annal. de Chimie. a) A. Simon, b) W. Jeitschko, c) J. Rossat-Mignod, d) S. Rundqvist, e) E. Parthé.
243. B. Oles and H.G. von Schnering, J.Phys.C.Sol.Stat. 14, 5559 (1981).
244. C.Y. Fong, C. Perlov and F. Wooten, J.Phys.C.Sol.Stat. 15, 2605 (1982).
245. H. Bilz, B. Büttner, A. Bussmann-Holder, W. Kress and U. Schroder, Phys.Rev.Lett. 48, 264 (1982).
246. G.M. Clark, 'The Structures of Non-molecular Solids - a co-ordinated polyhedron approach' (Wiley-Halsted 1972).
247. K. Adachi, M. Matsui, F. Kimura, Y. Omata, J.Phys.Soc.Jap. 49, 1629 (1980).

ELECTRON SPECTROSCOPY OF METALLIC SYSTEMS

P.J. Durham

SERC Daresbury Laboratory
Warrington
England

1. PREAMBLE

The electronic structure theories forming the topic of this School are aimed at making the observable behaviour of solids intelligible in terms of what is happening, on a microscopic scale, inside them. In practice this involves making careful calculations of a number of quantities, and discussing the physical effects underlying their success (or failure). From this point of view, band theory is the most important tool for understanding the detailed properties of a wide range of materials, as other lectures in this volume testify. Indeed, we know that if the density functional theory is applied sufficiently carefully, then band calculations give ground state properties (in particular, the charge density and total energy) very accurately. But what about the bands themselves ? In all but the simplest solids, the energy bands are richly complex objects, eminently meriting their nickname "spaghetti". It is hard not to wonder how real these states are, whether particular bands play an important role in some phenomena but not in others, and so on. Thus, we want to observe the bands directly, just as atomic physicists can observe individual states of their systems. One good way of doing this is to look at the Fermi surface, and well known techniques have been developed for this purpose. But to see states away from the Fermi level we have to apply some form of spectroscopy.

So, in these notes I will describe what information about the electronic structure can be obtained from experiments such as X-ray emission and absorption spectroscopy, photoemission etc. To make the most of this information one needs to know precisely what is being measured, otherwise effects intrinsic to the spectroscopic process might mask the underlying electronic structure. In other words one needs a proper theory of the spectroscopy. This is especially true of the complex systems under discussion at this School.

The most important spectroscopic experiment at present is angle-resolved photoemission. With this technique one does come remarkably close to observing individual bands, and I shall give a good deal of attention to it. There are, though, other experiments which provide complementary information, and so I will treat all of these techniques on the same theoretical footing.

Spectroscopic transitions produce a hole in a previously occupied state or an electron in a previously unoccupied state (at the least ; sometimes more complicated states are produced). These are excited states, and it is essentially the spectrum of such states which is measured. Here we hit a fundamental problem, because we know that density functional band structures do not represent the energies of quasi-particle excitations. On the other hand, as von Barth argues in this volume, the density functional eigenvalues for extended states are probably quite close to quasi-particle energies, and it is true that for metals at least most features of the spectra are found in 1-electron calculations. Moreover, the many-body effects which are undoubtedly present in the spectra have usually been discussed using very simplified models for the 1-electron part of the problem. This makes comparison with experiment less useful than it could be if the 1-electron problem were realistically treated. Thus my view here will be that one should do 1-electron calculations of spectroscopic transition rates as well as one can, and I will indicate in these notes what this can involve. In systems where the nature of the 1-electron states is in question (eg. disordered systems), one then stands to learn a lot about these states. In systems where the 1-electron structure is really well known (eg. Ni,Cu), comparison with experiment then bears directly on the validity of the 1-electron picture. For this programme to make sense, the 1-electron spectrum has at least to be a reasonable approximation to the quasi-particle spectrum and for this reason I will not discuss the spectroscopy of localised states (see von Barth in this volume). This means that I have to exclude core levels (eg. in core level X-ray photoemission) and insulators, and concentrate on the spectroscopy of band states in metallic systems.

These notes are not a review of the current theoretical, let alone experimental, state of the electron spectroscopy of metals. They are intended to convey the nature of the information spectroscopy can provide, and to describe the important ingredients of realistic theoretical calculations. I start by showing how the transition rates can be expressed in terms of 1-electron Green's functions for a number of different spectroscopies. To do this I have borrowed a diagram technique from many-body theory. This serves my purpose, not because I will do any real many-body theory (I will not), but (a) because it does remind one that many-body effects are always present to some extent (I will expand on this later) - in my opinion the 1-electron and many-body approaches should not be as divorced from one another as they often seem to be, and (b) because I find the 1-electron diagrams (i.e. those of zeroth order in the electron-electron interaction) rather neat objects with which to work. Next, I show how the Green's functions can be obtained conveniently from a particular version of multiple scattering theory, and then go on to describe in more detail what is involved in numerical calculations. An advantage of this multiple scattering approach is its flexibility and this will enable me to generalise the calculations to the case of random metallic alloys. The actual calculations I will show are quite few (more are referred to), but have been chosen to illustrate important points in the discussion. I will concentrate for the most part on X-ray absorption and emission, and on photoemission spectra, but I will conclude with a glipse of some of the possibilities of X-ray scattering.

2. BASICS

I will start by discussing soft X-ray emission (SXE). This is historically appropriate because SXE seems to have been the first spectroscopy to have been brought to bear on the problem of extended electron states in solids, in the work of Skinner et al.[1] in the 1930's. In their book Mott and Jones[2] gave a summary of the contemporary results for x-ray emission and absorption, and even an early account of EXAFS.

In the x-ray emission process a valence electron fills a previously created core hole in state i (say), thereby emitting a photon. To describe this I first represent the ground state of the electron system and the photon vacuum by the ket $|0\rangle$. At some time in the distant past, $t = -\infty$, the system starts off in state $c_i|0\rangle$, where c_i is the destruction operator for electron state i. Thereafter the system evolves according to the operator

$$u(t,t') = e^{-iH(t-t')}$$

with the Hamiltonian for the uncoupled electrons and photons given by

$$H = \sum_n \varepsilon_n c_n^+ c_n (+ \text{ e-e interaction terms}) + \sum_{\underline{q}\lambda} \omega_{\underline{q}} a^+_{\underline{q}\lambda} a_{\underline{q}\lambda}$$

in which $a_{\underline{q}\lambda}$ is the destruction operator for photon state $(\underline{q},\lambda)$, with energy $\omega_{\underline{q}}$, and ε_n is the energy of electron state n. At some instant t=t' there occurs an electron-phonon interaction governed by the Hamiltonian

$$H_{int} = \frac{e}{c} \int d\underline{r}\, \underline{j}(\underline{r}).\underline{A}(\underline{r}) + \frac{e^2}{2mc^2} \int d\underline{r}\, \rho(\underline{r}) A^2(\underline{r})$$

$(\rho(\underline{r}), \underline{j}(\underline{r})$ = number, current density respectively)

and at a later time $t = t_0$ the system is in state

$$|i,t_0> = -i \int_{-\infty}^{t_0} dt'\; u(t_0,t')H_{int}\; u(t',-\infty)c_i|0>$$

(to first order in H_{int}). To obtain the transition probability we count the number of photons in state $(\underline{q},\lambda)$, i.e.

$$P_i^{q\lambda}(t_0) = <i,t_0|a^+_{\underline{q}\lambda}a_{\underline{q}\lambda}|i,t_0>$$

$$= \int_{-\infty}^{t_0} dt \int_{-\infty}^{t_0} dt' <0|c_i^+\, u(-\infty,t)H^+_{int}\; u(t,t_0)a^+_{\underline{q}\lambda}a_{\underline{q}\lambda}\; u(t_0,t')$$
$$H_{int}\; u(t',-\infty)c_i|0> \qquad (1)$$

The transition rate is just

$$W_i^{q\lambda} = \lim_{t_0\to\infty} \frac{dP_i^{q\lambda}(t_0)}{dt_0}$$

Note that to get a finite transition rate in first order only the following part of H_{int} need be considered ($\underline{e}^\lambda(\underline{q})$ = polarisation ; Ω is normalisation volume)

$$H_{int} = \sum_{nn'} \left(\frac{2\pi\hbar e^2}{\Omega\omega_{\underline{q}}}\right)^{1/2} \underline{e}^\lambda(\underline{q}) \,.\, \int d\underline{r}\; j_{nn'}(\underline{r})e^{-i\underline{q}.\underline{r}}\; a^+_{\underline{q}\lambda}c^+_n c_{n'}$$

This is the only part of the interaction which does not destroy a photon ($u(t,t')$ conserves photons). Also, since we can easily identify transitions involving a given core state i by looking at the frequency of the emitted radiation, we can specialise further to the interaction term

$$H_{int} = \sum_{n} \left(\frac{2\pi\hbar e^2}{\Omega\omega_{\underline{q}}}\right)^{1/2} \underline{e}^\lambda(\underline{q}) \,.\, \int d\underline{r}\; \underline{j}_{ni}(\underline{r})e^{-i\underline{q}.\underline{r}}\; a^+_{\underline{q}\lambda}c^+_i c_n$$

From equation (1) we see that $P_i^{q\lambda}(t_0)$ contains a product of time-dependent operators in a specified order. It is not a time-ordered product. Hence we cannot apply the Wick theorem and the rest of the apparatus of the standard diagrammatic perturbation theory to this quantity. (Actually, I will qualify this statement later, but for the moment let it stand). But suppose time is considered to run not from $-\infty$ to $+\infty$ in the usual way, but rather along a path consisting of two branches[3], and 'upper' or forward branch running from $-\infty$ to $+\infty$, and a 'lower' or backward branch running from $+\infty$ to $-\infty$. (What we are doing is to regard time as a complex variable, and to define a contour in the complex t plane. If the Hamiltonian is an analytic function of time there is great freedom of choice for this contour, but the 2-branch path just defined is particularly convenient here). If we make the convention that any time on the upper branch is to be considered as 'earlier' than any time on the lower branch, we can see that the operators in $P_i^{q\lambda}(t_0)$ are ordered along the path or contour. All interactions in the ket $|i,t_0\rangle$ are taken to lie on the upper branch, while those in the bra$\langle i,t_0|$ occur on the lower branch. Keldysh[3] showed that this generalisation leads to a diagrammatic perturbation theory very similar to the standard one, but involving four Green's functions distinguished by the time branches on which their end points lie :

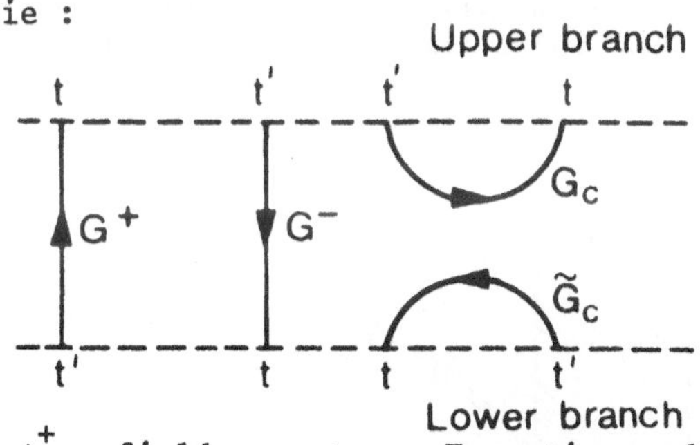

$$G_c = -i\langle 0|T_+\psi(t)\psi^+(t')|0\rangle$$
$$\tilde{G}_c = -i\langle 0|T_-\psi(t)\psi^+(t')|0\rangle$$
$$G^+ = i\langle 0|\psi^+(t')\psi(t)|0\rangle$$
$$G^- = -i\langle 0|\psi(t)\psi^+(t')|0\rangle$$

ψ,ψ^+ = field operators, T_+ = time-ordering, T_- = anti-time ordering.

Figure 1.

(This method of using two time branches is the original formulation of Keldysh. He[3] and others[4] then went on to define a 2 X 2 matrix Green's function

$$\begin{pmatrix} G_c & G^- \\ G^+ & \tilde{G}_c \end{pmatrix}$$

which simplifies some applications. However, in the applications I shall describe I find it easier to use the two branches explicitly. For further details of this approach see the book of Kadanoff and Baym[5] and the review of Langreth[6]).

The form of equation (1) shows that the general diagram for x-ray emission is

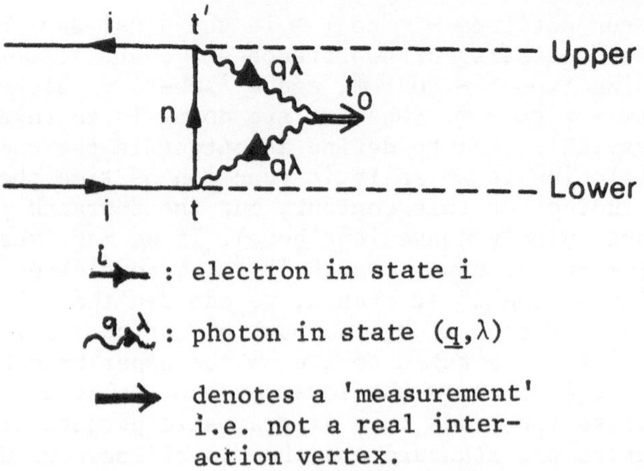

in which the hatched object conceals the usual daunting jungle of electron-electron interactions. The independent particle diagram looks much better

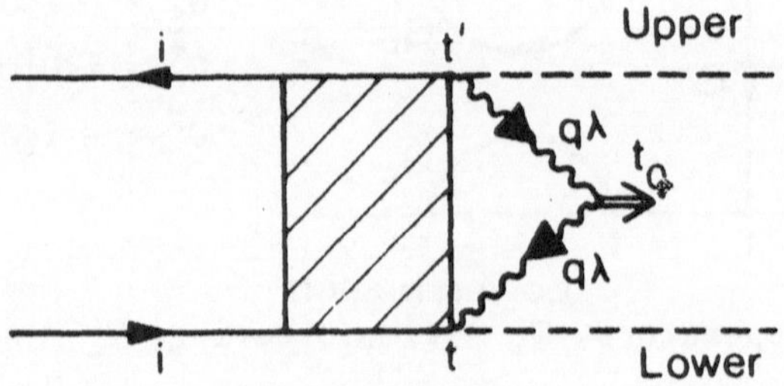

Figure 2.

Some general comments are now possible. First, note that these are diagrams for the full transition probability, and are therefore well suited to display interferences between different terms in the transition amplitude. Electron lines linking the two time branches correspond to real excitations created in the system, e.g. in this case a hole in state n. The spectrum of photons gives essentially the spectral density of such excitations. From now on I will not draw explicitly the dashed lines showing the time branches. Their implied location will be clear from the shape of the diagrams.

The time integrals in fig. 2 can easily be done if the usual Fourier transforms are introduced

$$G_c(\varepsilon) = \int_{-\infty}^{\infty} dt\ G_c(t)e^{i\varepsilon t}$$

and likewise for the other Green's functions. In this representation the transition rate is given by

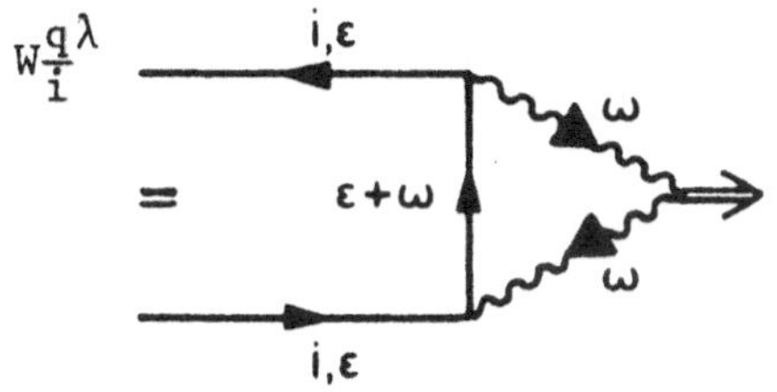

$$W_i^{q\lambda} = -i\int \frac{d\varepsilon}{2\pi} G_c^{(i)}(\varepsilon) H_{int} G_+(\varepsilon + \omega) H_{int}^{+} \tilde{G}_c^{(i)}(\varepsilon); \omega = \omega_{\underline{q}}$$

Figure 3.

In evaluating such diagrams it is useful to relate the Green's functions to the retarded Green's function which I denote by

$$G^+(\underline{r},\underline{r}';\varepsilon) = \langle\underline{r}|(\varepsilon - H_0 + i\eta)^{-1}|\underline{r}'\rangle$$

The formulae are

$$G_c(\underline{r},\underline{r}';\varepsilon) = \int \frac{d\varepsilon'}{2\pi} \frac{\mathrm{Im}\, G^+(\underline{r},\underline{r}';\varepsilon')}{\varepsilon' - \varepsilon - i\eta\,\mathrm{sign}(\varepsilon - \mu)}$$

$$\tilde{G}_c(\underline{r},\underline{r}';\varepsilon) = \int \frac{d\varepsilon'}{2\pi} \frac{\mathrm{Im}\, G^+(\underline{r},\underline{r}';\varepsilon')}{\varepsilon' - \varepsilon + i\eta\,\mathrm{sign}(\varepsilon - \mu)}$$

$$G_+(\underline{r},\underline{r}';\varepsilon) = -2i\,\mathrm{Im}\, G^+(\underline{r},\underline{r}';\varepsilon)\theta(\mu - \varepsilon)$$

$$G_-(\underline{r},\underline{r}';\varepsilon) = 2i\,\mathrm{Im}\, G^+(\underline{r},\underline{r}';\varepsilon)\theta(\varepsilon - \mu) \qquad (2)$$

(N.B. e-e interaction is excluded from H_0), (here μ is the Fermi level).

I want to continue this evaluation for a while to get to a more familiar looking result. If the core state has eigenvalue ε_i and width Γ then

$$G_c^{(i)}(\underline{r},\underline{r}';\varepsilon) = \frac{\phi_i(\underline{r})\phi_i^\star(\underline{r}')}{\varepsilon - \varepsilon_i - i\Gamma}\ ;\ \tilde{G}_c^{(i)}(\underline{r},\underline{r}';\varepsilon) = \frac{\phi_i(\underline{r})\phi_i^\star(\underline{r}')}{\varepsilon - \varepsilon_i + i\Gamma}$$

so that

$$W_i^{q\lambda} = -i\int \frac{d\varepsilon}{2\pi} \int d\underline{r} \int d\underline{r}'\, \phi_i^\star(\underline{r}) H_{int}(\underline{r}) \frac{G_+(\underline{r},\underline{r}';\varepsilon + \omega)}{(\varepsilon - \varepsilon_i)^2 + \Gamma^2} H_{int}^+(\underline{r}')\phi_i(\underline{r}')$$

$$= \frac{-i}{2\Gamma} \int d\underline{r} \int d\underline{r}'\, \phi_i^\star(\underline{r}) H_{int}(\underline{r}) G_+(\underline{r},\underline{r}';\varepsilon_i + \omega) H_{int}^+(\underline{r}')\phi_i(\underline{r}') \text{ as } \Gamma \to 0 \qquad (3)$$

or

$$W_i^{q\lambda} = -\frac{1}{\Gamma}\int d\underline{r}\int d\underline{r}'\phi_i^\star(\underline{r})X_{\underline{q}\lambda}(\underline{r})\,\mathrm{Im}G^+(\underline{r},\underline{r}';\varepsilon_i+\omega)X_{\underline{q}\lambda}^\star(\underline{r}')\phi_i(\underline{r}')\theta(\mu-\varepsilon_i+\omega) \tag{4}$$

where

$$X_{\underline{q}\lambda}(\underline{r}) = \left(\frac{2\pi\hbar e^2}{\Omega\omega_{\underline{q}}}\right)^{1/2} \underline{e}^\lambda(\underline{q}).\underline{j}(\underline{r})e^{-i\underline{q}.\underline{r}}$$

Now use the eigenfunction expansion for the retarded Green's function

$$G^+(\underline{r},\underline{r}';\varepsilon) = \sum_n \frac{\phi_n(\underline{r})\phi_n^\star(\underline{r}')}{\varepsilon-\varepsilon_n+i\eta}\ ;\ \phi_n(\underline{r}),\varepsilon_n \equiv n\text{'th eigenstate of } H_0$$

so that

$$W_i^{q\lambda} = \frac{\pi}{\Gamma}\sum_n |M_{in}|^2\ \delta(\varepsilon_i+\omega-\varepsilon_n)\theta(\mu-\varepsilon_i-\omega) \tag{5}$$

where

$$M_{in} = \int d\underline{r}\ \phi_i^\star(\underline{r})X_{\underline{q}\lambda}(\underline{r})\phi_n(\underline{r})$$

We have derived the Golden Rule !

Clearly, this is a fairly trivial example designed just to exhibit the way this method works out. It has defects from a physical point of view, however, because one starts the system off in an unstable excited state with a hole in core level i. The x-ray emission experiment is actually conducted by bombarding the sample with high energy electrons to excite the core hole and monitoring the emitted radiation - it is a scattering experiment in which photons are produced in the scattering event. We can describe this full process with precisely the same formalism. We just start off the system in a state consisting of the target in its ground state and an incident electron with high energy E_0. This excites by Coulomb interactions a core hole which then decays in the way just described. The rate of such processes is given by the diagram

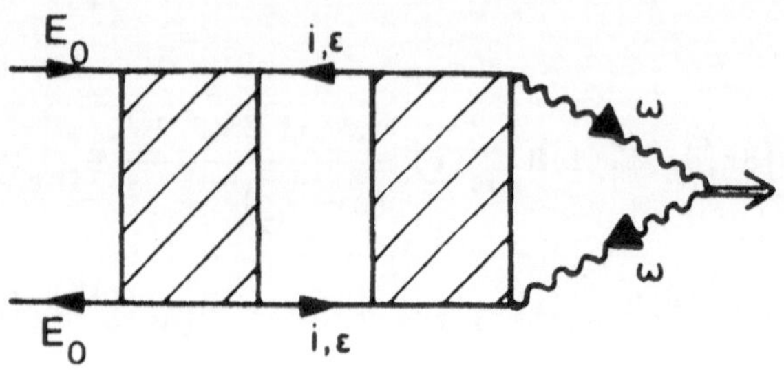

whose simplest components are

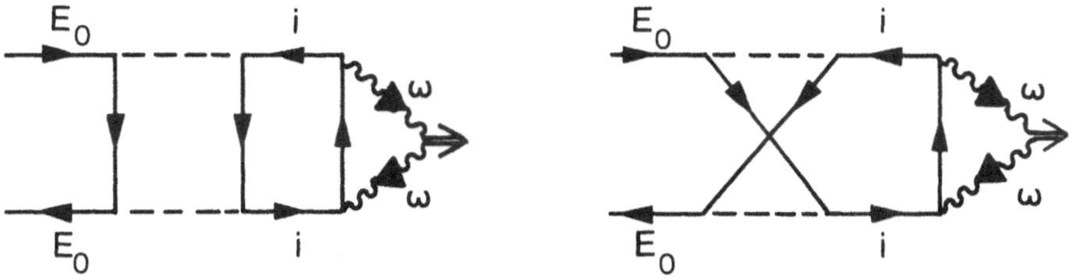

(------ = Coulomb interaction)

Figure 4.

In this way Yue and Doniach[7] were able to discuss the effects of the initial excitation process on the shape of the emission edge. We merely want to show how easy this formalism makes it to generalise the simple theory. Note also that in the long-lived hole approximation described above a diagrammatic analysis of the usual (i.e. non-Keldysh) type exists because[8]

$$W_i^{q\lambda} \propto \mathrm{Im}\chi_i(\omega)$$

where

$$\chi_i(\omega) =$$

(This is the formulation used by Nozieres and coworkers[9]). In general, the 2-point correlation functions encountered in linear response problems, while not containing time-ordered products of operators, are simply related to time-ordered susceptibilities, such as the one above, which do have a diagrammatic perturbation series. This is not so for higher order correlation functions such as the 4-point function occurring in the full scattering formalism for SXE. For such problems the Keldysh-type techniques are most useful since the resulting diagrams provide a systematic way of including interactions. In the independent particle approximation they are, of course, not essential. I introduce them not only to make contact with the many-body literature, but also because the diagrams provide a neat and easily appreciated representation of a number of different spectroscopic transitions. To show this, let me now give some more diagrams.

Consider first x-ray absorption, a process in which a photon is absorbed and a core hole in state i is "emitted". The transition rate is given by

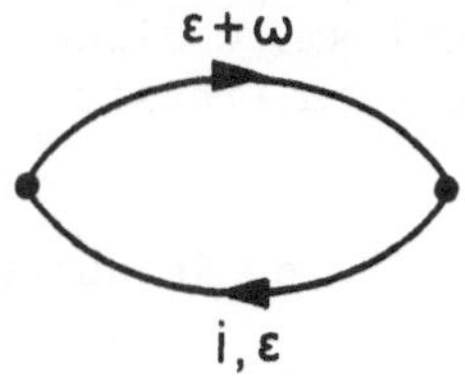

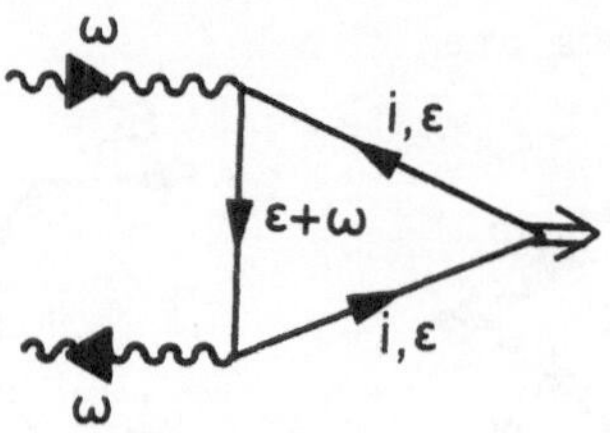

Figure 5.

which gives a Golden Rule expression like equation (5) but with a factor $\theta(\varepsilon_i + \omega - \mu)$. This is adequate to describe the transmission experiment. Often, though, one detects the core holes by monitoring their decay products. For example, one might measure the current of electrons ejected in an Auger transition involving the core hole :

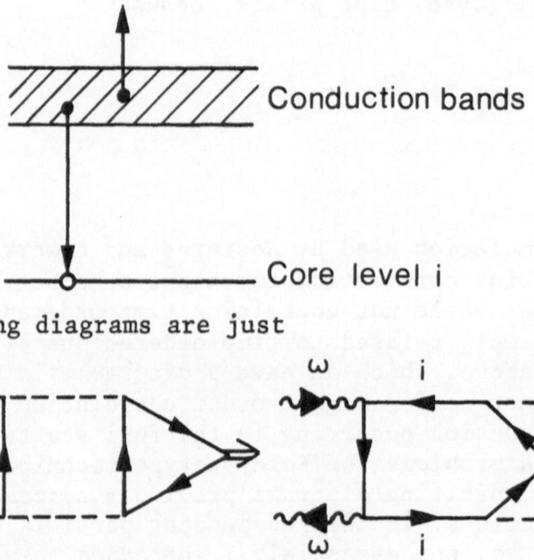

The corresponding diagrams are just

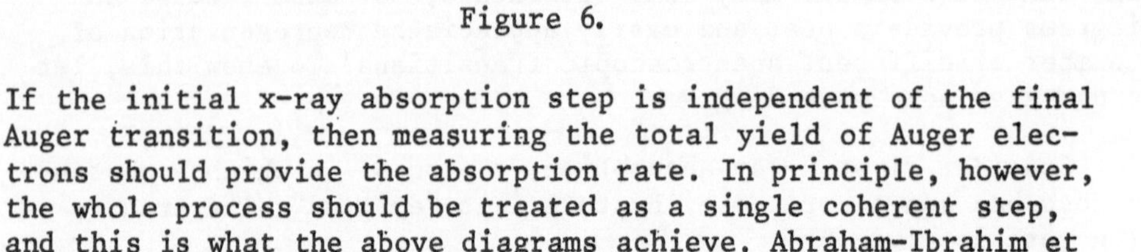

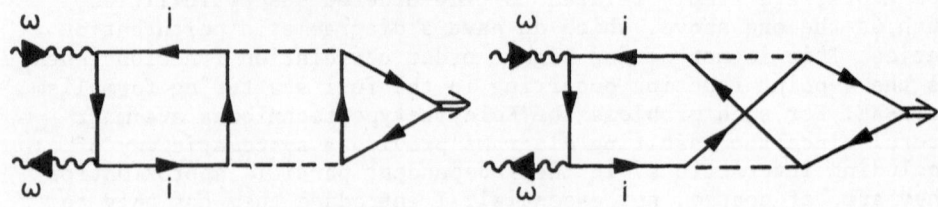

Figure 6.

If the initial x-ray absorption step is independent of the final Auger transition, then measuring the total yield of Auger electrons should provide the absorption rate. In principle, however, the whole process should be treated as a single coherent step, and this is what the above diagrams achieve. Abraham-Ibrahim et al.[10] have used this formalism to discuss the effect on Auger spectra of interferences between transitions involving degenerate holes on different sites.

In the last diagrams a photon is absorbed and an electron

emitted - they represent photoemission. The general diagram for photoemission is

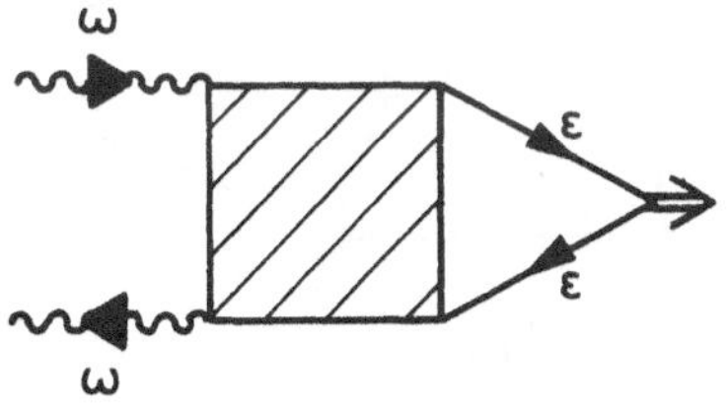

Figure 7.

of which the photo-excited Auger emission diagrams are a special case. The simplest diagram of this type, representing the independent particle approximation, is

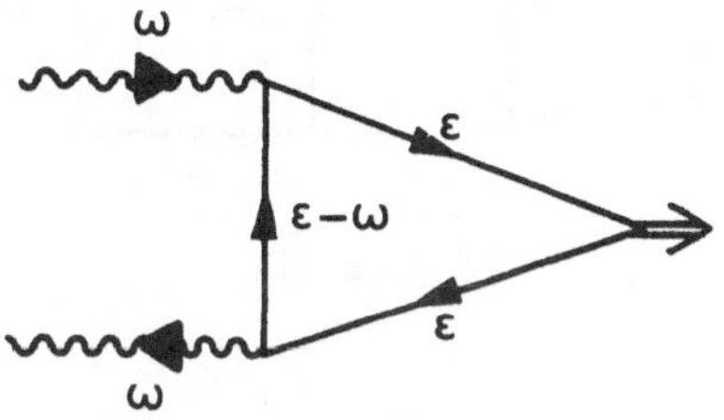

Figure 8.

containing the 3-current repsonse function (Schaich and Ashcroft)[11] This diagram shows that the system is left in an excited state with a hole in some formerly occupied core or valence level. It is the spectrum of such holes which is measured. By the same token, the diagrams in figure 6 show that the Auger current measures the 2-hole spectrum. Caroli et al.[12] have shown how to use such diagrams to give a systematic account of many-body effects in photoemission, and have treated in detail the electron-phonon interaction. Langreth and coworkers[13] have used the same technique to describe bulk and surface plasmon satellites in core level photoemission.

In figure 8 the photoelectron has energy ε. Many loss processes can be suffered by the photoelectron, causing it to emerge with energy < ε. For instance, an electron-hole pair might be excited

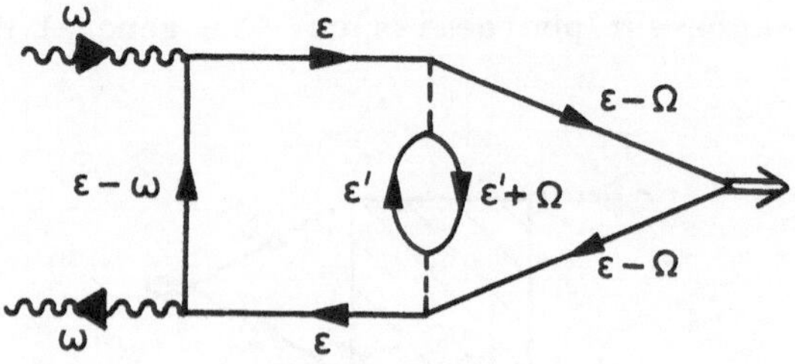

Figure 9.

The extra excited electron has energy $\varepsilon' + \Omega$, which might be sufficient for it to be able to escape from the sample and be externally observed. If I put the arrow of observation on this electron and deform the graph in a convenient way, I find a contribution to the secondary electron current

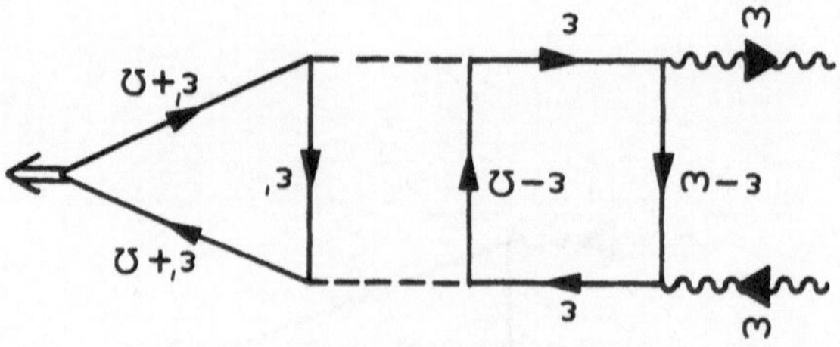

Figure 10.

Finally, I want to show some diagrams representing x-ray scattering[14]. Suppose the system starts in state $a^+_{\underline{q}\lambda}|0\rangle$, and evolves until some much later time t_0, at which the system is in state $|\underline{q}\lambda, t_0\rangle$, when we count the number of photons in state $\underline{q}'\lambda'$:

$$P_{\underline{q}\lambda,\underline{q}'\lambda'}(t_0) = \langle \underline{q}\lambda, t_0 | a^+_{\underline{q}'\lambda'} a_{\underline{q}'\lambda'} | \underline{q}\lambda, t_0 \rangle$$

The general diagram for the scattering rate is

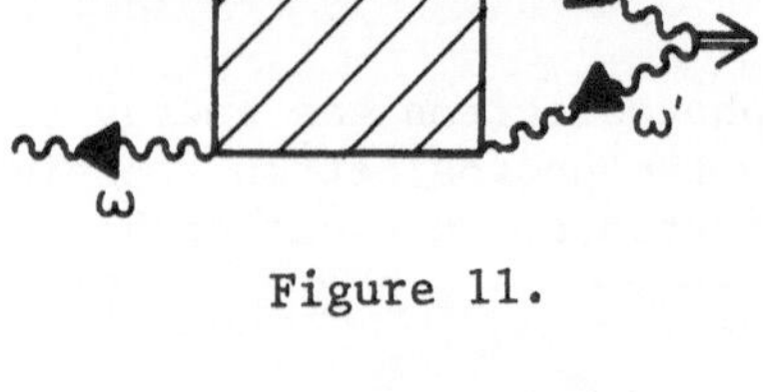

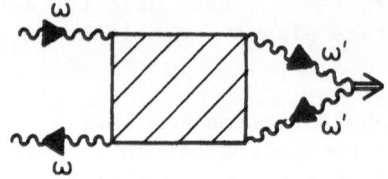

Figure 11.

of which the simplest types are

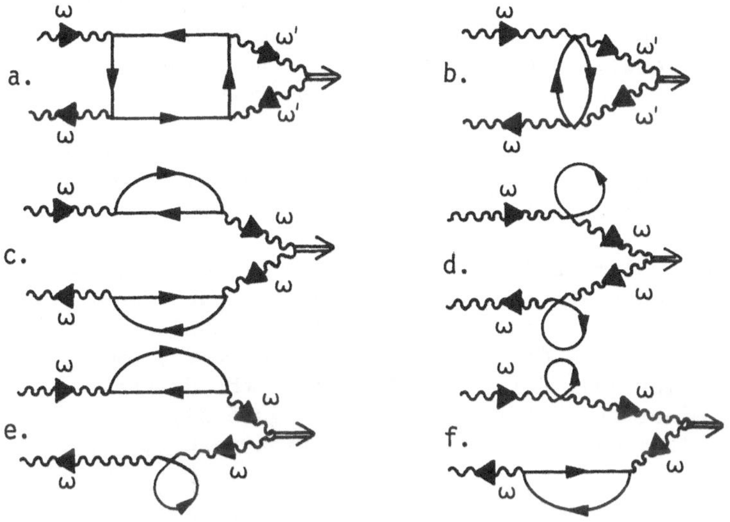

Figure 12.

Since two photons are involved, we have processes in which $\underline{j}.\underline{A}$ acts twice, or A^2 once. Figures 12a and b show inelastic scattering in which an excited electron and hole are left in the system. Figure 12a looks like a product of absorption and emission (see figures 3 and 5) and represents Raman or fluorescence scattering. Figure 12b represents Compton scattering. Figures 12c to f show elastic scattering - the electron system is left in its ground state. When a core state is involved, resonances leading to anomalous scattering can occur, and I shall return to this later. The amplitudes of these processes are second order in the vector potential and are therefore weak. But the rise of synchroton radiation now makes it feasible to exploit scattering experiments more fully than before, and here we have a formalism for treating them on the same footing as other spectroscopic techniques.

2. MULTIPLE SCATTERING THEORY

In the last section I obtained a Golden Rule formula for x-ray emission. This is a perfectly good starting point for a calculation if one already has the eigenfunctions ϕ_n and eigenvalues ε_n. In many ways, though, the preceeding formula, equation (4), involving the retarded Green's function, is more flexible. For one thing, as I shall show, it immediately allows one to apply the powerful technique of multiple scattering theory to the calculation of spectroscopic transition rates. Moreover, for disordered systems it is very difficult to work with wave functions because averages over the possible configurations of the system have to be made. While the configurationally averaged wave functions are perhaps useless, the averaged Green's function can easily be used to construct the average values of observables. I shall exemplify this by discussing random alloys. In this section, therefore, I want to show how multiple scattering theory can be used to get informative and calculationally useful formulae for the transition rates for x-ray emission and absorption and for photoemission within the independent particle approximation. This will enable me to make some general comments on the important ingredients of the formulae before going on, in the next section, to describe in more detail how numerical calculations can be made.

The general methodology is as follows. We first take the appropriate Keldysh diagram representing the independent particle approximation. Then, using equations (2) we evaluate the diagram in terms of retarded Green's functions. For x-ray emission this led to equation (4). For x-ray absorption the diagram (figure 5) is very similar to that for emission (figure 3), except that it contains $G_-(\varepsilon + \omega)$. So, for the absorption rate we get

$$W_i^{q\lambda} = -\frac{1}{\Gamma}\int d\underline{r}\int d\underline{r}'\phi_i^\star(\underline{r})X_{q\lambda}(\underline{r})\,\mathrm{Im}G^+(\underline{r},\underline{r}';\varepsilon_i+\omega)X^\star_{q\lambda}(\underline{r}')\phi_i(\underline{r}')\theta(\varepsilon_i+\omega-\mu) \tag{6}$$

The θ-functions in equations (4) and (6) just indicate that the emission process leaves a hole in an initially occupied state, while absorption results in an electron in an initially unoccupied state.

For photoemission figure 8 gives

$$I^{q\lambda}(\varepsilon) = \lim_{R\to\infty}\left(-\frac{\hbar}{m}\right)\left(\frac{\partial}{\partial R}-\frac{\partial}{\partial R'}\right)\int d\underline{r}\int d\underline{r}'G_c(\underline{R},\underline{r};\varepsilon)X_{q\lambda}(\underline{r})G_+(\underline{r},\underline{r}';\varepsilon-\omega) \times X^\star_{q\lambda}(\underline{r}')\tilde{G}_c(\underline{r}',\underline{R}';\varepsilon)\Big|_{\underline{R}'=\underline{R}} \tag{7}$$

Here we have recognised that the arrow of observation in this case means measuring the radially directed current of photoelectrons of energy ε at some distant external point $\underline{R}$. The first thing to note is that $G_c(\underline{R},\underline{r};\varepsilon)$ and $\tilde{G}_c(\underline{r}',\underline{R}';\varepsilon)$ give no current at $\underline{R}$ unless $\varepsilon > \mu + \Phi$ where Φ is the work function. Thus we can see from equations (2) that we can replace $G_c(\underline{R},\underline{r};\varepsilon)$ in equation (7) by the retarded Green's function $G^+(\underline{R},\underline{r};\varepsilon)$. Likewise $\tilde{G}_c(\underline{r}',\underline{R}';\varepsilon)$ becomes the advanced Green's function $G^-(\underline{r}',\underline{R}';\varepsilon)$. The standard development (see, for example, Feibelman and Eastman[15] or Liebsch[16]) notes that

$$G^+(\underline{R},\underline{r};\varepsilon) = G_0^+(\underline{R}-\underline{r};\varepsilon) + \int d\underline{r}_1 G_0^+(\underline{R}-\underline{r}_1;\varepsilon)V(\underline{r}_1)G^+(\underline{r}_1,\underline{r};\varepsilon)$$

where the free particle propagator is

$$G_0^+(\underline{R}-\underline{r};\varepsilon) = \frac{e^{i\kappa|\underline{R}-\underline{r}|}}{4\pi|\underline{R}-\underline{r}|} \cong \frac{1}{4\pi R}\,e^{i\kappa(R-\hat{\underline{R}}.\underline{r})}\ , \quad \kappa^2 = \varepsilon$$

since we can make $\underline{R}$ arbitrarily large. Thus the asymptotic formula ($\underline{R} \to \infty$) for the retarded Green's function is

$$G^+(\underline{R},\underline{r};\varepsilon) = \frac{e^{ikR}}{4\pi R}\ \psi_+(\underline{r},\varepsilon)$$

where the so-called "LEED state" ψ_+ is given by

$$\psi_+(\underline{r},\varepsilon) = e^{i\underline{k}.\underline{r}} + \int d\underline{r}_1\ e^{i\underline{k}.\underline{r}_1}\ V(\underline{r}_1)G^+(\underline{r}_1,\underline{r};\varepsilon)$$

and the wave vector $\underline{k} = -\kappa\hat{\underline{R}}$ points from the detector into the sample. The radial current at $\underline{R}$ is then easily found to be[17]

$$R^2 I^{\underline{q}\lambda}(\varepsilon) = -\frac{\hbar\kappa}{4\pi^2 m}\int d\underline{r}\int d\underline{r}'\psi_+(\underline{r},\varepsilon)X_{\underline{q}\lambda}(\underline{r})\,\mathrm{Im}G^+(\underline{r},\underline{r}';\varepsilon-\omega)$$
$$\times\ X^*_{\underline{q}\lambda}(\underline{r}')\psi^*_+(\underline{r}',\varepsilon)\theta(\mu-\varepsilon+\omega)\theta(\varepsilon-\mu-\Phi) \qquad (8)$$

where we have used equation (2) for $G_+(\underline{r},\underline{r}';\varepsilon+\omega)$.

The idea now is to calculate the retarded Green's functions in equations (4,6,8) by means of multiple scattering theory. To do this we have to assume that the electrons move independently in a 1-particle potential of non-overlapping muffin-tin form :

$$V(\underline{r}) = \sum_i v_i(\underline{r}-\underline{R}_i)\ ,\ \underline{R}_i \text{ is the position of site } i \qquad (9)$$

The scattering properties of each atom can then be summarized in a t-matrix for angular momentum $L \equiv \ell,m$

$$t_{i,L}(\varepsilon) = -\frac{1}{\kappa}\sin\delta_{i,L}(\varepsilon)e^{i\delta_{i,L}(\varepsilon)}$$

where $\delta_{i,L}(\varepsilon)$ is the l-wave phase-shift at energy ε of atom i. (Here I have assumed that the potentials $v_i(\underline{r})$ are spherically symmetrical. The multiple scattering theory can be generalised to handle non-spherical potentials[18], though it is not trivial to do so). As shown by Stocks in this volume, there is a convenient way of representing multiple scattering[19] through the scattering path matrix, or τ-matrix, whose on-shell matrix elements are given by

$$\tau^{ij}_{LL'}(\varepsilon) = t_{i,L}(\varepsilon)\delta_{ij}\delta_{LL'} + \sum_{n\neq i}\sum_{L''} t_{i,L}(\varepsilon)g^{in}_{LL''}(\varepsilon)\tau^{nj}_{L''L'}(\varepsilon) \quad (10)$$

where $g^{in}_{LL''}(\varepsilon)$ are the real space propagators defined by Stocks. $\tau^{ij}_{LL'}$ sums up the contribution to the total T-matrix of all multiple scattering paths beginning in channel L at site i and ending in channel L' at site j. This formulation is convenient because the retarded Green's function is given by the following neat formulae

$$G^+(\underline{r},\underline{r}';\varepsilon) = \sum_{LL'} Z_L^{(i)}(\underline{r}_i;\varepsilon)\tau_{LL'}^{ij}(\varepsilon)Z_{L'}^{(j)}(\underline{r}_j';\varepsilon) \quad \text{if } \underline{r} = \underline{r}_i + \underline{R}_i \text{ lies in muffin-tin } i, \text{ if } \underline{r}' = \underline{r}_j' + R_j \text{ lies in muffin-tin } j \neq i$$

$$= \sum_{LL'} Z_L^{(i)}(\underline{r}_i;\varepsilon)\tau_{LL'}^{ii}(\varepsilon)Z_{L'}^{(i)}(\underline{r}_i';\varepsilon) - \sum_L Z_L^{(i)}(\underline{r}_<,\varepsilon)J_L^{(i)}(\underline{r}_>,\varepsilon) \quad \text{if both } \underline{r} = \underline{r}_i + \underline{R}_i, \text{ and } \underline{r}' = \underline{r}_i' + \underline{R}_i \text{ lie in muffin-tin } i. \ \underline{r}_> \text{ is the greater of } \underline{r}_i, \underline{r}_i' \tag{11}$$

Here $Z_L^{(i)}(\underline{r},\varepsilon)$ and $J_L^{(i)}(\underline{r},\varepsilon)$ are respectively regular and irregular solutions of the Schrödinger equation containing only the single muffin-tin potential $v_i(\underline{r})$. They are real functions, and at the muffin-tin radius match smoothly on to the following exterior solutions :

$$Z_L^{(i)}(\underline{r},\varepsilon) = \kappa(n_\ell(\kappa r) - j_\ell(\kappa r)\cot\delta_{i,\ell}(\varepsilon))Y_L(\hat{\underline{r}})$$

$$J_L^{(i)}(\underline{r},\varepsilon) = j_\ell(\kappa r)Y_L(\hat{\underline{r}}) \qquad r > \text{muffin-tin radius}$$

where j_ℓ and n_ℓ are spherical Bessel and Neumann functions respectively, and $Y_L(\hat{\underline{r}})$ is a real spherical harmonic.

Now for x-ray absorption and emission, equations (4) and (6) show that, since the core state wave function $\phi_i(\underline{r})$ is, to a good approximation, localised within the muffin-tin sphere at site i (I use the label i to cover both the quantum numbers of the core state and its location - no ambiguities arise), we need $G^+(\underline{r},\underline{r}';\varepsilon_i + \omega)$ only for $\underline{r}$ and $\underline{r}'$ both lying within this sphere. Hence we can use equation (11) and find

$$W_i^{q\lambda} = -\frac{1}{\Gamma}\sum_{LL'} m_L^i(\varepsilon_i + \omega)\,\mathrm{Im}\tau_{LL'}^{ii}(\varepsilon_i + \omega)m_{L'}^{i\star}(\varepsilon_i + \omega) \quad \begin{matrix}\theta(\varepsilon_i+\omega-\mu)\text{absorption} \\ \theta(\mu-\varepsilon_i-\omega)\text{emission}\end{matrix} \tag{12}$$

where

$$m_L^i(\varepsilon_i + \omega) = \int d\underline{r}\ \phi_i^\star(\underline{r})X_{\underline{q}\lambda}(\underline{r})Z_L^{(i)}(\underline{r},\varepsilon_i + \omega) \tag{13}$$

For photoemission it is convenient to use the so-called acceleration form for the electron-phonon interaction

$$X_{\underline{q}\lambda}(\underline{r}) = -i\left(\frac{2\pi\hbar^3 e^2}{m^2\Omega\omega^3}\right)^{1/2} e^{-i\underline{q}\cdot\underline{r}}\, \underline{e}^{\lambda}(\underline{q})\cdot\nabla V(\underline{r})$$

$$= \sum_i \underline{c}\,(\omega)\cdot\nabla v_i(\underline{r} - \underline{R}_i)e^{i\underline{q}\cdot\underline{r}} = \sum_i X^{(i)}_{\underline{q}\lambda}(\underline{r}) \qquad (14)$$

where I have used the muffin-tin form, equation (9), for the total potential, and have lumped together all the fundamental constants, polarisation etc. in the quantity $\underline{c}(\omega)$, which I now regard as being given in atomic units. Since the interactions occur within muffin-tins we need to know the functions $\psi_+(\underline{r},\varepsilon)$, $\psi_+^{\star}(\underline{r}',\varepsilon)$ and $G^+(\underline{r},\underline{r}';\varepsilon - \omega)$ for $\underline{r}$ and $\underline{r}'$ lying within muffin-tins. For the retarded Green's function we can apply equations (11). For the LEED state it is straightforward to show that when $\underline{r}$ lies within muffin-tin i ($\underline{r} = \underline{r}_i + \underline{R}_i$) we can write

$$\psi_+(\underline{r},\varepsilon) = 4\pi \sum_n e^{i\underline{k}\cdot\underline{R}_n} \sum_{LL'} i^{-\ell}\, Y_L(-\hat{\underline{k}})\tau^{ni}_{LL'}(\varepsilon)Z^{(i)}_{L'}(\underline{r}_i,\varepsilon) \qquad (15)$$

Putting all this together we get the following expression for the photocurrent

$$R^2 I^{\underline{q}\lambda}(\varepsilon) = \frac{-\hbar\kappa}{4\pi^2 m} \sum_{\substack{n,n'\\ i,j'}} \sum_{\substack{L_1L_2\\ L_3L_4}} \sum_{LL'} \{i^{-\ell_1}\, e^{i\underline{k}\cdot\underline{R}_n}\, Y_{L_1}(-\hat{\underline{k}})\tau^{ni}_{L_1L_2}(\varepsilon)$$

$$\times M^{(i)}_{L_2L}(\varepsilon,\omega)\,\mathrm{Im}\tau^{ij}_{LL'}(\varepsilon - \omega)(M^{(j)}_{L'L_3}(\varepsilon,\omega)\tau^{jn'}_{L_3L_4}(\varepsilon)i^{-\ell_4}e^{i\underline{k}\cdot\underline{R}_{n'}}$$

$$\times Y_{L_4}(-\hat{\underline{k}}))^{\star}\}\theta(\mu - \varepsilon + \omega)\theta(\varepsilon - \mu - \Phi) \qquad (16)$$

where

$$M^{(i)}_{LL'}(\varepsilon,\omega) = \underline{c}(\omega)\cdot\int d\underline{r}\; Z^{(i)}_L(\underline{r}_i,\varepsilon)e^{i\underline{q}\cdot(\underline{r}_i + \underline{R}_i)}\nabla v_i(\underline{r}_i)Z^{(i)}_{L'}(\underline{r}_i,\varepsilon - \omega) \qquad (17)$$

Equations (12) and (16) are the general independent particle expressions for x-ray absorption and emission and for photoemission - general because they are valid for any arrangement of non-overlapping scatterers. At this stage we can make the following remarks.

1. X-ray emission or absorption is simpler than photoemission ! This is because the former involves a localised core state which (therefore) cannot propagate between different sites ; it is a local probe which depends only on the site-diagonal τ-matrix. In contrast, photoemission involves a conduction band hole which can propagate between sites. This brings in the non-site-diagonal

τ-matrix, $\tau^{ij}_{LL'}(\varepsilon - \omega)$, and gives the possibility of extracting $\underline{k}$-dependent information from the spectra.

2. The angular momentum sums can be truncated at quite low values. For example, for $\tau^{ij}_{LL'}(\varepsilon - \omega)$ in equation (16), $\ell = 2$ would be sufficient for a transition metal. For $\tau^{ni}_{L_1L_2}(\varepsilon)$, $\ell = 4$ is generally necessary since ε can be quite large.

3. The matrix elements $m^i_L(\varepsilon_i + \omega)$ and $M^{(i)}_{LL'}(\varepsilon,\omega)$ are atomic quantities in the sense that they depend only on the muffin-tin potential at site i. They obey the usual dipole selection rules. All the solid state effects are contained in the τ-matrices. What do these "atomic" matrix elements look like ?

Consider first the x-ray matrix element $m^i_L(\varepsilon_i + \omega)$. This is generally a smooth unstructured function of energy, as typified by the 2p → d matrix element for Mo shown in Figure 13. Note that if in equation (13) we had used instead of $Z^{(i)}_L(\underline{r},\varepsilon_i + \omega)$ the usual regular solution $R_L(\underline{r},\varepsilon)$, i.e. that which matches on to

$$[\, j_\ell(\kappa r)\cos\delta_{i,\ell}(\varepsilon) - n_\ell(\kappa r)\sin\delta_{i,\ell}(\varepsilon)]\, Y_L(\hat{\underline{r}})$$

at the muffin-tin radius, the matrix element would have shown a prominent peak at $\varepsilon_i + \omega$ = d-resonance energy. This is because the amplitude of $R^{(i)}_L(\underline{r},\varepsilon)$ within the muffin-tin peaks at a resonance. However, the normalisation of $Z^{(i)}_L(\underline{r},\varepsilon)$ is such that this effect is divided out, and in my experience the matrix elements $m^i_L(\varepsilon_i + \omega)$ never contribute significant structure to x-ray spectra. [Note that, as defined in equation (13), the matrix elements m^i_L contain information about the direction and polarisation of the absorbed or emitted photon. Thus there is a selection rule not only on ℓ but also on m - recall that L ≡ ℓ,m. In practice, experiments often do not discriminate in either photon angle or polarisation. However, one can easily sum over polarisations and integrate over angles by making use of the following formula[20]

$$\sum_{\lambda=1}^{2} (e^\lambda_z(\underline{q}))^2 = \sin^2\theta$$

where θ is the angle between $\underline{q}$ and the z-axis. The resulting formulae contain selection rules on ℓ, but the m's are simply summed over].

In the same way, the photoemission matrix elements $M^{(i)}_{LL'}(\varepsilon,\omega)$ show no peaks around partial wave resonances. Even so, quite spectacular atomic effects can arise in these matrix elements, as figure 14 shows. The very strong dips in the 4d → f matrix elements for Pd and Ag, known in atomic physics as Cooper minima[21] are associated with the node in the 4d radial wave-function. Such effects can certainly be seen in photoemission spectra, as figure

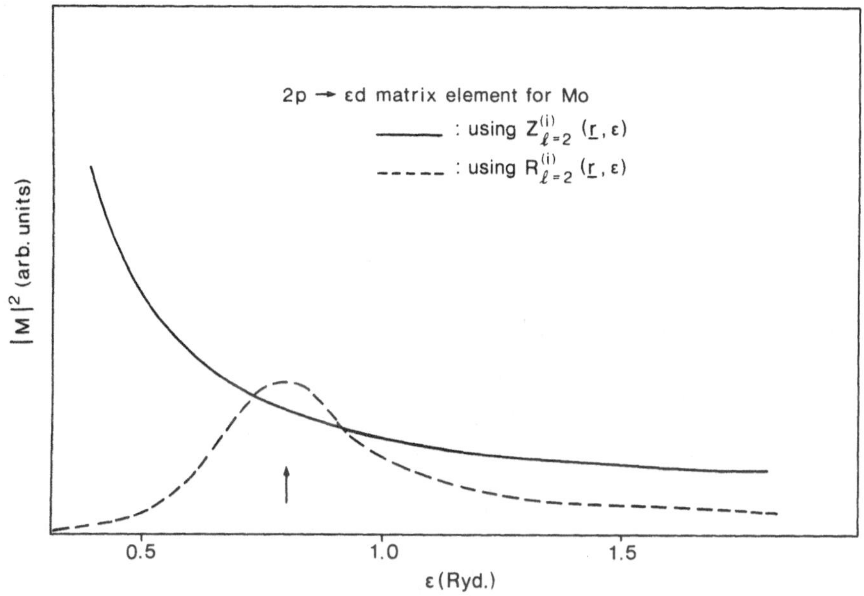

Figure 13. X-ray matrix element $m^{i}_{\ell=2}(E)$ for Mo. The arrow shows the energy of the 4d resonance.

15, taken from the work of Wehner et al.[22], shows. These data refer to photoemission from transition metals with adsorbed layers of CO, and show the intensity of d-band peaks relative to CO peaks as a function of photon energy.

It should be mentioned that all these matrix elements are very easily calculated since they only involve solutions of an atomic-like Schrödinger equation.

Equations (12) and (16) are the starting points we shall use to calculate x-ray and photoemission spectra, and in the next section I will indicate how they can be numerically evaluated. Let me conclude this section by showing somewhat schematically what kind of information the spectra contain.

First note that the local density of states at site i is given by

$$n^{(i)}(\varepsilon) = -\frac{2}{\pi}\int_{\text{cell}(i)} d\underline{r}\ \text{Im}\ G^{+}(\underline{r},\underline{r};\varepsilon)$$

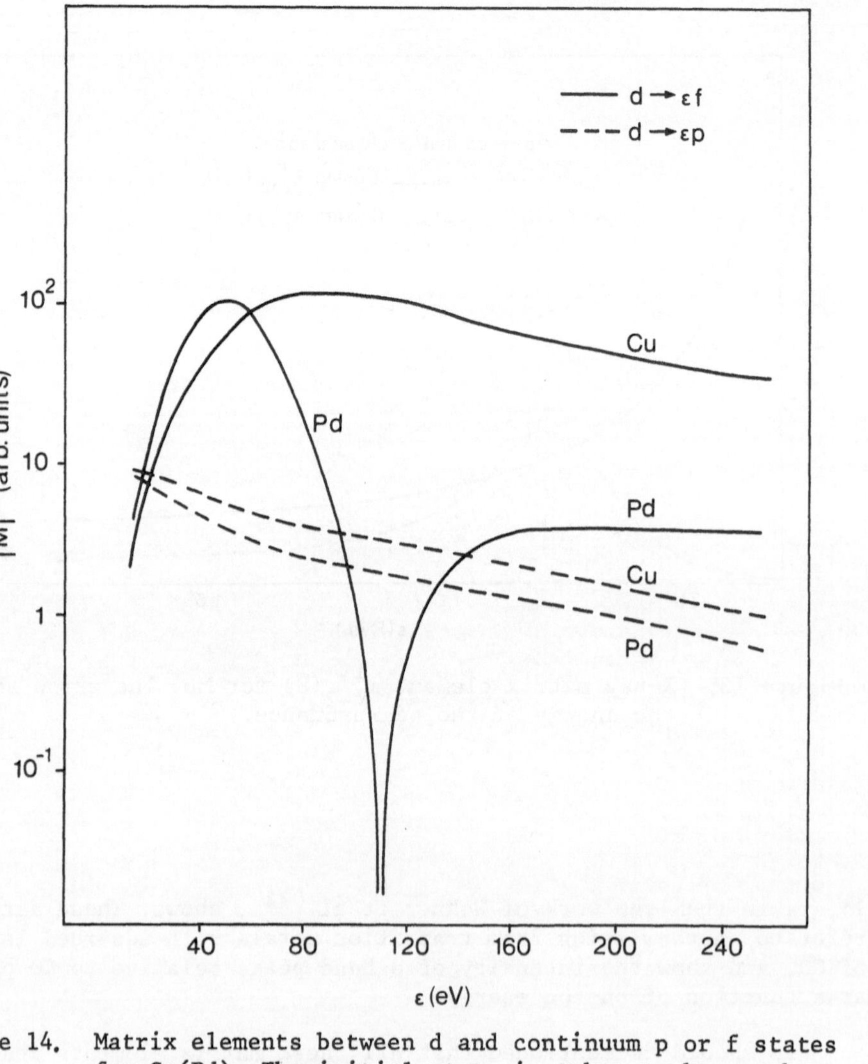

Figure 14. Matrix elements between d and continuum p or f states for Cu(Pd). The initial state is taken to be at the 3d(4d) resonance.

and that use of equation (11) immediately gives a decomposition into angular momenta

$$n^{(i)}(\varepsilon) = \sum_L n_L^{(i)}(\varepsilon) = \sum_L -\frac{2}{\pi}\int_{cell(i)} d\underline{r}\ (Z_L^{(i)}(\underline{r},\varepsilon))^2 \mathrm{Im}\tau_{LL}^{ii}(\varepsilon) \quad (18)$$

Thus, for systems in which $\tau_{LL'}^{ii}$ is diagonal in L, L' (e.g. for cubic systems, up to $\ell = 2$) we can write the x-ray transition

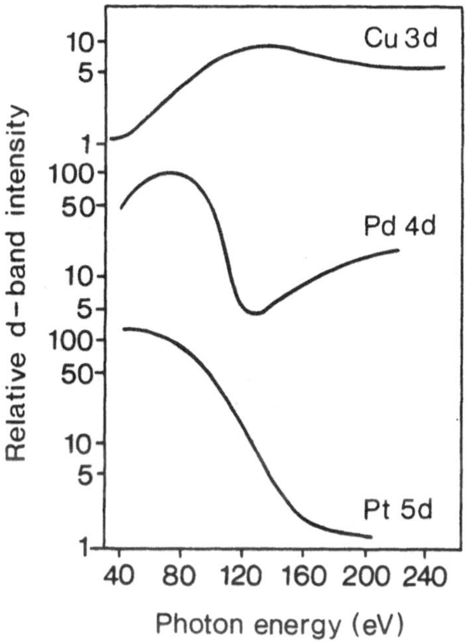

Figure 15. Taken from Wehner et al. (see text).

rate as

$$W_i^{q\lambda} = \frac{\pi}{2\Gamma} \sum_L |\bar{m}_L^i(\varepsilon_i + \omega)|^2 \, n_L^{(i)}(\varepsilon_i + \omega) \quad \begin{matrix} \theta(\varepsilon_i + \omega - \mu) \text{ absorption} \\ \theta(\mu - \varepsilon_i - \omega) \text{ emission} \end{matrix}$$

where

$$\bar{m}_L^i(\varepsilon_i + \omega) = \int d\underline{r}\; \phi_i^\star(\underline{r}) X_{\underline{q}\lambda}(\underline{r}) B_L^{(i)}(\underline{r}, \varepsilon_i + \omega)$$

and $B_L^{(i)}(\underline{r}) \propto Z_L^{(i)}(\underline{r})$ but is normalised to unity within cell i. Thus, x-ray emission (absorption) spectra measure the occupied (unoccupied) local density of states of a particular angular momentum, e.g. the p local density of states for K spectra (1s core state).

In x-ray absorption the photon energy can be made so high that the multiple scattering effects in $\tau_{LL'}^{ii}(\varepsilon_i + \omega)$ are weak. It then makes sense to truncate the t-matrix series for τ^{ii} at the single scattering term :

$$\tau^{ii}_{LL'}(\varepsilon) = t_{i,L}(\varepsilon)\delta_{LL'} + \sum_{j\neq i}\sum_{L''} t_{i,L}(\varepsilon)g^{ij}_{LL''}(\varepsilon)\tau^{ji}_{L''L'}(\varepsilon)$$

$$\cong t_{i,L}(\varepsilon)\delta_{LL'} + \sum_{j\neq i}\sum_{L''} t_{i,L}(\varepsilon)g^{ij}_{LL''}(\varepsilon)t_{j,L''}(\varepsilon)g^{ji}_{L''L'}(\varepsilon)t_{i,L'}(\varepsilon)$$

It is straightforward to show that if this truncated expression is used in equation (12), the standard formulae for EXAFS are obtained[23]. This approximate expression for $\tau^{ii}_{LL'}(\varepsilon)$ shows that the important information carried by the EXAFS part of the spectrum (i.e. a few hundreds of eV from the absorption edge) actually relates to the propagators $g^{ij}_{LL'}(\varepsilon)$; it is geometrical information about the arrangement of atoms in space. Indeed, EXAFS is now a highly developed technique for determining radial distribution functions in an enormous variety of materials[24].

To reveal what is measured in photoemission, I will first make some brutal assumptions. I will assume that the LEED state is a plane wave, $\psi_+(\underline{r},\varepsilon) \cong e^{i\underline{k}.\underline{r}}$, and I will forget about the surface. Then

$$R^2 I^{\underline{q}\lambda}(\varepsilon) \propto \sum_{i,j}\sum_{LL'} e^{i(\underline{k}-\underline{q}).(\underline{R}_i-\underline{R}_j)}\, \overline{M}^{(i)}_L \,\mathrm{Im}\tau^{ij}_{LL'}(\varepsilon-\omega)\overline{M}^{(j)\star}_{L'}$$

where

$$\overline{M}^{(i)}_L = \underline{c}(\omega).\int d\underline{r}_i\; e^{i(\underline{k}-\underline{q}).\underline{r}_i}\; \nabla v_i(\underline{r}_i)Z^{(i)}_L(\underline{r}_i,\varepsilon-\omega)$$

If the system is a perfect 1-component crystal then the matrix element is the same on each site and the positions $\underline{R}_i$, $\underline{R}_j$ fall on a regular lattice. Thus

$$R^2 I^{\underline{q}\lambda}(\varepsilon) \propto \sum_{LL'} \overline{M}_L \;\mathrm{Im}\;\tau_{LL'}(\underline{k}-\underline{q},\varepsilon-\omega)\overline{M}^{\star}_{L'}$$

where we have introduced the lattice Fourier transform of the τ-matrix

$$\tau_{LL'}(\underline{k},\varepsilon) = \frac{1}{N}\sum_{ij} e^{i\underline{k}.(\underline{R}_i-\underline{R}_j)}\;\tau^{ij}_{LL'}(\varepsilon) \tag{19}$$

The matrix $\underline{\tau}(\underline{k},\varepsilon)$ is a key quantity in KKR band theory[25]. As discussed by Stocks in this volume, for an ordered system its imaginary part has a number of sharp peaks defining the eigenvalues $\varepsilon_{\underline{k},\nu}$ for each band ν :

$$\mathrm{Im}\;\tau(\underline{k},\varepsilon) \sim \sum_{\nu} \delta(\varepsilon-\varepsilon_{\underline{k},\nu})$$

Thus the approximations used here imply that the photoemission spectrum will likewise contain sharp peaks

$$R^2 I^{\underline{q}\lambda}(\varepsilon) \sim \sum_{\nu} \delta(\varepsilon - \omega - \varepsilon_{\underline{k}-\underline{q},\nu})$$

which means that for a given photon energy and incidence angle (i.e. ω and $\underline{q}$) and for a given electron emission angle (i.e. $\underline{k}$), the current of photoelectrons peaks at energies for which $\varepsilon = \varepsilon_{\underline{k}-\underline{q},\nu} + \omega$, as illustrated in figure 16. These are the direct transitions (usually $q \ll k$) which figure so prominently in the interpretation of photoemission spectra, and which make it feasible to map out the band structure. (Actually, I have turned an oversimplification into a gross oversimplification by omitting to mention that measuring the electron emission angle only fixes the components of $\underline{k}$ parallel to the surface, $\underline{k}_{/\!/}$. Thus, direct transitions can be expected for any value of the normal component $k_\perp$ at which the condition $\varepsilon = \varepsilon_{\underline{k}-\underline{q},\nu} + \omega$ is satisfied).

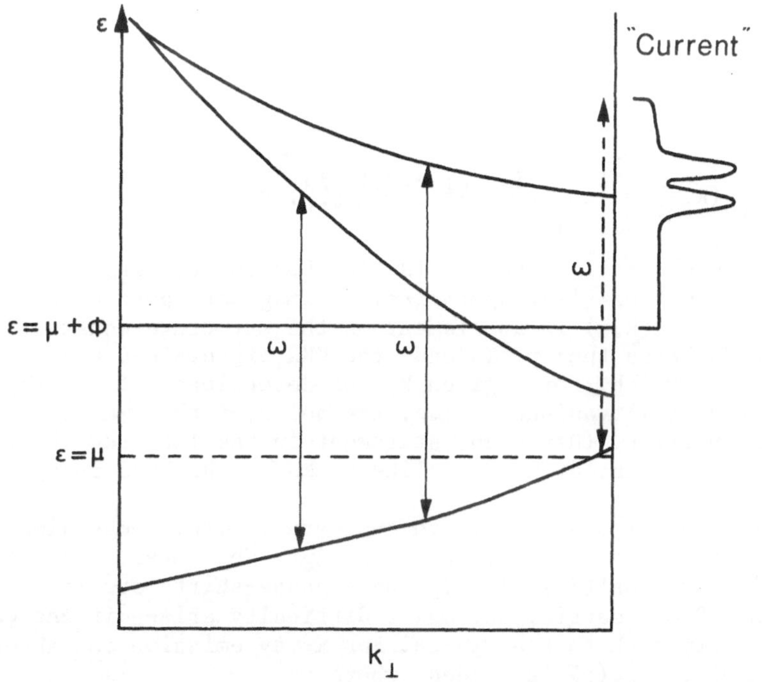

Figure 16. Schematic view of direct transitions.

The assumption of plane wave LEED states and neglect of the surface, while not totally ludicrous, are certainly not justified for angle-resolved UV photoemission. One must evaluate equation (16) in full to have a theory of photoemission worthy of the name.

However, it is useful to keep in mind : (1) that photoemission measures something in the nature of a spectral function, with peaks arising from individual bands, and (2) that x-ray emission and absorption spectroscopy measures particular angular momentum components of the local density of states at a given type of atom.

4. NUMERICAL CALCULATIONS

I now want to describe in more detail how realistic calculations of spectroscopic transition rates can be made, and to give some examples of such calculations. In the multiple scattering approach adopted in the last section, the key step is always the calculation of the scattering path matrix, $\tau^{ij}_{LL'}(\varepsilon)$, given by

$$\tau^{ij}_{LL'}(\varepsilon) = t_{i,L}(\varepsilon)\delta_{ij}\delta_{LL'} + \sum_{n\neq i}\sum_{L''} t_{i,L}(\varepsilon)g^{in}_{LL''}(\varepsilon)\tau^{nj}_{L''L'}(\varepsilon) \quad (20)$$

If the system is a perfect 1-component crystal (i.e. all t-matrices are the same and the vectors $\underline{R}_i$ lie on a regular lattice), then this equation reduces to

$$\tau^{-1}_{LL'}(\underline{k},\varepsilon) = t^{-1}_{L}(\varepsilon)\delta_{LL'} - G_{LL'}(\underline{k},\varepsilon) \quad (21)$$

where

$$G_{LL'}(\underline{k},\varepsilon) = \frac{1}{N}\sum_{ij} e^{i\underline{k}\cdot(\underline{R}_i - \underline{R}_j)}\, g^{ij}_{LL'}(\varepsilon)$$

and $\tau_{LL'}(\underline{k},\varepsilon)$ is likewise the lattice Fourier transform of $\tau^{ij}_{LL'}(\varepsilon)$, given in equation (19). Apart from an imaginary part equal to $\kappa = \varepsilon^{1/2}$, $G_{LL'}(\underline{k},\varepsilon)$ is an element of the structure constant matrix used in KKR band theory. Indeed, the KKR eigenvalues $\varepsilon_{\underline{k},\nu}$ are the energies at which, for a given $\underline{k}$, the determinant of the RHS of equation (21) vanishes ; i.e. the poles of the τ-matrix and/or Green's function. (Hence the statement in the last section that Im $\tau(\underline{k},\varepsilon)$ contains δ-function-like peaks at the band energies).

It is not too difficult to evaluate $\tau_{LL'}(\underline{k},\varepsilon)$-one simply (!) has to set up the KKR structure constants and invert the matrix given in equation (21). If s,p and d phase-shifts are included, this is a 9 X 9 matrix. The major difficulty arises if one wants to transform back to real space. For x-ray emission and absorption, for example, $\tau^{ii}_{LL'}(\varepsilon)$ is needed, where

$$\tau^{ii}_{LL'}(\varepsilon) = \Omega^{-1}_{BZ}\int_{BZ} d\underline{k}\; \tau_{LL'}(\underline{k},\varepsilon) \quad , \quad \Omega_{BZ} = \text{Brillouin zone volume} \quad (22)$$

If, as I have asserted, $\tau(\underline{k},\varepsilon)$ is a strongly peaked (indeed singular) function of $\underline{k}$, then it is clear that the Brillouin zone integral must be handled with care. (Of course, this is just the zone integral with which all band structure methods are faced in the course of calculating the density of states).

For local quantities like $\tau^{ii}_{LL'}(\varepsilon)$, however, it is feasible to make the calculation entirely in real space, using equation (20) directly. The idea is to represent the system by a finite cluster of atoms surrounding atom i. From equation (20) we have

$$\sum_{n}\sum_{L''}\left[t^{-1}_{i,L}(\varepsilon)\delta_{in}\delta_{LL''} - g^{in}_{LL''}(\varepsilon)(1-\delta_{in})\right]\tau^{nj}_{L''L'}(\varepsilon) = \delta_{ij}\delta_{LL'} \quad (23)$$

so that $\tau^{ij}_{LL'}(\varepsilon)$ is given by inverting a matrix with $t^{-1}_{i,L}$'s for each atom on the diagonal and $-g^{in}_{LL'}$'s off the diagonal. The greater the size of the cluster, the bigger the matrix, of course, but symmetry can be exploited to reduce its size[26]. The point of such cluster calculations is that although the truncation of the crystal at the cluster boundary is unphysical, a local quantity like $\tau^{ii}_{LL'}(\varepsilon)$ (giving, remember, the local density of states at site i) can be expected to be well represented if atom i lies at the centre of a sufficiently large cluster. Ideally, one would like to embed such clusters in a medium representing in some simple way the rest of the system, and several such schemes have been advanced[27]. However, one can just forget the region external to the cluster if one is careful to check the convergence of the calculation in shells of atoms. For a large enough cluster the correct answer is guaranteed. The great advantage of cluster calculations is that they can still be made when band structure calculations are either difficult (e.g. because of complicated unit cells) or impossible (e.g. because of broken translational symmetry).

A considerable number of calculations of x-ray emission and absorption spectra have now been made using k-space methods[28-33] the reader is referred to the original papers for details of the band structure calculations and zone integrals. A particularly detailed account of APW-based calculations for Ni has been given by Szmulowicz and Pease[34]. Real space cluster calculations have also been made for both crystalline[35] and disordered materials[36]. In figure 17 I show the results of cluster calculations[23] of the K shell (1s core state) absorption spectrum of Fe. This demonstrates the convergence of the calculation in atomic shells, and the final result for 5 or 6 shells compares well with both band theory calculations and experiment.

At this point I should mention some effects which, strictly speaking, fall outside the 1-electron theory, but which have to be considered when comparing theory with experiment. The first concerns the broadening due to the finite lifetime of the core hole and of conduction band holes (for emission) and electrons (for absorption). There is quite a lot of experimental and theoretical data on core hole widths. They usually arise mainly from

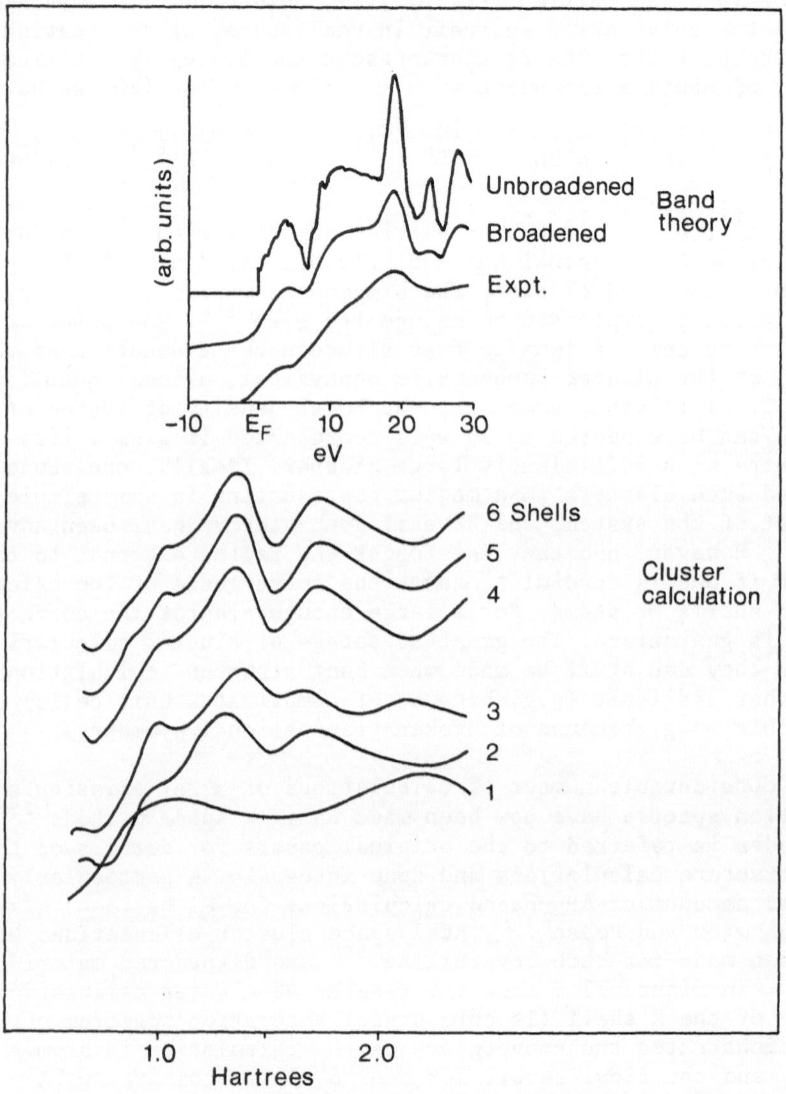

Figure 17. Cluster calculations of the K shell x-ray absorption spectrum of Fe. The inset is taken from Wakoh and Kubo.

Auger decay, and generally increase with the binding energy of the core level. Convenient tabulations exist[37] which show that the full width of the 1s state rises from about 0.8 eV for Z = 20 (Ca) to about 5 eV for Z = 40 (Zr). The 2p width rises from about 0.25 to 2 eV between Z = 20 and 40. Clearly these are very large broadenings, and will always obscure the fine details of the local density of states. This is perhaps not so bad for absorption spectra, which can span up to 10's of volts of interesting structure before the EXAFS limit is reached, but it certainly limits the amount of information in emission spectra to rather gross features of the electronic structure. As equation (3) shows, this broadening can be included in the calculations simply by folding the 1-particle spectrum with a Lorentzian of the appropriate width. The broadening due to finite conduction band lifetimes is usually included in the same way. These conduction band widths are less well understood, but can usually be expected to be a good deal smaller than core level widths. According to Fermi liquid theory the width should rise (quadratically) away from the Fermi level, and often a parameterisation along these lines of free electron calculations[38] is used. For transition metals, more sophisticated estimates based on the Hubbard model are becoming available, and these seem to give sensible results when incorporated into a realistic independent particle calculation[39]. In x-ray absorption the excited electron may have enough energy to excite a plasmon, and this decay mode no doubt affects its lifetime[40]. In general, quantitative results for conduction band lifetimes are scarce, and it is not unfortunate that in x-ray spectroscopy the broadening is usually dominated by the core level.

The second point is the effect of the potential due to the core hole on the spectra. The dynamical response of the conduction electrons to the core hole potential is a complex problem which has usually been tackled within the very simplified model used by Nozières and coworkers[9]. Very close to the threshold Nozières and De Dominicis[41] found an exact solution displaying the famous edge singularities, whose experimental significance has been thoroughly discussed by Citrin et al.[42]. Away from threshold the spectrum cannot be computed analytically even for simplified models, but numerical methods can be used. Out of such work the so-called final state rule has emerged[43], namely that the independent particle calculations which best agree with numerical solutions of the Nozières-De Dominicis equations are those which use the 1-electron potential appropriate to the final state (i.e. with a core hole for absorption and without for emission). It must be restated, though, that the calculations which motivate this rule use simplified models which may be appropriate for simple metals but are not obviously valid for more complex systems. In fact, band theoretical calculations of absorption spectra for transition metals seem to agree fairly satisfactorily with experiment without including the core hole potential at all[32,33]. On the other hand, for systems having localised unoccupied levels in the ground state (e.g. f levels in rare earths) the core hole potential is known to be important[44], and can give rise to new features in the spectrum. This problem is certainly worthy of further systematic study. The x-ray edge singularities, as such, seem only to be important in a few simple metals.

I now want to show how to apply the multiple scattering formalism to disordered systems, in particular the random substitutional alloys discussed at this school by Stocks. Consider an alloy A_xB_{1-x}, where x is the concentration of A atoms. This alloy does possess a regular crystalline lattice of sites, but each site is occupied by either an A or B atom with probability x and (1-x) respectively. One can still begin with equation (12), because it is valid for <u>any</u> arrangement of non-overlapping scatterers. The key point is that by identifying a particular core state one automatically specifies the type of atom in which it occurs. For example, the Cu 2p core states in a Cu-Ni alloy are well separated (by about 80 eV) from those of Ni. This means that to calculate the absorption or emission spectrum with core level c on atoms of type A (e.g. the $L_{2,3}$ spectra for Cu), equation (12) has to be averaged[45] over all configurations of the alloy which have an A atom at site i

$$\langle W_{A,c}(\omega)\rangle = -\frac{1}{\Gamma}\sum_{LL'} m^{A}_{L}(\varepsilon_c + \omega)\,\mathrm{Im}\langle\tau^{ii}_{LL'}(\varepsilon_c + \omega)\rangle_{Ai}\, m^{A\star}_{L'}(\varepsilon_c + \omega)$$

As indicated above, the matrix elements are atomic quantities in that they are completely determined by a single muffin-tin potential. By hypothesis there is just one such potential for all A atoms and one for all B atoms. So the main task is to calculate $\langle\tau^{ii}\rangle_{Ai}$, the 1-site restricted average of the τ-matrix over all configurations with an A atom at site i, and to do this we must invoke a theory of the electronic structure of random alloys. The best such theory appears at present to be the Korringa-Kohn-Rostoker coherent potential approximation, KKR-CPA for short. The way this approach works is described by Stocks in this volume, so I will just remark that the KKR-CPA very conveniently provides the 1-site restricted averages we need in the following form[46]

$$\langle\tau^{ii}_{LL'}(\varepsilon)\rangle_{Ai} = [\,\tau^{c,ii}(\varepsilon)(1 + (t_A^{-1} - t_c^{-1})\tau^{c,ii}(\varepsilon))^{-1}]_{LL'}$$

where $t_{A,L}(\varepsilon)$ is the A atom t-matrix, $t_{c,L}(\varepsilon)$ is the KKR-CPA effective t-matrix, and $\tau^{c,ii}_{LL'}(\varepsilon)$ is the τ-matrix for the (ordered) coherent lattice (i.e. that with $t_{c,L}(\varepsilon)$ on every site) :

$$\tau^{c,ii}_{LL'}(\varepsilon) = \Omega^{-1}_{BZ}\int_{BZ} d\underline{k}\,[\,t_c^{-1}(\varepsilon) - G(\underline{k},\varepsilon)]^{-1}_{LL'}$$

(cf. equations (21,22)). The hard part of this calculation is, as usual, the zone integration, and techniques for doing this have been detailed by Stocks et al.[47]. However, $\tau^{c,ii}_{LL'}(\varepsilon)$ has to be calculated anyway in the course of a KKR-CPA calculation to determine $t_{c,L}$, and so the evaluation of $\langle\tau^{ii}_{LL'}(\varepsilon)\rangle_{Ai}$ involves essentially no extra work.

This theory has been used quite successfully for Cu-Ni and Ag-Pd alloys[48]. X-ray spectroscopy is particularly useful in alloys (and multicomponent systems generally) because its local nature allows one to look at states on each type of atom separately. In figure 18 I show a nice example in which KKR-CPA-based calculations by Stocks[49] are compared with experimental x-ray emission data for $Li_{.8}Mg_{.2}$. Also shown are the s and p local densities of states, and their relationship to the emission spectra is clear. A relativistic extension of this approach has recently been made[50] and results for heavy systems are now becoming available.

In summary, as a tool for examining the extended states in solids x-ray spectroscopy has the following advantages : it's local, it's angular momentum resolved and it's relatively simple both experimentally and theoretically (at least at the 1-electron level). But it suffers from the major and intrinsic disadvantage that the involvement of a core hole broadens the spectra very considerably, and usually obliterates fine details of the density of states.

I now want to discuss calculations of angle-resolved photoemission spectra. I have no space to do justice to the current experimental situation in this very active field, and must refer the reader to recent review articles in the literature[51]. What I will do is to outline the salient features of the kind of calculation with which I have been involved, and show some results. I will also briefly describe how the multiple scattering approach may be applied to random substitutional alloys, and here the benefits of a full theory will be apparent.

The angle-resolved photoemission experiment requires a single crystal cleaved at a low index surface whose normal defines the z-direction in a system of coordinates. The photons are incident on this surface with specified energy, polarisation and direction, and the current of photoelectrons emitted with a certain energy and direction is measured. It is obvious that the surface is intrinsic to this experiment. Furthermore, since a number of inelastic processes impart to the photoelectron a finite mean free path, only a few atomic layers close to the surface can actually contribute to the photocurrent. (I have left the number of contributing layers deliberately vague, but it seems generally to be a number more like 10 than 100. A notorious "universal curve" of mean free path versus electron energy exists[52], but it is hard to say how universal or accurate it really is). The surface not only modifies

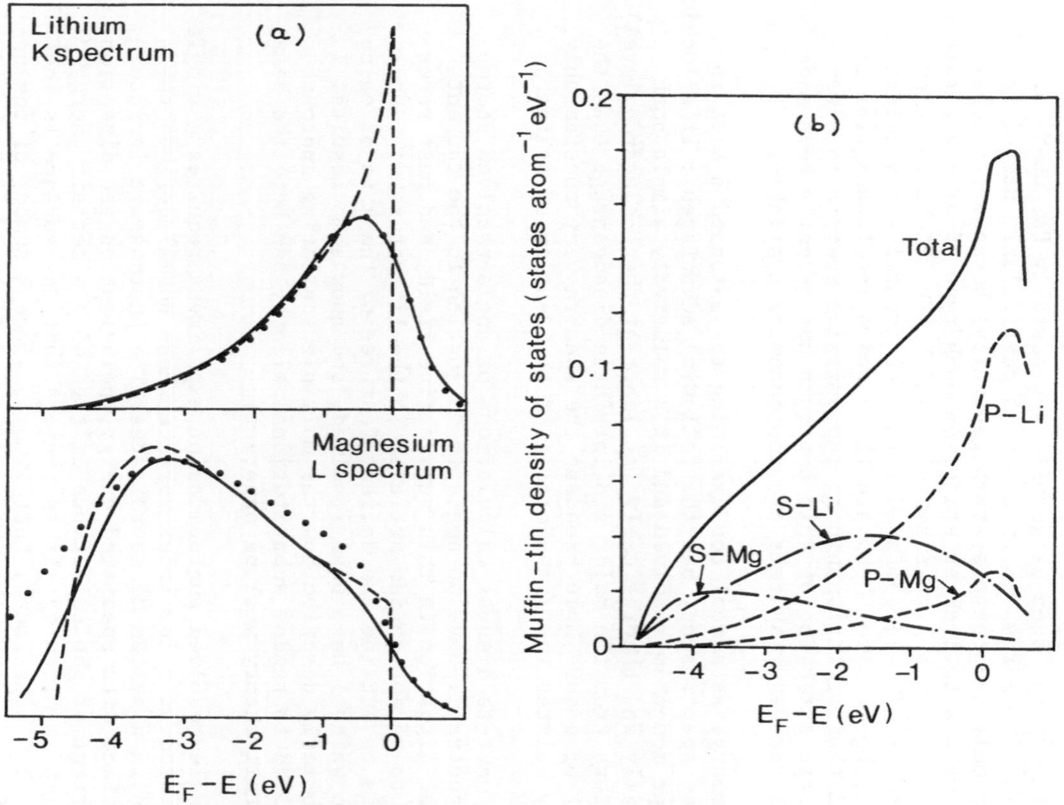

Figure 18. Taken from Stocks et al. (a) Calculated (full line) and experimental (dots) Li K and Mg L-soft x-ray emission spectra in $Li_{0.8}Mg_{0.2}$ alloys. The theoretical unbroadened spectra is given by the dashed lines. (b) Total and component muffin-tin densities of states in $Li_{0.8}Mg_{0.2}$ alloys.

the electronic structure in its vicinity (perhaps giving rise to surface states), but also acts as an extra source of photoemission, since it contributes an extra term, the surface barrier potential, to $V(\underline{r})$ in equation (14). This surface photoemission was investigated in model calculations by Schaich and Ashcroft[11] who found that it was not only comparable to but also interfered with the bulk photoemission originating in the underlying crystal. Thus the surface must be included in any realistic theory of photoemission, and in all the numerical calculations quoted below this is done in the following way. The crystal is described by a truncated (i.e. semi-infinite) array of non-overlapping muffin-tin potentials, augmented by a surface barrier potential which is flat in the x,y plane and a step function in the z-direction (see figure 19). The step is located so as to touch the top layer of muffin-tins, and its height is taken to be the sum of the Fermi energy and the work function.

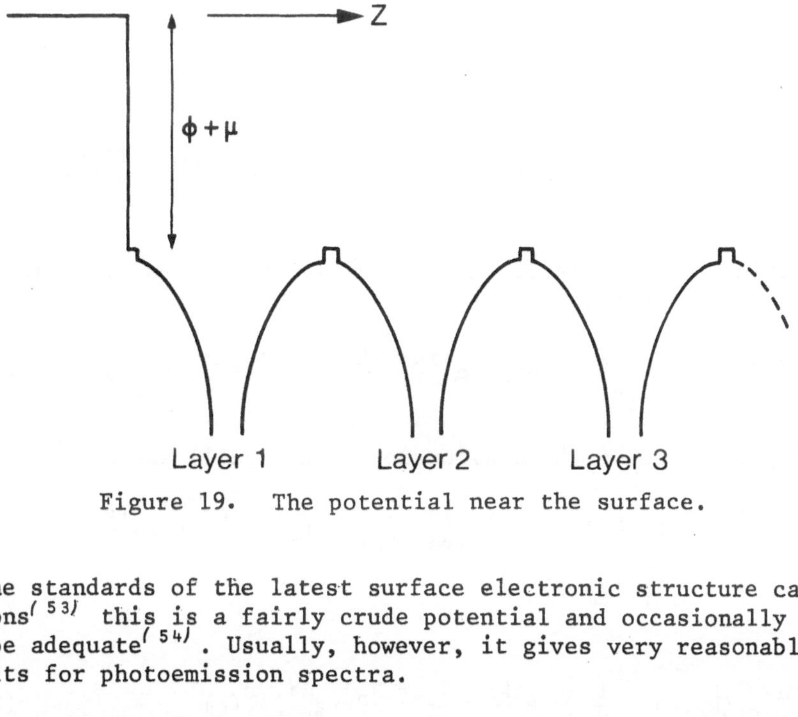

Figure 19. The potential near the surface.

By the standards of the latest surface electronic structure calculations[53] this is a fairly crude potential and occasionally may not be adequate[54]. Usually, however, it gives very reasonable results for photoemission spectra.

In equation (16) I gave a multiple scattering expression for the angle-resolved photocurrent which can be applied immediately to the potential just described. In fact, it gives only that part of the photocurrent which originates in the lattice, but the surface terms can be included in a straightforward way. For details of this and other aspects of a fairly complex calculation, I refer the reader to the original papers of Pendry[55], whose approach has been used or extended in all the work I shall subsequently quote. However, it is clear that the major difficulty in evaluating equation (16) for the above model is the breaking of translational symmetry in the z-direction, and I can show how this is handled relatively briefly.

The key ingredients in equation (16) are the τ-matrices ; thus, what we want is an economical way of finding $\tau^{ij}_{LL'}(\varepsilon)$ for a semi-infinite lattice. Of course, the real space equation (20) for it is still valid, but we should try to exploit the presumed lattice symmetry parallel to the layers. To do this, it is helpful to refine notation a little. Each layer will be labelled with a Greek letter α,β,γ etc., and within each layer the atomic positions will be labelled i,j, etc. Thus the position of atom (α,i) is $\underline{R}_\alpha + \underline{R}_i$, where $\underline{R}_\alpha$ is the origin of a coordinate system for layer α. The intralayer positions $\underline{R}_i$ are assumed to lie on a 2-dimensional regular lattice, and to be the same for each layer. This is the simplest case, but generalisations are easy to write down. We also suppose that all the muffin-tin potentials in a given layer are the same (but may differ from those in other layers), and are described by a t-matrix $t_{\alpha,L}$. The full τ-matrix of the system is thus

$$\tau^{\alpha i,\beta j}_{LL'} = t_{\alpha,L}\delta_{\alpha\beta}\delta_{ij}\delta_{LL'} + \sum_{\substack{(\gamma,n)\\ \neq(\alpha,i)}} \sum_{L''} t_{\alpha,L}\, g^{\alpha i,\gamma n}_{LL''}\, \tau^{\gamma n,\beta j}_{L''L'}$$

Now lattice Fourier transform parallel to the planes (i.e. in indices i,j) to obtain

$$\tau^{\alpha i,\beta j}_{LL'}(\varepsilon) = \frac{a}{(2\pi)^2}\int d\underline{k}_{/\!/}\; e^{-i\underline{k}_{/\!/}\cdot(\underline{R}_i - \underline{R}_j)}\; \tau^{\alpha\beta}_{LL'}(\underline{k}_{/\!/},\varepsilon) \tag{24}$$

(a being the area of the 2-dimensional unit cell) where

$$\tau^{\alpha\beta}_{LL'}(\underline{k}_{/\!/},\varepsilon) = \tau^{\alpha}_{LL'}(\underline{k}_{/\!/},\varepsilon)\delta_{\alpha\beta} + \sum_{\gamma\neq\alpha}\sum_{L_1L_2} \tau^{\alpha}_{LL_1}(\underline{k}_{/\!/},\varepsilon)\tilde{g}^{\alpha\gamma}_{L_1L_2}(\underline{k}_{/\!/},\varepsilon) \times \tau^{\gamma\beta}_{L_2L'}(\underline{k}_{/\!/},\varepsilon) \tag{25}$$

with

$$\tilde{g}^{\alpha\gamma}_{L_1L_2}(\underline{k}_{/\!/},\varepsilon) = \frac{1}{N}\sum_{in} e^{i\underline{k}_{/\!/}\cdot(\underline{R}_i - \underline{R}_n)}\; g^{\alpha i,\gamma n}_{L_1L_2}(\varepsilon) \tag{26}$$

and

$$\tau^{\alpha}_{LL'}(\underline{k}_{/\!/},\varepsilon) = \frac{1}{N}\sum_{ij} e^{i\underline{k}_{/\!/}\cdot(\underline{R}_i - \underline{R}_j)}\; \tau^{\alpha,ij}_{LL'}(\varepsilon) \tag{27}$$

$\tau^{\alpha,ij}_{LL'}$ is the τ-matrix for layer α alone. By analogy with equation

(21) we have

$$(\tau^{\alpha}_{LL'}(\underline{k}_{/\!/},\varepsilon))^{-1} = t^{-1}_{\alpha,L}(\varepsilon)\delta_{LL'} - G_{LL'}(\underline{k}_{/\!/},\varepsilon) \qquad (28)$$

Thus equations (24) to (28) show that we have a 2-dimensional KKR calculation within each layer, followed by a kind of 1-dimensional "cluster" calculation (i.e. in real space) for the scattering between the planes. This general structure is an inevitable consequence of breaking lattice symmetry in the z-direction. Actually, there is an alternative way of doing the 1-dimensional real space calculation which involves somewhat more physical quantities, and which is in fact used in the numerical calculations. Instead of describing the scattering by a single layer in terms of the τ-matrix $\tau^{\alpha,ij}_{LL'}(\varepsilon)$, we define transmission and reflection matrices which describe how the layer scatters plane waves. The basic formula which allows this to be done is ($h^{(+)}_{\ell}(\kappa r)$ = Hankel function)

$$\sum_i h^{(+)}_{\ell}(\kappa|\underline{r} - \underline{R}_i|)Y_L(\underline{r}-\underline{R}_i)e^{i\underline{k}_{/\!/}\cdot\underline{R}_i} = \begin{cases} \displaystyle\sum_{\underline{g}} \frac{-2\pi i^{-\ell}}{\kappa k_{\underline{g}z} a} Y_L(\hat{\underline{k}}^{+}_{\underline{g}})e^{i\underline{k}^{+}_{\underline{g}}\cdot\underline{r}} , & Z > 0 \\ \displaystyle\sum_{\underline{g}} \frac{2\pi i^{-\ell}}{\kappa k_{\underline{g}z} a} Y_L(\hat{\underline{k}}^{-}_{\underline{g}})\, e^{i\underline{k}_{\underline{g}}\cdot\underline{r}} , & Z < 0 \end{cases}$$

where $\underline{g}$ is a 2-dimensional reciprocal lattice vector of the layer, and the plane waves are defined as follows :

$$\underline{k}^{\pm}_{\underline{g}} = (\underline{k}_{/\!/} + \underline{g} , \pm k_{\underline{g}z}) \; ; \; k_{\underline{g}z} = (\kappa^2 - |\underline{k}_{/\!/} + \underline{g}|^2)^{1/2}$$

This equation describes how a set of spherical waves emerging with a Bloch-type phase factor from each atom of the layer adds up to give a set of diffracted plane waves, or beams. It is Huygens' principle for electrons. In the photoemission experiment, measuring the electron's emission angle means fixing its parallel momentum, $\underline{k}_{/\!/}$, and multiple scattering by the layers conserves $\underline{k}_{/\!/}$ modulo $\underline{g}$. Thus, the idea is to represent the waves scattered by a layer not by an angular momentum expansion about each atom, but by a plane wave expansion whose members have wave vectors $\underline{k}^{\pm}_{\underline{g}}$. Some of these waves are travelling in the +ve z direction ($\underline{k}^{+}_{\underline{g}}$), the rest in the -ve z direction ($\underline{k}^{-}_{\underline{g}}$). They all have energy $\varepsilon = \kappa^2$, so that $k_{\underline{g}z}$ can be real or imaginary, depending on $|\underline{k}_{/\!/} + \underline{g}|$. If $k_{\underline{g}z}$ is imaginary the propagation in the z-direction is damped, and this allows us to limit the number of $\underline{g}$'s involved in the expansion. We need all $\underline{g}$'s for which $k_{\underline{g}z}$ is real, and those for which $k_{\underline{g}z}$ is imaginary but small. For close-packed layers about 17 beams seems adequate, even to quite high photon energies. The scattering of these plane waves into one another by a layer is described by a transmission and a reflection matrix, as shown schematically in figure 20, and these are given in terms of the layer τ-matrices by

$$T^{\alpha}_{\underline{gg}'}(\underline{k}_{/\!/},\varepsilon) = \frac{8\pi^2 i}{a\, k_{\underline{g}z}} \sum_{LL'} i^{\ell}\, Y_L(\hat{\underline{k}}^{+}_{\underline{g}})\,\tau^{\alpha}_{LL'}(\underline{k}_{/\!/},\varepsilon)\,Y_{L'}(\hat{\underline{k}}^{+}_{\underline{g}})\,i^{-\ell'}$$
$$R^{\alpha}_{\underline{gg}'}(\underline{k}_{/\!/},\varepsilon) = -\frac{8\pi^2 i}{a\, k_{\underline{g}z}} \sum_{LL'} i^{\ell}\, Y_L(\hat{\underline{k}}^{-}_{\underline{g}})\,\tau^{\alpha}_{LL'}(\underline{k}_{/\!/},\varepsilon)\,Y_{L'}(\hat{\underline{k}}^{+}_{\underline{g}})\,i^{-\ell'} \qquad (29)$$

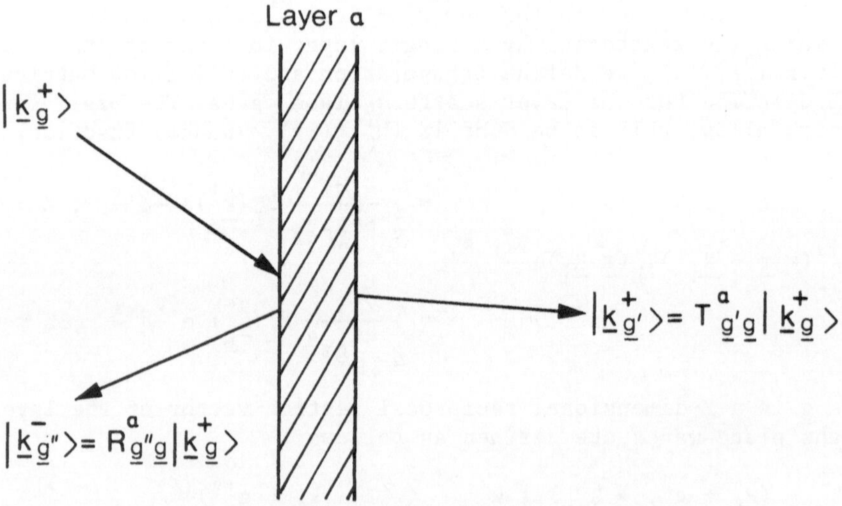

Figure 20. Schematic representation of layer transmission and reflection matrices.

Similarly one can define matrices $T^{\alpha\beta}_{\underline{gg}'}$ and $R^{\alpha\beta}_{\underline{gg}'}$, which give the amplitudes of plane waves $|\underline{k}^{+}_{\underline{g}'}\rangle$ emerging from layer β, given waves $|\underline{k}^{+}_{\underline{g}}\rangle$ incident on layer α. The equation for $T^{\alpha\beta}$ and $R^{\alpha\beta}$ in terms of T^{α} and R^{α} is as follows

$$\underline{\underline{M}}^{\alpha\beta}_{\underline{gg}'}(\underline{k}_{/\!/},\varepsilon) = \underline{\underline{M}}^{\alpha}_{\underline{gg}'}(\underline{k}_{/\!/},\varepsilon)\,\delta_{\alpha\beta} + \sum_{\underline{g}_1\underline{g}_2}\sum_{\gamma\neq\alpha} \underline{\underline{M}}^{\alpha}_{\underline{gg}_1}(\underline{k}_{/\!/},\varepsilon).\underline{P}^{\alpha\gamma}_{\underline{g}_1\underline{g}_2}(\underline{k}_{/\!/},\varepsilon)\,\underline{\underline{M}}^{\gamma\beta}_{\underline{g}_2\underline{g}'}(\underline{k}_{/\!/},\varepsilon) \qquad (30)$$

where

$$\underline{\underline{M}}^{\alpha\beta}_{\underline{gg}'} = \begin{pmatrix} T^{\alpha\beta}_{\underline{gg}'} & R^{\alpha\beta}_{\underline{gg}'} \\ R^{\alpha\beta}_{\underline{gg}'} & T^{\alpha\beta}_{\underline{gg}'} \end{pmatrix}, \quad \underline{\underline{M}}^{\alpha}_{\underline{gg}'} = \begin{pmatrix} T^{\alpha}_{\underline{gg}'} & R^{\alpha}_{\underline{gg}'} \\ R^{\alpha}_{\underline{gg}'} & T^{\alpha}_{\underline{gg}'} \end{pmatrix}$$

and

$$P^{\alpha\beta}_{\underline{gg}'} = \delta_{\underline{gg}'} \begin{pmatrix} e^{i\underline{k}^{+}_{\underline{g}}\cdot(\underline{R}_{\beta} - \underline{R}_{\alpha})} & 0 \\ 0 & e^{i\underline{k}^{-}_{\underline{g}}\cdot(\underline{R}_{\beta} - \underline{R}_{\alpha})} \end{pmatrix}$$

Thus $\underline{\underline{M}}^{\alpha\beta}_{\underline{gg}'}$ is again given by a kind of 1-dimensional cluster calculation in real space, with layer scattering matrices $\underline{\underline{M}}^{\alpha}_{\underline{gg}'}$ and propagators $P^{\alpha\beta}_{\underline{gg}'}$. The difference between equation (30) and equation (25) is that the propagators $\tilde{g}^{\alpha\beta}_{LL'}(\underline{k}_{/\!/},\varepsilon)$ require a little work to evaluate for each $\underline{k}_{/\!/}$, while $\underline{\underline{P}}^{\alpha\beta}_{\underline{gg}'}$ is trivial to set up. What I have given here, in fact, is a version of multiple scattering LEED theory[56]. The gist of what we do, then, is to recast the formula for the photocurrent, equation (16), in terms of $\tau^{\alpha\beta}_{LL'}(\underline{k}_{/\!/},\varepsilon)$ defined in equation (25). This is related, by the inverse of the transformation in equation (29), to $\underline{\underline{M}}^{\alpha\beta}_{\underline{gg}'}(\underline{k}_{/\!/},\varepsilon)$, which in turn is determined by equation (30). Once translational symmetry is broken, the inclusion of layers of different types of atoms (e.g. overlayers on clean metal surfaces) presents no further difficulty, so the method is quite flexible in this respect. For details of the convergence of the interlayer multiple scattering series, I refer the reader to the original work of Pendry.

This, in outline, is how the electronic part of the calculation is done. Now what about the photon field $\underline{A}(\underline{r})$? I have scarcely mentioned this so far, and have in fact tacitly assumed it to be unmodified from its form in the vacuum. Actually, the vector potential is modified in a solid, and in particular can undergo quite strong spatial variations at a metallic surface[57]. Feibelman and coworkers[58] have investigated this effect in Al, using calculations based on a jellium surface model, and have found that the intensity of surface state emission varies with photon energy in a way determined by the ω-dependence of the dielectric function of the metal. Such effects should presumably always be present in the relative intensities of "surface" to "bulk" transitions in photoemission, although they have only been seen (to my knowledge) in simple metals. They have not been taken into account in the calculations I shall describe.

Spectral broadening due to finite hole lifetimes must be included in photoemission calculations, however, just as in calculations of x-ray spectra. But the lifetimes of conduction band holes are generally much longer than those of core holes, and the broadening correspondingly smaller (often much less than 1 eV). The photoelectrons, on the other hand, can suffer quite large broadenings (4 eV or so), especially if they are sufficiently energetic to be able to excite plasmons. This final state effect does not show up as a broadening in energy, but rather in $\underline{k}$, i.e. in the angular structure of the spectra[59]. These broadenings can conveniently be included in the calculations by means of imaginary optical potentials, as is common practice in LEED analyses.

I can now illustrate the kinds of results produced by this approach. The simplest way of observing the dispersion of the initial state bands is to look at normal emission spectra for a number of different photon energies (see figure 16). Figures 21 and 22 show calculations, done by C.-G. Larsson, of such normal emission spectra for the (111) surfaces of Cu and Pd respectively, and peaks corresponding closely to bulk transitions in the Λ direction can clearly be seen. For Cu the peak close to the Fermi level is a surface state[60], and that at about 6 eV binding energy is associated with an initial state close to the L_2 state. The strong ω-dependence of this peak can be traced to a final state band which becomes accessible at $\sim$ 70 eV, and is not to be confused with a many-body satellite[61].

The dependence of the spectra on the electron emission angle is illustrated by figure 23, showing calculations and experimental data for the (100) face of Cu[62]. As the emission angle increases away from the normal a great wealth of structure unfolds, most of which can be correlated with bulk bands in the appropriate region of the Brillouin zone. It would take too long to demonstrate this in detail, but I have arrowed the peaks in the spectra corresponding to the upper part of the s-p band, and its characteristic behaviour as a function of emission angle can clearly be seen. This behaviour is summarised in figure 24, showing excellent agreement between theory and experiment (the movement of this peak is quite interesting - for a detailed explanation of why the curve turns around at $\theta_e \sim 30°$ see the paper by Durham and Kar[63].

This approach to photoemission can be generalised to the case of random substitutional alloys, as I shall now indicate. This generalisation is more tricky than for the x-ray emission and absorption calculations described above, because the general formula, equation (16), contains not one but three τ-matrices whose product has to be configurationally averaged. Thus, if this product is decoupled (i.e. the average of the product is approximated by the product of the averages), a vertex correction should be applied. In discussing this problem[64], I argued that because the disorder scattering felt by a high energy photoelectron is very much less than that felt by, say, a d-electron in a transition metal alloy, this vertex correction should be small, and I found that calculations which omit it do give reasonable results.

After making this decoupling the quantity to be averaged is of the form $\sum_{ij} M_i \,\mathrm{Im}\, \tau^{ij} M_j^*$, where M_i is a matrix element between

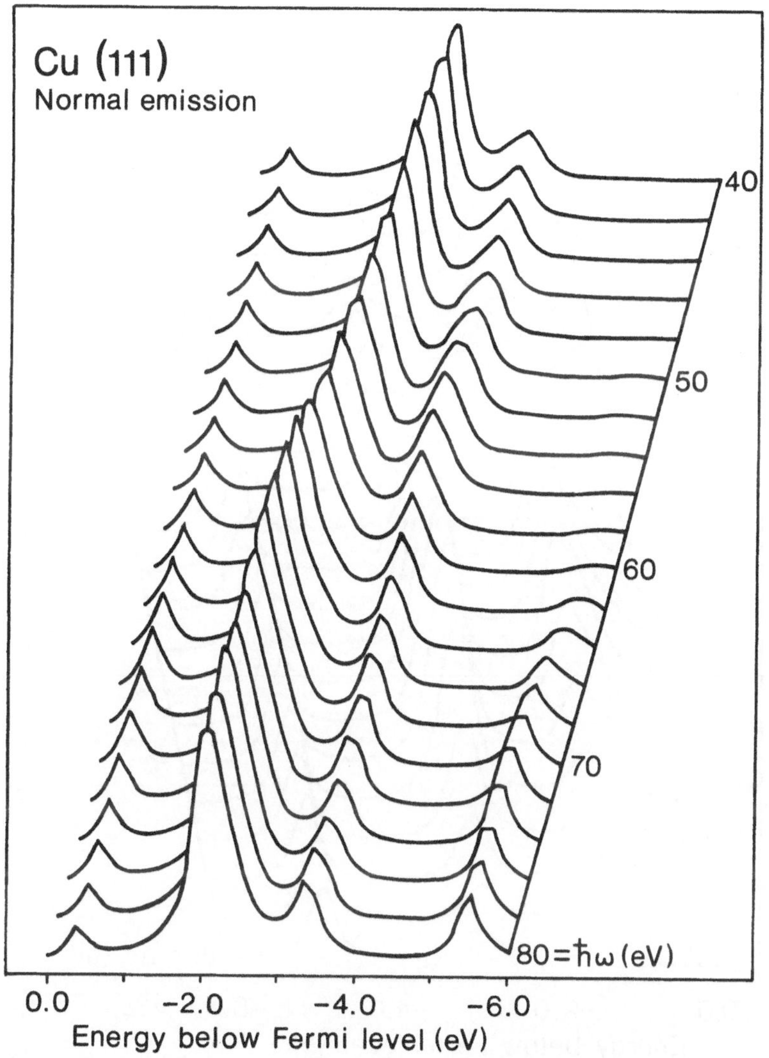

Figure 21 : ω-dependent photoemission calculations for Cu (111).

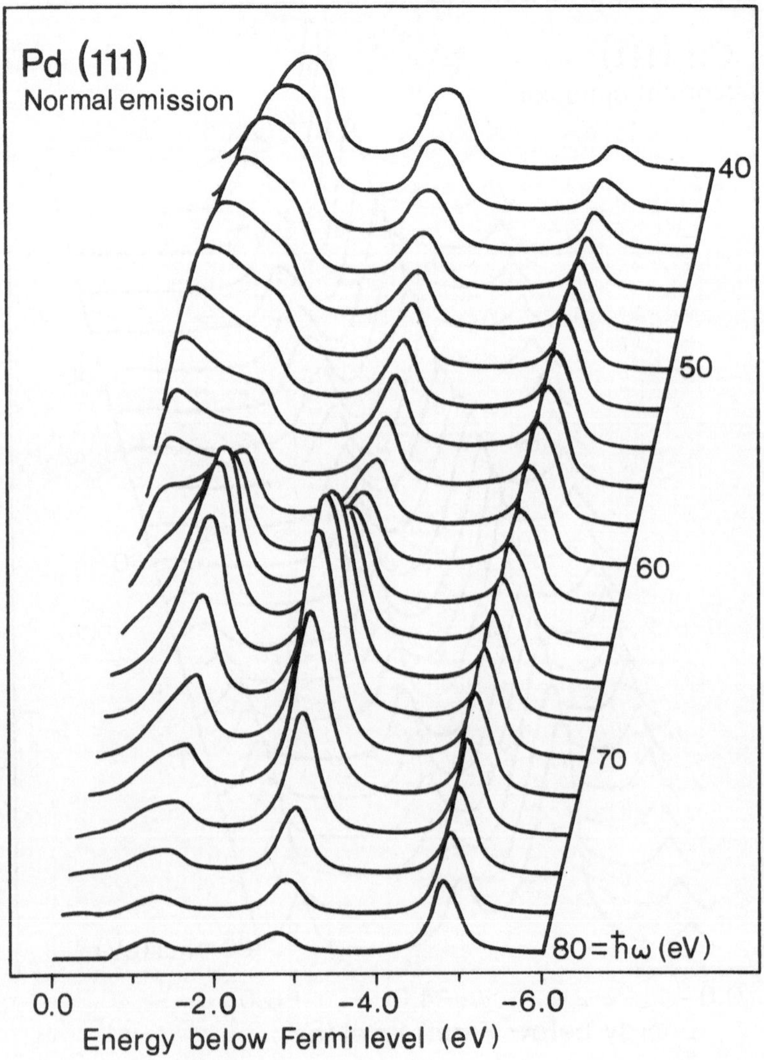

Figure 22. ω-dependent photoemission calculations for Pd(111).

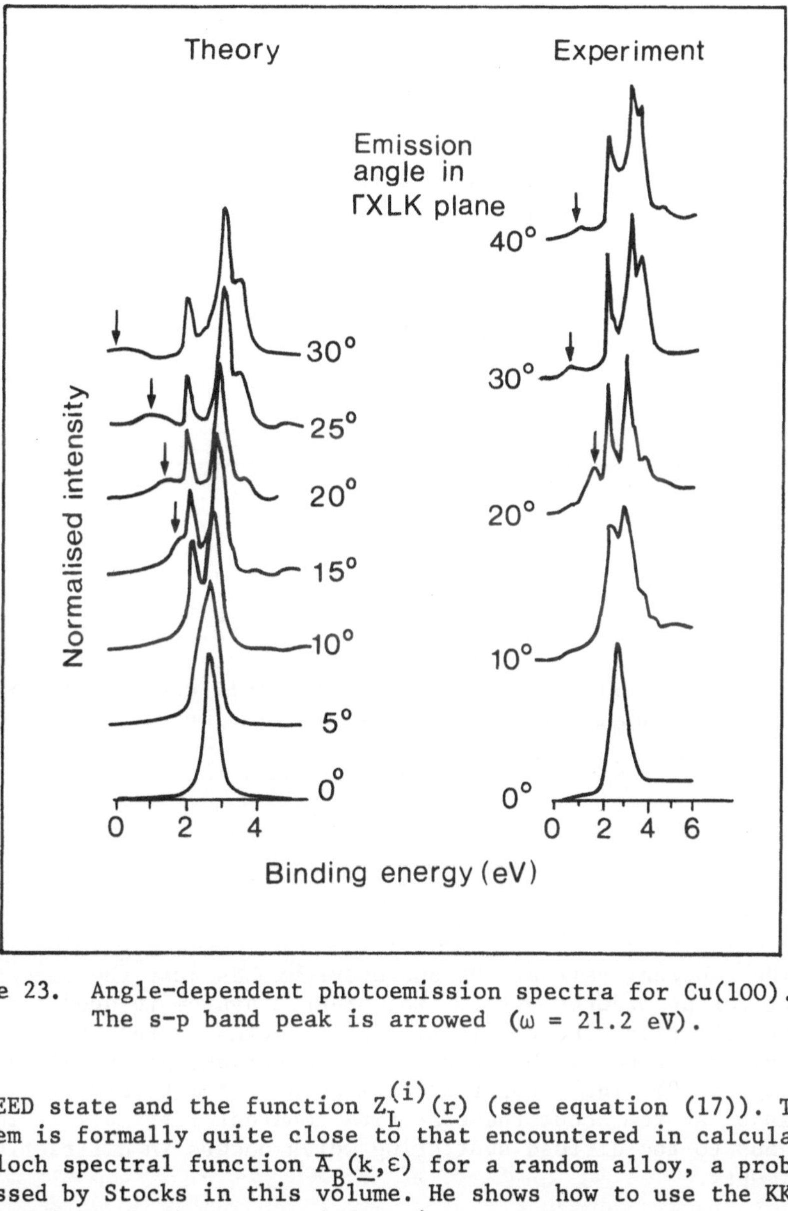

Figure 23. Angle-dependent photoemission spectra for Cu(100). The s-p band peak is arrowed (ω = 21.2 eV).

the LEED state and the function $Z_L^{(i)}(\underline{r})$ (see equation (17)). The problem is formally quite close to that encountered in calculating the Bloch spectral function $\bar{A}_B(\underline{k},\varepsilon)$ for a random alloy, a problem discussed by Stocks in this volume. He shows how to use the KKR-CPA to calculate the spectral function, and demonstrates the band-like features these calculations reveal. In the same way the KKR-CPA can be used to calculate the averaged photocurrent emitted from a random alloy surface, and since the analysis is similar to that for the spectral function[46], I will just quote the form of the result (details can be found in the original paper[64]) :

$$\langle I^{\underline{q}\lambda}(\varepsilon)\rangle = I^{\underline{q}\lambda}_{\text{coherent}}(\varepsilon) + I^{\underline{q}\lambda}_{\text{incoherent}}(\varepsilon)$$

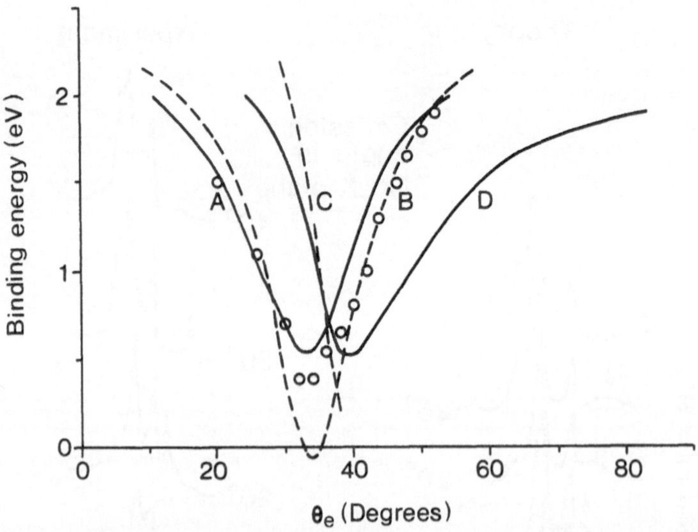

Figure 24. Binding energy of the s-p band peak (curve AB) seen in photoemission from Cu(100) ; θ_e is the emission angle in the ΓXULK plane. The points are experimental data (R.G. Jordan), the dashed curve shows full photocurrent calculations, and the full curve is the result of an optimised direct transition model.

The coherent part of the photocurrent has just the same form as that for an ordered system (the ordered system in question being the effective coherent lattice constructed by the KKR-CPA), and includes properly the effects of the surface potential barrier (see figure 19). The incoherent term arises from fluctuations of the random system about the effective coherent lattice ; it is analogous to the diffuse scattering which occurs in diffraction experiments.

Figure 25 shows the results of some calculations and experiments for the (100) surface of a $Cu_{.77}Ni_{.23}$ single crystal[62]. The extra broadening of the peaks due to disorder can be appreciated by comparing with figure 23, but individual peaks can still be resolved even in this concentrated alloy. The positions of these peaks correlate well with the Bloch spectral functions for this alloy, and constitute important evidence for the band-like features predicted by the KKR-CPA. Indeed, it was to examine the

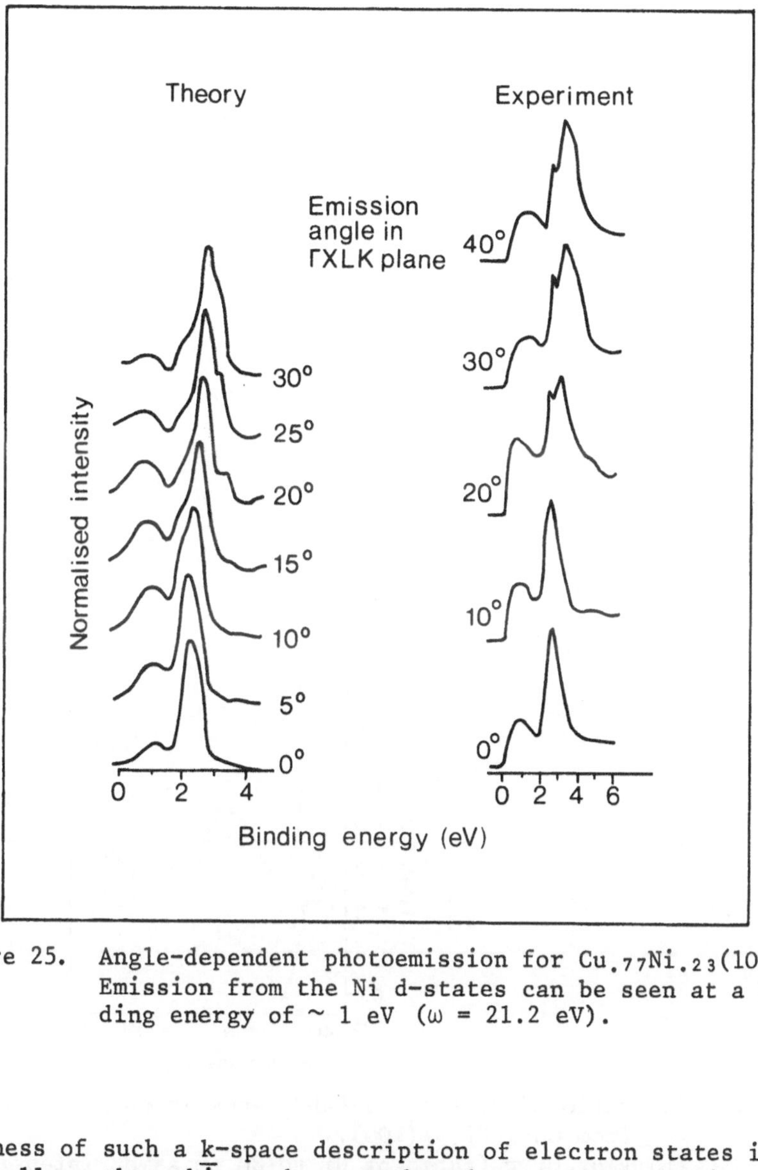

Figure 25. Angle-dependent photoemission for $Cu_{.77}Ni_{.23}(100)$. Emission from the Ni d-states can be seen at a binding energy of ~ 1 eV (ω = 21.2 eV).

usefulness of such a $\underline{k}$-space description of electron states in random alloys that this work was undertaken. Note that this correlation with the Bloch spectral function is not built into the photoemission calculation. The former describes the electronic structure of the bulk alloy, while the photoemission calculation contains a fairly realistic description of the surface. Figure 26 shows the calculated relative intensities of Ni and Cu d-band peaks in normal emission from the (100) of $Cu_{.62}Ni_{.38}$, as a function of photon energy[66]. An "oscillatory" behaviour like this was seen experimentally by Ling et al.[65], who ascribed it to a corresponding oscillation in the composition profile close to the surface.

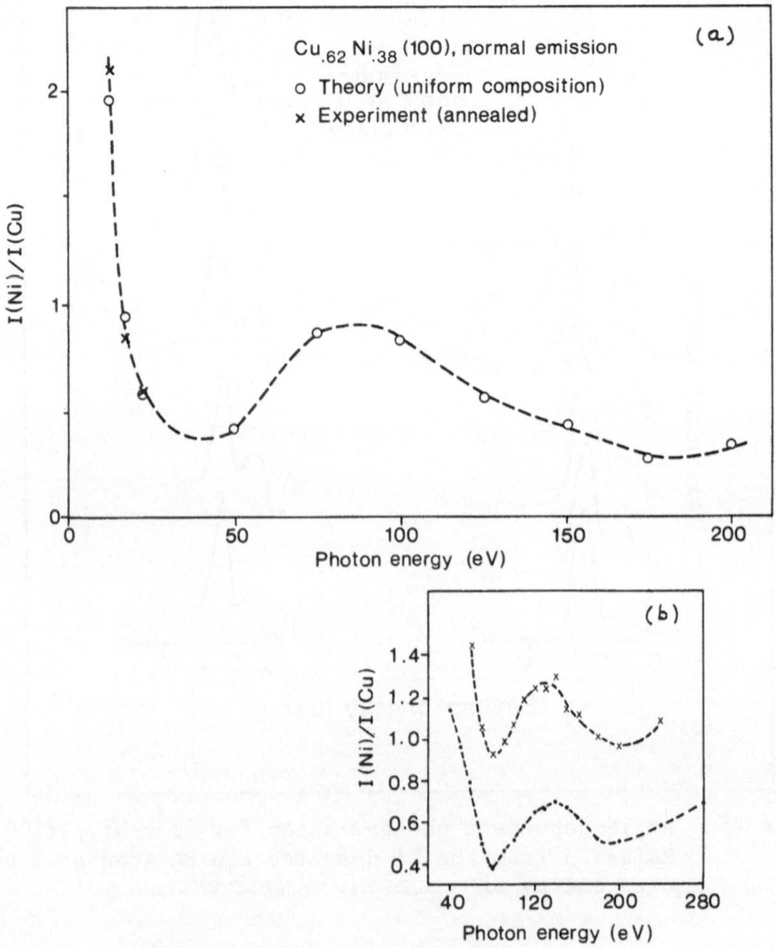

Figure 26. (a) Ratio of Ni to Cu d-band peaks in normal emission from $Cu_{.62}Ni_{.38}$(100).
(b) Plot of r (ratio of Ni to Cu emission intensities) versus hν (photon energy) for Cu/Ni (110). The dots are for the Cu/Ni I surface ; ~ 65% Cu, 35% Ni surface composition. The crosses are for the Cu/Ni II surface, ~ 65% Ni, 35% Cu. This figure is taken from Ling et al.

Cu is known to segregate to the surface of Cu-Ni alloys, and the suggestion was that this left a region correspondingly rich in Ni just below the Cu-rich surface. Since the escape depth of the photoelectrons varies with their energy, it is not implausible that an oscillation in the ratio of Ni to Cu photoemission peak heights might result. Our results show that this is unlikely to be the explanation. The oscillation in our calculation can be traced back to the $\underline{k}$-dependence of the disorder broadening in the Ni d-bands, an effect about which Ling et al. had no way of knowing. This illustrates the value of having a detailed theory of photoemission from alloys.

Surface-related features in alloy photoemission spectra can also be examined quantitatively. Figure 27a shows calculations(66) of the normally emitted photocurrent from the (111) surface of $Cu_{.77}Ni_{.23}$ (cf. figure 21). The feature close to the Fermi level is a surface state very similar to that found on pure Cu(111). The effects of simulating surface segregation by adding overlayers of pure Cu can be seen in the d-band transitions, but the surface state is almost unchanged. The effect of the bulk composition on the dispersion of the surface state is summarised in figure 27b ; the binding energy always follows a quasi-free-electron curve, whose composition dependence correlates closely with the radius of the Fermi surface neck around the L point. I have no space to analyse these features in full - I wanted mainly to give an indication of the capabilities of this kind of theory.

In conclusion, then, it appears that if the electronic structure is accurately calculated (including the surface), then 1-electron photoemission calculations can give a satisfactory account of quite complicated spectra, in transition metal systems at least. 1-electron theory cannot be the whole story, but the success it does achieve gives one hope that by careful comparison with experiment genuine many-body effects can be systematically isolated and quantified. Finally, I cannot leave this section without pointing out that the recently proposed and performed angle-resolved inverse photoemission experiment(67) (electrons in, photons out), can be described by almost exactly the same theory outlined above.

5. ELASTIC X-RAY SCATTERING

I will conclude these notes with a brief account of a topic of current interest in the synchrotron world; the anomalous scattering of X-rays. In section 2 I wrote down some diagrams corresponding to elastic X-ray scattering (figures 12c-f). In fact the corresponding scattering rate can be written

$$W^{\underline{q}_1\lambda_1,\underline{q}_2\lambda_2} = 2\pi\delta(\omega_{\underline{q}_1} - \omega_{\underline{q}_2})\, W_{\underline{q}_1\underline{q}_2}(\omega_{\underline{q}_1})$$

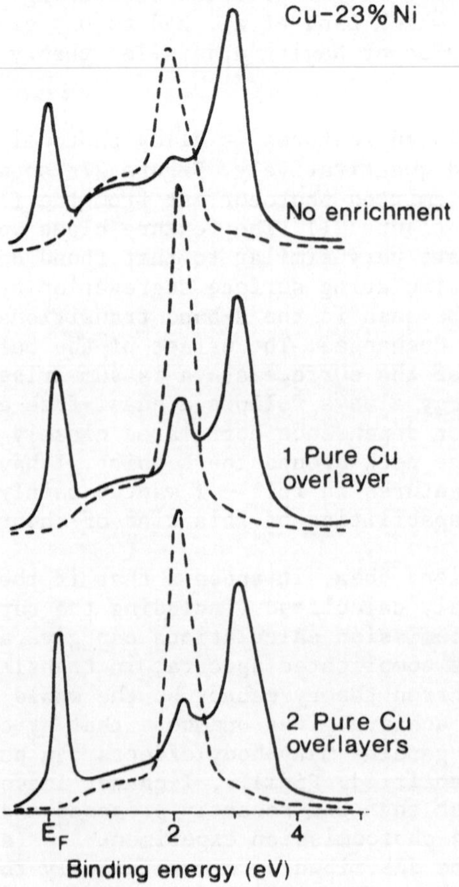

Figure 27a. Photoemission calculations for $Cu_{.77}Ni_{.23}(111)$, normal emission. The full curves refer to p-polarised radiation, the dashed curves to s-polarised.

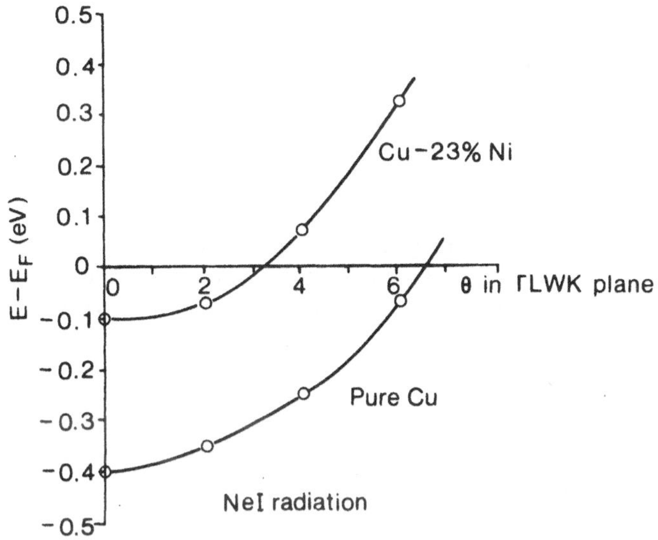

Figure 27b. Binding energy vs. emission angle for the (111) surface state.

where

$$W_{\underline{q}_1\underline{q}_2}(\omega) = |f_{\underline{q}_1\underline{q}_2}(\omega)|^2 \qquad (31)$$

The amplitude $f_{\underline{q}_1\underline{q}_2}(\omega)$ to scatter a photon $(\underline{q}_1,\lambda_1)$ into a photon $(\underline{q}_2,\lambda_2)$ with the same energy ω, has the following components

$$f_{\underline{q}_1\underline{q}_2} = f^0_{\underline{q}_1\underline{q}_2} + f^+_{\underline{q}_1\underline{q}_2} + f^-_{\underline{q}_1\underline{q}_2}$$

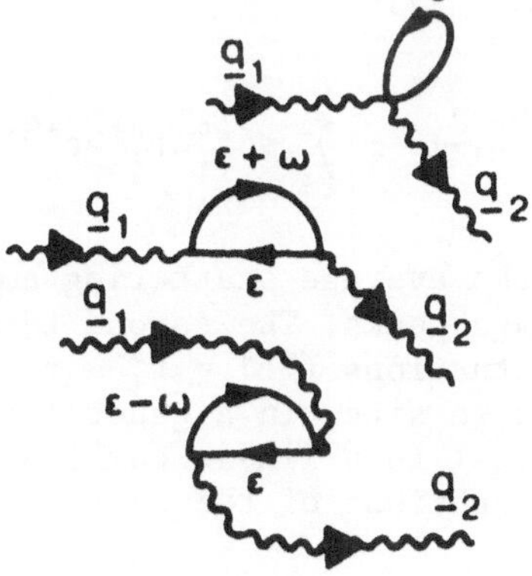

i.e. "A^2" acting once

i.e. "$\underline{j}.\underline{A}$" acting twice

(32)

It is straightforward to show that

$$f^0_{\underline{q}_1\underline{q}_2} = \left(\frac{\pi e^2 \hbar}{m\Omega\omega}\right) \underline{e}^{\lambda_1}(\underline{q}_1)\cdot\underline{e}^{\lambda_2}(\underline{q}_2)\rho(\underline{q}) \quad , \ \underline{q} \equiv \underline{q}_2 - \underline{q}_1$$

where

$$\rho(\underline{q}) = \int d\underline{r}\ e^{i\underline{q}\cdot\underline{r}}\ \rho(\underline{r})$$

This is the so-called normal part of the scattering amplitude, proportional to the Fourier transform of the charge density, $\rho(\underline{r})$. If $\rho(\underline{r})$ is decomposed into contributions from each site

$$\rho(\underline{r}) = \sum_i \rho_i(\underline{r} - \underline{R}_i)$$

then

$$f^0_{\underline{q}_1\underline{q}_2} = \sum_i f^0_{i,\underline{q}_1\underline{q}_2}\ e^{i\underline{q}\cdot\underline{R}_i} \tag{33}$$

where

$$f^0_{i,\underline{q}_1\underline{q}_2} \propto \int d\underline{r}\ \rho_i(\underline{r})e^{i\underline{q}\cdot\underline{r}}$$

(Note that the forward ($\underline{q} = 0$) scattering amplitude of atom i is just proportional to its atomic number). From equation (33) the usual kind of diffraction theory follows :

$$W_{\underline{q}_1\underline{q}_2} = \sum_{ij} f^0_i\ f^{0\star}_j\ e^{i\underline{q}\cdot(\underline{R}_i - \underline{R}_j)}$$

In an ordered system where all the f^0_i's are the same and the $\underline{R}_i$'s lie on a regular lattice, the diffracted X-rays form beams satisfying the Bragg condition. In a system with substitutional disorder like the random alloys discussed in the last section, $W_{\underline{q}_1\underline{q}_2}$ has to be averaged :

$$\begin{aligned} W_{\underline{q}_1\underline{q}_2} &= \sum_{ij} \langle f^0_i f^{0\star}_j\rangle\ e^{i\underline{q}\cdot(\underline{R}_i - \underline{R}_j)} \\ &= |\overline{f}^0|^2 \sum_{ij} e^{i\underline{q}\cdot(\underline{R}_i - \underline{R}_j)} + \sum_{ij} \langle\delta f^0_i \delta f^{0\star}_j\rangle e^{i\underline{q}\cdot(\underline{R}_i - \underline{R}_j)} \end{aligned} \tag{34}$$

The first term here involves the average scattering amplitude $\overline{f}^0$ of a site, and again gives Bragg peaks. The second term is the diffuse scattering due to fluctuations ($\delta f^0_i = f^0_i - \overline{f}^0$) in the scattering amplitude from site to site. In a general multicomponent system $W_{\underline{q}_1\underline{q}_2}$ is proportional to a linear combination of the partial pair correlation functions of the system. These partial correlation functions are a powerful way of describing the structure of a disordered system (e.g. alloys, glasses, etc.), and experimentalists have worked hard to extract them from diffraction data. The problem can be appreciated by looking at the diffuse scattering term in equation (34). This is the term which contains information about correlations in the occupancy of different sites, i.e. about the tendency of the atoms to order or cluster. If the scattering amplitudes of the different atoms are similar, then the fluctuations δf_i^0 are small, and the diffuse scattering is dominated by the coherent Bragg scattering. This situation can easily arise if the constituent atoms are near neighbours in the periodic table. If we could somehow make the scattering amplitudes of the atoms considerably different, this problem would go away. This is where the other terms in equation (32) come in.

In the same way that the X-ray emission/absorption and photoemission formulae were obtained in section 3, we can easily evalutate $f^{\pm}_{\underline{q}_1\underline{q}_2}$. The interesting case is that in which the electron at energy ε corresponds to a core state $\phi_i(\underline{r} - \underline{R}_i)$, ε_i on site i. Then

$$f^{\pm}_{\underline{q}_1\underline{q}_2} = \sum_i f^{\pm}_{i,\underline{q}_1\underline{q}_2} \; e^{i\underline{q}\cdot\underline{R}_i}$$

where

$$f^{\pm}_{i,\underline{q}_1\underline{q}_2} = -\int d\underline{r}\int d\underline{r}'\phi_i(\underline{r})X_{\underline{q}_1\lambda_1}(\underline{r})\int\frac{d\varepsilon}{\pi}\,\frac{\mathrm{Im}G^+(\underline{r},\underline{r}';\varepsilon \pm \omega)\theta(\varepsilon \pm \omega - \mu)}{\varepsilon - \varepsilon_i - i\Gamma}$$

$$\times\, X^{\star}_{\underline{q}_2\lambda_2}(\underline{r}')\phi_i^{\star}(\underline{r}') \tag{35}$$

Or, after using equation (11),

$$f^{\pm}_{i,\underline{q}_1\underline{q}_2}(\omega) = -\sum_{LL'}\int\frac{d\varepsilon}{\pi}\, m^i_{L\underline{q}_1}(E)\,\frac{\mathrm{Im}\tau^{ii}_{LL'}(E)}{E - \varepsilon_i \mp \omega - i\Gamma}\, m^{i\star}_{L'\underline{q}_2}(E)\theta(E - \mu) \tag{36}$$

where $m^i_{L\underline{q}}(E)$ is the same X-ray matrix element defined in equation (13). Note that the imaginary part of $f^+_{i,\underline{q}\underline{q}}$ is just the absorption rate, given by equation (12) ($\mathrm{Im}f^-$ vanishes), and so its real part is the Hilbert transform of the absorption spectrum. Thus $f^+_{i\,\underline{q}_1\underline{q}_2}(\omega)$ will change rapidly as ω crosses the absorption threshold $\omega = \mu - \varepsilon_i$. This is anomalous scattering. In the atomic case $\mathrm{Im}G^+(\underline{r},\underline{r}';\varepsilon + \omega)$ may well be dominated by a single discrete level $\phi_n(\underline{r})$, ε_n above μ. Then we find

$$f^+_i + f^-_i \propto \frac{|X_{in}|^2}{\omega^2 - \omega^2_{ni}} \, , \quad \omega_{ni} = \varepsilon_n - \varepsilon_i$$

X_{in} being the dipole matrix element between state i and n. This is the kind of formula atomic physicists write down[68], based on the atomic polarisability. The full scattering amplitude is usually written

$$f_i = f_i^0 + f_i' + if_i''$$

so that

$$f_i' = \mathrm{Re}(f_i^+ + f_i^-) \quad ; \quad f_i'' = \mathrm{Im}(f_i^+ + f_i^-)$$

Thus, in the above atomic example we could have the following picture :

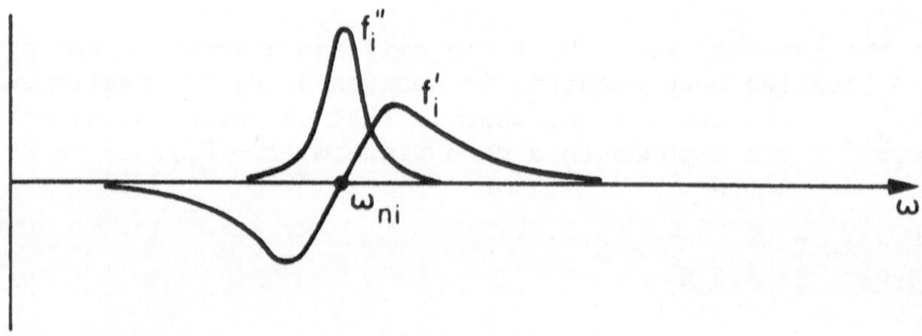

Figure 28.

and this shows the essence of the idea - by tuning ω across an absorption threshold f_i can be changed substantially. Furthermore, since the core states are localised, it is only the scattering from a particular type of atom which changes ; the scattering from all other types of atoms remains almost constant. It is just this kind of change in the relative scattering amplitudes which we argued would be a boon in diffraction studies on multicomponent systems.

Equation (36) indicates how the full electronic structure enters the scattering amplitude in condensed systems. In particular, it shows that band-related features should show up close to threshold in the real part of the scattering amplitude, and this is confirmed in the beautiful X-ray interferometry measurements of Hart and Siddons[69,70]. A good deal of work has yet to be done, theoretically and experimentally, to establish the usefulness of this technique, but its possibilities are exciting. The same is true of inelastic X-ray scattering, which in principle can probe many kinds of electronic excitation. The advent of very bright, stable, X-ray sources (i.e. synchrotron radiation from storage rings) makes it possible seriously to consider what can be done with such techniques, and the next few years could be interesting for X-ray physics.

REFERENCES

1. O'Bryan and Skinner, H., Phys.Rev. 45, 370 (1934).
2. Mott, N.F. and Jones, J., "Theory of the Properties of Metals and Alloys" (Oxford, Clarendon Press, 1936).
3. Keldysh, L.V., Soviet Physics JETP 20, 1018 (1965).
4. Caroli, C., Combescot, R., Nozières, P. and Saint-James, D., J.Phys.C. 4, 916 (1971).
5. Kadanoff, L.P. and Baym, G., "Quantum Statistical Mechanics" (Benjamin, New York, 1976).
6. Langreth, D.C. in "Linear and Nonlinear Electron Transport in Solids", ed. Devreese and Van Doren (Plenum, New York, 1976).
7. Yue, J.T. and Doniach, S., Phys.Rev. B8, 4578 (1973).
8. Mahan, G.D., "Many Particle Physics" (Plenum, New York, 1981).
9. Roulet, B., Gavrolet, J. and Nozières, P., Phys.Rev. 178, 1072 (1969).
10. Abraham-Ibrahim, S., Caroli, B., Caroli, C. and Roulet, B., Phys.Rev. B18, 6702 (1978) ; J.Physique 40, 861 (1979).
11. Schaich, W.L. and Ashcroft, N.W., Phys.Rev. B3, 2452 (1971).
12. Caroli, C., Lederer -Rozenblatt, D., Roulet, B. and Saint-James, D., Phys.Rev. B8, 4552 (1973).
13. Chang, J.J. and Langreth, D.C., Phys.Rev. B8, 4638 (1973).
14. Nozières, P. and Abrahams, E., Phys.Rev. B10, 3099 (1974).
15. Feibelman, P.E. and Eastman, D.E., Phys.Rev. B10, 4932 (1974).
16. Liebsch, A.R., in ref. 51.
17. Pendry, J.B., Surf.Sci. 57, 679 (1976) ; Mahan, G.D., Phys.Rev. B2, 4334 (1970) ; Adawai, I., Phys.Rev. 134, A788 (1964).
18. Faulkner, J.S., Phys.Rev. B19, 6186 (1979).
19. There are several monographs on the kind of scattering theory needed, in this context, to deal with a single muffin-tin potential (eg. those of Newton and Taylor). This is not so for multiple scattering theory. The version I use here can be pieced together from the articles of Lloyd P. and Smith P.V. (Adv. Physics 21, 69 (1975)) and of Faulkner, Gyorffy and Stocks (refs. 18, 25, 46). Someone should write a book on it !
20. Power, E.A., "Introductory Quantum Electrodynamics" (Longmans 1964).
21. Fano, U. and Cooper, J.W., Rev.Mod.Phys. 40, 441 (1968).
22. Wehner, P.S., Kevan, S.D., Williams, R.S., Davis, R.F. and Shirley, D.A., Chem.Phys.Letters 57, 334 (1978).
23. Durham, P.J., in "EXAFS" (ed. R. Prins and D. Koningsberger)1984.
24. eg. Teo BK and Joy, D.C. (eds.) "EXAFS spectroscopy : Techniques and Applications" (Plenum, New York, 1981).
25. Gyorffy, B.L. and Stocks, G.M., in "Electrons in Disordered Metals and at Metallic Surfaces", ed. Phariseau, Gyorffy and Scheire (Plenum, 1979).
26. Ries, G. and Winter, H., J.Phys.F. 9, 1589 (1979).
27. eg. Inglesfield, J.E., J.Phys.F. 11, 1287 (1981).
28. Goodings, D.A. and Harris, R., J.Phys.C. 2, 1808 (1969).

29. McMullen, T., J.Phys.C. 3, 2178 (1970).
30. Papaconstantopoulos, D.A., "Int.Conf. on Physics of X-ray Spectra", NBS (Gaithersborg 1976) p. 192.
31. Schwartz, K.H., Mohn, P. and Nowotny, H. in "Inner Shell and X-ray Physics of Atoms and Solids", ed. Fabian, Kleinpoppen and Watson (Plenum, 1981).
32. Wakoh, S. and Kubo, Y., Jap.J.Appl.Phys. 17 (S17-2) 193 (1978).
33. Mueller, J.E., Jepsen, O., Andersen, O.K. and Wilkins, J.W., Phys.Rev.Lett. 40, 720 (1978).
34. Szmulowicz, F. and Pease, D.M., Phys.Rev. B17, 3341 (1978).
35. House, D., Smith, P.V. and Gyorffy, B.L., J.Phys.F 3, 745 (1973).
36. Fairlie, R., Temmerman, W. and Gyorffy, B.L., J.Phys.F. (in press).
37. Keski-Rahkonen, O. and Krause, M.O., Atomic Data and Nuclear Data Tables 14, 139 (1974).
38. Blokhin, M.A. and Satchenko, V.P., Izv. Akad. Nauk. 24, 397 (1960).
39. Penn, D.R., Phys.Rev.Lett. 42, 921 (1979) ; Liebsch, A.R., Phys.Rev. B23, 5203 (1981) ; Treglia, G., Ducastelle, F. and Spanjaard, D., J.Physique 41, 281 (1980) ; Davis, L.C. and Feldkamp, L.A., Solid St. Commun. 34, 141 (1980) ; Durham, P.J., J.Phys.F. (in press).
40. Lundqvist, B.I., Phys.Stat.Sol. 32, 273 (1969).
41. Nozières, P. and De Dominicis, C.T., Phys.Rev. 178, 1097 (1969).
42. Citrin, P.H., Wartheim, G.K. and Schlüter, M., Phys.Rev. B20, 3067 (1980).
43. Von Barth, U. and Grossmann, G., Solid St. Commun. 32, 645 (1979) and Phys.Rev. B25, 5150 (1982) ; Mahan, G.D., Phys. Rev. B21, 1421 (1980).
44. Lang, N.D. and Williams, A.R., Phys.Rev. B16, 2408 (1977) ; Schönhammer, K. and Gunnarsson, Z.Physik B30, 297 (1978) ; Bianconi, A., Campagna, M. and Stizza, S., Phys.Rev. B (1982).
45. Durham, P.J., Ghaleb, D., Gyorffy, B.L., Hague, C.F., Mariot, J.M., Stocks, G.M. and Temmerman, W., J.Phys.F 9, 1719 (1979).
46. Faulkner, J.S. and Stocks, G.M., Phys.Rev. B21, 3222 (1980) ; Durham, P.J., Gyorffy, B.L. and Pindor, A.J., J.Phys.F. 10, 661 (1980).
47. Stocks, G.M., Temmerman, W. and Gyorffy, B.L. in "Electrons in Disordered Metals and at Metallic Surfaces" eds. Phariseau, Gyorffy and Scheire (Plenum, 1979).
48. Durham, P.J., Gyorffy, B.L., Hague, C.F., Mariot, J.M., Pindor, A.J. and Temmerman, W., "Physics of Transition Metals 1980" Inst.Phys.Conf.Ser. 55, p. 146 ; Munoz, C., Durham, P.J. and Gyorffy, B.L., J.Phys.F. (in press).
49. Stocks, G.M., Tagle, J.A., Calcott, T.A. and Arakawa, E.T. in "Inner Shell and X-ray Physics of Atoms and Solids" ed. Fabian, Kleinpoppen and Watson (Plenum 1981), p.619.
50. Weinberger, P., Staunton, J. and Gyorffy, B.L., to be published.

51. eg. "Photoemission and the Electronic Properties of Surfaces" ed. Feuerbacher, Fitton and Willis (Wiley 1978) ; Williams, R.H., Srivastava, G.P. and McGovern, I.T., Rep.Prog.Physics 43, 1357 (1980).
52. Helms, C.R. and Yu, K.Y., J.Vac.Sci.Tech. 12, 276 (1975).
53. for a recent review see Inglesfield, J.E., Rep.Prog.Physics 45, (1982).
54. eg. Surface states on W(100) seem to be poorly described (J.E. Inglesfield and N. Kar, private communication).
55. Pendry, J.B., ref. 17.
56. Pendry, J.B., "Low Energy Electron Diffraction" (Academic Press, 1974).
57. See the articles by Ashcroft N.W. and by Kliewer, K.L. in ref. 51.
58. Levinson, H.J., Plummer, E.W. and Feibelman, P.J., J.Vac.Sci. Tech. 17, 216 (1980).
59. Pendry, J.B. in ref. 51.
60. Gartland, P.O. and Slagsvold, B., Phys.Rev. B12, 4047 (1975).
61. Nilsson, P.O., Kanski, J. and Larsson, C.G., Solid St. Comm. 36, 111 (1980).
62. Allen, N.K., Durham, P.J., Gyorffy, B.L. and Jordan R.G., to be published.
63. Durham, P.J. and Kar, N., Surf.Sci. 111, L648 (1981).
64. Durham, P.J., J.Phys.F., 11, 2475 (1981).
65. Ling, D.T., Miller, J.N., Lindau, I., Spicer, W.E. and Stefan, P.M., Surf.Sci. 74, 612 (1978).
66. Wille, L.T., Durham, P.J. and Jordan, R.G., Solid St.Comm. (in press)
67. Pendry, J.B., Phys.Rev.Lett. 45, 1356 (1980) ; Woodruff, D.P. and Smith, N.V., Phys.Rev.Lett. 48, 283 (1982).
68. Wendin, G., Physica Scripta 21, 535 (1980).
69. Hart, M. and Siddons, D.P., Proc.R.Soc.Lond. A376, 465 (1981).
70. Siddons, D.P. (private communication).

ACKNOWLEDGEMENT

I would like to thank Drs. B.L. Gyorffy, G.M. Stocks, W.M. Temmerman, J.E. Inglesfield, N. Kar and L. Wille and Prof. J.B. Pendry for many illuminating discussions and collaborations. I also thank C.-G. Larsson, for permission to show his photoemission calculations.

FEATURES AND APPLICATIONS OF THE HAYDOCK RECURSION METHOD

Volker Heine

Cavendish Laboratory
Madingley Road
Cambridge CB3 OHE
England

1. SCOPE OF THE RECURSION METHOD

The subject of this talk has been reviewed in four articles by myself, Haydock, Bullett and Kelly constituting Solid State Physics, Vol. 35 to which reference will simply be made by page numbers.

The Haydock recursion method is applicable when one has (a) a Hamiltonian in terms of a discrete basis set such as atomic-like orbitals on a cluster of atoms (but see below for other applications), and (b) no assumptions about symmetry so that it is applicable at surfaces and in amorphous materials as well as in bulk crystals of course. In any situation of low symmetry and high degeneracy as in a solid, the eigenfunctions ψ_n of the Hamiltonian H are of relatively little use or significance (pp. 1-16). The reason is that the energy differences between consecutive eigenstates is of order 1/N where N is the number of atoms or orbitals, and a change in one atom is enough to scramble the nearly-degenerate eigenstates, without of course altering the physical properties of the piece of material significantly. Physically relevant quantities can usually be expressed in terms of the Green function, (pp. 6-9, 296-383) though last week I came across an exception. We mean the matrix element

$$G_{\chi\chi}(E) = \langle\chi|(E - H)^{-1}|\chi\rangle \tag{1}$$

with respect to some chosen state $|\chi\rangle$. The advantage of the method is that it calculates (1) directly without calculating the ψ_n. Whereas the calculation of the eigenstates and properties from them is a process of order N^3 in computer time required, the

Haydock recursion method computes (1) directly and requires a time of order N, an enormous saving by a factor N^2. In Cambridge we regularly use $N \sim 10^4$ orbitals on clusters of ~ 1000 atoms. One of the factors of N comes because H is a sparse matrix, i.e. consisting mostly of zeros. Each orbital is connected only to other orbitals on near atoms, but there is no restriction to nearest neighbours. The other factor of N comes from calculating a highly convergent approximation to the desired quantity (1) directly without a lot of unwanted information in the precise eigensolutions.

In most cases $|\chi\rangle$ is an atomic-like orbital on one atom so that from the imaginary part of (1) one obtains the local density of states. Early examples included amorphous materials, surfaces, Laves phases with larger unit cells, and cluster molecules (pp. 69-87, 296-383). But there are other applications such as photoemission from surfaces where one uses quite a different state for $|\chi\rangle$, in which case one talks about the projected density of states (pp. 86-7, 311). Another example is the degree of bonding between a pair of neighbouring atoms which depends on an off-diagonal matrix element of the Green function G - these have to be calculated as the difference between two diagonal elements (pp. 83, 302). Other applications include the precessional equation of motion of the spin on one atom in a metallic ferromagnet, e.g. above the Curie temperature: this allows us to calculate the coupling J_{ij} between pairs of atoms. Quite different but still depending on local basis sets are lattice vibrations (pp. 72, 356) and spin glasses (p. 379).

Let us turn to the actual mathematical technique. The answer to calculating (1) comes out in the form of a continued fraction

$$G_{\chi\chi}(E) = \cfrac{1}{E - a_0 - \cfrac{b_1^2}{E - a_1 - \cfrac{b_2^2}{E - \dots}}} \tag{2}$$

This may be unfamiliar mathematics to most of us quantum physicists but it is not very different and not very difficult. It arises by operating with H on $|u_0\rangle \equiv |\chi\rangle$ and then orthogonalising to $|u_0\rangle$ to obtain $|u_1\rangle$. In this way a sequence of new basis states, the recursion states $|u_n\rangle$, is built up from $|\chi\rangle$ by the recursion formula

$$b_{n+1}|u_{n+1}\rangle = H|u_n\rangle - a_n|u_n\rangle - b_n|u_{n-1}\rangle \tag{3}$$

The set of $|u_n\rangle$ are an orthonormal set in which H takes a simple tri-diagonal form (pp. 89, 222-8), i.e. H_{nm} is zero unless n = m or m $\pm$ 1. The Green function (1) is the zero-zero matrix element of the inverse of the matrix $[E - H_{mn}]$, which can be written down by elementary algebra due to the simple form of H_{mn} and turns out to be equal to (2) (pp. 90, 252).

Mathematically speaking the continued fraction (2) ought to be continued to level N to give (1) precisely but in practice we are dealing with a highly convergent formalism so that we normally stop at about level 15 or 20, though in special situations a much smaller number may suffice and in others a much larger number may be needed. If one stops at level r, one obtains roughly a resolution W/r in the density of states where W is the total band width. The reason for this good convergence can be seen in various ways. Mathematically the coefficients a_n, b_n relate to the (2r)th and (2r + 1)th moments of the projected density of states. Anything that one does with the continued fraction below some level r cannot alter the first 2r moments of the projected density of states. Physically we can see the convergence by taking for $|\chi\rangle \equiv |u_0\rangle$ a single atomic-like orbital on some atom, say atom 0. The matrix H contains hopping matrix elements connecting $|u_0\rangle$ to the near neighbours so that $|u_1\rangle$ is an orbital on the near neighbours surrounding atom 0 and the orthogonalisation to $|u_0\rangle$ ensures there is a hole in the middle. Similarly the $|u_n\rangle$ can be pictured as a set of expanding rings around atom 0, becoming more and more remote from it and hence influencing the local density of states on atom 0 less and less (p. 70). This picture is literally correct for a model with a single s-orbital on every atom with nearest neighbour hopping only. In other cases it is an oversimplification but the fact remains that the recursion basis orbitals $|u_n\rangle$ become progressively more remote from $|\chi\rangle$ as far as coupling through the Hamiltonian H is concerned, as is clear from its tridiagonal form.

As already remarked, the main but not only field of application lies in electronic structure with a tight-binding basis (Linear Combination of Atomic Orbitals = L.C.A.O.). One has to have H already in this form before starting with the recursion method. How do we get it ? The L.C.A.O. method was somewhat under a cloud for many years because of various ambiguities. In the last decade successful schemes have been set up for organic molecules (p. 147) and for transition metals (pp. 57, 192), and also a general scheme developed by P.W. Anderson and D.W. Bullett (pp. 47, 170). In the latter one sets up an equation which determines the best atomic-like set of orbitals, best in the sense of most localised and in spanning the desired function space. In practice the Hermann-Skillman atomic orbitals are often an adequate approximation, but one must insert them properly into the equations of the method which differ from simple old-fashioned L.C.A.O. procedures. Actually the method gives one $S^{-1}H$ where S is the overlap matrix (pp. 39, 177). The recursion method has now been formulated to deal properly with a non-orthogonal basis set with overlap matrix S and is quite simple to operate (pp. 225 - 8, 222). The previous approach (p. 311) with a double-sided recursion on the non-Hermitian $S^{-1}H$ is no longer recommended.

Self-consistency if required must also be expressed in atomic terms, for example through an intra-atomic Stoner-like exchange integral I such that the exchange splitting Δ on a given atom in a magnetic material is given by Im where m is the atomic moment. Similarly self-consistent charge transfers can be related to intra- and inter-atomic Coulomb integrals U_{ij} (p. 316). However to model the long range nature of the Coulomb interaction properly, for example to obtain exact screening in a metal, one would have to have U_{ij} falling off as $1/R_{ij}$ with distance R_{ij}.

Every good calculus should have its perturbation theory, and this one does : in fact it has several for doing different jobs e.g. applying a perturbation δV, changing the environment such as fcc to hcp, and disconnecting part of the system such as an adsorbed atom on a surface (pp. 99, 263, 446, 375).

2. TRUNCATION, TERMINATION AND CLUSTER SIZE

There are five related topics here, all concerned with obtaining maximum information for a given computing time, which must be considered together. Truncation means stopping the calculation after r recursions on (3), and termination is what you do with the continued fraction (2) having found r levels exactly. The method is inherently a discrete cluster method, and cluster size is clearly important, as are the boundary conditions to be applied at the edge of the cluster, and finally the shape of the cluster. We have already remarked that if one is calculating, say, the local density of states at the centre of the cluster, then successive iterations are calculating the electronic structure in the neighbourhood of the atom more and more accurately, so that in a general sense all five issues are concerned with the effect of 'distant' parts of the system, distant either in space or distant in the sense of linked only through a high order of H. The general invariance theorem of Green functions (pp. 9, 24) ensures reasonable convergence, but even so the issues are not trivial (pp. 215-294), and quite a sophisticated set of programs called the Cambridge Recursion Library has been developed by C. Nex to get the best out of the method. These programs are available by writing to him at the address at the head of this article with the equivalent of about ten dollars to cover the cost of the tape and copying. I have heard them described as a model of user-oriented programming and helpful documentation, and they are being up-dated as new procedures come along or replace old ones.

The main issue is terminating the continued fraction (2) having truncated it at some level r. A finite continued fraction is a rational function which can be simplified algebraically to a set of r poles

$$\sum_s w_s/(E - E_s) \tag{4}$$

with weights w_s at energies E_s. The projected density of states $n_\chi(E)$ is the imaginary part of this expression when E is given an infinitesimal imaginary part, i.e. it is a sum of delta functions

$$n_\chi(E) = \sum_s w_s \, \delta(E - E_s) \tag{5}$$

One procedure is to use this set of delta functions directly, and this is particularly recommended for carrying out a definite integral

$$\int_{-\infty}^{\infty} X(E) \, n_\chi(E) \, dE \tag{6}$$

over $n_\chi(E)$. It corresponds to "Gauss' n-point formula" in numerical integration because the positions and weights of the delta functions are 'chosen' to represent exactly the first 2r moments of $n_\chi(E)$. The programs for all this are contained in the Cambridge Recursion Library. The poles are found by noting that they are the eigenvalues of the truncated tridiagonal matrix and the weights by noting that the continued fraction truncated at level n is equal to the ratio of two polynomials

$$Q_n(E)/P_n(E) \tag{7}$$

where the $Q_n(E)$, $P_n(E)$ are calculated from a simple recurrence relation (pp. 254, 255-9).

For looking at $n_\chi(E)$ directly one wants a continuous function of E whereas simply a truncated continued fraction (4) gives the set of delta functions (5). The simplest course is to evaluate (2) which is equivalent to (4) at a complex energy E with a suitable finite imaginary part instead of an infinitesimal one. This broadens the delta functions (5) into Lorentzians, and is not recommended : if one takes the Lorentzians wide enough to smooth away the particular positions of the poles, then one also smoothes away much of one's information. A second approach is to continue the truncated fraction to infinity with constant coefficients, i.e. to set

$$a_n = a \quad , \quad b_n = b \quad \text{for all} \quad n > r \tag{8}$$

This tail can be summed into an analytic function called "the square root terminator" which gives $n_\chi(E)$ as a continuous function of E throughout the band extending from $E_{min} = a-2b$ to $E_{max} = a+2b$ (p. 314). This also is not recommended in general, though it has been widely used and sometimes gives good results for a continuous band. If the a,b correspond to too narrow a band then delta functions are split off at the edges, and if to too wide a band one gets strong spurious oscillations of $n_\chi(E)$ in the band. Recently Beer and Pettifor (this volume) have automated the calculation of a, b such as to just not give any split-off states, but this always gives a width slightly too narrow resulting in a small $E^{-1/2}$ singularity (instead of $E^{1/2}$) in n_χ at the band edges. However we have had also some bad experiences even when choosing the 'best' a,b. The answer seems to be that the more sharp structure $n_\chi(E)$ contains the worse the square root terminator becomes, as can be seen for example in the simple case of a rectangular density of states terminated with the a,b corresponding to the exact band width : this shows very strong oscillations even for quite high levels of truncation. We therefore use the routine which has been dubbed (rather inappropriately) "supersmooth" in the Cambridge Recursion Library (p. 258), which does a rather sophisticated processing of the recursion coefficients. We intend in future to call it "the method of bounds". If one truncates at level r, there are in fact upper and lower bounds on the integrated density of states $N_\chi(E)$ which cannot be exceeded no matter how one extends the continued fraction, and the "method of bounds" formula for $N_\chi(E)$ is the mean of these two bounds, which therefore retains all the information in the calculated a_n, b_n. The formula can be differentiated analytically to give a corresponding result for $n_\chi(E)$ which itself has no bounds. The evaluation is fast and numerically stable in terms of the $P_n(E)$, $Q_n(E)$ functions in (6). The method gives sharp band edges and even sharp internal band gaps with exponentially small ($\sim 10^{-16}$) density of states if these are not too narrow. Internal structure in n is resolved to about $(E_{max} - E_{min})/r$. The bounds are rigorous so that the method can safely be used under all circumstances. The method of bounds is also used if X(E) in (6) contains the Fermi function which effectively cuts off the integral at the Fermi level E_F. The Recursion Library contains routines for this, including the integrated density of states $N_\chi(E_F)$ and integrated energy $U_\chi(E_F)$ as special cases. Because the bounds are point-wise bounds at any E_F, the $N_\chi(E_F)$ and $U_\chi(E_F)$ etc. considered as functions of E_F through the band do not have quite the correct moments corresponding to the a_n, b_n.

Cluster size is the crucial factor : a cluster of diameter D has of order D^3 atoms in three dimensions so that the computing time increases as D^3 for a given truncation r. The trouble would appear to be that the continued application of the recursion (3) soon carries one outside the cluster. The a_n, b_n can be related to summing closed paths of length 2n on the lattice where each step corespond to a non-zero matrix elements of H in the original local basis. For the sake of discussion let us assume nearest neighbour interactions only and consider n_χ for the local density of states of the central atom, so that D/2 atomic step take one from the central atom to the cluster edge, giving a cubic cluster with about $D^3/2$ atoms for metallic structures. Thus D = 12 gives about 900 atoms, and in my magnetic calculations with 10 or 18 spin-orbitals per atom it means of order 10^4 orbitals. However only the first D/2 = 6 levels of the continued fraction correspond exactly to the infinite solid. This is not nearly as bad as it sounds. It is well known that random walks of length n (to the 'furthest point' of the walk) mostly curl around the origin and reach a distance which is Gaussian distributed with mean of order $n^{1/2}$, and provided this is less than D/2 most of the paths one is effectively counting are inside the cluster. This gives a limit $r \approx D^2/4$ for truncation, i.e. $r \approx 35$ for our case D = 12. For $6 < r < 35$ the method is making a small error because the cluster is not properly embedded in an infinite medium but at the same time it is solving exactly the Hamiltonian for the particular configuration of atoms inside the cluster which is far more important for $n_\chi(E)$. This argument applies particularly when there are several orbitals per atom because the paths are actually between orbitals rather than between atoms. Practical tests with fcc d-bands etc suggest one should go to a level at least

$$\tau \approx 3D/2 \tag{9}$$

i.e. about three times as far as the number of exact levels, i.e. about level 20 in our example. This cluster-truncation relation would appear to apply to continuous bands without a gap or other sharp structure, but its range of validity has not yet been fully researched. As a counter example Meek (p. 79) with his bond charge model required 100 levels to obtain adequate resolution in the physical part of his band which was only a fraction of the total band width. On the other hand Pettifor (private communication) with fcc d-bands finds that 5 levels with the knowledge that we have a continuous band, already contain a great deal of information. Suppose one has a large cluster representing, say, the interaction between two lattice defects, and suppose that for purposes of calculation total energy one needs the local density of states separately for each atom. The question arises, would 5 levels suffice ? Our rule (9) would suggest that it is inefficient to work with a large cluster and a small number of levels. Rather, to make best use of computing time one should take for each atom a surrounding sub-cluster of such a size and calculate such number of levels that (9) is satisfied or whatever relation is found best to substitute for (9) (pp. 231-3).

What about boundary conditions at the edge of the cluster ? If we are calculating the local density of states at the central atom, the simple answer is that if changing the boundary conditions makes a significant difference, then the cluster is not big enough for convergence. Although intuition might suggest that periodic boundary conditions would be better than simply cutting off H at the edge of the cluster, some tests by Nex and Kelly with small clusters and s-bands certainly show the reverse. I can produce theoretical arguments in both directions, so there may well be circumstances where one is better and other situations where the other is (pp. 231-3). If one is projecting onto a $|\chi\rangle$ which weights all atoms equally, as in a photoemission calculation, one certainly should use periodic boundary conditions to 'clothe' all atoms equally. Returning to the case of local density of states near the centre of the cluster, another technique which works well is to attach to each surface atom of the cluster a linear chain of 'atoms' of s-states with constant diagonal and hopping matrix elements chosen to match the band width. This is an effective embedding of the cluster in a medium with approximately the right behaviour, and gives very good results even for quite small clusters. The embedding chains contain far fewer orbitals and interactions than a corresponding enlargement of the cluster so that it is efficient in terms of computer time and storage. With non-constant matrix elements along the chains one can model a band gap. It is not clear whether just a single chain attached to all surface atoms (instead of one to each surface atom) would suffice.

Finally what about cluster shape ? If one is calculating a local density of states on one atom, it would appear sensible at first sight to surround it with a roughly spherical cluster to obtain maximum diameter for a given number of atoms. However this has a very bad effect : the corrections from cutting off at the cluster boundary are 'focussed' on the central atom and add up to sizeable corrections. It is better to use a cubic cluster shape to 'reflect' the corrections around the whole volume of the cube (pp. 75, 24, 233).

THE RECURSION METHOD AND THE ESTIMATION OF LOCAL DENSITIES OF STATES

N. Beer and D.G. Pettifor

Department of Mathematics
Imperial College
London

ABSTRACT

A technique is presented for terminating a continued fraction approximation to a local density of states. Results obtained by this method are compared with the results of a method due to Nex. It is found that the density of states and integrated quantities are more quickly convergent and that the moments of the density of states are preserved in the calculation, in contrast to what is found for the technique of Nex.

1. INTRODUCTION

The recursion method has been amply documented (Haydock 1980). Its use for a tight binding Hamiltonian leads to an expression for the local electronic density of states (LDOS) in the form of a continued fraction :

$$n(E) = -\frac{1}{\pi}\,\mathrm{Im}G_{00}(E)$$

$$G_{00}(E) = \cfrac{b_0^2}{E - a_0 - \cfrac{b_1^2}{E - a_1 - \cfrac{b_2^2}{E - a_2 - \ddots}}} \qquad (1)$$

Each successively deeper level in the fraction describes the contribution to the LDOS from increasingly distant atoms. The recursion method is a numerically stable technique for the calculation of the coefficients a_i, b_i^2. The continued fraction is, in fact, the solution to the classical moment problem (Akhiezer 1965) since the coefficients are related to the moments of the LDOS. In practical terms, however, we only ever calculate a finite number of coefficients (equivalently a finite number of moments) - the number determined by computer store/time, numerical stability or personal preference. Thus a decision must be made about how to approximate the effects of the uncalculated coefficients in the tail of the continued fraction.

The approximation investigated in this paper is based on the assumption that all the coefficients in the tail can be replaced by some constant coefficients a, b^2 ; this assumption leads to an analytic form for the tail of the continued fraction - the 'square root terminator' (Haydock, Heine and Kelly 1975). The tail of the fraction can be written

$$t(E) = \frac{b^2}{E - a - t(E)} \qquad (2)$$

So we find the square root terminator

$$t(E) = \frac{1}{2}\left[E - a - \sqrt{(E - a)^2 - 4b^2}\right] \qquad (3)$$

which gives a continuum of states in the range

$$a - 2b < E < a + 2b \qquad (4)$$

This method is only suitable when there are no gaps in the energy spectrum. If the band edges are known _a priori_, then since the coefficients a_i, b_i^2 tend to asymptotic values a_∞, b_∞^2 which are related to the band edges by (4), we could use these asymptotic values in (3). However, when the band edges are not known, we must make a reasonable choice of a, b^2. The effects of different choices will be discussed in the next section.

This termination technique will be compared with a different approximation due to Nex (1978) which was designed to handle totally general situations, such as a cluster that is localised or embedded. Nex showed that it is possible, using only the 2N + 1 coefficients a_i, b_i^2, i = 0,...,N to establish rigorous bounds, $N_\pm(E)$, on the _integrated_ LDOS. This technique, which is implemented in the recursion program package available from Cambridge, then uses the differential of the mean of these bounds as the approximation to the LDOS :

$$n(E) = \frac{1}{2}\frac{d}{dE}\left[N_+(E) + N_-(E)\right] \qquad (5)$$

2. DETERMINATION OF THE TERMINATING COEFFICIENTS

Experience with the use of a terminator has indicated a useful criterion to govern the choice of a, b^2. The terminating coefficients a, b^2 are related to the bandwidth by (4). It has been found that if the bandwidth chosen is too small, then the LDOS develops spurious peaks at the band edges, and weight may be lost from the band by delta functions splitting off from the band edges. These delta functions have an initially zero weight which increases with their distance from the band. On the other hand, if a bandwidth is chosen which is too large, tails with negligable weight are pulled from the LDOS out to the edges. But a more important effect is the emphasis of all the features in the LDOS - the LDOS tends to a spectrum of delta functions as the bandwidth tends to ∞.

A sensible physical criterion is thus : Given a finite number of coefficients, we must choose the a, b^2 that will give, for this set of coefficients, the minimum bandwidth consistent with no loss of weight from the band. We call these values a_c, b_c^2.

This criterion is easily translated into mathematical terms. The delta functions that would carry weight out of the band must be situated exactly at the band edges ; thus we demand that G_{00} diverge simultaneously at the top and bottom band edges.

We now note that the terminator (3) has a particularly simple form at the band edges

$$t(a \pm 2b) = \pm b \tag{6}$$

Then, making the substitution $E = a \pm 2b$ throughout the continued fraction we obtain the finite fraction

$$G_{00}(a \pm 2b) = \cfrac{1/2\, b_0^2}{\pm b - \frac{1}{2}(a_0 - a) - \cfrac{\frac{1}{4} b_1^2}{\pm b - \ddots \cfrac{}{} \cfrac{1/2\, b_N^2}{\pm b - (a_N - a)}}} \tag{7}$$

For given a, the N + 1 eigenvalues $\tilde{b}_0, \ldots, \tilde{b}_N$ of the corresponding finite tridiagonal matrix

$$\begin{bmatrix} \frac{1}{2}(a_0 - a) & \frac{1}{2} b_1 & 0 & & \\ \frac{1}{2} b_1 & \frac{1}{2}(a_1 - a) & \frac{1}{2} b_2 & & 0 \\ 0 & \frac{1}{2} b_2 & \ddots & & \\ & 0 & & \ddots & \frac{1}{\sqrt{2}} b_N \\ & & & \frac{1}{\sqrt{2}} b_N & (a_N - a) \end{bmatrix} \qquad (8)$$

are those values of b for which $G_{00}(a \pm 2b)$ diverges. The maximum eigenvalue b_{max} is that value of b for which no delta function has split off from the top of the band, and the minimum eigenvalue b_{min} is that b for which no delta function has split from the bottom of the band.

Since the terminator (3) only involves b^2 we must have

$$\tilde{b}_{max} = |\tilde{b}_{min}| = b_c \qquad (9)$$

This corresponds to the continued fraction diverging simultaneously at the top and bottom band edges. Satisfying condition (9) determines the values a_c, b_c^2 in the terminator.

The computation of the terminators a_c, b_c^2 by the procedure outlined above is extremely fast.

That the method gives the minimum bandwidth for no loss of weight from the band can be demonstrated by calculating the weight in the band as a function of b^2. If we use the value of a as given by the method above, then we obtain something of the form of the full line in figure 1. The arrowed value of b^2 is precisely b_c^2. If, however, we use a value of a different from a_c then we obtain the dashed line in figure 1 - to satisfy the condition of no loss of weight we would need a larger b_c^2.

3. RESULTS

In this section we present the LDOS obtained by both the termination method and the Cambridge method from the coefficients calculated for a canonical FCC tight binding d band (see O.K. Andersen, this volume). The FCC cluster used was of a sufficient size that 15 levels (30 coefficients $a_0 \ldots a_{14}$, $b_0^2 \ldots b_{14}^2$) are exact using first nearest neighbours only.

Figures 2(a) and (b) show the LDOS obtained using both methods for 5, 7 and 9 levels, and 2(c) and (d) compare the LDOS

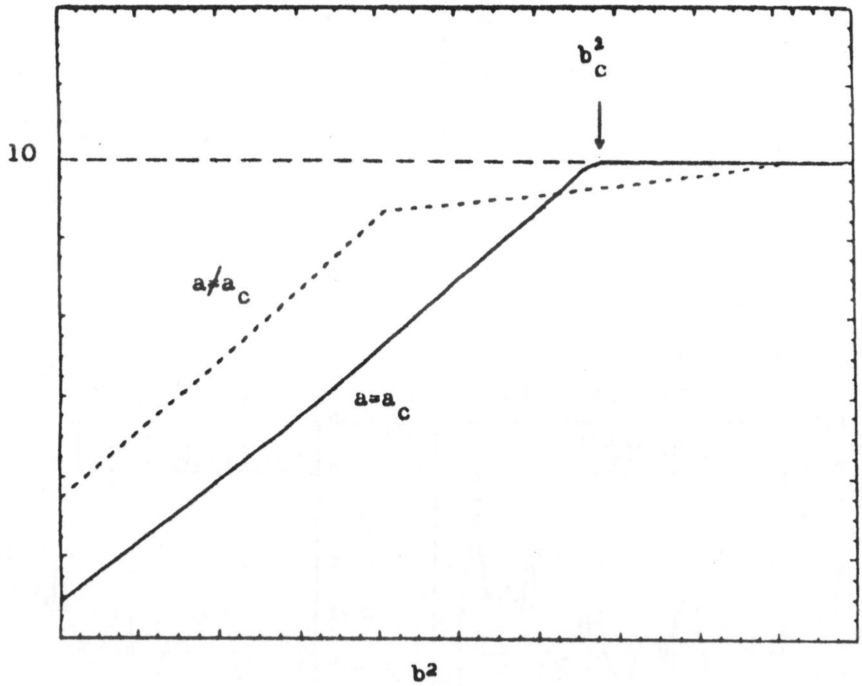

Fig. 1. Weight in FCC d band as a function of terminating coefficient b . Plotted for two values of a.

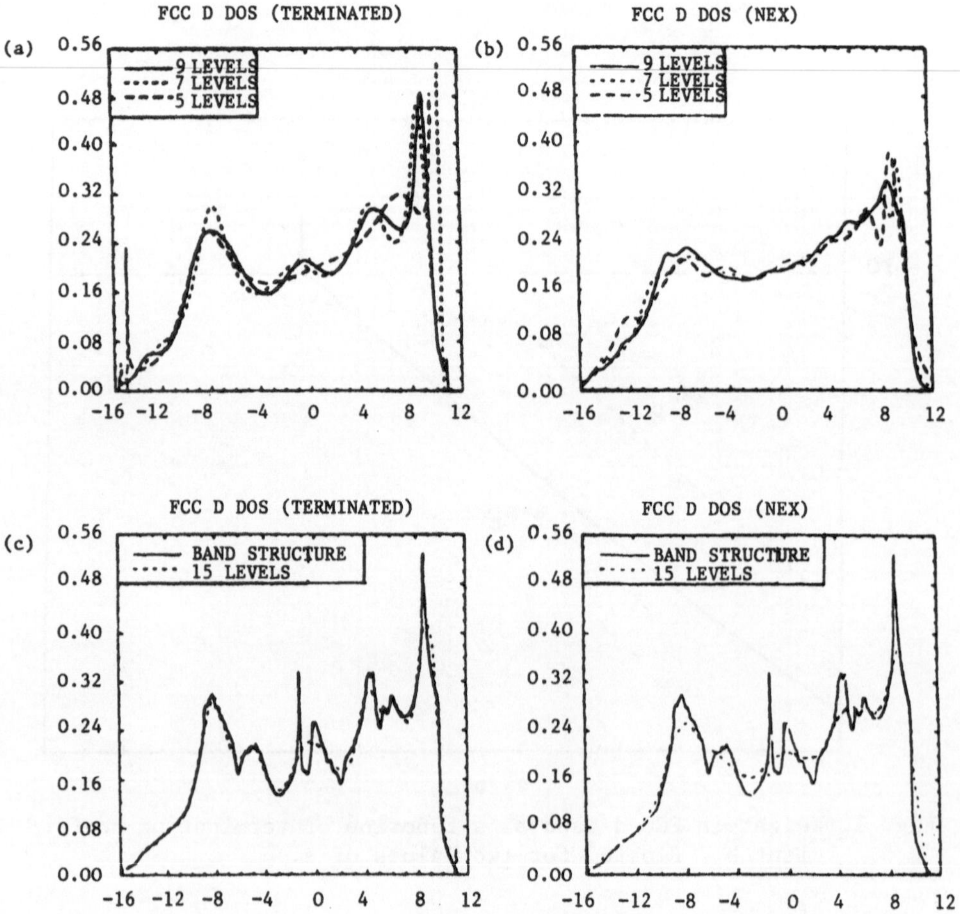

Fig. 2. Comparison of convergence of FCC d band DOS using termination method and Cambridge method.

calculated using all 15 exact levels with a tight binding band structure DOS (R. Muniz, private communication). As expected the more general Cambridge procedure gives a much slower convergence to the exact bulk DOS.

Although the LDOS themselves converge slowly, quantities integrated over LDOS, however, are expected to be more quickly convergent. We have investigated the structural energy

$$U = \int^{E_F} En(E)dE \tag{10}$$

as a function of band filling

$$N = \int^{E_F} n(E)dE \tag{11}$$

using the termination and the Cambridge procedures. In the Cambridge method bounds $U_\pm$ on U and $N_\pm$ on N are calculated and then averaged, and in the termination method n(E) is calculated from the terminated continued fraction and the integrations performed numerically. In order to emphasise the differences in the U's calculated using differing numbers of levels, a background has been subtracted. This background has been taken as the structural energy of a skew rectangular LDOS which was fitted to the first four FCC moments ($\mu_0,\ldots,\mu_3$). A simple theorem concerning moments of functions (Ducastelle and Cyrot-Lackmann 1971) tells us that our resulting structural energies <u>must</u> have at least two zeros distinct from those at N = 0 and N = 5.

The structural energies obtained with both methods are shown in figure 3. Those obtained for 7 and 9 levels with the Cambridge method have only one zero distinct from the band edges, indicating that in the course of the calculation the input information about the moments has been lost. The structural energies obtained by the terminator method do possess the correct number of zeros and are, moreover, almost identical for 5, 7 and 9 levels. The structural energies obtained with the Cambridge method do eventually approach the form of the terminated for a much greater number of levels (the difference in vertical scales in (a) and (b) should be noted).

The reason for the loss of information about the moments in the Cambridge method is straightforward to find in the details of how the LDOS and integrated quantities are calculated. Although the bounds $F_\pm(E)$ on the integral of f(E) are constructed using the coefficients (and thus using the moments), there is absolutely no guarantee that when we take the mean of these bounds all of these moments should be preserved. In the termination method, on the other hand, n(E) is calculated as a continued fraction in which the coefficients a_i, b_i^2, i = 0,...,N are unaffected by the (correct) terminator, and thus the moments are preserved.

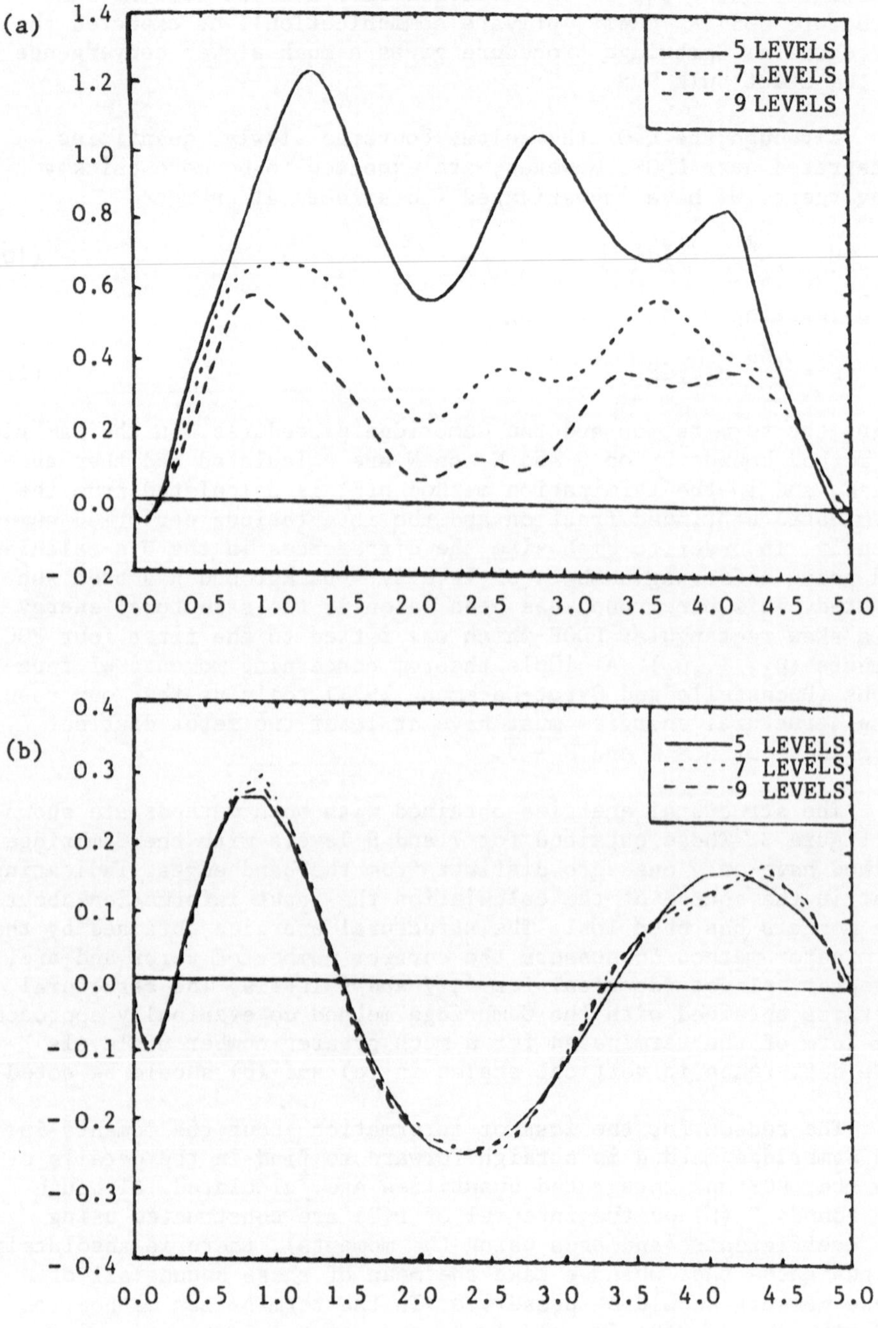

Fig. 3. Structural energy as a function of band filling calculated with 5, 7 and 9 levels.
(a) Cambridge method. (b) Termination method.

4. CONCLUSIONS

We have discussed two approximations used in estimating the LDOS written as a continued fraction, and shown that it is possible to find terminators a_c, b_c^2 that optimally match an infinite tail to a finite set of coefficients according to a sensible physical criterion. The results obtained using both the termination and the Cambridge method have been shown. As expected, for the case with no band gap, the LDOS obtained using the terminator procedure for a small number of levels are a considerably better representation of the bulk FCC DOS than those obtained with the Cambridge procedure. The scheme presented here can be generalized to the case of a single gap by using the terminator suggested by Turchi et al. (1982).

But the most important result presented here is that in the calculation of the LDOS and integrated quantities by the Cambridge method the moments are not preserved, even though they have been evaluated accurately to some level by the recursion method. The termination procedure preserves the moment information that is contained in the coefficients and, for the structural energy, gives results that are very quickly convergent, as can be seen in figure 3.

REFERENCES

Akhiezer, N.I., 1965, The Classical Moment Problem (Oliver and Boyd, Edinburgh - London).

Ducastelle, F. and Cyrot-Lackmann, F., 1971, J.Phys.Chem.Solids 32, 285-301.

Haydock, R., Heine, V. and Kelly, M.J., 1975, J.Phys.C. : Solid St.Phys. 8, 2591-605.

Haydock, R., 1980, Solid St. Phys. 35, 216-94.

Nex, C.M.M., 1978, J.Phys.A : Math.Gen. 11, 653-63.

Turchi, P., Ducastelle, F. and Treglia, G., 1982, J.Phys.C. : Solid St. Phys. 15, 2891-924.

COMPUTER EXPERIMENTS ON AMORPHOUS TRANSITION METALS

G.F. Weir and G.J. Morgan

Department of Physics
University of Leeds
LEEDS LS2 9JT, U.K.

INTRODUCTION

The calculation of the electronic properties of disordered systems is an extremely difficult one. Bloch's theorem, central to the understanding of crystalline solids, is no longer applicable and so any analytical attack on the problem becomes a formidable task even for the most simple case. Our particular interests lie in understanding the transport properties of amorphous transition metals and these are far from simple examples of disordered systems.

To tackle this problem we have used a series of simulation experiments on densely packed disorderd 'hard-sphere' models and what we hope to demonstrate is that by using this relatively simple approach considerable insight can be gained into the physics of these complicated systems.

METHOD

The simulations have been carried out using the equation of motion method on a typical amorphous transition metal of the 3d/4s series. These preliminary studies have been concerned with three major topics of current interest, namely :

(1) the calculation of the density of electronic states taking into account the hybridization between the s-band and all five d sub-bands
(2) the evaluation of the conductivity of unhybridised d electrons
(3) the calculation of the spectral function for the s-band, again including the effects of hybridisation.

We have used a tight binding (LCAO) scheme throughout which, as usual, assumes that the total wave function can be described as a sum of atomic orbitals, $\phi_L(\underline{r} - \underline{r}_j)$ (L = n,L,m), centered on the sites j

$$\Psi(\underline{r}) = \sum_L \sum_j a_{jL}\phi_L(\underline{r} - \underline{r}_j) \tag{1}$$

This is, of course, only an approximation and so the results we will present should be viewed with this in mind. However it must be said that it is well known that the tight binding approach gives a very good description of the bands in crystalline transition metals (comparison with more exact band calculations bears this out), and so even though we cannot hope to obtain precise answers, it is perfectly reasonable to expect that this treatment will successfully probe the underlying physics of these complex systems when sensibly applied. Furthermore it might also serve as a useful guide to the development of more exact theory, especially in the area of electron transport phenomena which is a particularly difficult problem.

The equation of motion method and its implementation have been explained elsewhere (Alben, Blume, Krakauer and Schwartz 1975, Weir and Morgan 1981). Suffice to say that the problem entails solving the time dependent Schrödinger equation for the site amplitudes with a set of starting conditions which depend on the particular property of interest.

$$i\hbar \frac{\partial a_{jL}(t)}{\partial t} - \sum_{L'} \sum_{\substack{j' \\ \neq j}} V_{LL'}(\underline{r}_j - \underline{r}_{j'})a_{j'L'}(t) = 0 \tag{2}$$

The summation is over 'near neighbours', which have been taken to be those that lie within three hard sphere radii of the site in question. This gives an average number of near neighbours of just over 13 for our 797 site model (Finney 1970) with periodic boundary conditions. Functional forms for the $V_{LL'}$ (V_{ss}, V_{sd}, V_{dd}) have been taken from Harrison's book (1980) as have his parameters for nickel.

CALCULATIONS AND RESULTS

(1) Density of States

The initial conditons are :

$$a_{jL}(0) = - ie^{i\theta_{jL}} \tag{3}$$

where θ_{jL} is a random phase between plus and minus π such that the average phase is essentially zero. To calculate the density of states we need the diagonal part of the Green's function and it can be shown quite easily that this is related to the site amplitudes as follows

$$G_{LL}(\underline{r}_j,\underline{r}_j;t) = a_{jL}(t)e^{-i\theta_{jL}} \tag{4}$$

Taking the imaginary part of the Fourier transform then gives the required result :

$$g(E) = -\frac{1}{\pi}\,\mathrm{Im}\,\{\int\sum_j G_{LL}(\underline{r}_j,\underline{r}_j;t)\,\exp(\frac{iEt}{\hbar})\,dt\} \tag{5}$$

In figure (1) we show the results for an 840 site fcc structure when there is no hybridization. Averaging over independent sets of phases will smooth out the data somewhat but the underlying features can already be discerned. Next we turn on the hybridisation and figure (2) shows the dramatic effect this has on the

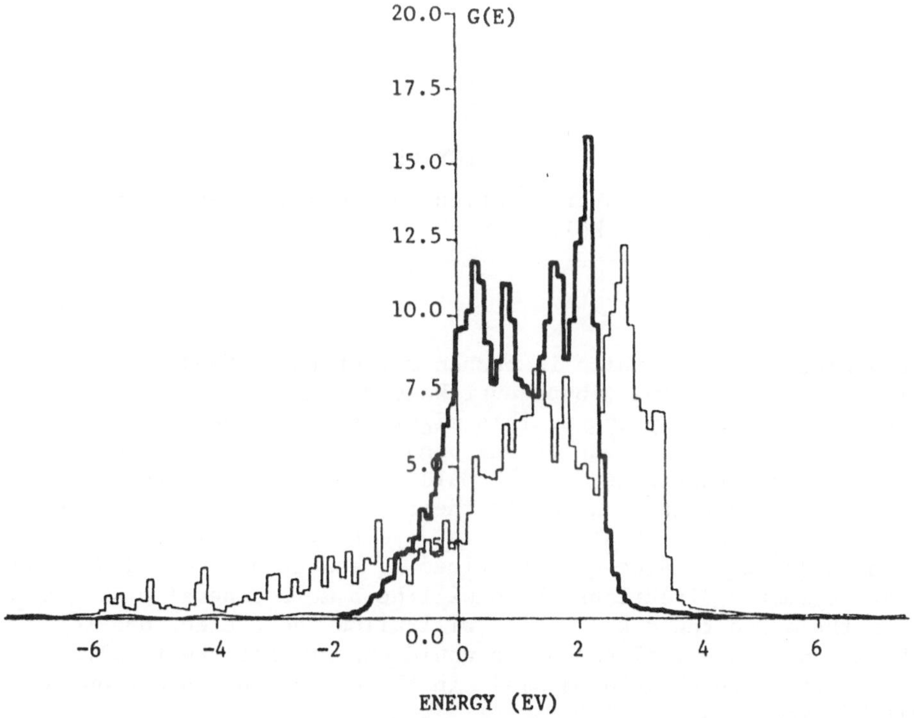

Fig. 1. The density of s and d states in crystalline nickel with no hybridisation. The density of states, G(E) (/atom, /spin, /ev), for d electrons is divided by five and corresponds to the heavier line in this figure and in figures 2, 3 and 4.

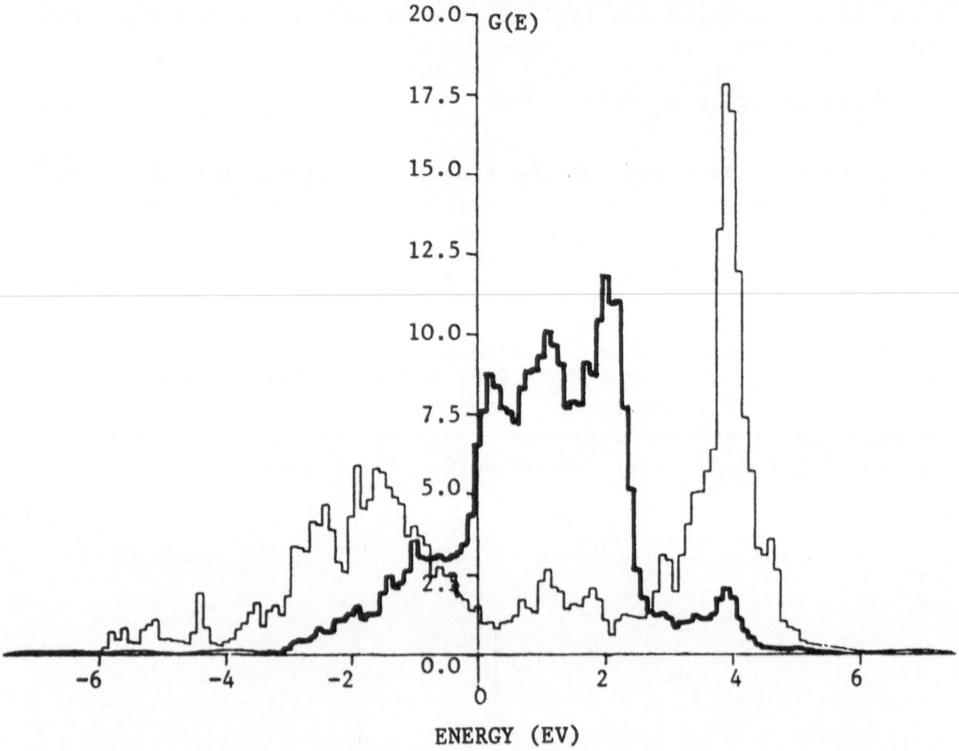

Fig. 2. The effect of hybridisation on the densities of states in crystalline nickel.

'free electron' band which is pushed out of the d band region resulting in a much lower, but nevertheless finite, density of s states in this energy range. This might at first seem surprising since an inspection of the energy bands along symmetry directions in any of the transition metals would lead one to believe that there should in fact be a gap in the s electronic density of states. However the point is that in special symmetry directions the mixing of states can be characterized by essentially one d sub-band. In off symmetry directions there will be a more general mixing between the s and the d band which will allow the s electrons to take up any energy, although one would expect that the total s density of states should be small in the d band region, as our results indicate.

Now we consider the amorphous case and, figure (3) shows the situation for no hybridisation. This, in essence, is a smoothed out version of the crystalline case and is exactly what one might predict since both models have the same density and approximately

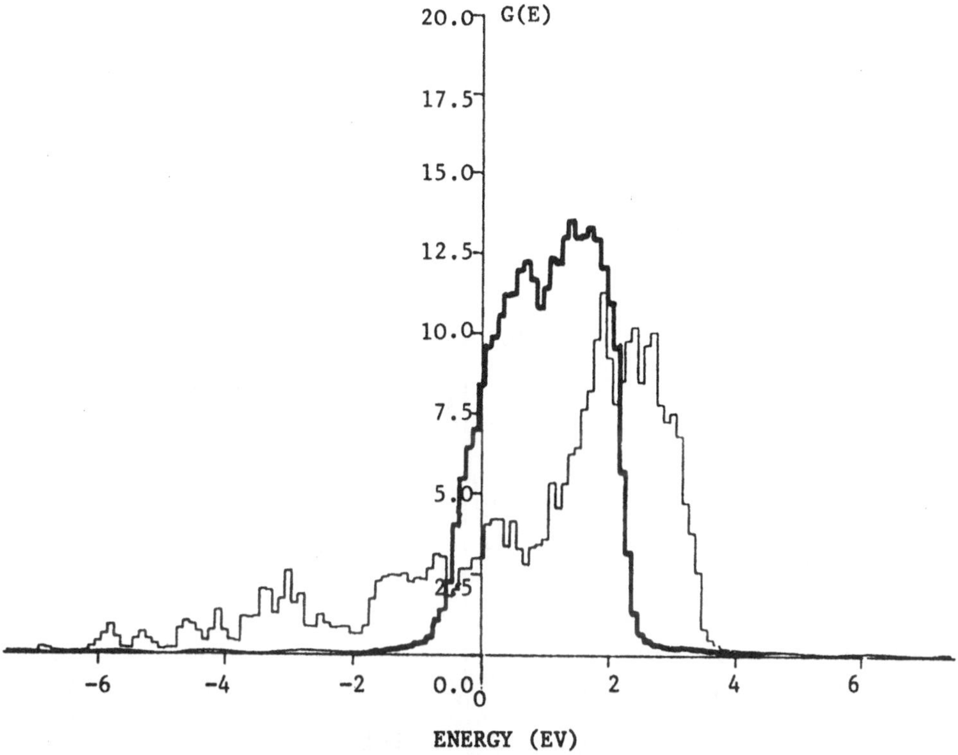

Fig. 3. The densities of states in 'amorphous' nickel with no hybridisation.

the same number of near neighbours. In figure (4) we have switched the hybridisation on and obtained results which are very similar to the crystalline case. This is quite surprising since the rule for mixing which applies to a periodic structure (only electrons of the same symmetry interact) is relaxed for the amorphous case. It would seem that the mixing in off symmetry directions in a crystal has approximately the same net effect, with regard to the density of states, as the completely general mixing in the amorphous state.

These results show how important hybridisation is in an amorphous solid and make one suspect that any formulation of realistic models to describe the transport mechanisms in such systems must give prime importance to this large effect. Most theories have tended to ignore the effects of hybridisation and assumed that all the current is carried by a free electron band. Conduction by the d electrons has been ruled out on the grounds that they are very tightly bound to the nuclei. However this is by no means an obvious consequence of their localised nature.

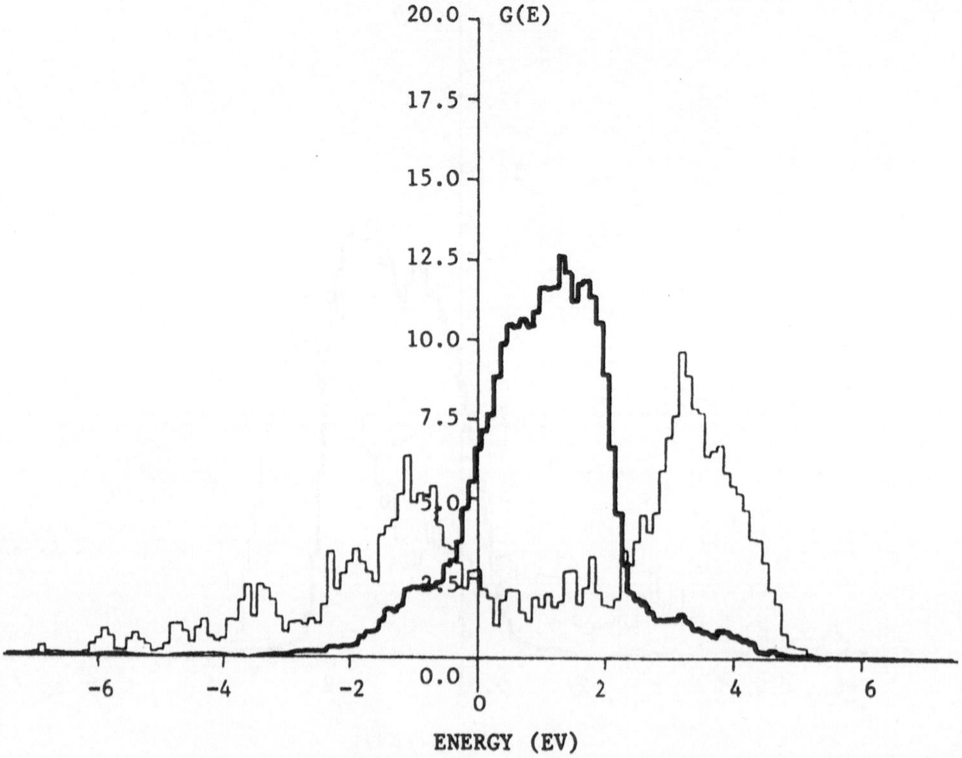

Fig. 4. The effect of hybridisation on the densities of states in 'amorphous' nickel.

In view of our results, that in energy regions spanning the positions of the Fermi energy there is a large depletion of s electrons for these systems, we were obviously interested to know how well the d electrons could conduct on their own.

(2) Conductivity

To simulate this on the computer we simply place a source term on the right hand side of the Schrödinger equation and allow it to inject electrons of a particular energy into one end of the sample for a certain time. When the source has been turned off we follow the centre of gravity of the pulse down the wire and a diffusion coefficient can be extracted by comparing this behaviour with a 'classical' experiment having the same injection rate. Then, using the density of states at that energy, we can evaluate the conductivity from the relationship :

$$\sigma(E) = 2e^2 g(E) D(E) \qquad (6)$$

where $\sigma(E)$ is the conductivity, $g(E)$ is the density of states/spin/volume/unit energy range and $D(E)$ is the diffusion coefficient.

In figure (5) we show the results for the unhybridised d band and compare them with the experimental value for liquid nickel. Although the agreement is very good it is, of course, the qualitative features which are important. It can be seen that the values for the conductivity are not orders of magnitude different from the measured values for the transition series, which are all typically $10^4(\Omega cm)^{-1}$. This shows quite clearly that an unhybridised d band can carry appreciable current. Turning on the hybridisation should not alter the picture in the central regions of the band since the d density of states is hardly modified by the interaction in that energy range. However in the 'wings' of the density of states, where the s density is starting to recover, the role of s-d hybridisation will be subtle. A model involving s and d electrons carrying current in parallel is probably more appropriate in these energy ranges.

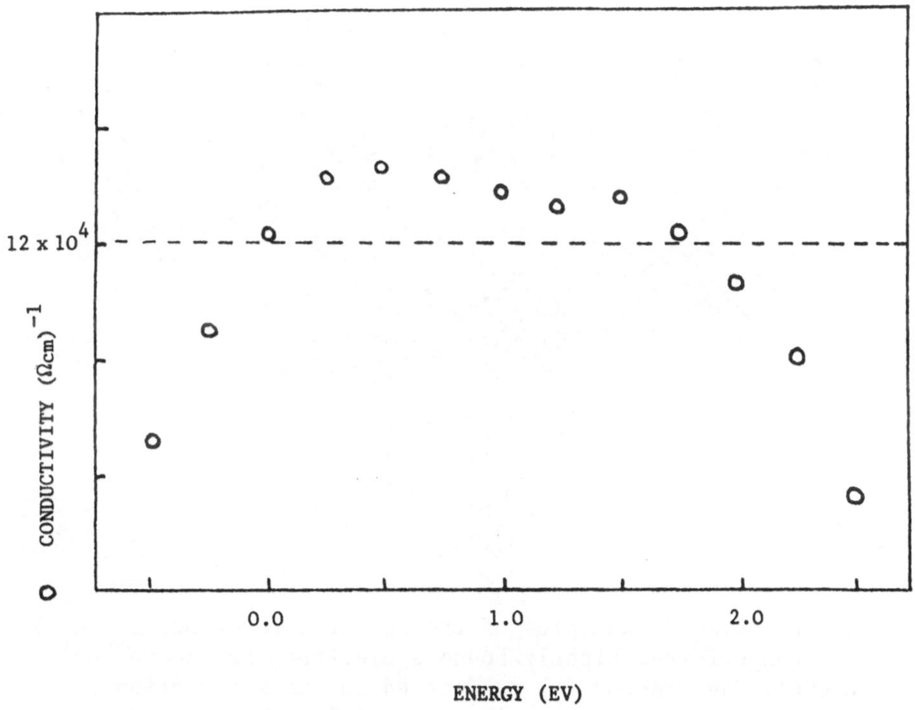

Fig. 5. The conductivity of unhybridised d electrons in 'amorphous' nickel. The dashed line refers to the experimental value for liquid nickel.

Another interesting point to note is that far out into the 'wings' diffusion becomes impossible (that is why the results in figure (5) do not span the whole d band width). This indicates the crossing of a mobility edge which is an important topic for future study.

(3) Spectral Functions

Our results so far have shown that a modified s band can persist throughout the d band region in an amorphous solid and so it seems reasonable to ask what the dispersion relationship for the s electrons looks like in such a system.

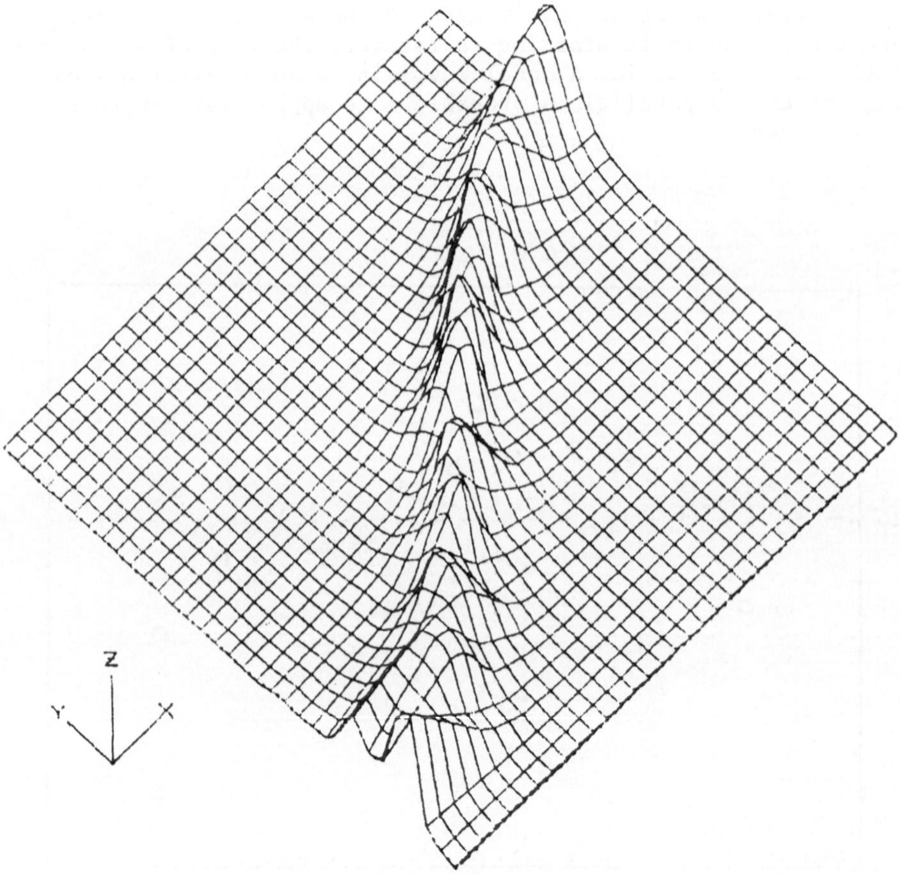

Fig. 6. A three-dimensional plot of the spectral function, $\rho_{ss}(k,E)$, for unhybridised tightly bound s electrons in 'amorphous' nickel. Wave vector (k) is plotted in the x-direction from k = 0.0 au to k = 0.8 au in steps of 0.025 au. Energy (E) is plotted in the y-direction from E = - 9.5 ev to E = + 6.5 ev in steps of 0.5 ev.

To do this we take the following initial conditions :

$$a_{jL}(0) = - i\phi_L(\underline{k})e^{i\underline{k}\cdot\underline{r}_j} \quad (7)$$

where $\phi_L(\underline{k})$ is the Fourier transform of the particular orbital, of interest (in this case an s type wave function) and k is the modulus of an allowed wave vector. The calculation is then performed (Morgan and Weir 1982) in exactly the same way as for the density of states and the spectral function is obtained from

$$\rho_{LL}(\underline{k},E) = -\frac{1}{\pi}\,\mathrm{Im}\{\int G_{LL}(\underline{k},\underline{k},t)\,\exp(\frac{iEt}{\hbar})dt\} \quad (8)$$

The spectral function for the unhybridised s band is presented in figure (6). If the peak in this function is taken to define the E(k) relationship then the shape of the dispersion curve would be very similar to the free electron band in a crystal. Figure (7)

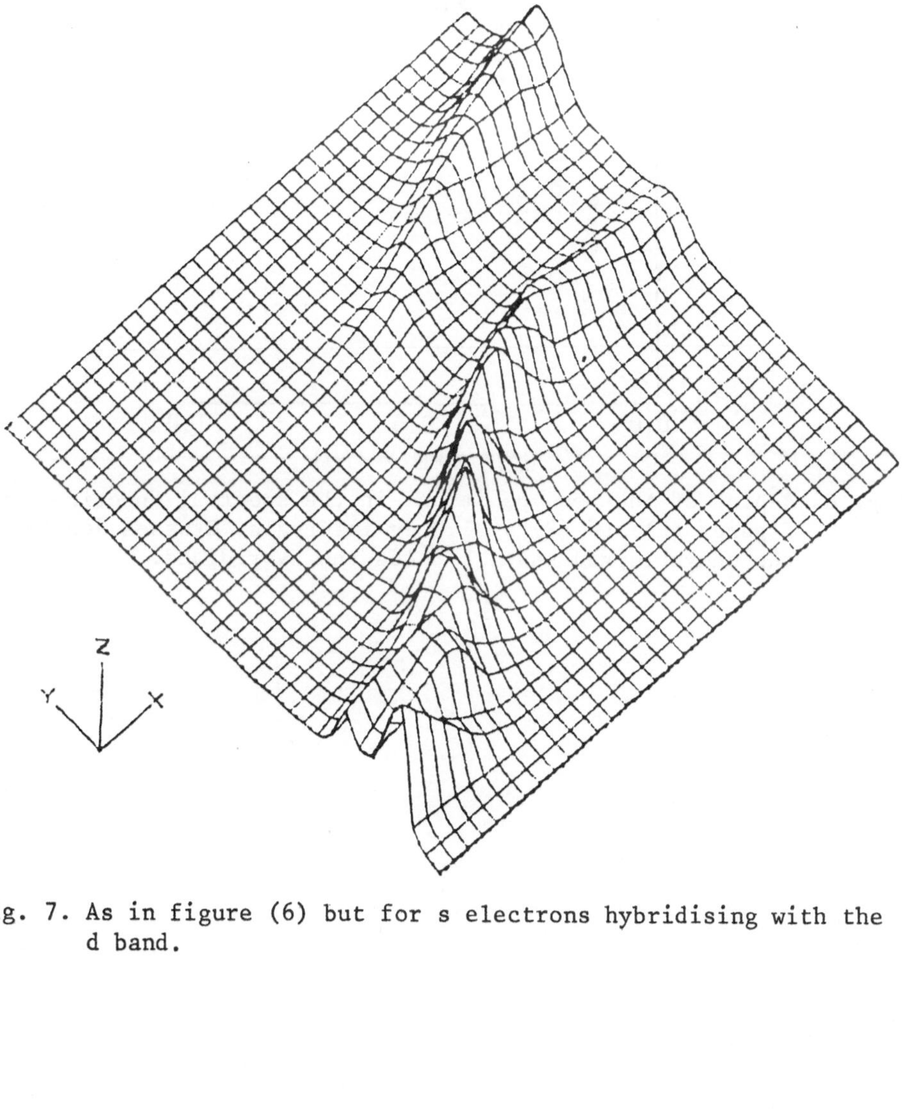

Fig. 7. As in figure (6) but for s electrons hybridising with the d band.

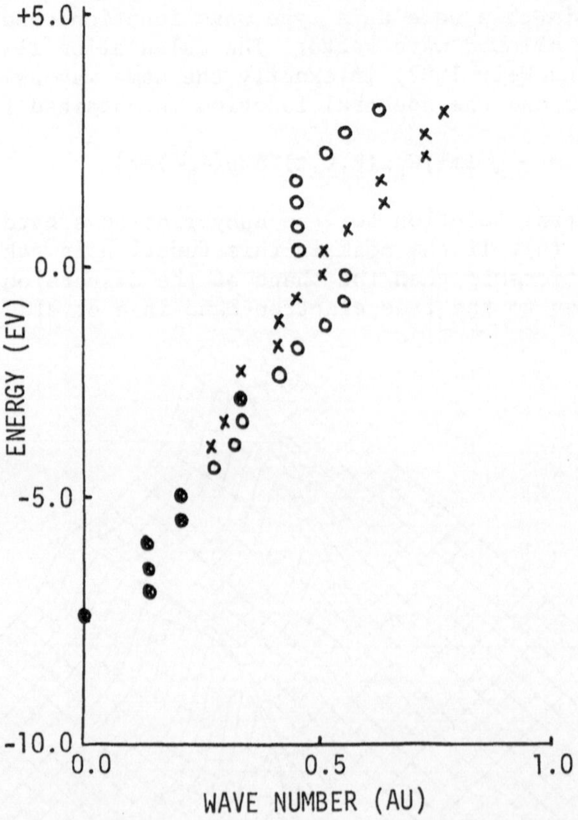

Fig. 8. A plot of the peak positions in the spectral functions in figures (6) and (7) at 1.0 ev intervals. Crosses denote the unhybridised case, circles the hybridised.

shows the situation when the hybridisation is included. Although the band has appeared to split as it crossed the d band region, there are still well defined peaks at all energies and these are shown, together with the unhybridised results, in figure (8). The hybridised case shows an S shaped anomally, familiar in anomalous dispersion in optics. It may be argued that since the s electronic lifetimes are becoming very short in the middle of the d band a dispersion relationship cannot be defined. However the finite density of states confirms that s electrons most certainly exist in this energy range. The fact that they have a large width is not altogether surprising and does not present any fundamental problem. The step-like nature of the results is only a consequence of using a discrete set of k points. A finer k scale would give smoother results but our findings are quite unambiguous and going to larger models would not alter the qualitative conclusions.

A very interesting consequence of this S shaped curve is that it predicts negative group velocities at certain energies. This would lead to positive Hall coefficients which are indeed observed in many liquid transition metals and amorphous transition metal alloys.

(An alternative approach using a perturbation theory gives very similar results for a free electron band hybridising with a d band. Details of this can be found in Weir, Howson, Gallagher and Morgan (1982)).

CONCLUDING REMARKS

At present we are extending the work on the conductivity to simulate the Hall effect directly. This involves looking for changes in the electronic distribution when a magnetic field is applied. The problem is that this is an extremely small effect. There are methods to get round some of the difficulties and it will be very interesting to see if our predictions can be realised in a computer experiment.

We hope that this short contribution has shown how useful the role of simulation studies can be in extracting the physics of these complex systems in a relatively straight forward and computationally inexpensive manner.

REFERENCES

Alben, R., Blume,M., Krakauer, H. and Schwartz, L., 1975, Phys. Rev. B12, 4090.
Finney, J.L., 1970, Proc.Roy.Soc. (London) A315, 479.
Harrison, W.A., 1980, 'Electronic Structure' (Freeman and Co.).
Morgan, G.J. and Weir, G.F., 1983, Phil.Mag.B. 47, 163-176.
Weir, G.F., Howson, M.A., Gallagher, B.L. and Morgan, C.J., 1983 Phil.Mag.B. 47, 177-181.
Weir, G.F. and Morgan, G.J., 1981, J.Phys.F. 11, 1833.

INDEX

Note - Additional compounds are shown in the Materials Index on pages 693 to 696